AF333767

Magnetic Methods and the Timing of Geological Processes

The Geological Society of London
Books Editorial Committee

Chief Editor
RICK LAW (USA)

Society Books Editors
JIM GRIFFITHS (UK)
DAVE HODGSON (UK)
HOWARD JOHNSON (UK)
PHIL LEAT (UK)
DANIELA SCHMIDT (UK)
RANDELL STEPHENSON (UK)
ROB STRACHAN (UK)
MARK WHITEMAN (UK)

Society Books Advisors
GHULAM BHAT (India)
MARIE-FRANÇOISE BRUNET (France)
JAMES GOFF (Australia)
MARIO PARISE (Italy)
SATISH-KUMAR (Japan)
MARCO VECOLI (Saudi Arabia)
GONZALO VEIGA (Argentina)
MAARTEN DE WIT (South Africa)

Geological Society books refereeing procedures

The Society makes every effort to ensure that the scientific and production quality of its books matches that of its journals. Since 1997, all book proposals have been refereed by specialist reviewers as well as by the Society's Books Editorial Committee. If the referees identify weaknesses in the proposal, these must be addressed before the proposal is accepted.

Once the book is accepted, the Society Book Editors ensure that the volume editors follow strict guidelines on refereeing and quality control. We insist that individual papers can only be accepted after satisfactory review by two independent referees. The questions on the review forms are similar to those for *Journal of the Geological Society*. The referees forms and comments must be available to the Society's Book Editors on request.

Although many of the books result from meetings, the editors are expected to commission papers that were not presented at the meeting to ensure that the book provides a balanced coverage of the subject. Being accepted for presentation at the meeting does not guarantee inclusion in the book.

More information about submitting a proposal and producing a book for the Society can be found on its website: www.geolsoc.org.uk.

It is recommended that reference to all or part of this book should be made in one of the following ways:

JOVANE, L., HERRERO-BERVERA, E., HINNOV, L. A. & HOUSEN, B. A. (eds) 2013. *Magnetic Methods and the Timing of Geological Processes*. Geological Society, London, Special Publications, **373**.

HERRERO-BERVERA, E. & CAÑÓN-TAPIA, E. 2013. On the directional geomagnetic signature of the Pringle Falls excursion recorded at Pringle Falls, Oregon, USA. *In:* JOVANE, L., HERRERO-BERVERA, E., HINNOV, L. A. & HOUSEN, B. A. (eds) *Magnetic Methods and the Timing of Geological Processes*. Geological Society, London, Special Publications, **373**, 261–278. First published online December 7, 2012, http://dx.doi.org/10.1144/SP373.12.

GEOLOGICAL SOCIETY SPECIAL PUBLICATION NO. 373

Magnetic Methods and the Timing of Geological Processes

EDITED BY

L. JOVANE
Universidade de São Paulo, Brazil

E. HERRERO-BERVERA
University of Hawaii at Manoa, USA

L.A. HINNOV
Johns Hopkins University, USA

and

B. HOUSEN
Western Washington University, USA

2013
Published by
The Geological Society
London

THE GEOLOGICAL SOCIETY

The Geological Society of London (GSL) was founded in 1807. It is the oldest national geological society in the world and the largest in Europe. It was incorporated under Royal Charter in 1825 and is Registered Charity 210161.

The Society is the UK national learned and professional society for geology with a worldwide Fellowship (FGS) of over 10 000. The Society has the power to confer Chartered status on suitably qualified Fellows, and about 2000 of the Fellowship carry the title (CGeol). Chartered Geologists may also obtain the equivalent European title, European Geologist (EurGeol). One fifth of the Society's fellowship resides outside the UK. To find out more about the Society, log on to www.geolsoc.org.uk.

The Geological Society Publishing House (Bath, UK) produces the Society's international journals and books, and acts as European distributor for selected publications of the American Association of Petroleum Geologists (AAPG), the Indonesian Petroleum Association (IPA), the Geological Society of America (GSA), the Society for Sedimentary Geology (SEPM) and the Geologists' Association (GA). Joint marketing agreements ensure that GSL Fellows may purchase these societies' publications at a discount. The Society's online bookshop (accessible from www.geolsoc.org.uk) offers secure book purchasing with your credit or debit card.

To find out about joining the Society and benefiting from substantial discounts on publications of GSL and other societies worldwide, consult www.geolsoc.org.uk, or contact the Fellowship Department at: The Geological Society, Burlington House, Piccadilly, London W1J 0BG: Tel. +44 (0)20 7434 9944; Fax +44 (0)20 7439 8975; E-mail: enquiries@geolsoc.org.uk.

For information about the Society's meetings, consult *Events* on www.geolsoc.org.uk. To find out more about the Society's Corporate Affiliates Scheme, write to enquiries@geolsoc.org.uk.

Published by The Geological Society from:
The Geological Society Publishing House, Unit 7, Brassmill Enterprise Centre, Brassmill Lane, Bath BA1 3JN, UK

The Lyell Collection: www.lyellcollection.org
Online bookshop: www.geolsoc.org.uk/bookshop
Orders: Tel. +44 (0)1225 445046, Fax +44 (0)1225 442836

The publishers make no representation, express or implied, with regard to the accuracy of the information contained in this book and cannot accept any legal responsibility for any errors or omissions that may be made.

© The Geological Society of London 2013. No reproduction, copy or transmission of all or part of this publication may be made without the prior written permission of the publisher. In the UK, users may clear copying permissions and make payment to The Copyright Licensing Agency Ltd, Saffron House, 6–10 Kirby Street, London EC1N 8TS UK, and in the USA to the Copyright Clearance Center, 222 Rosewood Drive, Danvers, MA 01923, USA. Other countries may have a local reproduction rights agency for such payments. Full information on the Society's permissions policy can be found at: www.geolsoc.org.uk/permissions

British Library Cataloguing in Publication Data

A catalogue record for this book is available from the British Library.
ISBN 978-1-86239-354-7
ISSN 0305-8719

Distributors
For details of international agents and distributors see:
www.geolsoc.org.uk/agentsdistributors

Typeset by Techset Composition India (P) Ltd, Bangalore and Chennai, India
Printed by Berforts Information Press Ltd, Oxford, UK

Contents

Palaeoclimatic changes from rock magnetic proxies

Magnetic methods and the timing of geological processes

L. JOVANE[1]*, L. HINNOV[2], B. A. HOUSEN[3] & E. HERRERO-BARVERA[4]

[1]*Instituto Oceanográfico da Universidade de São Paulo, Fisica, Geoquimica e Geologia, Praça do Oceanográfico 191, São Paulo, São Paulo 05508-120, Brazil*

[2]*John Hopkins University, Earth and Planetary Sciences, 3400 N. Charles Street, Baltimore, Maryland 21218, USA*

[3]*Department of Geology, Western Washington University, 516 High St., Bellingham, Washington 98225, USA*

[4]*University of Hawaii at Manoa, Hawaii Institute of Geophysics and Planetology, Honolulu, Hawaii 96822, USA*

**Corresponding author (e-mail: jovane@usp.br)*

Abstract: Magnetostratigraphy is best known as a technique that employs correlation among different stratigraphic sections using the magnetic directions that define geomagnetic polarity reversals as marker-horizons. The ages of the polarity reversals provide common tie points among the sections, allowing accurate time correlation. Recently, magnetostratigraphy has acquired a broader meaning, now referring to many types of magnetic measurements within a stratigraphic sequence. Many of these measurements provide correlation and age control not only for the older and younger boundaries of a polarity interval, but also within intervals. Thus, magnetostratigraphy no longer represents a dating tool based only on the geomagnetic polarity reversals, but comprises a set of techniques that includes measurements of all geomagnetic field parameters, environmental magnetism, rock magnetic and palaeoclimatic change recorded in sedimentary rocks, and key corrections to magnetic directions related to geodynamics, tectonics and diagenetic processes.

Discovery of geomagnetic reversals

Over the past century numerous methodologies have been developed to detect time variations of the geomagnetic field and environmentally significant magnetic properties in rocks. These methods comprise measurements of natural remanence, magnetic susceptibility, demagnetization and induced artificial magnetizations. Brunhes (1906) and Matuyama (1926) were among the first to recognize that old rocks have inclination values that are very different from today's values, and sometimes of opposite polarity to the present-day magnetic field. According to Matuyama (1926), these changes represent reversals in the polarity of the ancient geomagnetic field. Cox *et al.* (1963) recognized that these polarity reversals were global events, and that, by combining palaeomagnetic and geochronologic data, a sequence of geomagnetic field reversals could be constructed. This led directly to the development of the Geomagnetic Polarity Time Scale (GPTS).

Vine & Matthews (1963), using data acquired during marine cruises, recognized that magnetic anomalies had a symmetrical pattern with respect to the mid-ocean ridges, and that there was a relationship between geomagnetic field reversals and motions of the ocean floor. The oceanic magnetic anomalies are related to the magnetization of the oceanic basalts that cool down while spreading from mid-oceanic ridges, and can be used in combination with geomagnetic field polarity sequences from rocks found on land to further develop a globally extensive GPTS (Heirtzler *et al.* 1968). Opdyke (1972) first integrated magnetostratigraphy and biostratigraphy for Plio-Pleistocene marine sediments, and since that time biostratigraphic information has been increasingly used for correlation of the observed polarity sequences in sedimentary rocks with the appropriate part of the radioisotope-calibrated GPTS. Subsequently, Alvarez *et al.* (1977) recognized sets of magnetic polarity reversals within the Cenozoic Gubbio (Italy) sedimentary sequence. William Lowrie studied the geological and physical processes that permit pelagic sediments to keep magnetization and defined the magnetostratigraphy of those Italian sections (Lowrie *et al.* 1982), allowing the scientific community to build and further refine the GPTS (Cox *et al.* 1963; Heirtzler *et al.* 1968; LaBreque *et al.* 1977; Berggren *et al.* 1985; Cande & Kent 1992, 1995; Huestis & Acton 1997; Singer *et al.* 2002; Channell *et al.* 1995; Malinverno *et al.* 2012; Ogg 2012).

From: JOVANE, L., HERRERO-BERVERA, E., HINNOV, L. A. & HOUSEN, B. A. (eds) 2013. *Magnetic Methods and the Timing of Geological Processes*. Geological Society, London, Special Publications, **373**, 1–12. First published online April 25, 2013, http://dx.doi.org/10.1144/SP373.17 © The Geological Society of London 2013. Publishing disclaimer: www.geolsoc.org.uk/pub_ethics

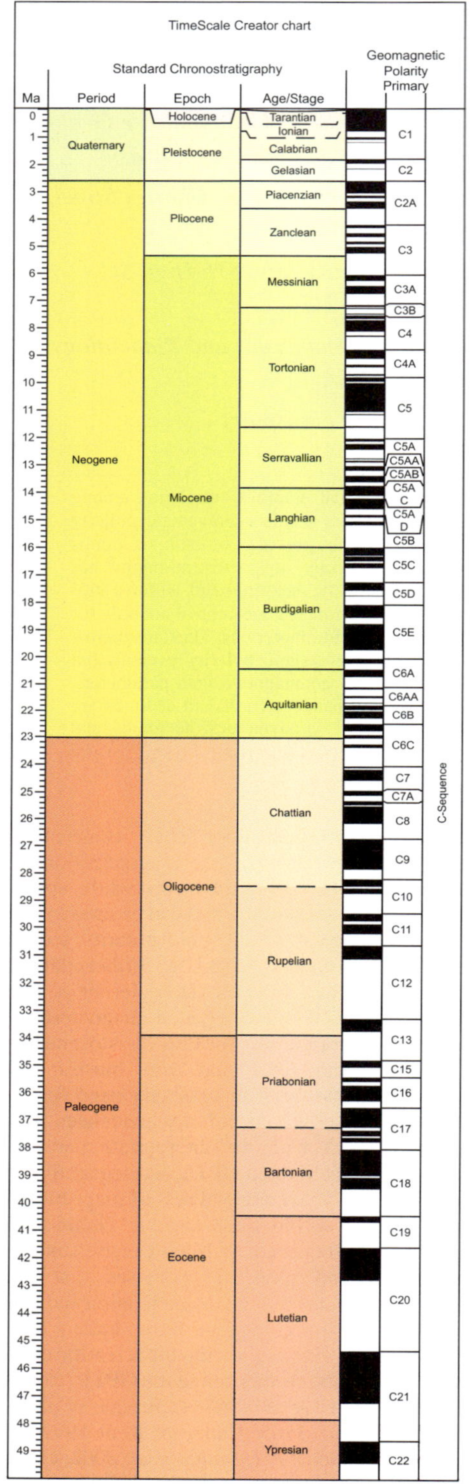

TimeScale Creator chart
Standard Chronostratigraphy
Geomagnetic Polarity Primary
Ma
Period
Epoch
Age/Stage
Holocene
Tarantian
Ionian
Calabrian
Gelasian
Quaternary
Pleistocene
Piacenzian
Zanclean
Pliocene
Messinian
Tortonian
Serravallian
Langhian
Burdigalian
Aquitanian
Neogene
Miocene
Chattian
Rupelian
Oligocene
Priabonian
Bartonian
Lutetian
Ypresian
Eocene
Paleogene
C1
C2
C2A
C3
C3A
C3B
C4
C4A
C5
C5A
C5AA
C5AB
C5AC
C5AD
C5B
C5C
C5D
C5E
C6A
C6AA
C6B
C6C
C7
C7A
C8
C9
C10
C11
C12
C13
C15
C16
C17
C18
C19
C20
C21
C22
C-Sequence

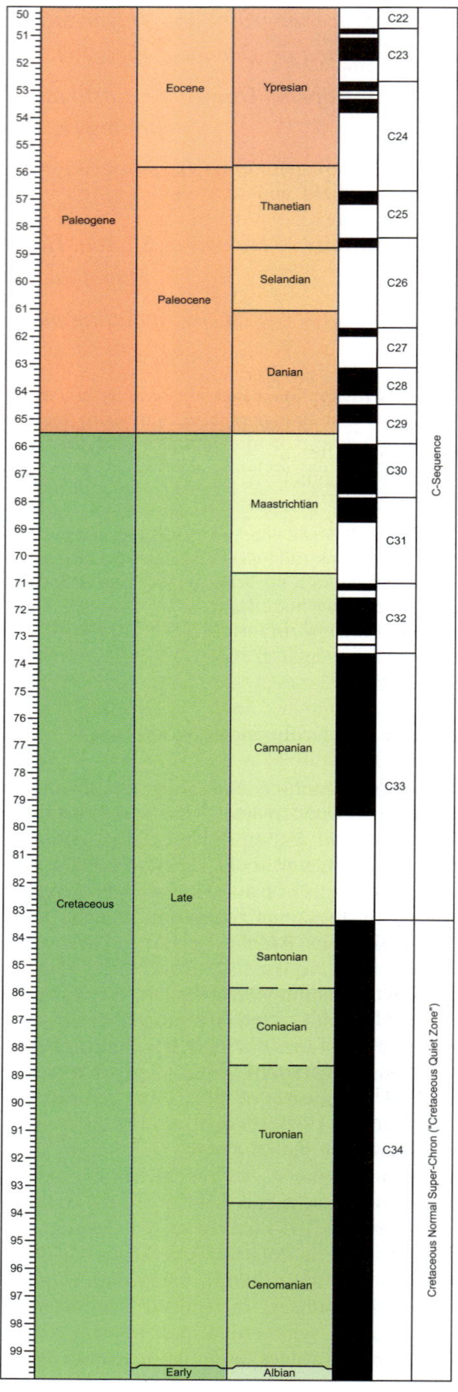

Eocene
Ypresian
Paleogene
Thanetian
Selandian
Paleocene
Danian
Cretaceous
Late
Maastrichtian
Campanian
Santonian
Coniacian
Turonian
Cenomanian
Early
Albian
C22
C23
C24
C25
C26
C27
C28
C29
C30
C31
C32
C33
C34
C-Sequence
Cretaceous Normal Super-Chron ("Cretaceous Quiet Zone")

Today, geomagnetic reversals are routinely recognized along stratigraphic sections composed of sedimentary (marine or continental) or volcanic materials. To determine a polarity stratigraphy, the sediment or rock must first contain a record of the geomagnetic field that is generally acquired at the time of emplacement. In order to measure the original magnetization that records the geomagnetic field polarity at the time of formation of the rock, sample direction must be measured in the field (i.e. *in situ*). The capacity of a rock to maintain its own magnetic field and resist demagnetization is related to the coercivity of its magnetic minerals. There are different ways by which rocks can record a natural remanent magnetization (NRM) in the presence of an external magnetic field (e.g. the geomagnetic field; Kodama 2012): (1) thermoremanent magnetization (TRM) is acquired when a rock cools down below the Curie temperature of its magnetic minerals; (2) chemical remanent magnetization (CRM) is acquired when a new magnetic mineral grows after the rock is formed and establishes its own magnetization; (3) viscous remanent magnetization (VRM) is attained in an ambient field for magnetic relaxation during time; (4) isothermal remanent magnetization (IRM) occurs in nature when rocks are struck by lightning and are submitted to a magnetic field larger that their coercivity; and, the most important in sediments, (5) detrital remanent magnetization (DRM) is acquired when depositional magnetic grains align themselves with the geomagnetic field as they are settling through the water column or are in unconsolidated sediment. Depositional magnetic grains deposited on the seafloor are then able to lock in and retain the original magnetization in the direction of the geomagnetic field during the initial consolidation of sediments (Tauxe *et al.* 2006; Tauxe & Yamazaki 2007). However, in some deep-sea sediments, a time delay of magnetization has been occasionally observed and is attributed to very low sedimentation rates and delayed lock-in below the sediment–water interface (Verosub 1977; Suganuma *et al.* 2011). This magnetization acquired near but not directly at the time of deposition is referred to as post-depositional remanent magnetization (pDRM). There are also some conditions in which magnetic crystals produced by magnetotactic bacteria may record the geomagnetic field (Petersen *et al.* 1986; Housen & Moskowitz 2006; Vasiliev *et al.* 2008; Yamazaki 2009; Roberts *et al.* 2011, 2012; Jovane *et al.* 2012).

The cause of the geomagnetic field reversals is still unknown, although geodynamical modelling demonstrates the occurrence of reversals (Glatzmaier & Roberts 1995; Kuang & Bloxham 1997). The main theory relates reversals to internal fluid instability of the Earth's outer core. In this region of the core, convective movements and complex vortices are created within tangent cylinders, that is, cylinders that are pretended coaxial movements in relation to the Earth's rotation axis, and tangent to the inner core/outer core boundary with their projection on Earth surface at 79.1° latitude. The thermal and compositional convection processes at the core-mantle boundary influence the geodynamo with long-term variations ($>10^5$ years) of intensity, inclination and declination (Olson *et al.* 2010). This organized motion evolves chaotically with the magnetic field produced by the electromagnetic dynamo growing, decaying and occasionally flipping in the opposite direction (Gubbins & Bloxham 1985, 1987; Bloxham & Jackson 1992; Olson & Aurnou 1999; Jackson *et al.* 2000; Hulot *et al.* 2002; Aurnou *et al.* 2003; Wardinski & Holme 2006). The geomagnetic reversals occur on the entire globe, also near the tangent cylinder and polar regions (Jovane *et al.* 2008). There are also other theories to explain reversals, for example, one in which geomagnetic reversals are linked to extraterrestrial impacts (Muller & Morris 1986).

Magnetostratigraphy and the geomagnetic polarity time scale

At its most fundamental level, magnetostratigraphy documents the geological record of polarity changes of the geomagnetic field. The individual normal (black) and reversed (white) (Fig. 1) polarity intervals are known as chrons and typically range in duration from 10 kyr to 10 Myr. The transition from a reversed-polarity chron to a normal-polarity chron and vice-versa is very short (± 5 kyr). This allows a numerical age to be assigned to each rock unit containing a polarity reversal within a stratigraphic succession. Since polarity reversals effectively occur simultaneously over the whole surface of the Earth, they can be used for global time correlation. Shorter periods of opposing polarity within a chron are called, depending on duration, 'subchrons,' 'microchrons' and 'cryptochrons' (Cande & Kent 1992). Longer periods with dominant single polarity are called 'superchrons' or 'megachrons,' also depending on duration (Opdyke & Channell 1996).

Chrons are conventionally labelled and named after the corresponding seafloor spreading magnetic

Fig. 1. Example of the GPTS for the past 100 million years, from *TSCreator 5.0* (www.tscreator.org). Courtesy of J. Ogg. (2012).

anomaly number (Cande & Kent 1992). Chron number is usually suffixed by the letter *n* or *r*, depending on whether the dominant magnetic polarity is normal or reversed, and is prefixed, for instance, by the letter C for Cenozoic, or M for Mesozoic (Gee & Kent 2007). The most recent reversal of the geomagnetic field occurred 781 kyr ago and is the boundary between Brunhes (C1n) and Matuyama (C1r) Chrons. Other normal and reversed chrons, Gauss (C2n) and Gilbert (C2r), contain distinctive subchrons, such as the Olduvai Subchron (C1r.2n) within the Matuyama Chron. The GPTS is continually updated to provide the most accurately known ages for the chron boundaries (e.g. Fig. 1).

Vector component diagrams (Zijderveld 1967) are used display stepwise demagnetization data. When a series of rock layers shows the same sign of inclination of the characteristic remanent magnetization (ChRM), it is called a magnetozone (or magnetostratigraphic unit), because within this interval a single polarity is constant. Magnetostratigraphy correlates magnetozones to the GPTS. Namely, inclination, declination and intensity of the ChRM for each sample are examined through a principal component analysis (PCA) of demagnetization steps (Kirschvink 1980). Demagnetization can be accomplished in different ways for different magnetic components. Stepwise thermal (TH) and alternating field (AF) demagnetizations are the most popular methods. Chemical and pressure demagnetization methods can also be applied to erase unwanted secondary components.

The calculated PCA inclination values sometimes do not represent the true inclination of the geomagnetic field at that time of the formation or deposition of the rock unit (Butler 1992; Tauxe 2010). This problem is related to geological factors that can be resolved using other magnetic methods. These factors may be related to diagenetic compaction or rotation of tectonic plates and tilt of structural blocks.

Beyond classical magnetostratigraphy

Dating with other recorded features of the geomagnetic field can also be undertaken, as follows.

Excursions and aborted reversals

It is very well known from palaeomagnetic records today that, in addition to polarity changes, the Earth's magnetic field has experienced changes from its regular near-axial configuration for brief periods of time without establishing, and perhaps not even approaching, a reversed state of the palaeofield. These types of behaviour are called geomagnetic excursions or 'aborted reversals'.

Such geomagnetic excursions have been reported in geological recorders such as lava flows of various ages in different parts of the world as well as from deep-sea, lake sediment and sedimentary rocks. This type of geomagnetic feature generally are observed to start with a sudden and often fairly smooth movement of the virtual geomagnetic poles (VGP) toward equatorial latitudes. The VGP may then return almost immediately, or it may cross the equator and move through latitudes in the opposite direction before travelling back again to resume a near-axial position. The term 'excursion' was defined to describe a VGP movement of more that $40°$ from the geographic pole for intermediate pole positions that end up with a return of the Earth's field to its pre-existing polarity (Barbetti & McElhinny 1976). During an excursion, the dynamo does not establish itself in the opposite polarity. Defined in this way, excursions are distinguished from secular variation (when the VGP colatitude is $10° < \theta < 40°$) and from short polarity episodes, which is a term applied when the opposite polarity (θ is $<40°$ or $>140°$) persists sufficiently long for at least one oscillation in the strength of the main dipole (about 104 years, Jacobs 1984).

Several short and almost complete changes in geomagnetic inclination have occurred within the present-day Brunhes Chron. These rapid and global geomagnetic events are called 'excursions' or 'aborted reversals' (Lund *et al.* 2006; Laj & Channell 2007). Among these excursions, the Laschamp (40–41 ka), Blake (*c.* 115–120 ka) and the Pringle Falls (*c.* 211–218 ka) are the most important geomagnetic events because they have been widely documented (e.g. Valet & Meynadier 1998; Guyodo & Valet 1999; Valet *et al.* 2008). Because excursion events have been recognized further back in time (e.g. Handschumacher *et al.* 1988; Sager *et al.* 1998; Tivey *et al.* 2006), we can infer that they occurred probably also during older chrons. These excursions, sometimes called tiny wiggles (e.g. Lanci & Lowrie 1997), cannot yet be used as time constrain.

Relative palaeointensity

The Earth's magnetic field can be simplified as a big dipole with the poles at the geographical poles, which is called the geocentric axial dipole (GAD). The behaviour of the geomagnetic field is complex and is related not only to the GAD but also to other components that change at different timescales: years (secular variation), millennia (excursions) and millions of years (reversals) (Gee & Kent 2007).

Several short and almost complete changes in geomagnetic inclination have occurred within the present-day Brunhes Chron. These rapid and global geomagnetic events are called 'excursions' or 'aborted reversals' (Lund *et al.* 2006; Laj & Channell 2007). An important long-term (10^5 year) variation of spherical parameters (inclination, declination and intensity) of the geomagnetic field is the secular variation (Valet *et al.* 2008), which includes geomagnetic jerks, westward drift and palaeointensity. Geomagnetic jerks are abrupt changes in one of the geomagnetic components related to inner core flow patterns or major earthquakes (Mandea *et al.* 2000; Florindo *et al.* 2005). The secular variation, which is not constant and uniform on the Earth, is mainly related today to a westward drift of about 0.2° per year since 1400 AD, and an eastward drift from 1000 to 1400 AD (Dumberry & Finlay 2007). The variation in intensity of the geomagnetic field through geological time is known as

palaeointensity. Measurements of palaeointensity in different geological sequences (e.g. lava flows, or marine or lake sediments) show a consistent pattern, allowing the reconstruction of a relative palaeointensity curve for the past 2 myr (Fig. 2; Sint-2000; Valet *et al.* 2005). Sint-2000 is a stack curve of independent palaeointensity records from various latitudes for the last 2 myr. It is possible to date a geological sequence by correlating the reference Sint-2000 intensity record to the palaeointensity record of the studied sequence. This is possible only when the magnetic minerals along the section are relatively uniform (King *et al.* 1983; Valet & Meynadier 1998), and when the magnetic intensity has been previously normalized for the concentration and grain-size of magnetic material (thus 'relative'), which may be climatically driven

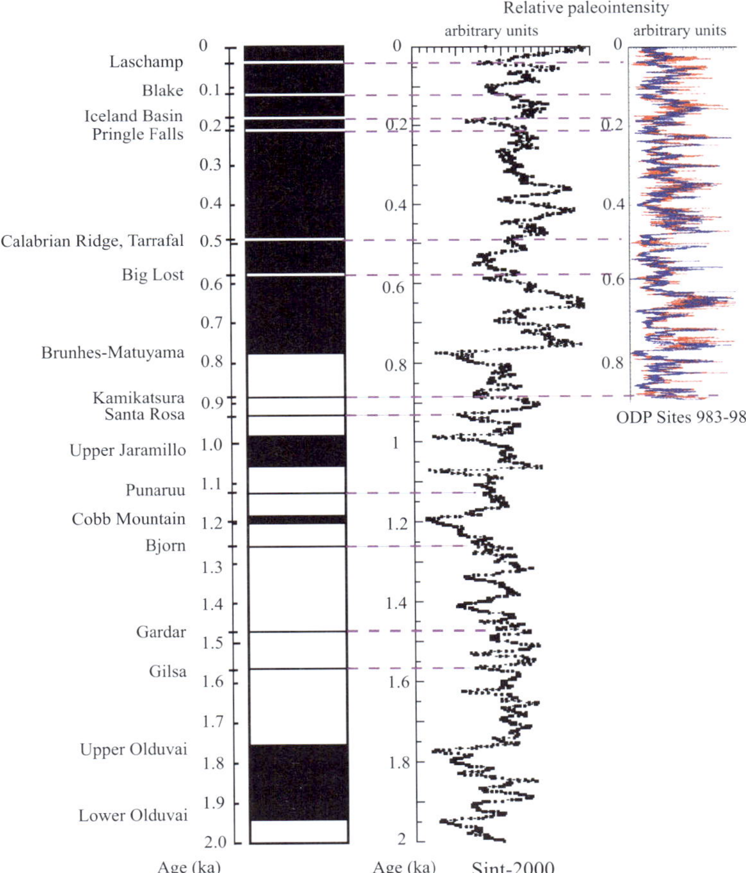

Fig. 2. Polarity column showing the position of the successive excursions and reversals with respect to the fluctuations of dipole field intensity derived from the composite Sint-2000 record and from ODP sites 983 (red line) and 984 (blue line) in the northeastern Atlantic Ocean (arbitrary units increasing toward right) (Guyodo & Valet 1999; Valet *et al.* 2008).

(e.g. Valet & Herrero-Barvera 2000; Channell *et al.* 2002; Yamazaki 2008). It is worth noting that severe declines in relative intensity coincide with geomagnetic reversals and excursions (Fig. 2; Valet *et al.* 2008).

Another application is the scatter of inclination and declination on a sphere, or angular dispersion (S) (e.g. Jovane *et al.* 2008). S is related to latitude, with higher values near the poles, and may indicate the geomagnetic stability of core vortices (Olson & Aurnou 1999; Aurnou *et al.* 2003).

Tectonics and deformation

While palaeomagnetic data have been used for a long time to examine plate-scale to regional/local tectonic and structural problems (e.g. Irving 1963, Beck 1981; Van der Voo 1988), a number of recent studies have combined magnetostratigraphic and palaeomagnetic analyses to examine timing and temporal variations in structural processes. This approach is made possible by the collection of more samples per site (or single geological bed) and more sites than is typical for standard magnetostratigraphic studies (e.g. Zhao *et al.* 2001; Liu *et al.* 2003; Titus *et al.* 2011). This technique has the potential to test velocity models of plate-margin and fault-zone deformation that are based primarily on Global Positioning System (GPS) data (e.g. McCaffrey 2005). GPS deformation studies utilize short (decades at most) time-series of GPS displacement data and their predictions of vertical-axis rotation can be compared with observations obtained from palaeomagnetic data (e.g. Titus *et al.* 2011).

Magnetic inclination and declination track the position of palaeopoles through geological time (Irving 1956). A reconstructed sequence of palaeopoles for a plate is known as an Apparent Polar Wander Path (APWP). We define the APWP through study of ChRM, which points to the VGP of the original position of a rock at the time it was formed. APWPs of the major tectonic plates have been reconstructed through much of Phanerozoic time (e.g. Besse & Courtillot 2002; Kearey *et al.* 2009), but there are still large uncertainties in the APWPs of minor plates and in reference poles (McElhinny & McFadden 2000).

Magnetic anisotropy studies can help define shallowing effects related to diagenetic compaction, flow direction during deposition or cooling and deformation of rocks owing to tectonic stresses. The larger datasets generated from these studies can be utilized to test and resolve other important aspects of how sedimentary rocks record the geomagnetic field. The inclination, declination and intensity of the DRM of sedimentary magnetic grains, and the TRM of lava flows, can be affected by geomagnetic components at the time of deposition or solidification (Tauxe 2010). Consequently, magnetic susceptibility may not be isotropic, giving different values when measured in different directions. In such cases, the anisotropy of magnetic susceptibility (AMS) is calculated as a tensor by comparing the magnetic susceptibility values in three perpendicular directions, which produce a matrix representing the ellipsoid of the magnetic susceptibility (e.g. Hrouda 1982). This ellipsoid of the magnetic fabric can be spherical, oblate (flattened) or prolate (cigar-shaped) providing information, for example, on the direction of a palaeocurrent (e.g. Ellwood & Whitney 1980) or lava flow (e.g. Stacey 1960; Ellwood 1978), strain patterns (e.g. Goldstein 1980), fabric/structure of granites (e.g. Ellwood & Whitney 1980) and the degree of compaction of strata studying the relation between inclination shallowing and oblation of the fabric (e.g. Tan & Kodama 2002).

Environmental magnetism

Environmental magnetism is used to identify and characterize magnetic mineralogy, and provides constraints for interpreting palaeomagnetic results and in understanding the factors that control environmental change. Environmental magnetic measurements are inexpensive, rapid and nondestructive, and provide fundamental information on the size, abundance and composition of magnetic minerals (Verosub & Roberts 1995; Kodama 2012; Liu *et al.* 2013). These measurements can be performed along a sedimentary sequence at high-resolution and include (1) magnetic susceptibility, and artificial remanences; (2) anhysteretic remanent magnetization (ARM); (3) isothermal remanent magnetization (IRM) and back-field isothermal remanent magnetization (BIRM); and (4) coercive-dependent parameters (S-ratio and 'hard' IRM referred to as HIRM).

Low-field magnetic susceptibility (χ = normalized for weight, κ = normalized for volume) represents how much magnetization (m) a material retains when a magnetic field (H) is applied (Table 1). Thus it is a complex parameter reflecting the sum of magnetic materials, such as ferromagnetic *sensu lato* (e.g. magnetite), antiferrimagnetic (e.g. hematite) and non-magnetic materials, such as paramagnetic (e.g. silicates or clays) and diamagnetic (e.g. quartz or carbonate). Results from low-field magnetic susceptibility must be interpreted with caution since it can be related to numerous processes.

Anhysteretic remanent magnetization activates only the finest magnetic minerals, frequently single domain (SD) grains, which do not have a domain wall and are uniformly magnetized. It is,

Table 1. *Units and conversions in the Système Internationale d'unités (SI) and the old CGS*

	CGS	SI
Energy	10^7 erg	1 joule (J)
Force (F)	10^5 dyne	1 Newton (N)
Current (I)	0.1 adamp	1 ampere (A)
Magentic Inducion (B)	10^4 gauss (G)	1 Tesla (T)
Magnetic Field (H)	$4\pi \times 10^{-3}$ oersted (Oe)	1 A m^{-1}
Magnetic Moment (M)	10^3 emu	1 A m^2
Magnetization/volume (m)	10^{-3} emu ($=$G cm^{-3})	1 A m^{-1}
Magnetization/mass (J)	1 emu g^{-1} or G cm^{-3} g^{-1}	1 A m^2 kg^{-1}
Magnetic susceptibility/volume (κ)	Dimensionless	Dimensionless
Magnetic susceptibility/mass (χ)	10^{-3} cm^3 g^{-1}	1 m^3 kg^{-1}

From Butler (1992) and Collinston (1983).

consequently, a proxy for the concentration of fine magnetite of eolian, biogenic or impactoclastic origin, or from other environmental processes. Isothermal remanent magnetization activates all magnetic grains up to saturation (also called SIRM or M_{rs}) that is usually at 1 T or 900 mT (Table 1). IRM, therefore, is the main proxy for magnetic concentration. The ratio between ARM and IRM (ARM/IRM$_{900}$) provides a proxy for magnetic grain size. Then, smaller IRMs can be applied backwards to IRM, which are called back-field IRM (BIRM), to produce IRM ratios that supply information on magnetic composition. These are the so-called S-ratios (King & Channell 1991) and HIRMs (imparted at 300 and 100 mT, respectively). They are calculated as, for example, S-ratio$_{300} =$ BIRM$_{300}$/IRM$_{900}$, S-ratio$_{100} =$ BIRM$_{100}$/IRM$_{900}$, HIRM$_{300} =$ (IRM$_{900} +$ BIRM$_{300}$)/2 or HIRM$_{100} =$ (IRM$_{900} +$ BIRM$_{100}$)/2 to indicate the different responses of high-coercivity magnetic minerals. The presence of hematite can be determined with the proxy IRM$_{900}$@AF120 mT   for the concentration of hematite by AF demagnetizing the IRM$_{900}$ step at 120 mT (Larrasoaña *et al.* 2003). Liu *et al.* (2007) also proposed the L-ratio (IRM$_{900} +$ BIRM$_{300}$)/(IRM$_{900} +$ BIRM$_{100}$) to define how the hardness of hematite can affect HIRM and the S-ratio.

Environmental magnetic records obtained by finely sampled measurements along sedimentary sequences can be evaluated as time series and related to palaeoclimatic and palaeoenvironmental change. In particular, palaeoclimate change at $10^5 - 10^6$ year scales – often resolved in the sedimentary record – has been strongly modulated by astronomically forced insolation (e.g. Berger 1988; Laskar *et al.* 2004, 2011; Table 2). The record of astronomical forcing frequencies in palaeomagnetic proxies can be used as a tool to perform astronomical tuning and develop continuous timescales along stratigraphic sequences (Hinnov & Hilgen 2012).

Organization of this special volume

Following three conference sessions entitled 'Magnetostratigraphy: not only a dating tool' and presented at the AGU Meeting of Americas (Foz de Iguaçu, Brazil) in 2010 and AGU Fall Meetings 2010, 2011 (San Francisco, USA), we decided to assemble the presentations into a Special Publication entitled *Magnetic Methods and the Timing of Geological Processes*. In this volume, we present research that includes innovations in classical magnetostratigraphy and emerging techniques that use sequential rock magnetic measurements to define chronology in stratigraphy. We introduce applications that use magnetic direction and geomagnetic polarity reversals to infer geodynamo processes, tectonics, diagenesis and climate change.

Table 2. *Periods and relative amplitudes of the main astronomical forcing cycles for 0–10 Ma (Laskar et al. 2004; Hinnov & Hilgen 2012). The frequencies and relative amplitudes change over different time intervals, and relative amplitudes change additionally as the result of climatic filtering*

Astronomical parameter	Period in kiloyears (relative amplitude)				
Orbital eccentricity	405 (1.0)	131 (0.43)	124 (0.56)	99 (0.48)	95 (0.74)
Obliquity (tilt)	53.04 (0.34)	40.80 (1.0)	40.11 (0.33)	39.45 (0.42)	29.75 (0.16)
Precession index	23.61 (1.0)	22.33 (0.87)	19.07 (0.45)	19.12 (0.69)	16.44 (0.08)

Finally, we present examples of astronomical forcing of environmental magnetism in the sedimentary record, and their utility in defining high-resolution timescales.

Part 1: integrated magnetostratigraphy

When integrated with biostratigraphy, cyclostratigraphy, chemostratigraphy and geochronology, magnetostratigraphy provides valuable timescale information. In this section, modern integrated magnetostratigraphy is demonstrated with marine Cenozoic studies. New Oligocene–Miocene magnetostratigraphy from the equatorial Pacific pinpoints the Oligocene–Miocene transition at the base of C6Cn.2n, and identifies three new excursions (**Guidry et al. 2012**). New work is presented on the Eocene climatic optimum interval (50–55 Ma) and Late Eocene–Oligocene transition into the Cenozoic icehouse (*c.* 33.5 Ma) from Integrated Ocean Drilling Progam sites and Italian sections (**Firth et al. 2012; Savian et al. 2013; Jovane et al. 2013**), and a composite integrated magnetostratigraphic sequence from Italy is provided for the entire Palaeogene Period (**Coccioni et al. 2012**).

Part 2: dating tectonic processes with magnetic methods

The evolution of palaeomagnetic declination in Chinese continental stratigraphy reveals block rotation, uplift and basin infilling related to the Cenozoic India–Asia collision, as discussed by three papers in this section (**Yan et al. 2012a, b; Fang et al. 2012**). The timing of these declination changes is constrained by magnetic reversal stratigraphy. A fourth paper (**Zhao et al. 2013**) analyses tectonic-driven sedimentation change in the Nankai Trough, Japan, with the assistance of integrated magnetostratigraphy.

Part 3: relative palaeointensity for dating geological sequences

In this section, the palaeomagnetic secular variation recovered from a Holocene Lake in Indonesia is compared with a recent geomagnetic model for 0–3 ka, and used to evaluate radiocarbon-based chronologies (**Haberzettl et al. 2012**). A new evaluation of the Pringle Falls lacustrine sequence confirms the clockwise loop of the VGP around the globe associated with the Pringle Falls Excursion (**Herrero-Bervera & Cañón-Tapia 2012**). The Laschamp, Skalamaelifell and Blake excursions, detected in sediments from offshore Queensland, Australia, are applied as chronostratigraphic constraints (**Herrero-Bervara & Jovane, this volume,**

in press). Finally, VGP trajectories measured on the Lower Cretaceous Serra Geral volcanic sequence are interpreted to indicate anisotropy in the Earth's interior (**Caminha-Maciel & Ernesto 2012**).

Part 4: palaeoclimatic change from rock magnetic proxies

The studies in this section resolve astronomical timescales of palaeoclimatic change using rock magnetism stratigraphy evaluated from a wide variety of marine environments: Plio-Pleistocene marginal marine, Cretaceous carbonate platform, Permian lower slope and basinal carbonates, Permo-Carboniferous glaciogenic rhythmites, and Ordovician shallow shelf limestone/mudstone. The Plio-Pleistocene study astronomically tunes a magnetic susceptibility series from a sedimentary section in northern Italy that otherwise exhibits no cyclicity (**Gunderson et al. 2012**). The Cretaceous study (**Hinnov et al., this volume, in press**) finds that ARM cyclicity is independent from host carbonate platform cycles in northeastern Mexico; the former is linked to a precession-forced eolian dust flux and the latter to much lower frequency sea-level fluctuations. **Ellwood et al. (2012a)** develop obliquity-tuned 'floating timescales' for magnetic susceptibility variations along Middle Permian sections, Texas, to assess sedimentation rates and durations of geological events. **Franco & Hinnov (2012)** evaluate anisotropy of magnetic susceptibility in Brazilian Permocarboniferous glacial rhythmites as a palaeoclimate indicator. Finally, the cyclic Ordovician Kope Formation, USA, is evaluated for astronomical frequencies through analysis of a composite high-resolution magnetic susceptibility series (**Ellwood et al. 2012b**).

We gratefully acknowledge the review of G. Acton, M. Ernesto, A. Malinverno and X. Zhao.

References

ALVAREZ, W., ARTHUR, M. A., FISCHER, A. G., LOWRIE, W., NAPOLEONE, G., PREMOLI-SILVA, I. & ROGGENTHEN, W. M. 1977. Upper Cretaceous-Paleocene magnetic stratigraphy at Gubbio, Italy. V. Type section for the late Cretaceous-Paleocene geomagnetic reversal time scale. *Geological Society of America Bulletin*, **88**, 383–389.

AURNOU, J. M., ANDREADIS, S., ZHU, L. & OLSON, P. L. 2003. Experiments on convection in Earth's core tangent cylinder. *Earth and Planetary Science Letters*, **212**, 119–134.

BARBETTI, M. F. & MCELHINNY, M. W. 1976. The Lake Mungo geomagnetic excursion. *Philosophical Transactions Royal Society of London*, **A281**, 515–540.

BECK, M. E., BURMESTER, R. F. & SCHOONOVER, R. 1981. Paleomagnetism and Tectonics of the

Cretaceous Mt. Stuart Batholith of Washington: translation or Tilt? *Earth and Planetary Science Letters*, **56**, 336–342.

BERGER, A. 1988. Milankovitch theory and climate. *Reviews of Geophysics*, **26**, 624–657.

BERGGREN, W. A., KENT, D. V., FLYNN, J. J. & VAN COUVERING, J. A. 1985. Cenozoic geochronology. *Geological Society of America Bulletin*, **96**, 1407–1418.

BESSE, J. & COURTILLOT, V. 2002. Apparent and true polar wander and the geometry of the geomagnetic field over the last 200 Myr. *Journal of Geophsical Research*, **107**, 2300.

BLOXHAM, J. & JACKSON, A. 1992. Time-dependent mapping of the magnetic field at the core–mantle boundary. *Journal of Geophysical Research*, **97**, 19 537–19 563.

BRUNHES, B. 1906. Recherches sur la direction d'aimantation des roches volcaniques. *Journal of Physics IV*, **5**, 705–724.

BUTLER, R. F. 1992. *Palaeomagnetism, Magnetic Domains to Geologic Terranes*. Blackwell Scientific, Boston, MA.

CAMINHA-MACIEL, G. & ERNESTO, M. 2013. Characteristic wavelengths in VGP trajectories from magnetostratigraphic data of the Early Cretaceous Serra Geral lava piles, southern Brazil. *In:* JOVANE, L., HERRERO-BERVERA, E., HINNOV, L. A. & HOUSEN, B. A. (eds) *Magnetic Methods and the Timing of Geological Processes*. Geological Society, London, Special Publications, 373, first published on March 25, 2013, doi: 10.1144/SP373.15.

CANDE, S. C. & KENT, D. V. 1992. A new geomagnetic timescale for the Late Cretaceous and Cenozoic. *Journal of Geophysical Research*, **97**, 13 917–13 951.

CANDE, S. C. & KENT, D. V. 1995. Revised calibration of the Geomagnetic Polarity Time Scale for the Late Cretaceous and Cenozoic. *Journal of Geophysical Research*, **100**, 6093–6095.

CHANNELL, J. E. T., ERBA, E., NAKANISHI, M. & TAMAKI, K. 1995. Late Jurassic–Early Cretaceous time scales and oceanic magnetic anomaly block models. *In:* BERGGREN, W. A. *ET AL.* (eds) *Geochronology, Timescales, and Stratigraphic Correlation*. Society for Sedimentary Geolists, Tulsa, OK, 51–63.

CHANNELL, J. E. T., MAZAUD, A., SULLIVAN, P., TURNER, S. & RAYMO, M. E. 2002. Geomagnetic excursions and paleointensities in the 0.9–2.15 Ma interval of the Matuyama Chron at ODP Site 983 and 984 (Iceland Basin). *Journal of Geophysical Research*, **107**, doi: 10.1029/2001JB000491.

COCCIONI, R., SIDERI, M. *ET AL.* 2012. Integrated stratigraphy (magneto-, bio- and chronostratigraphy) and geochronology of the Palaeogene pelagic succession of the Umbria–Marche Basin (central Italy). *In:* JOVANE, L., HERRERO-BERVERA, E., HINNOV, L. A. & HOUSEN, B. A. (eds) *Magnetic Methods and the Timing of Geological Processes*. Geological Society, London, Special Publications, 373, first published on November 16, 2012, doi: 10.1144/SP373.4.

COLLINSON, A. W. 1983. *Methods in Rock Magnetism and Paleomagnetism. Techniques and Instrumentation*. Chapman and Hall, New York.

COX, A., DOELL, R. R. & DALRYMPLE, G. B. 1963. Geomagnetic polarity epochs and Pleistocene geochronometry. *Nature*, **198**, 1049–1051.

DUMBERRY, M. & FINLAY, C. C. 2007. Eastward and westward drift of the Earth's magnetic field for the last three millennia. *Earth and Planetary Science Letters*, **254**, 146–157.

ELLWOOD, B. B. 1978. Flow and emplacement direction determined for selected basaltic bodies using magnetic susceptibility anisotropy measurements. *Earth Planetary Science Letters*, **41**, 254–264.

ELLWOOD, B. B. & WHITNEY, J. A. 1980. Magnetic fabric of the Elberton granite, Northeast Georgia. *Journal of Geophysical Research*, **85**, 1481–1486.

ELLWOOD, B. B., LAMBERT, L. L. *ET AL.* 2012*a*. Magnetostratigraphy susceptibility for the Guadalupian series GSSPs (Middle Permian) in Guadalupe Mountains National Park and adjacent areas in West Texas. *In:* JOVANE, L., HERRERO-BERVERA, E., HINNOV, L. A. & HOUSEN, B. A. (eds) *Magnetic Methods and the Timing of Geological Processes*. Geological Society, London, Special Publications, **373**, first published on August 14, 2012, doi: 10.1144/SP373.1.

ELLWOOD, B. B., BRETT, C. E., TOMKIN, J. H. & MACDONALD, W. D. 2012*b*. Visual identification and quantification of Milankovitch climate cycles in outcrop: an example from the Upper Ordovician Kope Formation, Northern Kentucky. *In:* JOVANE, L., HERRERO-BERVERA, E., HINNOV, L. A. & HOUSEN, B. A. (eds) *Magnetic Methods and the Timing of Geological Processes*. Geological Society, London, Special Publications, **373**, first published on August 14, 2012, doi: 10.1144/SP373.2.

FANG, X., LIU, D., SONG, C., DAI, S. & MENG, Q. 2012. Oligocene slow and Miocene–Quaternary rapid deformation and uplift of the Yumu Shan and North Qilian Shan: evidence from high-resolution magnetostratigraphy and tectonosedimentology. *In:* JOVANE, L., HERRERO-BERVERA, E., HINNOV, L. A. & HOUSEN, B. A. (eds) *Magnetic Methods and the Timing of Geological Processes*. Geological Society, London, Special Publications, **373**, first published on November 15, 2012, doi: 10.1144/SP373.5.

FIRTH, J. V., ELDRETT, J. S., HARDING, I. C., COXALL, H. K. & WADE, B. S. 2012. Integrated biomagnetochronology for the Palaeogene of ODP Hole 647A: implications for correlating palaeoceanographic events from high to low latitudes. *In:* JOVANE, L., HERRERO-BERVERA, E., HINNOV, L. A. & HOUSEN, B. A. (eds) *Magnetic Methods and the Timing of Geological Processes*. Geological Society, London, Special Publications, **373**, first published on October 1, 2012, doi: 10.1144/SP373.9.

FLORINDO, F., DE MICHELIS, P., PIERSANTI, A. & BOSCHI, E. 2005. Could the Mw = 9.3 Sumatra Earthquake Trigger a Geomagnetic Jerk? *EOS, Transactions of the American Geophysical Union*, **86**, 123–124.

FRANCO, D. R. & HINNOV, L. A. 2012. Anisotropy of magnetic susceptibility and sedimentary cycle data from Permo-Carboniferous rhythmites (Paraná Basin, Brazil): a multiple proxy record of astronomical and millennial scale palaeoclimate change in a glacial setting. *In:* JOVANE, L., HERRERO-BERVERA, E., HINNOV, L. A. & HOUSEN, B. A. (eds) *Magnetic Methods and the Timing of Geological Processes*. Geological Society, London, Special Publications, **373**, first published on November 6, 2012, doi: 10.1144/SP373.11.

GEE, J. S. & KENT, D. V. 2007. Source of oceanic magnetic anomalies and the geomagnetic polarity time scale. Geomagnetism. *In*: KONO, M. (ed.) *Treatise on Geophysics*. Elsevier, Amsterdam, **5**, 455–507.

GLATZMAIER, G. A. & ROBERTS, P. H. 1995. A three-dimensional convective dynamo solution with rotating and finitely conducting inner core and mantle. *Physics of the Earth and Planetary Interiors*, **91**, 63–75.

GOLDSTEIN, A. G. 1980. Magnetic susceptibility anisotropy of mylonites from the Lake Char mylonite zone, southeastern New England. *Tectonophysics*, **66**, 197–211.

GUBBINS, D. & BLOXHAM, J. 1985. Geomagnetic field analysis – III. Magnetic fields on the core–mantle boundary. *Geophysical Journal of the Royal Astronomical Society*, **80**, 695–713.

GUBBINS, D. & BLOXHAM, J. 1987. Morphology of the geomagnetic field and implications for the geodynamo. *Nature*, **325**, 509–511.

GUIDRY, E. P., RICHTER, C. *ET AL.* 2012. Oligocene–Miocene magnetostratigraphy of deep-sea sediments from the equatorial Pacific (IODP Site U1333). *In*: JOVANE, L., HERRERO-BERVERA, E., HINNOV, L. A. & HOUSEN, B. A. (eds) *Magnetic Methods and the Timing of Geological Processes*. Geological Society, London, Special Publications, **373**, first published on August 17, 2012, doi: 10.1144/SP373.7.

GUNDERSON, K. L., KODAMA, K. P., ANASTASIO, D. J. & PAZZAGLIA, F. J. 2012. Rock-magnetic cyclostratigraphy for the Late Pliocene–Early Pleistocene Stirone section, Northern Apennine mountain front, Italy. *In*: JOVANE, L., HERRERO-BERVERA, E., HINNOV, L. A. & HOUSEN, B. A. (eds) *Magnetic Methods and the Timing of Geological Processes*. Geological Society, London, Special Publications, **373**, first published on August 14, 2012, doi: 10.1144/SP373.8.

GUYODO, Y. & VALET, J.-P. 1999. Global changes in intensity of the Earth's magnetic field during the past 800 kyr. *Nature*, **399**, 249–252.

HABERZETTL, T., ST-ONGE, G., BEHLING, H. & KIRLEIS, W. 2012. Evaluating Late Holocene radiocarbon-based chronologies by matching palaeomagnetic secular variations to geomagnetic field models: an example from Lake Kalimpaa (Sulawesi, Indonesia). *In*: JOVANE, L., HERRERO-BERVERA, E., HINNOV, L. A. & HOUSEN, B. A. (eds) *Magnetic Methods and the Timing of Geological Processes*. Geological Society, London, Special Publications, **373**, first published on August 14, 2012, doi: 10.1144/SP373.10.

HANDSCHUMACHER, D. W., SAGER, W. W., HILDE, T. W. C. & BRACEY, D. R. 1988. Pre-Cretaceous tectonic evolution of the Pacific plate and extension of the geomagnetic polarity reversal time scale with implications for the origin of the Jurassic 'Quiet Zone'. *Tectonophysics*, **155**, 365–380.

HEIRTZLER, J. R., DICKSON, G. O., HERRON, E. M., PITMAN, W. C. III. & LE PICHON, X. 1968. Marine Magnetic Anomalies, geomagnetic field reversals, and motions of the ocean floor and continents. *Journal of Geophysical Research*, **73**, 2119–2136.

HERRERO-BERVERA, E. & CAÑÓN-TAPIA, E. 2012. On the directional geomagnetic signature of the Pringle Falls excursion recorded at Pringle Falls, Oregon, USA. *In*: JOVANE, L., HERRERO-BERVERA, E., HINNOV, L. A. & HOUSEN, B. A. (eds) *Magnetic Methods and the Timing of Geological Processes*. Geological Society, London, Special Publications, **373**, first published on December 7, 2012, doi: 10.1144/SP373.12.

HERRERO-BERVARA, E. & JOVANE, A. In press. On the palaeomagnetic and rock magnetic constraints regarding the age of IODP 325 Hole M0058A. *In*: JOVANE, L., HERRERO-BERVERA, E., HINNOV, L. A. & HOUSEN, B. A. (eds) *Magnetic Methods and the Timing of Geological Processes*. Geological Society, London, Special Publications, **373**, http://dx.doi.org/10.1144/SP373.19

HINNOV, L. A. & HILGEN, F. 2012. Chapter 4: cyclostratigraphy and astrochronology. *In*: GRADSTEIN, F. M., OGG, J. G., SCHMITZ, M. D. & OGG, G. M. (eds) *The Geologic Time Scale 2012*. Elsevier, Amsterdam, 63–83.

HINNOV, L., KODAMA, K. P., ANASTASIO, D. J., ELRICK, M. & LATTA, D. K. In press. Global Milankovitch cycles recorded in rock magnetism of the shallow marine lower Cretaceous Cupido Formation, northeastern Mexico. *In*: JOVANE, L., HERRERO-BERVERA, E., HINNOV, L. A. & HOUSEN, B. A. (eds) *Magnetic Methods and the Timing of Geological Processes*. Geological Society, London, Special Publications, **373**, http://dx.doi.org/10.1144/SP373.20

HOUSEN, B. A. & MOSKOWITZ, B. M. 2006. Depth distribution of magnetofossils in near surface sediments from the Blake/Bahama Outer Ridge, western North Atlantic Ocean, determined by low-temperature magnetism. *Journal of Geophysical Research*, **111**, G01005.

HROUDA, F. 1982. Magnetic anisotropy of rocks and its application in geology and geophysics. *Geophysical Surveys*, **5**, 37–82.

HUESTIS, S. P. & ACTON, G. D. 1997. On the construction of geomagnetic timescales from nonprejudicial treatment of magnetic anomaly data from multiple ridges. *Geophysical Journal International*, **129**, 176–182.

HULOT, G., EYMIN, C., LANGLAIS, B., MANDEA, M. & OLSEN, N. 2002. Small-scale structure of the geodynamo inferred from Oersted and Magsat satellite data. *Nature*, **416**, 620–623.

IRVING, E. 1956. Palaeomagnetic and palaeoclimatic aspects of polar wandering. *Geofisica pura e applicata*, **33**, 23–41.

IRVING, E. 1963. Paleomagnetism of the Narrabeen Chocolate Shales and the Tasmanian Dolerite. *Journal of Geophysical Research*, **68**, 2283–2287.

JACKSON, A., JONKERS, A. R. T. & WALKER, M. R. 2000. Four centuries of geomagnetic secular variation from historical records. *Philosophical Transactions of the Royal Society of London Series A*, **358**, 957–990.

JACOBS, J. A. 1984. *Reversals of the Earth's Magnetic field*. Adam Hilger, Bristol.

JOVANE, L., ACTON, G., FLORINDO, F. & VEROSUB, K. L. 2008. Geomagnetic field behavior at high latitudes from a paleomagnetic record from Eltanin Core 27–21 in the Ross Sea Sector, Antarctica. *Earth and Planetary Science Letters*, **263**, 435–443.

JOVANE, L., FLORINDO, F., BAZYLINSKI, D. A. & LINS, U. 2012. Prismatic magnetite magnetosomes from cultivated Magnetovibrio blakemorei strain MV-1: a magnetic fingerprint in marine sediments? *Environmental Microbiology Reports*, **4**, 664–668, doi: 10.1111/1758- 2229.12000

JOVANE, L., SAVIAN, J. ET AL. 2013. Integrated magneto-biostratigraphy of the middle Eocene-lower Oligocene interval from the Monte Cagnero section, central Italy. *In*: JOVANE, L., HERRERO-BERVERA, E., HINNOV, L. A. & HOUSEN, B. A. (eds) *Magnetic Methods and the Timing of Geological Processes*. Geological Society, London, Special Publications, **373**, first published on March 25, 2013, doi: 10.1144/SP373.13.

KEAREY, P., KLEPEIS, K. A. & VINE, F. J. 2009. *Global Tectonics*. 3rd edn. Wiley, Chichester.

KING, J. & CHANNELL, J. E. T. 1991. Sedimentary magnetism, environmental magnetism, and magnetostratigraphy. U.S. National Report to the International Union of Geodesy and Geophysics. *Reviews of Geophysics*, Supplement, **29**, 358–370.

KING, J. W., BANERJEE, S. K. & MARVIN, J. 1983. A new rock-magnetic approach to selecting sediments for geomagnetic paleointensity studies: application to paleointensity for the last 4000 years. *Journal of Geophysical Research*, **88**, 5911–5921.

KIRSCHVINK, J. L. 1980. The least-squares line and plane and the analysis of palaeomagnetic data. *Geophysical Journal of the Royal Astronomical Society*, **62**, 699–710.

KODAMA, K. 2012. *Paleomagnetism of Sedimentary Rocks: Process and Interpretation*. Wiley-Blackwell, Oxford.

KUANG, , W. & BLOXHAM, J. 1997. An Earth-like numerical dynamo model. *Nature*, **389**, 371–374.

LABREQUE, J. L., KENT, D. V. & CANDE, S. C. 1977. Revised magnetic polarity time scale for Late Cretaceous and Cenozoic time. *Geology*, **5**, 330–335.

LAJ, C. & CHANNELL, J. E. T. 2007. Geomagnetic excursions. *In*: KONO, M. (ed.) *Treatise in Geophysics: Volume 5, Geomagnetism*. Elservier, Amsterdam, Chapter 10, 373–416.

LANCI, L. & LOWRIE, W. 1997. Magnetostratigraphic evidence that 'tiny wiggles' in the oceanic magnetic anomaliy record represent geomagnetic paleointensity variations. *Earth and Planetary Science Letters*, **148**, 581–592.

LARRASOAÑA, J. C., ROBERTS, A. P., ROHLING, E. J., WINKLHOFER, M. & WEHAUSEN, R. 2003. Three million years of monsoon variability over the northern Sahara. *Climatic Dynamics*, **21**, 689–698.

LASKAR, J., ROBUTEL, P., JOUTEL, J., GASTINEAU, M., CORREIA, A. C. M. & LEVRARD, B. 2004. A numerical solution for the insolation quantities of the Earth. *Astronomy and Astrophysics*, **428**, 261–285.

LASKAR, J., FIENGA, A., GASTINEAU, M. & MANCHE, H. 2011. A new orbital solution for the long-term motion of the Earth. *Astronomy and Astrophysics.*, **532**, A89.

LIU, Z., ZHAO, X., CHENGSHAN, W., SHUN, L. & HAISHEN, Y. 2003. Magnetostratigraphy of Tertiary sediments from the Hoh Xil Basin: implications for the Cenozoic tectonic history of the Tibetan Plateau. *Geophysical Journal International*, **154**, 233–252.

LIU, Q. S., ROBERTS, A. P., TORRENT, J. & HORNG, C. S. 2007. What do the HIRM and S-ratio really measure in environmental magnetism? *Geochemistry Geophysics Geosystems*, **8**, Q09011.

LIU, Q. S., ROBERTS, A. P., LARRASOAÑA, J. C., BANERJEE, S. K., GUYODO, Y., TAUXE, L. & OLDFIELD, F. 2013. Environmental Magnetism: principles and applications. *Reviews of Geophysics*, **50**, RG4002.

LOWRIE, W., ALVEREZ, W., NAPOLEONE, G., PERCH-NIELSON, K., PREMOLI SILVA, I. & TOUMARINE, M. 1982. Paleogene magnetic stratigraphy in Umbrian pelagic carbonate rocks: the Contessa sections, Gubbio. *Geological Society of America Bulletin*, **93**, 414–432.

LUND, S., STONER, J. S., CHANNELL, J. E. T. & ACTON, G. 2006. A summary of Brunhes paleomagnetic field variability recorded in Ocean Drilling Program cores. *Physics of the Earth and Planetary Interiors*, **156**, 194–204.

MALINVERNO, A., HILDEBRANDT, J., TOMINAGA, M. & CHANNELL, J. E. T. 2012. M-sequence geomagnetic polarity time scale (MHTC12) that steadies global spreading rates and incorporates astrochronology constraints. *Journal of Geophysical Research*, **117**, B06104.

MANDEA, M., BELLANGER, E. & LE MOUL, J. L. 2000. A geomagnetic jerk for the end of the 20th century? *Earth and Planetary Science Letters*, **183**, 369–373.

MATUYAMA, M. 1926. On the direction of magnetization of basalt in Japan. *Proceedings of the Imperial Academy*, **5**, 203–205.

MCCAFFREY, R. 2005. Block kinematics of the Pacific–North America plate boundary in the southwestern US from inversion of GPS, seismological, and geologic data. *Journal of Geophysical Research*, **110**, B07401.

MCELHINNY, M. & MCFADDEN, P. 2000. *Paleomagnetism: Continents and Oceans*. International Geophysics Series, **73**, Academic Press, San Diego.

MULLER, R. A. & MORRIS, D. E. 1986. Geomagnetic reversals from impacts on the Earth. *Geophysical Research Letters*, **13**, 1177–1180.

OGG, J. G. 2012. Chapter 5: geomagnetic polarity time scale. *In*: GRADSTEIN, F. M., OGG, J. G., SCHMITZ, M. D. & OGG, G. M. (eds) *The Geologic Time Scale 2012*. Elsevier, Amsterdam, 85–113.

OLSON, P. & AURNOU, P. 1999. A polar vortex in the Earth's core. *Nature*, **402**, 170–173.

OLSON, P. L., COE, R. S., DRISCOLL, P. E., GLATZMAIER, G. A. & ROBERTS, P. H. 2010. Geodynamo reversal frequency and heterogeneous core-mantle boundary heat flow. *Physics of the Earth and Planetary Interiors*, **180**, 66–79.

OPDYKE, N. D. 1972. Palaeomagnetism of deep-sea cores. *Reviews of Geophysics and Space Physics*, **10**, 213–249.

OPDYKE, N. D. & CHANNELL, J. E. T. 1996. *Magnetic Stratigraphy*. Academic Press, San Diego, CA.

PETERSEN, N., VON DOBENECK, T. & VALI, H. 1986. Fossil bacterial magnetite in deep-sea sediments from the South Atlantic Ocean. *Nature*, **320**, 611–615.

ROBERTS, A. P., FLORINDO, F. ET AL. 2011. Magnetotactic bacterial abundance in pelagic marine environments is limited by organic carbon flux and availability of dissolved iron. *Earth and Planetary Science Letters*, **310**, 441–452.

ROBERTS, A. P., CHANG, L., HESLOP, D., FLORINDO, F. & LARRASOAÑA, J. C. 2012. Searching for single domain magnetite in the 'pseudo-single-domain' sedimentary haystack: implications of biogenic magnetite preservation for sediment magnetism and relative paleointensity determinations. *Journal of Geophysical Research*, **117**, B08104.

SAGER, W. W., WEISS, M. A., TIVEY, M. A. & JOHNSON, H. P. 1998. Geomagnetic polarity reversal model of deep-tow profiles from the Pacific Jurassic 'Quiet Zone'. *Journal of Geophysical Research*, **103**, 5269–5286.

SAVIAN, J., JOVANE, L., BOHATY, S. & WILSON, P. 2013. Middle Eocene to early Oligocene magnetostratigraphy of ODP Hole 711A (Leg 115), western equatorial Indian Ocean. *In*: JOVANE, L., HERRERO-BERVERA, E., HINNOV, L. A. & HOUSEN, B. A. (eds) *Magnetic Methods and the Timing of Geological Processes*. Geological Society, London, Special Publications, **373**, first published on March 25, 2013, doi: 10.1144/SP373.16.

SINGER, B. S., RELLE, M. R., HOFFMAN, K. A., BATTLE, A., GUILLOU, H., LAJ, C. & CARRACEDO, J. C. 2002. Ar/Ar ages of transitionally magnetized lavas on La Palma, Canary Islands, and the Geomagnetic Instability Timescale. *Journal of Geophysical Research*, **107**, 2307.

STACEY, F. D. 1960. Magnetic anisotropy of igneous rocks. *Journal of Geophysical Research*, **65**, 2429–2442.

SUGANUMA, Y., OKUNO, J., HESLOP, D., ROBERTS, A. P., YAMAZAKI, T. & YOKOYAMA, Y. 2011. Post-depositional remanent magnetization lock-in for marine sediments deduced from ^{10}Be and paleomagnetic records through the Matuyama-Brunhes boundary. *Earth and Planetary Science Letters*, **311**, 39–52, 2011.

TAN, X. & KODAMA, K. P. 2002. Magnetic anisotropy and paleomagnetic inclination shallowing in red beds: evidence from the Mississippian Mauch Chunk Formation, Pennsylvania. *Journal of Geophysical Research*, **107**, 2311.

TAUXE, L. 2010. *Essentials of Paleomagnetism*. University of California, Berkeley.

TAUXE, L. & YAMAZAKI, T. 2007. Paleointensities. *In*: SCHUBERT, G. (ed.) *Treatise on Geophysics*. Elsevier, Amsterdam.

TAUXE, L., STEINDORF, J. L. & HARRIS, A. J. 2006. Depositional remanent magnetization: toward an improved theoretical and experimental foundation. *Earth and Planetary Science Letters*, **244**, 515–529.

TITUS, S. J., CRUMP, S., McGUIRE, Z., HORSMAN, E. & HOUSEN, B. 2011. Using vertical axis rotations to characterize off-fault deformation across the San Andreas fault system, central California. *Geology*, **39**, 711–714.

TIVEY, M. A., SAGER, W. W., LEE, S.-M. & TOMINAGA, M. 2006. Origin of the Pacific Jurassic quiet zone. *Geology*, **34**, 789–792.

VALET, J.-P. & MEYNADIER, L. 1998. A comparison of different techniques for relative paleointensity. *Geophysical Research Letters*, **25**, 89–92.

VALET, J.-P. & HERRERO-BERVERA, E. 2000. Paleointensity experiments using alternating field demagnetization. *Earth and Planetary Science Letters*, **177**, 43–58.

VALET, J. P., MEYNADIER, L. & GUYODO, Y. 2005. Geomagnetic field strength and reversal rate over the past 2 Million years. *Nature*, **435**, 802–805.

VALET, J. P., PLENIER, G. & HERRERO-BERVERA, E. 2008. Geomagnetic excursions reflect an aborted polarity state. *Earth and Planetary Science Letters*, **274**, 472–478.

VAN DER VOO, R. 1988. Paleozoic paleogeography of North America, Gondwana, and intervening displaced terranes: comparisons of paleomagnetism with paleoclimatology and biogeographical patterns. *Geological Society of America Bulletin*, **100**, 311–324.

VASILIEV, I., FRANKE, C., MEELDIJK, J. D., DEKKERS, M. J., LANGEREIS, C. G. & KRIJGSMAN, W. 2008. Putative greigite magnetofossils from the Piocene epoch. *Nature Geoscience*, **1**, 782–786.

VEROSUB, K. L. 1977. Paleomagnetic record of Clear Lake, California. *Earth and Planetary Science Letters*, **36**, 219–230.

VEROSUB, K. L. & ROBERTS, A. P. 1995. Environmental magnetism: past, present, and future. *Journal of Geophysical Research*, **100**, 2175–2192.

VINE, F. J. & MATTHEWS, D. H. 1963. Magnetic anomalies over oceanic ridges. *Nature*, **199**, 947–949.

WARDINSKI, I. & HOLME, R. 2006. A time-dependent model of the Earth's magnetic field and its secular variation for the period 1980–2000. *Journal of Geophysical Research*, **111**, B12101, doi: 10.1029/2006JB004401.

YAMAZAKI, T. 2008. Magnetostatic inter actions in deep-sea sediments inferred from first-order reversal curve diagrams: implications for relative paleointensity normalization. *Geochemistry, Geophysics, Geosystems*, **9**, Q02005.

YAMAZAKI, T. 2009. Environmental magnetism of Pleistocene sediments in the North Pacific and Ontong-Java Plateau: temporal variations of detrital and biogenic components. *Geochemistry, Geophysics, Geosystems*, **10**, Q07Z04.

YAN, M., VAN DER VOO, R., FANG, X.-M. & SONG, C. 2012*a*. Magnetostratigraphy, fence diagrams and basin analysis. *In*: JOVANE, L., HERRERO-BERVERA, E., HINNOV, L. A. & HOUSEN, B. A. (eds) *Magnetic Methods and the Timing of Geological Processes*. Geological Society, London, Special Publications, **373**, first published on August 17, 2012, doi: 10.1144/SP373.3

YAN, M., FANG, X, VAN DER VOO, R., SONG, C. & LI, J. 2012*b*. Neogene rotations in the Jiuquan Basin, Hexi Corridor, China. *In*: JOVANE, L., HERRERO-BERVERA, E., HINNOV, L. A. & HOUSEN, B. A. (eds) *Magnetic Methods and the Timing of Geological Processes*. Geological Society, London, Special Publications, **373**, first published on November 16, 2012, doi: 10.1144/SP373.6.

ZHAO, X., LADNER, B., ROESSIG, K., WISE, S. & URQUHART, E. 2001. Magnetic stratigraphy and biostratigraphy of cenozoic sediments recovered from the Iberian Abyssal Plain. *In*: BESLIER, M.-O., WHITMARSH, R. B., WALLACE, P. J. & GIRARDEAU, J. (eds) *Proceedings of the Ocean Drilling Program, Scientific Results*, **172**, Ocean Drilling Program, Texas A & M University, College Station, TX, 1–73 [CD-ROM].

ZHAO, X., ODA, H. ET AL. 2013. Magnetostratigraphic results from sedimentary rocks of IODP's Nankai Trough Seismogenic Zone Experiment (NanTroSEIZE) Expedition 322. *In*: JOVANE, L., HERRERO-BERVERA, E., HINNOV, L. A. & HOUSEN, B. A. (eds) *Magnetic Methods and the Timing of Geological Processes*. Geological Society, London, Special Publications, **373**, first published on March 25, 2013, doi: 10.1144/SP373.14.

ZIJDERVELD, J. D. A. 1967. A.C. demagnetization of rocks. *In*: COLLISON, D. W., CREER, K. M. & RUNCORN, S. K. (eds) *Methods in Palaeomagnetism*. Elsevier, Amsterdam, 256–286.

Oligocene–Miocene magnetostratigraphy of deep-sea sediments from the equatorial Pacific (IODP Site U1333)

E. P. GUIDRY[1]*, C. RICHTER[1], G. D. ACTON[2], J. E. T. CHANNELL[3], H. F. EVANS[4], C. OHNEISER[5], Y. YAMAMOTO[6] & T. YAMAZAKI[7]

[1]*School of Geosciences, University of Louisiana at Lafayette, PO Box 44530, Lafayette, LA 70504, USA*

[2]*Department of Geology, University of California, Davis, One Shields Avenue, Davis, CA 95616, USA*

[3]*Department of Geological Sciences, University of Florida, PO Box 112120, Gainesville, FL 32611-2120, USA*

[4]*LDEO, PO Box 1000, Palisades, NY 10964-8000, USA*

[5]*Geology Department, University of Otago, PO Box 56, Dunedin 9054, New Zealand*

[6]*Center for Advanced Marine Core Research, Kochi University, B200 Monobe, Nankoku, Kochi 783-8502, Japan*

[7]*Geological Survey of Japan, AIST Tsukuba Central 7, Tsukuba 305-8567, Japan*

**Corresponding author (e-mail: emily.guidry@bg-group.com)*

Abstract: We present palaeomagnetic results from the Oligocene through Miocene part of the Integrated Ocean Drilling Program Site U1333 (1030.996′N, 138°25.159′W), which is located in 4853 m-deep water over seafloor with an estimated crustal age of 46 Ma. Detailed magnetostratigraphic investigations are essential to provide a sound age model for the study of the palaeoclimatic and palaeo-oceanographic history of the Cenozoic of the Equatorial Pacific and to improve the database of Pacific magnetostratigraphy. Rock magnetic measurements were carried out at 1 cm resolution on 81 U-channel samples from the spliced section with the goal of extracting a high-resolution record of the magnetostratigraphy. Stepwise demagnetization of the natural remanent magnetization yielded a well-defined magnetostratigraphy over a time interval of approximately 10 Ma between the base of Chron C6n (19.722 Ma) and the middle of Chron C11r (>29.9 Ma) and identification of the Oligocene–Miocene transition at the base of Subchron C6Cn.2n. The palaeomagnetic data are characterized by shallow inclinations, and by 180° alternations in declinations downhole, reflecting magnetic polarity zones. The relatively high temporal resolution allowed for the identification of three possible excursions previously not identified on the geomagnetic polarity time scale, which were recorded in Subchrons C8n.1r and C11n.2n and in Chron C11r.

The Pacific Equatorial Age Transect (PEAT) was cored by the Integrated Ocean Drilling Program (IODP) to provide material for palaeo-oceanographic, palaeoclimatic and other palaeo-environmental change analyses of the equatorial Pacific since the Eocene (Lyle *et al.* 2010; Pälike *et al.* 2010). The PEAT program was designed to investigate the interaction of the global climate with a major oceanic region over the entire Cenozoic era (65–0 Ma). Drill sites of the PEAT program (Fig. 1) were selected to recover a portion of the interval spanning the early Eocene through the Pleistocene, when the sediments were deposited near the palaeo-equator. The Pacific Ocean is important in the investigation of major changes in the global climate system because it represents nearly half of the world's ocean area and previous Deep Sea Drilling Project and Ocean Drilling Program (ODP) Legs 8, 9, 16, 85, 138 and 199 revealed that sediment recovered from the equatorial Pacific can be used to study changes through the Cenozoic in the carbon and climate cycles (van Andel *et al.* 1975; Pisias *et al.* 1995; Lyle *et al.* 2010; Pälike *et al.* 2010).

Key factors in successfully recovering relatively thick sections of pristine Cenozoic-age sediments are that the Pacific Plate has moved towards the NW throughout the Cenozoic and that a narrow equatorial upwelling zone produced a thick sediment

From: Jovane, L., Herrero-Bervera, E., Hinnov, L. A. & Housen, B. A. (eds) 2013. *Magnetic Methods and the Timing of Geological Processes*. Geological Society, London, Special Publications, **373**, 13–27.
First published online August 17, 2012, http://dx.doi.org/10.1144/SP373.7 © The Geological Society of London 2013.
Publishing disclaimer: www.geolsoc.org.uk/pub_ethics

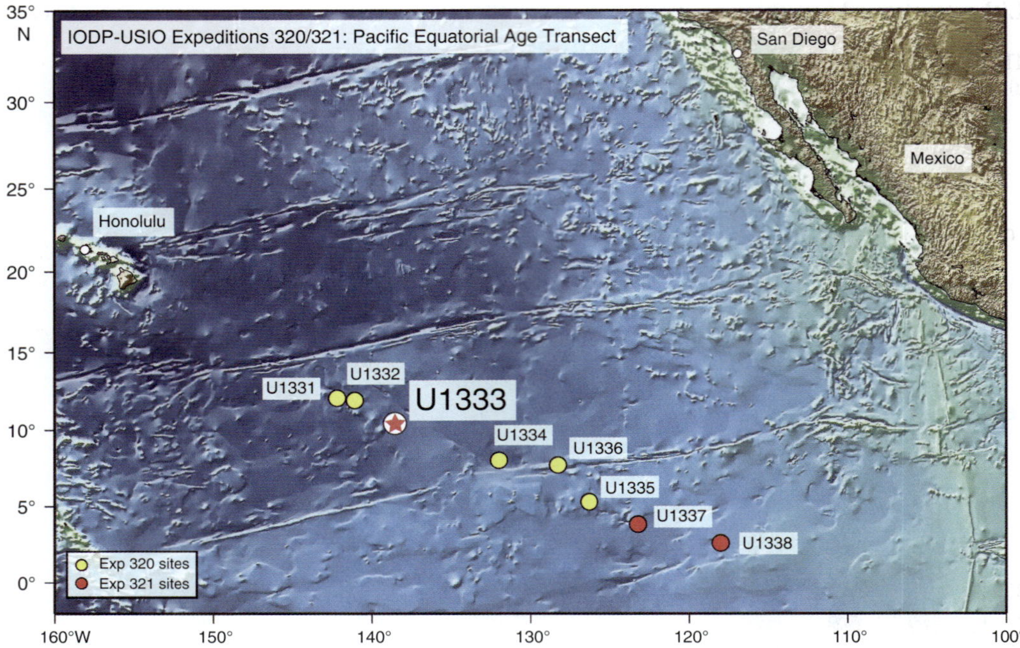

Fig. 1. Location map of sites drilled during IODP Expedition 320/321. Grey (yellow) circles, sites drilled during Expedition 320; dark grey (red) circles, sites drilled during Expedition 321; star, Site U1333 (after Pälike *et al.* 2010).

bulke. Therefore, sediments from the equatorial high-productivity zone were moved northward with the Pacific Plate, which enabled the recovery of older, diagenetically unaltered sediments at shallow depths by drilling along a transect (Lyle *et al.* 2002; Pares & Moore 2005; Lyle *et al.* 2010; Pälike *et al.* 2010). Transportation of the sediment depocentre to regions of lower sedimentation resulted in diminished overburden, which allowed for pristine preservation of sediments and microfossils, and minimized the potential for burial diagenesis. The recovered interval complements the Paleocene/Eocene and late Miocene to Recent intervals that were recovered by earlier drilling expeditions to the equatorial Pacific (Pisias *et al.* 1995; Lyle *et al.* 2002; Pälike *et al.* 2010).

At least two holes were cored at each site to ensure complete stratigraphic recovery. High-resolution hole-to-hole correlation allowed for the construction of a complete spliced section (Hagelberg *et al.* 1992; Pälike *et al.* 2010; Westerhold *et al.* 2012). The Expedition 320 composite stratigraphic section or splice for each site was constructed by aligning coeval features in cores from multiple adjacent holes to create a common composite depth scale for all holes at the site and then selecting the most representative and intact intervals, which were constructed into the complete continuous stratigraphic section. Sampling of the composite section at each site allows for the high-resolution measurement of any palaeo-environmental, chemical or physical property without any coring gaps or missing cycles.

Our study focuses on constructing a high-resolution magnetostratigraphy for the upper 85 m of the spliced stratigraphic section recovered from the three holes cored at Site U1333 (Expedition 320/321 Scientists 2010; Palmer *et al.* 2010). This includes the detection of potential brief excursions not documented in the standard geomagnetic polarity time scale (GPTS). Results from the initial shipboard investigations were promising (Expedition 320/321 Scientists 2010) and Oligocene and Miocene data from nearby Ocean Drilling Program Sites 1218 and 1219 (Lanci *et al.* 2004, 2005) were excellent, indicating that the sediments were likely to contain high-fidelity records of the palaeomagnetic field that could be extracted by an extensive high-resolution study. Detailed core descriptions, biostratigraphy and initial shipboard descriptions of magnetic and physical properties are presented in the Initial Reports of the expeditions (Expedition 320/321 Scientists 2010). The magnetic results of the lower part of Site U1333 will be published elsewhere by G. Acton and others.

Lithostratigraphy and biostratigraphy

The cores recovered at Site U1333 consist of middle Eocene to early Miocene-aged sediments, which are divided into four major lithological units comprising clay, clayey radiolarian ooze, clayey nannofossil ooze, nannofossil ooze and porcellanite (Expedition 320/321 Scientists 2010). The top 9 m

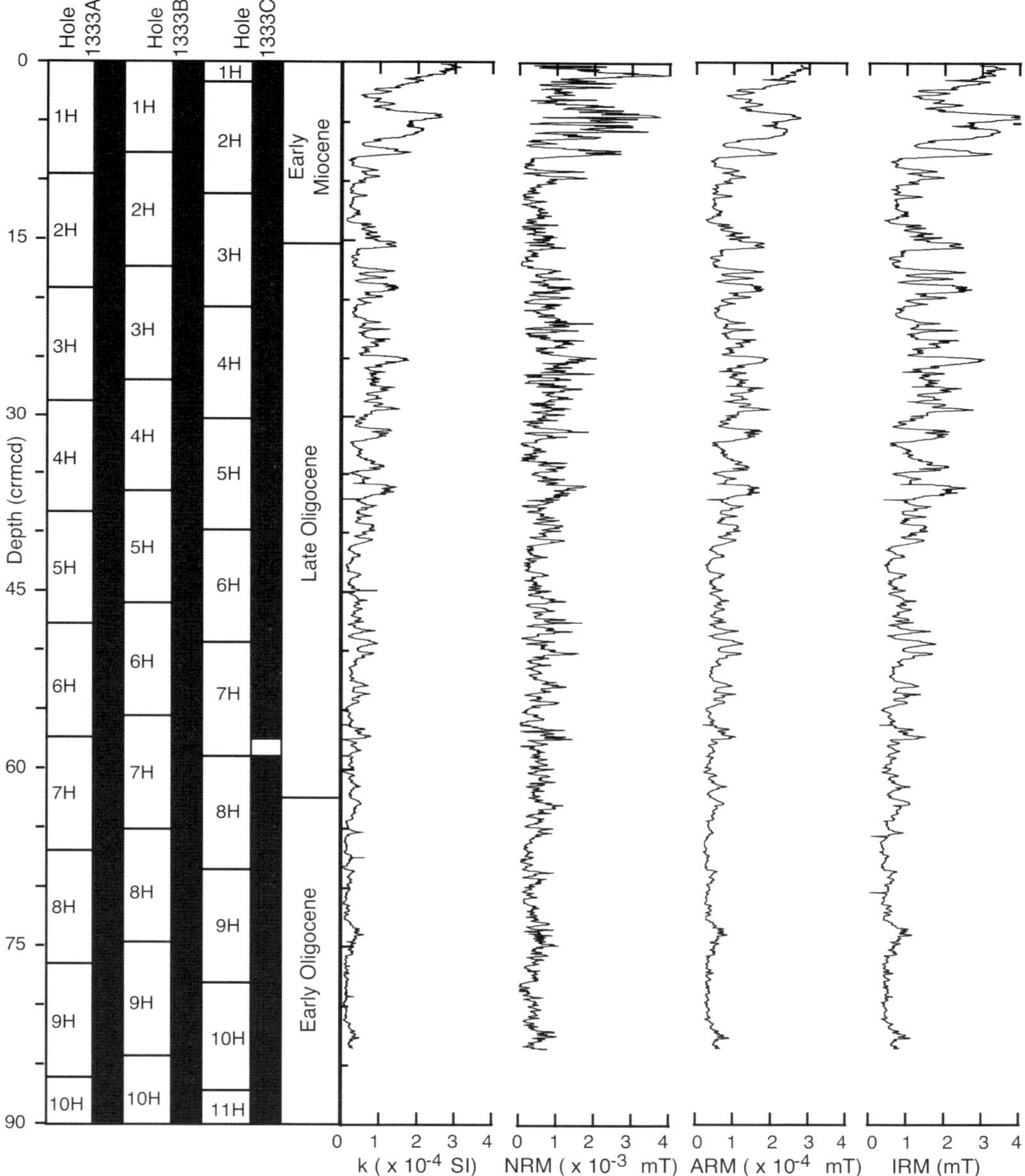

Fig. 2. Plot of core recovery from Holes U1333A, U1333B and U1333C along with composite splice records of downhole magnetic parameters of magnetic susceptibility (k), natural remanent magnetization (NRM) intensity after 25 mT demagnetization, anhysteretic remanent magnetization (ARM) and isothermal remanent magnetization (IRM). All four parameters exhibit variability and an overall decrease in the concentration of the magnetic carrier mineral downhole.

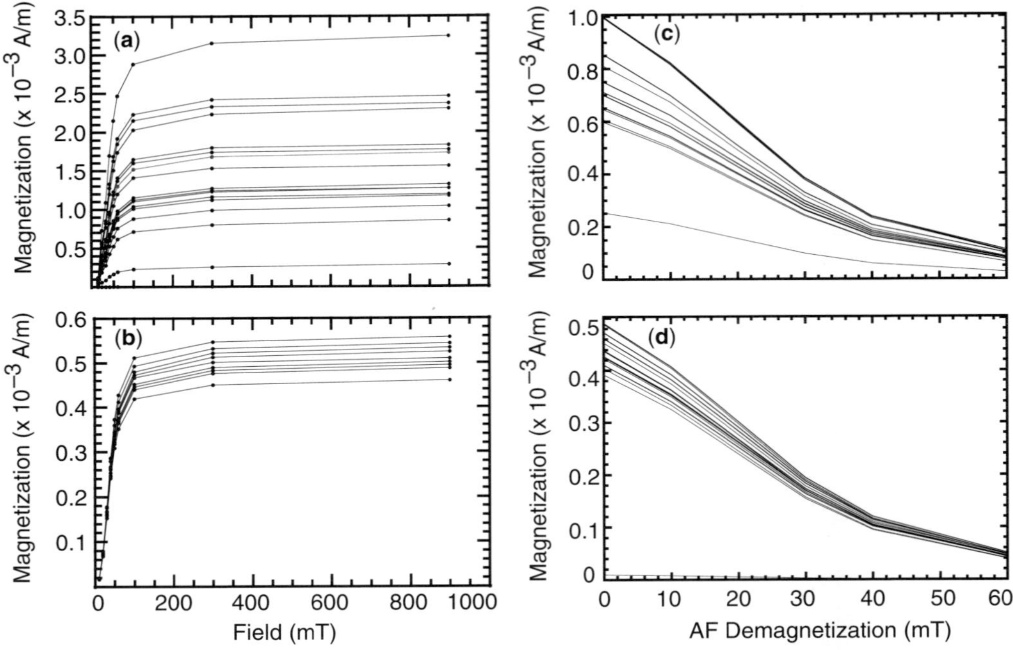

Fig. 3. (**a**) Acquisition behaviour of the IRM of U-channel 320-U1333A-01H-3; (**b**) acquisition behaviour of the IRM of U-channel 320-U1333B-09H-3; (**c**) demagnetization behaviour of the IRM of U-channel 320-U1333A-9H-3; and (**d**) demagnetization behaviour of the IRM of U-channel 320-U1333B-9H-6.

of the sequence comprises early Miocene-aged brown clays and oozes with varying amounts of nanno-fossils and radiolarians and minor bioturbation

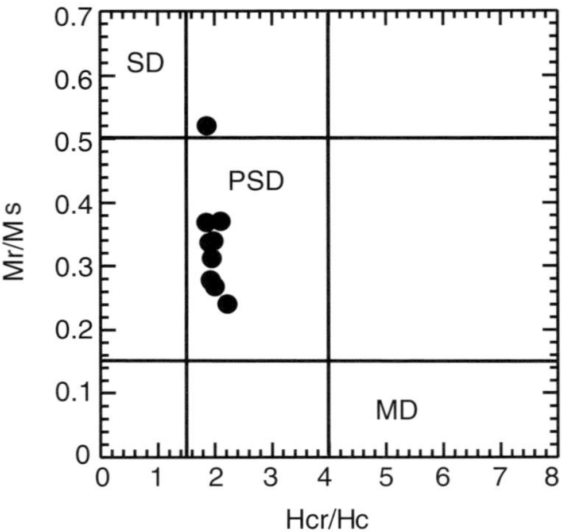

Fig. 4. Hysteresis ratios (Mr/Ms v. Hcr/Hc) indicate the pseudo-single domain (PSD) state of the magnetic carrier (Day *et al.* 1977) and vary within a relatively narrow range along the grain-size mixing line.

intensity. The boundary between the top litho-stratigraphic unit and the underlying older sediments is defined by the absence of clay as the main lithological component. The underlying unit consists of white to very pale brown nannofossil oozes with diatoms and radiolarians. Bioturbation is usually not visible or very minor. Volcanic ash particles and small pumice clasts (0.5–2 cm) are occasionally found.

The abundant microfossils in the sediments recovered at Site U1333 provide a high-resolution, concise biostratigraphic succession from the Eocene basement through the early Miocene (Expedition 320/321 Scientists 2010). A complete sequence of nannofossils zones, ranging from the lower Miocene Zone NN1 to the middle Eocene Zone NP15, was recovered. Radiolarians are present in most of the sediment section and provide a high-resolution biochronology ranging from Zones RN1 (lower Miocene) to RP13 (middle Eocene). Planktonic foraminifers are well preserved and abundant from the lower Oligocene to the lowest part of the Miocene. Benthic foraminifers occur throughout the entire sequence and indicate lower bathyal to abyssal palaeodepths. A detailed summary table of biostratigraphic results from Site U1333 is presented by Expedition 320/321 Scientists (2010) and is not reproduced here.

Rock magnetic properties

During Expedition 320, initial natural remanent magnetization (NRM) measurements were obtained along the archive-half core sections before and after alternating-field (AF) demagnetization up to 20 mT using the shipboard 2 G Enterprises long-core magnetometer (Expedition 320/321 Scientists 2010). The cores were subsequently transported to and stored at the IODP Core Repository at Texas A&M University, where we collected 81 U-channel samples (2 × 2 cm cross section and up to 150 cm long; Tauxe *et al.* 1983; Weeks *et al.* 1993). The samples span the top 85 m of the composite stratigraphic section, with some duplication to ensure continuity where the stratigraphic section is spliced from one drill hole to the next.

Magnetic parameters were measured every 1 cm along each U-channel sample as well as over the leader and trailer intervals (10 cm-long intervals before and after the samples passed through the SQUID sensors) with a 2 G Enterprises Model 755R cryogenic magnetometer housed in a shielded room at the palaeomagnetism laboratory of the

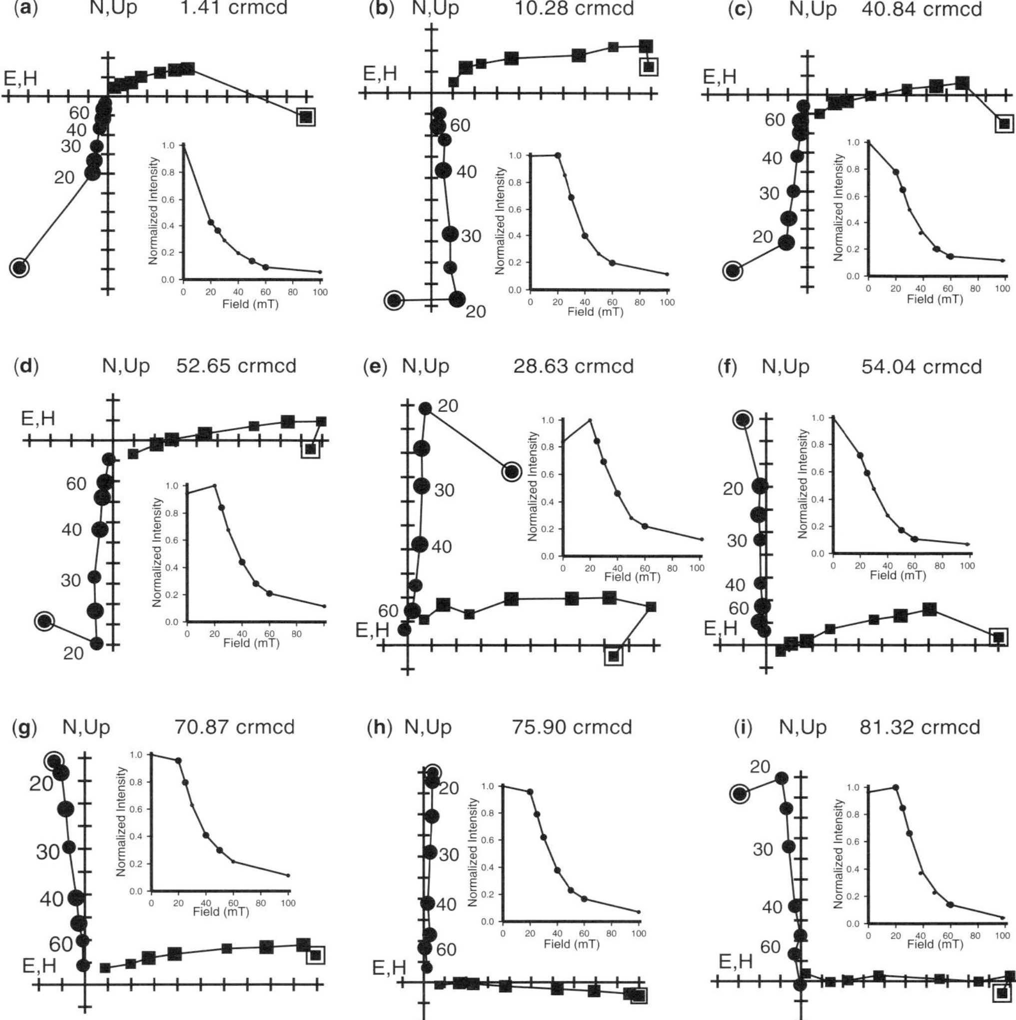

Fig. 5. Representative examples of vector component diagrams with normalized intensity decay plots of alternating field behaviour. Plots **a**–**d** are examples from reverse-polarity intervals and **e**–**i** from normal polarity intervals. Squares (circles) represent projections on the vertical (horizontal) plane.

University of California at Davis. The pickup coils of the magnetometer have a Gaussian-shaped response with a width of 4.17 cm for the *x*- and *y*-axes and 6.46 cm for the *z*-axis, resulting in an independent value roughly every fifth measurement. Step-wise AF demagnetization of the NRM was performed and measured in the pass-through magnetometer before and after 0, 20, 25, 30, 40, 50, 60 and 100 mT AF demagnetization steps (Fig. 2). An anhysteretic remanent magnetization (ARM) was imparted using a 100 mT peak alternating field and a 50 μT direct-current (DC) biasing field. The samples then were measured in the magnetometer following AF demagnetization at 0, 20, 30, 40, 50 and 60 mT. Isothermal remanent magnetizations (IRM) were imparted with a 900 mT field and measured following AF demagnetization at 0, 20, 30, 40, 50 and 60 mT. Selected samples were subjected to progressive IRM acquisition at 100, 200, 300, 500 600, and 900 mT prior to demagnetizing the saturation IRM. We do not use data from the leader, trailer or from within 4 cm of each end of the U-channels to avoid erroneous results related to sample edge effect.

Low-field magnetic susceptibilities were measured at the University of Florida at Gainesville using a U-channel track system in which the samples pass through a Sapphire Instruments S12B magnetic susceptibility meter with a 3.3 cm square coil (Thomas *et al.* 2003). The susceptibility measurement procedure was conducted on all U-channels twice to ensure accurate results; the mean value is used and displayed in Figure 2.

The magnetic susceptibility, ARM and IRM are concentration-dependent measurements that are sensitive to the amount of magnetic material present. The ARM is more sensitive to the fine-grained magnetic grains, whereas the magnetic susceptibility and the IRM measure coarser-grained magnetic particles. The magnetic susceptibility, ARM and IRM intensities correlate well with each other and to a lesser degree with the initial NRM intensity, which depends on magnetic concentration, palaeomagnetic field strength and any overprints that exist prior to magnetic cleaning. For example, the correlation coefficient R^2 between the magnetic susceptibility and the IRM intensity is 0.97, suggesting that the bulk of both signals is carried by the same magnetic mineral fraction.

Saturation IRM experiments were carried out on 13 U-channel samples to aid in the identification of the magnetic carrier mineral. Typical IRM acquisition curves of two U-channel samples (Fig. 3a, b)

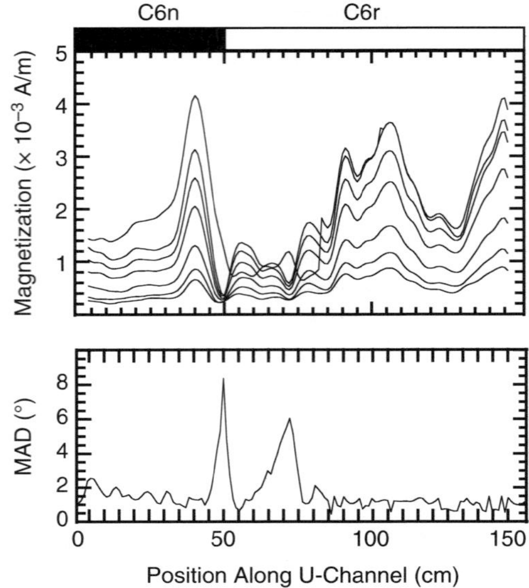

Fig. 6. Demagnetization behaviour of U-channel 320-U1333B-01H-3 at 0, 20, 25, 30, 40, 50, 60 and 100 mT. The associated maximum angular deviation (MAD) is usually very low (<3°), except at the magnetic reversal at 50 cm and the low-intensity interval between 65 and 75 cm.

exhibit very uniform behaviour and show that nearly full saturation is achieved in fields of 300 mT, which is consistent with magnetite being the dominant remanence carrier. The slight increase in IRM from 300 to 900 mT indicates the presence of a higher coercivity mineral (e.g. hematite). AF demagnetization of these samples (Fig. 3c, d) demonstrates a moderate to soft magnetization, with *c.* 90% of the intensity of magnetization lost by the 60 mT demagnetization step.

Hysteresis properties were determined on 10 typical samples using a Princeton Measurements alternating field gradient magnetometer at the University of Florida at Gainsville. The ratios of coercivity of remanence over coercivity (Hcr/Hc) and saturation remanence over saturation magnetization (Mr/Ms) are a function of variation in grain size (Day *et al.* 1977; Dunlop 1986). A plot of Mr/Ms v. Hcr/Hc (Day *et al.* 1977) shows little variation in magnetic grain size and indicates that the bulk magnetic grain sizes for all but one sample fall into the pseudo-single domain (PSD) field of magnetite (Fig. 4).

Fig. 7. Downhole plots of the virtual geomagnetic pole (VGP) in (**a**) the composite section, and Holes (**b**) U1333A, (**c**) U1333B and (**d**) U1333C. In addition, the results of the principal component analysis of discrete samples (solid circles) are shown for Hole U1333A.

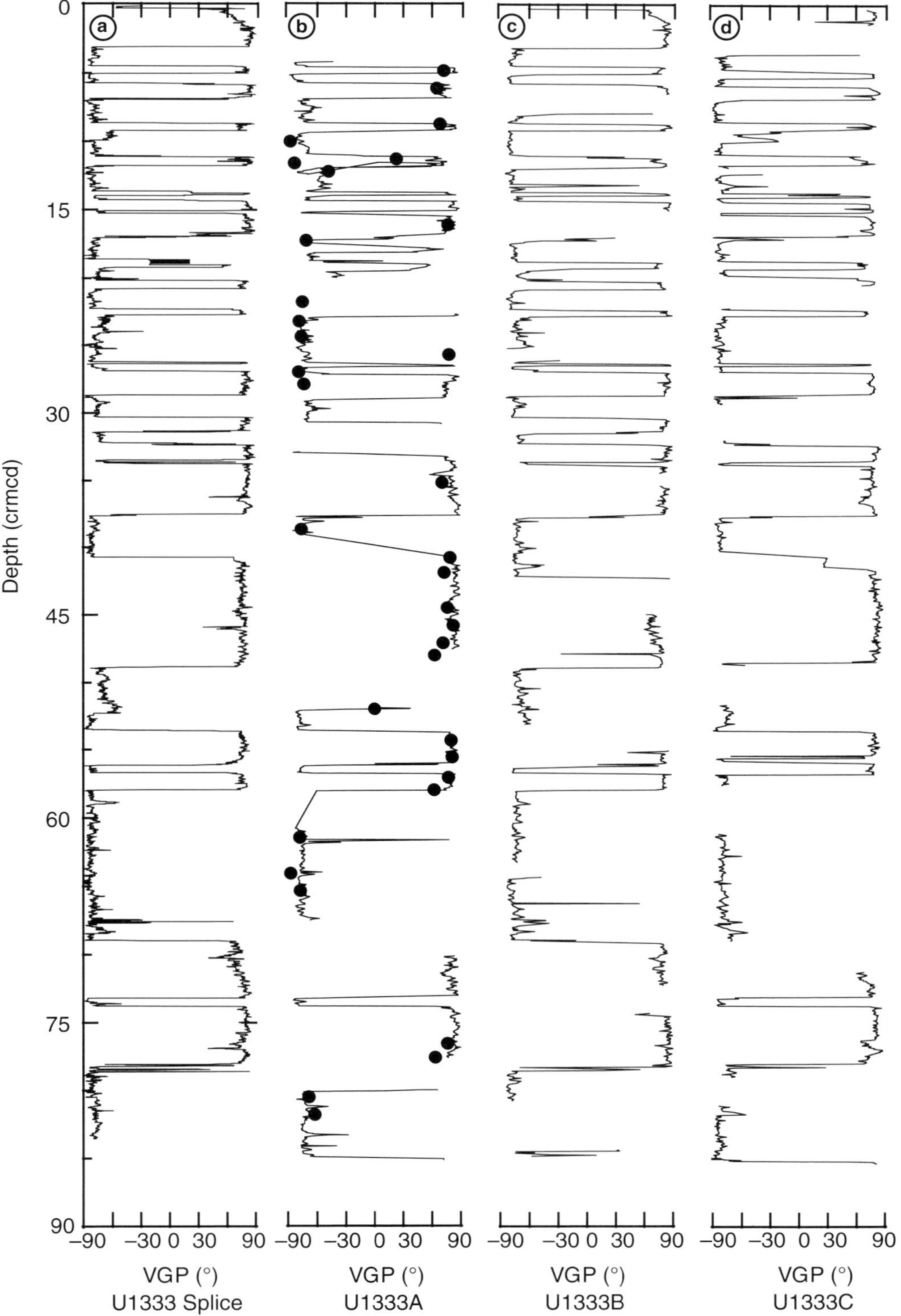
0
15
30
45
60
75
90
Depth (crmcd)
a
b
c
d
-90 -30 0 30 90
-90 -30 0 30 90
-90 -30 0 30 90
-90 -30 0 30 90
VGP (°)
VGP (°)
VGP (°)
VGP (°)
U1333 Splice
U1333A
U1333B
U1333C

Overall, the rock magnetic results indicate that the magnetic remanence of the sediments at Site U1333 is dominated by PSD magnetite. The excellent correlation between the magnetic susceptibility and the magnetic remanence indicates that the magnetic susceptibility, one of the cornerstones of the shipboard measurement and correlation efforts, is carried by magnetite and that changes within the susceptibility signal (Fig. 2) are probably caused by variations in the magnetite concentration.

Natural remanent magnetization

The raw NRM data were reformatted for use in the ZPLOTIT Software (Acton 2011). This program automatically estimated the principal component analysis (PCA) (Kirschvink 1980) at >9000 measurement positions to determine the characteristic remanent magnetization using the 'free' option of the PCA. The best-fit line that passes through the vector demagnetization data was determined for data between the 20 and 100 mT demagnetization steps omitting outliers as described in Acton (2011). Representative vector demagnetization diagrams of the demagnetization behaviour (Fig. 5) indicate the presence of a stable, well-defined characteristic remanence over the entire interval. The average maximum angular deviation (MAD) angle from the principal component analysis, a measurement for the quality of the demagnetization behaviour, is 2.03°.

Much of the sediment is affected by a secondary NRM component acquired during sampling, storage and/or transport (e.g. Acton *et al.* 2002; Richter *et al.* 2007). This overprint was readily removed in alternating fields of <25 mT (Fig. 5). In some cases, the linear demagnetization paths do not decay directly to the origin of the vector demagnetization diagrams, but instead end near the origin with a steep component remaining that is only a few per cent of the initial NRM. This high-coercivity component could be a small overprint acquired during coring or subsequently or by measurement artefacts. The demagnetization behaviour of a single U-channel sample (Fig. 6) exhibits decreasing intensity with increasing alternating fields and a significant increase in the MAD during the reversal at 50 cm, and in the low-intensity interval between 65 and 75 cm. This behaviour is expected during times of polarity reversal and/or low intensity because the magnetic intensity of the sample is low and the directions during demagnetization are therefore not well defined and yield larger MAD values.

The palaeomagnetic inclination is close to horizontal for normal and reverse-polarity intervals at near-equatorial latitudes and, therefore, the *c.* 180°

alternations in declination along the composite section are used for identifying changes in the magnetic polarity. The composite section provides a continuous magnetic reversal zonation, but not the absolute polarity because the Site U1333 cores were not azimuthally oriented. Determination of absolute polarity relies on correlating the magnetic polarity zonation sequence with the geomagnetic reversals of the GPTS. This correlation is sufficiently unambiguous on its own but can also be confirmed with the aid of biostratigraphic constraints (Expedition 320/321 Scientists 2010). Once the polarity had been determined, we computed the mean palaeomagnetic direction for each core from stable normal polarity and inverted reverse-polarity intervals. This 'core' mean declination was rotated to 0°, resulting in stable normal polarity interval having declinations near 0° and stable reversed-polarity intervals having declinations near 180° (Figs 5, 7 & 8).

Depth scale and composite section

Alignment of cores from multiple adjacent holes drilled at the same site can be used to establish a composite depth scale and for the construction of a composite spliced section of the stratigraphic interval (Hagelberg *et al.* 1992; Pälike *et al.* 2010; Westerhold *et al.* 2012). The depths of individual cores in a given hole are shifted by a constant amount that

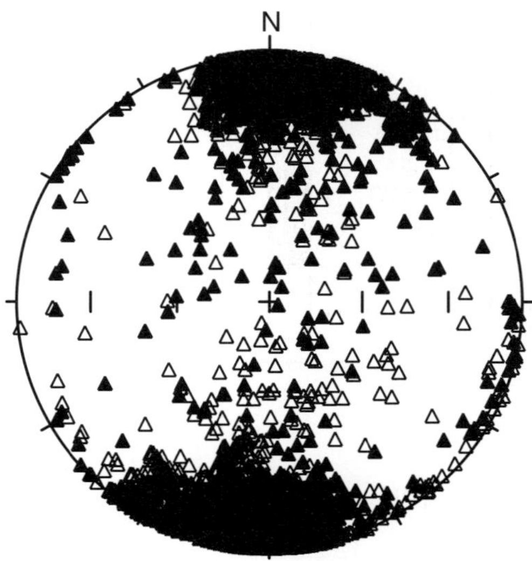

Fig. 8. Equal area stereographic projection of the characteristic remanent magnetization at Site U1333. Closed (open) symbols represent projection into the upper (lower) hemisphere.

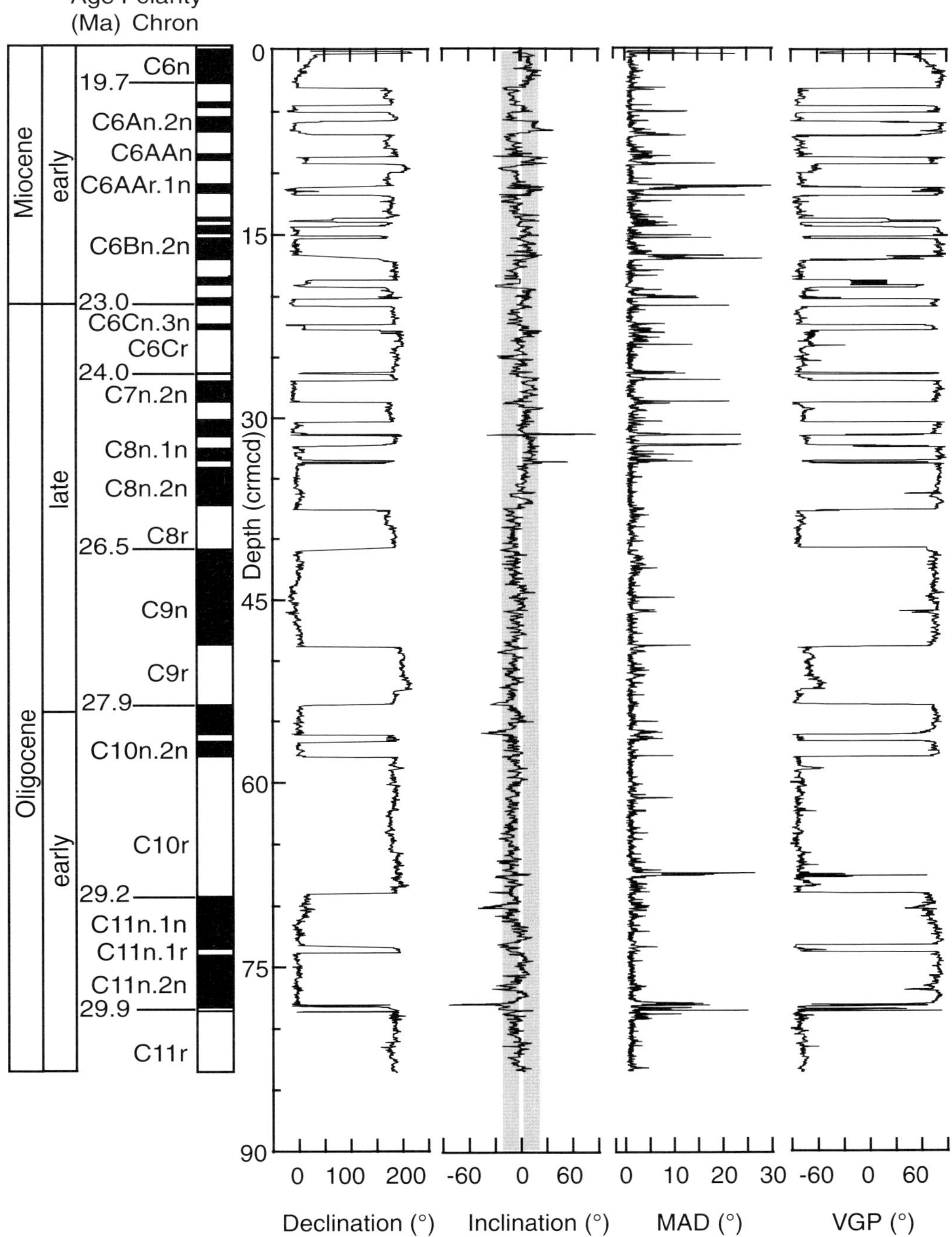

Fig. 9. Magnetostratigraphic interpretation, magnetic declination, inclination, MAD value and VGP of the spliced section at Site U1333.

best aligns the variations in physical properties (susceptibility, γ-ray density, colour reflectance and X-ray diffraction) to the equivalent cores in adjacent holes (Pälike *et al.* 2010; Westerhold *et al.* 2012). Data from each hole at Site U1333 were compiled into the composite splice using methods and offsets described by Pälike *et al.* (2010) and revised by Westerhold *et al.* (2012). The revised metres composite depth (rmcd) scale was then corrected for an estimated 15% core expansion and reported as a corrected rmcd (crmcd) scale to be comparable to the shipboard metres below seafloor (mbsf) depth scale and to yield realistic sedimentation rates (see Pälike *et al.* 2010; Westerhold *et al.* 2012 for detailed discussion).

Magnetostratigraphy

The composite record of magnetic declination and inclination expressed as virtual geomagnetic pole (VGP) latitude together with the shipboard VGP

Table 1. *Magnetic reversal data of the splice at Site U1333, with depth and corresponding age*

Sample	Polarity Chron	Age (Ma)	Depth (mbsf)	Depth (mcd)	Depth (crmcd)
U1333B-01H-3, 50 cm	C6n/C6r	19.722	3.50	3.50	3.04
U1333B-01H-4, 64 cm	C6r/C6An.1n	20.040	5.14	5.14	4.47
U1333B-01H-4, 128 cm	C6An.1n/C6An.1r	20.213	5.78	5.78	5.03
U1333A-01H-2, 50 cm	C6An.1r/C6An.2n	20.439	2.00	6.66	5.79
U1333A-01H-3, 23 cm	C6An.2n/C6Ar	20.709	3.23	7.89	6.86
U1333A-01H-4, 81 cm	C6Ar/C6AAn	21.083	5.31	9.97	8.67
U1333A-01H-4, 149 cm	C6AAn/C6AAr.1r	21.159	5.99	10.65	9.26
U1333B-02H-4, 56 cm	C6AAr.1r/C6AAr.1n	21.403	12.76	12.77	11.10
U1333B-02H-4, 136 cm	C6AAr.1n/C6AAr.2r	21.483	13.56	13.57	11.80
U1333C-03H-2, 44 cm	C6AAr.2r/C6AAr.2n	21.659	13.04	15.70	13.65
U1333C-03H-2, 81 cm	C6AAr.2n/C6AAr.3r	21.688	13.41	16.07	13.97
U1333C-03H-2, 122 cm	C6AAr.3r/C6Bn.1n	21.767	13.82	16.48	14.33
U1333C-03H-3, 58 cm	C6Bn.1n/C6Bn.1r	21.936	14.68	17.34	15.08
U1333C-03H-3, 82 cm	C6Bn.1r/C6Bn.2n	21.992	14.92	17.58	15.29
U1333C-03H-4, 113 cm	C6Bn.2n/C6Br	22.268	16.73	19.39	16.86
U1333C-03H-6, 22 cm	C6Br/C6Cn.1n	22.564	18.82	21.48	18.68
U1333B-03H-2, 138 cm	C6Cn.1n/C6Cn.1r	22.754	20.08	22.15	19.26
U1333B-03H-3, 95 cm	C6Cn.1r/C6Cn.2n	22.902	21.15	23.22	20.19
U1333B-03H-4, 19 cm	C6Cn.2n/C6Cn.2r	23.030	21.89	23.96	20.83
U1333B-03H-5, 37 cm	C6Cn.2r/C6Cn.3n	23.278	23.57	25.64	22.30
U1333B-03H-5, 94 cm	C6Cn.3n/C6Cr	23.340	24.14	26.21	22.79
U1333C-4H-5, 30 cm	C6Cr/C7n.1n	24.022	26.90	30.18	26.24
U1333C-4H-5, 47 cm	C7n.1n/C7n.1r	24.062	27.07	30.35	26.39
U1333B-04H-1, 106 cm	C7n.1r/C7n.2n	24.147	27.76	30.94	26.90
U1333B-04H-3, 12 cm	C7n.2n/C7r	24.459	29.82	33.00	28.70
U1333B-04H-4, 49 cm	C7r/C7An	24.756	31.69	34.87	30.32
U1333B-04H-5, 22 cm	C7An/C7Ar	24.984	32.92	36.10	31.39
U1333B-04H-5, 122 cm	C7Ar/C8n.1n	25.110	33.92	37.10	32.26
U1333C-05H-2, 35 cm	C8n.1n/C8n.1r	25.248	31.95	38.50	33.48
U1333C-05H-2, 64 cm	C8n.1r/C8n.2n	25.306	32.24	38.79	33.73
U1333C-05H-5, 52 cm	C8n.2n/C8r	26.032	36.62	43.17	37.54
U1333C-06H-1, 101 cm	C8r/C9n	26.508	40.61	47.14	40.99
U1333B-06H-4, 15 cm	C9n/C9r	27.412	50.35	56.19	48.86
U1333A-06H-3, 140 cm	C9r/C10n.1n	27.886	51.90	61.65	53.61
U1333B-07H-1, 134 cm	C10n.1n/C10n.1r	28.126	56.54	64.57	56.15
U1333B-07H-2, 46 cm	C10n.1r/C10n.2n	28.164	57.16	65.19	56.69
U1333B-07H-3, 38 cm	C10n.2n/C10r	28.318	58.58	66.61	57.92
U1333B-08H-4, 138 cm	C10r/C11n.1n	29.166	70.58	79.35	69.00
U1333A-8H-3, 146 cm	C11n.1n/C11n.1r	29.467	70.96	84.34	73.34
U1333C-09H-3, 54 cm	C11n.1r/C11n.2n	29.536	71.64	84.88	73.81
U1333B-09H-4, 132 cm	C11n.2n/C11r	29.957	80.02	89.95	78.22

Ages from Cande & Kent (1995), Lourens *et al.* (2004), and Pälike *et al.* (2006*a*, *b*).
mbsf, metres below seafloor; mcd, metres composite depth; crmcd, corrected revised mcd.

data for Holes U1333A, U1333B and U1333C are presented in Figure 7 on the crmcd scale. The shipboard VGPs were calculated from the magnetic inclination and declination after 20 mT AF demagnetization of the archive half, whereas the U-channel VGP data of the splice result from detailed principal component analysis after stepwise demagnetization. In addition, shipboard PCA results from discrete samples along Hole U1333A are displayed in Figure 7b (Expedition 320/321 Scientists 2010) and a stereographic projection of the characteristic remanent magnetization of the U-channel measurements is presented in Figure 8. All data show excellent agreement and support the use of the initial shipboard measurements and magnetostratigraphic interpretation of Pälike *et al.* (2010). However, the new magnetic record along the spliced section provides data of much higher resolution and quality control through PCA.

The interpretation of magnetic polarity from the composite declination record for Site U1333 (Fig. 9 and Table 1) is well supported by key nannofossil data spanning Biozones NN1 to NP22 and radiolarian zones RN2 to RP20 between 0 and 85 crmcd (Expedition 320/321 Scientists 2010). The magnetostratigraphy is based on correlation of the magnetic polarity zonation determined from the *c.* 180° alternations in declination with the GPTS as described in the 'Natural remanent magnetization' section. In this study, we used a GPTS that is a composite of the time scales of Cande and Kent (1995), Lourens *et al.* (2004), and Pälike *et al.* (2006*a*, *b*), as described in detail by Pälike *et al.* (2010).

The declination record for the section, when compared with the GPTS, provides a complete record from Chron C6n in the early Miocene through late and early Oligocene polarity intervals to the middle of Chron C11r (Fig. 9 and Table 1). The magnetic record is characterized by sharp 180° reversals of the magnetic declinations, which allows for the identification of magnetic chrons and subchrons. The palaeomagnetic inclination can also be indicative of polarity, with very shallow downward directions (positive inclinations) occurring preferentially in normal polarity intervals and very shallow upward directions (negative inclinations) occurring preferentially in reversed-polarity intervals. The inclination is, however, very shallow and has natural palaeosecular variability, making polarity determination based on inclination alone a precarious proposition (Fig. 9). The present-day geocentric axial dipole inclination at Site U1333 is 20.4°, but the expected palaeo-inclination is much shallower, with estimated values of 1.9° using the 30 Ma Pacific palaeomagnetic pole of Beaman *et al.* (2007) or 8.0° using the 32 Ma pole of Horner-Johnson and Gordon (2010). The shaded bars on the inclination record (Fig. 9, second panel) depict the

expected range of inclinations at the site given the motion of the Pacific Plate since the Oligocene. The magnetic polarity reversals and the correlation with the GPTS show that the sedimentary record is remarkably complete from Chron 11r to Chron 6n, and that sedimentation ceased or became insignificant at Site U1333 about 19 Ma ago. The Oligocene–Miocene boundary is defined as the reversed to normal magnetic transition represented by the base of magnetic Subchron C6Cn.2n (Shackleton *et al.* 2000). This transition, as well as the three short normal subchrons within C6Cn, is well resolved.

In the 85 m-long Oligocene–Miocene composite record of Site U1333, 41 chrons and subchrons of the GPTS are present. The record appears to be complete and uninterrupted by any significant hiatuses. A palaeomagnetic age model for Site U1333, listing depths and ages of magnetic polarity reversal events, is given in Table 1. An age–depth plot and sedimentation rates for Site U1333, using the GPTS reversal polarity ages (Cande & Kent *et al.* 1995; Lourens *et al.* 2004; Pälike *et al.* 2006*a*, *b*) are provided in Figure 10. Average sedimentation rates decrease from *c.* 11 m/Ma in the early Oligocene to *c.* 5 m/Ma in the late Oligocene, to *c.* 4 m/Ma in the Miocene. The trend to decreasing sedimentation rates results from Site U1333 moving out of the equatorial high-productivity zone as the Pacific Plate migrated northwestward throughout the Cenozoic.

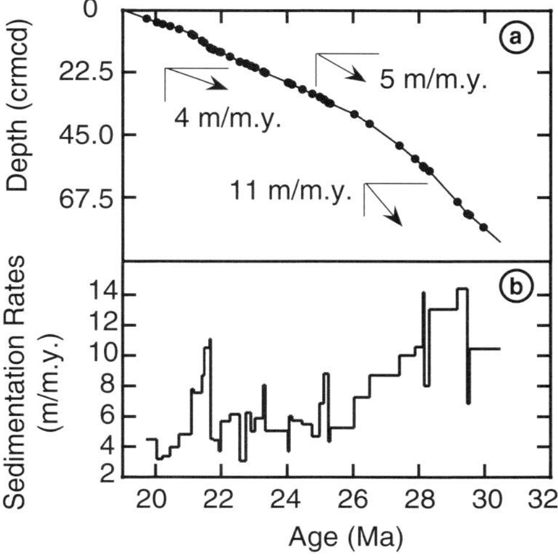

Fig. 10. (**a**) Plot of magnetic reversal depth v. age and (**b**) sedimentation rates v. age. Sedimentation rates decrease with time as the site is moved out of the equatorial zone of high productivity.

Possible geomagnetic excursions

Sedimentation rates vary between 4 and 14 m/Ma and are therefore high enough at the given measurement frequency to record brief polarity excursions. Figure 11a shows a possible excursion in Subchron C8n.1r (*c.* 33.64 crmcd), where a brief, previously undocumented, normal polarity interval is recorded in the reversed subchron. The declination and the VGP change by about 180°, the inclination steepens to values not typical for the equatorial position of the site, and the MAD increases. Vector demagnetization diagrams (Fig. 11a) exhibit well-defined magnetic demagnetization behaviour before, during and after the potential excursion. Visual inspection of the core images shows no coring disturbance or other apparent sedimentological anomalies in this interval. The duration of this possible excursion is estimated at 8 ka, which is well within the range of 30 ka or less defined by Cande & Kent (1992, 1995), and is similar to the duration of other geomagnetic events documented in the equatorial Pacific during the Miocene and Oligocene (Schneider 1995; Lanci *et al.* 2004, 2005). There are two more locations in the deeper part of the record, in Subchron C11n.2n and Chron C11r (Fig. 11b), which exhibit similar behaviour at 78.10 and 78.63 crmcd and show no apparent signs of coring disturbance

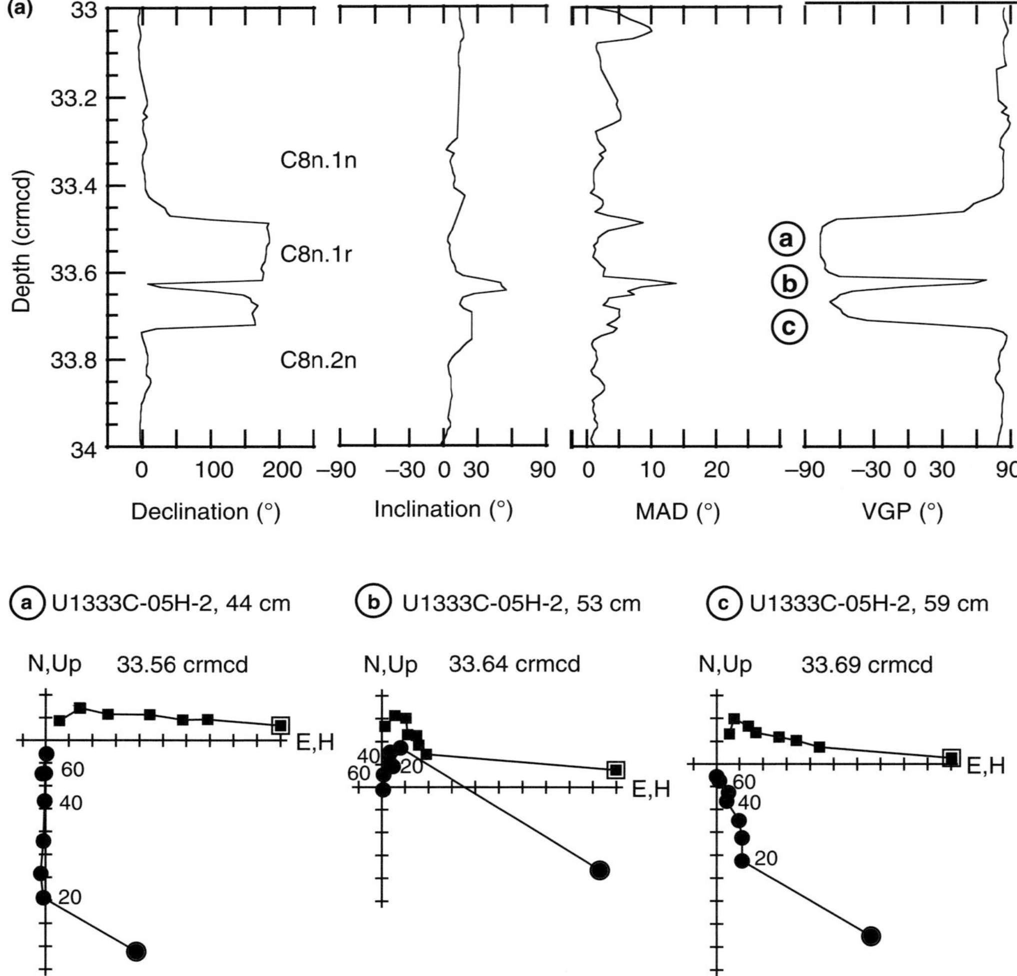

Fig. 11. Potential geomagnetic excursions not represented in the geomagnetic polarity time scale. Downhole plots of declination, inclination, MAD and VGP are depicted in addition to vector component diagrams from the locations marked with encircled letters.

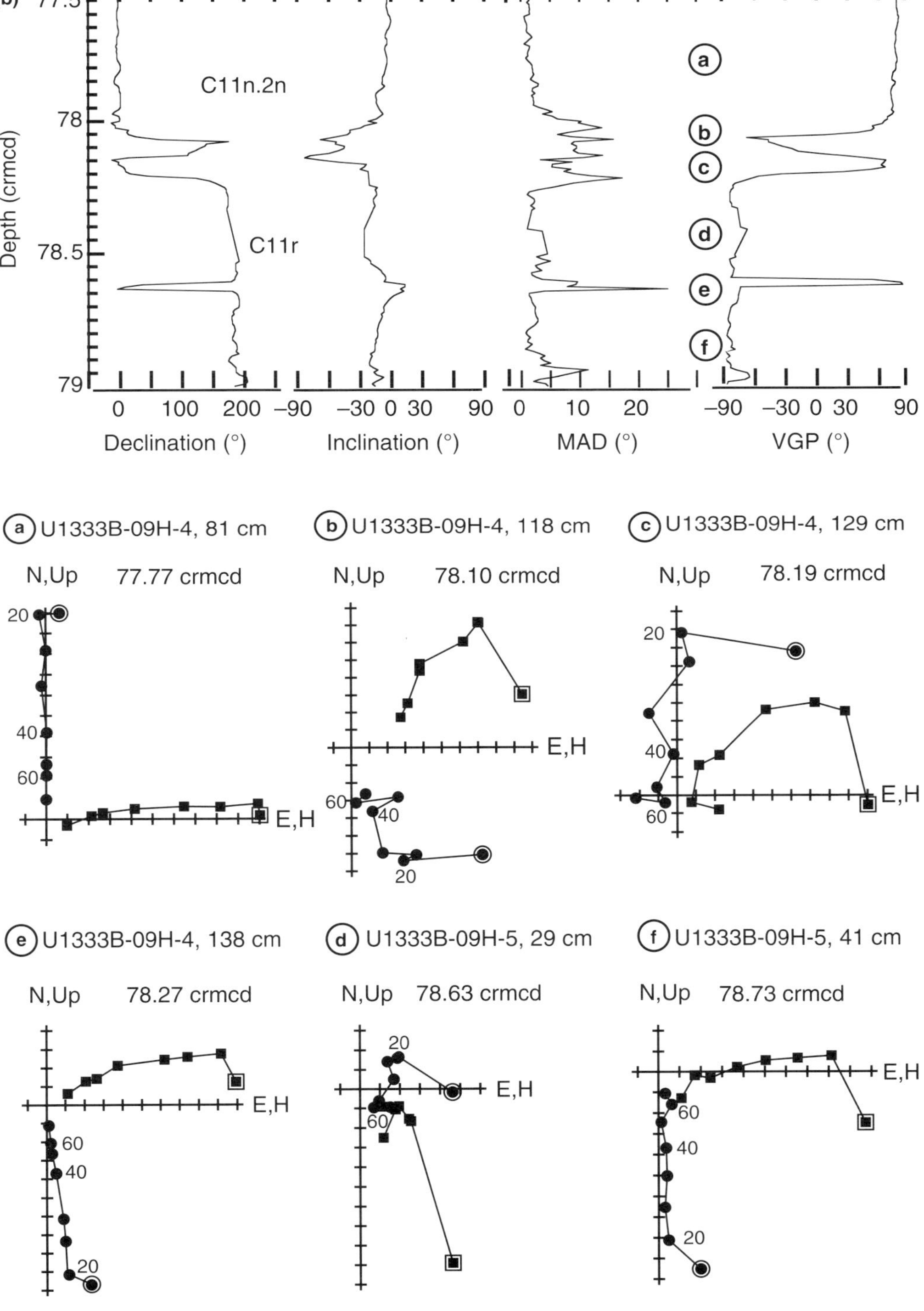

Fig. 11. (*Continued*)

or sedimentological anomalies. The estimated duration is 5 ka for the event in C11n.2n and 2 ka for the event in C11r. Our current data support the interpretation of these short intervals as geomagnetic excursions not previously reported; however, additional research on discrete samples, cores from outside the splice and comparison with results from adjacent sites needs to be conducted.

Summary and conclusions

Vector demagnetization diagrams and principal component analyses illustrate successful removal of a secondary magnetization and identification of the primary magnetization direction and intensity of the sediment samples. Alternating-field demagnetization behaviour, along with IRM acquisition curves that saturate in 300–400 mT fields, indicate the presence of magnetite as the main magnetic carrier. Hysteresis properties indicate PSD magnetite with a limited range of grain sizes. Furthermore, the characteristic remanent magnetization directions are shallow during stable polarity intervals, which are consistent with the expected direction based on the Pacific Plate apparent polar wander path, and reversed-polarity directions are antipodal to the normal polarity directions. Taken together, these observations indicate that the sediments at Site U1333 are high-fidelity recorders of the palaeomagnetic field.

Indeed, the top 85 crmcd of the sediment cored at IODP Site U1333 in the equatorial Pacific yielded an exceptional record of the direction of the geomagnetic field from the early Oligocene through the early Miocene. Stepwise demagnetization of the NRM revealed a very well resolved magnetostratigraphic record over a time interval of approximately 10 Ma between the base of Chron C6n (19.72 Ma) and the middle of Chron C11r (>29.9 Ma). The rarely documented magnetic transition of the Oligocene–Miocene boundary is identified as the reversed to normal magnetic transition represented by the base of magnetic Subchron C6Cn.2n. We also identified all three normal events of magnetic Chron C6Cn, which have only rarely been reported in the literature (Shackleton *et al.* 2000). In addition, we observed three potential excursions, occurring in Subchrons C8n.1r, C11n.2n and Chron C11r, which have not been reported previously. The duration of these events is estimated at 8, 5 and 2 ka, respectively.

This research used samples and data provided by the Integrated Ocean Drilling Program, which is sponsored by the US National Science Foundation and participating countries under the management of Ocean Leadership Inc. Funding for this work was provided by grants from Ocean Leadership (IUSSP410) and the US National Science Foundation (OCE 0961161, OCE 0961412 and OCE 0960999). Any opinions, findings and conclusions or recommendations expressed in this material are those of the author(s) and do not necessarily reflect the views of the National Science Foundation. Careful reviews by three anonymous reviewers greatly improved the manuscript. We acknowledge numerous discussions with the Expedition 320/321 Shipboard Scientific Party, help from H. Pälike and T. Westerhold with the ever evolving depth scale, and the professional support of the IODP technicians and the drill crew of the JOIDES Resolution.

References

ACTON, G. 2011. *ZPLOTIT Software Users' Guide Version, version 2011-01*, http://paleomag.ucdavis.edu/softwareZplotit.html

ACTON, G. D., OKADA, M., CLEMENT, B. M., LUND, S. P. & WILLIAMS, T. 2002. Paleomagnetic overprints in ocean sediment cores and their relationship to shear deformation caused by piston coring. *Journal of Geophysical Research*, **107**, 1.

BEAMAN, M., SAGER, W. W., ACTON, G. D., LANCI, L. & PARES, J. 2007. Improved Late Cretaceous and early Cenozoic paleomagnetic apparent polar wander path for the Pacific plate. *Earth and Planetary Science Letters*, **262**, 1–20.

CANDE, S. C. & KENT, D. V. 1992. Ultra-high resolution marine magnetic anomaly-profiles: a record of continuous paleointensity variations? *Journal of Geophysical Research*, **97**, 15 075–15 083.

CANDE, S. C. & KENT, D. V. 1995. Revised calibration of the geomagnetic polarity time scale for the Late Cretaceous and Cenozoic. *Journal of Geophysical Research*, **100**, 6093–6095.

DAY, R., FULLER, M. & SCHMIDT, V. A. 1977. Hysteresis properties of titanomagnetites: grain-size and compositional dependence. *Physics of the Earth and Planetary Interiors*, **13**, 260–267.

DUNLOP, D. J. 1986. Hysteresis properties of magnetite and their dependence of particle size: a test of pseudo-single-domain remanence models. *Journal of Geophysical Research*, **91**, 9569–9584.

EXPEDITION 320/321 SCIENTISTS. 2010. Site U1333. *In*: PÄLIKE, H., LYLE, M. ET AL. *Proceedings of the Integrated Ocean Drilling Program, 320/321*. Integrated Ocean Drilling Program Management International Inc., Tokyo, http://dx.doi.org/10.2204/iodp.proc.320321.105.2010

HAGELBERG, T., SHACKLETON, N. & PISIAS, N. Shipboard Scientific Party. 1992. Development of composite depth sections for Sites 844 through 854. *In: Proceedings of the Ocean Drilling Program*, Initial Reports, **138**. Ocean Drilling Program, College Station, TX, 79–85, http://dx.doi.org/10.2973/odp.proc.ir.138.105.1992

HORNER-JOHNSON, B. C. & GORDON, R. G. 2010. True polar wander since 32 Ma B.P.: a paleomagnetic investigation of the skewness of magnetic anomaly 12r on the Pacific plate. *Journal of Geophysical Research*, **115**, B09101, http://dx.doi.org/10.1029/2009jb006862

KIRSCHVINK, J. L. 1980. The least-squares line and plane and the analysis of paleomagnetic data. *Geophysical*

Journal of the Royal Astronomical Society, **62**, 699–718.

LANCI, L., PARES, J. M., CHANNELL, J. E. T. & KENT, D. 2004. Miocene magnetostratigraphy from Equatorial Pacific sediments (ODP Site 1218, Leg 199). *Earth and Planetary Science Letters*, **226**, 207–224.

LANCI, L., PARES, J. M., CHANNELL, J. E. T. & KENT, D. 2005. Oligocene magnetostratigraphy from Equatorial Pacific sediments (ODP Sites 1218 and 1219, Leg 199). *Earth and Planetary Science Letters*, **237**, 617–634.

LOURENS, L. J., HILGEN, F. J., LASKAR, J., SHACKLETON, N. J. & WILSON, D. 2004. The Neogene period. *In*: GRADSTEIN, F. M., OGG, J. *ET AL*. (eds) *A Geologic Time Scale 2004*. Cambridge Univ. Press, Cambridge, 409–440.

LYLE, M., WILSON, P. A. *ET AL*. 2002. *Proceedings of the Ocean Drilling Program*, Initial Reports, **199**. Ocean Drilling Program, College Station, TX, http://dx.doi.org/10.2973/odp.proc.ir.199.2002

LYLE, M., PÄLIKE, H. *ET AL*. 2010. The Pacific Equatorial Age Transect, IODP Expeditions 320 and 321: building a 50-million-year-long environmental record of the equatorial Pacific Ocean. *Scientific Drilling*, **9**, 4–15.

PÄLIKE, H., FRAZIER, J. & ZACHOS, J. C. 2006*a*. Extended orbitally forced palaeoclimatic records from the equatorial Atlantic Ceara Rise. *Quaternary Science Reviews*, **25**, 3463–3475.

PÄLIKE, H., NORRIS, R. D. *ET AL*. 2006*b*. The heartbeat of the Oligocene climate system. *Science*, **314**, 1894–1898, http://dx.doi.org/10.1126/science.1133822.

PÄLIKE, H., LYLE, M. *ET AL*. 2010. *Proceedings IODP, 320/321*. Integrated Ocean Drilling Program Management International Inc., Tokyo.

PALMER, E., RICHTER, C. *ET AL*. 2010. Paleomagnetic and environmental magnetic properties of sediments from IODP Site U1333 (Equatorial Pacific), Abstract GP13A-0761. *2010 Fall Meeting, AGU*, San Francisco, CA, 13–17 December.

PARES, J. M. & MOORE, T. C. 2005. New evidence for the Hawaiian hotspot plume motion since the Eocene. *Earth and Planetary Science Letters*, **237**, 951–959, http://dx.doi.org/10.1016/j.epsl.2005.06.012

PISIAS, N. G., MAYER, L. A., JANECEK, T. R., PALMER-JULSON, A. & VAN ANDEL, T. H. (eds) 1995. *In*: *Proceedings of the Ocean Drilling Program*, Scientific Results, **138**. Ocean Drilling Program, College Station, TX.

RICHTER, C., ACTON, G., ENDRIS, C. & RADSTED, M . 2007. *Handbook for Shipboard Paleomagnetists*. ODP Technical Note, 34. World Wide Web Address: http://www-odp.tamu.edu/publications/tnotes/tn34/INDEX.HTM

SCHNEIDER, D. A. 1995. Paleomagnetism of some Leg 138 sediments: detailing Miocene magnetostratigraphy. *In*: PISIAS, N. G., MAYER, L. A., JANECEK, T. R., PALMER-JULSON, A. & VAN ANDEL, T. H. (eds) *Proceedings of the Ocean Drilling Program*, Scientific Results, **138**. Ocean Drilling Program, College Station, TX, 59–72.

SHACKLETON, N. J., HALL, M. *ET AL*. 2000. Astronomical calibration age for the Oligocene-Miocene boundary. *Geology*, **28**, 447–450.

TAUXE, L., LABRECQUE, J. L., DODSON, R. & FULLER, M. 1983. U-channels – a new technique for paleomagnetic analysis of hydraulic piston cores. *EOS Transactions, American Geophysical Union*, **64**, 219.

THOMAS, R., GUYODO, Y. & CHANNELL, J. E. T. 2003. U-channel track for susceptibility measurements. *Geochemistry, Geophysics and Geosystems (G3)*, **4**, 1050.

VAN ANDEL, T. H., HEATH, G. R. & MOORE, T. C. 1975. *Cenozoic History and Paleoceanography of the Central Equatorial Pacific Ocean: A Regional Synthesis of Deep Sea Drilling Project Data*. The Geological Society of America, Boulder, CO, Memoirs, **143**.

WEEKS, R., LAJ, C. *ET AL*. 1993. Improvements in long-core measurement techniques: applications in palaeomagnetism and palaeoceanography. *Geophysical Journal International*, **114**, 651–662.

WESTERHOLD, T., RÖHL, U. *ET AL*. 2012. Revised composite depth scales and integration of IODP Sites U1331-U1334 and ODP Sites 1218–1220. *In*: *Proceedings of the Integrated Ocean Drilling Program, 320/321*. Integrated Ocean Drilling Program Management International Inc., Tokyo.

Integrated biomagnetochronology for the Palaeogene of ODP Hole 647A: implications for correlating palaeoceanographic events from high to low latitudes

J. V. FIRTH[1]*, J. S. ELDRETT[2], I. C. HARDING[3], H. K. COXALL[4] & B. S. WADE[5,6]

[1]*Integrated Ocean Drilling Program, 1000 Discovery Drive, College Station, TX 77845, USA*

[2]*Shell Exploration and Production Inc., 3737 Bellaire Boulevard, PO Box 481, Houston, TX 77001-0481, USA*

[3]*University of Southampton, National Oceanography Centre, Southampton, European Way, Southampton SO14 3ZH, UK*

[4]*School of Earth and Ocean Sciences, Cardiff University, Main Building, Park Place, Cardiff CF10 3YE, UK*

[5]*Department of Geology and Geophysics, Texas A&M University, College Station, TX 77843, USA*

[6]*Present address: School of Earth and Environment, University of Leeds, Woodhouse Lane, Leeds LS2 9JT, UK*

Corresponding author (e-mail: firth@iodp.tamu.edu)

Abstract: Lower Eocene to Oligocene microfossil-rich hemipelagic sediments in ODP Hole 647A, southern Labrador Sea, provide a strategic section for resolving the early history of high North Atlantic climates and ocean circulation, and for correlating with carbonate-poor lower Cenozoic sediments in the Arctic and Nordic seas. Our new, integrated palaeomagneto- and multigroup biostratigraphy (63 dinoflagellate cyst, calcareous nannofossil, planktonic foraminifer and diatom datums) significantly improves Site 647 chronostratigraphy and provides a framework for future studies. This new age model, coupled with provisional $\delta^{18}O$ analyses, provides greater confidence in the location of significant ocean-climate events at this site, including the Eocene–Oligocene transition and the Middle Eocene Climatic Optimum. Early Eocene hyperthermals may also be present near the base of the section. Palaeomagnetic age control is significantly improved in the Eocene, but not in the Oligocene. Revised estimates of sedimentation and biogenic flux indicate changes in supply and preservation that may be climatically controlled. A Lower to Middle Eocene hiatus is more precisely constrained, with a *c.* 4 million year duration. Age and depth errors quantify the age uncertainties throughout the section. Our revised age model will play an important role in stratigraphic correlation between very high latitude and lower latitude sites.

Supplementary material: All tables with ages and age-derived calculations based on the Gradstein *et al.* (2004) timescale used herein are reproduced as supplementary tables using both the Gradstein *et al.* (2004) and the Cande & Kent (1995) timescales (Tables DS1–DS6). Discrete sample and shipboard pass-through cryomagnetometer palaeomagnetic data, planktonic foraminifer and fine fraction (<20 μm) stable isotope data, raw and processed core GRA density data, and specifications and results of GRA density spectral analyses are also provided as supplementary tables (Tables DS7–DS11). These tables are available at www.geolsoc.org.uk/SUP18546.

Ocean Drilling Program (ODP) Hole 647A is the highest latitude (53°20′N, 45°16′W) Eocene to Oligocene continuously cored sediment sequence overlying oceanic crust that has significant biogenic carbonate, biosilica and organic sediment components suitable for detailed multiproxy palaeoceanographic and biostratigraphic analysis. Located south of Greenland in the southern Labrador Sea, sediments accumulating at this site contain an imprint of climatic and oceanic variation in a key region of the northern North Atlantic Ocean, where today the water column overturns and dense deepwater masses exit from the Labrador and Nordic Seas (Fig. 1). Preservation of microfossils, especially

From: JOVANE, L., HERRERO-BERVERA, E., HINNOV, L. A. & HOUSEN, B. A. (eds) 2013. *Magnetic Methods and the Timing of Geological Processes*. Geological Society, London, Special Publications, **373**, 29–78.
First published online October 1, 2012, http://dx.doi.org/10.1144/SP373.9 © The Geological Society of London 2013.
Publishing disclaimer: www.geolsoc.org.uk/pub_ethics

calcitic varieties, is also exceptionally good at Site 647, offering great potential for geochemical palaeoceanographic and palaeoclimate studies. In these respects the Eocene–Oligocene sequence from Site 647 is an important and strategic target for (i) linking the virtually carbonate-free, very high northern latitude Palaeogene sediments of the Arctic Ocean and Nordic Seas to Palaeogene carbonate-rich sediments of the North Atlantic Ocean and elsewhere, and (ii) unravelling the Early Cenozoic history of high North Atlantic climate and ocean circulation. An obstacle to advancing Site 647 studies is that previous chronological control has been limited, especially in the Palaeogene, such that the location and imprint of important global events in the Early Cenozoic, including the Eocene–Oligocene transition (EOT, *c.* 34 Ma) and the Middle Eocene Climatic Optimum (MECO; *c.* 40 Ma), have been unconstrained.

In this paper, we provide new palaeomagnetic and multigroup biostratigraphic data (dinoflagellate cysts, calcareous nannofossils and planktonic foraminifers) that provide important new constraints on Site 647 chronostratigraphy. Diatom biostratigraphic events of Baldauf & Monjanel (1989) are also used based on newer diatom/magnetostratigraphic correlations from other Oligocene deep-sea drill sites. Stable isotope data were produced to help constrain the EOT. Our revised age model allows us to estimate new sedimentation and mass accumulation rates that have palaeoceanographic

significance. We also performed an independent test of the age model over one segment of the Site 647 record by comparing cycles in down-core physical properties data with predicted Milankovitch periodicities. Results support orbital frequencies of 23, 41 and 97 ka in part of the Site 647 Middle Eocene.

Site location, geological setting and lithostratigraphy

ODP Site 647 is located on the flank of Gloria Drift in the southern Labrador Sea (53°19.878′N, 45°15.720′W, 3869 m water depth, Fig. 1). The drift is one of a series of North Atlantic drift deposits formed by pulses of strong bottom-flowing currents produced through cooling and sinking of Norwegian Sea Overflow Water, which spills over the Greenland–Scotland Ridge through major channels at the Denmark Strait and Faroe–Shetland channel (Srivastava *et al.* 1987; Srivastava & Arthur 1989). The sedimentary sequence of the Gloria Drift is well characterized by seismic profiling and coring during DSDP Leg 12 (Laughton *et al.* 1972) and ODP Leg 105 (Srivastava *et al.* 1987; Srivastava & Arthur 1989). The seismic stratigraphy includes a set of prominent age-calibrated reflectors identified as R2, R3 and R4 (Fig. 2) that can be traced regionally. Reflector R2 corresponds to an unconformity above

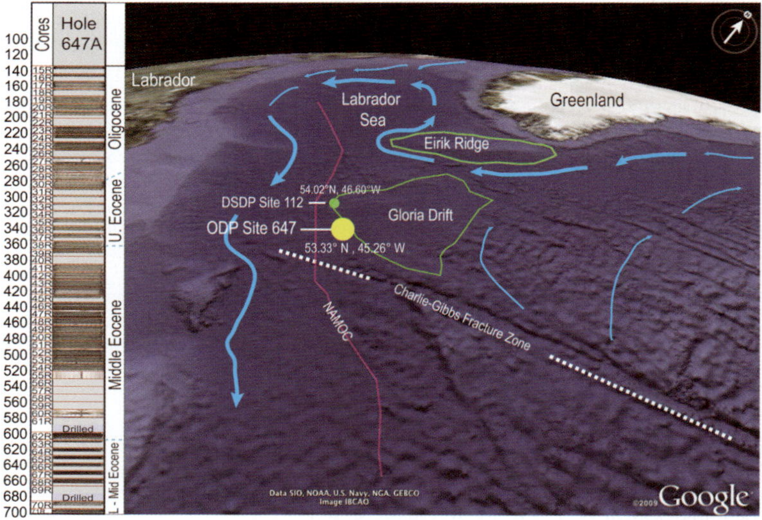

Fig. 1. Map of the NW Atlantic Ocean and Labrador Sea, showing locations of ODP Site 647 and DSDP Site 112, the Gloria Drift and Eirik Ridge Drift, the Charlie–Gibbs Fracture Zone and the Northwest Atlantic Mid-Ocean Channel (NAMOC). Modern deep water circulation patterns shown (blue arrows) are from Srivastava *et al.* (1987). Palaeogene core images, core recovery and major epoch boundaries for Hole 647A (Cores 15R–71R) are summarized on the left. Map courtesy of Google Earth, 2009.

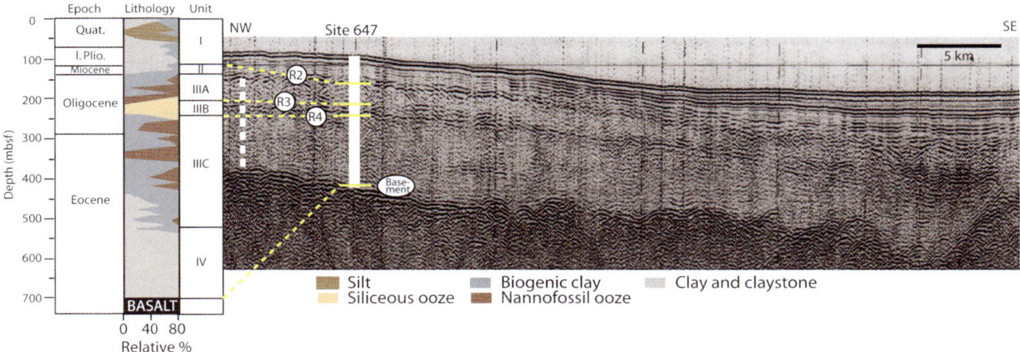

Fig. 2. Segment of Hudson single-channel seismic reflection line 8 (Hudson 84030-8S, original survey data, adapted from Arthur *et al.* 1989*a*, *b*) tied to the lithological section recovered at ODP Site 647. Time-calibrated lithological units are shown in relation to three prominent reflectors (R2, R3 and R4). Lithological contributions are determined from smear slides (Srivastava *et al.* 1987). White dashed line, Middle Eocene to Lower Oligocene part of the section.

upper Miocene strata that are overlain by the Plio-Pleistocene glaciomarine sequence (Srivastava *et al.* 1987; Srivastava & Arthur 1989). Shipboard and subsequent studies identified an expanded and continuous Middle Eocene to Oligocene sequence below the prominent regional seismic reflector identified as R4, which corresponds to the change from siliceous- to calcareous-biogenic claystones within the Lower Oligocene at about 240 m below the seafloor (mbsf; Firth *et al.* 1987).

Coring at Site 647 penetrated 699 m of sediment and 37 m of basaltic basement. Suboptimal drilling conditions resulted in variable core recovery, ranging from 100 to 10% in the Palaeogene, and only a single copy of the Eocene to Oligocene record (Fig. 1). The EOT itself appears to be intact. Plio-Pleistocene sediments, consisting of mud, clay, nannofossil clay and biosilica-bearing clay extend down to *c.* 116 mbsf (reflector R2), and unconformably overlie Miocene cores of similar lithology (13R and 14R; 116 to 135.4 mbsf; Srivastava *et al.* 1987; Fig. 2). These in turn unconformably overlie biogenic clay and clayey ooze of Oligocene to Middle Eocene age (135.4–530.3 mbsf; Fig. 1). Abundant biosilica occurs between 150 and 240 mbsf. From 530.3 to 699 mbsf (the sediment/basement basalt contact), the sediment consists of varicoloured claystones and foraminifer/nannofossil bearing claystones of Middle to Early Eocene age (Fig. 1). The basalt has a minimum age of 54–55 Ma based on the age of the oldest recovered sediments (Srivastava *et al.* 1987). A benthic foraminifer-based subsidence model (Kaminski *et al.* 1989) calculated an Eocene to Oligocene palaeowater depth of 2–3 km. The high clay content of Site 647 sediments, which results from the proximity to Labrador and Greenland continental sources, facilitated high sedimentation rates.

Other high-latitude North Atlantic Palaeogene sections

Spot coring at DSDP Site 112 to the NW of Site 647 revealed a similar succession of Palaeogene oozes and clays (Laughton *et al.* 1972). The Eocene–Oligocene boundary (EOB) is located at approximately *c.* 425 mbsf at this site compared with between 270 and 280 mbsf at Site 647. DSDP Site 407, also overlying ocean crust and containing carbonate sediments, is approximately 10° further north than Site 647 (*c.* 63°N), but it bottomed in Lower Oligocene sediments, and did not reach the anomaly C13-age basement (Luyendyk *et al.* 1978). Many other drill sites exist in the northeastern North Atlantic that contain Eocene and/or Oligocene sediments, with either spot or continuous coring. However, these all occur in shallower palaeowater depths and/or are positioned on top of continental crust (e.g. the Hatton–Rockall Plateau or Goban Spur). Re-studying Site 647 is a good starting point for improved correlation of all these North Atlantic sites to other regions of the world's oceans, where high-quality magnetostratigraphy and cyclostratigraphy are allowing unprecedented improvements in constraining the timing of global palaeoceanographic events (e.g. Pälike *et al.* 2006).

Site 647 microfossils and existing chronology

The hemipelagic clays that make up the *c.* 560 m of Palaeogene sediments at this site contain abundant and diverse assemblages of dinoflagellate cysts (dinocysts; Head & Norris 1989), calcareous nannofossils (nannofossils; Firth 1989), diatoms (Baldauf & Monjanel 1989), radiolarians (Lazarus & Pallant

1989) and benthic foraminifers (Kaminski *et al.* 1989). Planktonic foraminifers are fewer in number and of lower diversity than other microfossil groups in this site, and as compared with lower latitude Eocene and Oligocene sequences. High clay content, however, has resulted in good, typically excellent (glassy) preservation of foraminiferal and calcareous nannofossil calcite, of a quality comparable to Palaeogene material from Tanzania (Bown *et al.* 2008; Pearson & Burgess 2008). Believed to be virtually unaltered in its geochemical composition, isotopic data based on this glassy foraminifer material have resulted in a paradigm shift in our understanding of past warm climates (Pearson *et al.* 2001; Wilson *et al.* 2002; also new data in this study). The high North Atlantic location and fortuitous combination of abundant and diverse calcitic, siliceous and organic-walled microfossils make Hole 647A a strategic site for linking the virtually carbonate-free, very high northern latitude Palaeogene sediments of the Arctic Ocean (e.g. Clark 1974; Kitchel & Clark 1982; Bukry 1984; Mudie 1985; Mudie *et al.* 1986; Dell-Agnese & Clark 1994; Magavern *et al.* 1996; Backman *et al.* 2008; Sangiorgi *et al.* 2008; St. John 2008; Stickley *et al.* 2009; Suto *et al.* 2009) and of the Norwegian–Greenland Sea (N–G Sea; e.g. Manum 1976; Müller 1976; Schrader & Fenner 1976; Goll 1989; Manum *et al.* 1989; Firth 1996; Poulsen *et al.* 1996; Williams & Manum 1999; Eldrett *et al.* 2004, 2007; Brinkhuis *et al.* 2006) to Palaeogene carbonate-rich sediments of the North Atlantic Ocean (e.g. DSDP Legs 12, 41, 43, 47, 48, 49, 80, 82, 93, 95; ODP Legs 101, 149, 154, 171B, 207).

The original shipboard and post-cruise magnetostratigraphic studies carried out on Hole 647A yielded few well-constrained magnetic polarity reversal events (Srivastava *et al.* 1987; Clement *et al.* 1989). Palaeogene shipboard biostratigraphy was also minimal and mostly restricted to widely spaced shipboard sampling (Srivastava *et al.* 1987). Detailed post-cruise biostratigraphic studies for dinocysts (Head & Norris 1989), diatoms (Baldauf & Monjanel 1989), nannofossils (Firth 1989), radiolarians (Lazarus & Pallant 1989) and benthic foraminifers (Kaminski *et al.* 1989) provided valuable data on very diverse microfossil assemblages. However, the interpreted ages of diatom, radiolarian, dinocyst and benthic foraminifer bioevents were largely based on comparison to the Site 647 nannofossil zonation (Firth 1989) rather than on comparison with independent magnetostratigraphically calibrated events from other locations. For planktonic foraminifers, reduced diversity of these higher latitude sediments compared with low latitudes (where the biozonation schemes are better defined and calibrated) presented challenges to identifying useful bioevents such that much detail was lacking. Recent improvements in foraminifer taxonomic concepts and global datum

calibrations (Pearson *et al.* 2006; Wade *et al.* 2011), combined with increased sampling, offered opportunities to improve on this. Calibration of dinocyst biostratigraphy with magnetostratigraphy in the Norwegian–Greenland Sea (Eldrett *et al.* 2004) offered a similar opportunity for further biostratigraphic refinements, as well as better constraints for the interpretation of magnetic reversals.

Methods

Palaeomagnetics

(1) The original shipboard pass-through cryomagnetic data was reviewed and determined to be too disrupted by poor core recovery and core disturbance (biscuiting), owing to rotary coring, to be reliable for primary interpretations of polarity reversals. However, much of the 5 mT alternated field (AF) demagnetization of shipboard cryomagnetic data supports both old and new discrete palaeomagnetic data.

(2) Original shipboard and shore-based discrete palaeomagnetic data (Srivastava *et al.* 1987 and Clement *et al.* 1989) were reviewed and ranked in quality according to condition of core samples and level of demagnetization (0 (natural remanent magnetization, NRM), 3, 5 and 10 mT).

(3) New discrete palaeomagnetic data are presented here, with analysis of sample and data quality, and are integrated with the original discrete data (Srivastava *et al.* 1987; Clement *et al.* 1989). The new palaeomagnetic data are considered to be of better quality for this hole than most of the original data, because of the more complete demagnetization of each sample up to 80 mT.

Approximately 100 palaeomagnetic samples were obtained from the Eocene (Cores 647A–30R–54R) of Hole 647A with a sampling resolution of one sample per core section (i.e. one sample every 1–1.5 m). Palaeomagnetic measurements were made on discrete samples (7 cm^3 cubes) with a 2 G Enterprises narrow-access pass-through direct-current cryogenic magnetometer equipped with high-resolution pick-up coils. All measurements were made in the magnetically-shielded palaeomagnetic laboratory at the School of Ocean and Earth Science, University of Southampton. The NRM of the samples was subjected to stepwise AF demagnetization at peak alternating fields of 5, 10, 15, 20, 25, 30, 35, 40, 50, 60, 70 and 80 mT using an AF demagnetizer that is configured in-line with the cryogenic magnetometer. The stability of the NRM was investigated by inspection of vector component plots (Fig. 3). If the magnetization of the sediment decayed to the origin along a

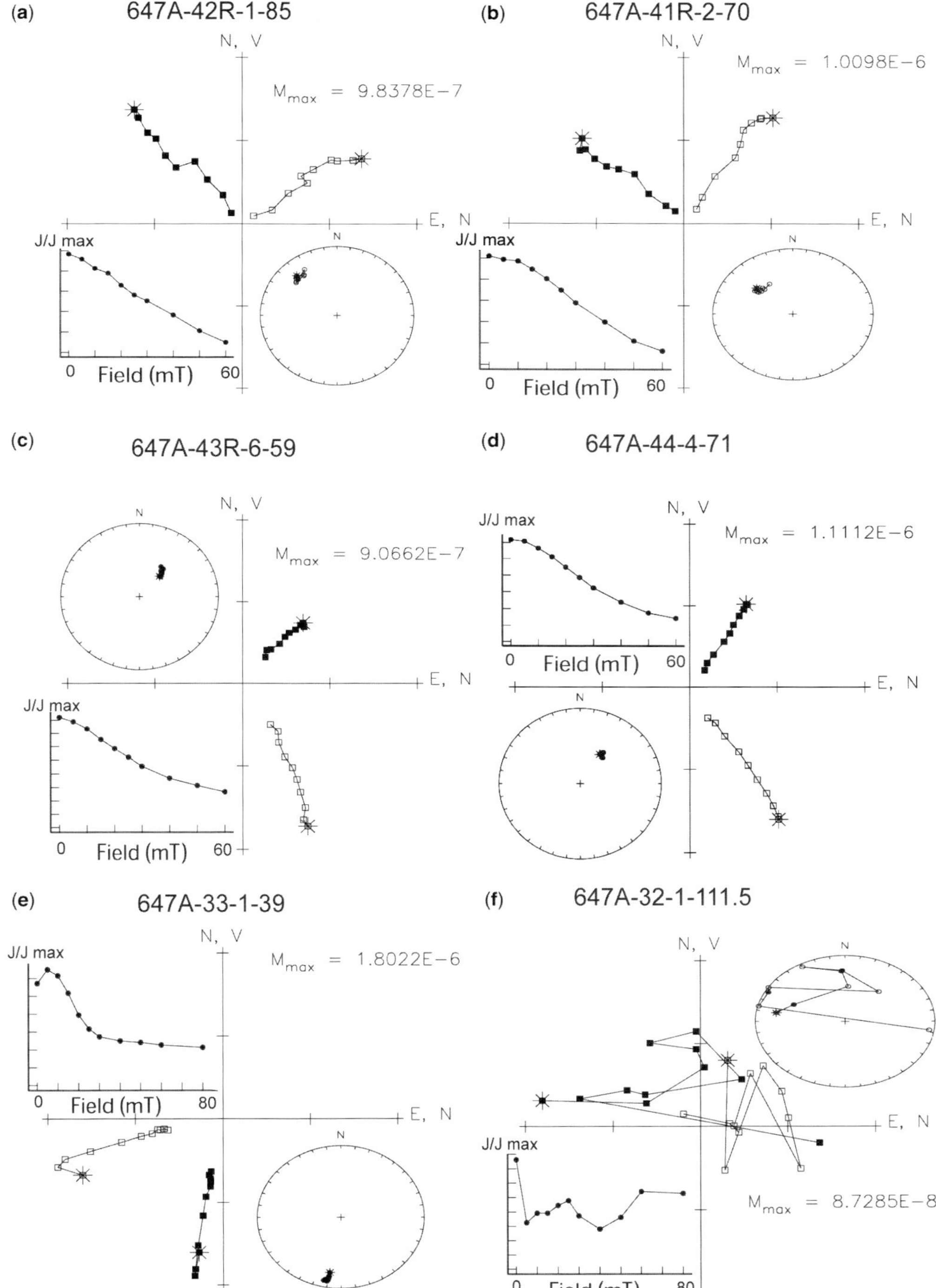

Fig. 3. Vector component diagrams of demagnetization behaviour for representative samples from Hole 647A. (**a**, **b**) Demagnetized to 60 mT, stable behaviour, reversed polarity. (**c**, **d**) Demagnetized to 60 mT, stable behaviour, normal polarity. (**e**) Demagnetized to 80 mT, stable behaviour, normal polarity. (**f**) Unstable magnetization.

straight path, the characteristic remanent magnetization (ChRM) was calculated using principal component analysis (Kirschvink 1980). The quality of fit of the best-fit line to the selected demagnetization steps was quantified by determining the maximum angular deviation for each measured sample (Kirschvink 1980). A minimum of four data points were used for principal component analysis and samples with maximum angular deviation values >15 were not considered when constructing the magnetic polarity stratigraphy. The studied cores were not azimuthally oriented. However, at the high latitude of the studied sites, the ambient geomagnetic field is dominated by a near-vertical component, which makes it possible to uniquely determine the polarity using the palaeomagnetic inclination alone, without azimuthal orientation of the cores. Unfortunately, ODP cores can sometimes be dominated by a near-vertical drilling-induced overprint (e.g. Roberts *et al.* 1996; Fuller *et al.* 1998; Acton *et al.* 2002) and it can be difficult to discriminate the ChRM component from the drilling-induced overprint. In such cases, a conservative interpretation was employed and samples suspected of a drilling-induced remagnetization were excluded from further analysis. Thermal demagnetization of an isothermal remanent magnetization was performed on a representative set of samples from different lithologies (Fig. 4). Isothermal remanent magnetizations were imparted using an inducing field of 0.9 T. Thermal demagnetization was performed at temperatures of 25, 50, 100, 150, 200, 250, 300, 350, 400, 450, 500, 550 and 580 °C.

We also cleaned the 5mT AF demagnetized shipboard pass-through cryomagnetic data by removing all values from the top and bottom 10–15 cm of each section, to avoid any end effects from measuring voids at the ends of each section. These data are plotted along with discrete sample data, but are not relied upon for our primary polarity interpretations. Although this higher-resolution dataset shows a lot of scatter, much of it shows similarity to the polarity values of the discrete samples.

Biostratigraphy

Since Site 647 was drilled in 1985, the age control on many biostratigraphic datums has been improved for most microfossil groups, based on direct correlation to polarity reversal stratigraphy, orbital cyclostratigraphy or both. Where possible, high-latitude bioevent datum calibrations were used (primarily sites of middle to high northern or southern latitudes) to reduce likely biases associated with latitudinal diachroneity and ecological exclusion. These are most common among diatoms and dinocysts, for which detailed northern high latitude studies exist (e.g. Eldrett *et al.* 2004). The resulting Site 647 age model presented here is used to test,

and also to provide new calibrations for, bioevents. The construction of age references for bioevents for each microfossil group is discussed below.

Dinoflagellate cysts. Eocene and Oligocene dinocysts are extremely diverse and abundant in Hole 647A (Head & Norris 1989), yet precise independent age control on Palaeogene dinocysts was lacking at the time of that publication. Damassa *et al.* (1990) calibrated 57 Palaeogene dinocyst ranges to standard nannofossil zones in 18 North Atlantic DSDP sites. However, these sites did not have good palaeomagnetic age control at that time. Eldrett *et al.* (2004) produced the first integrated dinocyst biomagnetostratigraphy from three sites (DSDP Site 338, ODP Sites 643 and 913) from the Norwegian–Greenland Sea. Many dinocyst datums recognized in these sites have also been found in the North Sea and northern Europe, which demonstrated the utility of this integrated palaeomagnetic/dinocyst stratigraphy for application to Hole 647A. The original slides of Martin Head and Geoff Norris were re-examined, as well as 41 additional samples, to identify any biostratigraphically useful dinocyst taxa with age calibrations from the Norwegian–Greenland Sea. Because Site 647 is further south, within the open North Atlantic, it may have experienced somewhat warmer or different palaeoceanographic conditions than the Norwegian–Greenland Sea sites. Therefore, the dinocyst datums in Hole 647A have preliminary age estimates that could be biased owing to the higher-latitude, restricted basinal setting of the Norwegian–Greenland Sea reference sites.

Calcareous nannofossils. Eocene and Oligocene nannofossils are also extremely diverse and abundant in Hole 647A (Firth 1989). Age control of Palaeogene nannofossil datums has developed in recent years, yet Site 647 is still fairly remote from any other Palaeogene sections and there is still no regional calibration as there is for dinocysts (Eldrett *et al.* 2004). Nannofossil age calibrations for this study, therefore, are derived from several middle- and high-latitude sites, primarily from the Southern Hemisphere, that are considered to resemble the cooler conditions at Site 647 (Table 1). Twenty additional smear slides were analysed to augment the initial research conducted by Firth (1989).

Diatoms. The Oligocene of Hole 647A has an abundant, diverse diatom assemblage (Baldauf & Monjanel 1989). As with the nannofossils, a compilation of palaeomagnetic age calibrations of diatoms derived from several middle- and high-latitude sites was used to provide new age calibrations of diatom taxa recorded by Baldauf & Monjanel (1989) in Hole 647A (Table 1).

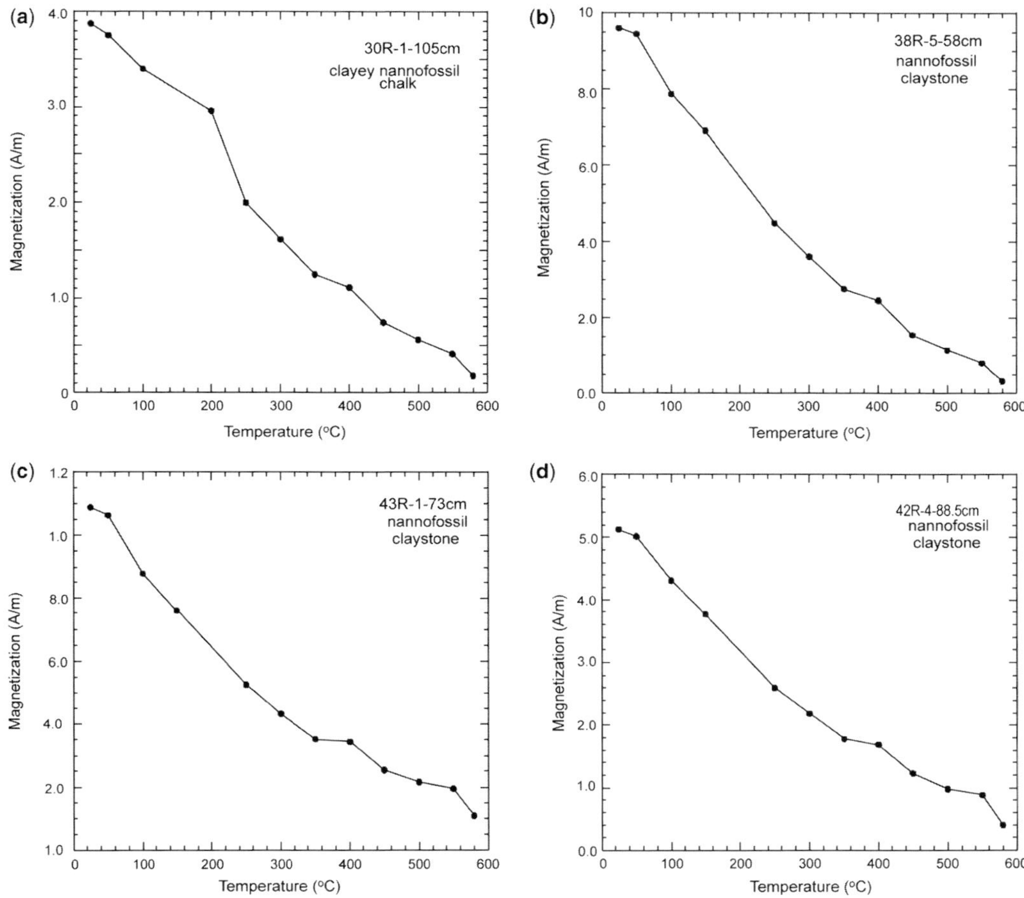

Fig. 4. Results of thermal demagnetization of an isothermal remanent magnetization for Hole 647A. In each case, the magnetization decreases to near zero values near 580 °C, which indicates that magnetite is the dominant magnetic mineral in the studied sediments.

Planktonic foraminifers. Species' evolutionary ranges and age calibration of Palaeogene planktonic foraminifers in the northern North Atlantic is poorly resolved. The distribution of species in Hole 647A was limited to the initial shipboard site report (Srivastava *et al.* 1987). Analysis of planktonic foraminifer stable isotopes was reported by Arthur *et al.* (1989*a*); however, they concluded that significant diagenesis had occurred, rendering the geochemical data from the foraminifers untrustworthy for palaeoceanographic interpretation. Their conclusions may have been affected by the sampling of variably preserved planktonic foraminifers in this hole. More recently, Pearson & Burgess (2008) showed that planktonic foraminifer preservation is variable in this hole, ranging from excellent 'glassy' preservation during the E–O sequence and parts of the Middle Eocene, to more poorly preserved in other parts of the section. A new treatment of the planktonic foraminifer biostratigraphy in Hole 647A is produced here, based on the taxonomy of Pearson *et al.* (2006).

For this study a total of 60 samples were examined for planktonic foraminifers from Cores 20R to 68R (*c.* 187.38–661.80 mbsf; Oligocene to Lower Eocene). An additional higher-resolution set of 67 samples from Cores 39R-3, 42–44 cm to 20R-3, 88–90 cm (369.52–187.38 mbsf) were also examined around the EOT. All samples were washed over a 63 μm mesh sieve and dried prior to microscopic examination. Oligocene Cores 23R–26R contain very few planktonic foraminifers, whereas specimens show signs of corrosion and recrystallization close to basement (Arthur *et al.* 1989*a, b*). Foraminiferal sized particles (>45 μm) make up <1% of the total sample weight as a consequence of heavy

Table 1. *Magnetostratigraphic calibrations of nannofossil datums from high and mid-latitude sites.* (converted to Gradstein et al. 2004, timescale)

Eocene and Oligocene Nannofossil Datums	Hsü et al. (1984), Site 522[†]	Parker et al. (1985), Site 558[‡]	Backman, (1986a), Site 528	Backman, (1986b), Site 577[§]	Backman (1987), Sites 522 and 523[¶]	Wei & Wise (1989) Site 516[1]	Thomas et al. (1990), Site 689	Thomas et al. (1990), Site 690	Marino & Flores (2002), Site 1090[2]	Roberts et al. (2003), Sites 744 and 748	Stickley et al. (2004); Site 1168	Stickley et al. (2004); Site 1170	Stickley et al. (2004); Site 1172	Agnini et al. (2007); Site 1262	Predicted mean ages in Hole 647A	Predicted age ranges in Hole 647A
Age references of nannofossil datums from middle and high latitudes	Age (Ma)	Age (Ma)	Age (Ma)	Age (Ma)	Age (Ma)	Age (Ma)	Age (Ma)	Age (Ma)	Age (Ma)	Age (Ma)	Age (Ma)	Age (Ma)	Age (Ma)	Age (Ma)	Age (Ma)	Age (Ma)
LO *S. ciperoensis*	28.7	27.4				27.1									27.9	27.1–28.7
HO *R. umbilicus* (>14 μm)					31.9	32.7	31.1	31.6		31.5	33.0				32.0	31.1–33.0
LO *S. distentus*						33.5									33.5	33.5
HO *I. recurvus*					32.9	32.8		33.9			33.0				33.4	32.8–33.9
HO *E. formosa*					32.9	33.1									33.0	32.9–33.1
HO *D. saipensis*					34.4	34.1			34.3						34.3	34.1–34.4
HO *D. barbadiensis*					34.6	34.4									34.5	34.4–34.6
HO *R. reticulate*					35.0	35.0			36.1	36.1					35.5	35.0–36.1
LO *R. bisecta* (>10 μm)					39.6	39.4			39.5–40.4						39.9	39.4–40.4
HO *C. solitus*						38.1	38.0			39.7		38.0	38.4		38.8	38.0–39.7
LO *R. reticulate*						40.5	39.7	41.0	40.4	41.1		38.0	39.3		39.6	38.0–41.1
HO *Nannotetrina cristata* (=HO *N.* spp.)					40.9	40.9									40.9	40.9
LO *R. umbilicus* (>14 μm)					41.2	41.5	41.1	41.8					42.9		42.0	41.1–42.9
HO *C. gigas*						41.5									41.5	41.5
LO *C. gigas*						44.0									44.0	44.0
HO *D. lodoensis*								48.0							48.0	48.0
HO *T. orthostylus*				50.9											50.9	51.0
LO *D. lodoensis*								52.3						53.5	52.9	52.3–53.5
LO *T. orthostylus*			53.9	53.9										54.2	54.0	53.9–54.2

HO, highest occurrence; LO, lowest occurrence.

[†]LO *S. ciperoensis* correlated to base C10n.2n.

[‡]Parker *et al.*'s (1985) more thorough analysis of nannofossils places the LO *S. ciperoensis* clearly at the base of Section 22-CC, not in Section 23-5 as reported by Bukry (1985) and Miller *et al.* (1985).

[§]Backman's (1986*a, b*) datum for the HO of *T. orthostylus* is from Site 577, which was in the tropics during the Paleocene to Early Eocene; however it is the only magnetostratigraphically correlated age for this datum.

[¶]HO *Nannotetrina* spp.; LO of *D. hesslandii* (equivalent to *R. bisecta*).

[1]HO *Nannotetrina fulgens.*

[2] LO *R. bisecta* between top and base of C18r.

Table 1. *Continued*

Thomas *et al.* (1990), Site 689 nannofossil data	Mid-point Depth of datum	Chrons surrounding datum[†]	Chron Depths surrounding datum	% Datum Depth > upper Chron Depth	Chron ages surrounding datum (Ma)	Datum Age (Ma)
HO *R. umbilicus* (>14 μm)	106.55	Top C12n	104.38	0.87	30.627	31.05
		Base C12n	106.88		31.116	
HO *C. solitus*	153.15	Top C17n.1n	135.77	0.97	36.512	37.99
		Top C18n.1n	153.70		38.032	
LO *R. reticulata*	158.96	Top C18n.1n	153.70	0.67	38.032	39.65
		Top C19n	161.54		40.439	
LO *R. umbilicus* (>14 μm)	164.37	Base C19n	163.16	0.51	40.671	41.14
		Top C20n	165.55		41.590	
Thomas *et al.* (1990), Site 690 nannofossil Data						
HO *I. recurvus*	89.91	Base C12n	84.01	0.50	31.116	33.72
		Base C16n.2n	95.70		36.276	
HO *R. umbilicus* (>14 μm)	84.92	Base C12n	84.01	0.08	31.116	31.52
		Base C16n.2n	95.70		36.276	
LO *R. reticulata*	105.85	Base C19n	105.65	0.32	40.671	40.97
		Top C20n	106.27		41.590	
LO *R. umbilicus* (>14 μm)	108.30	Top C20n	106.27	0.17	41.590	41.79
		Base C20n	118.23		42.774	
HO *D. lodoensis*	127.55	Base C21n	123.63	0.57	47.235	48.02
		Top C22n	130.48		48.599	
LO *D. lodoensis*	135.66	Base C23n.2n	133.18	0.60	51.901	52.35
		Top C24n.1n	137.33		52.648	
Parker *et al.* (1985), Site 558 nannofossil data						
LO S. ciperoensis	146.32	Top C9n	137.70	0.61	26.71	27.40
		Base C9n	151.77		27.83	

Oligocene diatom data	Schrader & Fenner (1976), Site 338[‡]	Roberts *et al.* (2003), Site 744	Roberts *et al.* (2003), Site 748	Gersonde *et al.* (1999); ODP Leg 177 IR	Predicted mean ages in Hole 647A	Predicted age ranges in Hole 647A
Age references of diatom datums from middle and high latitudes	Age (Ma)	Age (Ma)	Age (Ma)	Age (Ma)	Age (Ma)	Age (Ma)
LO *Rhizosolenia gravida/oligocaenica*		33.79	33.98	33.42	33.70	33.42–33.98
LO *Synedra jouseanus*		30.99	31.1	30.78	30.94	30.78–31.10
HO *Sceptroneis pupa*	31.12–33.27				32.19	31.12–33.27
LO *Sceptroneis pupa*	31.12–33.27				32.19	31.12–33.27

[*]Mean age values and errors based on maximum and minimum reported ages are calculated when data are available from multiple sites.
[†]Khan *et al.* (1985).
[‡]Calibrated to Eldrett *et al.* (2004).

clay dilution. Planktonic foraminifer assemblages are of lower diversity than tropical sites and many of the typical tropical–subtropical biozone markers are absent, as reported at nearby DSDP Site 112 (Berggren 1972). There is also a bias towards small test sizes with much of the assemblage represented in the 63–150 μm size fraction. Despite these limitations, some of the tropical–subtropical biozonal markers are present and provide constraints for the age model. The datums are largely those from Wade *et al.* (2011) with amendments for high-latitude regions where datums applicable to Site 647 cores have been calibrated or applied (Roberts *et al.* 2003; Cooper 2004; Huber & Quillévéré 2005).

Stable isotopes

Fine fraction (<20 μm) bulk sediment. Stable isotope analysis was conducted on 24 samples of the sediment fine fraction (<20 μm) from 36R-4, 113–115 cm to 29R-1, 2–3 cm. Samples were crushed with a mortar and pestle and analysed with a DeltaPlusXP mass spectrometer at Texas A&M University. Approximately 546 ± 170 μg of sample was reacted with concentrated phosphoric acid at 70 °C. Seven of the samples were analysed in duplicate, and two samples from Core 29R were run three or four times because of low wt% carbonate. The results were calibrated to PDB (Pee Dee Belemnite) through the NBS-19 reference standard ($\delta^{13}C = 1.95\%_o$, $\delta^{18}O = -2.20\%_o$). Standard deviations of the NBS-19 reference standard were 0.06‰ for $\delta^{18}O$. All values are reported in standard delta notation (δ) in parts per mil (‰) relative to Vienna Pee Dee Belemnite.

Benthic foraminifers. Stable isotopic analyses were also carried out using the shallow-infaunal species *Oridorsalis umbonatus* (>63 μm) in 17 samples across the E–O interval from 39R-3, 42–44 cm to 20R-3, 88–90 cm. Analyses were performed using a ThermoFinnigan MAT252 mass spectrometer equipped with an automated KIEL III carbonate preparation unit at Cardiff University. Stable isotope results were calibrated to the Vienna Pee Dee Belemnite scale using the international standard NBS19 and analytical precision was better than 0.05‰ for $\delta^{18}O$.

Physical properties and spectral analysis

Sediment physical property data was collected during the drilling of Site 647 on a multisensor track (Srivastava *et al.* 1987). This includes GRA density data from all the cores, collected at an average interval of 0.6 cm. Where core recovery and quality were good, spectral analysis was performed to look for cyclicity that might provide an independent test of the robustness of the age model. We cleaned the raw GRA data by comparison with core photographs, removing spurious data caused by cracks, drilling disturbance or concretions. We then ran a 35-point (*c.* 20 cm) smoothing on the raw cleaned data, and subtracted the smoothed values from the raw values to create a detrended dataset. We analysed both the smoothed data and the detrended data using the Blackman–Tukey, Periodogram and Maximum Entropy methods in AnalySeries 2.0.4.2© (Paillard *et al.* 1996). Since the data came from semi-lithified sediments with abundant cracks, we chose to use these three spectral analysis methods instead of just the most commonly used Blackman–Tukey method, to see if our results were reproducible. Spectral analyses were set for higher resolution rather than higher confidence, and therefore the results may contain some spurious peaks. The goal was to establish if any observed lithological cycles fit close to or within the cycle thickness ranges for orbital periodicities (precession, obliquity and eccentricity) predicted from our age model.

Constructing an age model with error boundaries

All data are presented on the Gradstein *et al.* (2004) timescale. The primary age model is based on linear interpolation between interpreted Chron boundaries, with possible alternative Chron boundaries showing uncertainties in our interpretation in some portions of the hole. Each interpreted Chron boundary occurs between two samples. We have used the midpoint depth between the top and bottom bounding samples for each event, and used the depths of the top and bottom bounding samples to represent the depth error for the event (Fig. 5). To determine the linear interpolation between two Chron boundaries, we connect the midpoint depth values of each event with a solid black line (Fig. 5). This represents the primary age model line of correlation (LOC). The age model error between two Chron boundaries is shown by connecting the top and bottom bounding samples of each event with solid red lines (Fig. 5).

Visual inspection of age-depth scatter graphs of our new compiled biostratigraphic datum list (Table 2) enabled us to subdivide the stratigraphic column into four depth intervals in which simple linear regression lines could be calculated for the biostratigraphic datums. These regression lines in turn were used to aid in the interpretation of the palaeomagnetic data to determine the final magnetostratigraphy for the new age model. We present and discuss these results from the oldest sediments (Core 71R) to youngest sediments (Core 15R) as Intervals 1–4 (Figs 6–9), and summarize the overall age model in Figure 10.

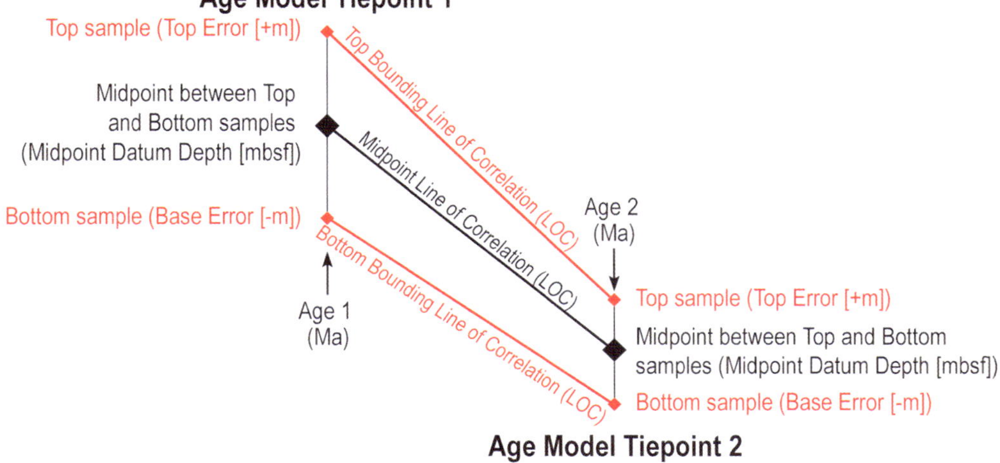

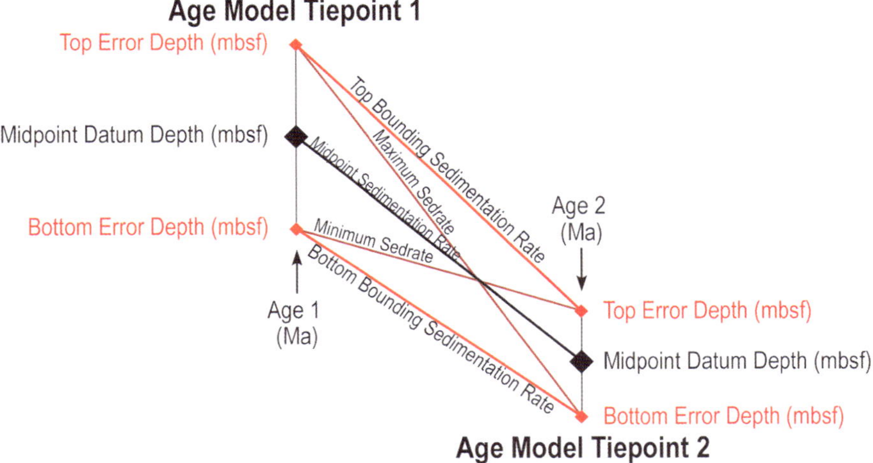

Fig. 5. (**a**) Illustration of a biostratigraphic or palaeomagnetic datum, including its depth error, and the 'mid-point', 'top' bounding and 'bottom' bounding lines of correlations (LOCs) between two datums. (**b**) Illustration of the method used to calculate these horizons allowing calculation of, maximum and minimum sedimentation rates between two age control datums. The primary age model is based on interpreted palaeomagnetic Chron boundary datums, guided by the distribution of biostratigraphic datums in the hole. Alternative possible LOCs are plotted with dashed lines instead of solid lines in Figures 6–10.

Results

Biostratigraphy

The stratigraphic ranges of all 63 age-diagnostic taxa from the four microfossil groups are summarized in Tables 2–6, and most of these are illustrated in Figures 11–13. Their age/depth distribution in the hole, even without palaeomagnetic data, show a significantly more robust dataset for determining an age model than the original research from Leg 105. Their lowest and highest occurrences were instrumental in

Table 2. *Depths, ages and error ranges of all stratigraphically useful microfossil bioevents from Hole 647A*

Fossil datum number	Top sample	Bottom sample	Age references	Fossil group	Datum type	Taxon	Datum depth Top (mbsf)	Datum depth Base (mbsf)	Datum depth Mean (mbsf)	Depth error (±) (m)	Minimum age (Ma)	Maximum age (Ma)	Mean age (Ma)	Age error (±) (Myr)
63	17R-1, 107–112	17R-1, 116–118	1	CN	LO	*Sphenolithus ciperoensis*	155.87	155.96	155.92	0.05	27.1	28.7	27.9	0.8
62	20R-3, 52–54	20R-5, 50–52	1	D	LO	*Synedra jouseana*	187.02	190.00	188.51	1.49	30.6	33.6	32.1	1.5
61	21R-1, 92–94	21R-3, 77–79	2	PF	HO	*Turborotalia ampliapertura*	194.02	196.87	195.45	1.43	30.4	30.4	30.4	0.0
60	22R-1, 11–13	23R1–118–120	1	CN	HO	*Isthmolithus recurvus*	202.81	213.48	208.15	5.33	32.8	33.9	33.4	0.5
59	23R-4, 91–93	23R-5, 94–96	1	CN	HO	*Reticulofenestra umbilicus* (>14 μm)	213.48	214.90	214.19	0.71	31.1	33.0	32.0	1.0
58	23R-5, 102–104	23R-CC	1	D	HO	*Sceptroneis pupa*	219.32	221.82	220.57	1.25	31.1	33.3	32.2	1.1
57	24R-2, 83–85	24R-3, 86–88	1	D	LO	*Sceptroneis pupa*	224.23	225.76	225.00	0.77	31.1	33.3	32.2	1.1
56	21R-1, 92–94	21R-3, 77–79	3	PF	HO	*Chiloguembelina ototara*	227.07	241.40	234.24	7.17	29.6	29.6	29.6	0.0
55	25R-3, 112–114	25R-3, 112–114	4	DC	HO	*Spiniferites manumii*	235.62	235.62	235.62	0.00	31.1	32.2	31.6	0.6
54	25R-5, 61–63	27R-1, 51–53	1	CN	HO	*Ericsonia formosa*	238.11	250.91	244.51	6.40	32.9	33.1	33.0	0.1
53	25R-5, 61–63	27R-1, 51–53	1	CN	LO	*Sphenolithus distentus*	238.11	250.91	244.51	6.40	33.5	33.5	33.5	0.0
52	26R-1, 30–32	27R-1, 47–49	2	PF	HO	*Pseudohastigerina naguewichiensis*	241.40	250.87	246.14	4.74	32.2	32.2	32.2	0.0
51	26R-CC	27R-CC	1	D	LO	*Rhizosolenia gravida/ oligocaenica*	242.33	259.62	250.98	8.65	33.8	34.0	33.9	0.1
50	28R-CC	29R-1, 9–11	2	PF	HO	*Pseudohastigerina micra*	269.70	269.79	269.75	0.05	33.9	33.9	33.9	0.0
49	30R-2, 62–65	30R-3, 100–104	4	DC	LO	*Chiropteridium galea*	281.52	283.40	282.46	0.94	33.4	33.8	33.6	0.2
48	30R-6, 30–33	30R-7, 30–32	4	DC	LO	*Spiniferites manumii*	287.20	288.70	287.95	0.75	33.4	33.8	33.6	0.2
47	31R-1, 4–6	31R-2, 4–6	1	CN	HO	*Discoaster saipanensis*	289.04	290.54	289.79	0.75	34.1	34.4	34.3	0.1
46	31R-1, 56–58	31R-CC	2	PF	HO	*Turborotalia cerroazulensis*	289.56	298.60	294.08	4.52	34.0	34.0	34.0	0.0
45	32R-2, 19–21	33R-1, 10–12	4	DC	HO	*Areosphaeridium diktyoplokum*	300.29	308.40	304.35	4.05	33.2	33.6	33.4	0.2
44	32R-2, 11–13	33R-1, 138–140	1	CN	HO	*Reticulofenestra reticulata*	300.21	309.68	304.95	4.74	35.0	36.1	35.5	0.5
43	33R-1, 10–12	33R-1, 91–94	4	DC	HO	*Areosphaeridium michoudii*	308.40	309.21	308.81	0.41	35.4	35.5	35.4	0.0

42	33R-1, 10–12	33R-1, 91–94	4	DC	HO	*Heteraulacacysta porosa*	308.40	309.21	308.81	0.41	36.2	36.3	36.2	0.1
41	33R-1, 138–140	35R-1, 36–38	1	CN	HO	*Discoaster barbadiensis*	309.68	327.96	318.82	9.14	34.4	34.6	34.5	0.1
40	35R-2, 39.5–41.0	35R-2, 74.5–76.0	2	PF	HO	*Globigerinatheka index*	328.57	329.85	329.21	0.64	34.5	34.5	34.5	0.0
39	36R-1, 10–13	37R-2, 88–90	4	DC	HO	*Cribroperidinium* sp. 2 (Damassa *et al.* 1990)	337.20	349.08	343.14	5.94	35.5	35.9	35.7	0.2
38	39R-2, 77–80	39R-3, 108–112	4	DC	HO	*Phthanoperidinium distinctum*	368.37	370.20	369.29	0.91	37.3	39.2	38.3	0.9
37	43R-2, 120–124	44R-4, 95–97	4	DC	HO	*Diphyes colligerum*	407.50	419.95	413.73	6.22	39.1	40.6	39.8	0.7
36	47R-6, 42–44	47R-7, 23–28	1	CN	LO	*Reticulofenestra bisecta* (>10 μm)	451.42	452.73	452.08	0.66	39.4	40.4	39.9	0.5
35	47R-5, 49–51	48R-3, 60–62	2	PF	LO	*Globigerinatheka index*	449.99	456.80	453.40	3.41	41.9	41.9	41.9	0.0
34	47R-7, 23–28	48R-1, 91–93	1	CN	HO	*Chiasmolithus solitus*	452.73	454.11	453.42	0.69	38.0	39.7	38.8	0.9
33	49R-1, 55–57	49R-6, 40–42	2	PF	HO	*Acarinina mcgowrani*	463.35	470.70	467.03	3.67	37.7	37.7	37.7	0.0
32	49R-1, 55–57	49R-6, 40–42	2	PF	HO	*Acarinina bullbrooki*	463.35	470.70	467.03	3.67	39.8	39.8	39.8	0.0
31	50R-2, 76–78	50R-6, 59–61	5	PF	HO	*Acarinina collactea*	474.76	480.59	477.68	2.91	37.4	37.4	37.4	0.0
30	50R-2, 76–78	50R-6, 59–61	2	PF	HO	*Turborotalia frontosa*	474.76	480.59	477.68	2.91	38.8	38.8	38.8	0.0
29	51R-1, 87–89	51R-5, 71–73	4	DC	LO	*Svalbardella cooksoniae*	482.97	488.81	485.89	2.92	40.6	41.3	40.9	0.4
28	51R-1, 87–89	51R-5, 71–73	4	DC	HO	*Wetzeliella ovalis*	482.97	488.81	485.89	2.92	39.3	42.9	41.1	1.8
27	51R-1, 87–89	51R-5, 71–73	4	DC	LO	*Cribroperidinium* sp. 2 (Damassa *et al.* 1990)	482.97	488.81	485.89	2.92	41.4	42.3	41.9	0.4
26	52R-4, 25–28	53R-4, 121–125	4	DC	LO	*Rhombodinium rhomboideum*	496.55	507.11	501.83	5.28	39.3	39.4	39.4	0.1
25	52R-4, 25–28	53R-4, 121–125	4	DC	LO	*Wetzeliella ovalis*	496.55	507.11	501.83	5.28	40.5	46.1	43.3	2.8
24	54R-6, 15–17	56R-1, 11–16	1	CN	LO	*Reticulofenestra reticulata*	518.75	530.41	524.58	5.83	38.0	41.1	39.6	1.5
23	56R-1, 11–16	57R-CC, 0–2	1	CN	HO	*Nannotetrina* spp. (*cristata*)	530.41	549.50	539.96	9.55	40.9	40.9	40.9	0.0
22	59R-1, 35–37	61R-1, 29–33	2	PF	LO	*Turborotalia pomeroli*	559.75	579.09	569.42	9.67	41.5	41.5	41.5	0.0
21	60R-4, 70–77	61R-1, 12–14	4	DC	LO	*Phthanoperidinium distinctum*	574.30	578.92	576.61	2.31	40.5	43.2	41.8	1.4
20	61R-1, 12–14	62R-1, 0–2	4	DC	LO	*Phthanoperidinium comatum*	578.92	598.10	588.51	9.59	41.4	41.4	41.4	0.0
19	62R-1, 0–2	62R-2, 92–94	4	DC	LO	*Enneadocysta arcuata*	598.10	600.52	599.31	1.21	40.6	42.4	41.5	0.9

(*Continued*)

Table 2. *Continued*

Fossil datum number	Top sample	Bottom sample	Age references	Fossil group	Datum type	Taxon	Datum depth Top (mbsf)	Datum depth Base (mbsf)	Datum depth Mean (mbsf)	Depth error ($\pm$) (m)	Minimum age (Ma)	Maximum age (Ma)	Mean age (Ma)	Age error ($\pm$) (Myr)
18	62R-2, 92–94	62R-5, 88–91	4	DC	HO	*Diphyes ficusoides*	600.52	604.98	602.75	2.23	44.5	46.2	45.3	0.9
17	62R-5, 87–89	63R-1, 36–38	1	CN	OCC/ HO (>)	*Chiasmolithus gigas*	604.97	608.16	606.57	1.59	41.5	41.5	41.5	0.0
16	63R-1, 44–46	63R-1, 105–108	4	DC	LO	*Heteraulacacysta porosa*	608.24	608.85	608.55	0.31	40.4	44.3	42.4	2.0
15	63R-1, 44–46	63R-1, 105–108	4	DC	HO	*Cerebrocysta magna*	608.24	608.85	608.55	0.31	44.6	46.2	45.4	0.8
14	63R-1, 139–142	63R-2, 36–38	1	CN	LO	*Reticulofenestra umbilicus (>14 µm)*	609.20	609.66	609.43	0.23	41.1	42.9	42.0	0.9
13	63R-1, 36–38	63R-2, 36–38	1	CN	OCC/ LO (<)	*Chiasmolithus gigas*	608.16	609.66	608.91	0.75	44.0	44.0	44.0	0.0
12	64R-1, 30–32	64R-2, 78–80	4	DC	HO	*Eatonicysta ursulae*	617.80	619.78	618.79	0.99	48.0	48.6	48.3	0.3
11	64R-2, 78–80	64R-3, 28–30	4	DC	HO	*Dracodinium pachydermum*	619.78	620.78	620.28	0.50	46.0	46.5	46.3	0.3
10	64R-3, 88–91	65R-1, 27–29	4	DC	HO	*Charlesdowniea columna*	621.38	627.47	624.43	3.05	48.3	48.8	48.6	0.2
9	65R-1, 117–121	66R-1, 15–18	4	DC	LO	*Cerebrocysta magna*	628.37	636.95	632.66	4.29	48.0	48.6	48.3	0.3
8	65R-1, 117–121	66R-1, 15–18	4	DC	LO	*Diphyes ficusoides*	628.37	636.95	632.66	4.29	49.2	50.0	49.6	0.4
7	65R-1, 117–121	66R-1, 15–18	4	DC	LO	*Dracodinium pachydermum*	628.37	636.95	632.66	4.29	49.2	49.4	49.3	0.1
6	65R-3, 133–138	66R-1, 52–54	1	CN	HO	*Discoaster lodoensis*	631.53	637.32	634.43	2.90	48.0	48.0	48.0	0.0
5	67R-3, 50–52	68R-1, 47–49	1	CN	HO	*Tribrachiatus orthostylus*	650.00	656.67	653.34	3.33	51.0	51.0	51.0	0.0
4	68R-4, 110–112	68R-4, 110–112	6	PF	LO	*Acarinina primitiva*	661.80	661.80	661.80	0.00	51.6	51.6	51.6	0.0
3	68R-4, 110–112	68R-4, 110–112	6	PF	LO	*Acarinina collactea*	661.80	661.80	661.80	0.00	48.9	48.9	48.9	0.0
2	69R-CC, 14–16	70R-1, 47–49	1	CN	LO	*Discoaster lodoensis*	666.24	685.67	675.96	9.71	52.3	53.5	52.9	0.6
1	71R-2, 41–41	71R-2, 41–41	1	CN	LO (<)	*Tribrachiatus orthostylus* (base of sediment)	696.71	696.71	696.71	0.00	53.9	54.2	54.0	0.2

Fossil group: CN, Calcareous nannofossil; D, Diatom; DC, Dinoflagellate cyst; PF, Planktonic foraminifer.

Datum type: HO, highest occurrence; LO, lowest occurrence; OCC, Occurrence; >, minimum depth; <, maximum depth.

Top and bottom sample IDs: Core/type-section, interval (cm).

Age references: 1, Firth, this paper; 2, Wade *et al.* (2011); 3, Cooper (2004); 4, Eldrett *et al.* (2004); 5, Roberts *et al.* (2003); 6, Huber & Quillévéré (2005).

Table 3. *Stratigraphic ranges of key Eocene and Oligocene dinocyst marker species in Hole 647A. Taxonomy follows that of Eldrett et al. 2004*

Core	Type	Section	Interval Top (cm)	Interval Bottom (cm)	Datum depth Top (mbsf)	Areosphaeridium diktyoplokum	Areosphaeridium michoudii	Cerebrocysta magna	Charlesdowniea columna	Chiropteridium galea complex	Cribroperidinium sp. 2	Diphyes colligerum	Diphyes ficusoides	Dracodinium pachydermum	Eatonicysta ursulae	Enneadocysta arcuata	Heteraulacacysta porosa	Phthanoperidinium comatum	Phthanoperidinium distinctum	Rhombodinium rhomboideum	Spiniferites manumii	Svalbardella cookoniae	Wetzeliella ovalis
25	R	3	112	114	235.62																**X**		
25	R	CC			240.05													X*					
28	R	2	105	108	262.65					X*						X					X	X	
28	R	4	101	103	265.61													X			X		
30	R	1	30	32	279.97													X			X	X	
30	R	2	62	65	281.25					X								X			X		
30	R	3	100	104	283.40											X		X			X	X	
30	R	6	30	33	287.20													X			**X***		
30	R	7	30	32	288.70													X					
31	R	1	17	20	289.17											X							
32	R	2	19	21	300.29																		
33	R	1	10	12	308.40	**X**																	
33	R	1	91	94	309.21	X	**X**									X	**X**	X			X		
35	R	1	135	138	328.95	X										X	X*	X				X	
35	R	2	74.5	76	329.87												X	X					
35	R	CC	0	2	332.70	X	X*									X	X					X*	
36	R	1	0	2	337.20	X	X										X					X	
37	R	2	88	90	349.08	X	X				**X**												
38	R	1	0	2	356.40											X	X	X					
38	R	4	86	89	361.76											X	X						
38	R	CC	7.5	10.5	366.08											X	X					X	
39	R	2	77	80	368.37												X	X					
39	R	3	108	112	370.18												X	X	X				
41	R	3	58	61	388.98	X											X	X		X			
42	R	2	98	102	397.58	X											X			X			
43	R	2	120	124	407.50																		
44	R	4	95	97	419.95						X	**X***					X	X				X	
45	R	1	139	143	425.49	X										X	X					X	

(Continued)

Table 3. *Continued*

Core	Type	Section	Interval Top (cm)	Interval Bottom (cm)	Datum depth Top (mbsf)	Areosphaeridium diktyoplokum	Areosphaeridium michoudii	Cerebrocysta magna	Charlesdowniea columna	Chiropteridium galea complex	Cribroperidinium sp. 2	Diphyes colligerum	Diphyes ficusoides	Dracodinium pachydermum	Eatonicysta ursulae	Enneadocysta arcuata	Heteraulacacysta porosa	Phthanoperidinium comatum	Phthanoperidinium distinctum	Rhombodinium rhomboideum	Spiniferites manumii	Svalbardella cookoniae	Wetzeliella ovalis
46	R	5	60	63	440.40	X					X							X					
47	R	2	92	95	445.92													X		X*			
48	R	2	107	110	455.77						X						X	X	X	X		X	
50	R	3	91	94	476.41						X							X	X	X			
50	R	6	95	98	480.95						X	X						X	X	X		X	
51	R	1	87	89	482.97						**X***					X	X		X	X		**X**	X*
51	R	5	71	73	488.81											X*			X	X			X
52	R	4	25	28	496.72								X			X	X		X	**X**			**X**
53	R	4	121	125	507.11	X	X											X					
54	R	6	31	34	518.91											X			X*				
56	R	1	8	10	530.38		X									X							
60	R	4	70	77	574.30	X											X			**X**			
61	R	1	12	14	578.00											X							
62	R	1	0	2	578.92											**X**							
62	R	2	92	94	600.52																		
62	R	5	87	89	604.98	X*							**X***			X	X						
63	R	1	44	46	608.24	X						X	X				**X**						
63	R	1	105	108	608.85			**X**				X											
64	R	1	30	32	617.80	X		X					X		**X**								
64	R	2	78	80	619.78			X							X								
64	R	3	28	30	620.78			X*					X	**X**	X								
64	R	3	88	91	621.38	X		X							X								
65	R	1	27	29	627.47	X		X	**X***			X	X		X								
65	R	1	117	121	628.27			**X**	X			X	**X**	**X***	X				**X**				
66	R	1	15	18	636.95			X	X						X*			**X**					

Bold represents biostratigraphic datum.
*Specimens photographed from that sample.

Table 4. *Stratigraphic ranges of key Eocene and Oligocene calcareous nannofossil marker species in Hole 647A. Taxonomy follows that of Firth (1989)*

Core	Type	Section	Interval Top (cm)	Interval Bottom (cm)	Depth (mbsf)	*Sphenolithus ciperoensis*	*Sphenolithus distentus*	*Reticulofenestra umbilicus* (<14 μm)	*Isthmolithus recurvus*	*Ericsonia formosa*	*Discoaster saipanensis*	*Discoaster barbadiensis*	*Reticulofenestra reticulata*	*Chiasmolithus solitus*	*Reticulofenestra bisecta* (<10 μm)	*Nannotetrina* spp. (*N. cristata*)	*Chiasmolithus gigas*	*Discoaster lodoensis*	*Tribrachiatus orthostylus*
15	R	1	68	70	136.08										C				
15	R	2	66	68	137.56		P								C				
15	R	3	3	5	138.43										C				
16	R	1	117	119	146.27										C				
16	R	2	90	92	147.50		P								C				
16	R	3	36	38	148.46	P	R								C				
17	R	1	107	112	155.87	**P***	R								C				
17	R	1	116	118	155.96		R		r						C				
17	R	2	21	23	156.51										C				
17	R	4	111	113	160.41		R*								C				
17	R	5	111	113	161.91		R								C				
18	R	1	32	34	164.42		F		r						C				
18	R	2	28	30	165.56		R		r						A				
19	R	1	9	11	173.89		R		r						C				
19	R	4	14	16	178.44		R		r		r				C				
19	R	6	24	26	181.54		R		r						A				
20	R	1	106	108	184.56		R								A				
20	R	3	75	76	187.25		R	r	r						C				
20	R	5	75	76	190.25										C				
21	R	1	67	69	193.77		R								C				
21	R	2	69	71	195.29		P	r	r						C				
21	R	3	69	71	196.79		R								C				
22	R	1	11	13	202.81			r							C				
23	R	1	118	120	213.48				**R**						F				
23	R	2	110	112	214.90		P	**R**	R		r				F				
23	R	3	111	113	216.41		R	R	R						C				
23	R	5	94	96	219.24			R	R						C				
24	R	4	23	25	226.63		P	R	P						F				
25	R	1	58	60	232.08			R							C				
25	R	4	61	63	236.61		R	F	R						C				
25	R	5	61	63	238.11		**P**	R	R						F				
27	R	1	51	53	250.91			A	C	**R**					F				
27	R	7	17	22	259.57			C	F						C				
28	R	1	71	73	260.81			A	A	R					C				
28	R	4	72	74	265.32			A	C	R					C				
29	R	1	2	3	269.72			C*		R*									
30	R	1	16	18	279.56			C	C	R					C				
30	R	1	80	82	280.20				F*						C				
30	R	2	15	17	281.05			A	F	R					C				
30	R	6	18	20	287.08			C	C	R					C				
31	R	1	4	6	289.04			C	C	R					C				
31	R	2	4	6	290.54			A	C	F	**P**				A				
32	R	1	61	63	299.21			C	C	F					C				
32	R	2	11	13	300.21			C	C	F			R		F				
33	R	1	138	140	309.68			A	C	R			R			**R**			
35	R	1	36	38	327.96			C	C	F			R	R	C				
35	R	1	77	80	328.37			C							C*				

(Continued)

Table 4. *Continued*

Core	Type	Section	Interval Top (cm)	Interval Bottom (cm)	Depth (mbsf)	Sphenolithus ciperoensis	Sphenolithus distentus	Reticulofenestra umbilicus (<14 μm)	Isthmolithus recurvus	Ericsonia formosa	Discoaster saipanensis	Discoaster barbadiensis	Reticulofenestra reticulata	Chiasmolithus solitus	Reticulofenestra bisecta (<10 μm)	Nannotetrina spp. (N. cristata)	Chiasmolithus gigas	Discoaster lodoensis	Tribrachiatus orthostylus
35	R	3	35	37	330.95			C	F	F	F	R	C		C				
36	R	1	108	110	338.28			F	F	P	R		F		F				
36	R	4	108	110	342.78			C	F	F	F	R	C		C				
37	R	2	47	49	348.67			C	F	R	F	R	C		C				
37	R	4	47	49	351.67			C	C	F	F	R	C		C				
38	R	1	101	103	357.41			R	F	R	F		C		F				
38	R	6	101	103	364.91			C	F	R	F	R	C		C				
39	R	1	75	77	366.85			C	F	R	F	R	C		C				
39	R	3	75	77	369.85			F	F	R	F		C		F				
41	R	1	24	26	385.64			C	C	R	F	R	C		C				
41	R	3	24	26	388.64			C	F	R	F	R	C*		C				
41	R	3	58	61	388.98			C			F*	R*	C		C				
42	R	1	19	21	395.29			C	F		R	R	A		C				
42	R	7	19	21	404.29			C	C	R	R	R	F		A				
43	R	1	24	26	405.04			F	F	R	F		C		C				
43	R	6	15	17	412.45			C	F	R	F	P	C		C				
43	R	7	17	19	413.97				F	F	F		C		C				
44	R	1	103	105	415.53			C	F	R	F		C		C				
44	R	6	100	102	423.00			C	R	R	C	P	A		C				
45	R	1	42	44	424.52			C			F	R	C		C				
45	R	2	28	30	425.88			C	P	F	R		C		C				
46	R	3	33	35	437.13			C	R	F	F	R	C		F				
46	R	6	101	103	442.31			C		F	F	F	C		F				
47	R	2	63	65	445.63			F		R	R	R	C		R				
47	R	5	43	45	449.93			C		F	F	R	A		F				
47	R	6	42	44	451.42			F					F		**R**				
48	R	1	91	93	454.11			F		F	F	R	C	**R**					
48	R	2	91	93	455.61			C		F	R	R	F	P		P?			
48	R	5	90	92	460.10			C		F	C	F	R	R					
49	R	1	91	93	463.71			C		F	F	R	R	R					
49	R	6	90	92	471.20			F		R	R		R	C					
50	R	1	109	111	473.59			F		F		F	R	P					
50	R	6	109	111	481.09			C		F	R	F	R	R					
51	R	1	110	112	483.20			F		F	F		F						
51	R	5	114	116	489.24			C		F		R	R	R					
52	R	2	122	124	494.52			C		F		P	F	C					
52	R	6	122	124	500.52			F		F	R	R	R	C					
53	R	1	36	38	501.76			C		F		R	P	A					
53	R	5	39	41	507.79					F		R	P	A					
54	R	1	34	36	511.44			C			C	P	R	A		R?			
54	R	5	18	20	517.28					F		R	P	C					
54	R	6	15	17	518.75			R		R		R	**P**	C					
56	R	1	11	16	530.41			F		R		R		C					
57	R	CC	0	2	549.50			C		R		P		C			**P**		
59	R	CC	0	2	559.45									C*					
60	R	CC	5	7	578.71			F		F		R		A					
62	R	1	82	84	598.92						R	R	R	F				P*	
62	R	2	83	85	600.43			R			R	F	R	C					

(Continued)

Table 4. *Continued*

Core	Type	Section	Interval Top (cm)	Interval Bottom (cm)	Depth (mbsf)	*Sphenolithus ciperoensis*	*Sphenolithus distentus*	*Reticulofenestra umbilicus* (<14 μm)	*Isthmolithus recurvus*	*Ericsonia formosa*	*Discoaster saipanensis*	*Discoaster barbadiensis*	*Reticulofenestra reticulata*	*Chiasmolithus solitus*	*Reticulofenestra bisecta* (<10 μm)	*Nannotetrina* spp. (*N. cristata*)	*Chiasmolithus gigas*	*Discoaster lodoensis*	*Tribrachiatus orthostylus*
62	R	5	87	89	604.97			P		R	P	R		F		P			
63	R	1	36	38	608.16			R		F	R	R		C			**P***		
63	R	1	139	142	609.20			**R**											
66	R	1	52	54	637.32						R			F				**F**	
66	R	2	49	51	638.79					R		R		P				F	
67	R	1	140	142	647.90					F		R						F	r
67	R	3	50	52	650.00							P		R				F	
68	R	1	47	49	656.67					P	P	R		P				F	**R**
68	R	2	48	50	658.18					F		P		F				F	F*
68	R	4	36	38	661.06					R				R				F	F
69	R	CC	14	16	666.24												**C***		F
70	R	1	47	49	685.67														F
70	R	2	47	49	687.17									C					R
70	R	3	47	49	688.67									F					P
71	R	1	145	146	696.25					R				F					R
71	R	2	24	26	696.54							R		F					
71	R	2	41	41	696.71							R		F					**R**

Bold represents biostratigraphic datum.
*Specimens photographed from that sample.

deciphering several difficult portions of the stratigraphic section, including the presence of a major unconformity, and showing relatively unchanging sedimentation rates in a couple of intervals where core recovery was low and palaeomagnetic data was sparse. Their main contribution in this study was to improve the identification of significantly more palaeomagnetic polarity reversal datums with which we have constructed our new primary age model.

Magnetic polarity reversal stratigraphy

In general, the Eocene part of Hole 647A still has better quality palaeomagnetic data for interpreting polarity reversal stratigraphy than the Oligocene, because the latter has intervals of poor core recovery, greater core disturbance and a biosiliceous rich interval (about 150–250 mbsf) with very low magnetic intensity. The Eocene polarity reversal stratigraphy is improved over Clement *et al.* (1989), whereas the Oligocene palaeomagnetic data above Core 30R is still based solely on shipboard (Srivastava *et al.* 1987) and initial post-cruise (Clement *et al.* 1989) data. The Oligocene palaeomagnetic data

cannot be overinterpreted, because most of it is 0 mT demagnetized NRM data. The best we are able to do is to present 'speculative chron equivalents' to depth intervals based on comparison with biostratigraphic data. In other words, such intervals can only be considered as 'might be consistent' with a Chron assignment of '*x*'.

Besides ranking palaeomagnetic data quality, we first separated our polarity reversal interpretations and chron assignments into three types (Figs 6–10), before comparison with biostratigraphic data:

Chron datums with only one possible (reasonable) depth: these are recorded as the midpoints between positive and negative inclination values, with depth errors of plus or minus ($\pm$) one-half the distance between adjacent positive or negative inclination samples. These are plotted as black diamonds in our figures (Figs 6–10). Solid black lines connect the midpoint depths of these Chron datums, whereas solid red lines connect the top and bottom bounding errors of these Chron datums.

Chron datums with greater than one possible depth assignment or age assignment: each possibility is recorded as additional midpoint values, each with ($\pm$) errors. These are plotted as red diamonds

Table 5. *Stratigraphic ranges of key Eocene and Oligocene planktonic foraminifera biostratigraphic marker species in Hole 647A. Taxonomy follows Pearson* et al. *(2006)*

Core	Type	Section	Top (cm)	Bottom (cm)	Depth (mbsf)	Chiloguembelina ototara	Globigerinatheka index	Turborotalia frontosa	Turborotalia pomeroli	Turborotalia cerroazulensis	Turborotalia ampliapertura	Pseudohastigerina micra	Pseudohastigerina naguewichiensis	Paragloborotalia opima	Acarinina collactea	Acarinina bullbrooki	Acarinina primitiva	Acarinina mcgowrani
20	R	5	95	97	190.45													
21	R	1	92	94	194.02									R				
21	R	3	77	79	196.87						R							
23	R	1	134	136	213.64						R							
23	R	3	54	56	215.84													
25	R	1	58	60	232.08													
25	R	4	91	93	236.91									R				
26	R	1	30	32	241.40						R?			F				
27	R	1	29	30	250.69									X*				
27	R	1	47	49	250.87	C							R					
27	R	CC			260.10	X												
28	R	1	103	105	261.13	F					R		R					
28	R	2	35	37	261.95						F							
28	R	4	95	97	265.55	C												
28	R	CC			269.70								X					
29	R	1	9	11	269.79	C					C	R	R					
30	R	1	73	75	280.13	A					R	R	R	F				
30	R	3	67	69	283.07	A*					C*	R	R	F				
31	R	1	56	58	289.56	C					C	R						
31	R	CC			298.60					X								
32	R	1	83	85	299.43	F						R						
33	R	1	103	106	309.33	C					R	R						
35	R	1	91	93	328.51	C												
35	R	1	97	99	328.57	C												
35	R	2	40	41	329.50					X								
35	R	2	75	76	329.85	C	R							C				
35	R	3	62	64	331.22	A	F			R				C				
35	R	4	41	43	332.51	A	C*			R								
35	R	CC			337.20	X												
36	R	1	69	71	337.89	C					R							
36	R	3	82	84	341.02					R*								
36	R	4	76	78	342.46	C	R			R		F						
37	R	2	57	59	348.77	A												
37	R	4	85	87	352.05	A					F	F						
38	R	1	25	27	356.65	A					F	F						
38	R	4	44	46	361.34	F					F							
38	R	7	26	28	365.66	C												
39	R	1	88	90	366.98	C					R	R						
39	R	3	42	44	369.52	F					F	F						
41	R	6	54	56	393.44	A	R											
42	R	5	56	58	401.66													
43	R	6	62	64	412.92	C	C											
44	R	3	50	52	418.00													
45	R	1	60	62	424.70	C	C											
46	R	2	78	80	436.08													
47	R	5	49	51	449.99	R	R											
48	R	3	60	62	456.80													
49	R	1	55	57	463.35	F			A			C						
49	R	6	40	42	470.70				A								F	C

(Continued)

Table 5. *Continued*

Core	Type	Section	Top (cm)	Bottom (cm)	Depth (mbsf)	*Chiloguembelina ototara*	*Globigerinatheka index*	*Turborotalia frontosa*	*Turborotalia pomeroli*	*Turborotalia cerroazulensis*	*Turborotalia ampliapertura*	*Pseudohastigerina micra*	*Pseudohastigerina naguewichiensis*	*Paragloborotalia opima*	*Acarinina collactea*	*Acarinina bullbrooki*	*Acarinina primitiva*	*Acarinina mcgowrani*
50	R	2	76	78	474.76	R			A									
50	R	6	59	61	480.59	R		**F**	A			F			**R**			
51	R	2	115	117	484.75	F			A*			F						
51	R	6	74	76	490.34	F		F	A			F						
52	R	4	70	72	497.00			F	A			R			C			
52	R	6	125	127	500.55	R		F	A			F			C			
53	R	3	77	79	505.17			C	F			F						F
54	R	3	60	62	514.70			A	F									F
59	R	1	35	37	559.75			A	**F**			R			C			C
61	R	1	29	33	579.09			A				C						R
62	R	2	9	11	599.69			A				C				C*		
62	R	6	24	26	605.84													
63	R	1	65	67	608.45			C				A			C			F
64	R	3	32	34	620.82													
65	R	3	55	57	630.75													
66	R	3	31	33	640.11			C				F			C			C
67	R	3	56	58	650.06													
68	R	1	119	121	657.39			C				C			F	F	C	F
68	R	4	110	112	661.80	X		F				F*			**C***		**X***	C

Bold represents biostratigraphic datum.
*Specimens photographed from that sample.

Table 6. *Stratigraphic ranges of key Oligocene diatom marker species in Hole 647A*

Core	Type	Section	Interval Top (cm)	Interval Bottom (cm)	Depth (mbsf)	*Rhizosolenia oligocaenica*	*Synedra jouseana*	*Sceptroneis pupa*
15	R	1	125	127	136.65		R	
15	R	2	124	126	138.14		R	
15	R	CC			139.62		R	
16	R	1	49	51	145.59		R	
18	R	1	50	52	164.60	R	R	
18	R	CC			166.52		R	
19	R	5	124	126	181.04	R		
20	R	1	48	50	183.98	R		
20	R	3	52	54	187.02	R	**R**	
20	R	5	50	52	190.00	R		
20	R	CC			191.49	R		
21	R	1	124	126	194.34	R		
21	R	2	124	126	195.84	R		
21	R	3	124	126	197.34	R		
23	R	CC			221.82			R
24	R	2	83	85	224.23			R
24	R	4	86	88	227.26	R		
24	R	CC			227.50	R		
26	R	CC			242.33	**R**		

Taxonomy follows Baldauf & Monjanel (1989).
Bold represents biostratigraphic datum.

on all figures. Dashed red lines connect these alternative Chron datums.

Speculative Chron intervals interpreted mainly from cleaned 5 mT AF demagnetized shipboard cryomagnetic data are plotted as purple dashed lines for the Lower to Middle Eocene, to show possible consistency with biostratigraphic datums where discrete palaeomagnetic data does not exist, and core recovery is low (Fig. 6). Speculative Chron datums were interpreted mainly from 0 mT AF discrete samples in the Oligocene, which are of very low fidelity, and are plotted only to show possible consistency with biostratigraphic datums. These

are plotted as purple diamonds connected by purple dashed lines (Fig. 9).

Interval 1: Lower Eocene to lower Middle Eocene; Cores 647A–71R to 62R (700–600 mbsf); Chrons C24n–C21n or C20n? (Fig. 6)

Interval 1 comprises claystones at the base of the Site 647 sedimentary section recovered in Cores 647A–71R to 62R (700–600 mbsf) (Fig. 6). Radiometric age dating of basalts from Hole 647A (Roddick

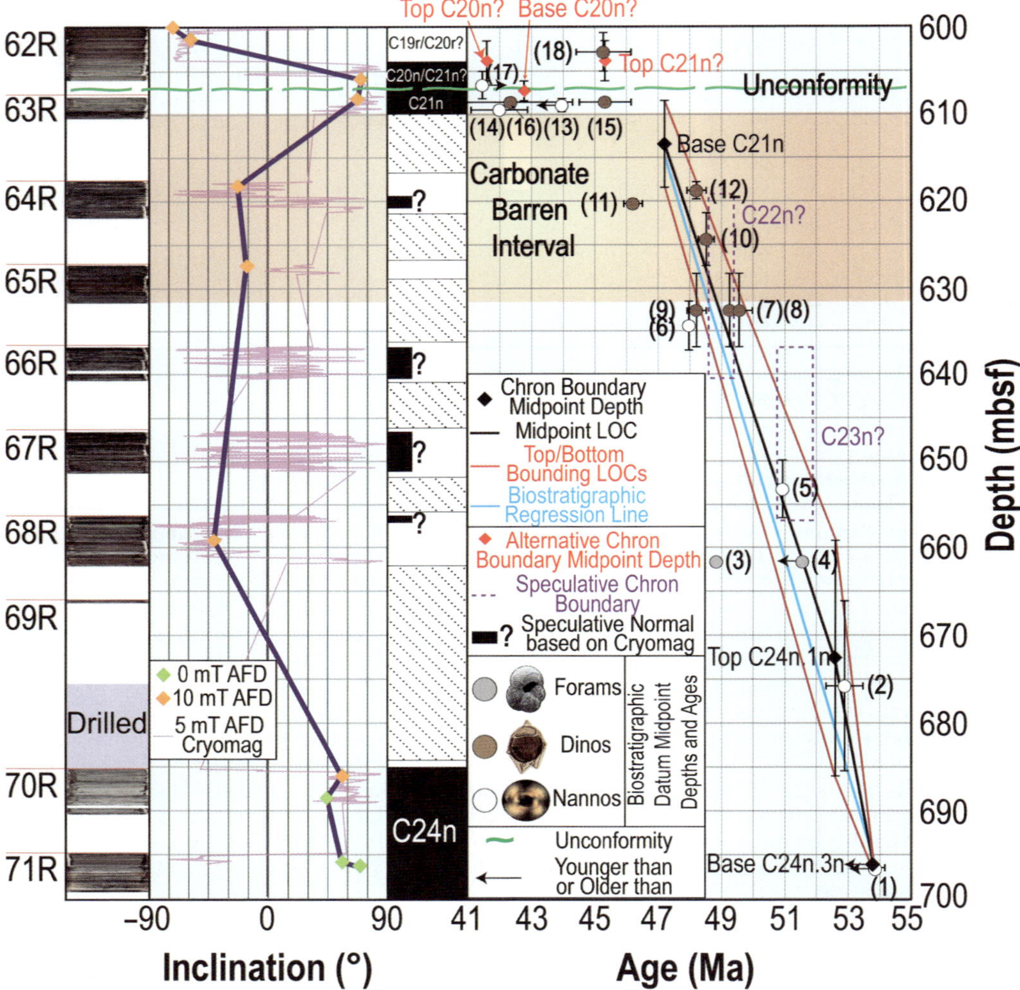

Fig. 6. Integrated magnetobiochronological summary for ODP Hole 647A Interval 1 (700–600 mbsf, Chron C24n–C21n or C20n?). Shown are magnetic inclination values, biostratigraphic datums, biostratigraphic regression line, interpreted primary LOCs and Chron boundaries, tentative LOCs, and interpreted unconformity from Cores 647A–71R to 62R. See Table 2 for key to biostratigraphic datums.

1989) could not constrain the age of basement because of extensive alteration of the basalts. Therefore, the age of basal sediments is used to interpret the age of the oceanic crust at this site.

Core 647A–70R and Sections 71R-1 and -2 contain four discrete positive inclination values. These are consistent with the mostly positive inclination values from cryomagnetic data for Cores 70R and the top of 71R (Srivastava *et al.* 1987). We interpret these positive inclinations to correspond to Chron C24n based on the lowest occurrence of *Tribrachiatus orthostylus* in Core 71R (datum no. 1, Table 2, Fig. 6), consistent with the interpreted age of crust based on magnetic anomaly patterns (Srivastava 1978). Mid-point, top-bounding and bottom-bounding LOCs are therefore constructed between the base of C24n.3n and top of C24n.1n (Fig. 6).

Poor core recovery above Core 70R, up through Core 62R, severely limits the number of interpretable magnetostratigraphic datum events in the upper part of Interval 1. The change from two discrete negative inclinations in Cores 64R and 65R to one discrete positive inclination sample near the top of Core 63R is interpreted to mark the Chron C21n/C21r boundary. This is supported by the biostratigraphic linear regression line, as well as by most of the cryomagnetic data in the same intervals as the discrete values. We therefore constructed our primary mid-point, top and bottom bounding LOCs between the top of C24n.1n and the base of C21n. The mid-point LOC is very close to the biostratigraphic linear regression line, and the top and bottom bounding LOCs encompass most of the biostratigraphic datums between Cores 64R and 70R. Five new dinocyst datums (numbers 7–10, and 12; Table 2, Fig. 6) occur close to the mid-point LOC as well as the calculated biostratigraphic regression line within and adjacent to an interval labeled the Carbonate Barren Interval (Core 65R up to the bottom half of Core 63R; see also Fig. 10 for percentage carbonate for this hole). These datums lend support to the interpretation that sedimentation rates are roughly constant between approximately 670 and 615 mbsf.

Cores 68R–64R contain 5 mT AF cryomagnetic data outside of three discrete palaeomagnetic samples. These data suggest there could be intervals of normal polarity within these cores. We have drawn speculative boundaries for Chrons C22n and C23n based on these possible normal polarity intervals. These indicate that Chrons C22n and C23n could possibly fall close to the primary mid-point LOC between top C24n.1n and base C21n. Further high-resolution discrete sample palaeomagnetic analyses are worth analysing in these cores to determine if their magnetic reversal stratigraphy can be further refined.

A cluster of five bioevents covering an age range of roughly 4 Ma occur within Section 63R-1. This suggests the presence of a hiatus at approximately 608 mbsf (Fig. 6). Both discrete and cryomagnetic inclination measurements indicate an up-section switch from positive to negative polarity in the lower part of Core 62R. One interpretation is that this reversal represents the top of Chron C21n. This would result in only a small change in sedimentation rate from the mid-point for base Chron C21n to mid-point of top Chron C21n. The highest occurrence of *Diphyes ficusoides* (datum no. 18, Table 2, Fig. 6), which plots within the depth of error of this possible top of C21n, in Core 62R, supports this interpretation. The distribution of the biostratigraphic data above 600 mbsf (see below), combined with the cluster of datums at *c.* 608 mbsf, supports a more abrupt interruption to sedimentation. Consequently, we interpret the rare occurrence of *D. ficusoides* in Core 62R as being reworked, and interpret Core 63R as being bounded above by an unconformity. This requires Core 62R to be significantly younger than Core 63R, probably within the age range of Chron C20n, which provides an alternative, and we believe more likely, interpretation of the Core 63R–62R interval (see discussion below).

Interval 2: Middle Eocene; Cores 647A–63R to 42R (610–395 mbsf); Chrons C21n?–C18n (Fig. 7)

Interval 2 unconformably overlies Interval 1 and is represented by claystone and nanno/foraminifer-bearing claystones. It includes an interval of extremely poor recovery through Cores 61R–55R; however, five biostratigraphic datums do occur between Cores 61R and 55R, and help to constrain the age model using the calculated biostratigraphic regression line (Fig. 7). Interval 2 extends upwards to a magnetic reversal within Core 42R. The simple biostratigraphic linear regression line (Fig. 7) between 600 and 395 mbsf is based on 19 microfossil datums. The top of this line falls close to, and is therefore consistent with, a relatively well-defined normal polarity interval (discussed below), whereas the base of the line falls close to the three youngest of the five biostratigraphic datums that strongly indicate an unconformity between Cores 63R and 62R.

The one discrete positive inclination value at the base of Core 62R, along with positive cryomagnetic inclinations in the same interval, most likely represents a normal-polarity interval younger than C21n. The shift in discrete and cryomagnetic data from positive to negative inclination within Core

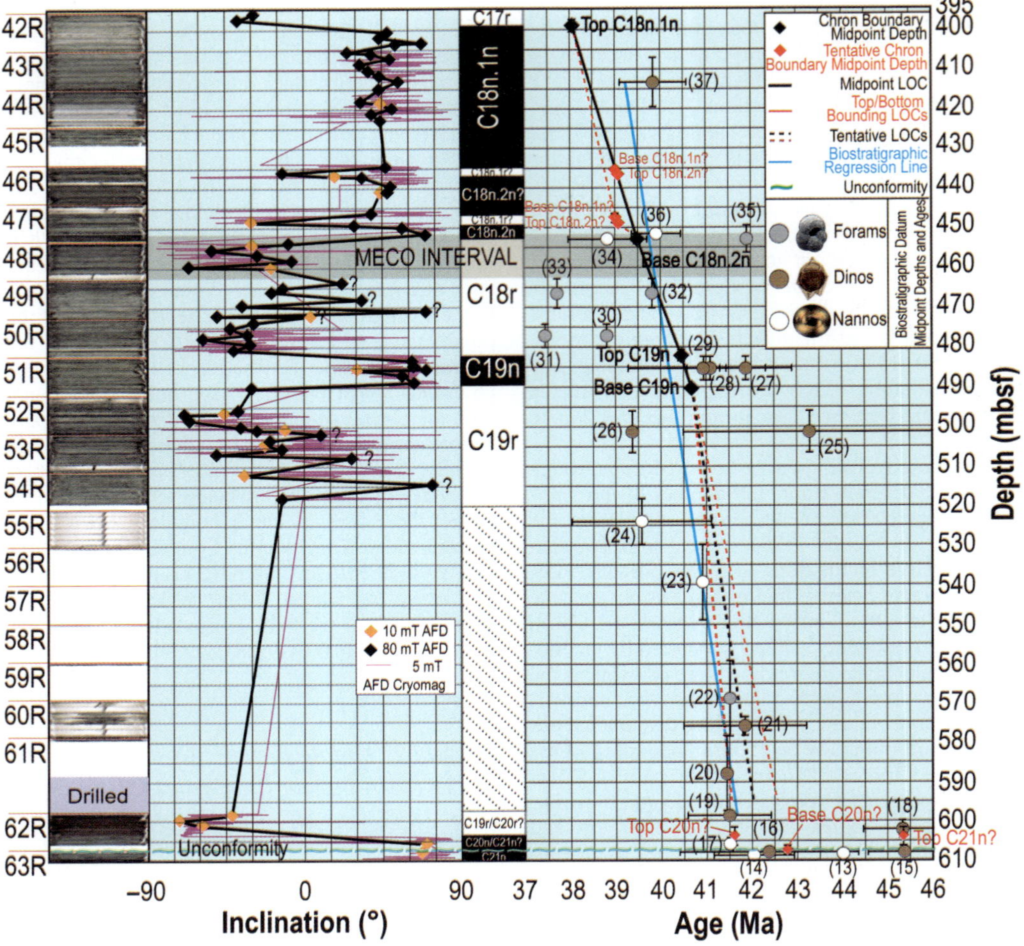

Fig. 7. Integrated magnetobiochronological summary for ODP Hole 647A Interval 2 (610–395 mbsf, Chron C20n or C21n? to C17r). Format as for Figure 6. Includes the depth of interpreted unconformity between Cores 647A–63R and 62R. Our age model predicts the MECO event to occur in Core 48R up to the base of Core 47R. See Table 2 for key to biostratigraphic datums.

62R also indicates that this represents the top of a normal polarity interval. Based on correlations to the ages of the youngest biostratigraphic datums in Core 63R, and to the regression line trend of biostratigraphic datums from Cores 62R up to 51R, this shift from normal to reversed polarity most closely fits that of the top of Chron C20n. Since the normal polarity interval at the base of Core 62R is short, and because there is also some missing section at the base of Core 62R, we cannot be absolutely sure where within C20n this normal interval may reside. Considering this, we have drawn only a tentative LOC for our primary age model from the base of Chron C19n to the mean age between the top and base of Chron C20n (black dashed line in Fig. 7). We have also drawn tentative top and bottom bounding LOCs (red dashed lines in Fig. 7) from the base of C19n to the base C20n and top C20n, respectively, to provide an envelope of error.

Cores 61R–55R contain very low to no core recovery and therefore contain sparse biostratigraphic datums and no palaeomagnetic data. Both discrete and cryomagnetic data in Cores 54R and 53R contain widely fluctuating magnetic inclination values, and do not show consistency of values between the two datasets. Core 52R has mostly consistent negative inclination values in both the discrete and cryomagnetic data. A distinct shift to positive inclination values occurs in the base of Core 51R and another distinct shift to reversed polarity occurs in the base of overlying Core 50R.

Clement *et al.* (1989) previously interpreted Chron C19n using shipboard cryomagnetic data (Srivastava *et al.* 1987) and additional discrete data for the interval between *c.* 490 and 481 mbsf (lower Core 51R to lower Core 50R). Our new discrete magnetic polarity data support their interpretation, with small depth errors for both the top and base of C19n. Three biostratigraphic datums (the lowest occurrence of *Svalbardella cooksoniae*, and highest occurrences of *Wetzeliella ovalis* and *Cribroperidinium* sp. 2.) within Core 51R are not inconsistent with this interpretation. In addition, Core 52R can be interpreted with some confidence to represent the top of C19r.

From Cores 50R–42R there is a robust pattern of magnetic polarity reversals but sparse biostratigraphic control provides little help in their interpretation and identification. Above Chron C19n, Cores 50R and 48R have predominantly negative discrete inclination values and predominantly negative cryomagnetic inclination values. Core 49R has a scatter of four negative and four positive discrete inclination values, but does not have any cryomagnetic data. Above Core 48R is a thick interval of predominantly positive discrete and cryomagnetic inclination values. Therefore we interpret the interval from the base of Core 50R to the top of Core 48R to be of overall reversed polarity, and assign it to Chron C18r.

From the base of Core 47R (*c.* 453 mbsf) upwards to near the top of Core 42R (*c.* 400 mbsf), both discrete and cryomagnetic inclinations are predominantly positive, which we interpret to represent the normal polarity interval Chron C18n. 1n–C18n.2n. The very short time interval represented by Chron C18n.1r (66 ka, Gradstein *et al.* 2004) might be represented by either of two discrete negative inclination values, in Core 46R (*c.* 436 mbsf) or 47R (*c.* 449 mbsf). These discrete values are also closely associated with thin cryomagnetic negative inclination excursions. Both possible locations of C18n.1r are plotted on Figure 7, and show possible alternative LOCs that are close to, or only slightly different than, our primary LOC. Our interpretation of the base of Chron C18n.2n is the same as the original interpretation of Clement *et al.* (1989).

Interval 3: Upper Eocene to Lower Oligocene;
647A Core 42R to Core 27R (400–250 mbsf);
Chrons C18n–C13n (Fig. 8)

Interval 3 comprises 150 m of biogenic clay and nannofossil ooze in which bio- and magnetostratigraphic data produce relatively well-constrained

LOCs (Fig. 8), despite some fairly considerable gaps in core recovery.

Above the top of C18n.1n, the upper part of Core 42R and all of Core 41R (except for one positive inclination sample at about 392 mbsf) have negative inclination values. We interpret this mostly negative inclination interval as C17r. Very low core recovery in Cores 40R and 39R precludes any interpretation of magnetic polarity. This large gap in data creates a large depth error in the polarity reversal between the interpreted C17r interval and the overlying positive inclination interval that occurs from the base of Core 39R (*c.* 370 mbsf) up to the top of Core 38R (*c.* 355 mbsf). This positive inclination interval is interpreted to represent part or all of C17n. Another data gap in Core 37R creates a second large depth error between the interpreted C17n and the overlying negative inclination interval in Cores 36R and 35R. We interpret top of C17n.1n as occurring between the top of Core 38R and the base of Core 36R. Since Chron C17n contains three normal polarity subchrons, we plotted the single positive inclination sample at *c.* 392 mbsf as occurring within C17n.3n, and plotted two alternative age interpretations (the bases of subchrons C17n.1n and C17n.2n) for the polarity reversal centred at *c.* 378 ± 8 mbsf (Fig. 8). This shows that the placement of the bases of either C17n.1n or C17n.2n, and placement of C17n.3n, shows only slight deviations from our primary interpreted LOC, which is between the top of C18n.1n and the top of C17n.1n. We have therefore used the mid-points of base C17n.2n and base C17n.1n alternative age interpretations to define our top and bottom bounding LOCs between the top of C17n.1n and the top of C18n.1n. The slightly greater age errors are shown in Figure 8 with dashed red lines. Only one biostratigraphic datum, the highest occurrence of *Phthanoperidinium distinctum* at *c.* 370 mbsf, occurs between 400 and 350 mbsf, and has a fairly large age calibration error. Therefore, the interpretation of the lower 50 m of Interval 3 is based exclusively on the palaeomagnetic inclination pattern.

In contrast, the upper 100 m of Interval 3 contain 13 biostratigraphic datums, which produce a simple linear regression line that helps us to interpret a mostly low recovery interval with sparse palaeomagnetic data, which extends up to the EOT (Fig. 8).

Discrete sample negative inclinations in Cores 36R and 35R are interpreted as C16r, overlying C17n. Poor core recovery in Cores 34R–32R creates another large data gap. However biostratigraphic data indicate that the age of this interval should be somewhere between the age of C16r and C13R. Core 33R contains one negative inclination sample overlain by one positive inclination sample. Core 32R also has one positive inclination

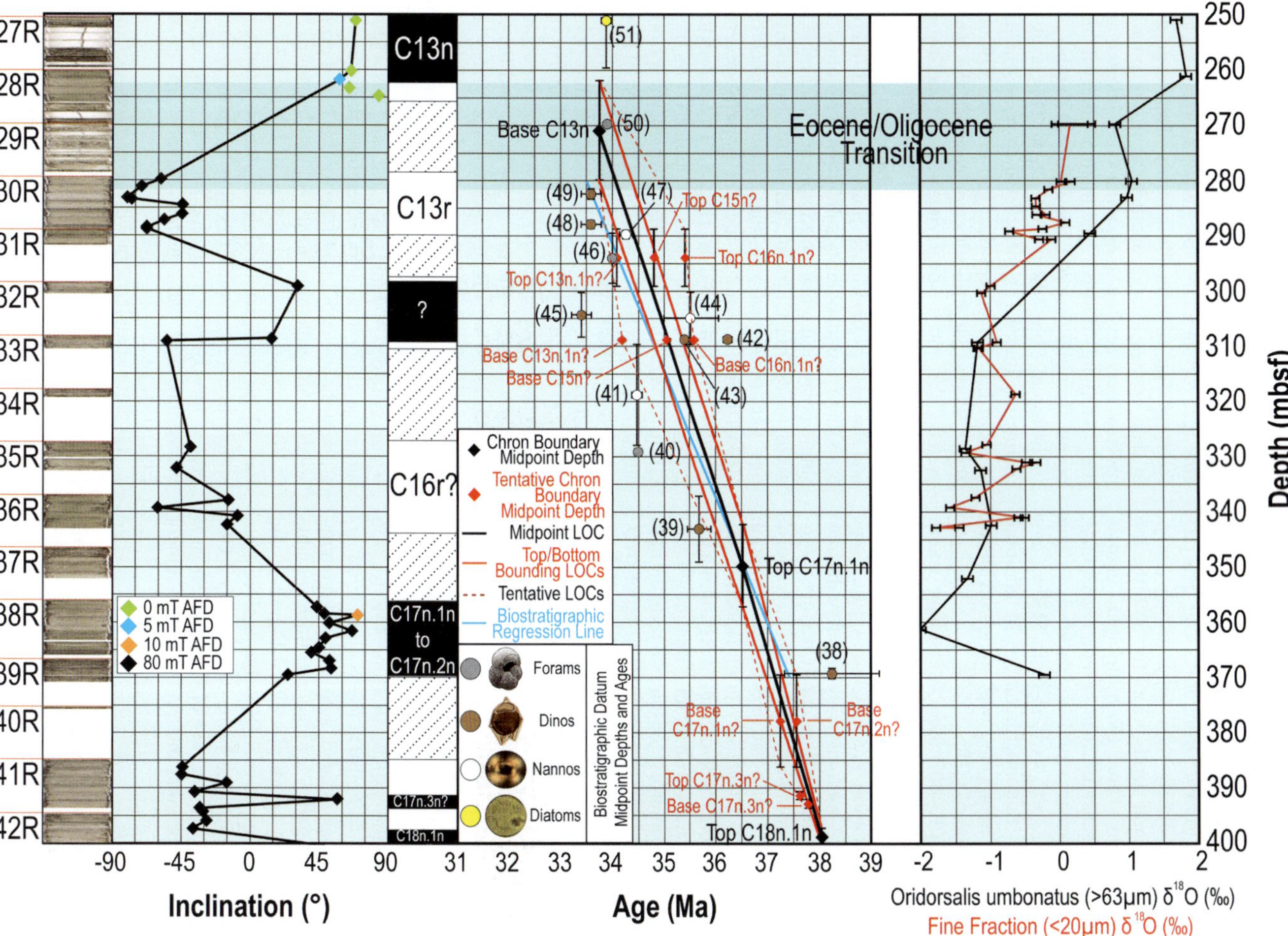

Fig. 8. Integrated magnetobiochronological summary for ODP Hole 647A Interval 3 (400–250 mbsf, Chron C18n–C13n, Cores 42R–27R). Format as for Figure 6. On our age model the EOB (*c.* 33.9 Ma, Gradstein *et al.* 2004) is predicted to fall within Core 29R. This is supported by the sediment fine fraction and benthic foraminifera δ^{18}O chemostratigraphy (i.e. the EOB is *c.* half-way through the plateau in benthic δ^{18}O). See Table 2 for key to biostratigraphic datums.

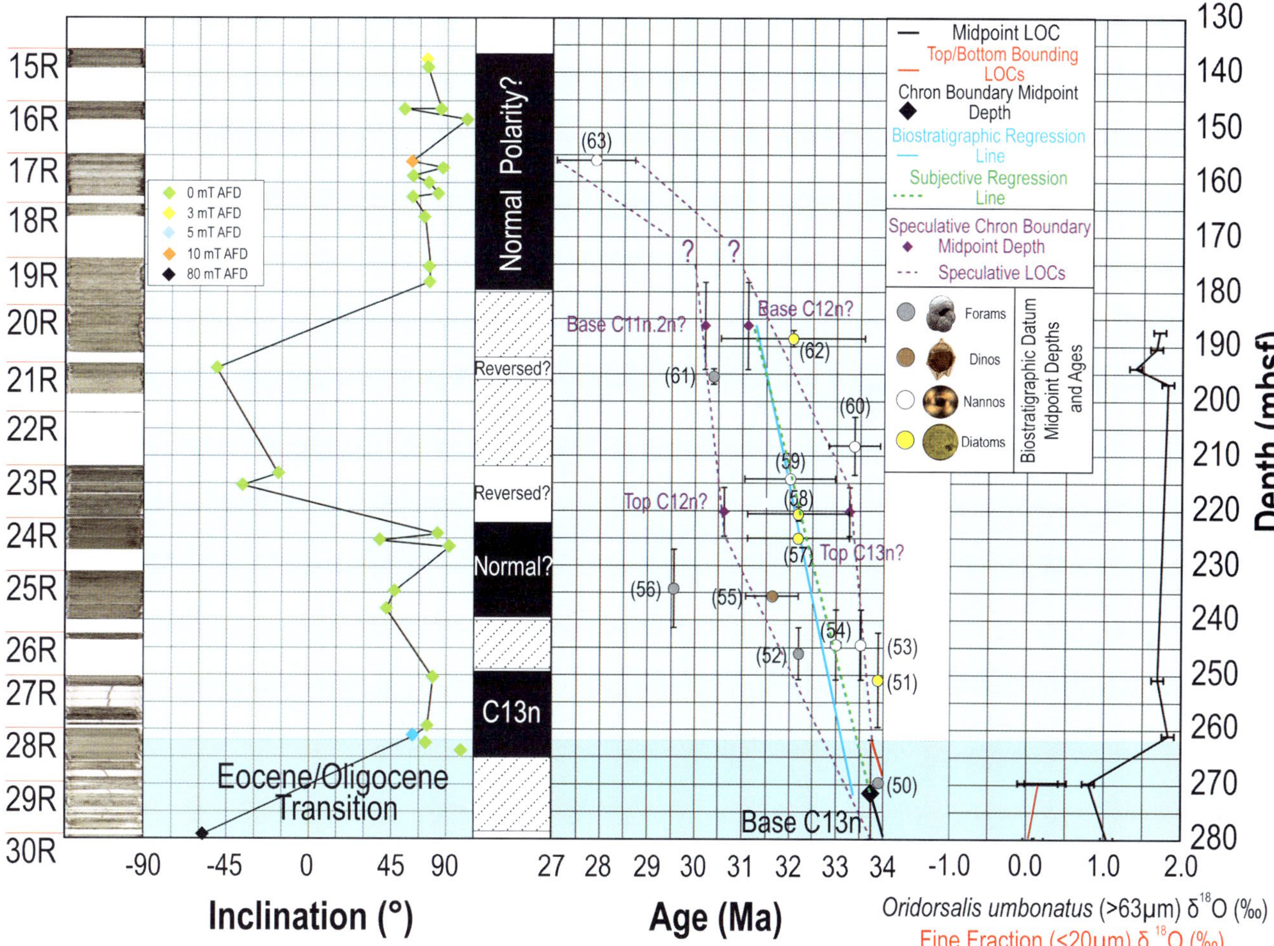

Fig. 9. Integrated magnetobiochronologic summary for ODP Hole 647A Interval 4 (280–130 mbsf, Chrons C13r to C10–C9?, Cores 30R–15R). Format as for Figures 6–8. See Table 2 for key to biostratigraphic datums.

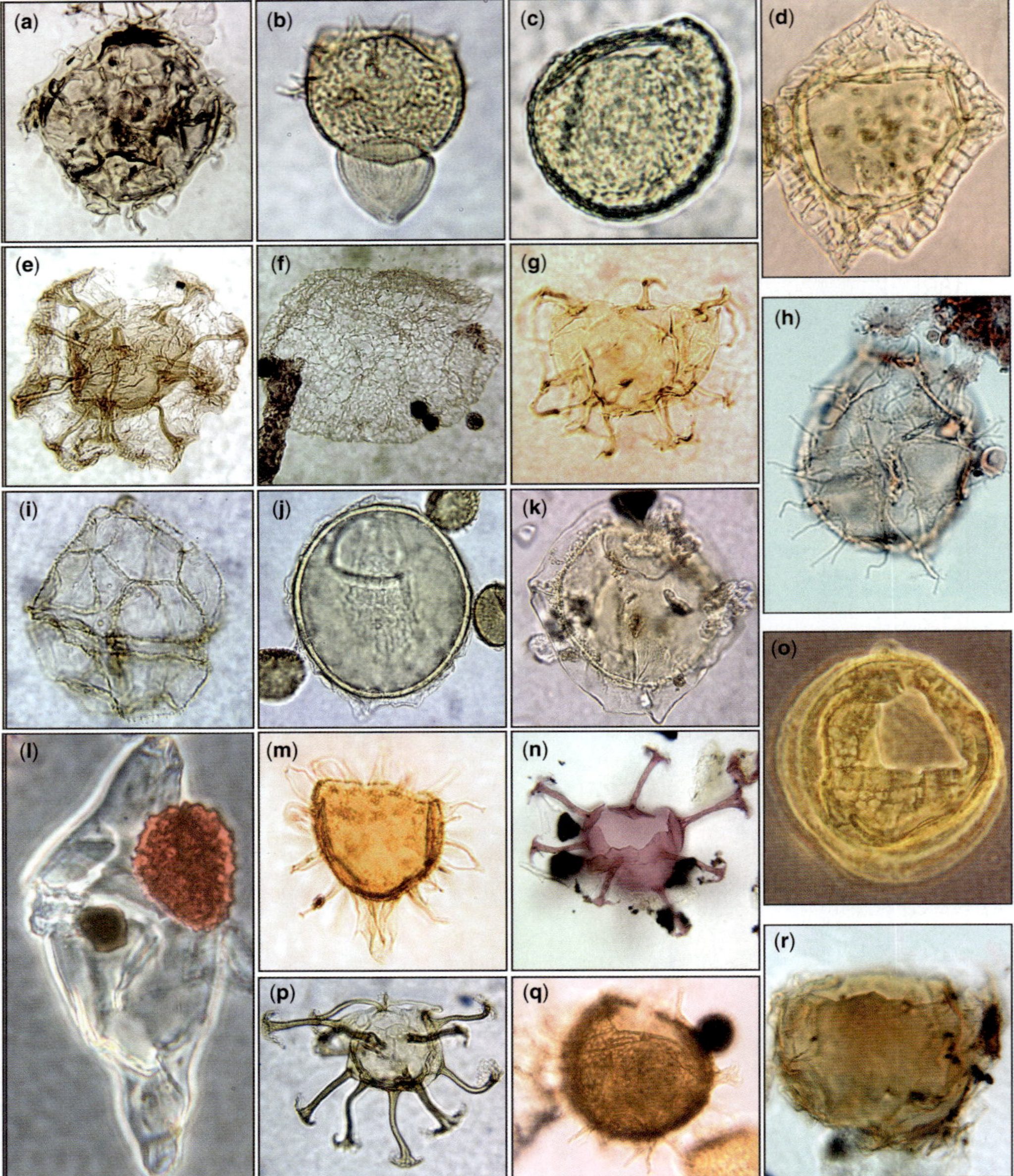

Fig. 11. Plane transmitted light (Pl) and phase contrast (Ph) photomicrographs of Eocene and Oligocene dinocyst biostratigraphic marker species used in construction of the Site 647 age model. Locations of specimens on slides are recorded using the England Finder coordinate system. (**a**) *Dracodinium pachydermum*, sample 105-647A−65R-1, 117−121 cm, G22-3, Pl, 80 μm long. (**b**) *Diphyes ficusoides*, sample 105-647A-62R-5, 87−89 cm, Q23, Pl, 55 μm long. (**c**) *Cerebrocysta magna*, sample 105-647A-64R-3, 28−30 cm, Q21, Pl, 49 μm long. (**d**) *Charlesdowniea columna*, sample 105-647A-65R-1, 27−29 cm, G-28, Pl, 108 μm long. (**e**) *Eatonicysta ursulae*, sample 105-647A-66R-1, 15−18 cm, M23-3, Pl, 80 μm long. (**f**) *Heteraulacacysta porosa*, sample 105-647A-35R-1, 135−138 cm, P22-1, Pl, 75 μm wide. (**g**) *Enneadocysta arcuata*, sample 105-647A-51R-5, 71−73 cm, O42, Pl, 59 μm wide. (**h**) *Phthanoperidinium comatum*, sample 105-647A-25R-CC, J45−4, Pl, 51 μm long. (**i**) *Phthanoperidinium distinctum*, sample 105-647A-54R-6, 31-34 cm, W33-1, Pl, 53 μm long. (**j**) *Wetzeliella ovalis*, sample 105-647A-51R-1, 87−89 cm, P27−3, Pl, 110 μm long. (**k**) *Rhombodinium rhomboideum*, sample 105-647A-47R-2, 92−95 cm, G32-3, Pl, 110 μm long. (**l**) *Svalbardella cooksoniae*, sample 105-647A-35R-CC, Q21-3, Ph, 107 μm long.

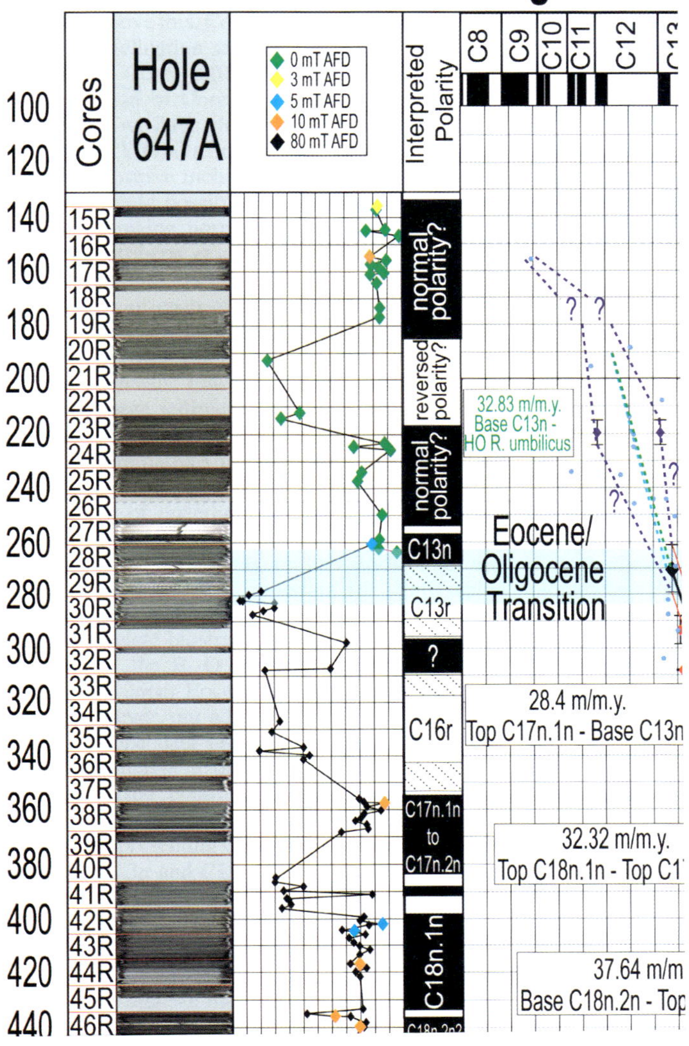

Magnetic R
Hole 647A
Cores
Interpreted Polarity
0 mT AFD
3 mT AFD
5 mT AFD
10 mT AFD
80 mT AFD
C8
C9
C10
C11
C12
C13
100
120
140
160
180
200
220
240
260
280
300
320
340
360
380
400
420
440
15R
16R
17R
18R
19R
20R
21R
22R
23R
24R
25R
26R
27R
28R
29R
30R
31R
32R
33R
34R
35R
36R
37R
38R
39R
40R
41R
42R
43R
44R
45R
46R
normal polarity?
reversed polarity?
normal polarity?
C13n
C13r
?
C16r
C17n.1n to C17n.2n
C18n.1n
C18n.2n2
32.83 m/m.y.
Base C13n -
HO R. umbilicus
Eocene/
Oligocene
Transition
28.4 m/m.y.
Top C17n.1n - Base C13n
32.32 m/m.y.
Top C18n.1n - Top C1
37.64 m/m
Base C18n.2n - Top
?
?
?
?
?

Table 7. *Primary age model tie-point depths, ages, errors, and sedimentation rates for each LOC (using Gradstein* et al. *2004*)*

Tie-point ranking	Tie-point type	Chron/ biostratigraphic event	Chron depth top (mbsf)	Chron depth base (mbsf)	Chron depth mid–point (mbsf)	Depth error ($\pm$) m	Age (Ma)
Speculative tie-point	Base?	C11n.2n?	178.21	194.16	186.19	7.98	30.217
Speculative tie-point	Base?	C12n?	178.21	194.16	186.19	7.98	31.116
Speculative tie-point	Top?	C12n?	215.74	224.63	220.19	4.45	30.627
Tentative Primary	**HO**	*Reticulofenestra umbilicus* **(>14 μm)**	**213.48**				**31.05**
Tentative Primary	**HO**	*Reticulofenestra umbilicus* **(>14 μm)**			**214.19**	**0.71**	**32.01**
Tentative Primary	**HO**	*Reticulofenestra umbilicus* **(>14 μm)**		**214.90**			**32.97**
Speculative tie-point	Top?	C13n?	215.74	224.63	220.19	4.45	33.266
Primary	**Top**	**C13r**	**261.92**	**279.93**	**270.93**	**9.01**	**33.738**
Secondary possible	Top	C13n.1n (= C13r [0.31], Pälike *et al.* 2006)	288.83	299.18	294.01	5.18	34.066
Secondary possible	Top	C15n	288.83	299.18	294.01	5.18	34.782
Secondary possible	Top	C16n.1n	288.83	299.18	294.01	5.18	35.404
Secondary possible	Base	C13n.1n (= C13r [0.41], Pälike *et al.* 2006)	308.68	309.16	308.92	0.24	34.164
Secondary possible	Base	C15n	308.68	309.16	308.92	0.24	35.043
Secondary possible	Base	C16n.1n	308.68	309.16	308.92	0.24	35.567
Primary	**Top**	**C17n.1n**	**342.35**	**357.27**	**349.81**	**7.46**	**36.512**
Secondary possible	Base	C17n.1n	369.59	386.30	377.95	8.35	37.235

Secondary possible	Base	C17n.2n	369.59	386.30	377.95	8.35	37.549
Secondary possible	Top	C17n.3n	390.75	392.16	391.46	0.71	37.610
Secondary possible	Base	C17n.3n	392.16	393.73	392.95	0.78	37.771
Primary	**Top**	**C18n.1n**	**397.38**	**400.49**	**398.94**	**1.56**	**38.032**
Secondary possible	Base	C18n.1n	434.77	436.12	435.45	0.68	38.975
Secondary possible	Top	C18n.2n	436.12	437.05	436.59	0.47	39.041
Secondary possible	Base	C18n.1n	446.35	448.65	447.50	1.15	38.975
Secondary possible	Top	C18n.2n	448.65	449.38	449.02	0.36	39.041
Primary	**Base**	**C18n.2n**	**451.67**	**453.99**	**452.83**	**1.16**	**39.464**
Primary	**Top**	**C19n**	**480.63**	**483.39**	**482.01**	**1.38**	**40.439**
Primary	**Base**	**C19n**	**490.27**	**490.36**	**490.32**	**0.05**	**40.671**
Secondary possible	Top	C20n	601.46	606.11	603.79	2.32	41.590
Tentative Primary	**Midpoint Age Between Top and Base C20n**	**C20n**	**601.46**	**606.11**	**603.79**	**2.32**	**42.182**
Secondary possible	Base	C20n	606.11	608.38	607.25	1.13	42.774
Secondary possible	Top	C21n	601.46	606.11	603.79	2.32	45.346
Secondary possible	Base	C20n	608.38	618.43	613.41	5.03	42.774
Primary	**Base**	**C21n**	**608.38**	**618.43**	**613.41**	**5.03**	**47.235**
Speculative tie-point	Top?	C22n?	619.55	619.65	619.60	0.05	48.599
Speculative tie-point	Base?	C22n?	640.55	640.60	640.58	0.02	49.427
Speculative tie-point	Top?	C23n.1n?	636.85	636.90	636.88	0.02	50.730
Speculative tie-point	Base?	C23n.2n?	657.10	657.15	657.13	0.02	51.901
Primary	**Top**	**C24n.1n**	**659.38**	**686.24**	**672.81**	**13.43**	**52.648**
Primary	**Base (<)**	**C24n.3n**	**696.43**	**696.43**	**696.43**	**0.00**	**53.808**

*Ages of Chron boundaries from Gradstein *et al.* 2004 do not include age uncertainties.
Primary chron tie-points are used for the primary age model.
Tentative primary interpreted chron or biostratigraphic tie-points are used for the primary age model.
Secondary possible tie-points are not used in the age model.
Speculative tie-points from NRM discrete or from cryomagnetic data are not used in age model.
<, Younger than, for datum at base of studied section.

Table 7. *Continued*

Tie-point Top	Tie-point Base	Midpoint LOC sedimentation rate (m Ma^{-1})	Top bounding LOC sedimentation rate (m Ma^{-1})	Bottom bounding LOC sedimentation rate (m Ma^{-1})	Maximum sedimentation rate (m Ma^{-1})	Minimum sedimentation rate (m Ma^{-1})
HO *Reticulofenestra umbilicus* (>14 μm)	Top C13r	32.83	63.07	24.19	86.52	17.49
Top C13r	Top C17n.1n	28.44	28.99	27.88	34.37	22.50
Top C17n.1n	Top C18n.1n	32.32	36.20	28.43	38.25	26.39
Top C18n.1n	Base C18n.2n	37.64	37.91	37.36	39.53	35.74
Base C18n.2n	Top C19n	29.93	29.70	30.15	32.53	27.32
Top C19n	Base C19n	35.80	41.55	30.04	41.94	29.66
Base C19n	Midpoint age between Top and Base C20n	75.10	55.08	125.95	126.05	55.04
Base C21n	Top C24n.1n	10.97	9.42	12.53	14.38	7.57
Top C24n.1n	Base C24n.3n	20.36	31.94	8.78	31.94	8.78

all the biostratigraphic datums are equally robust for determining age. However, biostratigraphic data show a large spread of ages, suggesting that the calibrations used are not appropriate for this locality, which is remote from the Southern Ocean and low-latitude calibration sites. Therefore, we have also drawn a subjective best-fit LOC from the mid-point of top C13r through the highest occurrences of *Ericsonia formosa* and *Reticulofenestra umbilicus* (Fig. 9, green dashed line) to compare with the biostratigraphic linear regression line. This LOC relies on two commonly used nannofossil datums in Lower Oligocene sediments throughout the world. The resultant LOC falls within the age ranges of the lowest and highest occurrences of *Sceptroneis pupa*, and would extrapolate upwards close to the lowest occurrence of *Synedra jouseana* at *c.* 185 mbsf. Although it is entirely subjective, it is very similar to the biostratigraphic linear regression line, and still suggests that Cores 28R–24R may be age-equivalent to Chrons C13n, or Chrons 13n–C12n. The age error of *R. umbilicus* is also quite large, giving age uncertainties for this subjective LOC that are almost as large as those suggested by the palaeomagnetic data or the linear regression line. Therefore we cannot determine any LOC within the Oligocene without a very large amount of uncertainty. For the purpose of calculating a tentative sedimentation rate and new ages for biostratigraphic datums within the Oligocene of Hole 647A, we use this subjective LOC between the base C13n and highest occurrence of *R. umbilicus* (see below).

Except for the lowest occurrence of *Sphenolithus ciperoensis*, no other biostratigraphic control exists for the upper part of the Oligocene, above 180 mbsf. This datum suggests that a change in sedimentation rate may exist between Cores 21R and Core 17R. The top of the section in Hole 647A is therefore interpreted to be as young as *c.* 28.7–27 Ma, or Chron C10n–C9n equivalent.

The dramatic change in composition from almost completely calcareous claystones to diatom oozes, and then a gradual change back to calcareous claystones occurs from Cores 26R–17R. Percentage carbonate decreases over this interval (Fig. 10) and planktonic foraminifers become extremely scarce and poorly preserved. The preservation of nannofossils remains moderate to good, indicating that nannofossil production continued alongside biosiliceous production. Biogenic carbonate preservation (primarily nannofossils) does deteriorate significantly up-section, but only above most of the biosiliceous-rich interval, from Cores 17R–15R. This alludes to the interplay of dissolution v. recrystallization at the seafloor on biogenic carbonate preservation, suggesting that nannofossils were deposited in deep water with low carbonate ion concentration (i.e. below the carbonate compensation depth) where they were more likely to dissolve and recrystallize in carbonate rich pore-waters compared with carbonate poor pore-waters. Considering the large sediment compositional changes, it seems likely that a simple linear age model for the Oligocene is unsuitable and a model involving variable sedimentation rates is required to account for the contributions from biosilica. For example, the possible change in sedimentation rate between Cores 21R and Core 17R suggested by the lowest

Table 8. *Simple linear regressions (http://mikescosmos.com/LinReg.php) between primary age model tie-points ($r^2 = 1$)*

Top tie-point	Base tie-point	Chron depth mid-point LOC	Chron depth top bounding LOC	Chron depth bottom bounding LOC	
HO *Reticulofenestra umbilicus* ($>$14 μm)	Top C13r	$y = 0.0304547x + 25.4869$	$y = 0.0158547x + 29.5853$	$y = 0.0413348x + 22.1672$	
Top C13r	Top C17n.1n	$y = 0.0351673x + 24.2101$	$y = 0.0344896x + 24.7045$	$y = 0.0358676x + 23.6976$	
Top C17n.1n	Top C18n.1n	$y = 0.0309383x + 25.6895$	$y = 0.0276213x + 27.0558$	$y = 0.0351689x + 23.9472$	
Top C18n.1n	Base C18n.2n	$y = 0.0265726x + 27.4311$	$y = 0.0263769x + 27.5504$	$y = 0.0267664x + 27.3123$	
Base C18n.2n	Top C19n	$y = 0.0334133x + 24.3335$	$y = 0.0336671x + 24.2576$	$y = 0.0331633x + 24.4082$	
Top C19n	Base C19n	$y = 0.0279182x + 26.9822$	$y = 0.0240664x + 28.872$	$y = 0.0332855x + 24.3491$	
Base C19n	Base C20n	–	$y = 0.0189221x + 31.3931$	–	
Base C19n	Mean age between top and Base C20n	$y = 0.0133163x + 34.1418$	–	–	
Base C19n	Top C20n	–	–	$y = 0.00793678x + 36.7794$	
Base C21n	Top C24n.1n	$y = 0.0911279x - 8.66379$	$y = 0.106137x - 17.3368$	$y = 0.079826x - 2.13178$	
Top C24n.1n	Base C24n.3n	$y = 0.0491109x + 19.6057$	$y = 0.031309x + 32.0034$	$y = 0.113837x - 25.4716$	
HO *Reticulofenestra umbilicus* ($>$14 μm) mid-point depth: 218.48 mbsf	Top C13r mid-point depth: 270.93 mbsf	Mean age from mid-point LOC (Ma) ($y = 0.0304547 - 25.4869$)	Maximum age from top bounding LOC (Ma) ($y = 0.0158547 - 29.5853$)	Minimum age from bottom bounding LOC (Ma) ($y = 0.0413348x + 22.1672$)	Age error ($\pm$) (Myr)
HO	*Reticulofenestra umbilicus* ($>$14 μm)	32.1	33.0	31.2	0.9
HO	*Sceptroneis pupa*	32.2	33.1	31.3	0.9
LO	*Sceptroneis pupa*	32.3	33.1	31.5	0.8
HO	*Chiloguembelina* 'ototara'	32.6	33.2	32.1	0.5
HO	*Spiniferites manumii*	32.7	33.3	31.9	0.7
HO	*Ericsonia formosa*	32.9	33.6	32.0	0.8
LO	*Sphenolithus distentus*	32.9	33.6	32.0	0.8
HO	*Pseudohastigerina naguewichiensis*	33.0	33.4	32.5	0.4
LO	*Rhizosolenia gravida/ oligocaenica*	33.1	33.4	32.9	0.3
HO	*Pseudohastigerina micra*	33.7	33.9	33.3	0.3

(Continued)

Table 8. *Continued*

Top tie-point	Base tie-point	Chron depth mid-point LOC	Chron depth top bounding LOC	Chron depth bottom bounding LOC	
Top C13r mid-point depth: 270.93 mbsf	Top C17n.1n mid-point depth: 349.81 mbsf	Mean age from mid-point LOC (Ma) ($y = 0.0351673x + 24.2101$)	Maximum age from top bounding LOC (Ma) ($y = 0.0344896x + 24.7045$)	Minimum age from bottom bounding LOC (Ma) ($y = 0.0358676x + 23.6976$)	Age error ($\pm$) (Myr)
LO	*Chiropteridium galea*	34.1	34.4	33.9	0.3
LO	*Spiniferites manumii*	34.3	34.6	34.1	0.3
HO	*Discoaster saipensis*	34.4	34.7	34.1	0.3
HO	*Turborotalia cerroazulensi*	34.6	34.7	34.4	0.1
HO	*Areosphaeridium diktyoplokum*	34.9	35.1	34.8	0.2
HCO	*Reticulofenestra reticulata*	34.9	35.1	34.8	0.1
HO	*Areosphaeridium michoudii*	35.1	35.3	34.8	0.3
HO	*Heteraulacacysta porosa*	35.1	35.3	34.8	0.3
HO	*Discoaster barbadiensis*	35.4	36.0	34.8	0.6
HO	*Globigerinatheka index*	35.8	36.1	35.5	0.3
HO	*Cribroperidinium* sp. 2	36.3	36.7	35.8	0.5
Top C17n.1n mid-point depth: 349.81 mbsf	Top C18n.1n mid-point depth: 398.94 mbsf	Mean age from mid-point LOC (Ma) ($y = 0.0309383 - 25.6895$)	Maximum age from top bounding LOC (Ma) ($y = 0.0276213x + 27.0558$)	Minimum age from bottom bounding LOC (Ma) ($y = 0.0351689x + 23.9472$)	Age error ($\pm$) (Myr)
HO	*Phthanoperidinium distinctum*	37.1	37.3	36.9	0.2
Top C18n.1n mid-point depth: 398.94 mbsf	Base C18n.2n mid-point depth: 452.83 mbsf	Mean age from mid-point LOC (Ma) ($y = 0.0265726x + 27.4311$)	Maximum age from top bounding LOC (Ma) ($y = 0.0263769x + 27.5504$)	Minimum age from bottom bounding LOC (Ma) ($y = 0.0267664x + 27.3123$)	Age error ($\pm$) (Myr)
HO	*Diphyes colligerum*	38.4	38.3	38.6	0.1
LO	*Reticulofenestra bisecta* (>10 μm)	39.4	39.5	39.4	0.0
LO	*Globigerinatheka index*	39.5	–	39.4	0.1

Base C18n.2n mid-point depth: 452.83 mbsf	Top C19n mid-point depth: 482.01 mbsf	Mean age from mid-point LOC (Ma) ($y = 0.0265726x + 27.4311$)	Maximum age from top bounding LOC (Ma) ($y = 0.0336671x + 24.2576$)	Minimum age from bottom bounding LOC (Ma) ($y = 0.0267664x + 27.3123$)	Age error ($\pm$) (Myr)
LO	*Globigerinatheka index*	39.5	39.6	–	0.1
HO	*Chiasmolithus solitus*	39.5	39.5	39.5	0.0
HO	*Acarinina mcgowrani*	39.8	40.1	39.7	0.2
HO	*Acarinina bullbrooki*	39.8	40.1	39.7	0.2
HO	*Acarinina collactea*	40.1	40.4	40.0	0.2
HO	*Turborotalia frontosa*	40.1	40.4	40.0	0.2

Top C19n mid-point depth: 482.01 mbsf	Base C19n mid-point depth: 490.32 mbsf	Mean age from mid-point LOC (Ma) ($y = 0.0279182x + 26.9822$)	Maximum age from top bounding LOC (Ma) ($y = 0.0240664x + 28.872$)	Minimum age from bottom bounding LOC (Ma) ($y = 0.0332855x + 24.3491$)	Age error ($\pm$) (Myr)
LO	*Svalbardella cooksoniae*	40.5	40.6	40.4	0.1
HO	*Wetzeliella ovalis*	40.5	40.6	40.4	0.1
LO	*Cribroperidinium* sp. 2	40.5	40.6	40.4	0.1

Base C19n mid-point depth: 490.32 mbsf	Mean age between Top and Base C20n mid-point depth: 603.79 mbsf	Mean age between Top and Base C20n Midpoint Depth (Ma) ($y = 0.0133163x + 34.1418$)	Maximum Age from Base C20n, depth Top (Ma) ($y = 0.0189221x + 31.3931$)	Minimum Age from Top C20n, Depth Base (Ma) ($y = 0.00793678x + 36.7794$)	Age error ($\pm$) (Myr)
LO	*Rhombodinium rhomboideum*	40.8	41.0	40.7	0.1
LO	*Wetzeliella ovalis*	40.8	41.0	40.7	0.1
LO	*Reticulofenestra reticulata*	41.1	41.4	40.9	0.3
HO	*Nannotetrina* spp. (*cristata*)	41.3	41.8	41.0	0.4
LO	*Turborotalia pomeroli*	41.7	42.4	41.2	0.6
LO	*Phthanoperidinium distinctum*	41.8	42.3	41.3	0.5
LO	*Phthanoperidinium comatum*	42.0	42.7	41.4	0.7
LO	*Enneadocysta arcuata*	42.1	42.8	41.5	0.6

(*Continued*)

J. V. FIRTH *ET AL.*

Table 8. *Continued*

Top tie-point	Base tie-point	Chron depth mid-point LOC	Chron depth top bounding LOC	Chron depth bottom bounding LOC	
Base C21n mid-point depth: 613.41 mbsf	Top C24n.1n mid-point depth: 672.81 mbsf	Mean age from mid-point LOC (Ma) ($y = 0.0911279x - 8.66379$)	Maximum age from top bounding LOC (Ma) ($y = 0.1061370x - 17.3368$)	Minimum age from bottom bounding LOC (Ma) ($y = 0.079826x - 2.13178$)	Age error ($\pm$) (Myr)
HO	*Eatonicysta ursulae*	47.7	48.4	47.2	0.6
HO	*Dracodinium pachydermum*	47.9	48.6	47.3	0.6
HO	*Charlesdowniea columna*	48.2	49.3	47.5	0.9
LO	*Cerebrocysta magna*	49.0	50.3	48.0	1.1
LO	*Diphyes ficusoides*	49.0	50.3	48.0	1.1
LO	*Dracodinium pachydermum*	49.0	50.3	48.0	1.1
HO	*Discoaster lodoensis*	49.2	50.3	48.3	1.0
HO	*Tribrachiatus orthostylus*	50.9	52.4	49.8	1.3
LO	*Acarinina primitiva*	51.6	52.9	50.7	1.1
LO	*Acarinina collactea*	51.6	52.9	50.7	1.1
Top C24n.1n mid-point depth: 672.81 mbsf	Base C24n.1n mid-point depth: 696.43 mbsf	Mean age from mid-point LOC (Ma) ($y = 0.0491109x + 19.6057$)	Maximum age from top bounding LOC (Ma) ($y = 0.0313090x + 32.0034$)	Minimum age from bottom bounding LOC (Ma) ($y = 0.113837x - 25.4716$)	Age error ($\pm$) (Myr)
LO	*Discoaster lodoensis*	52.8	53.5	50.4	1.5
LO	*Tribrachiatus orthostylus*	<53.8	<53.8	<53.8	0.00

HO, highest occurrence; LO, lowest occurrence; LOC, line of Correlation.
Recalculated ages and age errors for all biostratigraphic datums in Hole 647A (Gradstein *et al.* 2004).

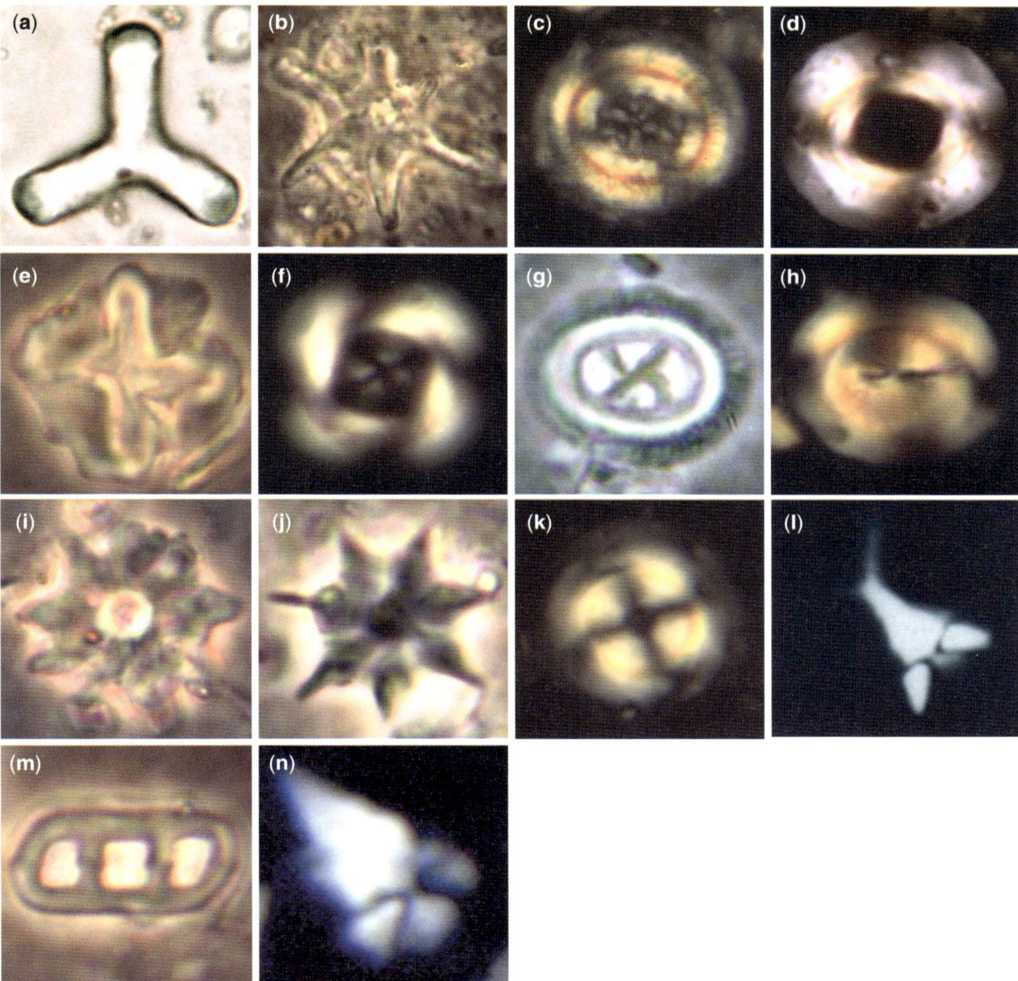

Fig. 12. Plane-polarized (Pl-pol), cross-polarized (x-pol), phase contrast (Ph) and plane transmitted light (Pl) photomicrographs of Eocene and Oligocene nannofossil and diatom biostratigraphic marker species used in construction of the Site 647 age model. (**a**) *Tribrachiatus orthostylus*, sample 105-647A-68R-2, 48–50 cm, J20-1, Pl-pol, 12 μm wide.(**b**) *Discoaster lodoensis*, sample 105-647A-69R-CC, J21-3, Ph, 26.5 μm wide. (**c**) *Chiasmolithus gigas*, sample 105-647A-63R-1, 36–38 cm, N20-1, x-pol, 18 μm long. (**d**) *Reticulofenestra umbilicus* (>14 μm), sample 105-647A-29R-1, 2–3 cm, S44, x-pol, 18 μm long. (**e**) *Nannotetrina cristata*, sample 105-647A-62R-1, 82–84 cm, R29-4, Ph, 11.5 μm wide. (**f**) *Reticulofenestra reticulata*, sample 105-647A-41R-3, 58-61 cm, R27, x-pol, 8.5 μm wide. (**g**) *Chiasmolithus solitus*, sample 105-647A-59R-CC, R-25, Ph, 9 μm long. (**h**) *Reticulofenestra bisecta* (≥10 μm), sample 105-647A-35R-1, 77-80 cm, M23, x-pol, 10 μm long. (**i**) *Discoaster barbadiensis*, sample 105-647A-41R-3, 58–61 cm, L38, Ph, 11.5 μm wide. (**j**) *Discoaster saipanensis*, sample 105-647A-41R-3, 58–61 cm, L29-1, Ph, 14 μm wide. (**k**) *Ericsonia formosa*, sample 105-647A-29R-1, 2-3 cm, K24-4, xpol, 16 μm wide. (**l**) *Sphenolithus distentus*, sample 105-647A-17R-4, 111-113 cm, 45° to x-pols, 6 μm long. (**m**) *Isthmolithus recurvus*, sample 105-647A-29R-1, 2-3 cm, R46-2, Ph, 8 μm long. (**n**) *Sphenolithus ciperoensis*, sample 105-647A-17R-1, 107-112 cm, N33, 45° to x-pols, 4.5 μm long.

occurrence of *S. ciperoensis* in Core 17R coincides with the upsection increase in percentage carbonate and the decline in percentage biosilica. Better palaeomagnetic data and more suitable calibrations for the various microfossil groups are needed, but this is beyond the scope of this study.

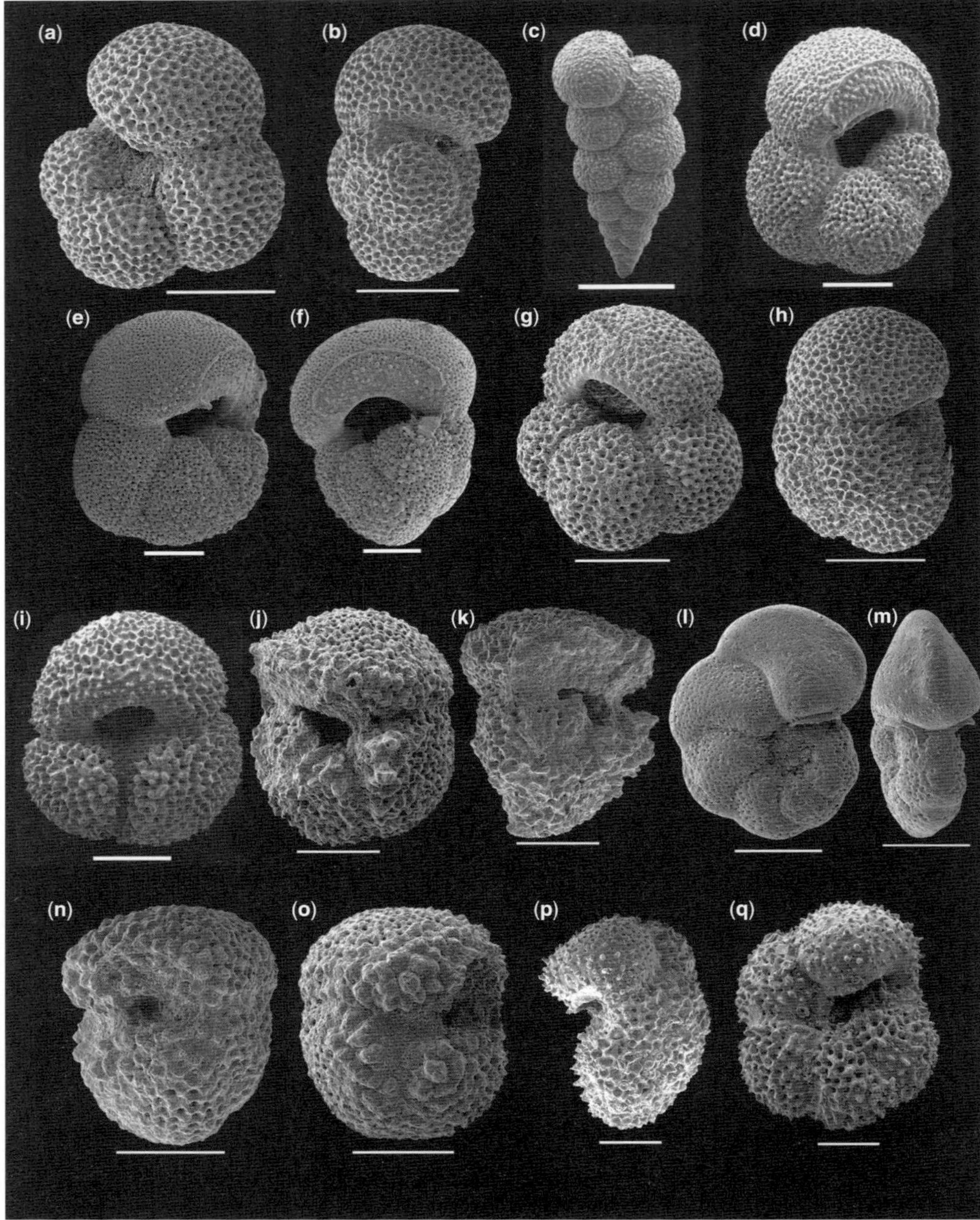

Fig. 13. Scanning electron micrographs of Eocene and Oligocene planktonic foraminifer biostratigraphic marker species used in construction of the Site 647 age model. Edge views of same specimen. Scale bars a–o, 100 μm; p and q, 50 μm. Taxonomy follows Pearson *et al.* (2006) and the Oligocene Planktonic Foraminifera Working Group (Wade *et al.* in preparation). (**a, b**) *Paragloborotalia nana*, 647A-27R-1, 28.5-30, Zone O1.(**c**) *Chiloguembelina ototara*, 647A-30R-3, 67-69 cm, EOT interval. (**d**) *Turborotalia ampliapertura*, 647A-30R-3, 67–69 cm; EOT interval. (**e, f**) *Turborotalia cerroazuelensis*, 647A-36R-3, 82–84 cm, Late Eocene. (**g, h**) *Turborotalia pomeroli* 647A-51R-2, 115–117 cm, Middle Eocene. (**i**) *Globigerinatheka index*, 647A-35R-4, 41–43 cm, Late Eocene. (**j, k**) *Acarinina bullbrooki* 647A-62R-2, 9–11 cm, Middle Eocene. (**l, m**) *Pseudohastigerina micra*

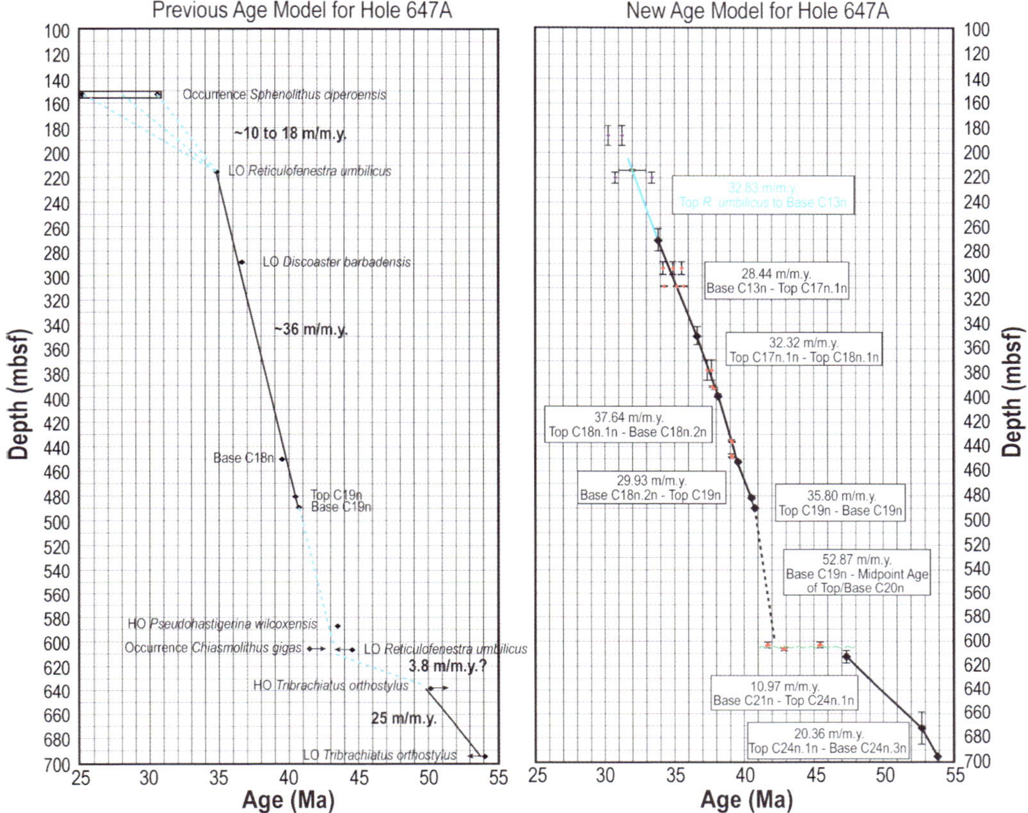

Fig. 14. Comparison between the previous age model of Baldauf *et al.* (1989) and Arthur *et al.* (1989*a*, *b*) and the new age model presented herein. The new age model has more than twice the number of age/depth tie-points, based mostly on palaeomagnetic data, than the previous age model, which relied primarily on nannofossil tie-points.

Comparison of age models and new calibrated biostratigraphic datum ages for Hole 647A

A comparison between our new age model and the original shipboard age model is presented in Figure 14. The previous age model (Baldauf *et al.* 1989; Arthur *et al.* 1989*a*, *b*) was constrained by five nannofossil tie-points and three palaeomagnetic tie-points, and consisted of four linear segments with four estimated sedimentation rates for the entire 565 m Palaeogene section. Our new age model is constrained by nine palaeomagnetic tie-points and one biostratigraphic tie-point (Table 7). The entire Palaeogene section in Hole 647A is now subdivided into nine segments of different sedimentation rates, with quantitative estimates of the error on each segment and each sedimentation rate. This significant increase in age resolution leads to more accurate estimates of bulk and component sedimentation rates and accumulation rates that have palaeoceanographic significance. It also will enable future research to focus on specific intervals where important palaeoceanographic events may be recorded.

Our new age calibrations for the biostratigraphic datums in Hole 647A, with age error estimates, are compiled in Table 8. Our calibrated high northern latitude bioevents provide new constraints on the diachrony of evolutionary and extinction events

Fig. 13. (*Continued*) 647A-68R-4, 110–112 cm, late Early Eocene. (**n, o**) *Acarinina primitiva* 647A-68R-4, 110–112 cm, late Early Eocene. (**p, q**) *Acarinina collactea* 647A-68R-4, 110–112 cm, late Early Eocene.

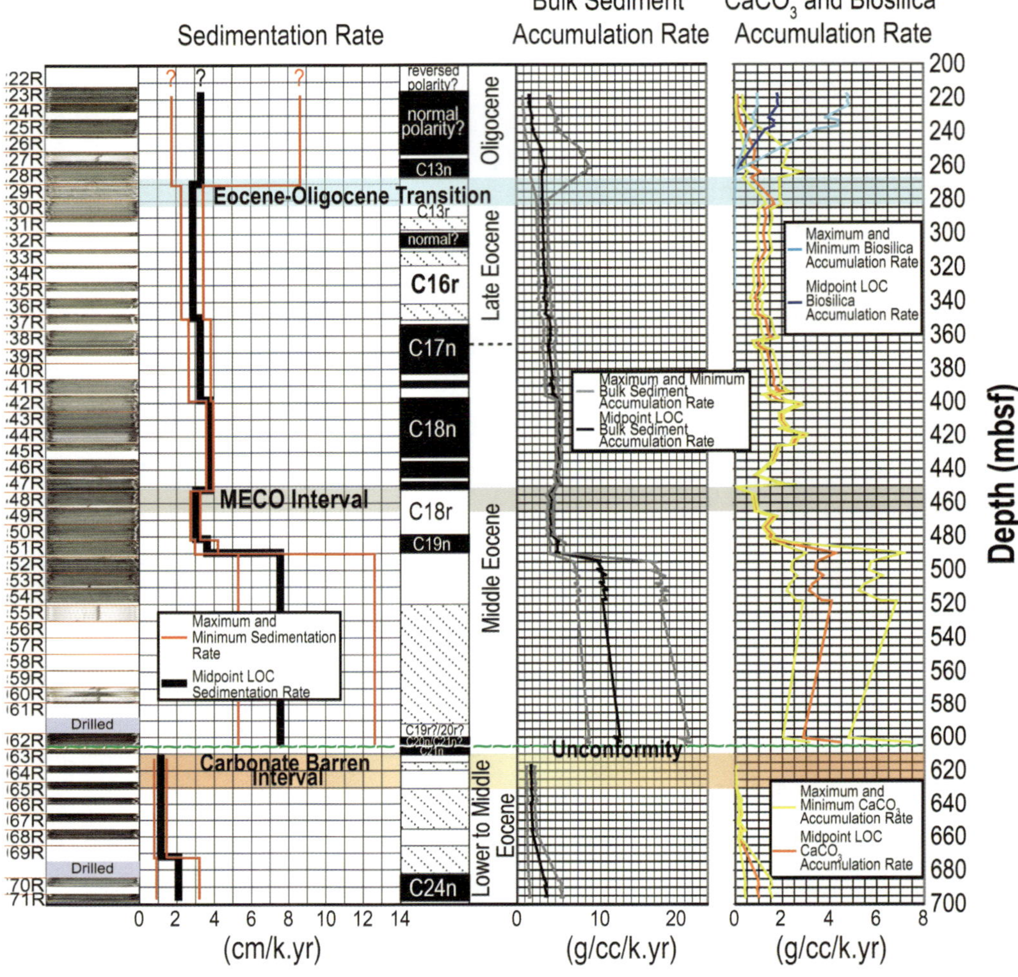

Fig. 15. Sedimentation rates, bulk sediment, $CaCO_3$, and biosilica accumulation rates for Hole 647A, from the top age model datum (highest occurrence of *R. umbilicus*) to basement (Table 7, plotted v. depth (mbsf). Large uncertainty is apparent between Base C19n and C20n? (*c.* 604 mbsf). Also, above the base of C13r, uncertainty is very large because of the age range error of the highest occurrence of *R. umbilicus*. An interval of very low to no sediment accumulation occurs between Cores 62R and 63R. Changes in sedimentation rates cannot be discerned across the Carbonate Barren Interval or the EOT. A decline in carbonate accumulation rate, but not sedimentation rate, is associated with the predicted level of the MECO interval.

between tropical and the high northern latitudes. Biostratigraphic datums that were not recalculated are those that occur within Core 63R (containing an unconformity), the highest occurrence of *D. ficusoides* in Core 62R, which is interpreted to be reworked, and those that occur above the highest occurrence of *R. umbilicus*, because of the increasingly large age uncertainty that results from extrapolating above that level. The age of *S. ciperoensis* is also not recalculated, owing to the lack of any other age control in Cores 16R–19R.

Sedimentation rates and mass accumulation rates

Our new age model enables us to recalculate the sedimentation rates and mass accumulation rates of bulk sediment, biogenic $CaCO_3$, and biosilica after the original Leg 105 studies. These were calculated using shipboard Dry Bulk Density and percentage $CaCO_3$ data (Srivastava *et al.* 1987) additional percentage $CaCO_3$ data (Stein *et al.* 1989) and percentage biosilica data from Bohrmann & Stein (1989).

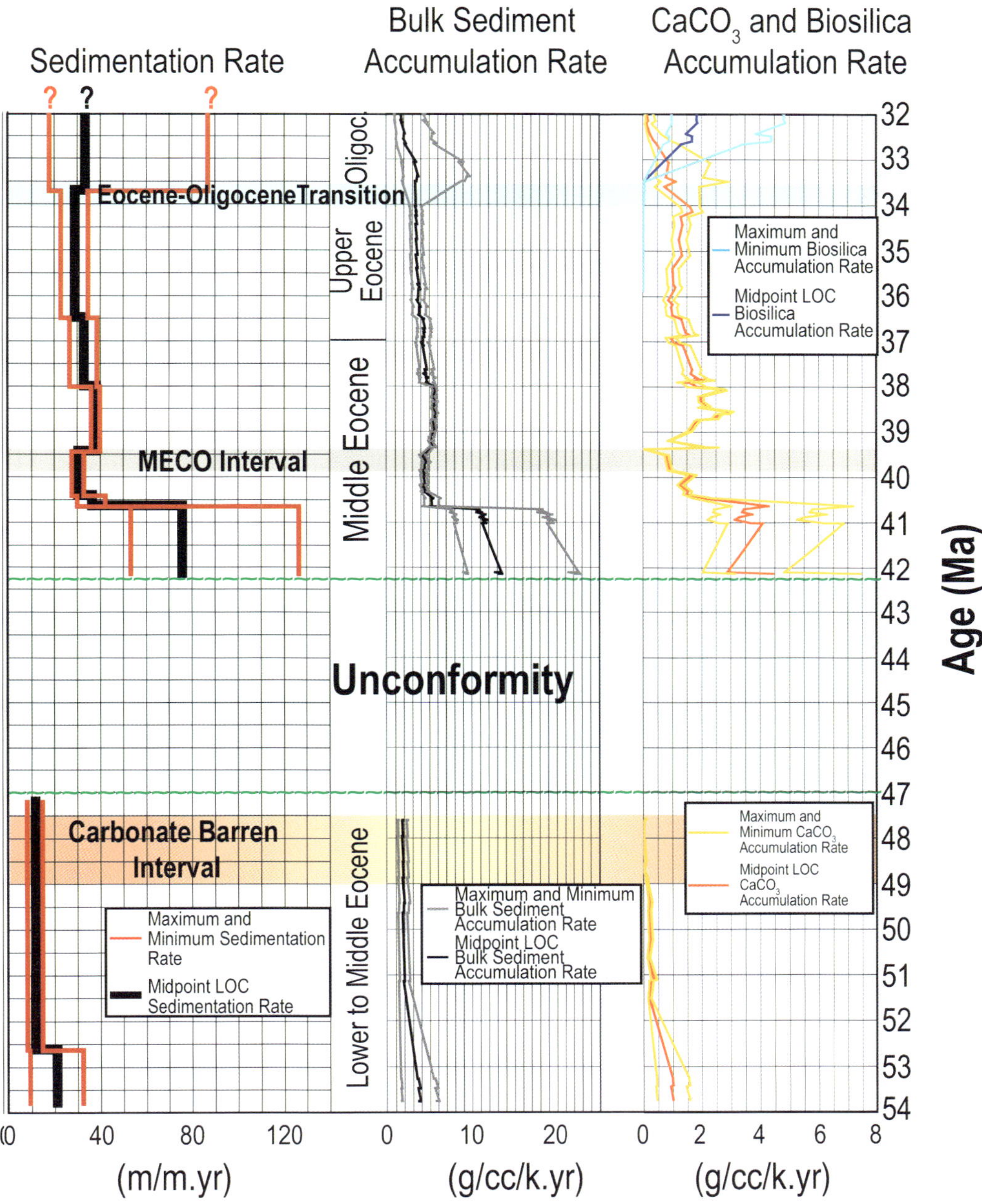

Fig. 16. Sedimentation rates, bulk sediment, CaCO₃, and biosilica accumulation rates for Hole 647A, from the top Age Model datum (highest occurrence of *R. umbilicus*) to basement (Table 7), plotted v. age (Ma). Format as for Figure 15. This interpretation emphasizes the >4 Ma unconformity in the Middle Eocene at Hole 647A between *c.* 42.2 and 47.0 Ma.

Percentage biogenic CaCO₃ and biosilica are plotted by depth in the summary age model (Fig. 10), whereas the sedimentation and accumulation rates are plotted both by depth and by age (Figs 15 & 16, respectively). This provides the opportunity to examine the influence of two major lithological changes on sedimentation in Hole 647A, that is (i) the interval of carbonate-free claystones (610–630 mbsf, Cores

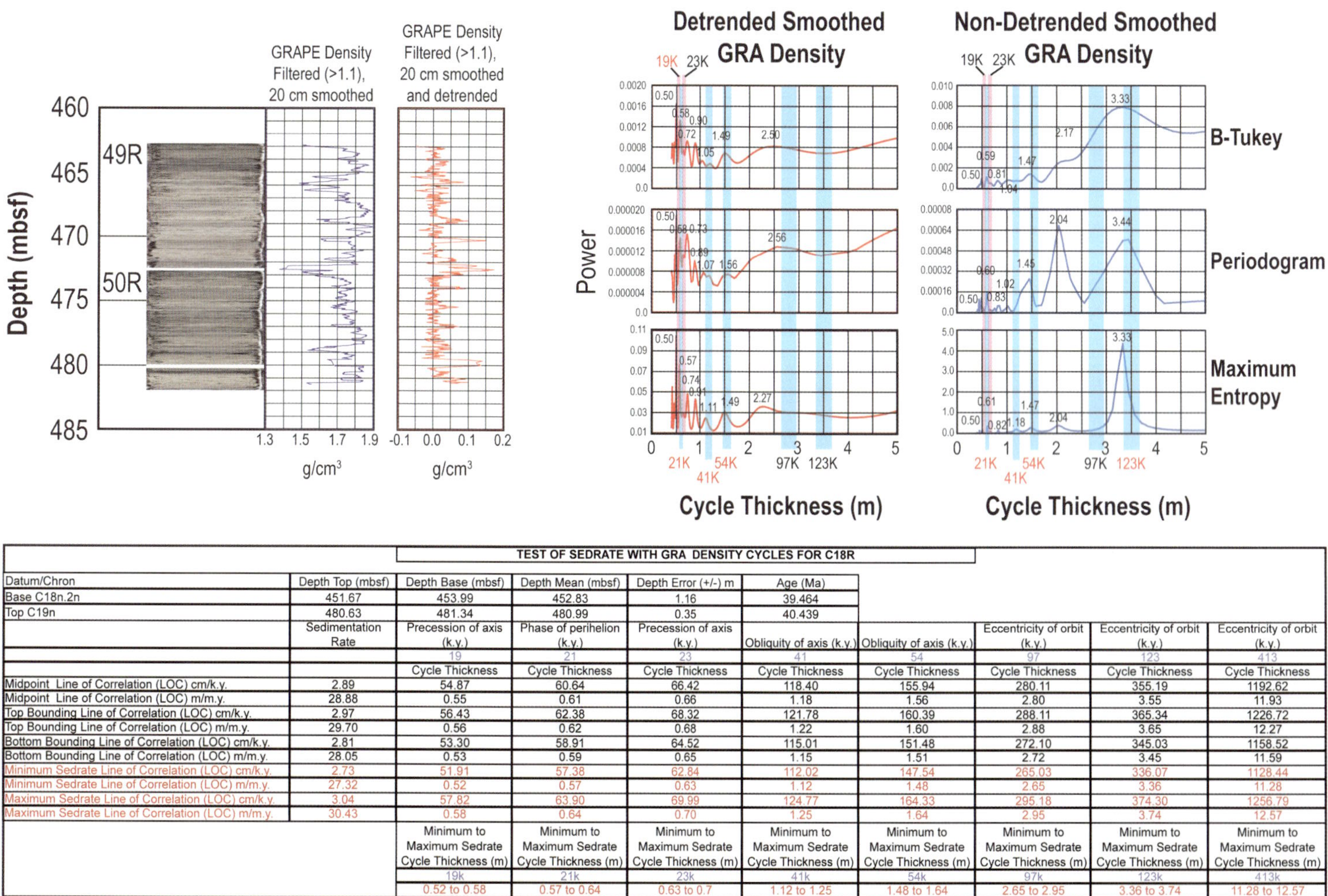

TEST OF SEDRATE WITH GRA DENSITY CYCLES FOR C18R									
Datum/Chron	Depth Top (mbsf)	Depth Base (mbsf)	Depth Mean (mbsf)	Depth Error (+/-) m	Age (Ma)				
Base C18n.2n	451.67	453.99	452.83	1.16	39.464				
Top C19n	480.63	481.34	480.99	0.35	40.439				
	Sedimentation Rate	Precession of axis (k.y.)	Phase of perihelion (k.y.)	Precession of axis (k.y.)	Obliquity of axis (k.y.)	Obliquity of axis (k.y.)	Eccentricity of orbit (k.y.)	Eccentricity of orbit (k.y.)	Eccentricity of orbit (k.y.)
		19	21	23	41	54	97	123	413
		Cycle Thickness	Cycle Thickness	Cycle Thickness	Cycle Thickness	Cycle Thickness	Cycle Thickness	Cycle Thickness	Cycle Thickness
Midpoint Line of Correlation (LOC) cm/k.y.	2.89	54.87	60.64	66.42	118.40	155.94	280.11	355.19	1192.62
Midpoint Line of Correlation (LOC) m/m.y.	28.88	0.55	0.61	0.66	1.18	1.56	2.80	3.55	11.93
Top Bounding Line of Correlation (LOC) cm/k.y.	2.97	56.43	62.38	68.32	121.78	160.39	288.11	365.34	1226.72
Top Bounding Line of Correlation (LOC) m/m.y.	29.70	0.56	0.62	0.68	1.22	1.60	2.88	3.65	12.27
Bottom Bounding Line of Correlation (LOC) cm/k.y.	2.81	53.30	58.91	64.52	115.01	151.48	272.10	345.03	1158.52
Bottom Bounding Line of Correlation (LOC) m/m.y.	28.05	0.53	0.59	0.65	1.15	1.51	2.72	3.45	11.59
Minimum Sedrate Line of Correlation (LOC) cm/k.y.	2.73	51.91	57.38	62.84	112.02	147.54	265.03	336.07	1128.44
Minimum Sedrate Line of Correlation (LOC) m/m.y.	27.32	0.52	0.57	0.63	1.12	1.48	2.65	3.36	11.28
Maximum Sedrate Line of Correlation (LOC) cm/k.y.	3.04	57.82	63.90	69.99	124.77	164.33	295.18	374.30	1256.79
Maximum Sedrate Line of Correlation (LOC) m/m.y.	30.43	0.58	0.64	0.70	1.25	1.64	2.95	3.74	12.57
		Minimum to Maximum Sedrate Cycle Thickness (m)	Minimum to Maximum Sedrate Cycle Thickness (m)	Minimum to Maximum Sedrate Cycle Thickness (m)	Minimum to Maximum Sedrate Cycle Thickness (m)	Minimum to Maximum Sedrate Cycle Thickness (m)	Minimum to Maximum Sedrate Cycle Thickness (m)	Minimum to Maximum Sedrate Cycle Thickness (m)	Minimum to Maximum Sedrate Cycle Thickness (m)
		19k	21k	23k	41k	54k	97k	123k	413k
		0.52 to 0.58	0.57 to 0.64	0.63 to 0.7	1.12 to 1.25	1.48 to 1.64	2.65 to 2.95	3.36 to 3.74	11.28 to 12.57

64R and 65R) within the carbonate-rich Lower and Middle Eocene sequence, and (ii) the interval of high biosilica content between *c.* 150 and 240 mbsf (Cores 17R–26R) in the Lower Oligocene sediments. It also allows us to see whether the MECO event and the EOT affected the bulk and compositional accumulation rates.

The carbonate barren interval in the Lower Eocene sediments (Cores 64R and 65R) has the lowest accumulation rate interval for the entire section. The bulk sediment and $CaCO_3$ accumulation rates below this level, however, are not significantly higher when the error range is considered (Figs 12 & 13), and when compared with the accumulation rates of the Middle Eocene to Oligocene deposits. Within the Eocene, the highest bulk and $CaCO_3$ accumulation rates occurred in the Middle Eocene between 42 and 40.5 Ma, which corresponds to the period of highest nannofossil species richness for Hole 647A (Firth 1989). This occurs below the level of the MECO event as estimated by our age model, which falls within a *c.* 1 Ma trough in $CaCO_3$ accumulation (*c.* 485–455 mbsf; *c.* 40.4–39.4 Ma; Figs 12 & 13). Another smaller increase in $CaCO_3$ accumulation is recorded between 38.7 and 37.8 Ma, after which it declines and remains at a steady rate until the Early Oligocene biosilica bloom occurred. Oligocene biosilica accumulation appears to have increased rapidly between 33.5 and 32.5 Ma, after the EOT, with a roughly coeval decrease in $CaCO_3$ accumulation rate. Age model error is large within this segment of Hole 647A, producing large uncertainties in the calculated sedimentation and accumulation rates. It is therefore not possible to draw clear conclusions concerning the relationship between biosilica and $CaCO_3$ fluxes, or the overall bulk sediment accumulation rate, which includes clays of terrestrial origin.

Using spectral analysis of core physical properties to test the age model

Cores 49R and 50R have almost complete recovery, with relatively little coring disturbance (mostly cracks of competent semi-lithified calcareous claystones, as opposed to biscuiting or rubble), and are constrained chronologically between the base of Chron C18n.2n and the top of Chron C19n. Core 48R has slightly less than 100% recovery, and the

single hole coring at Site 647 in rough weather would suggest that an even greater coring gap between Cores 48R and 49R could exist. They provide the two best adjacent cores for using an independent test of a portion of our age model. GRA density data with a resolution of 0.6 cm was the only non-destructive, high-resolution physical property data collected on these cores during Leg 105. Showing a pattern of GRA density fluctuations, this continuous record was suitable for spectral analysis to determine possible orbital cyclicity patterns (Fig. 14). The results provide an independent test of sedimentation rates calculated by our age model.

First, we calculated predicted thicknesses, with errors, for eight different orbital cyclicities that might occur in Cores 48R and 49R (axial precession −19 and 23 kyr; perihelion precession −21 kyr; obliquity of axis −41 and 54 kyr; eccentricity of orbit −97, 123 and 413 kyr) derived from the midpoint, maximum and minimum sedimentation rates between the base C18n.2n and top C19n. Then we performed multiple spectral analyses on the smoothed and detrended GRA density data from these two cores (see Methods section, above). We performed separate spectral analyses on both the raw smoothed and detrended data (Fig. 17). The Blackman–Tukey, Periodogram and Maximum Entropy methods produced different results between the detrended and non-detrended datasets. The non-detrended smoothed data show the following peaks: (a) a small peak *c.* 0.60 m thick; (b) a second small peak *c.* 1.08 m thick; (c) a third, broader and more prominent peak *c.* 1.46 m thick; (d) a fourth, more prominent broad peak *c.* 2.1 m thick; and (e) a fifth, broad, most prominent peak at *c.* 3.4 m thick. These appear to match up closely with the following predicted orbital cycle thicknesses: 0.60 m = 21 kyr, 1.08 m = 41 kyr, 1.46 m = 54 kyr and 3.4 m = 123 kyr. The 2.1 m-thick peak does not match any predicted orbital cycle thickness.

The detrended data show the following peaks: (a) a narrow, most prominent peak 0.5 m thick; (b) a second narrow, prominent peak *c.* 0.58 m thick; (c) a third smaller peak *c.* 1.08 m thick; (d) a fourth smaller peak *c.* 1.51 m thick; and (e) a broad prominent peak *c.* 2.44 m thick. These appear to match closely with the following predicted orbital cycle thicknesses: 0.5 m = 19 kyr, 0.58 m = 21 kyr, 1.08 m = 41 kyr, 1.51 m = 54 kyr. The 2.44 m thick peak is slightly less thick than the predicted 97 kyr peak,

Fig. 17. Comparison of predicted orbital cycle thicknesses in Cores 48R and 49R (C18r) v. calculated cycle thicknesses using core GRA density data (cleaned, 20 cm smoothed, detrended), and using the Blackman–Tukey, Periodogram, and Maximum Entropy methods (Paillard *et al.* 1996). The apparent GRA density-based cycle thicknesses within Cores 48R and 49R fall mostly within minimum to maximum predicted cycle thickness ranges of several orbital cycles, based on the sedimentation rate from the Base C18n.2n to the Top C19n.

and has a very broad range of thickness, which we cannot explain. Therefore, the multiple spectral analysis methods show reasonable reproducibility within each dataset (detrended and non-detrended), but emphasize different spectral peaks and have different peak thicknesses. Taken together, they appear to show prominent precession (19 and 21 kyr), obliquity (41 and 54 kyr) and eccentricity (123 kyr) cyclicity.

Coring imperfections, including some disturbance and missing material, although minimal compared with much of the hole, may have influenced the GRA density spectral signal (Hagelberg *et al.* 1992). This test of an independent dataset with the predicted cyclicities from our age model within Chron C18r indicates that (a) they are reasonably close, supporting this portion of the age model, and (b) precession, obliquity and eccentricity cycles occur in this middle part of the Eocene for Hole 647A. Additional spectral peaks also occur in each spectral analysis that do not correspond to any of the orbital cyclicities listed above. These could be spurious peaks possibly caused by selecting higher resolution/lower confidence analyses instead of higher confidence/lower resolution analyses.

Implications for palaeoceanography

The Labrador Sea Site 647 Palaeogene sedimentary sequence is of great importance, both as the most northerly and continuous microfossil-rich record (including calcitic groups) of Early Cenozoic northern hemisphere polar cooling, and as a monitor of North Atlantic deep water history, sampling the exchange of surface waters between the North Atlantic and Arctic Ocean and the Norwegian–Greenland Sea, since the Eocene (Miller *et al.* 2005; Cramer *et al.* 2011; O'Regan *et al.* 2011). Through our revised Site 647 age model and microfossil studies we provide improvements in age control that help identify several significant ocean-climate events during this critical phase of climate evolution, some for the first time.

Our palaeomagnetic record combined with low-resolution stable isotopic data help considerably in constraining the position of the EOB and associated climatic transition at Site 647. Correlation of global Eocene–Oligocene $\delta^{18}O$ (Zachos *et al.* 1996; Coxall *et al.* 2005; Katz *et al.* 2008; Coxall & Wilson 2011) to planktonic foraminifer records (Coxall & Pearson 2007; Pearson *et al.* 2008) reveals that the extinction of *Hantkenina* spp., which denotes the EOB worldwide (Nocchi *et al.* 1988), occurs roughly midway through a plateau in $\delta^{18}O$ values separating two rapid $\delta^{18}O$ increases in magnetochron C13r. If the *O. umbonatus* $\delta^{18}O$ increase of *c.* 1.5‰ through C13r to C13n (*c.* 290–260 mbsf)

in our records is equivalent to this step feature, and the single data point at *c.* 270 mbsf is within the plateau, then we predict the EOB to occur in the middle of Core 29R at *c.* 275 mbsf. This is *c.* 15 m above the estimated position of the EOB based on a turnover in benthic foraminifers in the original Leg 105 Site 647 age model (290 mbsf, Kaminski *et al.* 1989). A value of 285 mbsf corresponds (within the resolution of our record) to the 'first step' of the isotopic increase, which we know to be correlated with ocean cooling and widespread associated benthic and planktonic assemblage turnover (Pearson *et al.* 2008). The HO of *Turborotalia cerroazulensis* (datum no. 47, Table 2, Fig. 8), occurring at *c.* 294 mbsf (± 4.52 m) in Core 31R, and *c.* 65 ka below the extinction of *Hantkenina* spp. at its Massignano calibration site (Berggren & Pearson 2005), is consistent with this interpretation. The trend of increasing $\delta^{18}O$ prior to the EOT *sensu-stricto* alludes to local deep-ocean cooling and/or circulation changes at this high northern latitude site.

These constraints are important for establishing the response of the high North Atlantic to the major temperature changes, Antarctic ice growth and sea level change that characterize this interval (Shackleton & Kennett 1975; Zachos *et al.* 1996; Coxall *et al.* 2005; Coxall & Pearson 2007; Schulte *et al.* 2009; Wade *et al.* 2012). Higher resolution analyses of this interval are already proceeding (Coxall *et al.* 2011).

In the Middle Eocene, we predict the peak warming event of the MECO interval, which has been calibrated globally to occur at *c.* 40 Ma (around magnetic Chrons C18r to C18n.2n) on the Berggren *et al.* (1995) time scale (Jovane *et al.* 2007; Bohaty *et al.* 2009) or 39.35 Ma on the Gradstein *et al.* (2004) time scale, to occur in the base of Core 47R at about 453 mbsf (Figs 7 & 10). The highest occurrence of *Acarinina bullbrooki* (39.8 Ma, Wade *et al.* 2011) is consistent at Site 647 (Fig. 7) with the calibration of this bioevent from the SW Atlantic Ocean (Rio Grande Rise; Pujol 1983), suggesting isochronous extinction between the south and north Atlantic. The LO of *Reticulofenestra bisecta* (>10 μm) also falls nicely on the LOC, supporting its previously calibrated age (Wei & Wise 1989). Our age model places the base of C18n.2n and the top of C19n at the same location as Clement *et al.* (1989).

Jovane *et al.* (2007) documented the first Northern Hemisphere stable isotopic record across the MECO event in the Contessa Highway Section in Italy, correlating it roughly to the base of Chron C18n.2n. Bohaty *et al.* (2009) compiled a comprehensive stable isotope and carbonate accumulation record of many southern ocean DSDP/ODP sites, and two North Atlantic ODP sites, recalibrated to a common age model (Berggren *et al.* 1995; Cande & Kent 1995). This work enabled them to compare these data over a large portion of the oceans across the MECO, and to define the onset of warming to occur roughly 0.5 Myr before the brief peak warming event, which is correlated to slightly above the base of C18n.2n.

This allows us to predict the stratigraphic location of the onset of warming at roughly 462 mbsf (± 2 m), within Sections 49R-1 to -3, and the peak warming at roughly 453 mbsf, at the base of Core 47R. We have used these values to plot the MECO interval on our age model and sediment accumulation rate plots (Figs 7 & 10–12). Current research on this event within Hole 647A (Polling *et al.* 2011) supports our predicted location of the event. This will lead to comparisons with the magnitude and timing of this event in other regions (Jovane *et al.* 2007; Bohaty *et al.* 2009).

Core recovery and stratigraphic control are generally poor in the lowest 100 m of the hole (Cores 62R–71R) and it was not possible to closely constrain the level of the Early Eocene hypothermals, which have been recognized as globally significant events through the results of recent scientific drilling on Shatsky Rise, ODP Leg 198 (Bralower *et al.* 2002); Demerara Rise, ODP Leg 207 (Erbacher *et al.* 2004) and on Walvis Ridge, ODP Leg 208 (Zachos *et al.* 2004). If the lowest sedimentary magnetic normal corresponds to C24n, as interpreted here, then Cores 70R and 71R might contain a record of the Eocene Thermal Maximum '3' (ETM3), as calibrated by Galeotti *et al.* (2010). Greenishgrey laminations in Section 647–71R-2, a feature that has been found to be associated with ETM3 elsewhere, supports this prediction.

Revised estimates of sedimentation and biogenic flux provide indications of changes in sediment supply and preservation that may be climatically controlled. This includes intervals of very low carbonate accumulation during the Middle Eocene that might correspond to shallowing of the carbonate compensation depth and lysocline, as suggested by Kaminski *et al.* (1989). It is possible that these correspond to global changes in the ocean carbon system as recognized elsewhere in the Atlantic and Pacific during the Middle Eocene (Tripati *et al.* 2005; Edgar *et al.* 2007). Early versions of the Hole 647A age model (Srivastava *et al.* 1987; Baldauf *et al.* 1989; Srivastava & Arthur 1989) speculated on the presence of a Lower to lower Middle Eocene hiatus or interval of reduced sedimentation between 600–630 mbsf (Srivastava & Arthur 1989). Aubry (1995) suggested that an unconformity was present within Core 63R, between a dinocyst datum and a nannofossil datum. Middle and Lower Eocene stratigraphic control, was limited and our new multigroup biostratigraphic and palaeomagnetic information, including a revised (deeper) depth constraint for the LO of *Areosphaeridium diktyoplokum*, provides

robust support for a hiatus separating Cores 62R and 63R. The length of the hiatus, with age error determinations, is thus now better constrained at about 4 Ma duration. The cause of the hiatus is unknown but may reflect some combination of erosion by deep-water currents related to changes in spreading rates, basin geometry and sea level, or climate related overturning (Kaminski *et al.* 1989; Browning *et al.* 1996; Cramer *et al.* 2011; Hohbein *et al.* 2012). The new age constraints will allow more accurate comparisons with other North Atlantic sequences, which will help determine the causal mechanisms. For example, DSDP Site 401 in the northeastern North Atlantic also contains unconformities that overlap parts of Chrons C20n and C21n (Aubry 1995), although it has more sediment accumulation within Chron C20r than is apparent at Site 647. Extensive comparison and re-evaluation of North Atlantic unconformities is beyond the scope of this paper.

Conclusions

We have produced a refined age model for ODP Hole 647A from the Early Eocene to Oligocene that provides a reliable framework for interpreting the dynamics of environmental change during the Early Cenozoic in this strategic and climatically sensitive region of the high North Atlantic. Capitalizing on the abundance of most important oceanic microfossil groups: calcareous nannofossils, planktonic foraminifers, benthic foraminifers, dinoflagellate cysts, diatoms and radiolarians, we have improved palaeomagnetic constraints, which results in an integrated biomagnetochronology. This is augmented by a low resolution upper Eocene to Lower Oligocene benthic foraminifer $\delta^{18}O$ stratigraphy to provide additional constraints on the EOT, and spectral analysis over the least disturbed core interval in Chron C18r to test for orbital cyclicity as an independent assessment of the new chronology.

This study has yielded 63 biostratigraphic datums from four microfossil groups. We have also identified 9 Polarity Chron reversal boundaries, and 10 additional tentative Chron boundaries that fit within or very close to the error range of the primary age model. Almost all biostratigraphic datums now have recalibrated ages and age errors for Hole 647A. Therefore we have produced a template for future biomagnetostratigraphic control for many other Northern Hemisphere sites. The magnetostratigraphic calibration of dinocysts from the Norwegian–Greenland Sea (Eldrett *et al.* 2004) worked well in Hole 647A, and is now tied to nannofossil and foraminifer datums in the same hole. The results from Hole 647A have the potential for re-interpretation of many other North Atlantic DSDP and ODP sites in a similar manner, and to assess the diachroneity of bioevents between tropical and high latitude locations.

Samples were provided by the Integrated Ocean Drilling Program (IODP). The IODP is sponsored by the US National Science Foundation (NSF), and other participating countries. We thank K. Tao, A. Kasson and J. Becker for assistance with stable isotope analyses. BW acknowledges support from NSF CAREER grant EAR-0847300. We thank J. Backman for discussion and M. O'Regan for providing an improved lithostratigraphic/seismic stratigraphic summary figure. We are grateful to M. Head and G. Norris for providing their original palynological samples for review. We thank N. Pressling for double-checking discrete palaeomagnetic sample data and reviewing the palaeomagnetic methods section. We appreciate the thorough reviews by F. Frontalini, L. Jovane, and one anonymous reviewer, that helped considerably in improving this paper.

References

ACTON, G. D., OKADA, M., CLEMENT, B. M., LUND, S. P. & WILLIAMS, T. 2002. Paleomagnetic overprints in ocean sediment cores and their relationship to shear deformation caused by piston coring. *Journal of Geophysical Research*, **107**, 2067–2081. http://dx.doi.org/10.1029/2001JB000518

AGNINI, C., FORNACIARI, E., RAFFI, I., RIO, D., ROEHL, U. & WESTERHOLD, T. 2007. High-resolution nannofossil biochronology of middle Paleocene to early Eocene at ODP Site 1262: implications for calcareous nannoplankton evolution. *Marine Micropaleontology*, **64**, 215–248.

ARTHUR, M. A., DEAN, W. E., ZACHOS, J. C., KAMINSKI, M., RIEG, S. H. & ELMSTROM, K. 1989a. Geochemical expression of early diagenesis in middle Eocene–Lower Oligocene pelagic sediments in the southern Labrador Sea, Site 647, ODP Leg 105. *In: Proceedings of the Ocean Drilling Program, Scientific Results.* Ocean Drilling Program, **105**. Texas A&M University, College Station, Texas, 111–135.

ARTHUR, M. A., SRIVASTAVA, S. P., KAMINSKI, M. A., JARRARD, R. & OSIER, J. 1989b. Seismic stratigraphy and history of deep circulation and sediment drift development in Baffin Bay and the Labrador Sea. *In: Proceedings of the Ocean Drilling Program, Scientific Results*, **105**. Ocean Drilling Program. Texas A&M University, College Station, Texas, 957–988.

AUBRY, M.-P. 1995. From chronology to stratigraphy: interpreting the Lower and Middle Eocene stratigraphic record in the Atlantic Ocean. *In: BERGGREN, W. A. (ed.) Geochronology, Time Scales and Global Stratigraphic Correlation.* Society for Sedimentary Geology, Tulsa, OK, Special Publications, **54**, 213–274.

BACKMAN, J. 1986a. Accumulation patterns of Tertiary calcareous nannofossils around extinctions. *Geologische Rundschau*, **75**, 185–196.

BACKMAN, J. 1986b. Late Paleocene to Middle Eocene calcareous nannofossil biochronology from the Shatsky Rise, Walvis Ridge and Italy. *Palaeogeography, Palaeoclimatology, Palaeoecology*, **57**, 43–59.

BACKMAN, J. 1987. Quantitative calcareous nannofossil biochronology of Middle Eocene through Early Oligocene sediment from DSDP sites 522 and 523. *Abhandlungen Geologische Bundesanstalt*, **39**, 21–32.

BACKMAN, J., JAKOBSSON, M. ET AL. 2008. Age model and core-seismic integration for the Cenozoic Arctic Coring Expedition sediments from the Lomonosov Ridge. *Paleoceanography*, **23**, PA1S03, http://dx.doi.org/10.1029/2007PA001476

BALDAUF, J. G. & MONJANEL, A.-L. 1989. An Oligocene diatom biostratigraphy for the Labrador Sea: DSDP Site 112 and ODP Hole 647A. *In: SRIVASTAVA, S. P., ARTHUR, M. A. ET AL. (eds) Proceedings of the Ocean Drilling Program, Scientific Results.* Ocean Drilling Program, **105**. Texas A&M University, College Station, Texas, 323–347.

BALDAUF, J. G., CLEMENT, B. ET AL. 1989. Magnetostratigraphic and biostratigraphic synthesis of ocean drilling program leg 105: Labrador Sea and Baffin Bay. *In: SRIVASTAVA, S. P., ARTHUR, M. A. ET AL. (eds) Proceedings of the Ocean Drilling Program, Scientific Results.* Ocean Drilling Program, **105**. Texas A&M University, College Station, Texas, 935–956.

BERGGREN, W. A. 1972. Cenozoic Biostratigraphy and Paleobiogeography of the North Atlantic. *In: LAUGHTON, A. S., BERGGREN, W. A. ET AL. (eds) Initial Reports of the Deep Sea Drilling Project*, **12**. US Government Printing Office, Washington, DC, 965–1001.

BERGGREN, W. A. & PEARSON, P. 2005. A revised tropical to subtropical Paleogene planktonic foraminiferal zonation. *Journal of Foraminiferal Research*, **35**, 279–298.

BERGGREN, W. A., KENT, D. V., SWISHER, C. C. III & AUBRY, M.-P. 1995. A revised Cenozoic geochronology and chronostratigraphy. *In: BERGGREN, W. A., KENT, D. V., AUBRY, M.-P. & HARDENBOL, J. (eds) Geochronology, Time Scales and Global Stratigraphic Correlation: A Unified Temporal Framework for an Historical Geology.* Society for Sedimentary Geology, Tulsa, OK, Special Publications, **54**, 129–212.

BOHATY, S. M., ZACHOS, J. C., FLORINDO, F. & DELANEY, M. L. 2009. Coupled greenhouse warming and deep-sea acidification in the middle Eocene. *Paleoceanography*, **24**, PA2207, http://dx.doi.org/2210.1029/2008PA001676.

BOHRMANN, G. & STEIN, R. 1989. Biogenic silica at ODP Site 647 in the Southern Labrador Sea: occurrence diagenesis, and paleoceanographic implications. *In: SRIVASTAVA, S. P., ARTHUR, M. A. ET AL. (eds) Proceedings of the Ocean Drilling Program, Scientific Results.* Ocean Drilling Program, **105**. Texas A&M University, College Station, Texas, 155–170.

BOWN, P. R., JONES, T. D. ET AL. 2008 A Paleogene calcareous microfossil Konservat-Lagerstätte from the Kilwa Group of coastal Tanzania. *Bulletin of the Geological Society of America*, **120**, 3–12.

BRALOWER, T. J., PREMOLI SILVA, I., MALONE, M. J. ET AL. (eds) 2002. *Proceedings of the Ocean Drilling Program, Initial Reports.* Ocean Drilling Program, **208**. Texas A&M University, College Station, Texas.

BRINKHUIS, H., SCHOUTEN, S. ET AL. & THE EXPEDITION SCIENTISTS. 2006. Episodic fresh surface waters in the Eocene. *Nature*, **441**, 606–609.

BROWNING, J. V., MILLER, K. G. & PAK, D. K. 1996. Global implications of lower to middle Eocene sequence boundaries on the New Jersey coastal plain: the icehouse cometh. *Geology*, **24**, 639–642.

BUKRY, D. 1984. Paleogene paleoceanography of the Arctic Ocean is constrained by the middle or late Eocene age of USGS Core Fl-422: evidence from silicoflagellates. *Geology* **12**, 199–201.

BUKRY, D. 1985. Mid-Atlantic ridge coccolith and silicoflagellate biostratigraphy, deep sea drilling project sites 558 and 563. *In*: BOUGAULT, H., CANDE, S. C. ET AL. (eds) *Initial Reports of the Deep Sea Drilling Project*, **82**. US Goverment Printing Office, Washington, DC, 591–603.

CANDE, S. C. & KENT, D. V. 1995 Revised calibration of the geomagnetic polarity timescale for the Late Cretaceous and Cenozoic. *Journal of Geophysical Research*, **100**, 6093–6095. http://dx.doi.org/10.1029/94JB03098

CLARK, D. L. 1974. Late Mesozoic and early Cenozoic sediment cores from the Arctic Ocean. *Geology*, **2**, 41–44.

CLEMENT, B. M., HALL, F. J. & JARRARD, R. D. 1989. The magnetostratigraphy of ocean drilling program leg 105 sediments. *In*: SRIVASTAVA, S. P., ARTHUR, M. A. ET AL. (eds) *Proceedings of the Ocean Drilling Program, Scientific Results*. Ocean Drilling Program. Texas A&M University, College Station, Texas, **105**, 583–595.

COOPER, R. A. 2004. *The New Zealand Geological Timescale*. Institute of Geological and Nuclear Science, Wellington, Monographs, **22**.

COXALL, H. K. & PEARSON, P. N. 2007. The Eocene–Oligocene transition. *In*: WILLIAMS, M., HAYWOOD, A. M., GREGORY, F. J. & SCHMIDT, D. N. (eds) *Deep Time Perspectives on Climate Change: Marrying the Signal from Computer Models and Biological Proxies*. The Micropaleontological Society, The Geological Society, London, Special Publications, 351–387.

COXALL, H. K. & WILSON, P. A. 2011. Early Oligocene glaciation and productivity in the eastern equatorial Pacific; insights into global carbon cycling. *Paleoceanography*, **26**, A2221.

COXALL, H. K., WILSON, P. A., PÄLIKE, H., LEAR, C. H. & BACKMAN, J. 2005. Rapid stepwise onset of Antarctic glaciation and deeper calcite compensation in the Pacific Ocean. *Nature*, **433**, 53–57, http://dx.doi.org/10.1038/nature03135.

COXALL, H., BACKMAN, J. & PEARSON, P. 2011. Glassy foram stable isotope records of Eocene–Oligocene climate change from two latitudinal extremes: the high north Atlantic and the Indo-Pacific warm pool. *In*: EGGER, H. (ed.) Climate and Biota of the Early Paleogene, Conference Program and Abstracts, 5–8 June 2011, Salzburg. *Berichte der Geologischen Bundesanstalt*, **85**, 60.

CRAMER, B. S., MILLER, K. G., BARRETT, P. J. & WRIGHT, J. D. 2011. Late Cretaceous–Neogene trends in deep ocean temperature and continental ice volume: reconciling records of benthic foraminiferal geochemistry (and Mg/Ca) with sea level history. *Journal of Geophysical Research*, **116**, http://dx.doi.org/10.1029/2011JC007255.

DAMASSA, S. P., GOODMAN, D. K., KIDSON, E. J. & WILLIAMS, G. L. 1990. Correlation of Paleogene dinoflagellate assemblages to standard nannofossil zonation in North Atlantic DSDP Sites. *Review of Palaeobotany and Palynology*, **65**, 331–339.

DELL'AGNESE, D. J. & CLARK, D. L. 1994. Siliceous microfossils from the warm Late Cretaceous and Early Cenozoic Arctic Ocean. *Journal of Paleontology*, **68**, 31–47.

EDGAR, K. M., WILSON, P. A., SEXTON, P. F. & SUGANUM, Y. 2007. No extreme bipolar glaciation during the main Eocene calcite compensation shift. *Nature*, **448**, 908–911.

ELDRETT, J. S., HARDING, I. C., FIRTH, J. V. & ROBERTS, A. P. 2004. Magnetostratigraphic calibration of Eocene–Oligocene dinoflagellate cyst biostratigraphy from the Norwegian–Greenland Sea. *Marine Geology*, **204**, 91–127.

ELDRETT, J. S., HARDING, I. C., WILSON, P. A., BUTLER, E. & ROBERTS, A. P. 2007. Continental ice in Greenland during the Eocene and Oligocene. *Nature*, **446**, 176–179.

ERBACHER, J., MOSHER, D. C., MALONE, M. J. ET AL. 2004. *Proceedings of the Ocean Drilling Program, Scientific Reports*. Ocean Drilling Program, **207**. Texas A&M University, College Station, Texas. http://dx.doi.org/10.2973/odp.proc.ir.207.2004

FIRTH, J. V. 1989. Eocene and Oligocene calcareous nannofossils from the Labrador Sea, ODP Leg 105. *In*: SRIVASTAVA, S. P., ARTHUR, M. A. ET AL. (eds) *Proceedings of the Ocean Drilling Program, Scientific Results*. Ocean Drilling Program, **105**. Texas A&M University, College Station, Texas, 263–286.

FIRTH, J. V. 1996. Upper middle Eocene to Oligocene dinoflagellate biostratigraphy and assemblage variations in Hole 913B, Greenland Sea. *In*: THIEDE, J., MYHRE, A. M., FIRTH, J. V., JOHNSON, G. L. & RUDDIMAN, W. F. (eds) *Proceedings of the Ocean Drilling Program, Scientific Results*. Ocean Drilling Program, **151**. Texas A&M University, College Station, Texas, 203–242.

FIRTH, J. V. and LEG 105 SHIPBOARD SCIENTIFIC PARTY. 1987. Paleontologic and geophysical correlations in Baffin Bay and the Labrador Sea: ODP Leg 105. *Cushman Foundation for Foraminiferal Research, Special Publication*, **24**, 1–6.

FULLER, M., HASTEDT, M. & HERR, B. 1998. Coring-induced magnetization of recovered sediment. *In*: WEAVER, P. P. E., SCHMINCKE, H.-U., FIRTH, J. V. & DUFFIELD, W. (eds) *Proceedings of the Ocean Drilling Program, Scientific Results*. Ocean Drilling Program, **157**. Texas A&M University, College Station, Texas, 47–56.

GALEOTTI, S., KRISHNAN, S. ET AL. 2010. Orbital chronology of Early Eocene hyperthermals from the Contessa Road section, central Italy. *Earth and Planetary Science Letters*, **290**, 192–200.

GERSONDE, R., HODELL, D. A., BLUM, P. ET AL. (eds) 1999. *Proceedings of the Ocean Drilling Program, Initial Reports*. Ocean Drilling Program, **177**. Texas A&M University, College Station, Texas.

GOLL, R. M. 1989. A synthesis of Norwegian-Greenland Sea biostratigraphies: ODP Leg 104 on the Vøring Plateau. *In*: ELDHOLM, O., THIEDE, J. & TAYLOR, E. (eds) *Proceedings of the Ocean Drilling Program, Scientific Results*. Ocean Drilling Program, **104**. Texas A&M University, College Station, Texas, 777–826.

GRADSTEIN, F. M., OGG, J. G. *ET AL.* 2004. *A Geological Time Scale*. Cambridge University Press, Cambridge.

HAGELBERG, T., SHACKLETON, N., PISIAS, N. & SHIPBOARD SCIENTIFIC PARTY 1992. Development of composite depth sections for sites 844 through 854. *In*: MAYER, L., PISIAS, N., JANECEK, T. *ET AL.* (eds) *Proceedings of the Ocean Drilling Program, Initial Reports*. Ocean Drilling Program, **138**. Texas A&M University, College Station, Texas, 79–85.

HEAD, M. J. & NORRIS, G. 1989. Palynology and dinocyst stratigraphy of the Eocene and Oligocene in ODP Leg 105, Hole 647A, Labrador Sea. *In*: SRIVASTAVA, S. P., ARTHUR, M. A. *ET AL.* (eds) *Proceedings of the Ocean Drilling Program, Scientific Results*. Ocean Drilling Program, **105**. Texas A&M University, College Station, Texas, 515–550.

HOHBEIN, M. W., SEXTON, P. F. & CARTWRIGHT, J. A. 2012. Onset of North Atlantic Deep Water production coincident with inception of the Cenozoic global cooling trend. *Geology*, **40**, 255–258.

HSÜ, K. J., PERCIVAL, S. F., WRIGHT, R. C. & PETERSEN, N. P. 1984. Numerical ages of magnetostratigraphically calibrated biostratigraphic zones. *In*: HSU, K. J., LA BRECQUE, J. L. *ET AL.* (eds) *Initial Reports of the Deep Sea Drilling Project*, **73**. US Government Printing Office, Washington, DC, 623–635.

HUBER, B. T. & QUILLÉVÉRÉ, F. 2005. Revised Paleogene planktonic foraminiferal biozonation for the Austral Realm. *Journal of Foraminiferal Research*, **35**, 299–314.

JOVANE, L., FLORINDO, F. *ET AL.* 2007. Evidence of the middle Eocene climatic optimum (MECO) event from the Contessa Highway section, Umbrian Apennines, Italy. *Geological Society of America Bulletin*, **119**, 413–427, http://dx.doi.org/10.1130/B25917.1.

KAMINSKI, M. A., GRADSTEIN, F. M. & BERGGREN, W. A. 1989. Paleogene benthic foraminifer biostratigraphy and paleoecology at Site 647, Southern Labrador Sea. *In*: SRIVASTAVA, S. P., ARTHUR, M. A. *ET AL.* (eds) *Proceedings of the Ocean Drilling Program, Scientific Results*. Ocean Drilling Program, **105**. Texas A&M University, College Station, Texas, 705–730.

KATZ, M. E., MILLER, K. G., WRIGHT, J. D., WADE, B. S., BROWNING, J. V., CRAMER, B. S. & ROSENTHAL, Y. 2008. Stepwise transition from the Eocene greenhouse to the Oligocene icehouse. *Nature Geoscience*, **1**, 329–334.

KHAN, M. J., KENT, D. V. & MILLER, K. G. 1985. Magnetostratigraphy of Oligocene to Pleistocene sediments, Sites 558 and 563. *In*: BOUGAULT, H., CANDE, S. C. *ET AL.* (eds) *Initial Reports of the Deep Sea Drilling Project*, **82**. US Government Printing Office, Washington, DC, 385–392.

KIRSCHVINK, J. L. 1980. The least-squares line and plane and the analysis of palaeomagnetic data. *Geophysical Journal of the Royal Astronomical Society*, **62**, 699–718.

KITCHELL, J. A. & CLARK, D. L. 1982. Late Cretaceous–Paleogene paleogeography and paleocirculation: evidence of north polar upwelling. *Palaeogeography, Palaeoclimatology, Palaeoecology*, **40**, 135–165.

LAUGHTON, A. S., BERGGREN, W. A. *ET AL.* (eds) 1972. *Initial Reports of the Deep Sea Drilling Project*, **12**. US Government Printing Office, Washington, DC.

LAZARUS, D. & PALLANT, A. 1989. Oligocene and Neogene radiolarians from the Labrador Sea: ODP Leg 105. *In*: SRIVASTAVA, S. P., ARTHUR, M. A. *ET AL.* (eds) *Proceedings of the Ocean Drilling Program, Scientific Results*. Ocean Drilling Program, **105**. Texas A&M University, College Station, Texas, 349–380.

LUYENDYK, B. P., CANN, J. R. *ET AL.* (eds) 1978. *Initial Reports of the Deep Sea Drilling Project*, **49**. US Government Printing Office, Washington, DC.

MAGAVERN, S., CLARK, D. L. & CLARK, S. L. 1996. $^{87/86}$Sr, phytoplankton, and the nature of the Late Cretaceous and Early Cenozoic Arctic Ocean. *Marine Geology*, **133**, 183–192.

MANUM, S. 1976. Dinocysts in Tertiary Norwegian–Greenland Sea sediments (Deep Sea Drilling Project 38), with observations on palynomorphs and palynodebris in relation to environment. *In*: TALWANI, M., UDINTSEV, G. *ET AL.* (eds) *Initial Reports of the Deep Sea Drilling Project*, **38**. US Government Printing Office, Washington, DC, 897–919.

MANUM, S. B., BOULTER, M. C., GUNNARSDOTTIR, H., RANGNES, K. & SCHOLZE, A. 1989. Eocene to Miocene palynology of the Norwegian-Greenland Sea (ODP Leg 104). *In*: ELDHOLM, O., THIEDE, J. & TAYLOR, E. (eds) *Proceedings of the Ocean Drilling Program, Scientific Results*, **104**. Ocean Drilling Program. Texas A&M University, College Station, Texas, 611–662.

MARINO, M. & FLORES, J.-A. 2002. Middle Eocene to early Oligocene calcareous nannofossil stratigraphy at Leg 177 Site 1090. *Marine Micropaleontology*, **45**, 383–398.

MILLER, K. M., AUBRY, M-P., KHAN, M. J., MELILLO, A. J., KENT, D. V. & BERGGREN, W. A. 1985. Oligocene-Miocene biostratigraphy, magnetostratigraphy, and isotopic stratigraphy of the western North Atlantic. *Geology*, **13**, 257–261.

MILLER, K. M., WRIGHT, J. D. & BROWNING, J. V. 2005. Visions of ice sheets in a greenhouse world. *Marine Geology*, **217**, 215–231.

MUDIE, P. J. 1985. Palynology of the CESAR Cores, Alpha Ridge. *Geological Survey of Canada, Paper*, **84–22**, 149–174.

MUDIE, P. J., STOFFYN-EGLI, P. & VAN WAGONER, N. A. 1986. Geological constraints for tectonic models of the Alpha Ridge. *In*: JOHNSON, G. L. & KAMINUMA, K. (eds) *Polar Geophysics. Journal of Geodynamics*, **6**, 215–236.

MÜLLER, C. 1976. Tertiary and Quaternary calcareous nannoplankton in the Norwegian-Greenland Sea, DSDP, Leg 38. *In*: TALWANI, M., UDINTSEV, *ET AL.* (eds) *Initial Reports of the Deep Sea Drilling Project*, **38**. US Government Printing Office, Washington, DC, 823–841.

NOCCHI, M., MONECHI, S. *ET AL.* 1988. The extinction of Hantkeninidae as a marker for defining the Eocene/Oligocene boundary: a proposal. *In*: PREMOLI SILVA,

I., COCCIONI, R. & MONTANARI, A. (eds) *The Eocene/ Oligocene Boundary in the Marche-Umbria Basin (Italy)*. International Union of Geological Sciences Commission on Stratigraphy, The International Subcommission of Paleogene Stratigraphy, Ancona, Monte Cònero, 249–252.

O'REGAN, A. M., WILLIAMS, C. J., FREY, K. E. & JAKOBSSON, M. 2011. A synthesis of the long-term paleoclimatic evolution of the Arctic. Special Issue on the International Polar Year (2007–2009). *Oceanography, The Changing Arctic Ocean*, **24**, 66–80.

PAILLARD, D., LABEYRIE, L. & YIOU, P. 1996. Macintosh program performs time-series analysis. *EOS, Transactions. American Geophysical Union*, **77**, 379.

PÄLIKE, H., NORRIS, R. D. *ET AL.* 2006. The heartbeat of the Oligocene climate system. *Science*, **314**, 1894–1898.

PARKER, M. E., CLARK, M. & WISE, S. W. JR. 1985. Calcareous nannofossils of Deep Sea Drilling Project Sites 558 and 563, North Atlantic Ocean: biostratigraphy and the distribution of Oligocene braarudosphaerids. *In*: BOUGAULT, H., CANDE, S. C. *ET AL.* (eds) *Initial Reports of the Deep Sea Drilling Project*, **82**. US Government Printing Office, Washington, DC, 559–589.

PEARSON, P. N. & BURGESS, C. E. 2008. Foraminifer test preservation and diagenesis: comparison of high latitude Eocene sites. *In*: AUSTIN, W. E. N. & JAMES, R. H. (eds) *Biogeochemical Controls on Palaeoceanographic Environmental Proxies*. Geological Society, London, Special Publications, **303**, 59–72.

PEARSON, P. N., DITCHFIELD, P. W., SINGANO, J., HARCOURT-BROWN, K. G., NICHOLAS, C. J., SHACKLETON, N. J. & HALL, M. A. 2001. Warm tropical sea surface temperatures in the Late Cretaceous and Eocene epochs. *Nature*, **413**, 481–487.

PEARSON, P. N., OLSSON, R. K., HEMLEBEN, C., HUBER, B. T. & BERGGREN, W. A. (eds) 2006. *Atlas of Eocene Planktonic Foraminifera. Cushman Foundation of Foraminiferal Research*. Allen Press, Lawrence, KS, Special Publications.

PEARSON, P. N., MCMILLAN, I. K., WADE, B. S., DUNKLEY JONES, T., COXALL, H. K., BOWN, P. R. & LEAR, C. H. 2008. Extinction and environmental change across the Eocene–Oligocene boundary in Tanzania. *Geology*, **36**, 179–182.

POLLING, M., HOUBEN, A. J. P. *ET AL.* 2011. The Middle Eocene Climatic Optimum (MECO) in the high latitudes of the North Atlantic: temperature and biotic change. *In*: EGGER, H. (ed.) Climate and Biota of the Early Paleogene, Conference Program and Abstracts, 5–8 June 2011, Salzburg. *Berichte der Geologischen Bundesanstalt*, **85**, 20.

POULSEN, N. E., MANUM, S. B., WILLIAMS, G. L. & ELLEGAARD, M. 1996. Tertiary dinoflagellate biostratigraphy of sites 907, 908 and 909 in the Norwegian–Greenland Sea. *In*: THIEDE, J., MYHRE, A. M., FIRTH, J. V., JOHNSON, G. L. & RUDDIMAN, W. F. (eds) *Proceedings of the Ocean Drilling Program, Scientific Results*. Ocean Drilling Program, **151**. Texas A&M University, College Station, Texas, 255–287.

PUJOL, C. 1983. Cenozoic planktonic foraminiferal biostratigraphy of the southwestern Atlantic (Rio Grande Rise): Deep Sea Drilling Project Leg 72. *In*: BARKER,

P. E., CARLSON, R. L., JOHNSON, D. A. *ET AL.* (eds) *Initial Reports of the Deep Sea Drilling Project*, **72**. US Government Printing Office, Washington, DC, 623–673.

ROBERTS, A. P., STONER, J. S. & RICHTER, C. 1996. Coring-induced magnetic overprints and limitations of the long-core paleomagnetic measurement technique: some observations from ODP Leg 160, Eastern Mediterranean Sea. *In*: EMEIS, K.-C., ROBERTSON, A. H. F., RICHTER, C. *ET AL.* (eds) *Proceedings of the Ocean Drilling Program, Scientific Results*. Ocean Drilling Program, **160**. Texas A&M University, College Station, Texas, 497–505.

ROBERTS, A. P., BICKNELL, S. J., BYATT, J., BOHATY, S. M., FLORINDO, F. & HARWOOD, D. M. 2003. Magnetostratigraphic calibration of Southern Ocean diatom datums from the Eocene–Oligocene of Kerguelen Plateau (Ocean Drilling Program sites 744 and 748). *Palaeogeography, Palaeoclimatology, Palaeoecology*, **198**, 145–168.

RODDICK, J. C. 1989. K–Ar dating of basalts from Site 647, ODP Leg 105. *In*: SRIVASTAVA, S. P., ARTHUR, M. A. *ET AL.* (eds) *Proceedings of the Ocean Drilling Program, Scientific Results*. Ocean Drilling Program, **105**. Texas A&M University, College Station, Texas, 885–887.

SANGIORGI, F., VAN SOELEN, E. E. *ET AL.* 2008. Cyclicity in the middle Eocene central Arctic Ocean sediment record: orbital forcing and environmental response. *Paleoceanography*, **23**, PA1S04, http://dx.doi.org/ 10.1029/2007PA001477.

SCHRADER, H-J. & FENNER, J. 1976. Norwegian Sea Cenozoic diatom biostratigraphy and taxonomy. *In*: TALWANI, M., UDINTSEV, G. *ET AL.* (eds) *Initial Reports of the Deep Sea Drilling Project*, **38**. US Government Printing Office, Washington, DC, 897–919.

SCHULTE, P., WADE, B. S., KONTNY, A. & SELF-TRAIL, J. M. 2009. The Eocene–Oligocene sedimentary record in the Chesapeake Bay impact structure: implications for climate and sea level changes on the western Atlantic margin. *In*: GOHN, G. S., KOEBERL, C., MILLER, K. G. & REIMOLD, W. U. (eds) *The ICDP-USGS Deep Drilling Project in the Chesapeake Bay Impact Structure: Results from the Eyreville Core Holes*. Geological Society of America, Boulder, CO, Special Papers, **458**, 839–865. http://dx.doi.org/10.1130/ 2009.2458(35)

SHACKLETON, N. J. & KENNETT, J. P. 1975. Paleotemperature history of the Cenozoic and the initiation of Antarctic glaciation: oxygen and carbon isotope analysis in DSDP Sites 277, 279 and 281. *In*: KENNETT, J. P., HOUTZ, R. E. *ET AL.* (eds) *Initial Reports of the Deep Sea Drilling Project*, **29**. US Government Printing Office, Washington, DC, 881–884.

SRIVASTAVA, S. P. 1978. Evolution of the Labrador Sea and its bearing on the early evolution of the North Atlantic. *Geophysical Journal of the Royal Astronomical Society*, **52**, 313–357.

SRIVASTAVA, S. P. & ARTHUR, M. A. 1989. Tectonic evolution of the Labrador Sea and Baffin Bay: constraints imposed by regional geophysics and drilling results from Leg 105. *In*: SRIVASTAVA, S. P., ARTHUR, M. A. *ET AL.* (eds) *Proceedings of the Ocean Drilling Program,*

Scientific Results. Ocean Drilling Program, **105**. Texas A&M University, College Station, Texas, 989–1009.

SRIVASTAVA, S. P., ARTHUR, M. A. *ET AL.* (eds) 1987. *Proceedings of the Ocean Drilling Program, Initial Reports*. Ocean Drilling Program, **105**. Texas A&M University, College Station, Texas.

STEIN, R., LITTKE, R., STAX, R. & WELTE, D. H. 1989. Quantity, provenance, and maturity of organic matter at ODP Sites 645, 646, and 647: implications for reconstruction of paleoenvironments in Baffin Bay and Labrador Sea during Tertiary and Quaternary Time. *In*: SRIVASTAVA, S. P., ARTHUR, M. A. *ET AL.* (eds) *Proceedings of the Ocean Drilling Program, Scientific Results*. Ocean Drilling Program, **105**. Texas A&M University, College Station, Texas, 185–208.

STICKLEY, C. E., BRINKHUIS, H. *ET AL.* 2004. Late Cretaceous–Quaternary biomagnetostratigraphy of ODP Sites 1168, 1170, 1171, and 1172, Tasmanian Gateway. *In*: EXON, N. F., KENNETT, J. P. & MALONE, M. J. (eds) *Proceedings of the Ocean Drilling Program, Scientific Results*. Ocean Drilling Program, **189**. Texas A&M University, College Station, Texas, 1–57.

STICKLEY, C. E., ST. JOHN, K., KOÇ, N., JORDAN, R. W., PASSCHIER, S., PEARCE, R. & KEARNS, L. E. 2009. Evidence for middle Eocene Arctic sea ice from diatoms and ice-rafted debris. *Nature*, **460**, 376–380.

ST JOHN, K. 2008. Cenozoic ice-rafting history of the Central Arctic Ocean: terrigenous sands on the Lomonosov ridge. *Paleoceanography*, **23**, PA1S05, http://dx.doi.org/10.1029/2007PA001483.

SUTO, I., JORDAN, R. W. & WATANABE, M. 2009. Taxonomy of middle Eocene diatom resting spores and their allied taxa from the central Arctic Basin. *Micropaleontology*, **55**, 259–312.

THOMAS, E., BARRERA, E. *ET AL.* 1990. Upper Cretaceous-Paleogene stratigraphy of Sites 689 and 690, Maud Rise (Antarctica). *In*: BARKER, P. F., KENNETT, J. P. *ET AL. Proceedings of the Ocean Drilling Program,*

Scientific Results. Ocean Drilling Program, **113**. Texas A&M University, College Station, Texas, 901–914.

TRIPATI, A., BACKMAN, J., ELDERFIELD, H. & FERRETTI, P. 2005. Eocene bipolar glaciation associated with global carbon cycle changes. *Nature*, **436**, 341–346.

WADE, B. S., PEARSON, P. N., BERGGREN, W. A. & PÄLIKE, H. 2011. Review and revision of Cenozoic tropical planktonic foraminiferal biostratigraphy and calibration to the geomagnetic polarity and astronomical time scale. *Earth-Science Reviews*, **104**, 111–142.

WADE, B. S., HOUBEN, A. J. P. *ET AL.* 2012. Multiproxy record of abrupt sea-surface cooling across the Eocene–Oligocene transition in the Gulf of Mexico. *Geology*, **40**, 159–162.

WEI, W. & WISE, S. W. JR. 1989. Paleogene calcareous nannofossil magnetobiochronology: results from South Atlantic DSDP Site 516. *Marine Micropaleontology*, **14**, 119–152.

WILLIAMS, G. & MANUM, S. 1999. Oligocene–Early Miocene Dinocyst Stratigraphy of Hole 985A (Norwegian Sea). *In*: RAYZMO, M. E., JANSEN, E., BLUM, P. & HERBERT, T. D. (ed.) *Proceedings of the Ocean Drilling Program, Scientific Results*. Ocean Drilling Program, **162**. Texas A&M University, College Station, Texas, 99–109.

WILSON, P. A., NORRIS, R. D. & COOPER, M. J. 2002. Testing the mid-Cretaceous greenhouse hypothesis using glassy foraminiferal calcite from the core of the Turonian tropics on Demerara Rise. *Geology*, **30**, 607–610.

ZACHOS, J. C., QUINN, T. M. & SALAMY, K. A. 1996. High-resolution (104 years) deep-sea foraminiferal stable isotope records of the Eocene–Oligocene climate transition. *Paleoceanography*, **11**, 251–266.

ZACHOS, J. C., KROON, D., BLUM, P. *ET AL.* (eds) 2004. *Proceedings of the Ocean Drilling Program, Initial Reports*. Ocean Drilling Program, **208**. Texas A&M University, College Station, Texas.

Integrated magnetobiostratigraphy of the middle Eocene–lower Oligocene interval from the Monte Cagnero section, central Italy

LUIGI JOVANE[1,2]*, JAIRO F. SAVIAN[1,3], RODOLFO COCCIONI[4],
FABRIZIO FRONTALINI[4], GIUSEPPE BANCALÀ[4], RITA CATANZARITI[5],
VALERIA LUCIANI[6], STEVEN M. BOHATY[1], PAUL A. WILSON[1] & FABIO FLORINDO[7]

[1]*Ocean and Earth Science, University of Southampton, National Oceanography Centre Southampton, European Way, Southampton SO14 3ZH, UK*

[2]*Instituto Oceanográfico, Universidade de São Paulo, Rua do Matão 1226, 05508-090, São Paulo, Brazil*

[3]*Instituto de Astronomia, Geofísica e Ciências Atmosféricas, Universidade de São Paulo, Rua do Matão 1226, 05508-090, São Paulo, Brazil*

[4]*Dipartimento di Scienze dell'Uomo, dell'Ambiente e della Natura, Università degli Studi di Urbino 'Carlo Bo,' Campus Scientifico, Località Crocicchia, 61029 Urbino, Italy*

[5]*Istituto di Geoscienze e Georisorse CNR, 56124 Pisa, Italy*

[6]*Università degli Studi, Ferrara, Italy*

[7]*Istituto Nazionale di Geofisica e Vulcanologia, Via di Vigna Murata 605, 00143 Rome, Italy*

Corresponding author (e-mail: luigijovane@gmail.com)

Abstract: The Monte Cagnero sedimentary section, which crops out in the northeastern Apennines near Urbania in the Umbria–Marche Basin (Italy), contains well-exposed strata spanning the middle Eocene to lower Oligocene interval. We use an integrated magnetobiostratigraphic approach to generate a high-resolution age model for the Monte Cagnero section, with the goal of obtaining a reliable chronostratigraphic framework for studying Eocene–Oligocene palaeoceanographic changes during the switch from greenhouse to icehouse conditions. The studied sediments consist of alternating reddish and greenish limestones and marlstones. A new integrated age model for the section is based on high-resolution palaeomagnetic analyses, combined with detailed planktonic foraminiferal and calcareous nannofossil biostratigraphic results. Rock magnetic measurements show that the magnetic mineralogy is dominated by a mixture of high- and low-coercivity minerals, probably representing a combination of hematite and magnetite. A robust magnetostratigraphic signal, together with the identification of key planktonic foraminiferal and nannofossil biostratigraphic events, allows construction of a detailed age model for the section. Based on these results, we infer that the section spans a continuous interval (within magnetochron resolution) from the middle Eocene to lower Oligocene (*c.* 41–27 Ma; Chrons C18r–C12r). The Monte Cagnero section, therefore, represents a sequence that is suitable for studying the impact of the Neo-Tethyan gateway closure on subtropical Eocene circulation and determining the nature and timing of palaeoceanographic changes in the Tethys through the late middle Eocene to early Oligocene interval.

The Cenozoic Era, which represents the last 65 million years (myr) from the end of the Cretaceous to the present, is a transformational time in the evolution of Earth's palaeoclimate. Benthic foraminiferal oxygen isotope ($\delta^{18}O$) records from all major ocean basins indicate a cooling trend over the past 50 myr, since the early Eocene climatic optimum (*c.* 53–51 Ma; e.g. Zachos *et al.* 2001, 2008). The middle Eocene to early Oligocene was a key interval for Earth's transition from warmer (greenhouse) to cooler (icehouse) conditions. This interval is characterized by the expansion of Antarctic large ice sheets to ≥50% the size of the present day volume (e.g. Zachos *et al.* 1996; Lear *et al.* 2000; Coxall *et al.* 2005; Katz *et al.* 2008; Coxall & Wilson 2011), culminating in the Eocene–Oligocene

From: JOVANE, L., HERRERO-BERVERA, E., HINNOV, L. A. & HOUSEN, B. A. (eds) 2013. *Magnetic Methods and the Timing of Geological Processes*. Geological Society, London, Special Publications, **373**, 79–95.
First published online March 25, 2013, http://dx.doi.org/10.1144/SP373.13 © The Geological Society of London 2013.
Publishing disclaimer: www.geolsoc.org.uk/pub_ethics

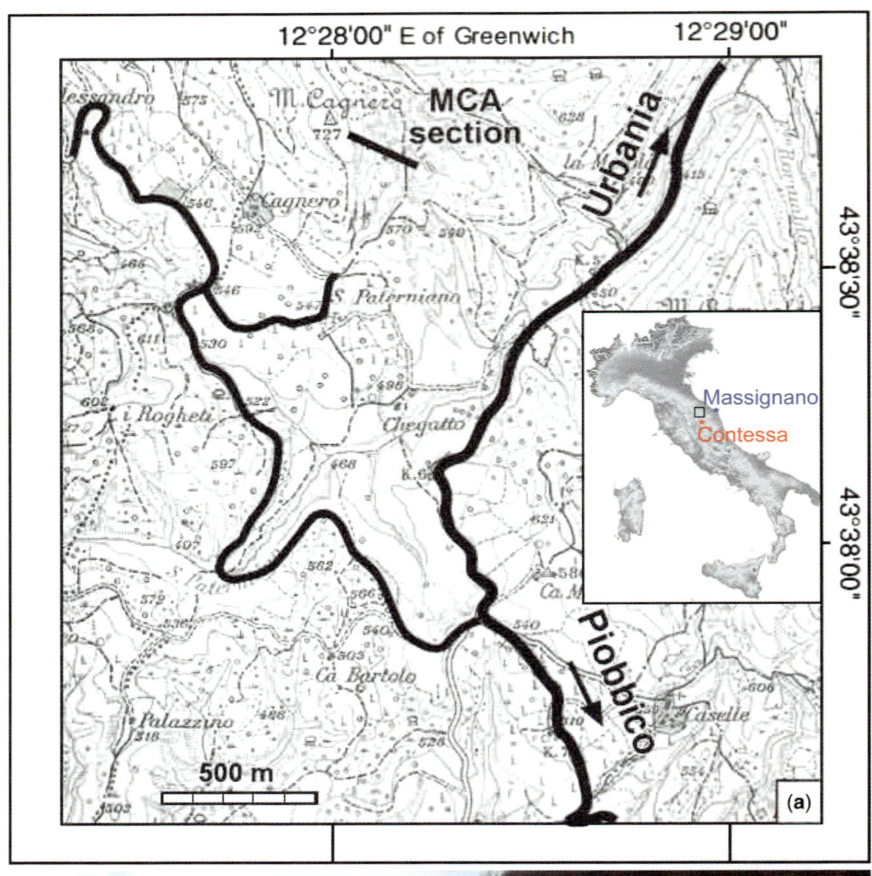

12°28'00" E of Greenwich
12°29'00"
MCA section
Urbania
Piobbico
Massignano
Contessa
500 m
43°38'30"
43°38'00"
(a)

(b)

(E–O) climate transition (Oi-1; *c.* 33.5 Ma; Miller *et al.* 1991).

The long-term cooling trend through the Eocene did not occur as a single gradual change. There are a series of distinct cooling and warming episodes superimposed on the cooling trend (e.g. Miller *et al.* 1991; Jovane *et al.* 2007*a*, *b*; Zachos *et al.* 2008; Cramer *et al.* 2011). The most important climatic warming event in this interval is the Middle Eocene Climatic Optimum (MECO) – an event that is characterized by a large negative $\delta^{18}O$ excursion at *c.* 40 Ma (Bohaty & Zachos 2003; Jovane *et al.* 2007*b*; Bohaty *et al.* 2009; Edgar *et al.* 2010). The causes of this event, which has been identified at multiple deep-sea drill sites around the globe and in outcrop sections in the Tethys region (Jovane *et al.* 2007*b*; Bohaty *et al.* 2009; Edgar *et al.* 2010; Luciani *et al.* 2010; Spofforth *et al.* 2010; Dawber & Tripati 2011), are still unknown.

The Eocene–Oligocene transition represents an abrupt climatic threshold reflecting ice growth in Antarctica (Katz *et al.* 2008; Merico *et al.* 2008; Cramer *et al.* 2011). The main hypothesized causes for the Oi-1 abrupt cooling at the E–O transition include the opening of Southern Ocean gateways (e.g. Livermore *et al.* 2005), decreasing concentrations of greenhouse gases in the atmosphere and orbital forcing (e.g. DeConto & Pollard 2003; Coxall *et al.* 2005; Lyle *et al.* 2005; Pagani *et al.* 2005, 2011; Pearson *et al.* 2009), and closure of the Neo-Tethys gateway (e.g. Jovane *et al.* 2007*a*, 2009). Concerning the last hypothesis, new data from the Tethys Ocean are needed to better assess the effects of the gradual closure of the Neo-Tethys gateway, which was driven by collision between Arabia and Eurasia (Jovane *et al.* 2007*a*, 2009; Allen & Armstrong 2008). The gradual nature of the closure resulted in palaeoceanographic oscillations, particularly clear evidence in environmental magnetic properties, in the Neo-Tethys region prior to the Oi-1 event (Jovane *et al.* 2007*a*), perhaps as a result of variations in the Sub Tropical Eocene Neo-Tethys current (Jovane *et al.* 2009).

The Monte Cagnero section (Fig. 1), near Urbania, central Italy, is an important sedimentary sequence for study of Eocene and Oligocene palaeoclimate events. Biostratigraphic studies of several microfossil groups have been carried out for this section, including dinoflagellates (Brinkhuis & Biffi 1993), calcareous nannofossils (Baumann & Roth 1969; Baumann 1970; Maiorano & Monechi 2006) and foraminifera (Baumann & Roth 1969; Baumann 1970; Parisi *et al.* 1988; Verducci & Nocchi 2004). An integrated stratigraphy of the Oligocene pelagic sequence at Monte Cagnero has been constructed by Coccioni *et al.* (2008). The integrated stratigraphy indicates a stratigraphic equivalence between metres stratigraphic level (msl) 114.1 at Monte Cagnero and msl 19 at the section at Massignano, which is near Ancona City (Monte Conero) and represents the Global boundary stratotype section and point (GSSP) for the Eocene–Oligocene boundary (Premoli Silva & Jenkins 1993; Jovane *et al.* 2004; Coccioni *et al.* 2008; Hyland *et al.* 2009). A previous magnetostratigraphic interpretation (Hyland *et al.* 2009) covers mainly the Oligocene part of the section, from 100 to 146.5 msl, and in the lower part the Priabonian–Rupelian boundary. The Monte Cagnero section is a unique and representative section for the Eocene and Oligocene epochs in the Umbria–Marche Basin, and offers the possibility of correlation with the Contessa Highway section (e.g. Jovane *et al.* 2007*b*), *c.* 40 km south, and with the Massignano stratotype section (e.g. Jovane *et al.* 2009), *c.* 100 km east (Fig. 1).

In this study, we present a high-resolution magneto-, bio- and chronostratigraphic study of the Monte Cagnero section spanning the middle Eocene to lower Oligocene interval. We extend the stratigraphic study of the section from 128 msl down to 58 msl. The main objectives of this work were (1) to obtain an accurate magnetostratigraphy for the middle Eocene–lower Oligocene interval, and (2) to refine the age calibration of biostratigraphic events in the same interval. This work has resulted in better documentation of the relationship between the gradual closure of the Neo-Tethys gateway and the resulting variation of the Sub-Tropical Eocene Neo-Tethys (Jovane *et al.* 2009). A robust age model forms the framework for future palaeoceanographic studies at this site, which will be used in studies of important climatic events, such as the MECO and the Oi-1 events.

Geological setting and sampling

The Monte Cagnero section (43°38′50′N, 12°28′05′E and 727 m above sea-level) lies within the Umbria–Marche Basin (northeastern Apennines, Chiarabba *et al.* 2005) near Urbania, Italy (Fig. 1). The study section ranges from 58 to 128 msl (70 m thick) and comprises the Priabonian–Rupelian boundary at 114.1 msl, which corresponds to 19 msl in the Massignano Section (Coccioni *et al.* 2008; Hyland *et al.* 2009).

Fig. 1. (**a**) Location map of the Monte Cagnero Section. (**b**) Representative picture from Scaglia Cinerea part of the Monte Cagnero section (taken in March 2008).

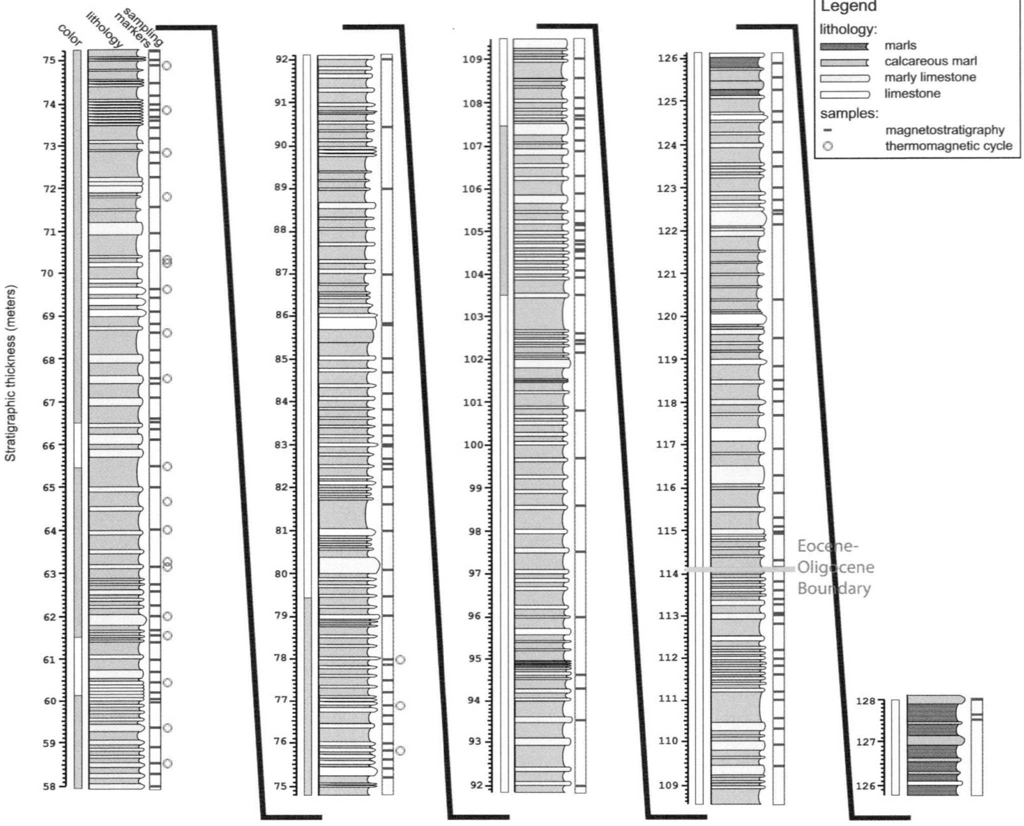

Fig. 2. High-resolution lithostratigraphic sequence of the Monte Cagnero section from metres stratigraphic level (msl) 58 to 128 following the legend. Purple diamonds and blue lines indicate samples used for thermomagnetic curves and magnetostratigraphy. At 114.1 msl the Eocene–Oligocene boundary is marked (Coccioni *et al.* 2008; Hyland *et al.* 2009).

The Monte Cagnero section has been proposed as the Rupelian/Chattian GSSP (Coccioni *et al.* 2008). The lithologies of the entire Eocene and Oligocene section consist of alternating reddish/greenish grey calcareous marl and marly limestones from the Scaglia Variegata and Scaglia Cinerea formations, which largely represent pelagic carbonate successions (Fig. 2) (e.g. Jovane *et al.* 2007a, b, 2009), The lower part of the study section consists of *c.* 21 m (58.00–79.50 msl) of 0.15–1.00 m-thick red-violet limestone and 0.20–2.00 m-thick violet marls with intercalations of white marl corresponding to the lower member of Scaglia Variegata Formation. We recognize the middle member of the Scaglia Variegata Formation as *c.* 24 m (79.50–103.50 msl) of alternating 0.15–0.35 m-thick marly limestones and green/grey marls. The upper member of the Scaglia Variegata Formation (103.50–107.50 msl) is represented by a series of 0.15 m-thick beds of red calcareous marls. The Scaglia Variegata Formation comprises a time interval from the end of the early Eocene to the end of the late Eocene (*c.* 47–*c.* 35 Ma; e.g. Jovane *et al.* 2007a). Above 107.50 msl, the Scaglia Variegata Formation grades into grey marl and marly limestone of the mainly Oligocene Scaglia Cinerea Formation (e.g. Jovane *et al.* 2007a). Benthic foraminiferal data from Guerrera *et al.* (1988) and Parisi *et al.* (1988) interpret lower bathyal palaeo-depths (1000–2000 m) during the middle Eocene, which gradually shoaled in the early late Eocene to mid-bathyal depths (800–1000 m). In the early Oligocene, upper bathyal palaeo-depths (400–600 m) are estimated. Between 103.5 and 103.30 msl lies a 0.2 m layer of slightly deformed marl of the Scaglia Cinerea Formation associated with slumping.

Methods

Palaeomagnetism

A total of 700 samples were collected at *c.* 0.1 m intervals from the Monte Cagnero section and used for all analyses reported here. Palaeomagnetic sampling consisted of 241 oriented block samples for magnetostratigraphic analysis at *c.* 0.3 m resolution. The oriented block samples were cut into two or three paired cubes (named A and B), in order to apply both alternating field (AF) (A samples) and thermal demagnetization (B samples) methods. Before analysis, each dry sample was weighed with a high precision balance, and, owing to the irregular volume of the samples, all data were normalized relative to mass.

Because of the low intensity of the natural remanent magnetization (NRM) observed in a pilot study, samples were collected between 78 and 102 msl from two parallel transects at a distance of 20 m from the central main transect. The purpose of this sampling strategy was to increase the number of samples in this key interval. We designated these two sections as Y and Z, in addition to the central section called X. All the materials studied are archived in the laboratory of the Dipartimento di Scienze della Terra, della Vita e dell'-Ambiente, Università di Urbino, Italy.

All palaeomagnetic measurements were carried out at the University of Southampton's palaeomagnetic laboratory of the National Oceanography Centre Southampton, UK. We measured the low-field mass-specific bulk magnetic susceptibility (χ) in all samples using a KLY-4 Kappabridge (AGICO) instrument. Measurements were performed using a three-axis 2-G Enterprises cryogenic magnetometer (model 755R), housed in a magnetically shielded room. Progressive AF demagnetization steps (5, 10, 15, 20, 25, 30, 35, 40, 45, 50, 60, 70 and 80 mT) were applied to all A specimens, and stepwise heating to the B specimens (25, 100, 150, 200, 250, 300, 350, 400, 450, 500, 520, 540, 560, 580, 600, 620, 640, 660, 680 and 700 °C). The 241 discrete A samples from the Monte Cagnero section were subjected to progressive AF demagnetization to remove possible magnetic overprints due to secondary magnetization and to discern characteristic remanent magnetization (ChRM) directions.

Magnetic components were identified using stereographic projections, orthogonal and demagnetizing intensity plots. The ChRM directions were calculated using principal component analysis (Kirschvink 1980) using the palaeomagnetic data software package REMASOFT 3.0. The palaeomagnetic results were rectified removing the tectonic tilt of Monte Cagnero section, which dips between 20 and 22° to the WNW (mean strike N30°E).

To obtain information on the magnetic mineralogy of the study samples, thermomagnetic analyses were performed using a CS-3 Kappabridge at the palaeomagnetic laboratory of the IAG, University of São Paulo, São Paulo (Brazil). Thermomagnetic analyses are produced by heating samples and then measuring susceptibility as a function of temperature, revealing features typical of individual magnetic minerals or groups of minerals (e.g. Hrouda 1994; Dunlop & Özdemir 1997; Hrouda 2003). The temperature dependence of magnetic susceptibility was measured up to 700 °C (high temperature) on selected samples (22). The thermomagnetic curves were measured in argon atmosphere in order to eliminate oxidation and alteration of magnetic minerals during heating. For the same set of samples, low-temperature thermomagnetic curves were also performed down to -192 °C for the same samples. These measurements were performed under air atmosphere. The thermomagnetic curves were interpreted and evaluated using the software package CUREVAL (AGICO).

Biostratigraphy

Planktonic foraminifera. Marly and soft marly-limestone samples for planktonic foraminiferal analysis were disaggregated using dilute hydrogen peroxide and Desogen (a surfactant). Hard marly limestone and limestone samples were prepared using the cold acetolyse technique of Lirer (2000), sieving through a 63 μm mesh, and drying at 50 °C. This method enabled extraction of identifiable foraminifera even from indurated limestones. The combined preparation techniques allowed accurate taxonomic determination and detailed analysis of foraminiferal assemblages through the entire Monte Cagnero section. The residues were studied with a binocular microscope to characterize assemblages and identify biostratigraphic marker species, following the taxonomic criteria of Pearson *et al.* (2006). The two planktonic foraminiferal zonations of Berggren *et al.* (1995) and Wade *et al.* (2011) were followed.

Calcareous nannofossils. Samples for calcareous nannofossil analysis were prepared from unprocessed material as simple smear slides using standard preparation methods (Bown & Young 1998). Smear slides were studied using a Leitz Laborlux 12 Pol light microscope both under crossed-polarized and transmitted light at a magnification of 1250×. Nannofossil taxa were identified according to the taxonomic concepts summarized by Perch-Nielsen (1985). The standard calcareous nannofossil zonations of Martini (1971) and Okada & Bukry (1980) are widely used for low- and mid-latitude

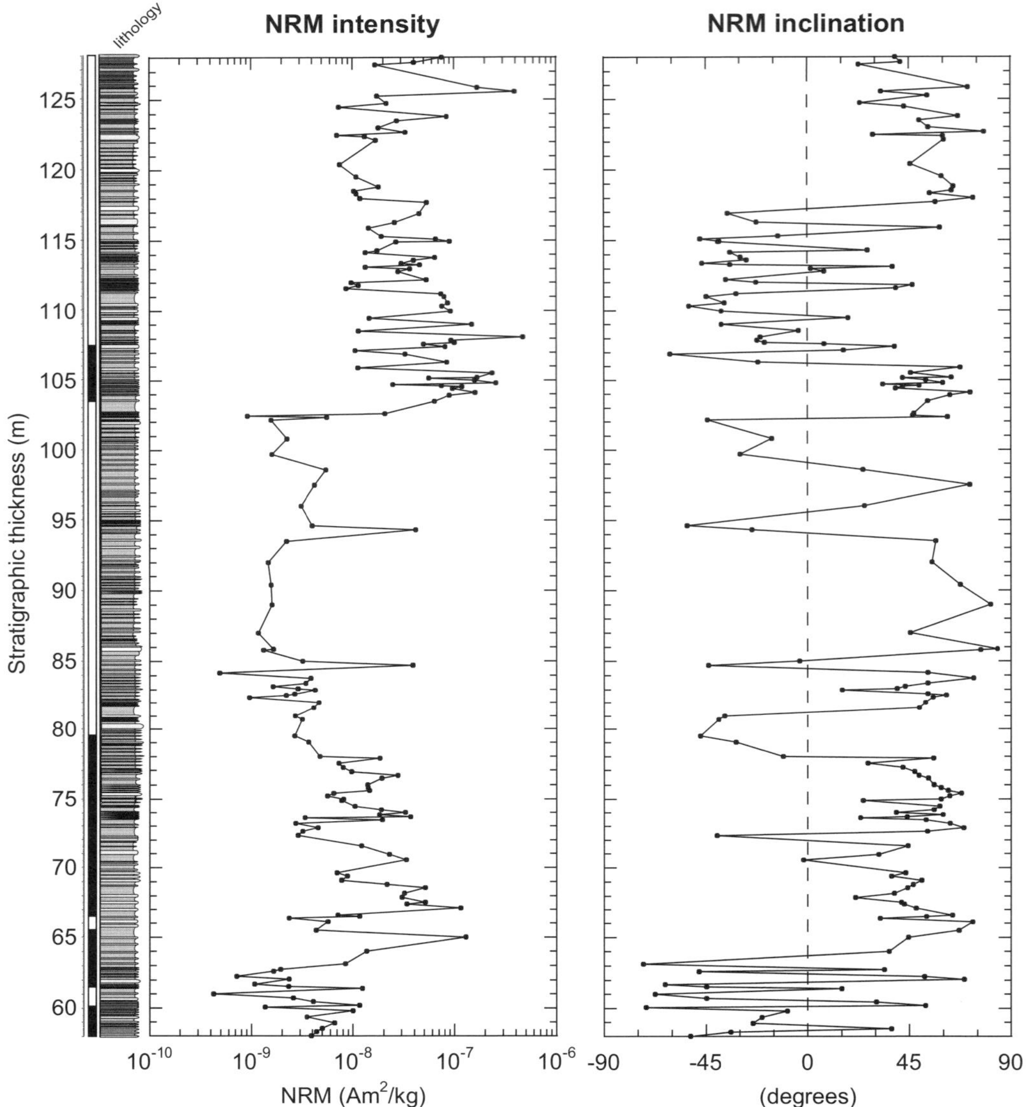

Fig. 3. Intensity and inclination of the NRM along the Monte Cagnero section from 58 to 128 msl compared with lithostratigraphic column.

biostratigraphic studies in the Palaeogene, and these were adopted in this paper.

Results

Palaeomagnetic behaviour and polarity zonation

The NRM of the Monte Cagnero sample set varies from 4.27×10^{-10} to 4.71×10^{-7} A m^2 kg^{-1}, with mean values of 3.49×10^{-8} A m^2 kg^{-1} (Fig. 3). The NRM intensities are significantly lower in the interval between 78 and 104 msl and the magnetic behaviour during demagnetization is not completely resolvable.

AF demagnetization was more effective in removing the secondary magnetization than thermal demagnetization at the Monte Cagnero section. In most samples, a low-coercivity component was removed after AF demagnetization between 10 and

80 mT and after thermal demagnetization at 200–540 °C. Stable palaeomagnetic behaviour (with maximum angular deviation (MAD) values below 15°) was obtained for 105 samples (44% of the total analysed samples). Thermal treatment performed on all 241 B samples showed stable behaviour for 65 samples (27%). For a total of 170 samples (70%), based on the observed demagnetization behaviour (Fig. 4), we isolated the ChRM component recognizing both normal and reverse polarity directions (Fig. 5).

The magnetic polarity record from the Monte Cagnero section was divided into 13 magnetozones (Fig. 5). A high-resolution record was obtained between 58 and 79 msl and between 102 and 124 msl (Fig. 5). Between these intervals (79–102 msl), a much lower-resolution record was obtained owing to the complex behaviour of the samples during demagnetization. Many of these samples with poor demagnetization behaviour have MAD values >15° and were not used to define the magnetic polarity record. The magnetozones were defined using multiple and consecutive samples with the same polarities along the section.

The arithmetic mean inclination and declination for the normal polarity samples used in the magnetozones are 50.8 and 245.8° (colatitudes = 48.8°; $N = 94$; $\alpha_{95} = 3.40$), respectively. For the reverse polarity samples, the means are −42.8 and 83.2° (colatitudes= − 48.9°; $N = 69$; $\alpha_{95} = 3.52$), respectively. Samples with ChRM inclinations below 20° were not considered when calculating the means. At this location, the expected average values are 56 for inclination and 6° for declination (Besse & Courtillot 1991). Thus, published values of inclination and declination are close to the cone of confidence determined here. For the samples that retained a weak magnetization (in several cases we were not able to remove the secondary components), we believe the reversal test and results to be acceptable.

The thermomagnetic heating and cooling curves for selected samples from Monte Cagnero section are shown in Figure 6. For all samples, we observe magnetic mineralogical transformations during thermal treatment detected by the variation of magnetic susceptibility v. temperature changes. Samples at 58.60 and 68.65 msl (Fig. 6a, d) show initially low values of magnetic susceptibility corresponding to a presence of dia- and/or paramagnetic minerals. Samples at 63.35 msl (Fig. 6b) and 64.70 msl (Fig. 6c) show an increase in the magnetic susceptibility at around 300–400 °C, and then a major fall after 580 °C, indicating the presence of magnetite with minimal presence of hematite. The sample at 64.70 msl shows reversible behaviour after heating; however, after heating at 700 °C under open-air conditions, the samples at 58.60, 68.65 and 63.35 msl suffered an irreversible mineralogical transformation, indicated by the higher susceptibility of the cooling curve. For samples at 58.60 and 68.65 msl, the susceptibility increases rapidly during cooling between 600 and 500 °C, indicating that magnetite is product of clays during thermal treatment. In samples from 63.35 and 64.70 msl, the increase begins at approximately 650 °C, suggesting that new hematite was also produced. These mixtures of different magnetic phases are often found in carbonate samples, and the formation of secondary (authigenic) iron oxides is thought to result from the presence of pre-existing iron-bearing minerals (Channell *et al.* 1982; Channell & McCabe 1994; Jovane *et al.* 2007*a*, *b*).

Biostratigraphy

The biostratigraphy of the Monte Cagnero section was initially documented by Baumann & Roth (1969). They established a correlation between planktonic foraminiferal and calcareous nannofossil biozonations from the Eocene–Oligocene boundary interval to the uppermost Oligocene and identified seven nannofossil biozones corresponding to Zones NP19/20–NP24 of Martini (1971). Compared with several other sections of the Umbria–Marche Scaglia succession, the Monte Cagnero section is the most expanded and continuous section straddling the E–O boundary (Baumann 1970). Further biostratigraphic refinement of the Monte Cagnero section, encompassing *c.* 55 m of the Eocene (Scaglia Variegata Formation) and *c.* 30 m of the overlying Oligocene section (Scaglia Cinerea Formation), was completed by Guerrera *et al.* (1988) and Parisi *et al.* (1988). A further refinement of the middle–late Eocene planktonic foraminiferal biostratigraphy at the Monte Cagnero section was performed by Verducci & Nocchi (2004), who also attempted to identify significant palaeoclimatic and palaeoceanographic events across this time interval. More recently, Coccioni *et al.* (2008) proposed the Monte Cagnero section as a potential candidate for the Chattian GSSP, and Hyland *et al.* (2009) nominated this section as a parastratotype for the Eocene–Oligocene GSSP at Massignano.

Planktonic foraminifera. Planktonic foraminifera are abundant and diverse throughout the study interval of the Monte Cagnero section, with high species richness and assemblages typical of late Eocene–early Oligocene low-latitude pelagic environments. Preservation is mostly moderate to good, therefore offering the possibility of accurate taxonomic.

The following main biostratigraphic events (bioevents) were identified and correlated with the planktonic foraminiferal biozones (from bottom to top): (1) lowest occurrence (LO) of *Orbulinoides*

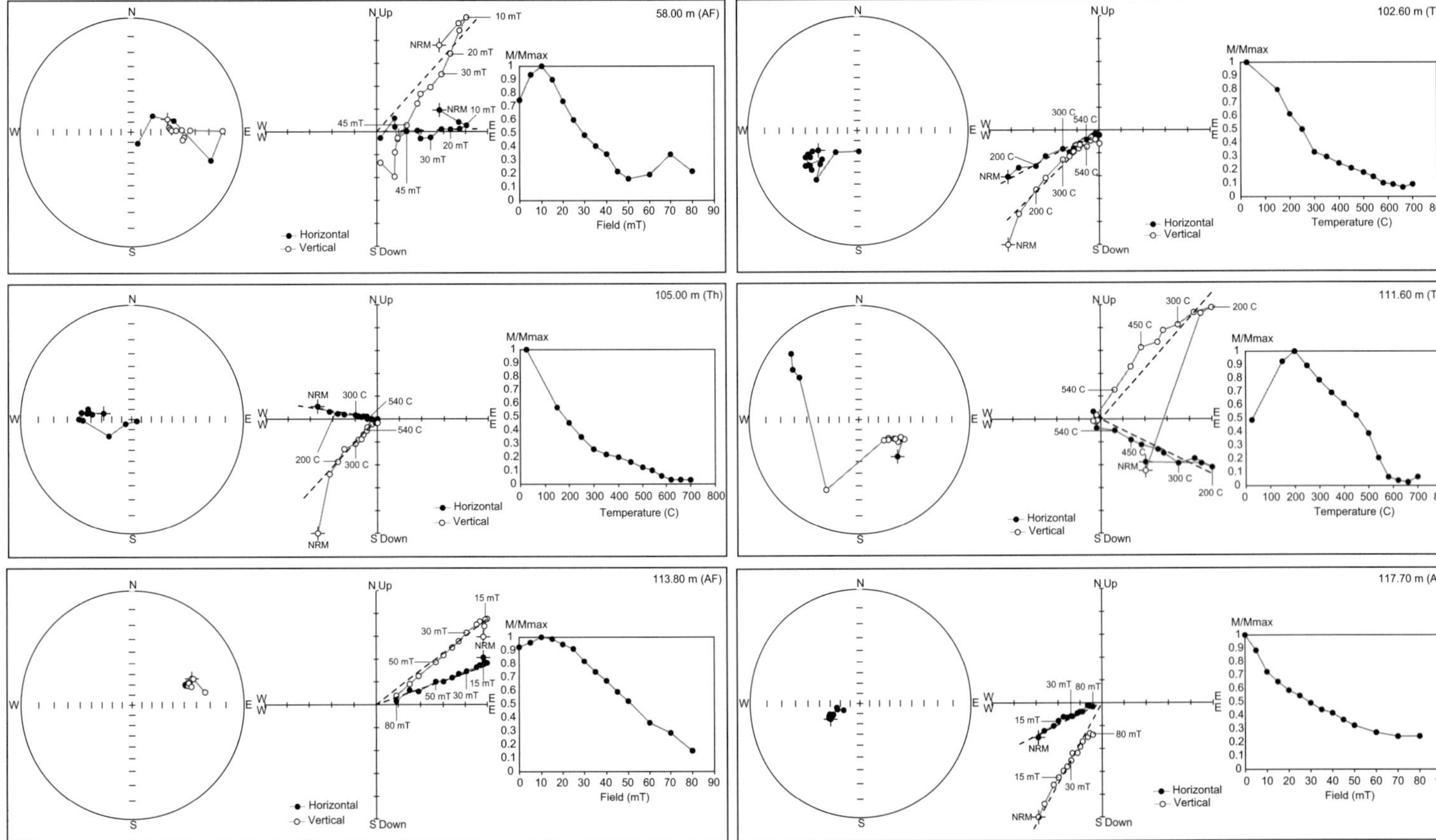

Fig. 4. Selected representative orthogonal plots, stereograms and intensity decay plots. Solid (open) squares represent the projection on the horizontal (vertical) plane. The demagnetization levels are represented in milli-Tesla.

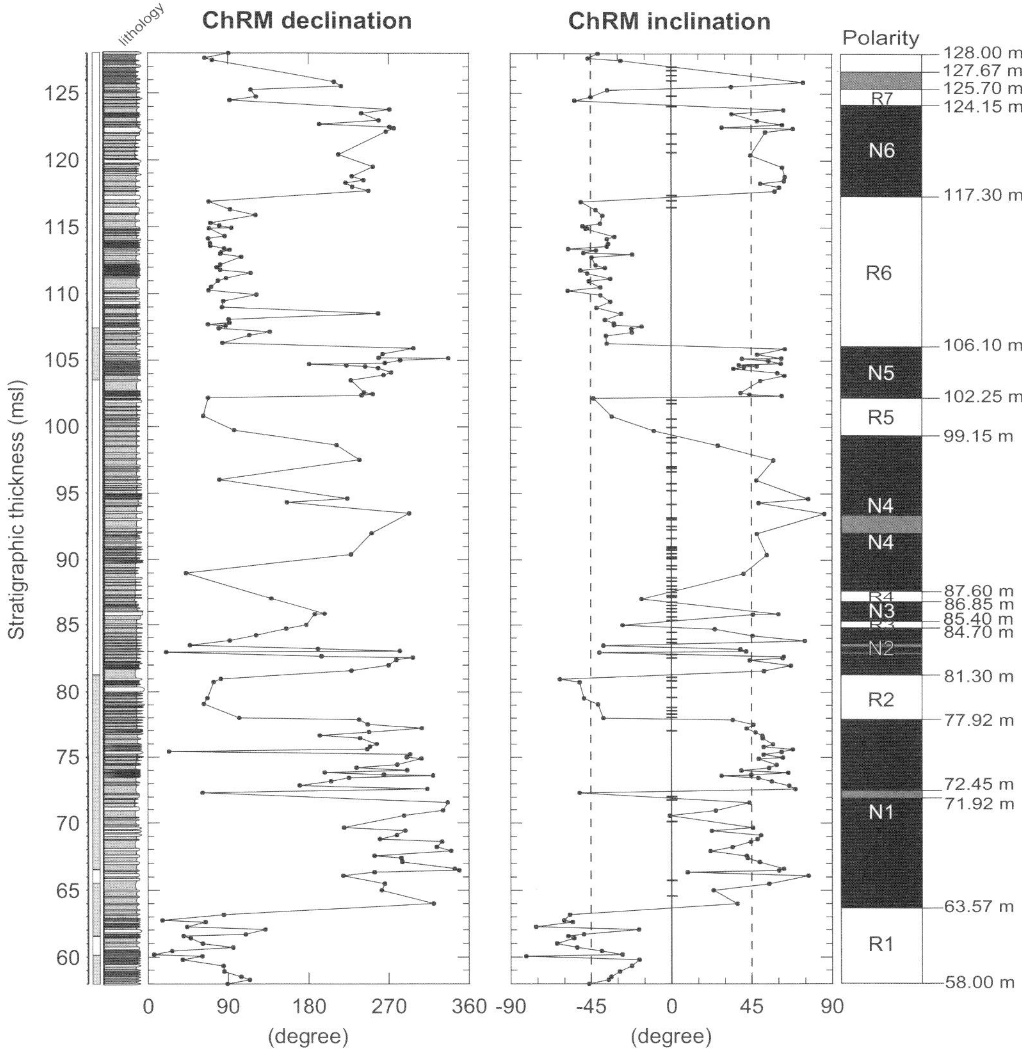

Fig. 5. Intensity and inclination of the ChRM along the Monte Cagnero section from 58 to 128 msl compared with lithostratigraphic column. On the right site, we also include the magnetozones and the boundary between them in metres stratigraphic level. Black horizontal lines at 0° in the inclination of ChRM represent samples for which we have not been able to calculate the ChRM because of high MAD values. The dashed line represents the GAD (geocentric axial dipole) based on the reconstruction of Besse & Courtillot (1991).

beckmanni at 63.2 msl; (2) highest occurrence (HO) of *O. beckmanni* at 65.5 msl; (3) HO of *Morozovelloides crassatus* at 85.0 msl; (4) LO of *Globigerinatheka semiinvoluta* at 86.5 msl; (5) HO of *Globigerinatheka semiinvoluta* at 101.1 msl; (6) LO of *Turborotalia cunialensis* at 102.50 msl; (7) HO of *Globigerinatheka index* 107.52 msl; (8) HO of *Cirbrohantkenina inflata* 112.90 msl; (9) HO of *Turborotalia cerroazulensis* 113.6 msl; and (10) HO of *Hantkenina alabamensis* at 114.1 msl. The

lower part of the study section is assigned to Zones P12 of Berggren *et al.* (1995) and E11 of Wade *et al.* (2011), based on the absence of *O. beckmanni*, which defines the lower and the upper boundaries of Zone P13 (*Globigerapsis beckmanni* Total Range Zone) of Berggren *et al.* (1995) and Zone E12 (*O. beckmanni* Taxon-range Zone) of Wade *et al.* (2011), respectively (Fig. 7). Following Berggren *et al.* (1995), the LOs of *G. semiinvoluta* and *T. cunialensis* and the HOs of *C. inflata* and

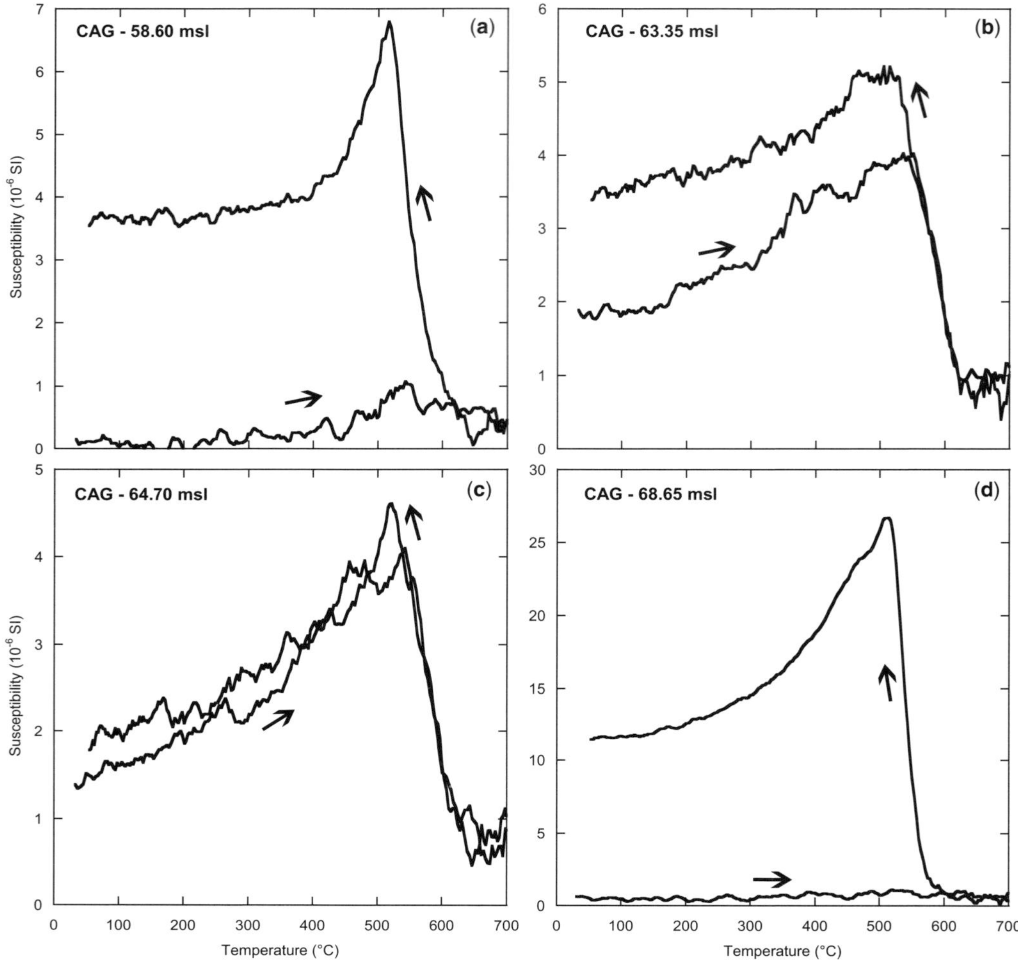

Fig. 6. Temperature v. magnetic susceptibility for four samples from the Monte Cagnero section. Details on the significance of magnetic mineral associations from heating progression and the cooling return are given in the text. The furnace diamagnetism has been subtracted.

T. cerroazulensis are used to mark the P14/P15, P15/P16, P16/P17 and P17/P18 zonal boundaries, respectively (Fig. 7). According to Wade *et al.* (2011), the HOs of *M. crassatus*, *G. semiinvoluta* and *G. index* and the LO of *H. alabamensis* define the E13/E14, E14/E15, E15/E16 and El6/O1 zonal boundaries, respectively (Fig. 7). Following Coccioni *et al.* (1988), Nocchi *et al.* (1988) and Premoli Silva & Jenkins (1993), the E–O boundary is placed at 114.1 msl, at the extinction level of the planktonic foraminiferal Family Hantkeninidae (Fig. 7).

Calcareous nannofossils. In all study samples from the Monte Cagnero section, calcareous nannofossils are abundant, and preservation varies from moderate to poor. The following primary bioevents were identified in the Monte Cagnero section (from bottom to top; Fig. 7): (1) HO of *Chiasmolithus solitus* at 70.0 msl; (2) LO of *Chiasmolithus oamaruensis* at 84.0 msl; (3) HO of *Chiasmolithus grandis* at 85.50 msl; (4) HO of *Chiasmolithus oamaruensis* at 98.5 msl; (5) HO of *Discoaster saipanensis* at 108.5 msl; and (6) acme base (AB) of *Clausicoccus obrutus* >5.7 μm at 114.5 msl. The lower part of the section is assigned to zones NP16 and CP14a of Martini (1971) and Okada & Bukry (1980), respectively, based on the presence of *Reticulofenestra umbilica* and the absence of *Blackites gladius* (Fig. 7). The HO of

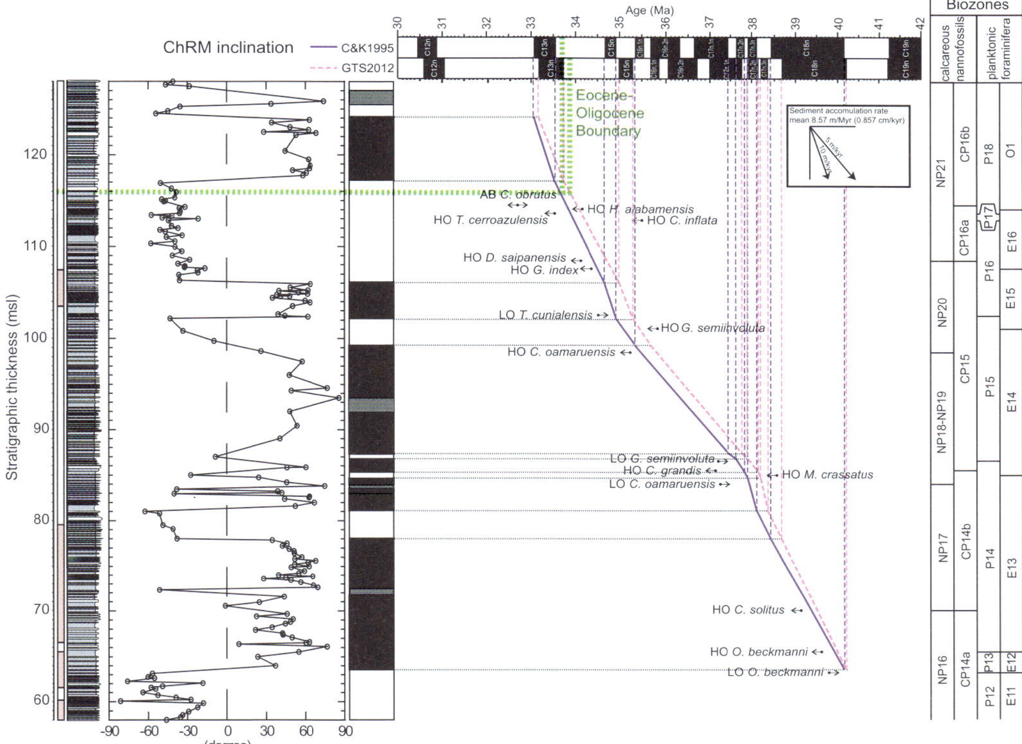

Fig. 7. Correlation between age of the geomagnetic polarity time scale (C&K1995, Cande & Kent 1995; Gee & Kent, 2007; GTS2012, Vandenberghe *et al.* 2012) and ChRM inclination of the Monte Cagnero section polarity zonations (purple solid line). Calcareous nannofossil and planktonic foraminiferal data were used to constrain the interpretation. Planktonic foraminiferal zones are after (P) Berggren *et al.* (1995) and (E) Wade *et al.* (2011); calcareous nannofossil zones are after (NP) Martini (1971) and (CP) Okada & Bukry (1980). Correlation between chron ages at Monte Cagnero and the geological time scale (GTS2012; Vandenberghe *et al.* 2012) and magnetozones is also shown (pink dashed line). LO, Lowest occurrence; HO, Highest occurrence; AB, Acme base. Sedimentation rates are tabulated for each chron interval. The Eocene–Oligocene boundary is at114.1 msl (Coccioni *et al.* 2008; Hyland *et al.* 2009).

C. solitus marks the NP16/NP17 and CP14a/CP14b zonal boundaries of Martini (1971) and Okada & Bukry (1980), respectively. Following the biostratigraphical scheme of Martini (1971), we used the LO of *C. oamaruensis* to define the base of NP17/NP18–NP19 zonal boundary, and the HOs of *C. oamaruensis* and *D. saipanensis* to mark NP18–19/NP20 and NP20/NP21 zonal boundaries, respectively (Fig. 7). Since the LO of *Isthmolithus recurvus*, which defines both NP18/NP19 and CP15a/CP15b zonal boundaries of Martini (1971) and Okada & Bukry (1980), respectively, was not identified, zones NP18 and NP19 have been combined, as well as subzone CP15a (Fig. 7). According to the biostratigraphical biozonations of Okada & Bukry (1980), the HOs of *C. grandis* and *D. saipanensis* were used to define the CP14b/CP15, and CP15/CP16a zonal boundaries, respectively

(Fig. 7). The AB of *C. obrutus* >5.7 μm was proposed by Hyland *et al.* (2009) to define the CP16a/CP16b zonal boundary instead of the AE of *Clausicoccus* as suggested by Backman (1987). This event is placed immediately above the E–O boundary and falls in the middle-upper part of Chron C13r (Fig. 7).

Discussion

Correlation to the geomagnetic polarity time scale

A series of 13 magnetozones are recognized within the Monte Cagnero section in a sequence of six normal polarity intervals and seven reverse polarity intervals (Fig. 5). We interpreted the magnetic polarity record based on the geomagnetic polarity time scale (GPTS) calibrations of Cande & Kent (1995; C&K1995) and Vandenberghe *et al.* (2012; GTS2012). Magnetozones R1 (58.00–63.57 msl), N1 (63.57–77.92 msl), R2 (77.92–81.30 msl), N2 (81.30–84.70 msl), R3 (84.70–85.40 msl), N3 (85.40–86.42 msl), N3 (86.42–88.00 msl), N4 (88.00–99.15 msl), R5 (99.15–102.25 msl), N5 (102.25–106.10 msl), R6 (106.10–117.30 msl), N6 (117.30–124.15) and R7 (124.15–128.00 msl) are interpreted to correspond respectively to Chrons C18r, C18n, C17r, C17n.3n, C17n.2r, C17n.2n, C17n.1r/C17n.1n/C16n, C15r, C15n, C13r, C13n and C12r (Figs 5 & 7). Our magnetostratigraphy is based on a straightforward interpretation for most of the section, with some uncertainties in the depth interval between *c.* 90 and 95 msl – an interval which is characterized by low concentration of magnetic minerals. Through linear interpolation between chron boundary ages (Cande & Kent 1995; Vandenberghe *et al.* 2012), we have constructed a detailed age–depth model for the Monte Cagnero study section (Fig. 7). The age–depth model for the interval between *c.* 90 and 95 msl is mainly based on the biostratigraphic results because the magnetostratigraphy in this interval is less reliable than the rest of the section (Fig. 7).

The top of the study section has an interpreted age of *c.* 32.6 Ma within Chron 12r (C&K1995, 30.939–33.058 Ma; GTS2012, 31.034–33.157), while the bottom has an age of *c.* 40.8 Ma within Chron 18r (40.130–41.257 Ma) (Fig. 7). No faults, condensed layers or hiatuses have been recognized along the section that might indicate a time discontinuity in the section. Consequently, the entire Monte Cagnero section spans a time interval of at least 13 myr (8 myr in our segment and the 5 myr through the Oligocene section documented by Hyland *et al.* (2009) and Coccioni *et al.* (2008)) in one continuous and undisturbed sedimentary sequence. The Chron 12r interval was identified in previous work on this section (Hyland *et al.* 2009), which is in agreement with our polarity and age interpretations.

Eocene–Oligocene sedimentation rates at Monte Cagnero

Based on the GPTS calibrations of Cande & Kent (1995) (confirmed in Gee & Kent 2007), sedimentation rates throughout the Monte Cagnero section vary from 14.07 m myr^{-1} (Chron 13n) to 5.34 m myr^{-1} (Chron 17n.1r). The average sedimentation rate, which is calculated by linear interpolation between consecutive chron boundaries, is 8.57 m myr^{-1} (0.857 cm kyr^{-1}) (Fig. 8a).

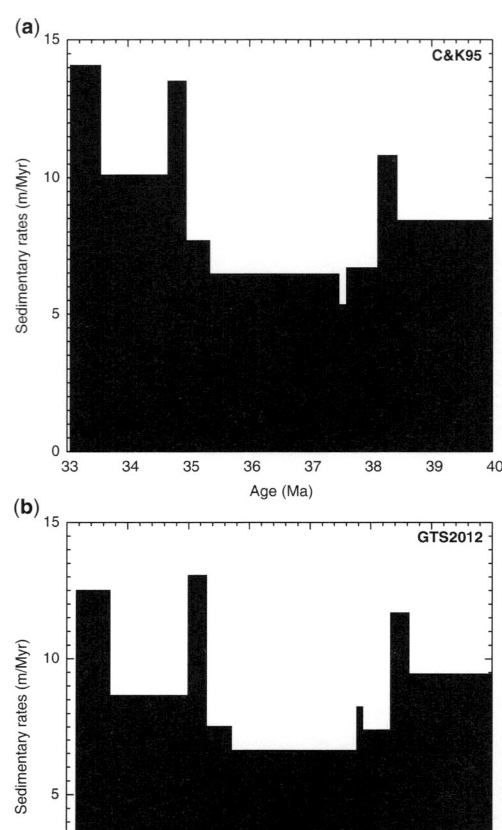

Fig. 8. Variation of sedimentation rates at Monte Cagnero v. age for the middle Eocene–early Oligocene (**a**) The average sedimentation rate is 8.57 m Myr^{-1} using Cande & Kent (1995), Gee & Kent (2007) and Jovane *et al.* (2010). (**b**) The average sedimentation rate is 8.68 m Myr^{-1} using the GTS from Vandenberghe *et al.* (2012). Depths are the metres stratigraphic level (msl).

Appling the most recent GTS chron ages from Vandenberghe *et al.* (2012), the calculated sedimentation rates for the Monte Cagnero section are very similar to those obtained using the GPTS of Cande & Kent (1995) (Fig. 8b) with an average sedimentation rate of 8.68 m myr^{-1}. These values confirm the duration of the chrons for this period from Cande & Kent (1995), Gee & Kent (2007) and Jovane *et al.* (2010), and GTS2012 (Vandenberghe *et al.* 2012). Furthermore, the calculated sedimentation rates for the Monte Cagnero section are comparable with sedimentation rates of coeval sections at Massignano (upper Eocene to lower Oligocene) and Contessa Highway (middle Eocene), which have average rates of 7.3 and 6.4 m myr^{-1}, respectively (Jovane *et al.* 2007*a*, *b*).

Refinement of age calibrations for biostratigraphic events

All of the biostratigraphic marker taxa that define the standard planktonic foraminiferal zones of Berggren *et al.* (1995) and Wade *et al.* (2011), through the middle–late Eocene and early Oligocene, occur in the analysed material from the Monte Cagnero section. Additionally, most of the key markers used to define the standard calcareous nannofossil zonations of Martini (1971) and Okada & Bukry (1980) occur together with several additional marker species. Our integrated biostratigraphic study provides evidence that the 58–127 msl interval at the Monte Cagnero section is continuous, within biostratigraphic and magnetostratigraphic resolution, between planktonic foraminiferal Zones P12 and P18 of Berggren *et al.* (1995) and between Zones E11 and O1 of Wade *et al.* (2011). Also, a complete record of calcareous nannofossil Zones NP16–NP21 of Martini (1971) and CP14a–CP16b of Okada & Bukry (1980) is present (Fig. 7).

At Monte Cagnero, however, some bioevents that mark the zonal boundaries are placed at different stratigraphic levels compared with the current standard magneto-biochronological timescale (Luterbacher *et al.* 2004, with updates from Ogg

Table 1. *Planktonic foraminiferal and calcareous nannofossil events and interpretation of chrons boundaries for the Monte Cagnero section*

Planktonic foraminiferal and calcareous nannofossil events and Chron boundaries	Depth (msl)	Age (Ma)	Astronomical Age (Ma)
LO *Orbulinoides beckmanni*	63.20	40.50[1]	40.330[6]
C18r/C18n.2n	63.57	40.130[3]	40.120[6]
HO *Orbulinoides beckmanni*	65.50	40.00[1]	39.860[6]
HO *Chiasmolithus solitus*	70.00	40.40[2]	
C18n.2n/C18n.1r	71.92	39.756[4]–39.631[3]	
C18n.1r/C18n.1n	72.45	39.686[4]–39.552[3]	
C18n.1n/C17r	77.92	38.668[4]–38.426[3]	
C17r/C17n.3n	81.30	38.380[4]–38.113[3]	
LO *Chiasmolithus oamaruensis*	84.00	37.00[2]	
C17n.3n/C17n.2r	84.70	36.100[4]–7.920[3]	
HO *Morozovelloides crassatus*	85.00	38.00[1]	
C17n.2r/C17n.2n	85.40	38.135[4]–37.848[3]	
HO *Chiasmolithus grandis*	85.50	37.10[2]	
LO *Globigerinatheka semiinvoluta*	86.50	38.00[1]	
C17n.2n/C17n.1r	86.85	37.908[4]–37.604[3]	
C17n.1r/C17n.1n	87.60	37.785[4]–37.473[3]	
C17n.1n/C16r?	92.00	36.969[4]–36.618[3]	
C16r?/C16n	93.40	36.700[4]–36.341[3]	
HO *Chiasmolithus oamaruensis*	98.50	35.40[2]	35.404[5]
C16n/C15r	99.15	36.706[4]–35.343[3]	35.340[5]
HO *Globigerinatheka semiinvoluta*	101.10	35.80[1]	35.520[5]
C15r/C15n	102.25	35.294[4]–34.940[3]	34.960[5]
LO *Turborotalia cunialensis*	102.50	35.80[1]	35.185[5]
C15n/C13r	106.10	34.999[4]–34.655[3]	34.728[5]
HO *Globigerinatheka index*	107.52	34.30[1]	34.400[5]
HO *Discoaster saipanensis*	108.50	34.20[2]	
HO *Cribrohantkenina inflata*	112.90	34.00[2]	34.221[5]
HO *Turborotalia cerroazulensis*	113.60	33.80[1]	
HO *Hantkenina alabamensis*	114.10	33.70[1]	33.703[5]
AB *Clausicoccus obrutus*	114.50	n.a.	
C13r/C13n	117.30	33.705–33.545[3]	33.545[5]
C13n/C12r?	126.67	33.157–33.058[3]	

Bioevent ages of Wade *et al.* (2011) (1), and Berggren *et al.* (1995) (2), Magnetochron ages from Cande & Kent (1995) (3), and Vandenberghe *et al.* (2012) (4). Astronomical ages from Jovane *et al.* (2006) (5), and Jovane *et al.* (2010) (6).

et al. 2008). The main discrepancies, from bottom to top, are as follows:

(1) the HO of *C. solitus* (NP16/NP17 and CP14a/CP14b zonal boundaries) occurs higher in the upper part of Chron C18n.2n;

(2) the LO of *C. oamaruensis* (NP17/NP18–NP19 zonal boundary) is found lower in Chron C17n.3n;

(3) the HO of *C. grandis* (CP14b/CP15 zonal boundary) occurs lower at the base of Chron C17n.2n;

(4) the HO of *G. semiinvoluta* (E14/E15 zonal boundary) is identified higher in the middle part of Chron C15r;

(5) the HO of *D. saipanensis* (NP20/NP21 and CP15/CP16a zonal boundaries) is recognized higher in the lower part of Chron C13r.

These discrepancies are probably due to the fact that the markers adopted in standard calcareous nannofossil zonations of Martini (1971) and Okada & Bukry (1980) for the Palaeogene Period might be latitudinally restricted and facies-dependent (Fornaciari *et al.* 2010). Moreover, the biostratigraphic resolution of the calcareous nannofossil and planktonic foraminiferal schemes for the middle Eocene–early Oligocene is low and some bioevents have rarely been calibrated to high-resolution magnetostratigraphic records. Accordingly, correlations of the standard calcareous nannofossil and planktic foraminiferal biohorizons, to the GPTS (Cande & Kent 1995; CK95) remain disputed for Palaeogene times. The data presented above and summarized in Table 1 expand the existing database of calcareous nannofossil and planktonic foraminiferal biochronology during the middle–late Eocene and early Oligocene.

Conclusions

The Monte Cagnero section located in the northeastern Apennines near Urbania in the Umbria–Marche Basin (Italy) provides the most complete and continuous stratigraphic sequence representing the middle Eocene to lower Oligocene interval (*c.* 55–28 Ma) currently identified in central Italy, spanning a time interval of at least *c.* 13 myr. In this study, we have obtained new insight into the nature and age of these strata based on high-resolution magnetostratigraphic analyses and detailed calcareous nannofossil and planktonic foraminiferal analyses. These data form the basis of a new, robust age-depth model for the middle Eocene–lower Oligocene interval.

The Monte Cagnero section consists of limestone and marlstone of reddish and greenish colour. The magnetic mineralogy of these rocks is represented mainly by low-coercivity minerals (most likely magnetite or maghemite) in a mixture in some intervals of high-coercivity minerals (most likely hematite). The high-resolution definition of the occurrence of those intervals along the section has important implications for climate studies and palaeoceanographic events in the Neo-Tethys Realm during from the middle to the late Eocene. Our new magneto-biostratigraphy for the Monte Cagnero section also provides a precise age model across several crucial climate events. The refined age information for this section, in addition to multidisciplinary climatic proxy studies, that are currently ongoing, will allow reconstruction of environmental change across the greenhouse-icehouse transition. The Monte Cagnero section probably records important middle Eocene–early Oligocene climate events, such as the E–O climate transition at *c.* 34 Ma and the MECO event at *c.* 40 Ma. New insights into these events and related regional climate changes in the Tethys region will be gained through future palaeoclimatic studies of the Monte Cagnero section and application of the high-resolution integrated age-model presented here.

We appreciate constructive comments from two anonymous reviewers and the editor L. Hinnov. Financial support for this research is provided from the European Community through Marie Curie Action (FP7-PEOPLE-IEF-2008 proposal no. 236311) inside the NEO-TETHYS project. J. F. S. is supported by Conselho Nacional de Desenvolvimento Científico (Process 201508/2009-5) and Conselho de Desenvolvimento de Pessoal do Nível Superior.

References

ALLEN, M. B. & ARMSTRONG, H. A. 2008. Arabia–Eurasia collision and the forcing of mid-Cenozoic global cooling. *Palaeogeography, Palaeoclimatology, Palaeoecology*, **265**, 52–58, doi: 10.1016/j.palaeo.2008.04.021.

BACKMAN, J. 1987. Quantitative calcareous nannofossil biochronology of middle Eocene through early Oligocene sediment from DSDP Sites 522 and 523. *Abhandlungen der Geologischen Bundesanstalt, Vienna*, **39**, 21–31.

BAUMANN, P. 1970. Micropalaontologische und stratigraphische Untersuchungen der Obereozanen–Oligozanen Scaglia im zentralen Apennin (Italien). *Eclogae Geologicae Helvetiae*, **63**, 1133–1211.

BAUMANN, P. & ROTH, P. H. 1969. Zonierung des Obereozaens und Oligozaens des Monte Cagnero (Zentralapennin) mit planktonischen Foraminiferen und Nannoplankton. *Eclogae Geologicae Helvetiae*, **62**, 303–323.

BERGGREN, W. A., KENT, D. V., SWISHER, C. C. III & AUBRY, M.-P. 1995. A revised Cenozoic geochronology and chronostratigraphy. *In*: BERGGREN, W. A., KENT, D. V., AUBRY, M. P. & HARDENBOL, J. (eds) *Geochronology, Time Scales and Global Stratigraphic Correlation*. Society of Economic Paleontologists and

Mineralogists, Tulsa, OK, Special Publications, **54**, 129–212.

BESSE, J. & COURTILLOT, V. 1991. Revised and synthetic apparent polar wander paths of the African, Eurasian, North American and Indian plates, and true polar wander since 200 Ma. *Journal of Geophysical Research*, **96**, 4029.

BOHATY, S. M. & ZACHOS, J. C. 2003. Significant Southern Ocean warming event in the late middle Eocene. *Geology*, **31**, 1017–1020.

BOHATY, S. M., ZACHOS, J. C., FLORINDO, F. & DELANEY, M. L. 2009. Coupled greenhouse warming and deep-sea acidification in the middle Eocene. *Paleoceanography*, **24**, PA2207, doi: 10.1029/2008PA00 1676.

BOWN, P. R. & YOUNG, J. R. 1998. Techniques. *In*: BOWN, P. R. (ed.) *Calcareous Nannofossil Biostratigraphy*. Kluwer Academic, Dordrecht, 16–28.

BRINKHUIS, H. & BIFFI, U. 1993. Dinoflagellate cyst stratigraphy of the Eocene/Oligocene transition in central Italy. *Marine Micropaleontology*, **22**, 131–183.

CANDE, S. C. & KENT, D. V. 1995. Revised calibration of the geomagnetic polarity timescale for the late Cretaceous and Cenozoic. *Journal of Geophysical Research*, **100**, 6093–6095.

CHANNELL, J. E. T. & MCCABE, C. 1994. Comparison of magnetic hysteresis parameters of unremagnetized and remagnetized limestones. *Journal of Geophysical Research*, **99**, 4613–4623.

CHANNELL, J. E. T., FREEMAN, R., HELLER, F. & LOWRIE, W. 1982. Timing of diagenetic haematite growth in red pelagic limestones from Gubbio (Italy). *Earth and Planetary Science Letters*, **58**, 189–201.

CHIARABBA, C., JOVANE, L. & DI STEFANO, R. 2005. A new view of Italian seismicity using 20 years of instrumental recordings. *Tectonophysics*, **395**, doi: 10.1016/j.tecto.2004.09.013.

COCCIONI, R., MONACO, P., MONECHI, S., NOCCHI, M. & PARISI, G. 1988. Biostratigraphy of the Eocene/Oligocene boundary at Massignano (Ancona, Italy). *In*: PREMOLI SILVA, I., COCCIONI, R. & MONTANARI, A. (eds) *International Subcommission on Paleogene Stratigraphy*. Industrie Grafiche F.lli Aniballi, Ancona, Special Publications, 59–80.

COCCIONI, R., MARSILI, A. *ET AL.* 2008. Integrated stratigraphy of the Oligocene pelagic sequence in the Umbria–Marche basin (northeastern Apennines, Italy): a potential Global Stratotype Section and Point (GSSP) for the Rupelian/Chattian boundary. Geological Society of America Bulletin, **120**, 487–511.

COXALL, H. K. & WILSON, P. A. 2011. Early Oligocene glaciation and productivity in the eastern equatorial Pacific: insights into global carbon cycling. *Paleoceanography*, **26**, PA2221, doi: 10.1029/2010PA002021

COXALL, H. K., WILSON, P. A., PÄLIKE, H., LEAR, C. H. & BACKMAN, J. 2005. Rapid stepwise onset of Antarctic glaciations and deeper calcite compensation in the Pacific Ocean. *Nature*, **433**, 53–57.

CRAMER, B. S., MILLER, K. G., BARRETT, P. J. & WRIGHT, J. D. 2011. Late Cretaceous–Neogene trends in deep ocean temperature and continental ice volume: reconciling records of benthic foraminiferal geochemistry

(δ^{18}O and Mg/Ca) with sea level history. *Journal of Geophysical Research*, **116**, C12023, doi: 10.1029/2011JC007255.

DAWBER, C. & TRIPATI, A. 2011. Constraints on glaciation in the Middle Eocene (46–37 Ma) from Ocean Drilling Program (ODP) Site 1209 in the tropical Pacific Ocean. *Paleoceanography*, **26**, PA2208, doi: 10.1029/2010PA002037.

DECONTO, R. M. & POLLARD, D. 2003. Rapid Cenozoic glaciations of Antarctic induced by declining atmospheric CO_2. *Nature*, **421**, 245–249.

DUNLOP, D. J. & ÖZDEMIR, Ö. 1997. *Rock Magnetism: Fundamentals and Frontiers*. Cambridge University Press, New York.

EDGAR, K. M., WILSON, P. A., SEXTON, P. F., GIBBS, S. J., ROBERTS, A. P. & NORRIS, R. D. 2010. New biostratigraphy, magnetostratigraphy and isotopic insights into the Middle Eocene Climatic Optimum in low latitudes. *Palaeogeography, Palaeoclimatology, Palaeoecology*, **297**, 670–682.

FORNACIARI, E., AGNINI, C., CATANZARITI, R., RIO, D., BOLLA, E. M. & VALVASONI, E. 2010. Mid-Latitude calcareous nannofossil biostratigraphy and biochronology across the middle to late Eocene transition. *Stratigraphy*, **7**, 229–264.

GEE, J. S. & KENT, D. V. 2007. Source of oceanic magnetic anomalies and the geomagnetic polarity time scale. Geomagnetism. *In*: KONO, M. (ed.) *Treatise on Geophysics*. Elsevier, Amsterdam, **5**, 455–507.

GUERRERA, F., MONACO, P., NOCCHI, M., PARISI, G., FRANCHI, R., VANNUCCI, S. & GIOVANNINI, G. 1988. La Scaglia Variegata Eocenica nella sezione di Monte Cagnero (bacino marchigiano interno): studio litostratigrafico, petrografico e biostratigrafico. *Bollettino Società Geologica Italiana*, **107**, 81–99.

HROUDA, F. 1994. A technique for the measurement of thermal changes of magnetic susceptibility of weakly magnetic rocks by the CS-2 apparatus and KLY-2 Kappabridge. *Geophysical Journal International*, **118**, 604–612, doi: 10.1111/j.1365-246X.1994.tb03987.

HROUDA, F. 2003. Indices for numerical characterization of the alteration processes of magnetic minerals taking place during investigation of temperature variation of magnetic susceptibility. *Studia Geophysica et Geodaetica*, **47**, 847–861.

HYLAND, E., MURPHY, B. *ET AL.* 2009. Integrated stratigraphic and astrochronologic calibration of the Eocene–Oligocene transition in the Monte Cagnero section (northeastern Apennines, Italy): a potential parastratotype for the Massignano global stratotype section and point (GSSP). *In*: KOEBERL, C. & MONTANARI, A. (eds) *The late Eocene Earth – Hothouse, Icehouse, and Impacts*. Geological Society of America, Boulder, CO, Special Papers, **452**, 303–322.

JOVANE, L., FLORINDO, F. & DINARÈS-TURELL, J. 2004. Environmental magnetic record of paleoclimate change: revisiting the Eocene–Oligocene stratotype section, Massignano, Italy. *Geophysical Research Letters*, **31**, doi: 10.1029/2004GL020554.

JOVANE, L., FLORINDO, F., SPROVIERI, M. & PALIKE, H. 2006. Astronomic calibration of the late Eocene/early Oligocene Massignano section (Central Italy). *Geochemistry, Geophysics and Geosystems*, **7**, doi: 10.1029/2005GC001195.

JOVANE, L., SPROVIERI, M., FLORINDO, F., ACTON, G., COCCIONI, R., DALL'ANTONIA, B. & DINARÈS-TURELL, J. 2007a. Eocene–Oligocene paleoceanography changes in the stratotype section, Massignano, Italy: clues from rock magnetism and stable isotopes. *Journal of Geophysical Research*, **112**, B11101.

JOVANE, L., FLORINDO, F. ET AL. 2007b. The middle Eocene climatic optimum event in the Contessa Highway section, Umbrian Apennines, Italy. Geological Society of America Bulletin, **119**, 413–427.

JOVANE, L., COCCIONI, R., MARSILI, A. & ACTON, G. 2009. The late Eocene greenhouse–icehouse transition: observations from the Massignano global stratotype section and point (GSSP). *In*: KOEBERL, C. & MONTANARI, A. (eds) *The Late Eocene Earth-Hothouse, Icehouse, and Impacts*. Geological Society of America, Boulder, CO, Special Papers, **452**, 149–168.

JOVANE, L., SPROVIERI, M., COCCIONI, R., FLORINDO, F., MARSILI, A. & LASKAR, J. 2010. Astronomic calibration of the middle Eocene Contessa Highway section (Gubbio, Italy). *Earth and Planetary Science Letters*, **298**, 77–88.

KATZ, M. E., MILLER, K. G., WRIGHT, J. D., WADE, B. S., BROWNING, J. V., CRAMER, B. S. & ROSENTHAL, Y. 2008. Stepwise transition from the Eocene greenhouse to the Oligocene icehouse. *Nature Geoscience*, **1**, 329–334.

KIRSCHVINK, J. L. 1980. The least-squares line and plane and the analysis of paleomagnetic data. *Geophysical Journal of the Royal Astronomical Society*, **62**, 699–718.

LEAR, C. H., ELDERFIELD, H. & WILSON, P. A. 2000. Cenozoic deep-sea temperatures and global ice volumes from Mg/Ca in benthic foraminiferal calcite. *Science*, **287**, 269–272.

LIRER, F. 2000. A new technique for retrieving calcareous microfossils from lithified lime deposits. *Micropaleontology*, **46**, 365–369.

LIVERMORE, R., NANKIVELL, A., EAGLES, G. & MORRIS, P. 2005. Paleogene opening of Drake Passage. *Earth and Planetary Science Letters*, **236**, 459–470, doi: 10.1016/j.epsl.2005.03.027.

LUCIANI, V., GIUSBERTI, L., AGNINI, C., FORNACIARI, E., RIO, D., SPOFFORTH, D. J. A. & PÄLIKE, H. 2010. Ecological and evolutionary response of Tethyan planktonic foraminifera to the middle Eocene climatic optimum (MECO) from the Alano section (NE Italy). *Palaeogeography, Palaeoclimatology, Palaeoecology*, **292**, 82–95.

LUTERBACHER, H. P., ALI, J. R. ET AL. 2004. The Paleogene period. *In*: GRADSTEIN, F. M., OGG, J. G. & SMITH, A. G. (eds) *A Geologic Time Scale*. Cambridge University Press, Cambridge, 384–408.

LYLE, M. W., OLIVAREZ LYLE, A., BACKMAN, J. & TRIPATI, A. 2005. Biogenic sedimentation in the Eocene equatorial Pacific: the stuttering greenhouse and Eocene carbonate compensation depth. *In*: LYLE, M., WILSON, P. & FIRTH, J. (eds) *Proceedings of the Ocean Drilling Program, Scientific Results*, **199**. Ocean Drilling Program, College Station TX, 1–35, doi: 10.2973/odp.proc.sr.199.219.2005; http://www.odp.tamu.edu/publications/199_SR/219/219.htm

MAIORANO, P. & MONECHI, S. 2006. Early to late Oligocene calcareous nannofossil bioevents in the Mediterranean (Umbria–Marche Basin, central Italy). *Rivista Italiana di Paleontologia e Stratratigrafia*, **112**, 261–273.

MARTINI, E. 1971. Standard Tertiary and Quaternary calcareous nannoplankton zonation. *In*: FARINACCI, A. (eds) *Proceedings of the Second Planktonic Conference, Rome 1970*. Tecnoscienza, Rome, **2**, 739–785.

MERICO, A., TYRRELL, T. & WILSON, P. A. 2008. Eocene/Oligocene ocean de-acidification linked to Antarctic glaciations by sea-level fall. *Nature*, **452**, 979–983.

MILLER, K. G., WRIGHT, J. D. & FAIRBANKS, R. G. 1991. Unlocking the ice house: Oligocene–Miocene oxygen isotopes, eustasy, and margin erosion. *Journal of Geophysical Research*, **96**, 6829–6948.

NOCCHI, M., MONECHI, S. ET AL. 1988. The extinction of Hantkeninidae as a marker for recognizing the Eocene–Oligocene boundary: a proposal. *In*: PREMOLI SILVA, I., COCCIONI, R. & MONTANARI, A. (eds) *The Eocene/ Oligocene Boundary in the Marche-Umbria Basin (Italy)*. International Subcommission on Paleogene Stratigraphy, Ancona, Industrie Grafiche Fratelli Aniballi, Special Publications, 249–252.

OGG, J. G., OGG, G. & GRADSTEIN, F. M. 2008. *The Concise Geologic Time Scale*. Cambridge University Press, Cambridge.

OKADA, H. & BUKRY, D. 1980. Supplementary modification and introduction of code numbers to the low-latitude coccolith biostratigraphic zonation (Bukry, 1973, 1975). *Marine Micropaleontology*, **5**, 321–325, doi: 10.1016/0377-8398(80)90016-X.

PAGANI, M., ZACHOS, J. C., FREEMAN, K. H., TIPPLE, B. & BOHATY, S. 2005. Marked decline in atmospheric carbon dioxide concentrations during the Paleogene. *Science*, **309**, 600–603.

PAGANI, M., HUBER, M. ET AL. 2011. The role of carbon dioxide during the onset of Antarctic glaciation. *Science*, **334**, 1261–1264.

PARISI, G., GUERRERA, F., MADILE, M., MAGNONI, G., MONACO, P., MONECHI, S. & NOCCHI, M. 1988. Middle Eocene to early Oligocene calcareous nannofossil and foraminiferal biostratigraphy in the Monte Cagnero section, Piobbico (Italy). *In*: PREMOLI SILVA, I., COCCIONI, R. & MONTANARI, A. (eds) *The Eocene/Oligocene Boundary in the Marche-Umbria basin (Italy)*. International Subcommission on Paleogene Stratigraphy, Ancona.

PEARSON, P. N., OLSSON, R. K., HEMLEBEN, C., HUBER, B. T. & BERGGREN, W. A. 2006. *Atlas of Eocene Planktonic Foraminifera*. Cushman Foundation of Foraminiferal Research, Fredericksburg, USA, Special Publications, **41**, 514.

PEARSON, P. N., FOSTER, G. L. & WADE, B. S. 2009. Atmospheric carbon dioxide through the Eocene–Oligocene climate transition. *Nature*, **461**, 1110–1113, doi: 10.1038nature 08447.

PERCH-NIELSEN, K. 1985. Mesozoic calcareous nannofossils. *In*: BOLLI, H. M., SAUNDERS, J. B. & PERCH-NIELSEN, K. (eds) *Plankton Stratigraphy*. Cambridge University Press, Cambridge, 329–426.

PREMOLI SILVA, I. & JENKINS, D. G. 1993. Decision on the Eocene–Oligocene boundary stratotype. *Episodes*, **16**, 379–382.

SPOFFORTH, D. J. A., AGNINI, C. ET AL. 2010. Organic carbon burial following the middle Eocene climatic optimum in the central western Tethys. *Paleoceanography*, **25**, PA3210.

VANDENBERGHE, N., HILGEN, F. J. *ET AL.* 2012. The paleogene period, chapter 28. *In*: GRADSTEIN, F. M., OGG, J. G. & SMITH, A. G. (eds) *A Geologic Time Scale*. Cambridge University Press, Cambridge, 855–941.

VERDUCCI, M. & NOCCHI, M. 2004. Middle to Late Eocene main planktonic foraminiferal events in the Central Mediterranean area (Umbria-Marche basin) related to paleoclimatic changes. *Neues Jahrbuch für Geologie und Paläontologie, Abhandlungen*, **234**, 361–413.

WADE, B. S., PEARSON, P. N., BERGGREN, W. A. & PÄLIKE, H. 2011. Review and revision of Cenozoic tropical planktonic foraminiferal biostratigraphy and calibration to the geomagnetic polarity and astronomical time scale. *Earth − Science Reviews*, **104**, 111–142, doi: 10.1016/j.earscirev.2010.09.003.

ZACHOS, J. C., QUINN, T. M. & SALAMY, K. A. 1996. High-resolution (10^4 years) deep-sea foraminiferal stable isotope records of the Eocene–Oligocene climate transitions. *Palaeoceanography*, **11**, 251–266.

ZACHOS, J. C., PAGANI, M., SLOAN, L., THOMAS, E. & BILLUPS, K. 2001. Trends, rhythms, and aberrations in global climate 65 Ma to present. *Science*, **292**, 686–693.

ZACHOS, J. C., DICKENS, G. R. & ZEEBE, R. E. 2008. An early Cenozoic perspective on greenhouse warming and carbon-cycle dynamics. *Nature*, **451**, 279–283.

Middle Eocene to early Oligocene magnetostratigraphy of ODP Hole 711A (Leg 115), western equatorial Indian Ocean

JAIRO F. SAVIAN[1,2]*, LUIGI JOVANE[2,3], STEVEN M. BOHATY[2] & PAUL A. WILSON[2]

[1]*Departamento de Geofísica, Instituto de Astronomia, Geofísica e Ciências Atmosféricas, Universidade de São Paulo, São Paulo 05508-090, Brazil*

[2]*School of Ocean and Earth Science, National Oceanography Centre, University of Southampton, Southampton SO14 3ZH, UK*

[3]*Instituto Oceanográfico, Universidade de São Paulo, São Paulo 05508–120, Brazil*

**Corresponding author (e-mail: savian@iag.usp.br)*

Abstract: Ocean Drilling Program (ODP) Site 711, located in the western equatorial Indian Ocean near the Seychelles Archipelago on Madingley Rise, is an important site for studying middle Eocene to early Oligocene climatic evolution. This site is ideal for studying the impact of Neo-Tethyan gateway closure on Indian Ocean currents and circulation to further understand global climate changes through the greenhouse to icehouse transition. Middle Eocene-to-lower Oligocene strata recovered within Hole 711A (Cores 711A-14X to 21X) primarily consist of clay-bearing nannofossil oozes/chalks, with layers rich in radiolarians. Here, we report a high-resolution magnetostratigraphic record and a new integrated age model for the middle Eocene-to-lower Oligocene section of Hole 711A. Correlation of the polarity pattern to the geomagnetic polarity timescale provides a record from Chron C19r (middle Eocene) to C12r (early Oligocene). Our results extend the existing polarity record down into the middle Eocene and confirm published results from the lower Oligocene section of the hole. Overall, these new results from Hole 711A have important implications for identifying and dating global climate change events, and for reconstructing calcite compensation depth history at this site.

Supplementary material: Magnetostratigraphic data used for construction of age models for Hole 711A included in this study are available at: www.geolsoc.org.uk/SUP18595

Ocean Drilling Program (ODP) Leg 115 in the western equatorial Indian Ocean was designed to study the Cenozoic evolution of the Réunion hotspot and Neogene sedimentation and dissolution history in the Indian Ocean Basin (Duncan *et al.* 1990). The main objective for coring at Site 711 was to obtain a complete Neogene sediment sequence to study time-dependent vertical changes in the position of the calcite compensation depth (CCD) and variability of carbonate preservation within the sublysocline transition zone (Backman *et al.* 1988). Moreover, accurate age models were to be developed by a combination of bio- and magnetostratigraphy (Backman *et al.* 1988). Hole 711A is located at a water depth at 4430 m on Madingley Rise (Fig. 1); a 250 m-thick sedimentary section was recovered. Sediments from Hole 711A span the time period between the Pleistocene and middle Eocene. Palaeodepth estimates for the site during the middle Eocene and early Oligocene interval (50–30 Ma) are approximately 3450 m at 42 Ma and 3750 m at 30 Ma (Peterson & Backman 1990).

Previous palaeomagnetic studies from the Oligocene interval of Hole 711A (Schneider & Kent 1990*a*; Touchard *et al.* 2003) indicated that the sediments retain a good record of the Earth's past magnetic field; however, the inclination data appear scattered because of magnetic overprints associated with the coring processes (e.g. Roberts *et al.* 1996; Fuller *et al.* 1998; Acton *et al.* 2002). Schneider & Kent (1990*a*) studied the magnetostratigraphy of Hole 711A down to the top of Chron C9n, recognizing the magnetostratigraphic Chrons C18 and C19 based on nannofossil Zone CP19. Touchard *et al.* (2003) obtained a implemented magnetostratigraphy from 214 samples that spanned the section between 98.8 and 157.38 mbsf, which corresponded to the time interval from Chron C13r to C19n (Schneider & Kent 1990*a*) and zonal boundary NP19-20/NP21 (Okada 1990).

The Eocene and Oligocene represent a critical time interval for palaeoclimate and palaeoceanography evolution. Deep-sea oxygen and carbon stable isotope records (e.g. Miller *et al.* 1991; Zachos *et al.* 2001, 2008) indicate that the climate system experienced progressive high-latitude cooling from the early Eocene climatic optimum (*c.* 50–52 Ma), which culminated in the oxygen isotopic (Oi-1)

From: JOVANE, L., HERRERO-BERVERA, E., HINNOV, L. A. & HOUSEN, B. A. (eds) 2013. *Magnetic Methods and the Timing of Geological Processes*. Geological Society, London, Special Publications, **373**, 97–110.
First published online March 25, 2013, http://dx.doi.org/10.1144/SP373.16 © The Geological Society of London 2013.
Publishing disclaimer: www.geolsoc.org.uk/pub_ethics

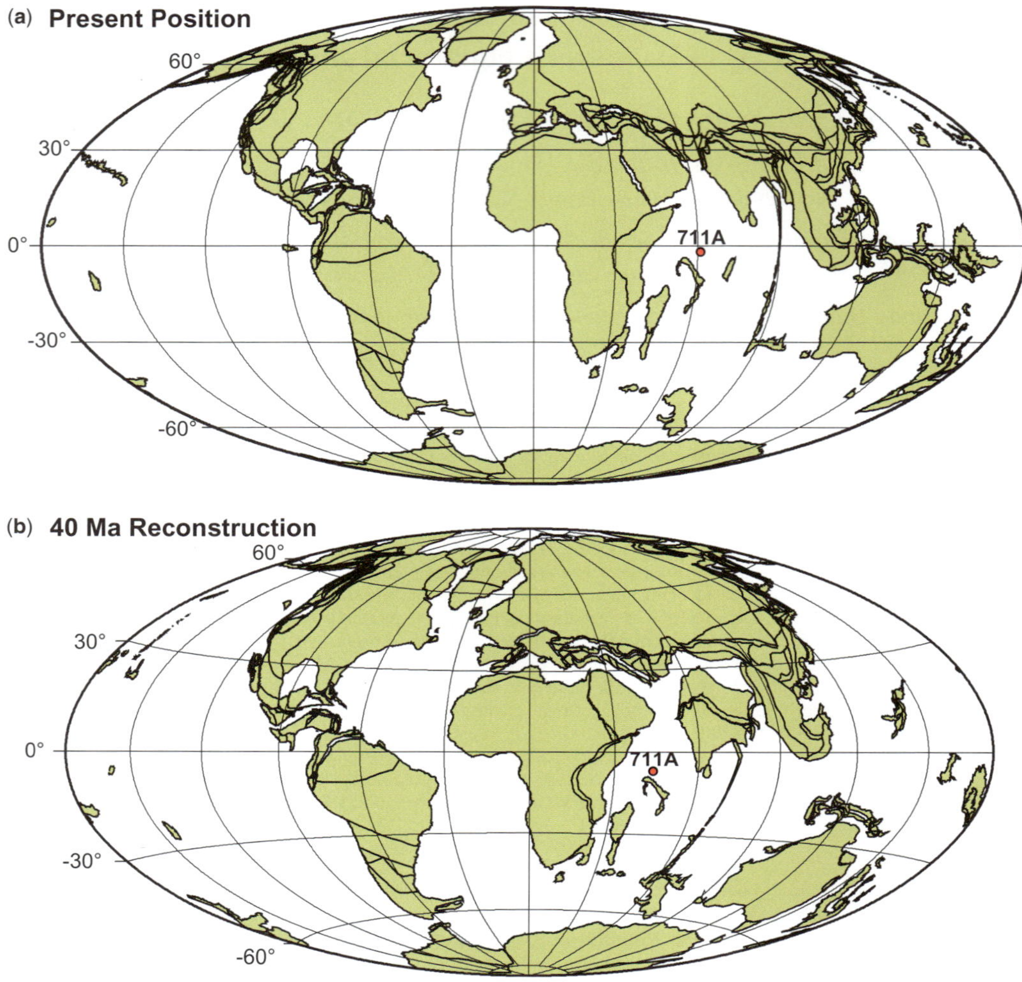

Fig. 1. (**a**) Location of ODP Site 711 in the western equatorial Indian Ocean. (**b**) Palaeogeographic reconstruction showing location of ODP Site 711 at 40 Ma. The maps were generated from the Ocean Drilling Stratigraphic Network (GEOMAR, Kiel, Germany).

excursion (*c.* 34 Ma; Miller *et al.* 1991). Superimposed on this long-term cooling trend are climatic instabilities associated with warming (e.g. Bohaty & Zachos 2003; Jovane *et al.* 2007; Bohaty *et al.* 2009; Sexton *et al.* 2011) and cooling events (e.g. Tripati *et al.* 2005). The largest warming event is the middle Eocene climatic optimum (MECO; Bohaty & Zachos 2003; Jovane *et al.* 2007; Edgar *et al.* 2010; Luciani *et al.* 2010; Spofforth *et al.* 2010), which is interpreted as a global warming event by Bohaty *et al.* (2009). The MECO event is characterized by a progressive long-term increase (*c.* 1.0‰) in benthic foraminiferal $\delta^{13}C$ values in several sites in the southern Ocean. This long-term

trend began at *c.* 40.7 Ma and ended at *c.* 39.6 Ma (Bohaty *et al.* 2009). The large negative $\delta^{18}O$ shift at 40 Ma from several sites indicates that MECO was a global warming event (Bohaty *et al.* 2009). Bijl *et al.* 2010 reconstructed the pCO_2 and sea surface temperatures during the MECO, suggesting a warming of *c.* 5 °C of the sea surface temperatures.

The Oi-1 event at the Eocene–Oligocene boundary was a pivotal event in the shift from greenhouse to icehouse climate state. The Oi-1 event was associated with an abrupt stepwise onset of Antarctic glaciation with global shifts in distribution of marine biogenic sediments, an overall increase in ocean fertility, a major drop in the CCD, and an onset of Atlantic thermohaline circulation (e.g. Coxall *et al.* 2005; Coxall & Wilson 2011). The Oi-1 cooling event was associated with an abrupt increase (*c.* 1.5‰) in benthic foraminiferal $\delta^{18}O$ values, which is attributed to development of ice sheets on Antarctica, which grew to $\geq 50\%$ near present day (e.g. Shackleton & Kennett 1975; Coxall *et al.* 2005). Variations in the Earth's climate system are caused by changes in ocean circulation owing to the opening of Southern Ocean gateways (e.g. Kennett 1977; Lyle *et al.* 2007), reduction of atmospheric greenhouse gases (DeConto & Pollard 2003; Pagani *et al.* 2005; Merico *et al.* 2008; Liu *et al.* 2009), superimposed orbital forcing (Coxall *et al.* 2005) and closure of the Neo-Tethys ocean gateway at *c.* 35 Ma (late Eocene; e.g. Allen & Armstrong 2008; Jovane *et al.* 2009).

In order to identify and provide new insight into the major Eocene and Oligocene climatic events, we developed a integrated high-resolution age model for Hole 711A. The middle Eocene-to-lower Oligocene interval is well represented in sediments form this hole, and in this paper, we present new high-resolution magnetostratigraphy for this section. Previously published biostratigraphic results (Okada 1990) from this site are used to constrain the correlation to the GPTS (Table 1). Magnetostratigraphic results have been previously reported for Hole 711A by Schneider & Kent (1990*a*) and Touchard *et al.* (2003). However, these earlier studies do not cover the lower part of the section owing to the proximity of the site to the equator and the difficulty envisioned in obtaining polarity data from these azimuthally unoriented samples. Our new results from Hole 711A have important implications for identification and dating global climate change events (MECO, Oi-1), and for reconstructing CCD history of this site.

Geological setting, lithology, and sampling

ODP Hole 711A ($2°44.56'S$, $61°09.78'E$) was drilled near the Seychelles Archipelago, between Madingley Rise and Carlsberg Ridge at a water depth of 4430 m, northern Indian Ocean (Fig. 1). Hole 711A was advanced to 104.5 m below seafloor (mbsf) using the advanced piston corer system, recovering 95.1 m (91%) and advanced to 249.6 mbsf using the extended core barrel system, recovering 108.7 m (80%) (Backman *et al.* 1988). We studied the depth interval of Cores 711A-14X to -21X, which encompass sediments from the middle Eocene to lower Oligocene interval. The middle Eocene-to-lower Oligocene strata within Hole 711A primarily consist of carbonate-rich sediments

Table 1. *Calcareous nannofossil events and interpretation of Chrons boundaries for the Hole 711A*

Calcareous nannofossil events and chrons boundaries	Depth (mbsf)	Age (Ma)
FO *Reticulofenestra umbilica*	198.1	42.67 (C20n) (1)
C20n/C19r	197.3	42.356 (2)
C19r/C19n	193.5	41.521 (2)
LO *Chiasmolithus solitus*	192.1	40.4 (C18r) (3)
C19n/C18r	191.3	41.257 (2)
FO *Dictyococcites scrippsae*	188.49	39.86 (C18n.2n) (1)
C18r/C18n	187.3	40.13 (2)
C18n/C17r	182.7	38.426 (2)
C17r/C17n	180.07	38.113 (2)
LO *Chiasmolithus grands*	177.22	37.1 (C17n.1n) (3)
C17n/C16n	169	36.341 (2)
FO *Isthmolithus recurvos*	163.12	36 (C16n.2n) (3)
C16n/C15r	163.2	35.343 (2)
C15r/C15n	161.4	34.94 (2)
C15n/C13r	159.3	34.655 (2)
LO *Discoaster barbadiensis*	158.5	34.3 (C13r) (3)
C13r/C13n	153.2	33.545 (2)
LO *Ericsonia Formosa*	151.2	32.8 (C12r) (3)
C13n/C12r	148.4	33.058 (2)
LO *Reticulofenestra umbilica*	140	32.3 (C12r) (3)
FO *Sphenolithus distentus*	125.8	31.5–33.1 (C12r) (3)
C12r/C12n	129.8	30.939 (2)

Calcareous nannofossil zonations of Jovane *et al.* (2010) (1) and Berggren *et al.* (1995) (3) and references therein. Magnetochron ages from Gee & Kent (2007) (2).

characterized as clay-bearing nannofossil oozes/chalks (68–173 mbsf; Lithostratigraphic Unit III), and clay-bearing nannofossil chalks with interbedded radiolarian oozes and radiolarian-bearing nannofossil chalks with carbonate content comprise about 70–80% (173–249.6 mbsf; Lithostratigraphic Unit IV; Backman *et al.* 1988; Okada 1990).

Previous biostratigraphic work in the Eocene–Oligocene section of Hole 711A includes studies of planktonic foraminifera, diatoms and nannofossils (Fenner & Mikkelsen 1990; Fornaciari *et al.* 1990; Johnson 1990; Mikkelsen 1990; Okada 1990; Rio *et al.* 1990). Moreover, $CaCO_3$, opal and magnetic susceptibility parameters have also been measured (Hampel & Bohrmann 1990; Peterson & Backman 1990; Robinson 1990). Palaeomagnetic measurements (progressive alternating-field demagnetization) have been measured in all Leg 115 sites (Schneider & Kent 1990*a*), and a more detailed palaeomagnetic study was performed by Touchard *et al.* (2003) in the upper 100–160 mbsf interval of Hole 711A. In this present study, a total of 375 discrete samples were taken by International ODP personnel in Kochi (Japan) pressing 7 cm^3 plastic samples boxes into the working half of the cores at an average *c.* 20 cm spacing.

Methods

Measurements were carried out in the palaeomagnetic laboratory of the National Oceanography Centre Southampton, University of Southampton, UK. All measurements were made using a 2G-Enterprises superconducting quantum interference device (SQUID) magnetometer (model 755R). To minimize sources of noise, the magnetometer is situated in a magnetically shielded room, which reduces the intensity of the ambient magnetic field. The sensitivity of the instrument corresponds to a rock magnetization of *c.* 10^{-6} A m^{-1}. We measured the low-field mass magnetic susceptibility (χ) of all samples using a KLY-4 Kappabridge (Agico Ltd), prior to measurement of the natural remanent magnetization (NRM).

In order to recognize and remove secondary NRM components we use alternating field (AF) demagnetization to isolate the characteristic remanent magnetization (ChRM). This process effectively randomizes the magnetization of grains with coercivities below the peak AF intensity, which allows the remanence of higher coercivity grains to be isolated. The NRM of all samples from Hole 711A was measured after progressive stepwise AF demagnetization at 5, 10, 15, 20, 25, 30, 35, 40, 45, 50, 60, 70 and 80 mT.

In order to display stepwise demagnetization data, we used vector component diagrams (Zijderveld 1967; Dunlop 1979). The ChRM direction were determined using stereographic projections, orthogonal and demagnetizing intensity plots of stepwise demagnetization data, allowing the various remanence components that make up the NRM to be separately identified. The directions of these different components were then calculated using principal component analysis (Kirschvink 1980). This method analyses the best-fit line through straight-line, single component line. This linearity can be assessed by calculating the maximum angular derivation (MAD). A good line fit is indicated by a MAD $< 15°$. This method is only valid where linear components of the demagnetization path can be isolated. Principal component analyses were performed using the palaeomagnetic data software package REMASOFT 3.0 (Chadima & Hrouda 2006).

The rock-magnetic measurements consisted of hysteresis curves and IRM (isothermal remanence magnetization) performed at room temperature using a vibrating sample magnetometer (VSM MicroMagTM 3900) and a Pulse Magnetizer (MMPM-9) at the National Oceanography Centre Southampton, University of Southampton, UK. Hysteresis loops were measured in maximum field of 1 T and average time 500 ms. IRM acquisition curves were carried out on representative samples along the section. The samples were submitted by applying incrementally increasing fields to initially demagnetized samples until a maximum value of 1.5 T was reached.

Results

Palaeomagnetic polarity zonation

A total of 375 discrete samples were analysed using the SQUID magnetometer. Of the 375 analysed samples, 248 (66%) had no reliable magnetic directions (MAD $\geq 15°$) after AF demagnetization. The other 127 samples (34%) displayed stable palaeomagnetic behaviour upon stepwise AF demagnetization (Fig. 2). The ChRM directions with MAD $\leq 15°$ displayed normal and reverse polarities defining a magnetic polarity zonation. In particular, several samples had inclination values that were intermediate between normal and reverse directions. For azimuthally unoriented samples from sedimentary rocks that formed in low latitudes, determining the polarity of sedimentary units may be difficult. The polarity ambiguity arises because the samples are azimuthally unoriented, and the declination cannot be used to determine polarity; the inclination is shallow near the equator and the angular distance between reversed and normal polarity inclinations is small; and the

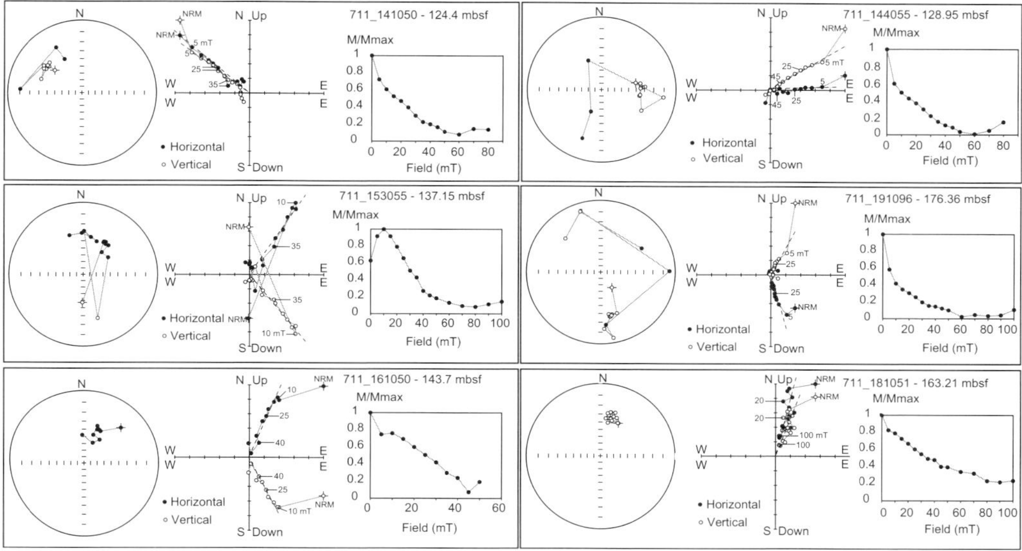

Fig. 2. Orthogonal and vector component diagrams (with normalized intensity decay plots) for AF demagnetization data for six representative specimens from the Hole 711A. Open circles indicate projections onto the vertical plane, and solid black circles indicate projections onto the horizontal plane.

palaeomagnetic inclinations from any samples will have some degree of dispersion about their mean inclination. Thus it is likely that, when the mean inclination is shallow, the sign of the inclination will not be indicative of the polarity (e.g. McFadden & Reid 1982; Cox & Gordon 1984). In our case, only two samples (163.7 and 193.7 mbsf) presented inclinations less than $10°$. The sign of the inclination of these two samples cannot be used as a definitive estimate of magnetic polarity. The samples with MAD values $\geq 15°$ were discarded. Some of them were showing low intensity owing to low magnetic concentration; the others can be interpreted as drilling-induced overprint.

Figure 3 illustrates the lithologic units of the middle Eocene–lower Oligocene sedimentary interval of Hole 711A. The abundance and preservation of the calcareous nannofossils vary from whole dissolved to moderate at Hole 711A (Backman *et al.* 1988; Okada 1990; Wei *et al.* 1992). The NRM demagnetization paths of the 375 samples from Hole 711A have considerable variations that are attributed to the relatively low intensity of magnetizations (5.23×10^{-5} to 1.77×10^{-1} A m^{-1} with an arithmetic average of 5.27×10^{-3} A m^{-1}; Fig. 3). Several samples are characterized by high NRM intensity, which is associated with dished lithologies (186.92 and 187.01 mbsf; e.g. very dark greyish brown clay-bearing radiolarian ooze). Arithmetic mean directions were computed separately for normal and reverse polarity populations and calculated

following Jovane *et al.* (2008). The normal polarity samples have an arithmetic mean inclination of $29.0°$ ($N = 59$; $\alpha_{95} = 2.5$) and the reverse polarity samples have an arithmetic mean inclination of $-30.7°$ ($N = 37$; $\alpha_{95} = 2.6$). These values show that our magnetostratigraphic record passes the reversal test. Our interpretation was compared with predicted palaeolatitudes of Site 711A according to the absolute plate motion model based on fixed African hotspots (e.g. Duncan 1981; Kidd *et al.* 1992). Based on these mean directions, the calculated mean palaeolatitude of Hole 711A is $16°$S. However, based on the reconstructed position of Hole 711A (see Fig. 1b), the 40 Ma palaeolatitude would be $9.5°$S, and the expected inclination would be $19°$. The observed mean inclination is approximately $10°$ steeper than the expected inclination. These inclinations are steeper than the present-day geocentric axial dipole inclination and it is attributed to three systematic error sources: the method of calculating the mean inclination; drift of the hotspots; and a large non-dipole field (Schneider & Kent 1990a). This result is compatible with previous palaeoreconstructions for the Eocene–Oligocene period (Schneider & Kent 1990a; Royer & Coffin 1992; Zachos *et al.* 1992). However, our mean inclination is steeper than expected (according to the fixed-hotspot models), and this could be due to some unresolved combination of unaccounted for motion of the plate, long-term non-dipole fields and effects of hotspot

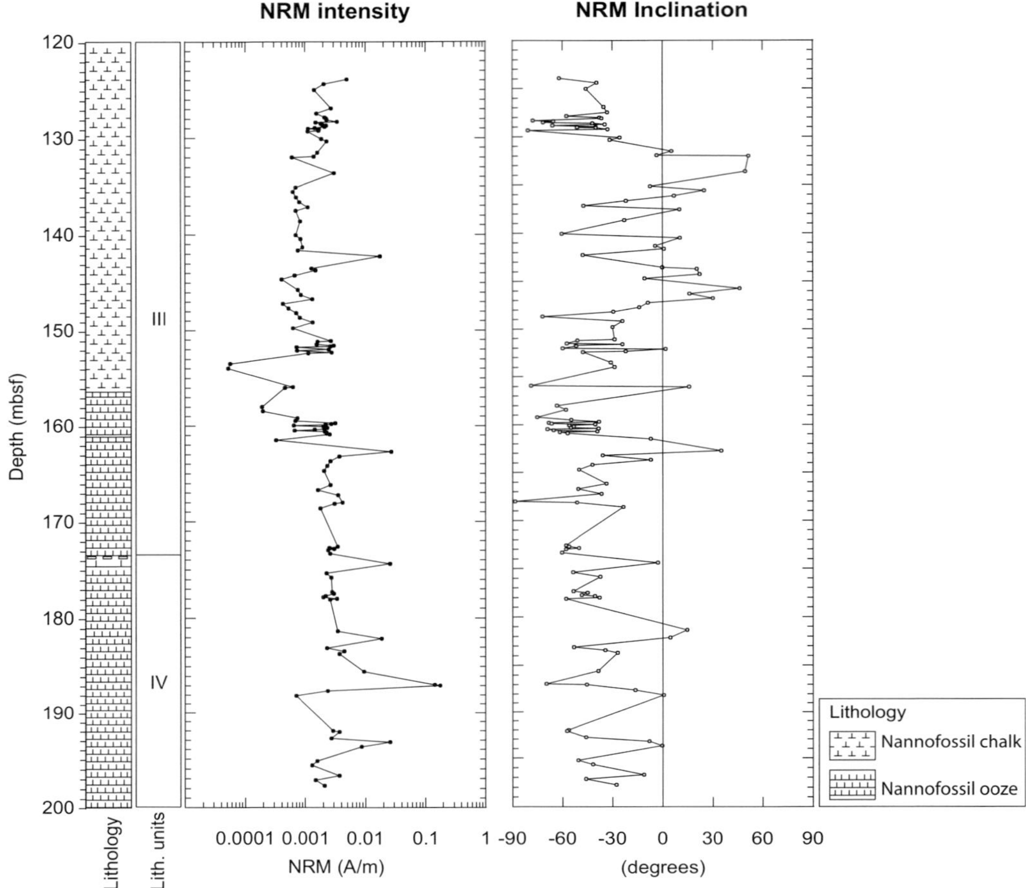

Fig. 3. Detailed lithostratigraphic column for ODP Hole 711A in the interval between 120 and 200 mbsf. The second column shows the lithostratigraphic units as defined in shipboard work (Backman *et al.* 1988), including Unit III (nannofossil oozes, clay-bearing nannofossil oozes, clay-bearing nannofossil chalks) and Unit IV (distinguished from unit III by the consistent occurrence of radiolarians). The third panel shows the stratigraphic variations of intensity and fourth panel shows the inclination of the NRM.

motion on the plate circuit models (Schneider & Kent 1990*b*). Based on palaeocenographic and palaeomagnetic records, the globally tectonic implications of Site 711 have been reconstructed. The drill site has moved northward since the formation these sediments by approximately 600 km.

The magnetic polarity record of the studied portion of Hole 711A can be subdivided into 14 magnetozones (Fig. 4). The magnetozones are defined as intervals with multiple, consecutive samples with polarities that are distinctly different from neighbouring intervals. In only one case, at 142.23 mbsf, the sample has a polarity opposite that of the rest of the magnetozone, but it is not used to determine the polarity.

Rock magnetic properties

Isothermal remanent magnetization acquisition curves were obtained for five representative samples of Hole 711A at fields up to 1.5 T (Fig. 5). IRM acquisition experiments indicated that saturation magnetization (M_s) occurs at a field between 60 and 80 mT, which typifies low- to medium-coercivity remanence carriers such as magnetite and/or titanomagnetite (Ti-poor magnetite; e.g. Dunlop & Özdemir 1997). These results confirm previous published results from the lower Oligocene section of the Hole 711A (Touchard *et al.* 2003). The presence of a low-coercivity mineral is also confirmed from the rapid decrease in

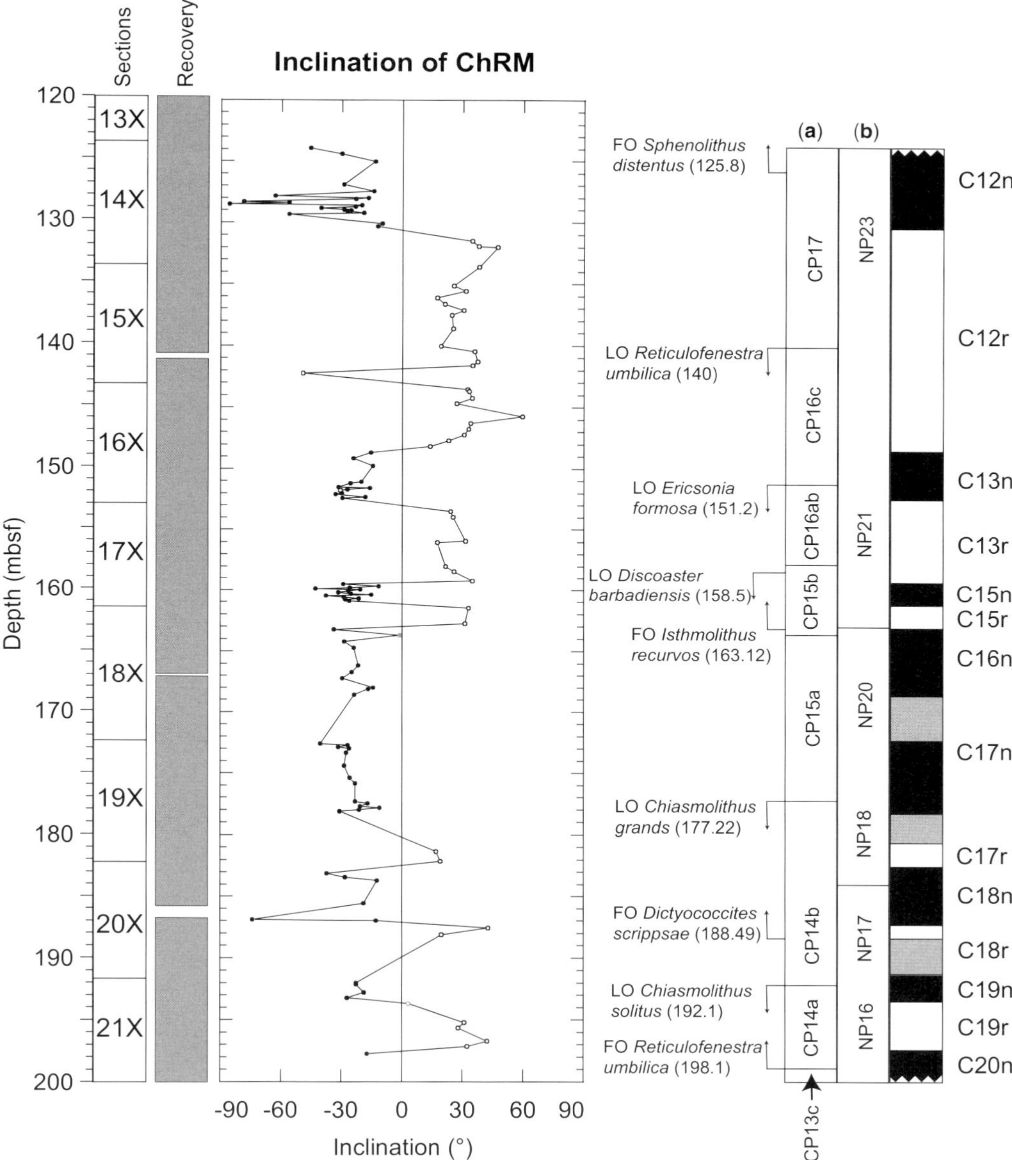

Fig. 4. Core number (left), core recovery (grey bars), and variations in the declination and inclination in Hole 711. The magnetostratigraphic interpretation is shown on the right side, which is constrained by the nannofossil biostratigraphy of Okada (1990). Nannofossil zones, using the zonal of both Okada & Bukry (1980) (**a**) and Martini (1971) (**b**), are defined in the section. Inclination values steeper than 10° are black while values below 10° are grey points.

intensity of remanence during AF demagnetization (Fig. 2).

Hysteresis cycles for the 20 representative samples along the 80 m of section show a low-coercivity behaviour (Fig. 6a), which probably indicates the presence of magnetite and/or titan-omagnetite (e.g. Dunlop & Özdemir 1997; Krása *et al.* 2011). The hysteresis standard parameters like M_s, saturation remanent magnetization (M_{rs}), coercive force (H_c) and coercivity of remanence

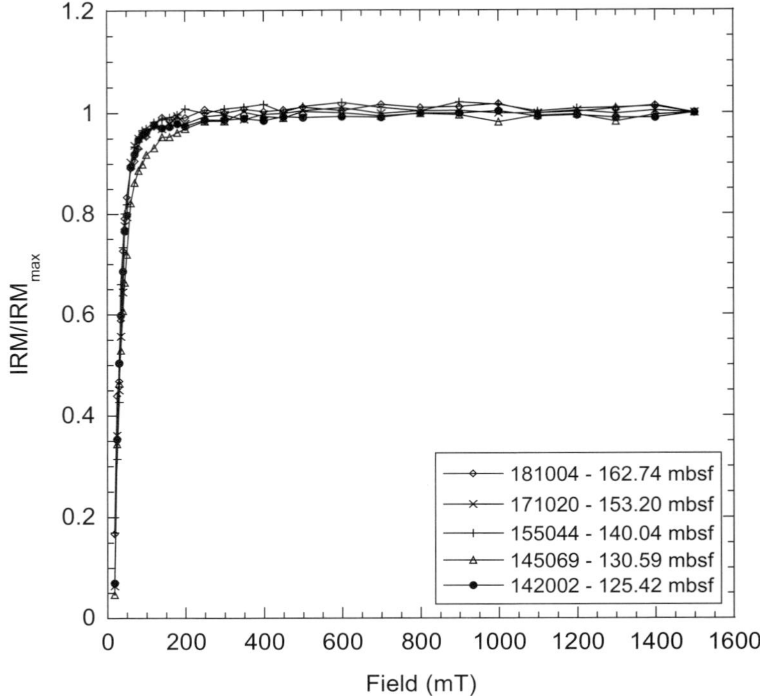

Fig. 5. Isothermal remanent acquisition curves for five representative samples within the ODP Hole 711A study section (normalized intensities v. applied magnetic field).

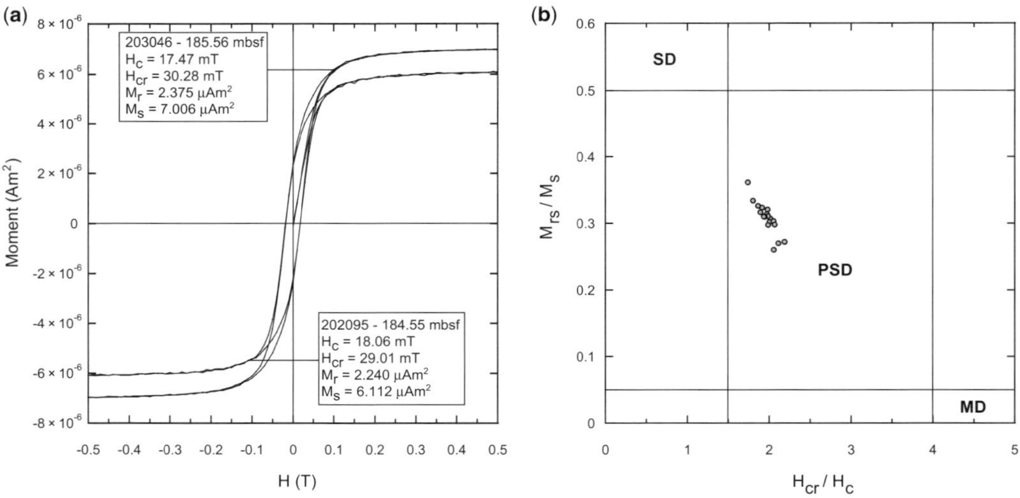

Fig. 6. (a) Hysteresis curves for two representative samples from Hole 711A indicating a similar magnetic behaviour (low-coercivity behaviour, magnetite and/or titanomagnetite). (b) M_{rs}/M_s v. $H_c r/H_c$ plot for sediments from the Hole 711A. The magnetic grain size shows pseudo-single domain particles.

(H_{cr}) are combined as ratio and approaching pseudo-single domain values. Magnetic hysteresis data (Fig. 6a) are characterized by stably magnetized samples that have moderate H_{cr}/H_c (1.74–2.10) and M_{rs}/M_s (2.61–3.61). Hysteresis ratios for samples from 711A indicate a pseudo-single domain assemblage (Fig. 6b; Day *et al.* 1977) with the dominance of magnetite and/or titanomagnetite particles (e.g. Dunlop & Özdemir 1997).

Biostratigraphy

In order to obtain a high-resolution integrated age model, magnetostratigraphic data need to be combined with biostratigraphic data in the same interval. Okada (1990) and Wei *et al.* (1992) conducted a biostratigraphic investigation on the Palaeogene interval of Site 711, studying calcareous nannofossil using biozones of Martini (1971) and Okada & Bukry (1980) (Fig. 4).

The base of the studied section (198.1 mbsf), which defines Zone CP14a, near the CP13c–CP14a zonal boundary of Okada & Bukry (1980), coinciding with the lower part of Zone NP16–17 of Martini (1971), was characterized by the first occurrence (FO) of *Reticulofenestra umbilica* (Okada 1990; Table 1). The FO of *Dictyococcites scrippsae* at 188.49 mbsf occurs approximately in the CP14a–CP14b of Okada & Bukry (1980), which is defined by last occurrence (LO) of *Chiasmolithus solitus* (192.1 mbsf). The LO of *Chiasmolithus grands* (177.22 mbsf) occurs near the CP14b–CP15a zonal boundary of and NP18–NP20. The FO of *Isthmolithus recurvos* (163.12 mbsf) and the LO of *Discoaster barbadiensis* occur at approximately the CP15a–CP15b boundary and

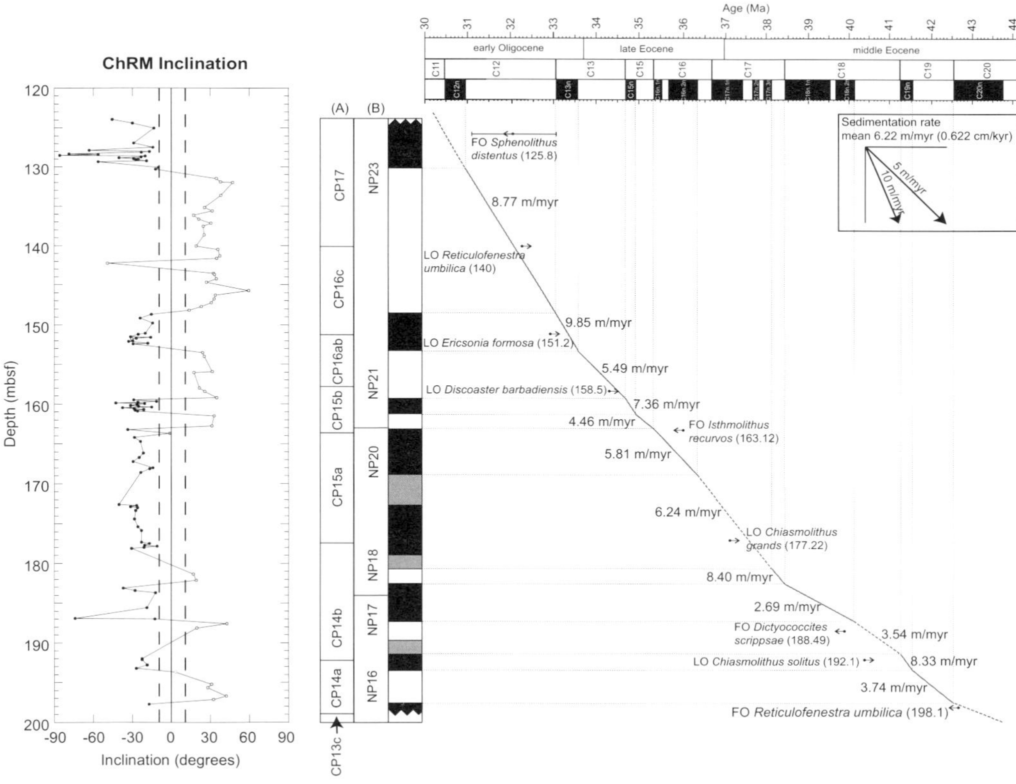

Fig. 7. Inclination of ChRM and age v. depth plot with correlation of the ODP Hole 711A polarity zonation to the geomagnetic polarity time scale of Gee & Kent (2007) and Jovane *et al.* (2010). Inclination values steeper than 10° are black while values below 10° are grey points and limited by dashed lines. Calcareous nannofossil and planktonic foraminiferal datums are used to constrain the interpretation. The biostratigraphic events from Okada (1990) are based on the calcareous nannofossil zonal schemes of Martini (1971) and Okada & Bukry (1980). The grey bars in the geomagnetic polarity zonation represent the uncertainties in the definition of polarity zonation in relationship with GPTS.

the CP15b–CP16ab boundary, respectively. The CP16ab–CP16c zonal boundary is characterized by the LO of *Ericsonia formosa* (151.2 mbsf). The LO of *Reticulofenestra umbilica*, which marks the CP16c–CP17 zonal boundary, occurs at 140 mbsf. The FO of *Sphenolithus distentus* (125.8 mbsf) occurs in the CP17 and NP23.

Discussion

Correlation to the geomagnetic polarity time scale

In this work, we present a new magnetic polarity record between 123.93 and 198.91 mbsf of Hole 711A. We recognize 14 magnetozones in the middle Eocene–lower Oligocene interval following the published biostratigraphic of Okada (1990) (C12n–C20n; Fig. 4). The new interpretation of the magnetic polarity pattern provides a correlation with geomagnetic polarity time scale (GPTS) of Gee &

Kent (2007) and Jovane *et al.* (2010) between the top of Chron C12r (30.939–33.058 Ma) and the top of Chron C20n (42.536–43.789 Ma; Fig. 7). From 160 to 120 mbsf (late Eocene to early Oligocene) our interpretation is in good agreement with previous magnetostratigraphic results reported for Hole 711A (Touchard *et al.* 2003). However, below 160 mbsf our interpretation is completely new, spanning approximately 8 myr.

Our new age interpretation following ages of Gee & Kent (2007) differs significantly of previously published data collected during the shipboard work (Backman *et al.* 1988) as well as subsequent works (Okada 1990; Rio *et al.* 1990; Schneider & Kent 1990*a*). We use the GPTS of Gee & Kent (2007) and Jovane *et al.* (2010) – use of the Gradstein *et al.* (2004) GPTS results in small changes in (<1 million years) in the age–depth relationships, and minor changes in the calculated sedimentation rate of the section (see Figures S1 and S2 in the Supplementary Material).

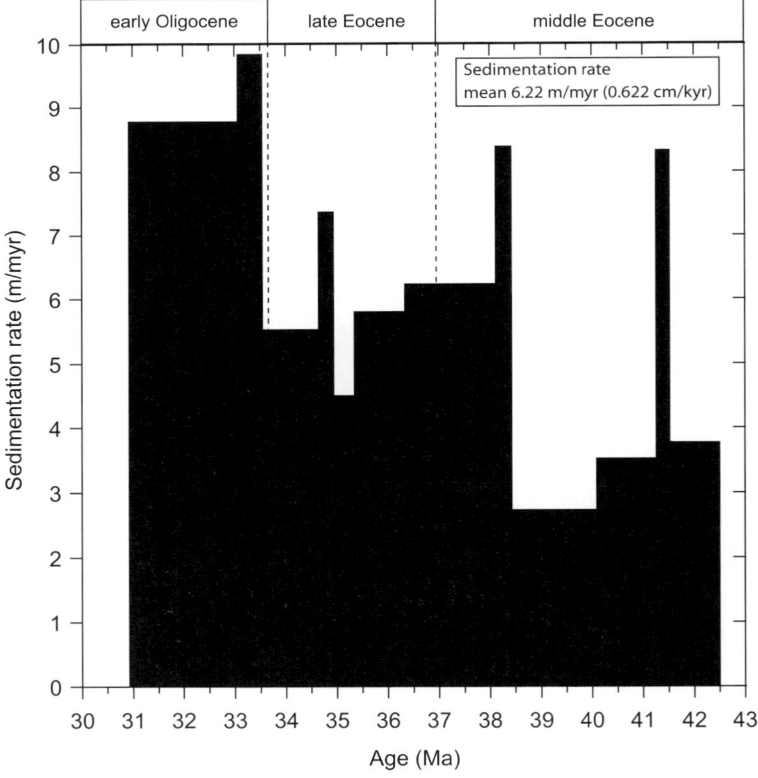

Fig. 8. Variation in sedimentation rate along the Middle Eocene–early Oligocene interval are shown as linear interpolation between the chron's boundary at Site 711. The average sedimentation rate is 6.22 m myr^{-1}. Age is from the Gee & Kent (2007) and Jovane *et al.* (2010) and depths are the metres below sea-floor (mbsf).

Age–depth model

With the aim of perform a new age model, we correlated the 12 magnetozones with GPTS (Gee & Kent 2007; Jovane *et al.* 2010). To constrain the magnetostratigraphic interpretation for the interval between 123.93 and 198.91 mbsf, we used the nannofossil stratigraphy presented by Okada (1990). Overall the linear correlation between magnetostratigraphy and biostratigraphy for Hole 711A (Fig. 7) shows small uncertainties.

The age–depth curve for the middle Eocene to the lower Oligocene interval of Hole 711A is presented in Figure 7. The resulting sedimentation rates are relatively uniform and are consistent with biostratigraphic data. The average sedimentation rate for the interval between 197.43 mbsf (base of Chron C19r) and 130.91 mbsf (top of Chron C12r) is 6.22 m myr^{-1} (0.622 cm kyr^{-1}; Fig. 8).

The new age model obtained in this study improves the dating of the bio-chronostratigraphical events during Eocene–Oligocene period. In this study we show that the Eocene–Oligocene boundary is placed at approximately 33.7 Ma (*c.* 155 mbsf), at the top of the Chron C13r, which is in agreement with Touchard *et al.* (2003) and Jovane *et al.* (2006). Hole 711A represents the most complete and least disturbed palaeomagnetic record from the northern Indian Ocean for the interval from the middle Eocene to the early Oligocene. The defined age model for this section will allow further development of Eocene–Oligocene palaeoceanographic records at this site with reliable age constraints. In particular, we are now able to define the Chrons C18n and C18r in Hole 711A, which span the interval of the MECO event (Bohaty & Zachos 2003; Jovane *et al.* 2007; Bohaty *et al.* 2009; Edgar *et al.* 2010). We observe a decrease in the sedimentation rates along the Chron C18n, which can be easily related to the carbonate concentration data for the same interval (Peterson & Backman 1990), showing that in this interval there is a decrease in carbonate mass accumulation rates (Bohaty *et al.* 2009). We also observe an increase in sedimentation rates along Chrons 13n and 12r, which is attributed to deepening of the CCD (Coxall *et al.* 2005; Katz *et al.* 2008).

Conclusions

We present a new magnetostratigraphic record and redefined age model for the middle Eocene–lower Oligocene interval of Indian Ocean ODP Hole 711A. The new magnetostratigraphic results are integrated with published biostratigraphic results, and the integrated magnetozones are correlated to Chrons C19r to C12r, (*c.* 42.5 and 30.9 Ma). Our magnetostratigraphy provides the basis for further improvement based on cyclostratigraphy and astrochronology.

Isothermal remanence curves and hysteresis curves provide information on the magnetic mineralogy of Hole 711A. The primary magnetic mineral carrier in pelagic sediments at this site is interpreted to be (titano)magnetite. Similar magnetic properties are observed throughout the study section.

Using our newly developed age–depth model for the Eocene–Oligocene section of Hole 711A, it will be possible to document palaeoceanographic variability within the middle Eocene–early Oligocene interval at this site. The Eocene–Oligocene transition and the MECO event, in particular, are present within continuous sections.

Our data indicate a decrease in the sedimentation rates at Site 711 during the MECO event, which may be related to the CCD shoaling inferred by Bohaty *et al.* (2009). In contrast, a considerable increase in the sedimentation rate during the Eocene–Oligocene transition is also observed in our magnetostratrigraphic data. This period is characterized by an abrupt stepwise onset of Antarctic glaciations, with a major deepening in the CCD (Coxall *et al.* 2005).

The study was carried out within the framework of the NEO-TETHYS project, which is sponsored by the European Community through Marie Curie Actions (FP7-PEOPLE-IEF-2008 proposal no. 236311). We acknowledge the Conselho Nacional de Desenvolvimento Científico (Process 201508/2009 5) and Conselho de Desenvolvimento de Pessoal do Nível Superior. We also thank the Integrated Ocean Drilling Program (IODP) for the samples used in this research. The IODP is sponsored by the US National Science Foundation and participating countries under the management of the Joint Oceanographic Institutions Inc. We also thank A. P. Roberts for providing comments on the manuscript.

References

ACTON, G. D., OKADA, M., CLEMENT, B. M., LUND, S. P. & WILLIAMS, T. 2002. Paleomagnetic overprints in ocean sediment cores and their relationship to shear deformation caused by piston coring. *Journal of Geophysics Research*, **107**, 1–15.

ALLEN, M. B. & ARMSTRONG, H. A. 2008. Arabia–Eurasia collision and the forcing of mid-Cenozoic global cooling. *Palaeogeography, Palaeoclimatology, Palaeoecology*, **265**, 52–58.

BACKMAN, J., DUNCAN, R. *ET AL.* 1988. *Shipboard Scientific Party, Proceedings of the Ocean Drilling Program*, **115**. Ocean Drilling Program, Texas A & M University, College Station, TX.

BERGGREN, W. A., KENT, D. V., SWISHER, C. & AUBRY, M.-P. 1995. A revised Cenozoic geochronology and chronostratigraphy. *In*: BERGGREN, W. A., KENT, D. V., AUBREY, M.-P. & HARDENBOL, J. (eds)

Geochronology, Time-Scales and Global Stratigraphic Correlation. Society of Economic Paleontologists and Mineralogists, Tulsa, OK, Special Publications, **54**, 129–212.

BIJL, P. K., HOUBEN, A. J. P. *ET AL.* 2010. Transient middle Eocene atmospheric CO_2 and temperature variations. *Science*, **330**, 819–821.

BOHATY, S. M. & ZACHOS, J. C. 2003. Significant Southern Ocean warming event in the late middle Eocene. *Geology*, **31**, 1017–1020.

BOHATY, S. M., ZACHOS, J. C., FLORINDO, F. & DELANEY, M. L. 2009. Coupled greenhouse warming and deep-sea acidification in the middle Eocene. *Paleoceanography*, **24**, PA2207, doi: 10.1029/2008PA001676

CHADIMA, M. & HROUDA, F. 2006. Remasoft 3.0: a user-friendly paleomagnetic data browser and analyzer. *Travaux Géophysiques*, **XXVII**, 20–21.

COX, A. & GORDON, R. G. 1984. Paleolatitudes determined from paleomagnetic data from vertical cores. *Reviews of Geophysics and Space Physics*, **22**, 47–72.

COXALL, H. K. & WILSON, P. A. 2011. Early Oligocene glaciation and productivity in the eastern equatorial Pacific: insights into global carbon cycling. *Paleoceanography*, **26**, PA2221.

COXALL, H. K., WILSON, P. A., PÄLIKE, H., LEAR, C. H. & BACKMAN, J. 2005. Rapid stepwise onset of Antarctic glaciations and deeper calcite compensation in the Pacific Ocean. *Nature*, **433**, 53–57.

DAY, R., FULLER, M. & SCHMIDT, V. A. 1977. Hysteresis properties of titanomagnetites: grain-size and compositional dependence. *Physics of the Earth and Planetary Interiors*, **13**, 260–267.

DECONTO, R. M. & POLLARD, D. 2003. Rapid Cenozoic glaciations of Antarctic induced by declining atmospheric CO_2. *Nature*, **421**, 245–249.

DUNCAN, R. A. 1981. Hotspots in the southern oceans-an absolute frame of reference for motion of the Gondwana continents. *Tectonophysics*, **74**, 29–42.

DUNCAN, R. A., BACKMAN, J. & PETERSON, L. C. 1990. *Proceedings of the Ocean Drilling Program, Scientific Results*, **115**. Ocean Drilling Program, Texas A & M University, College Station, TX.

DUNLOP, D. J. 1979. On the use of Zijderveld vector diagrams in multicomponent paleomagnetic studies. *Physics of the Earth and Planetary Interior*, **20**, 12–24.

DUNLOP, D. J. & ÖZDEMIR, Ö. 1997. *Rock Magnetism: Fundamentals and Frontiers.* Cambridge University Press, Cambridge.

EDGAR, K. M., WILSON, P. A., SEXTON, P. F., GIBBS, S. J., ROBERTS, A. P. & NORRIS, R. D. 2010. New biostratigraphy, magnetostratigraphy and isotopic insights into the Middle Eocene Climatic Optimum in low latitudes. *Palaeogeography, Palaeoclimatology, Palaeoecology*, **297**, 670–682.

FENNER, J. & MIKKELSEN, N. 1990. Eocene–Oligocene diatoms in the western Indian Ocean: taxonomy, stratigraphy, and paleoecology. *Proceedings of the Ocean Drilling Program, Scientific Results*, **115**. Ocean Drilling Program, Texas A & M University, College Station, TX, 433–466.

FORNACIARI, E., RAFFI, I., RIO, D., VILLA, G., BACKMAN, J. & OLAFSSON, G. 1990. Quantitative distribution patterns of Oligocene and Miocene calcareous nannofossils from the western equatorial Indian Ocean. *Proceedings of the Ocean Drilling Program, Scientific Results*, **115**. Ocean Drilling Program, Texas A & M University, College Station, TX, 237–254.

FULLER, M., HASTEDT, M. & HERR, B. 1998. Coring induced magnetization of recovered sediments. *In*: WEAVER, P. P. E., SCHMINKE, H.-U., FIRTH, J. V. & DUFFIELD, W. (eds) *Proceedings of the Ocean Drilling Program*, **157**. Ocean Drilling Program, Texas A & M University, College Station, TX, 47–56.

GEE, J. S. & KENT, D. V. 2007. Source of oceanic magnetic anomalies and the geomagnetic polarity time scale. Geomagnetism. *In*: KONO, M. (ed.) *Treatise on Geophysics*, **5**, 455–507, Elsevier, Amsterdam.

GRADSTEIN, F. M., OGG, J. G. & SMITH, A. G. 2004. *A Geological Time Scale.* Cambridge University Press, Cambridge.

HEMPEL, P. & BOHRMANN, G. 1990. Carbonate-free sediment components and aspects of silica diagenesis at Sites 707, 709, and 711 (Leg 115, western Indian Ocean). *Proceedings of the Ocean Drilling Program, Scientific Results*, **115**. Ocean Drilling Program, Texas A & M University, College Station, TX, 677–698.

JOHNSON, D. A. 1990. Radiolarian biostratigraphy in the central Indian Ocean, Leg 115. *Proceedings of the Ocean Drilling Program, Scientific Results*, **115**. Ocean Drilling Program, Texas A & M University, College Station, TX, 395–410.

JOVANE, L., FLORINDO, F., SPROVIERI, M. & PÄLIKE, H. 2006. Astronomic calibration of the late Eocene/early Oligocene Massignano section (central Italy). *Geochemistry, Geophysics, Geosystems*, **7**, Q07012, doi: 10.1029/2005GC001195.

JOVANE, L., FLORINDO, F. *ET AL.* 2007. The middle Eocene climatic optimum event in the Contessa Highway section, Umbrian Apennines, Italy. *GSA Bulletin*, **119**, 413–427.

JOVANE, L., ACTON, G., FLORINDO, F. & VEROSUB, K. L. 2008. Geomagnetic field behavior at high latitudes from a paleomagnetic record from Eltanin core 27–21 in the Ross Sea sector, Antarctica. *Earth and Planetary Science Letters*, **267**, 435–443.

JOVANE, L., COCCIONI, R., MARSILI, A. & ACTON, G. 2009. The late Eocene greenhouse–icehouse transition: observations from the Massignano global stratotype section and point (GSSP). *In*: KOEBERL, C. & MONTANARI, A. (eds) *The Late Eocene Earth – Hothouse, Icehouse, and Impacts.* Geological Society of America, Boulder, CO, Special Papers, **452**, 149–168

JOVANE, L., SPROVIERI, M., COCCIONI, R., FLORINDO, F., MARSILI, A. & LASKAR, J. 2010. Astronomical calibration of the middle Eocene Contessa Highway section (Gubbio, Italy). *Earth and Planetary Science Letters*, **298**, 77–88.

KATZ, M. E., MILLER, K. G. *ET AL.* 2008. Stepwise transition from the Eocene greenhouse to the Oligocene icehouse. *Nature Geoscience*, **1**, 329–334.

KENNETT, J. P. 1977. Cenozoic evolution of Antarctic glaciation, the Circum-Antarctic Ocean, and their impact on global paleoceanography. *Journal of Geophysical Research*, **82**, 3843–3860.

KIDD, R. B., RAMSAY, A. T. S., SYKES, T. J. S., BALDAUF, J. G., DAVIES, T. A., JENKINS, D. G. & WISE, S. W. JR. 1992. *An Indian Ocean Framework for Paleoceanographic Synthesis Based on DSDP and ODP Results*. Synthesis of Results from Scientific Drilling in the Indian Ocean, Geophysical Monograph, **70**, 403–422.

KIRSCHVINK, J. L. 1980. The least-squares line and plane and the analysis of palaeomagnetic data. *Geophysical Journal of the Royal Astronomical Society*, **62**, 699–710.

KRÁSA, D., MUXWORTHY, A. R. & WILLIAMS, W. 2011. Room and low-temperature magnetic properties of 2-D magnetite particles arrays. *Geophysical Journal International*, **185**, 167–180, doi: 10.1111/j.1365-246X.2011.04956.x.

LIU, Z., PAGANI, M. *ET AL.* 2009. Global cooling during the Eocene–Oligocene climate transition. *Science*, **323**, 1187–1190.

LUCIANI, V., GIUSBERT, L., AGNINI, C., FORNACIARI, E., RIO, D., SPOFFORTH, D. J. A. & PÄLIKE, H. 2010. Ecological and evolutionary response of Tethyan phanktonic foraminifera to the middle Eocene climatic optimum (MECO) from the Alano section (NE Italy). *Palaeogeography, Palaeoclimatology, Palaeoecology*, **292**, 82–95.

LYLE, M., GIBBS, S., MOORE, T. C. & REA, D. K. 2007. Late Oligocene initiation of the Antarctic Circumpolar Current: evidence from the South Pacific. *Geology*, **35**, 691–694.

MARTINI, E. 1971. Standard Tertiary and Quaternary calcareous nannoplankton zonation. *In*: FARINACCI, A. (ed.) *Proceedings of the Second Planktonic Conference, Rome 1970*. Tecnoscienza, Rome, **2**, 739–785.

MCFADDEN, P. L. & REID, A. B. 1982. Analysis of palaeomagnetic inclination data. *Geophysical Journal of the Royal Astronomical Society*, **69**, 307–319.

MERICO, A., TYRRELL, T. & WILSON, P. A. 2008. Eocene/Oligocene ocean de-acidification linked to Antarctic glaciations by sea-level fall. *Nature*, **452**, 979–983.

MIKKELSEN, N. 1990. Cenozoic diatom biostratigraphy and paleoceanography of the western equatorial Indian Ocean. *Proceedings of the Ocean Drilling Program, Scientific Results*, **115**. Ocean Drilling Program, Texas A & M University, College Station, TX, 411–432.

MILLER, K. G., WRIGHT, J. D. & FAIRBANKS, R. G. 1991. Unlocking the ice house: Oligocene–Miocene oxygen isotopes, eustasy, and margin erosion. *Journal of Geophysical Research*, **96**, 6829–6948.

OKADA, H. 1990. Quaternary and paleogene calcareous nannofossils, Leg 115. *Proceedings of the Ocean Drilling Program, Scientific Results*, **115**. Ocean Drilling Program, Texas A & M University, College Station, TX, 129–174.

OKADA, H. & BUKRY, D. 1980. Supplementary modification and introduction of code numbers to the low-latitude coccolith biostratigraphic zonation (Bukry, 1973; 1975). *Marine Micropaleontology*, **5**, 321–325.

PAGANI, M., ZACHOS, J. C., FREEMAN, K. H., TIPPLE, B. & BOHATY, S. 2005. Marked decline in atmospheric carbon dioxide concentrations during the Paleogene. *Science*, **309**, 600–603.

PETERSON, L. C. & BACKMAN, J. 1990. Late Cenozoic carbonate accumulation and the history of the carbonate compensation depth in the western equatorial Indian Ocean. *Proceedings of the Ocean Drilling Program, Scientific Results*, **115**. Ocean Drilling Program, Texas A & M University, College Station, TX, 467–508.

RIO, D., FORNACIARI, E. & RAFFI, I. 1990. Late Oligocene through early Pleistocene calcareous nannofossils from western equatorial Indian Ocean (Leg 115). *Proceedings of the Ocean Drilling Program, Scientific Results*, **115**. Ocean Drilling Program, Texas A & M University, College Station, TX, 175–236.

ROBERTS, A. P., STONER, J. S. & RICHTER, C. 1996. Coring-induced magnetic overprints and limitations of the long-core paleomagnetic measurement technique some observations from Leg 160 Eastern Mediterranean Sea. *In*: EMEIS, K.-C., ROBERTSON, A. H. F. & RICHTER, C. (eds) *Proceedings of the Ocean Drilling Program, Initial Reports*, **160**. Ocean Drilling Program, Texas A & M University, College Station, TX, 497–505.

ROBINSON, S. G. 1990. Applications for whole-core magnetic susceptibility measurements of deep-sea sediments: Leg 115 results. *Proceedings of the Ocean Drilling Program, Scientific Results*, **115**. Ocean Drilling Program, Texas A & M University, College Station, TX, 737–772.

ROYER, J.-Y. & COFFIN, M. F. 1992. Jurassic to Eocene plate tectonic reconstructions in the Kerguelen Plateau Region. *In*: WISE, S. W. JR., JULSON, A. A. P., SCHLICH, R. & THOMAS, E. (eds) *Proceedings of the Ocean Drilling Program, Scientific Results*, **120**. Ocean Drilling Program, Texas A & M University, College Station, TX, 917–928.

SCHNEIDER, D. A. & KENT, D. V. 1990*a*. Paleomagnetism of Leg 115 sediments: implications for Neogene magnetostratigraphy and paleolatitude of the Reunion Hotspot. *Proceedings of the Ocean Drilling Program, Scientific Results*, **115**. Ocean Drilling Program, Texas A & M University, College Station, TX, 717–736.

SCHNEIDER, D. A. & KENT, D. V. 1990*b*. Testing models of the Tertiary paleomagnetic field. *Earth and Planetary Science Letters*, **101**, 260–271.

SEXTON, P. F., NORRIS, R. D. *ET AL.* 2011. Eocene global warming events driven by ventilation of oceanic dissolved organic carbon. *Nature*, **471**, 349–353.

SHACKLETON, N. J. & KENNETT, J. P. 1975. *Paleotemperature history of the Cenozoic and the initiation of Antarctic glaciation: oxygen and carbon isotope analyses in DSDP sites 277, 279, and 281*. Initial Report of Deep Sea Drilling Project, **29**, 743–755.

SPOFFORTH, D. J. A., AGNINI, C. *ET AL.* 2010. Organic carbon burial following the middle Eocene climatic optimum in the central western Tethys. *Paleoceanography*, **25**, PA3210.

TOUCHARD, Y., ROCHETTE, P., AUBRY, M. P. & MICHARD, A. 2003. High-resolution magnetostratigraphic and biostratigraphic study of Ethiopian traps-related products in Oligocene sediments from the Indian Ocean. *Earth and Planetary Science Letters*, **206**, 293–508.

Tripati, A., Backman, J., Elderfield, H. & Ferretti, P. 2005. Eocene bipolar glaciation associated with global carbon cycle changes. *Nature*, **436**, 341–346.

Wei, W., Villa, J. & Wise, S. W. Jr. 1992. Paleoceanographic implications of Eocene–Oligocene calcareous nannofossils from sites 711 and 748 in the Indian Ocean. *Proceedings of the Ocean Drilling Program, 120*. Ocean Drilling Program, Texas A & M University, College Station, TX, 979–999.

Zachos, J. C., Rea, D. K., Seto, K., Nomura, R. & Niitsuma, N. 1992. *Paleogene and Early Neogene Deep Water Paleoceanography of the Indian Ocean as Determined from Benthic Foraminifer Stable Carbon and Oxygen Isotope Records.* Synthesis of Results from Scientific Drilling in the Indian Ocean, Geophysical Monographs, **70**, 351–385.

Zachos, J. C., Pagani, M., Sloan, L., Thomas, E. & Billups, K. 2001. Trends, rhythms, and aberrations in global climate 65 Ma to present. *Science*, **292**, 686–693.

Zachos, J. C., Dickens, G. R. & Zeebe, R. E. 2008. An early Cenozoic perspective on greenhouse warming and carbon-cycle dynamics. *Nature*, **451**, 279–283.

Zijderveld, J. D. A. 1967. A.C. demagnetization of rocks. *In*: Collison, D. W., Creer, K. M. & Runcorn, S. K. (eds) *Methods in Palaeomagnetism*, Elsevier, Amsterdam, 256–286.

Integrated stratigraphy (magneto-, bio- and chronostratigraphy) and geochronology of the Palaeogene pelagic succession of the Umbria–Marche Basin (central Italy)

RODOLFO COCCIONI[1]*, MARIANNA SIDERI[1], GIUSEPPE BANCALÀ[1], RITA CATANZARITI[2], FABRIZIO FRONTALINI[1], LUIGI JOVANE[3], ALESSANDRO MONTANARI[4] & JAIRO SAVIAN[5]

[1]*Dipartimento di Scienze della Terra, della Vita e dell'Ambiente, Università degli Studi di Urbino 'Carlo Bo', Campus Scientifico 'E. Mattei', Località Crocicchia, 61029 Urbino, Italy*

[2]*Istituto Geoscienze e Georisorse CNR, 56124 Pisa, Italy*

[3]*Instituto Oceanográfico, Universidade de São Paulo, 05508–090 São Paulo, Brazil*

[4]*Osservatorio Geologico di Coldigioco, 62021 Apiro, Italy*

[5]*Departamento de Geofísica, Instituto de Astronomia, Geofísica e Ciências Atmosféricas, Universidade de São Paulo, 05508–090 São Paulo, Brazil*

**Corresponding author (e-mail: rodolfo.coccioni@uniurb.it)*

Abstract: Extensive outcrops in the Umbria–Marche Basin of central Italy include some of the most complete successions of Palaeogene sediments known from the Tethyan Realm. Owing to the continuous deposition in a pelagic setting, a rather modest tectonic overprint, the availability of excellent age control through magneto-, bio-, chemo- and tephrostratigraphy, and direct radioisotopic dates from interbedded volcaniclastic layers, these sediments have played a prominent role in the establishment of standard Palaeogene time scales. We present here a complete and well-preserved Palaeogene pelagic composite succession of the Umbria–Marche Basin, which provides the means for an accurate and precise calibration of the Palaeogene time scale. As a necessary step towards the compilation of a more robust database on a wide scale so as to improve the magneto-, bio- and chronostratigraphic framework of the classical southern Tethyan zonations, enabling regional and supraregional correlations, we have constructed a record of reliable Palaeogene planktonic foraminifera, calcareous nannofossil and dinocyst biohorizons commonly used in tropical to subtropical Cenozoic zonations. In addition, an age model is provided for the Palaeogene pelagic composite succession based on magnetostratigraphy, planktonic foraminifera and calcareous nannofossils, which contributes to an integrated chronology for the Palaeogene Tethyan sediments from *c.* 65.5 to 23 Ma.

Supplementary material: Tables 1 to 13 which provide further details of the Palaeogene pelagic succession of the Umbria–Marche Basin (central Italy) are available at http://www.geolsoc.org.uk/SUP18539

The Palaeogene carbonate succession located in the Umbria–Marche (U–M) Basin in the northeastern Apennines of Italy is probably the most thoroughly studied in the Tethyan Realm, thanks to the accessibility of numerous well-exposed outcrops throughout this region, and its remarkable record of many crucial aspects of Earth's history from the Cretaceous–Palaeogene (K–Pg) boundary through to the uppermost Oligocene (Coccioni *et al.* 1986, 1988, 1989, 1994, 2000, 2008, 2009, 2010, 2012; Cresta *et al.* 1989; Bellagamba & Coccioni 1990; Mattias *et al.* 1992; Brinkhuis & Biffi 1993; Premoli Silva & Jenkins 1993; Lowrie & Lanci 1994; Montanari *et al.* 1994, 1997; Galeotti *et al.* 2000, 2004, 2010; Montanari & Koeberl 2000; Spezzaferri *et al.* 2002; Coccioni & Galeotti 2003; Bodiselitsch *et al.* 2004; Jovane *et al.* 2004, 2006, 2007a, b, 2009, 2010; Brown *et al.* 2009; Giusberti *et al.* 2009; Hyland *et al.* 2009; Pross *et al.* 2010, and references therein).

We present here a complete U–M Palaeogene pelagic composite succession (PPCS) that we constructed from six well-exposed and thoroughly studied sections magnetobiostratigraphically calibrated and radioisotopically dated from several interbedded volcanosedimentary layers, which provides an improved integrated chronostratigraphy for the Palaeogene time scale, from the K–Pg boundary at 65.5 Ma to the uppermost Oligocene at 23 Ma.

From: JOVANE, L., HERRERO-BERVERA, E., HINNOV, L. A. & HOUSEN, B. A. (eds) 2013. *Magnetic Methods and the Timing of Geological Processes*. Geological Society, London, Special Publications, **373**, 111–131.
First published online October 16, 2012, http://dx.doi.org/10.1144/SP373.4 © The Geological Society of London 2013.
Publishing disclaimer: www.geolsoc.org.uk/pub_ethics

Geological and stratigraphic setting

The U–M Apennines is a foreland fold-and-thrust belt (Chiarabba *et al.* 2005), which was formed in the latest phase of the Alpine–Himalayan orogenesis. These mountains are entirely made from marine sedimentary rocks of the so-called U–M succession, which represent a continuous record of the geotectonic evolution of an epeiric sea from the Late Triassic to the Pleistocene. During the Late Triassic and Early Jurassic, at the time of the opening of the North Atlantic, the rifting between Europe and Africa took place within a formerly continuous continental crust at the southern margin of Europe (Centamore *et al.* 1980; Bosellini 1989). This rifting formed oceanic ancestral basins, which are now preserved in the present Alpine mountain chains, including the Pennine–Liguride Ocean. This new ocean, perhaps connected to part of the ancient Tethyan Ocean to the east, outlined a northward-pointing promontory of African continental crust, commonly referred to as Adria or the Adriatic Promontory (Channell *et al.* 1979). The Adriatic Promontory was isolated from inputs of clastic sediments. As a large, and nearly isolated, passive continental margin, Adriatic Promontory underwent extensional faulting. Normal faults defined a complex of subsiding blocks leading to an irregular topography of structural highs (horsts) and adjacent depocentres (grabens and/or half-grabens). Where shallow water carbonate deposition could keep up with subsidence, and where the faulted blocks were large enough to support productive carbonate platform environments, very thick sequences of shallow water carbonates developed on the Adria's crust. In other regions, subsidence and complex block faulting carried the seafloor down, below the photic zone and out of the zone of shallow-water carbonate production. Areas with this history became pelagic basins, such as the U–M Basin, in which the Palaeogene carbonate succession was deposited recording the geologic, biologic and oceanographic evolution of this region, with remarkable continuity and completeness.

The Palaeogene portion of the U–M pelagic succession is represented by three distinct formations (from bottom to top): Scaglia Rossa (early Turonian to Middle Eocene), Scaglia Variegata (Middle to late Eocene) and Scaglia Cinerea (late Eocene to earliest Miocene). The Italian terms 'Rossa', 'Variegata' and 'Cinerea' refer to the red, variegated and ashy colour, respectively, and 'Scaglia' to the scaly, conchoidal fracture typical of these thinly bedded rocks. The Palaeogene Scaglia Rossa corresponds to the R3 and R4 members of Alvarez and Montanari (1988) and is made up of well-bedded, pink to red limestones with red marly interbeds in the Middle–Lower Danian, at the Selandian–Thanetian

transition and from the Upper Thanetian to Middle Ypresian. Radiolarian chert occurs from the Middle Ypresian to the Lower Lutetian portion of this formation. The Palaeogene Scaglia Rossa varies in thickness between 80 and 100 m. Beginning in the Middle Eocene, terrigenous clay input became increasingly important. The Scaglia Variegata represents a transitional interval between the Scaglia Rossa and Scaglia Cinerea and consists of an alternation of white, greenish-grey and pink to red marly limestones and calcareous marls. Its lower boundary is marked by the last occurrence of nodular chert (Lowrie *et al.* 1982). The top of the Scaglia Variegata is conventionally placed at the top of the uppermost reddish interval (Monaco *et al.* 1987; Coccioni *et al.* 1988: Odin & Montanari 1988). The Scaglia Variegata varies from 30 and 95 m in thickness. The Scaglia Cinerea consists mainly of calcareous marls and marls and subordinate marly limestones greenish-grey and grey in colour. This unit varies from 100 to 200 m in thickness and its top is set at the base of the Raffaello Level (Coccioni *et al.* 1989, 1994), a volcanoclastic bentonite that represents a regional marker bed easily recognizable in the field, which marks the base of the overlying early to Middle Miocene Bisciaro Formation. Biotite-rich layers emplaced as air-fall volcanic ash are found at different stratigraphic levels in the uppermost Scaglia Variegata and throughout the Scaglia Cinerea.

The Palaeogene pelagic sediments the U–M Basin were deposited well above the CCD at Middle to lower bathyal depths (1000–1500 m; Galeotti *et al.* 2004; Jovane *et al.* 2007*a*; Giusberti *et al.* 2009; Coccioni *et al.* 2010) and at *c.* 30°N palaeolatitude (Jovane *et al.* 2007*a*).

Materials and methods

This study focuses on six key sections and some distinct portions of them from three different localities (Contessa, Monte Cagnero and Massignano; Figs 1–7), which we used to construct, on the basis of magnetostratigraphic correlation, a complete and well-preserved stratigraphic composite section spanning the whole Palaeogene Period (Fig. 8).

In the Contessa Valley, which cuts a NE-dipping, half-anticlinal structure a few kilometres NW of the Medieval city of Gubbio, the Scaglia Rossa, Scaglia Variegata and Scaglia Cinerea are exposed in continuous fresh road-cuts and quarry faces several metres high. There, we have taken into consideration three discrete portions of the Paleocene–Middle Eocene Contessa Highway section together with the Paleocene–Lower Eocene Contessa Road, the Upper Oligocene Contessa Barbetti Road and the Upper Oligocene–Lower Miocene Contessa

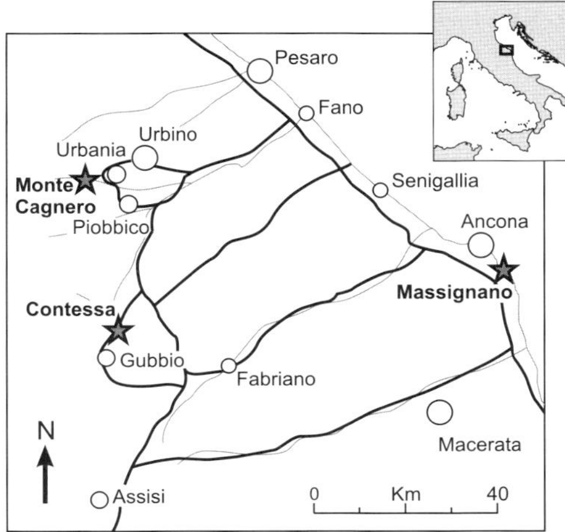

Fig. 1. Location map of the Contessa, Monte Cagnero and Massignano sections.

Quarry sections (Lowrie *et al.* 1982; Cresta *et al.* 1989; Montanari *et al.* 1997; Galeotti *et al.* 2000, 2004, 2010; Jovane *et al.* 2007*a*, 2010; Giusberti *et al.* 2009; Coccioni *et al.* 2010, 2012; Pross *et al.* 2010). At the Middle Eocene–Lower Oligocene Monte Cagnero section, which belongs to the Monte Montiego anticline and is located near Piobbico, three discrete portions have been selected (Parisi *et al.* 1988; Verducci & Nocchi 2004; Coccioni *et al.* 2008; Hyland *et al.* 2009; Pross *et al.* 2010; Jovane *et al.* 2012). Finally, we have considered the Upper Eocene–Lower Oligocene Massignano section, presently the official Global Stratotype Section and Point (GSSP) for the Eocene–Oligocene boundary, which is located on the west limb of the Monte Conero anticline, 10 km south of the city of Ancona (Bice & Montanari 1988; Coccioni *et al.* 1988, 2000, 2009; Brinkhuis & Biffi 1993; Premoli Silva & Jenkins 1993; Spezzaferri *et al.* 2002; Jovane *et al.* 2006, 2007*b*, 2009; Brown *et al.* 2009).

Biostratigraphic determinations are based on planktonic foraminifera, calcareous nannofossils and dinoflagellate cysts (here referred to as dinocyst), which are the main components of the microfossil assemblages in the Palaeogene pelagic sediments of the U–M Basin. In this study the planktonic foraminiferal standard zonations of Berggren *et al.* (1995), Berggren and Pearson (2005, 2006) and Wade *et al.* (2011), the calcareous nannofossil standard zonations of Martini (1971) and Okada and Bukry (1980), and the dinoflagellate cysts standard zonations of Brinkhuis and Biffi (1993) and Pross *et al.* (2010) were applied. The recognized bioevents, including some that were never reported in the U–M Basin in previous studies, and standard biozones are correlated to the magnetostratigraphy

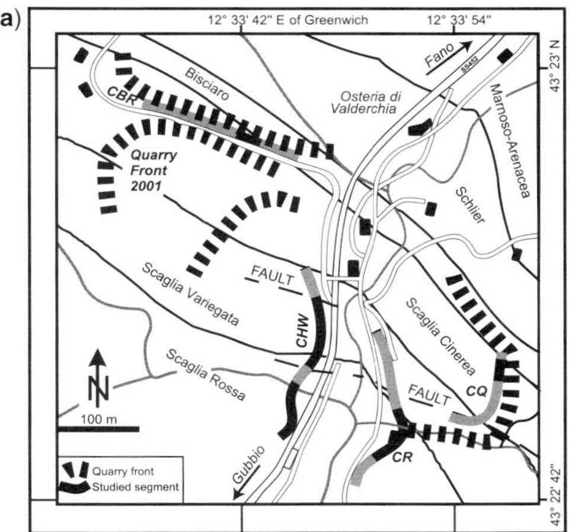

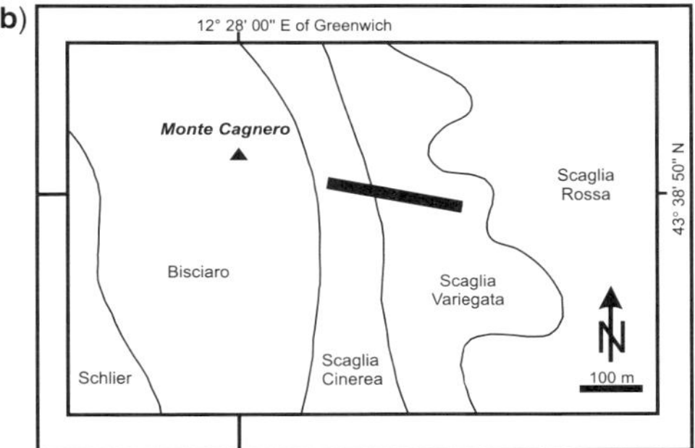

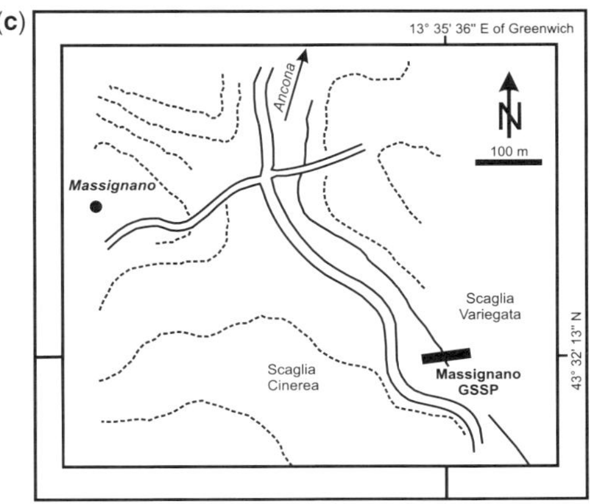

Fig. 2. (**a**) Location map of the Contessa sections (CHW, Contessa Highway; CR, Contessa Road; CBR, Contessa Barbetti Road; CQ, Contessa Quarry) on a simplified map that gives the formation boundaries in the Contessa Valley. (**b**) Location map of the Monte Cagnero section on a simplified geological map that gives the formation boundaries. (**c**) Location map of the Massignano section on a simplified geological map that gives the formation boundaries.

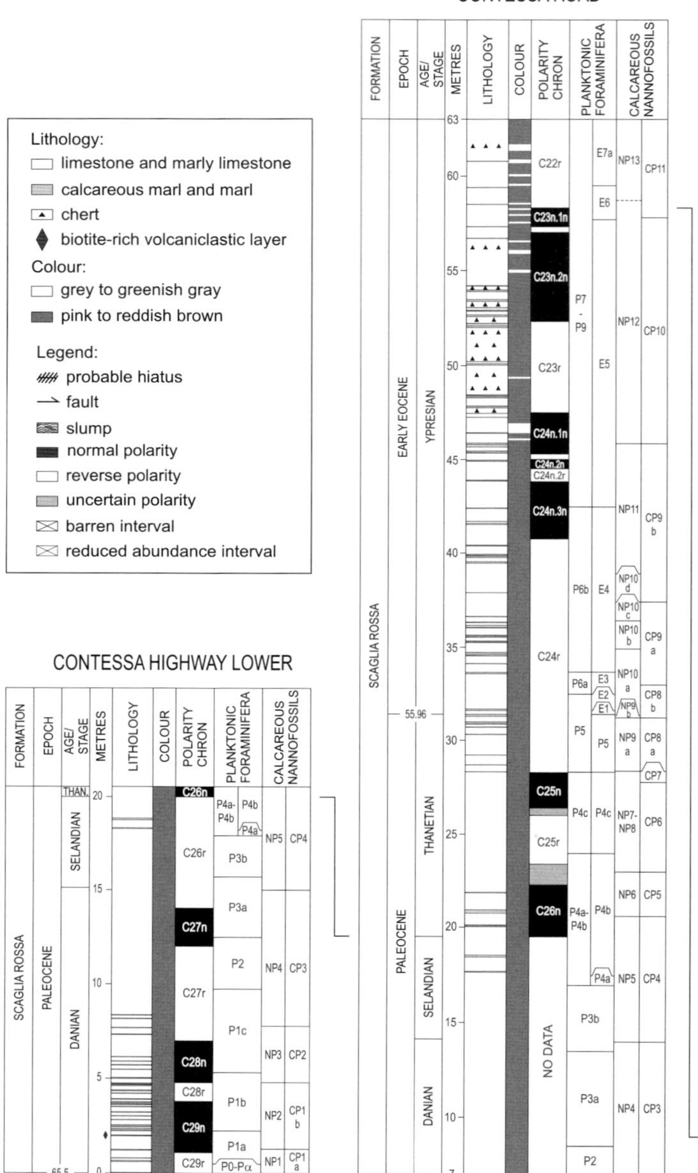

Fig. 3. Integrated stratigraphy of the Contessa Highway Lower and Contessa Road sections. Chronostratigraphy is after Ogg *et al.* (2008). Correlation between these sections is tied to the base of Chron C26n. Planktonic foraminiferal Zones and Subzones are after Berggren *et al.* (1995), Berggren and Pearson (2005) and Wade *et al.* (2011) (codified as P). Calcareous nannofossil Zones and Subzones are after Martini (1971) (codified as NP) and Okada and Bukry (1980) (codified as CP).

as it was determined in the selected sections in previous studies (Figs 3–7).

The key sections along with their related magneto-, bio- and chronostratigraphic and geochronological data were used to construct a complete and well-preserved PPCS of the Umbria–Marche Basin (Fig. 8).

Age–depth models for the PPCS (Figs 9–13) were built with a line of correlation relying on magnetochron boundaries.

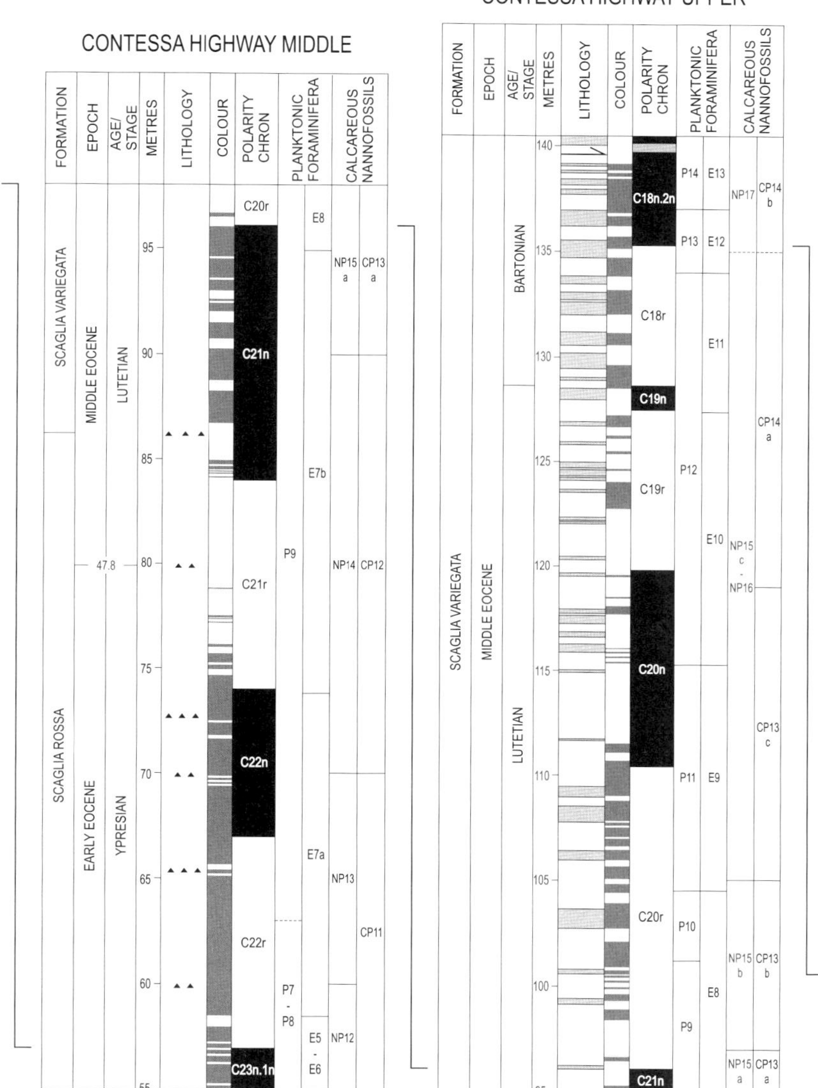

Fig. 4. Integrated stratigraphy of the Contessa Highway Middle and Contessa Highway Upper sections. Chronostratigraphy is after Molina *et al.* (2011), Ogg *et al.* (2008), Jovane *et al.* (2010), Ogg *et al.* (2008) and Molina *et al.* (2011). Correlation between Contessa Road and Contessa Highway Middle sections is tied to the top of Chron C23n.1n and that between Contessa Highway Middle and Contessa Highway Upper sections to the top of Chron C21n. Planktonic foraminiferal Zones and Subzones are after Berggren *et al.* (1995) (codified as P) and Berggren and Pearson (2006) and Wade *et al.* (2011) (codified as E). Calcareous nannofossil Zones and Subzones are after Martini (1971) (codified as NP) and Okada and Bukry (1980) (codified as CP). Legend as in Figure 3.

Results and discussion

The key sections

Contessa Highway Lower section. This section (lat. 43°22′47′″N; long. 12°33′45″E) starts at the K–Pg boundary, which is defined at metre level 0 (Figs 2a & 3). The Contessa Highway Lower section extends from the Danian to the lowermost Thanetian and is *c.* 21 m thick. It consists of well-bedded pink to red limestones with red marly interbeds in the

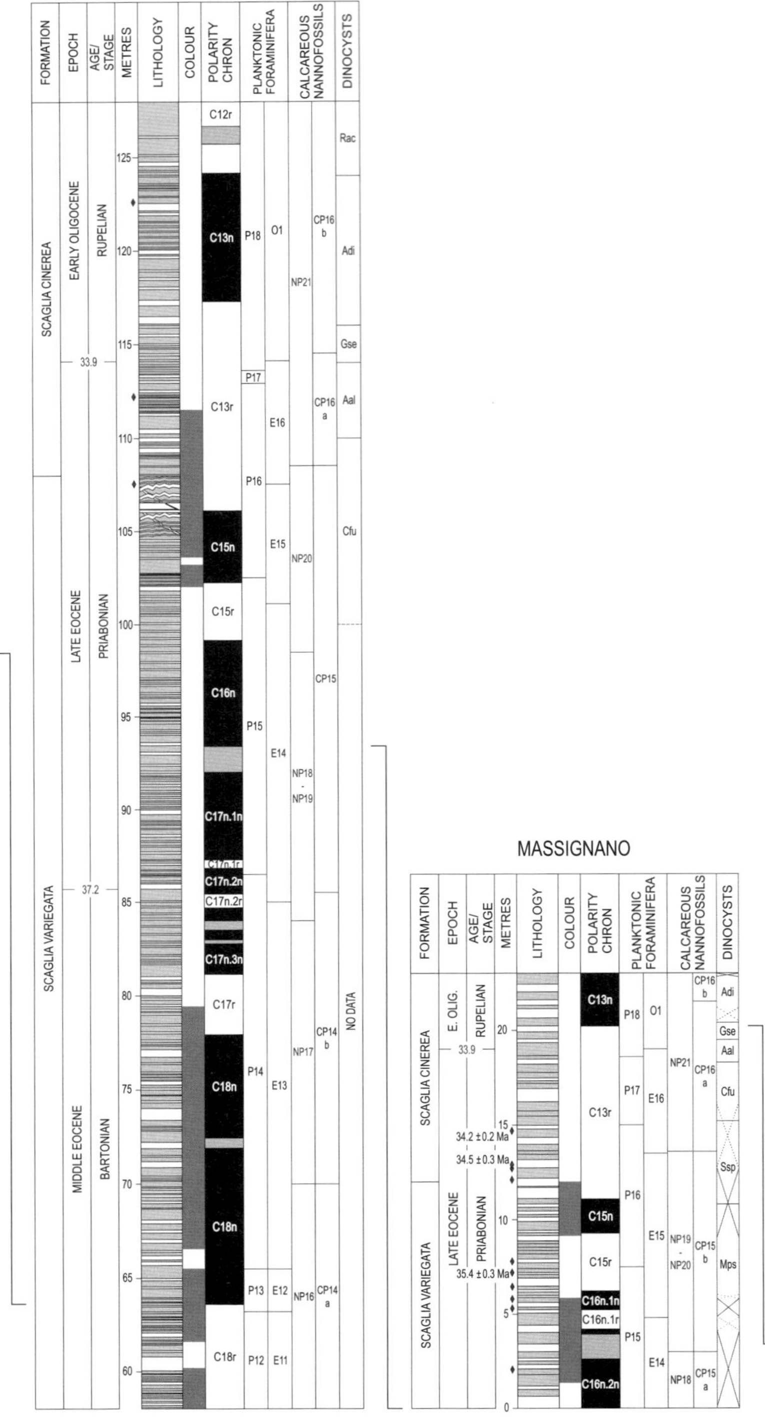
MONTE CAGNERO LOWER
MASSIGNANO

Middle–Lower Danian. A biotite-rich layer, the so-called Alessandro Level, is found at 1.98 m in the lowermost Danian.

Thirty-six reliable biohorizons were recognized, 14 of which are based on planktonic foraminifera and 22 on calcareous nannofossils.

Following Lowrie *et al.* (1982), Cresta *et al.* (1989), Fornaciari *et al.* (2007) and Coccioni *et al.* (2010), and according to our unpublished data, the Contessa Highway Lower section spans calcareous nannofossil Zones NP1–NP5 of Martini (1971) and CP1a–CP4 of Okada and Bukry (1980), planktonic foraminiferal Zones P0–Pα to P4a–P4b of Berggren *et al.* (1995) and Zones P0–Pα to P4b of Berggren and Pearson (2005) and Wade *et al.* (2011), and the upper part of Chron C29r to the lowermost part of Chron C26n (Fig. 3).

Contessa Road section. The base of this section (lat. 43°22′47‴ N; long. 12°33′50‴ E) is placed 7 m above the K–Pg boundary (Figs 2a & 3). The 56 m-thick Contessa Road extends from the Upper Danian to the Middle Ypresian. It consists of well-bedded pink to red, occasionally white, limestones and marly limestones with red marly interbeds at the Selandian–Thanetian transition and from the Upper Thanetian to the Middle Ypresian. Chert occurs in the Middle–Lower Ypresian from 48 m upwards.

Sixty reliable biohorizons were identified, 19 of which were based on planktonic foraminifera and 41 on calcareous nannofossils.

Following Lowrie *et al.* (1982), Aubry (1995), Galeotti *et al.* (2000, 2004, 2010), Raffi *et al.* (2005), Angori *et al.* (2007) and Giusberti *et al.* (2009), and according to Agnini *et al.* (2006) and our unpublished data, the Contessa Road section spans calcareous nannofossil Zones NP4–NP13 of Martini (1971) and CP3–CP11 of Okada and Bukry (1980), planktonic foraminiferal Zones P2 to P7–P9 of Berggren *et al.* (1995) and Berggren and Pearson (2005) and Zones E1–E7a of Berggren and Pearson (2006) and Wade *et al.* (2011), and the base of Chron C26n to the Middle part of Chron C22r (Fig. 3). The correlation between the Contessa Highway Lower section and the Contessa Road section is tied to the base of Chron C26n (Fig. 3).

Contessa Highway Middle section. This section (lat. 43°22′51‴N; long. 12°33′44″E) starts 55 m above the K–Pg boundary (Figs 2a & 4). The Contessa Highway Middle section extends from the Middle Ypresian to the Middle Lutetian and is 43 m thick. It consists of well-bedded pink to red and subordinately white and greenish-grey limestones and marly limestones. Chert occurs between 60 and 86 m, that is, from the Middle Ypresian to the Lower Lutetian interval. The uppermost cherty bed marks the base of the Scaglia Variegata.

Seven reliable biohorizons were recognized in this section, four of which were based on planktonic foraminifera and three on calcareous nannofossils.

Following Lowrie *et al.* (1982), Cresta *et al.* (1989) and Jovane *et al.* (2007*a*), and according to Agnini *et al.* (2006) and our unpublished data, the Contessa Highway Middle section spans calcareous nannofossil Zones and Subzones NP12–NP15a of Martini (1971) and CP11–CP13a of Okada and Bukry (1980), planktonic foraminiferal Zones P7–P8 to P9 of Berggren *et al.* (1995) and Zones E5–E6 to E8 of Berggren and Pearson (2006) and Wade *et al.* (2011), and Chron C23n.1n to lowermost part of Chron C20r (Fig. 4). The correlation between the Contessa Road section and the Contessa Highway Middle section is tied to the top of Chron C23n.1n (Figs 3 & 4).

Contessa Highway Upper section. The Contessa Highway Upper section (lat. 43°22′51″N; long. 12°33′44″E), which has been proposed by Jovane *et al.* (2010) as GSSP for the Lutetian–Bartonian boundary, extends from the Middle Lutetian to the Lower Bartonian and is 45 m thick (Figs 2a & 4). The base of this section is placed 95 m above the K–Pg boundary. The Contessa Highway Upper section consists of well-bedded white, greenish-grey and pink to red marly limestones and calcareous marls. A normal fault interrupts the section at its top.

Twenty-one reliable biohorizons were identified, 12 of which are based on planktonic foraminifera and nine on calcareous nannofossils.

Following Lowrie *et al.* (1982), Cresta *et al.* (1989) and Jovane *et al.* (2007*a*, 2010), and according to Aubry (1991) and our unpublished data, the Contessa Highway Upper section spans calcareous nannofossil Zones and Subzones NP15a–NP17 of Martini (1971) and CP13a–CP14b of Okada and Bukry (1980), planktonic foraminiferal Zones P9–P14 of Berggren *et al.* (1995) and Zones E8–E13 of Berggren and Pearson (2006) and Wade *et al.*

Fig. 5. Integrated stratigraphy of the Monte Cagnero Lower and Massignano sections. Chronostratigraphy is after Ogg *et al.* (2008). Correlation between Contessa Highway Upper and Monte Cagnero Lower sections is tied to the base of Chron C18n and that between Monte Cagnero Lower and Massignano sections to the base of Chron C16n. Planktonic foraminiferal Zones and Subzones are after Berggren *et al.* (1995) (codified as P) and Berggren and Pearson (2006) and Wade *et al.* (2011) (codified as E and O). Calcareous nannofossil Zones and Subzones are after Martini (1971) (codified as NP) and Okada and Bukry (1980) (codified as CP). Legend as in Figure 3.

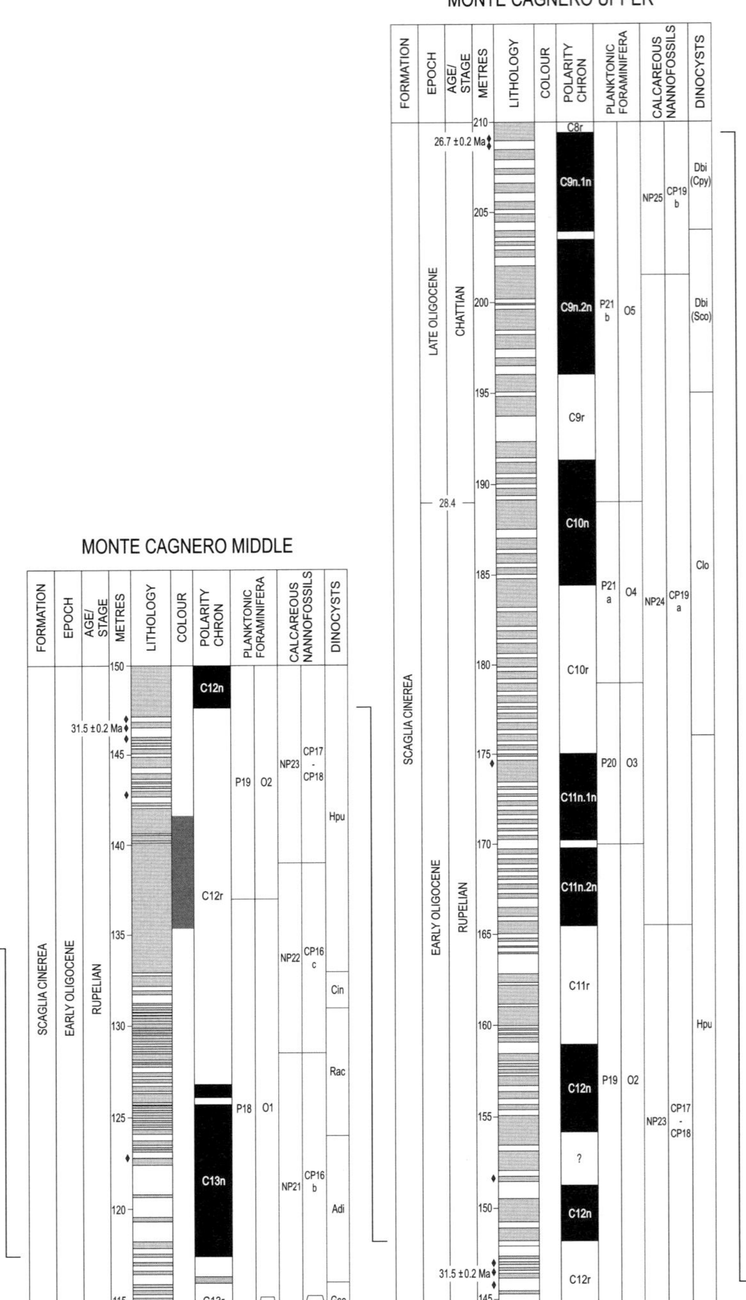

Fig. 6. Integrated stratigraphy of the Monte Cagnero Middle and Monte Cagnero Upper sections. Chronostratigraphy is after Coccioni *et al.* (2008) and Ogg *et al.* (2008). Correlation between Massignano and Monte Cagnero Middle sections is tied to the base of Chron C13n and that between Monte Cagnero Middle and Monte Cagnero Upper sections to the base of Chron C12n. Planktonic foraminiferal Zones and Subzones are after Berggren *et al.* (1995) (codified as P) and Berggren and Pearson (2006) and Wade *et al.* (2011) (codified as E and O). Calcareous nannofossil Zones and Subzones are after Martini (1971) (codified as NP) and Okada and Bukry (1980) (codified as CP). Legend as in Figure 3.

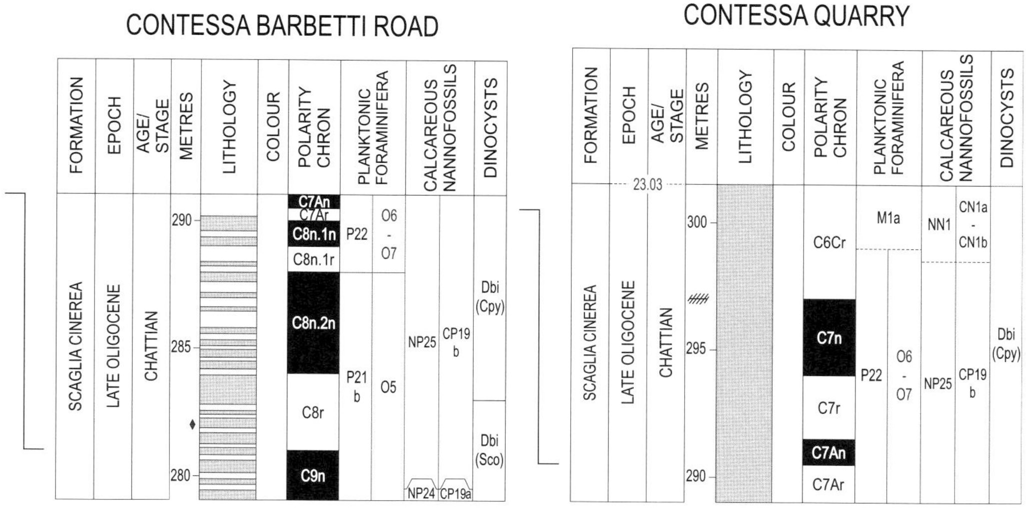

Fig. 7. Integrated stratigraphy of the Cava Barbetti Road and Contessa Quarry sections. Chronostratigraphy is after Ogg *et al.* (2008). Correlation between Monte Cagnero Upper and Cava Barbetti Road sections is tied to the base of Chron C8r and that between Cava Barbetti Road and Contessa Quarry sections to the base of Chron C7An. Planktonic foraminiferal Zones and Subzones are after Berggren *et al.* (1995) (codified as P and M) and Berggren and Pearson (2006) and Wade *et al.* (2011) (codified as O and M). Calcareous nannofossil Zones and Subzones are after Martini (1971) (codified as NP) and Okada and Bukry (1980) (codified as CP). Legend as in Figure 3.

(2011), and the uppermost part of Chron C21n to Chron C18n.2n (Fig. 4). The correlation between the Contessa Highway Middle section and the Contessa Highway Upper section is tied the top of Chron C21n (Fig. 4).

Monte Cagnero Lower section. The Monte Cagnero Lower section (lat. 43°38′51″N; long. 12°28′24″E) extends from the Lower Bartonian to the Lower Rupelian and is 70 m thick (Figs 2b & 5). It consists of calcareous marls and marls with subordinate marly limestones white, greenish-grey, grey and pink to red in colour. Discrete biotite-rich layers occur at 107.55, 112.2 and 122.8 m (Parisi *et al.* 1988; Hyland *et al.* 2009). The transition between the Scaglia Variegata and the Scaglia Cinerea contains deformations possibly related to soft sediment slumping and/or faulting.

Twenty reliable biohorizons were recognized, 10 of which are based on planktonic foraminifera, six on calcareous nannofossils and four on dinocysts.

Following Parisi *et al.* (1988), Brinkhuis and Biffi (1993), Verducci and Nocchi (2004), Hyland *et al.* (2009) and Jovane *et al.* (2012), the Monte Cagnero Lower section spans calcareous nannofossil Zones and Subzones NP16–NP21 of Martini (1971) and CP14a–CP16b of Okada and Bukry (1980), planktonic foraminiferal Zones P12–P18 of Berggren *et al.* (1995) and Zones E11 to O1 of Berggren and Pearson (2006) and Wade *et al.* (2011), dinocyst

Zones Cfu to Rac of Brinkhuis and Biffi (1993), and the middle part of Chron C18r to the lowermost part of Chron C12r (Fig. 5). The correlation between the Contessa Highway Upper section and the Monte Cagnero Lower section is tied to the base of Chron C18n (Figs 4 & 5).

Massignano section. The GSSP for the Eocene/Oligocene boundary at Massignano (lat. 43°32′13′N; long. 13°35′36′ E) extends from the Middle Priabonian to the lowermost Rupelian and is 23 m thick (Figs 2c & 5). It consists of calcareous marls and marls with subordinate marly limestones greenish-grey, grey and reddish in colour. The boundary between Scaglia Variegata and Scaglia Cinerea falls at 12 m. Discrete biotite-rich layers occur at 2.04, 5.25, 5.8, 6.5, 7.25, 7.75, 12.1, 12.7, 12.9 and 14.7 m. Biotite from layers at 7.25, 12.9 and 14.7 m provided ages of 35.4 ± 0.3, 34.5 ± 0.3 and 34.2 ± 0.2 Ma, respectively (Montanari *et al.* 1985; Odin *et al.* 1991; Oberli & Meier 1991).

Twenty-two reliable biohorizons were identified, nine of which were based on planktonic foraminifera, eight on calcareous nannofossils and five on dinocysts.

According to Bice and Montanari (1988), Coccioni *et al.* (1988, 2000, 2009), Brinkhuis and Biffi (1993), Premoli Silva and Jenkins (1993), Lowrie and Lanci (1994), Spezzaferri *et al.* (2002), Jovane *et al.* (2004, 2006, 2007*b*, 2009), Tori and Monechi

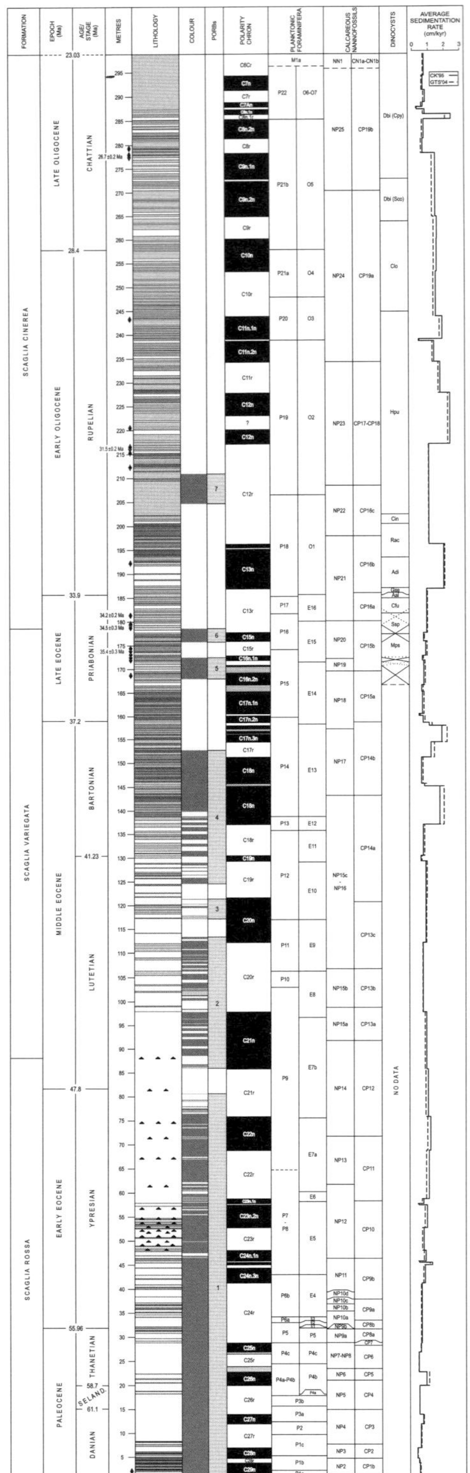

(2007) and Brown *et al.* (2009), the Massignano section spans calcareous nannofossil Zones and Subzones NP18–NP21 of Martini (1971) and CP15a–CP16b of Okada and Bukry (1980), planktonic foraminiferal Zones P15–P18 of Berggren *et al.* (1995) and Zones E14 to O1 of Berggren and Pearson (2006) and Wade *et al.* (2011), dinocyst Zones Mps to Adi of Brinkhuis and Biffi (1993) and Chron C16n.2n to the lower part Chron C13n (Fig. 5). The correlation between the Monte Cagnero Lower section and the Massignano section is tied to the base of Chron C16n (Fig. 5).

Monte Cagnero Middle section. The Monte Cagnero Middle section (lat. 43°38′51′N; long. 12°28′24′E) extends within the Lower to Middle Rupelian (Figs 2b & 6). For this study, we have re-measured the section previously studied by Coccioni *et al.* (2008) and Hyland *et al.* (2009) adding 1 m between 124 and 125 m. Accordingly, the Monte Cagnero Middle section is currently 36 m thick and consists of calcareous marls and marls with subordinate marly limestones greenish-grey, grey and reddish-violet in colour. Discrete biotite-rich layers occur at 122.8, 142.8, 145.8, 146.6 and 147 m. Volcaniclastic biotite at 146.6 m provided an age of 31.5 ± 0.2 Ma (Coccioni *et al.* 2008).

Fourteen reliable biohorizons were recognized, two of which were based on planktonic foramini-fera, eight on calcareous nannofossils and four on dinocysts.

According to Coccioni *et al.* (2008), Hyland *et al.* (2009) and Pross *et al.* (2010), the Monte Cagnero Middle section spans calcareous nannofossil Zones and Subzones NP21–NP23 of Martini (1971) and CP16a to CP17–CP18 of Okada and Bukry (1980), planktonic foraminiferal Zones P18–P19 of Berggren *et al.* (1995) and Zones E16 to O2 of Berggren and Pearson (2006) and Wade *et al.* (2011), dinocyst Zones Gse to Hpu of Brinkhuis and Biffi (1993), and the upper part of Chron C13r to the upper Chron C12n (Fig. 6). The correlation between the

Fig. 8. Integrated stratigraphy of the Palaeogene pelagic composite succession of the Umbria–Marche Basin together changes in the sedimentation rate according to the geomagnetic polarity time scales of Cande and Kent (1995) (CK′95) and Gradstein *et al.* (2004) (GTS′04). Chronostratigraphy is after Coccioni *et al.* (2008), Ogg *et al.* (2008), Jovane *et al.* (2010) and Molina *et al.* (2011). Planktonic foraminiferal Zones and Subzones are after Berggren *et al.* (1995) (codified as P and M) and Berggren and Pearson (2006) and Wade *et al.* (2011) (codified as O and M). Calcareous nannofossil Zones and Subzones are after Martini (1971) (codified as NP) and Okada and Bukry (1980) (codified as CP). Legend as in Figure 3. PORBs, Palaeogene oceanic red beds.

Massignano section and the Monte Cagnero Middle section is tied to the base of Chron C13n (Figs 5 & 6).

Monte Cagnero Upper section. This section (lat. 43°38′51′N; long. 12°28′24′E) has been proposed by Coccioni *et al.* (2008) as the GSSP for the Rupelian–Chattian boundary (Figs 2b & 6). According to the re-measurement of the underlying Monte Cagnero Middle section, the Monte Cagnero Upper section is currently 66 m thick and the Rupelian–Chattian boundary falls at 189 m instead of at 188 m as previously published by Coccioni *et al.* (2008) (Figs 2b & 6). This section consists of calcareous marls and marls with subordinate marly limestones greenish-grey and grey in colour. Discrete biotite-rich layers occur at 145.8, 146.6, 147, 151.65, 174.5, 208.7 and 209.1 m. Biotite from the layer at 146.6 m provided an age of 31.5 ± 0.2 Ma and those from layers at 208.7 and 209.1 m provided an age of 26.7 ± 0.2 Ma (Coccioni *et al.* 2008).

Twenty-one reliable biohorizons were identified, five of which were based on planktonic foraminifera, 10 on calcareous nannofossils and six on dinocysts.

Following Coccioni *et al.* (2008), Hyland *et al.* (2009) and Pross *et al.* (2010), the Monte Cagnero Upper section spans calcareous nannofossil Zones and Subzones NP23–NP25 of Martini (1971) and CP17–CP18 to CP19b of Okada and Bukry (1980), planktonic foraminiferal Zones and Subzones P19–P21b of Berggren *et al.* (1995) and Zones O2–O5 of Berggren and Pearson (2006) and Wade *et al.* (2011), dinocyst Zones and Subzones Hpu to Dbi (Cpy) of Pross *et al.* (2010), and the upper part of Chron C12r to the lowermost part of Chron C8r (Fig. 6). The correlation between the Monte Cagnero Middle section and the Monte Cagnero Upper section is tied to the base of Chron C12n (Fig. 6).

Contessa Barbetti Road section. The Contessa Barbetti Road section (lat. 43°23′01′N; long. 12°33′36′E) extends within the Upper Chattian and is 12 m thick (Figs 2a & 7). It consists of calcareous marls and marls with subordinate marly limestones, greenish-grey and grey in colour. A discrete biotite-rich layer occurs at 282 m (Coccioni *et al.* 2008).

Five reliable biohorizons were recognized, one of which was based on planktonic foraminifera, one on calcareous nannofossils and three on dinocysts.

According to Lowrie *et al.* (1982), Coccioni *et al.* (2008) and Pross *et al.* (2010), the Contessa Barbetti Road section spans calcareous nannofossil Zones and Subzones NP24–NP25 of Martini (1971) and CP19a–CP19b of Okada and Bukry (1980), planktonic foraminiferal Zones and Subzones P21b–P22 of Berggren *et al.* (1995) and Zones O5 to O6–O7 of Berggren and Pearson (2006) and Wade *et al.* (2011), dinocyst Zones and Subzones Dbi (Sco) to Dbi (Cpy) of Pross *et al.* (2010), and the upper part of Chron C9n to the middle of C7An (Fig. 7). The correlation between the Monte Cagnero Upper section and the Contessa Barbetti Road section is tied to the top of Chron C9n (Figs 6 & 7).

Contessa Quarry section. The Contessa Quarry section (lat. 43°22′50′N, long. 12°33′54′E) extends from the uppermost Chattian to the lowermost Aquitanian and is 12 m thick (Figs 2a & 7). It consists of greenish-grey and grey calcareous marls and marls.

Five reliable biohorizons were identified, one of which were based on planktonic foraminifera and four to calcareous nannofossils.

Following Lowrie *et al.* (1982), Montanari *et al.* (1997) and Pross *et al.* (2010), the Contessa Quarry section spans calcareous nannofossil Zones and Subzones NP25–NN1 of Martini (1971) and CP19b to CN1a–CN1b of Okada and Bukry (1980), planktonic foraminiferal Zones and Subzones P22 of Berggren *et al.* (1995) and Zones O6–O7 of Berggren and Pearson (2006) and Wade *et al.* (2011) to M1a of Berggren *et al.* (1995) and Wade *et al.* (2011), dinocyst Zone and Subzone Dbi (Cpy) of Pross *et al.* (2010) and Chron C7Ar to the top of Chron C6Cr (Fig. 7). The correlation between the Contessa Barbetti Road section and the Contessa Quarry section is tied to the base of Chron C7An (Fig. 7).

The PPCS of the Umbria–Marche Basin

The key sections along with their magneto-, bio- and chronostratigraphic and geochronological data were used to construct a 298.8 m-thick PPCS for the U–M Basin via correlations based on magnetostratigraphy (Fig. 8).

Magneto-, bio- and chronostratigraphy. Eighty-six reliable bioevents are identified through the succession, 40 of which were based on planktonic foraminifera, 35 on calcareous nannofossils and 11 on dinocysts.

Our study provides evidence that the U–M PPCS is complete and spans from the upper part of Chron C29r to the top of Chron C6Cr and from planktonic foraminiferal Zones P0–Pα to M1a and E1 to M1a, and calcareous nannofossil Zones and Subzones NP1 to NN1 and CP1a to CN1b (Fig. 8).

However, some bioevents used here to mark zonal boundaries occur at different stratigraphic levels with respect to the current standard magneto-,

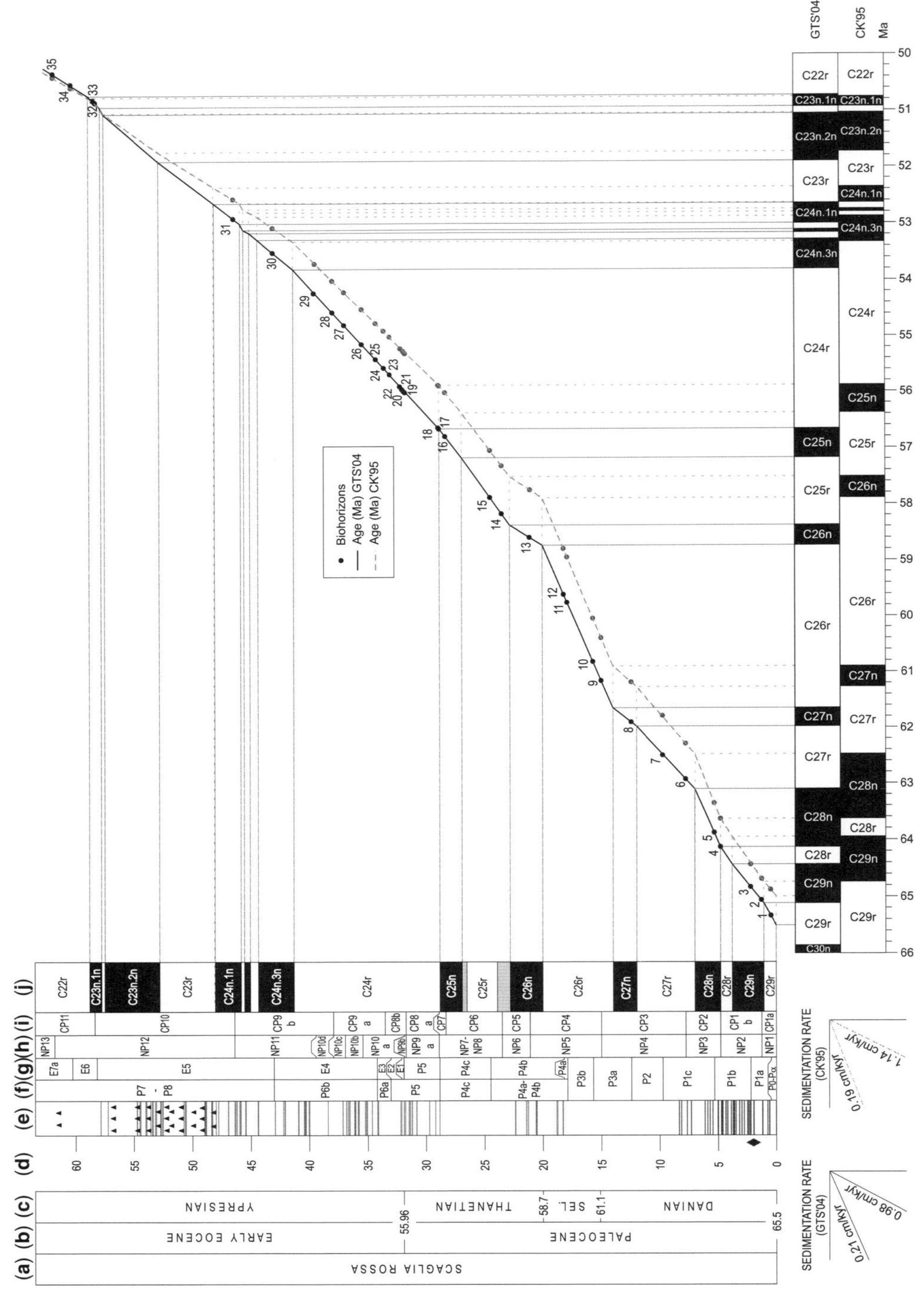

bio- and chronological time scales of Gradstein *et al.* (2004) and Ogg *et al.* (2008) and to the recent tropical Cenozoic planktonic foraminiferal zonation of Wade *et al.* (2011). In particular, the lowest occurrence (LO) of *Morozovella aragonensis* (P6b/P7 or E4/E5 zonal boundary), which has been placed by Wade *et al.* (2011) at the base of Chron C23r, is recognized in the middle of Chron C24n.3n in our U–M PPCS (i.e. *c.* 0.8 Ma older). The highest occurrence (HO) of this species (P11/P12 or E9/E10 zonal boundary) occurs higher in Chron C20n (*c.* 0.5 Ma younger). The HO of *Guembelitrioides nuttalli* (E10/E11 zonal boundary) is recognized higher in Chron C19r (*c.* 0.8 Ma younger). The HO of *Turborotalia ampliapertura* (P19/P20 or O3/O3 zonal boundary), which according to Wade *et al.* (2011) is placed in middle Chron C11r.2r, occurs in the middle of Chron C11r.1r (*c.* 0.65 Ma younger). The HO of *Paragloborotalia opima* (P21b/P22 or O5/O6 zonal boundary), which has been placed by Wade *et al.* (2011) in the middle of Chron C9n, is recognized at the top of Chron C8n.2n (*c.* 1.5 Ma younger). The LO of *Reticulofenestra umbilica* >14 μm (CP13c/CP14a zonal boundary) is recognized much higher in Chron C20n (*c.* 1.1 Ma younger). The LOs of *Chiasmolithus oamaruensis* (NP17/NP18 zonal boundary) and *Sphenolithus ciperoensis* (NP23/NP24 or CP18/CP19a zonal boundary) are lower in Chron C17n.2n (*c.* 0.7 Ma older) and in Chron C11n.2n (*c.* 0.2 Ma older), respectively.

The chronostratigraphy applied for the U–M PPCS follows Aubry *et al.* (2007), Coccioni *et al.* (2008), Ogg *et al.* (2008) and Jovane *et al.* (2010) and Molina *et al.* (2011).

Geochronology. Twenty-one biotite-rich layers emplaced as air fall volcanic ashes occur in our U–M PPCS, (Figs 5–7). As yet, K–Ar, Ar–Ar and Rb–Sr dates on pristine biotite separates have provided the geochronologic calibration of six discrete stratigraphic levels, contributing to the improvement of the accuracy of the Palaeogene time scale (Montanari *et al.* 1985; Odin & Montanari, 1988; Oberli & Meier 1991; Odin *et al.* 1991; Coccioni *et al.* 2008; Fig. 8).

They are as follows, from bottom to top:

- Level 1 – 173.9 m: uppermost part and lower part of planktonic foraminiferal Zones P15 and E15, respectively; lower part of calcareous nannofossil Zone NP20 and Subzone CP15b; lower part of Chron C15r: 35.4 ± 0.3 Ma.
- Level 2 – 179.6 m: middle part and uppermost part of planktonic foraminiferal Zones P16 and E15, respectively; uppermost part of calcareous nannofossil Zone NP20 and Subzone CP15b; lower part of Chron C13r: 34.5 ± 0.3 Ma.
- Level 3 – 181.4 m: uppermost part and lower part of planktonic foraminiferal Zones P16 and E16, respectively; lowermost part of calcareous nannofossil Zone NP21 and Subzone CP16a; middle part of Chron C13r; 34.2 ± 0.2 Ma.
- Level 4 – 216.1 m: lower part of planktonic foraminiferal Zones P19 and O2; lower part of calcareous nannofossil Zones NP23 and CP17–CP18; uppermost part of Chron C12r; 31.5 ± 0.2 Ma.
- Level 5 – 277.6 m: upper part of planktonic foraminiferal Subzone P21b and Zone O5; lower part of calcareous nannofossil Zone NP25 and Subzone CP19b; uppermost part of Chron C9n.1n; 26.7 ± 0.2 Ma.
- Level 6 – 278 m: upper part of planktonic foraminiferal Subzone P21b and Zone O5; lower part of calcareous nannofossil Zone NP25 and Subzone CP19b; uppermost part of Chron C9n.1n; 26.7 ± 0.2 Ma.

Fig. 9. Age–depth plot for the Paleocene–Lower Eocene of the Palaeogene pelagic composite succession of the Umbria–Marche Basin with correlation to the geomagnetic polarity times scales of Cande & Kent (1995) (CK′95) and Gradstein *et al.* 2004 (GTS′04). Chronostratigraphy is after Ogg *et al.* (2008). Legend: (**a**) formation; (**b**) epoch; (**c**) age/stage; (**d**) metres; (**e**) lithology; (**f**) planktonic foraminiferal Zones and Subzones of Berggren *et al.* (1995); (**g**) planktonic foraminiferal Zones and Subzones of Berggren & Pearson (2006) and Wade *et al.* (2011); (**h**) calcareous nannofossil Zones and Subzones of Martini (1971); (**i**) calcareous nannofossil Zones and Subzones of Okada & Bukry (1980); (**j**) magnetochron. Numbers correspond to planktonic foraminiferal and calcareous nannofossil biohorizons: (1) HO *Parvularugoglobigerina eugubina*; (2) LO *Cruciplacolithus intermedius*; (3) LO *Subbotina triloculinoides*; (4) LO *Chiasmolithus danicus*; (5) LO *Globanomalina compressa*; (6) LO *Ellipsolithus macellus*; (7) LO *Praemurica uncinata*; (8) LO *Morozovella angulata*; (9) LO *Fasciculithus tympaniformis*; (10) LO *Igorina albeari*; (11) LO *Globanomalina pseudomenardii*; (12) HO *Parasubbotina variospira*; (13) LO *Heliolithus kleimpellii*; (14) LO *Discoaster mohleri*; (15) LO *Acarinina soldadoensis*; (16) LO *Discoaster nobilis*; (17) HO *Globanomalina pseudomenardii*; (18) LO *Discoaster multiradiatus*; (19) LO Calcareous Nannofossil Extinction Taxa; (20) LO *Tribrachiatus bramlettei*; (21) LO *Acarinina sibaiyaensis*; (22) LO *Pseudohastigerina wilcoxensis*; (23) HO *Morozovella velascoensis*; (24) LO *Discoaster diastypus*; (25) LO *Morozovella formosa*; (26) LO *Tribrachiatus digitalis*; (27) HO *Tribrachiatus digitalis*; (28) LO *Tribrachiatus contortus*; (29) HO *Tribrachiatus contortus*; (30) LO *Morozovella aragonensis*; (31) LO *Discoaster lodoensis*; (32) HO *Morozovella subbotinae*; (33) LO *Crepidolithus crassus*; (34) LO *Acarinina cuneicamerata*; (35) HO *Tribrachiatus orthostylus*. The biohorizons are delineated as Lowest Occurrence (LO) and Highest Occurrence (HO).

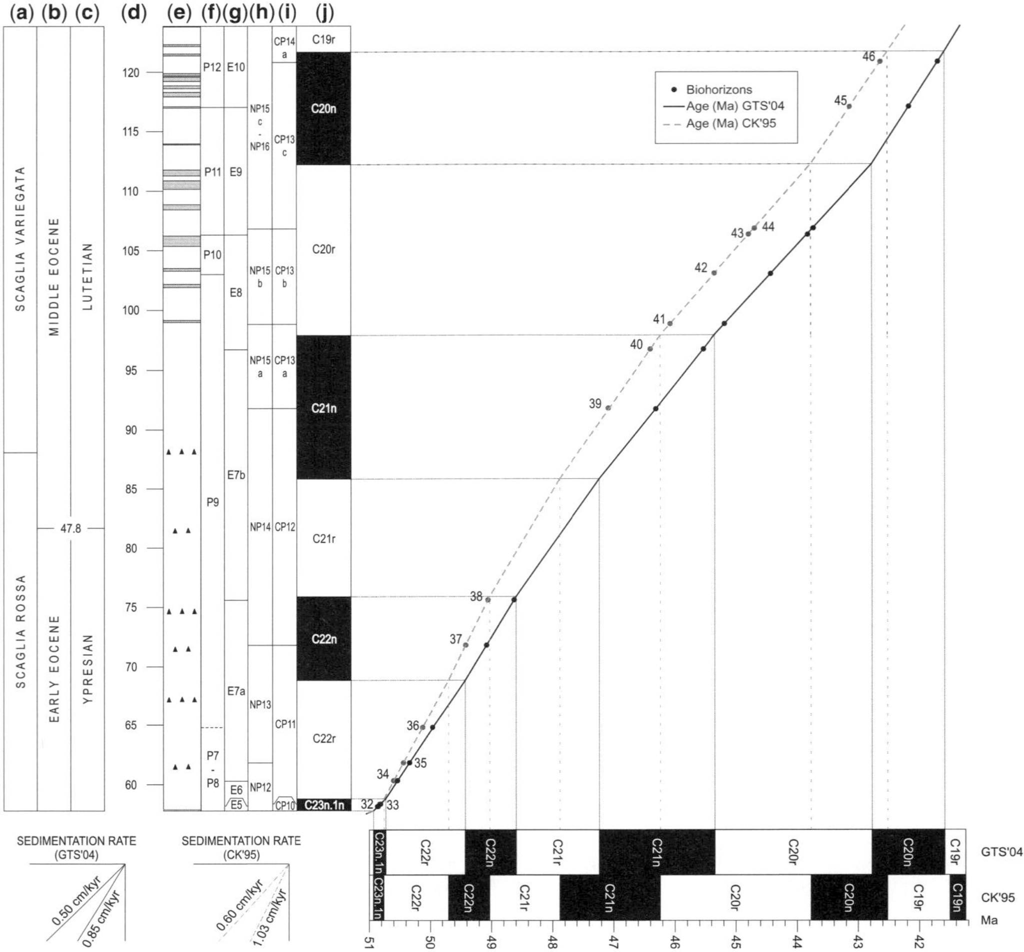

Fig. 10. Age–depth plot for the Lower–Middle Eocene of the Palaeogene pelagic composite succession of the Umbria–Marche Basin with correlation to the geomagnetic polarity times scales of Cande and Kent (1995) (CK′95) and Gradstein *et al.* 2004 (GTS′04). Chronostratigraphy is after Ogg *et al.* (2008) and Molina *et al.* (2011). Numbers correspond to planktonic foraminiferal and calcareous nannofossil biohorizons: (32) HO *Morozovella subbotinae*; (33) LO *Crepidolithus crassus*; (34) LO *Acarinina cuneicamerata*; (35) HO *Tribrachiatus orthostylus*; (36) LO *Planorotalites palmerae*; (37) *LO Discoaster sublodoensis*; (38) LO *Turborotalia frontosa*; (39) LO *Nannotetrina fulgens*; (40) LO *Guembelitrioides nuttalli*; (41) LO *Chiasmolithus gigas*; (42) LO *Hantkenina nuttalli*; (43) LO *Globigerinatheka kugleri*; (44) HO *Chiasmolithus gigas*; (45) HO *Morozovella aragonensis*; (46) LO *Reticulofenestra umbilica* >14μm. The biohorizons are delineated as Lowest Occurrence (LO) and Highest Occurrence (HO). Legend as in Figure 9.

Age model for the PPCS of the Umbria–Marche Basin. Our pelagic composite succession shows an excellent magnetostratigraphic and calcareous plankton biostratigraphical record for the Palaeogene time interval, allowing the construction of a western Tethyan reference age model. The age–depth model for the U–M PPCS and age assignment of biohorizons (Figs 9–13) were built with a line of correlation tied to the available magnetostratigraphy. The age model allows estimation of variations in sedimentation rates and provides tightly interpolated age assignments for 75 planktonic foraminiferal and calcareous nannofossil biohorizons, 40 of which were based on planktonic foraminifera and 35 on calcareous nannofossils (Figs 9–13). In addition, this reference age model can be used to estimate the time duration of calcareous nannofossil and planktonic foraminiferal Zones and Subzones.

Throughout the U–M PPCS, the mean sedimentation rate ranges from 0.19 cm/ka (Chron C28n,

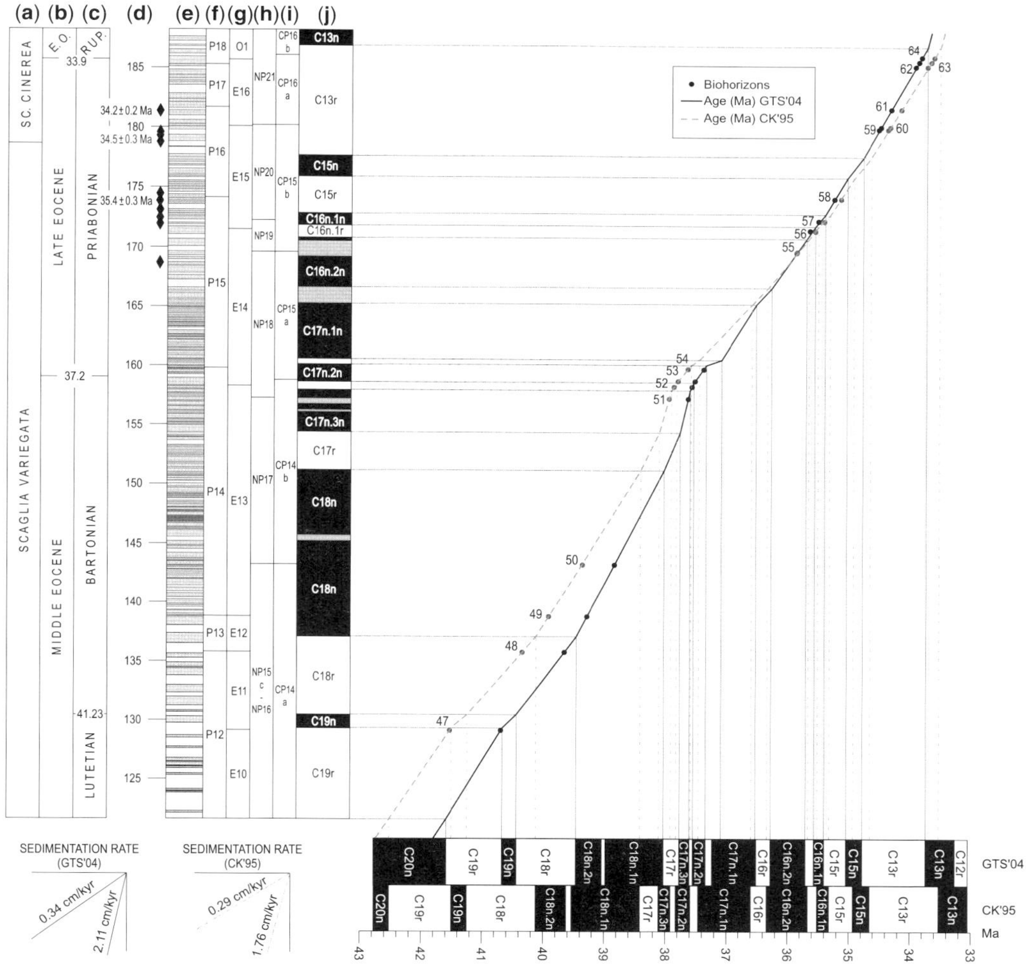

Fig. 11. Age–depth plot for the Middle–Upper Eocene of the Palaeogene pelagic composite succession of the Umbria–Marche Basin with correlation to the geomagnetic polarity times scales of Cande & Kent (1995) (CK'95) and Gradstein *et al.* 2004 (GTS'04). Chronostratigraphy is after Ogg *et al.* (2008) and Jovane *et al.* (2010). Legend as in Figure 9. Numbers correspond to planktonic foraminiferal and calcareous nannofossil biohorizons: (47) HO *Guembelitrioides nuttalli*; (48) LO *Orbulinoides beckmanni*; (49) HO *Orbulinoides beckmanni*; (50) HO *Chiasmolithus solitus*; (51) LO *Chiasmolithus oamaruensis*; (52) HO *Morozovelloides crassatus*; (53) HO *Chiasmolithus grandis*; (54) LO *Globigerinatheka semiinvoluta*; (55) LCO *Isthmolithus recurvus*; (56) HO *Globigerinatheka semiinvoluta*; (57) HO *Chiasmolithus oamaruensis*; (58) LO *Turborotalia cunialensis*; (59) HO *Globigerinatheka index*; (60) HO *Discoaster barbadiensis, Discoaster saipanensis*; (61) HO *Cribrohantkenina inflata*; (62) HO *Turborotalia cocoaensis, Turborotalia cunialensis*; (63) HO Hantkeninidae; (64) AB *Clausicoccus obrutus*. The biohorizons are delineated as Lowest Occurrence (LO), Highest Occurrence (HO), and Lowest Consistent Occurrence (LCO).

Lower Danian) to 2.44 cm/ka (Chron C8n.1r, Upper Chattian) and from 0.21 cm/ka (Chron C28n, Lower Danian) to 2.2 cm/ka (Chron C12n, Middle Rupelian) according to the polarity time scales of Cande and Kent (1995) and Gradstein *et al.* (2004), respectively. Important pulses of increased sedimentation rate took place during the Middle–Late Bartonian and early Rupelian and from the Middle Rupelian to the Middle Chattian (Figs 8–13).

The synthetic age model of calcareous microfossil biohorizons provides an excellent Palaeogene reference for the U–M Basin and may be extended to the whole southern margin of the Tethyan Realm, introducing new biostratigraphically useful

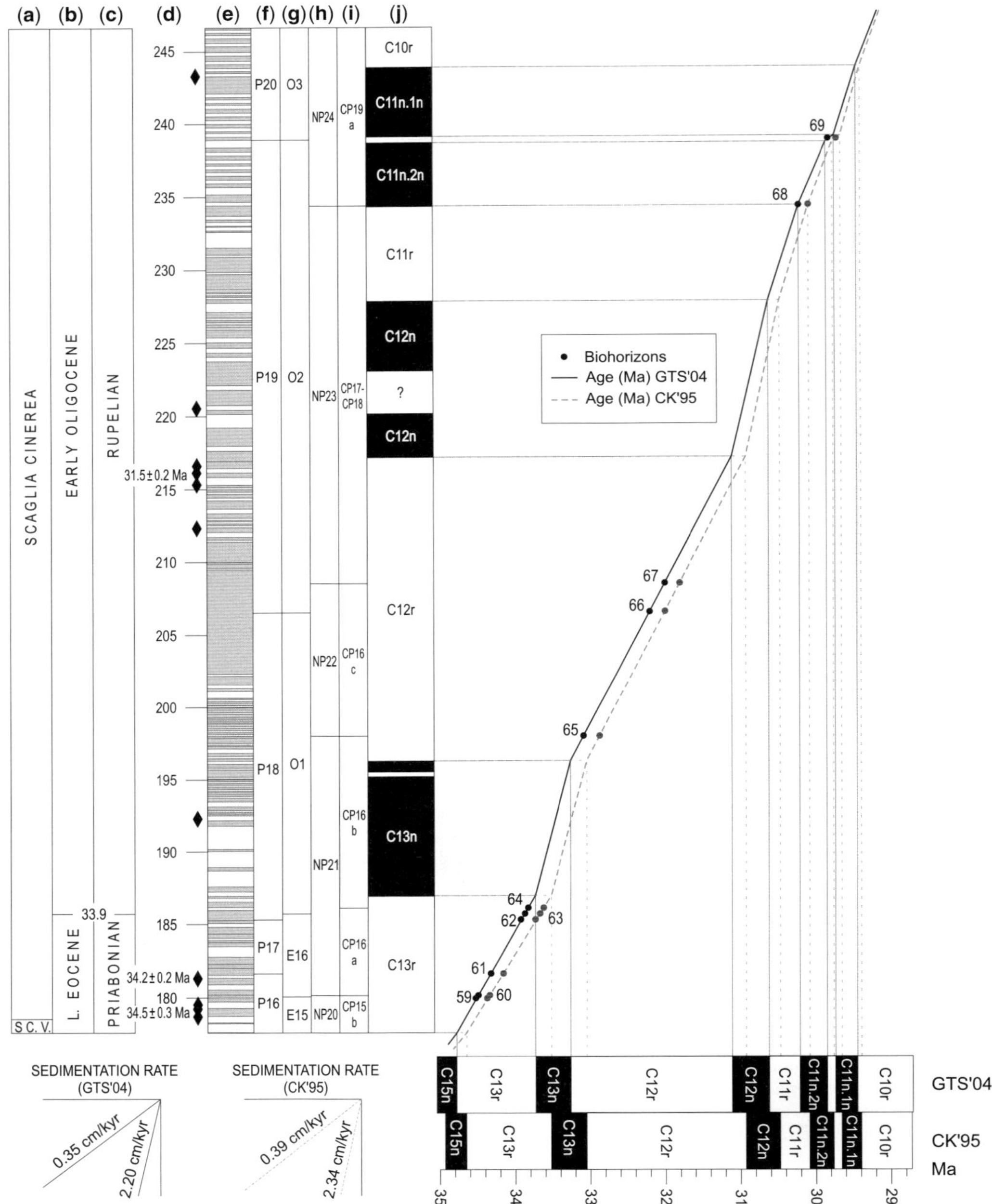

Fig. 12. Age–depth plot for the Upper Eocene–Lower Oligocene of the Palaeogene pelagic composite succession of the Umbria–Marche Basin with correlation to the geomagnetic polarity times scales of Cande & Kent (1995) (CK′95) and Gradstein *et al.* 2004 (GTS′04). Chronostratigraphy is after Ogg *et al.* (2008). Legend as in Figure 9. Numbers correspond to planktonic foraminiferal and calcareous nannofossil biohorizons: (59) HO *Globigerinatheka index*; (60) HO *Discoaster barbadiensis, Discoaster saipanensis*; (61) HO *Cribrohantkenina inflata*; (62) HO *Turborotalia cocoaensis, Turborotalia cunialensis*; (63) HO Hantkeninidae; (64) AB *Clausicoccus obrutus*; (65) HO *Ericsonia formosa*; (66) HO *Pseudohastigerina* spp., *Pseudohastigerina naguewichiensis*; (67) HO *Reticulofenestra umbilica*; (68) LO *Sphenolithus ciperoensis*; (69) HO *Turborotalia ampliapertura*. The biohorizons are delineated as Lowest Occurrence (LO), Highest Occurrence (HO), and Acme Beginning (AB).

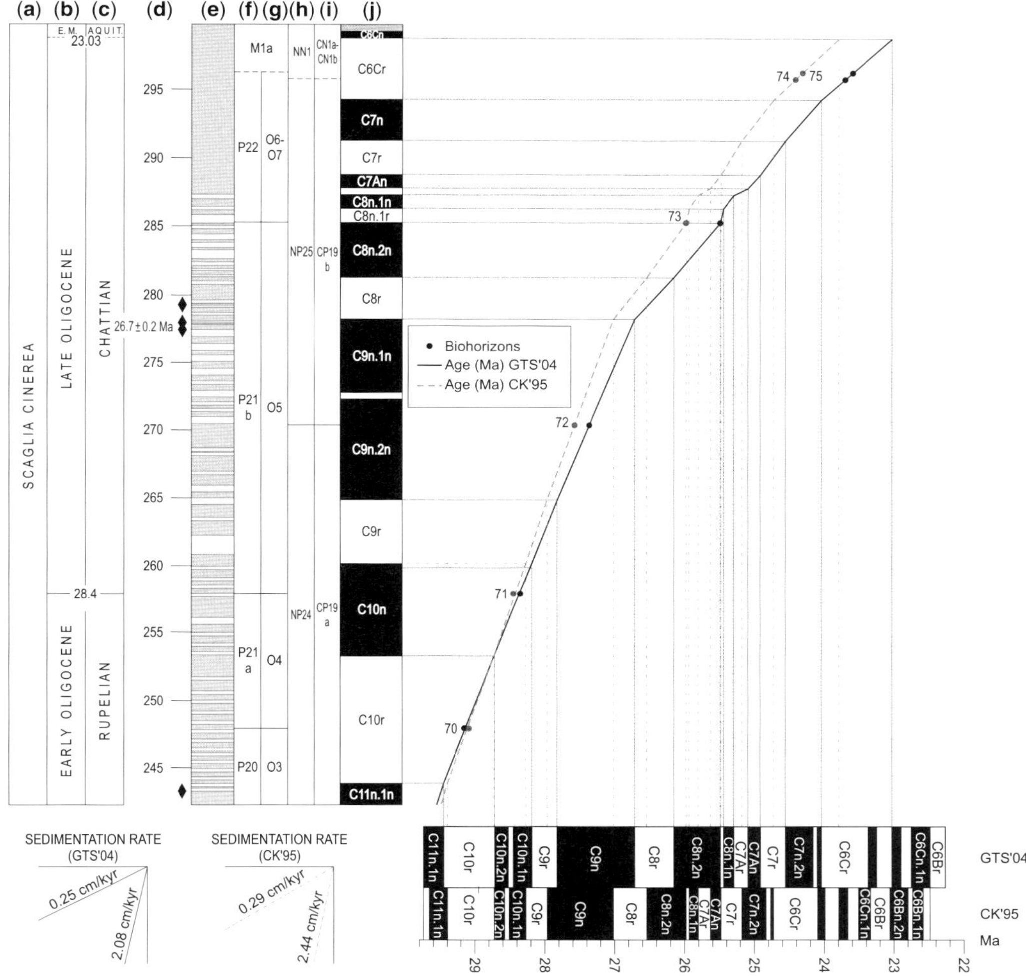

Fig. 13. Age-depth plot for the Lower Oligocene–lowermost Miocene of the Palaeogene pelagic composite succession of the Umbria–Marche Basin with correlation to the geomagnetic polarity times scales of Cande & Kent (1995) (CK'95) and Gradstein *et al.* 2004 (GTS'04). Chronostratigraphy is after Coccioni *et al.* (2008) and Ogg *et al.* (2008). Legend as in Figure 9. Numbers correspond to planktonic foraminiferal and calcareous nannofossil biohorizons: (70) LO *Globigerina angulisuturalis*; (71) HCO *Chiloguembelina cubensis*; (72) HCO *Sphenolithus distentus*; (73) HO *Paragloborotalia opima*; (74) HCO *Dictyococcites bisectus*; (75) LO *Paragloborotalia kugleri*. The biohorizons are delineated as Lowest Occurrence (LO), Highest Occurrence (HO), and Highest Consistent Occurrence (HCO).

biohorizons and testing their reproducibility within and beyond this region. However, application for the global scale or for the overall tropical to subtropical Realms should be cautioned owing to the evident diachroneity of several calcareous nannofossil and planktonic foraminiferal biohorizons among distinct oceanic basins and across latitudes for this time interval. The usefulness of all these biohorizons at the supra-regional and global scales as well as their synchroneity/diachroneity should be thoroughly tested and quantified by comparison with other reference sections, possibly with a good magnetostratigraphic record, with the help of other high-resolution stratigraphic tools such as chemostratigraphy and cyclostratigraphy. The difficult applicability of the available 'standard' calcareous nannofossil and planktonic foraminiferal biozonations for the Palaeogene time interval of the U–M

Basin calls for an in-depth re-evaluation of the use of global 'standard' v. the use of regional, well-correlated zonations.

Palaeogene Oceanic Red Beds. Seven reddish horizons are recognizable at different stratigraphic positions in our PPCS (Fig. 8), with the red colour arising from the early diagenetic oxidation of clastic iron particles (i.e. magnetite) leading to the authigenic formation of strongly pigmenting ferric oxide (haematite). On the analogy of the Cretaceous oceanic red beds (Hu *et al.*, 2005*a, b*; Neuhuber *et al.* 2007; Jansa & Hu 2009; Wang *et al.* 2009, 2011), these horizons are here referred to as Palaeogene oceanic red beds (PORBs). PORBs range from 2.8 (PORB 6) to 80.7 m (PORB 1) in thickness (Fig. 8). According to the mean sedimentation rate, their estimated duration varies from 400 ka to 16.2 Ma.

The lithological variability in the U–M PPCS may reflect a local response of the sedimentary system to the variation of local environmental conditions occurring in the context of significant palaeoclimatic and palaeoceanographical changes. Often primary carbonate productivity and terrigenous input changed in the U–M Basin and the resulting combination of oxygen demand in the sediments and sedimentation rates favoured the formation of authigenic haematite in the sediments which led to their reddening (Neuhuber *et al.* 2007; Cai *et al.* 2009; Hu *et al.* 2009; Trabucho-Alexandre *et al.* 2011). In this context, the change in sediment colour from white, greenish-grey and grey to red might reflect a change to more vigorous circulation, lower trophic levels and sedimentation rate, and better oxygenated bottom waters.

Concluding remarks

Our complete and well-preserved PPCS of the U–M Basin of central Italy represents the means for an accurate and precise calibration of the Palaeogene time scale. The record of Palaeogene planktonic foraminifera, calcareous nannofossil and dinocyst biohorizons identified in the composite sequence provides a refined, magnetostratigraphically calibrated age framework for future analyses of the Palaeogene in the western Tethys and a robust database on a wide scale. An age model is provided for the U–M PPCS based on magnetostratigraphy, planktonic foraminifera and calcareous nannofossils, which contributes to an integrated chronology for Palaeogene Tethyan sediments from *c.* 65.5 to 23 Ma. Ultimately, the temporally closely constrained biohorizons provided by our study may provide an improved basis for regional and supra-regional correlations and contribute to an improved understanding of the palaeoclimatic and palaeoceanographic dynamics of the Palaeogene world.

We thank R. Van der Voo, W. Lowrie and the Volume Editor E. Herrero-Bervera for their helpful comments.

References

AGNINI, C., MUTTONI, G., KENT, D. V. & RIO, D. 2006. Eocene biostratigraphy and magnetic stratigraphy from Possagno, Italy: the calcareous nannofossil response to climate variability. *Earth and Planetary Science Letters*, **241**, 815–830.

ALVAREZ, W. & MONTANARI, A. 1988. Geologic framework of the Northern Apennines pelagic carbonate sequence. *In*: PREMOLI SILVA, I., COCCIONI, R. & MONTANARI, A. (eds) *The Eocene–Oligocene Boundary in the Marche–Umbria Basin (Italy)*. International Subcommission on Paleogene Stratigraphy Special Publication. Industrie Grafiche Fratelli Aniballi, Ancona, 13–30.

ANGORI, E., BERNAOLA, G. & MONECHI, S. 2007. Calcareous nannofossils assemblages and their response to the Paleocene–Eocene Thermal Maximum event at different latitudes: ODP Site 690 and Tethyan sections. *In*: MONECHI, S., COCCIONI, R. & RAMPINO, M. R. (eds) *Large Ecosystem Perturbations: Causes and Consequences*. Geological Society of America, Boulder, CO, Special Papers, **424**, 69–85.

AUBRY, M. P. 1991. Sequence stratigraphy: eustasy or tectonic imprint? *Journal of Geophysical Research*, **96**, 6641–6679.

AUBRY, M. P. 1995. Towards an upper Paleocene–lower Eocene high resolution stratigraphy based on calcareous nannofossil stratigraphy. *Israel Journal of Earth-Sciences*, **44**, 239–253.

AUBRY, M. P., OUDA, K. ET AL. 2007. The Global Standard Stratotype-section and Point (GSSP) for the base of the Eocene Series in the Dababiya section (Egypt). *Episodes*, **30**, 271–286.

BELLAGAMBA, M. & COCCIONI, R. 1990. Deep-water agglutinated foraminifera from the Massignano section (Ancona, Italy): a proposed stratotype for the Eocene–Oligocene boundary. *In*: HEMLEBEN, C. H., KAMINSKI, M. A., KHUNT, W. & SCOTT, D. (eds) *Paleoecology, Biostratigraphy, Paleoceanography and Taxonomy of Agglutinated Foraminifera*. Kluwer, Dordrecht, NATO ASI Series, **327**, 883–921.

BERGGREN, W. A. & PEARSON, P. N. 2005. A revised tropical to subtropical Paleogene planktonic foraminiferal zonation. *Journal of Foraminiferal Research*, **35**, 279–298.

BERGGREN, W. A. & PEARSON, P. N. 2006. Tropical to subtropical planktonic foraminiferal zonation of the Eocene and Oligocene. *In*: PEARSON, P. N., OLSSON, R. K., HUBER, B. T., HEMLEBEN, Ch. & BERGGREN, W. A. (eds) *Atlas of Eocene Planktonic Foraminifera*. Cushman Foundation, Special Publications, Fredricksburg, Virginia, USA, **41**, 29–40.

BERGGREN, W. A., KENT, D. V., SWISHER, C. C. & AUBRY, M. P. 1995. A revised cenozoic geochronology and chronostratigraphy. *In*: BERGGREN, W. A., KENT, D. V., SWISHER, C. C. III. & AUBRY, M. P. (eds) *Geo-Chronology, Time and Global Stratigraphic Correlation*. Society of Economic Paleontologist and Mineralogist (Society for Sedimentary Geology), Tulsa, OK, Special Publications, **54**, 129–213.

BICE, D. M. & MONTANARI, A. 1988. Magnetic stratigraphy of the Massignano section across the Eocene–Oligocene boundary. *In*: PREMOLI SILVA, I., COCCIONI, R. & MONTANARI, A. (eds) *The Eocene/Oligocene Boundary in the Marche–Umbria Basin (Italy)*. International Subcommission on Paleogene Stratigraphy Special Publication. Industrie Grafiche Fratelli Aniballi, Ancona, 111–117.

BODISELITSCH, B., MONTANARI, A., KOEBERL, C. & COCCIONI, R. 2004. Delayed climate cooling in the late Eocene caused by multiple impacts: high resolution geochemical studies at Massignano, Italy. *Earth and Planetary Science Letters*, **223**, 283–302.

BOSELLINI, A. 1989. Dynamics of Tethyan carbonate platforms. *In*: CREVELLO, P., WILSON, J. L., SARG, R. & REED, R. (eds) *Controls of Carbonate Platform and Basin Development*. Society for Sedimentary Geology, Tulsa, OK, Special Publications, **44**, 3–14.

BRINKHUIS, H. & BIFFI, U. 1993. Dinoflagellate cyst stratigraphy of the Eocene/Oligocene transition in central Italy. *Marine Micropaleontology*, **22**, 131–183.

BROWN, R. E., KOEBERL, C., MONTANARI, A. & BICE, D. M. 2009. Evidence for a change in Milankovitch forcing caused by extraterrestrial events at Massignano, Italy, Eocene–Oligocene Boundary GSSP. *In*: KOEBERL, C. & MONTANARI, A. (eds) *The Late Eocene Earth – Hothouse, Icehouse, and Impacts*. Geological Society of America, Boulder, CO, Special Papers, **452**, 1–19.

CAI, Y., LI, X., HU, X., CHEN, X. & PAN, Y. 2009. Paleoclimatic approach to the origin of the coloring of Turonian pelagic from the Vispi Quarry section (Cretaceous, central Italy). *Cretaceous Research*, **30**, 1205–1216.

CANDE, S. C. & KENT, D. V. 1995. Revised calibration of the geomagnetic polarity timescale for the Late Cretaceous and Cenozoic. *Journal of Geophysical Research*, **100**, 6093–6095.

CENTAMORE, E., CHIOCCHINI, M., JACOBACCI, A., MANFREDINI, M. & MANGANELLI, P. 1980. The evolution of the Umbrian–Marchean Basin in the Apennine section of the Alpine orogenic belt (Central Italy). *In*: COGNE, J. & SLANSKY, M. (eds) *Mémoire du Bureau de Recherches Géologiques et Minières*. Editions Technip, Paris, **108**, 289–305.

CHANNELL, J. E. T., D'ARGENIO, B. & HORWATH, F. 1979. Adria, the African Promontory, in Mesozoic Mediterranean Palaeogeography. *Earth Science Review*, **15**, 213–292.

CHIARABBA, C., JOVANE, L. & DI STEFANO, R. 2005. A new view of Italian seismicity using 20 years of instrumental recordings. *Tectonophyics*, **395**, 251–268.

COCCIONI, R. & GALEOTTI, S. 2003. Deep-water benthic foraminiferal events from the Massignano Eocene/Oligocene boundary stratotype section and point (central Italy): Biostratigraphic, paleoecologic, and paleoceanographic implications. *In*: PROTHERO, D. R., IVANT, L. C. & NESBITT, E. (eds) *From Greenhouse to Icehouse: The Marine Eocene–Oligocene Transition*. Columbia University Press, New York, 438–452.

COCCIONI, R., MONACO, P., MONECHI, S., NOCCHI, M. & PARISI, G. 1986. The Eocene/Oligocene boundary at Massignano (Ancona, Italy): biostratigraphy based on calcareous nannofossils and planktonic foraminifera. *Bulletin of Liaison and Information, International Geological Correlation Program Project 196*, **6**, 37–44.

COCCIONI, R., MONACO, P., MONECHI, S., NOCCHI, M. & PARISI, G. 1988. Biostratigraphy of the Eocene/Oligocene boundary at Massignano (Ancona, Italy). *In*: PREMOLI SILVA, I., COCCIONI, R. & MONTANARI, A. (eds) *The Eocene–Oligocene Boundary in the Marche–Umbria Basin (Italy)*. International Subcommission on Paleogene Stratigraphy Special Publication. Industrie Grafiche Fratelli Aniballi, Ancona, 59–80.

COCCIONI, R., FRANCHI, R., NESCI, O., WEZEL, F. C., BATTISTINI, F. & PALLECCHI, P 1989. Stratigraphy and mineralogy of the Selli Level (Early Aptian) at the base of the Marne a Fucoidi in the Umbro–Marchean Apennines, Italy. *In*: WIEDMANN, J. (ed.) *Cretaceous of the Western Tethys. Proceedings 3rd International Cretaceous Symposium, Tübingen, 1987*. E. Schweizerbart'sche Verlagsbuchhandlung, Stuttgart, 563–584.

COCCIONI, R., MORANDI, N. & TATEO, F. 1994. The 'Livello Raffaello' (early Miocene) in the Umbria–Marche Apennines, Italy: stratigraphy, paleontology, mineralogy, and geochemistry. *In*: COCCIONI, R., MONTANARI, A. & ODIN, G. S. (eds) *Miocene Stratigraphy of Italy and Adjacent Regions. Giornale di Geologia*, **56**, 55–78.

COCCIONI, R., BASSO, D. *ET AL.* 2000. Marine biotic signal across a late Eocene impact layer at Massignano, Italy: evidence for long-term environmental perturbations? *Terra Nova*, **12**, 258–263.

COCCIONI, R., MARSILI, A. *ET AL.* 2008. Integrated stratigraphy of the Oligocene pelagic sequence in the Umbria–Marche basin (northeastern Apennines, Italy): a potential Global Stratotype Section and Point (GSSP) for the Rupelian/Chattian boundary. *Geological Society of America Bulletin*, **120**, 487–511.

COCCIONI, R., FRONTALINI, F. & SPEZZAFERRI, S 2009. *Late Eocene Impact-induced Climate and Hydrological Changes: Evidence from the Massignano Global Stratotype Section and Point (Central Italy)*. Geological Society of America, Boulder, CO, Special Papers, **452**, 97–118.

COCCIONI, R., FRONTALINI, F., BANCALÀ, G., FORNACIARI, E., JOVANE, L. & SPROVIERI, M. 2010. The Dan-C2 hyperthermal event at Gubbio (Italy): global implications, environmental effects, and cause(s). *Earth and Planetary Science Letters*, **297**, 298–305.

COCCIONI, R., BANCALA, G., CATANZARITI, R., FORNACIARI, E., FRONTALINI, F., GIUSBERTI, L., JOVANE, L., LUCIANI, V., SAVIAN, J. & SPROVIERI, M. 2012. An integrated stratigraphic record of the Paleocene–lower Eocene Gubbio (Italy): new insights into the early–middle Paleogene hyperthermals and global implications. *Terra Nova*, **24**, 380–386.

CRESTA, S., MONECHI, S. & PARISI, G. 1989. Stratigrafia del Mesozoico e Cenozoico nell'area umbro–marchigiana, Itinerari geologici sull'Appennino Umbro–Marchigiano (Italia). *Memorie Descrittive della Carta Geologica d'Italia*, **39**, 1–185.

FORNACIARI, E., GIUSBERTI, L. *ET AL.* 2007. An expanded Cretaceous–Tertiary transition in a pelagic setting of the Southern Alps (central-western Tethys). *Palaeogeography, Palaeoclimatology, Palaeoecology*, **255**, 98–131.

GALEOTTI, S., ANGORI, E. ET AL. 2000. Integrated stratigraphy across the Paleocene/Eocene boundary in the Contessa Road section, Gubbio (central Italy). *Bulletin de la Société Géologique de France*, **171**, 355–365.

GALEOTTI, S., KAMINSKI, M. A., SPEIJER, R. & COCCIONI, R. 2004. High resolution Deep Water Agglutinated Foraminiferal record across the Paleocene/Eocene transition in the Contessa Road section (central Italy). *In*: BUBIK, M. & KAMINSKI, M. A. (eds) *Proceedings of the 6th International Workshop on Agglutinated Foraminifera*. Arti Grafiche Editoriali, Urbino. Grzybowski Foundation Special Publication, **8**, 83–103.

GALEOTTI, S., KRISHNAN, S. ET AL. 2010. Orbital Chronology of Early Eocene hyperthermals from the Contessa Road section, central Italy. *Earth and Planetary Science Letters*, **290**, 192–200.

GIUSBERTI, L., COCCIONI, R., SPROVIERI, M. & TATEO, F. 2009. Perturbation at the sea floor during the Paleocene–Eocene Thermal Maximum: evidence from benthic foraminifera at Contessa Road, Italy. *Marine Micropaleontology*, **70**, 102–119.

GRADSTEIN, F. M., OGG, J. G. & SMITH, A. G. 2004. *A Geologic Time Scale 2004*. Cambridge University Press, Cambridge.

HU, X., JANSA, L. & SARTI, M. 2005a. Mid-Cretaceous oceanic red beds in the Umbria–Marche Basin, central Italy: constraints on paleoceanography and paleoclimate. *Palaeogeography, Palaeoclimatology, Palaeoecology*, **233**, 163–186.

HU, X., JANSA, L. ET AL. 2005b. Upper Cretaceous oceanic red beds (CORBs) in the Tethys: occurrences, lithofacies, age, and environments. *Cretaceous Research*, **26**, 3–20.

HU, X., CHENG, W. & JI, J. 2009. Origin of Cretaceous oceanic red beds from the Vispi Quarry, Central Italy: visible reflectance and inorganic geochemistry. *In*: HU, X., WANG, C., SCOTT, W., WAGREICH, M. & JANSA, L. (eds) *Oceanic Red Beds: Stratigraphy, Composition, Origins, and Paleoceanographic and Paleoclimatic Significance*. Society for Sedimentary Geology, Tulsa, OK, Special Publications, **91**, 183–197.

HYLAND, E., MURPHY, B. ET AL. 2009. Integrated stratigraphy and astrochronologic calibration of the Eocene–Oligocene transition in the Monte Cagnero section (Northeastern Apennines, Italy): a potential parastratotype for the Massignano GSSP. *In*: KOEBERL, C. & MONTANARI, A. (eds) *The Late Eocene Earth – Hothouse, Icehouse and Impacts*. Geological Society of America, Boulder, CO, Special Papers, **452**, 303–322.

JANSA, L. & HU, X. 2009. Cretaceous pelagic black shales and red beds in western Tethys: origins, paleoclimate, and paleoceanographic implications. *In*: HU, X., WANG, C., SCOTT, W., WAGREICH, M. & JANSA, L. (eds) *Oceanic Red Beds: Stratigraphy, Composition, Origins, and Paleoceanographic and Paleoclimatic Significance*. Society for Sedimentary Geology, Tulsa, OK, Special Publications, **91**, 59–72.

JOVANE, L., FLORINDO, F. & DINARÈS-TURELL, J. 2004. Environmental magnetic record of paleoclimate change: revisiting the Eocene–Oligocene stratotype section, Massignano, Italy. *Geophysical Research Letters*, **31**, http://dx.doi.org/10.1029/2004GL020554

JOVANE, L., FLORINDO, F., SPROVIERI, M. & PÄLIKE, H. 2006. Astronomic calibration of the late Eocene/early Oligocene Massignano section (central Italy). *Geochemistry Geophysics Geosystems*, **7**, Q07012, http://dx.doi.org/10.1029/2005GC001195

JOVANE, L., FLORINDO, F. ET AL. 2007a. The middle Eocene climatic optimum (MECO) event in the Contessa Highway section, Umbrian Apennines, Italy. *Geological Society of America Bulletin*, **119**, 413–427.

JOVANE, L., SPROVIERI, M. ET AL. 2007b. Eocene–Oligocene paleoceanographic changes in the stratotype section, Massignano, Italy: clues from rock magnetism and Supplementary Material isotopes. *Journal of Geophysical Research*, **112**, B11101, http://dx.doi.org/10.1029/2007JB004963

JOVANE, L., COCCIONI, R., MARSILI, A., FLORINDO, F., ACTON, G. & SPROVIERI, M. 2009. The late Eocene greenhouse–icehouse transition: observations from the Massignano Global Stratotype Section and Point (GSSP). *In*: KOEBERL, C. & MONTANARI, A. (eds) *The Late Eocene Earth – Hot House, Ice House, and Impacts*. Geological Society of America, Boulder, CO, Special Papers, **452**, 149–168.

JOVANE, L., SPROVIERI, M., COCCIONI, R., FLORINDO, F., MARSILI, A. & LASKAR, J. 2010. Astronomical calibration of the middle Eocene Contessa Highway section (Gubbio, Italy). *Earth and Planetary Science Letters*, **298**, 77–88.

JOVANE, L., SAVIAN, J. F. ET AL. 2012. Integrated magnetobiostratigraphy of the middle Eocene–lower Oligocene interval from the Monte Cagnero section, central Italy. *In*: JOVANE, L., HERRERO-BERVERA, E., HINNOV, L. A. & HOUSEN, B. A. (eds) *Magnetic Methods and the Timing of Geological Processes*. Geological Society, London, Special Publications, **373**, http://dx.doi.org/10.1144/SP373.xx

LOWRIE, W. & LANCI, L. 1994. Magnetostratigraphy of Eocene–Oligocene boundary sections in Italy: no evidence for short subchrons within chrons 12R and 13R. *Earth and Planetary Science Letters*, **126**, 247–258.

LOWRIE, W., ALVAREZ, W., NAPOLEONE, G., PERCH-NIELSEN, K., PREMOLI SILVA, I. & TOUMARKINE, M. 1982. Paleogene magnetic stratigraphy in Umbrian pelagic carbonate rocks: the Contessa sections, Gubbio. *Geological Society of America Bulletin*, **93**, 414–432.

MARTINI, E. 1971. Standard Tertiary and Quaternary Calcareous Nannoplankton Zonation. *In*: FARINACCI, A. (ed.) *Proceedings of the Second Planktonic Conference, Rome 1970*. Tecnoscienza, **2**, 739–785.

MATTIAS, P., CROCETTI, G. ET AL. 1992. Caratteristiche mineralogiche e litostratigrafiche della sezione eo-oligocenica di Massignano (Ancona, Italia) comprendente il limite Scaglia Variegata-Scaglia Cinerea. *Studi Geologici Camerti*, **12**, 93–103.

MOLINA, E., ALEGRET, L. ET AL. 2011. The Global Stratotype Section and Point (GSSP) for the base of the Lutetian Stage at the Gorrondatxe section, Spain. *Episodes*, **34**, 86–108.

MONACO, P., NOCCHI, M. & PARISI, G. 1987. Analisi stratigrafica e sedimentologica di alcune sequenze pelagiche dell'Umbria sud-orientale dall'Eocene inferiore all'Oligocene inferiore. *Bollettino della Società Geologica Italiana*, **106**, 71–91.

MONTANARI, A. & KOEBERL, C. 2000. *Impact Stratigraphy – The Italian Record*. Lecture Notes in Earth Sciences. **93**, Springer, Berlin.

MONTANARI, A., DRAKE, R. *ET AL.* 1985. Radiometric time scale for the upper Eocene and Oligocene based on K–Ar and Rb–Sr dating of volcanic biotites from the pelagic sequence of Gubbio, Italy. *Geology*, **13**, 596–599.

MONTANARI, A., CAREY, S., COCCIONI, R. & DEINO, A. 1994. Early Miocene tephra in the Apennine pelagic sequence: an inferred Sardinian provenance and implications for western Mediterranean tectonics. *Tectonics*, **13**, 1120–1134.

MONTANARI, A., BICE, D. M. *ET AL.* 1997. Integrated stratigraphy of the Chattian to Mid-Burdigalian pelagic sequence of the Contessa Valley (Gubbio, Italy). *In*: MONTANARI, A., ODIN, G. S. & COCCIONI, R. (eds) *Miocene Stratigraphy – An Integrated Approach. Developments in Palaeontology and Stratigraphy*. Elsevier, Amsterdam, **15**, 249–277.

NEUHUBER, S., WAGREICH, M., WENDLER, I. & SPÖTL, C. 2007. Turonian oceanic red beds in the Eastern Alps: concepts for palaeoceanographic changes in the Mediterranean Tethys. *Palaeogeography, Palaeoclimatology, Palaeoecology*, **251**, 222–238.

OBERLI, F. & MEIER, M. 1991. Age of Eocene–Oligocene boundary in the Marche–Umbria Basin, Italy, by high resolution U/Th–Pb dating. *TERRA Abstracts*, **3**, 286.

ODIN, G. S. & MONTANARI, A. 1988. The Eocene–Oligocene boundary at Massignano (Ancona, Italy): A potential boundary stratotype. *In*: PREMOLI SILVA, I., COCCIONI, R. & MONTANARI, A. (eds) *The Eocene–Oligocene Boundary in the Marche–Umbria Basin (Italy)*. International Subcommission on Paleogene Stratigraphy Special Publication. Industrie Grafiche Fratelli Aniballi, Ancona, 253–263.

ODIN, G. S., MONTANARI, A. *ET AL.* 1991. Reliability of volcano–sedimentary biotite ages across the Eocene–Oligocene boundary. *Chemical Geology*, **86**, 203–224.

OGG, J. G., OGG, G. & GRADSTEIN, F. M. 2008. *The Concise Geologic Time Scale*. Cambridge University Press, Cambridge.

OKADA, H. & BUKRY, D. 1980. Supplementary modification and introduction of code numbers to the low-latitude coccoliths biostratigraphic zonation (Bukry, 1973; 1975). *Marine Micropaleontology*, **5**, 321–325.

PARISI, G., GUERRERA, F. *ET AL.* 1988. Middle Eocene to early Oligocene calcareous nannofossil and foraminiferal biostratigraphy in the Monte Cagnero section, Piobbico (Italy). *In*: PREMOLI SILVA, I., COCCIONI, R. & MONTANARI, A. (eds) *The Eocene–Oligocene boundary in the Marche–Umbria Basin (Italy)*. International Subcommission on Paleogene Stratigraphy Special Publication. Industrie Grafiche Fratelli Aniballi, Ancona, 119–135.

PREMOLI SILVA, I. & JENKINS, D. G. 1993. Decision on the Eocene–Oligocene boundary stratotype. *Episodes*, **16**, 379–382.

PROSS, J., HOUBEN, A. J. P. *ET AL.* 2010. Umbria–Marche revisited: a refined magnetostratigraphic calibration of dinoflagellate cyst events for the Oligocene of the Western Tethys. *Review of Palaeobotany and Palynology*, **158**, 213–235.

RAFFI, I., BACKMAN, J. & PÄLIKE, H. 2005. Changes in calcareous nannofossil assemblages across the Paleocene/Eocene transition from the paleo-equatorial Pacific Ocean. *Palaeogeography, Palaeoclimatology, Palaeoecology*, **226**, 93–126.

SPEZZAFERRI, S., BASSO, D. & COCCIONI, R. 2002. Late Eocene planktonic foraminiferal response to an extraterrestrial impact at Massignano GSSP (northeastern Apennines, Italy). *Journal of Foraminiferal Research*, **32**, 188–199.

TORI, F. & MONECHI, S. 2007. Paleoecological and paleoclimatical significance of calcareous nannofossils at the E/O transition: new data from the Cagnero and Massignano sections, Umbria–Marche area (Italy). *In*: MONTANARI, A., KOEBERL, C., COCCIONI, R. & HILGEN, F. (eds) *The Late Eocene Earth: Hot House, Ice House, and Impacts – Abstract with Program and Field Trip Guide: Ancona, Italy. Geological Society of America Penrose Conference*, Osservatorio Coldigioco, Frontale di Apiro (Italy), 77.

TRABUCHO-ALEXANDRE, J., NEGRI, A. & DE BOER, P. 2011. Early Turonian pelagic sedimentation at Moria (Umbria–Marche, Italy: primary and diagenetic controls on lithological oscillations. *Palaeogeography, Palaeoclimatology, Palaeoecology*, **311**, 200–214.

VERDUCCI, M. & NOCCHI, M. 2004. Middle to late Eocene main planktonic foraminiferal events in the Central Mediterranean area (Umbria–Marche Basin) related to paleoclimatic changes. *Neues Jahrbuch für Geologie und Palaontologie Abhandlungen*, **234**, 361–413.

WADE, B. S., PEARSON, P. N., BERGGREN, W. A. & PÄLIKE, H. 2011. Review and revision of Cenozoic tropical planktonic foraminiferal biostratigraphy and calibration to the geomagnetic polarity and astronomical time scale. *Earth-Science Reviews*, **104**, 111–142.

WANG, C., HU, X., HUANG, Y., SCOTT, R. W. & WAGREICH, M. 2009. Overview of Cretaceous Oceanic Red Beds (CORBs): a window on global oceanic and climate change. *In*: HU, X., WANG, C., SCOTT, W., WAGREICH, M. & JANSA, L. (eds) *Oceanic Red Beds: Stratigraphy, Composition, Origins, and Paleoceanographic and Paleoclimatic Significance*. Society for Sedimentary Geology, Tulsa, OK, Special Publications, **91**, 13–33.

WANG, C., HU, X., HUANG, Y., WAGREICH, M., SCOTT, R. & HAY, W. 2011. Cretaceous oceanic red beds as possible consequence of oceanic anoxic events. *Sedimentary Geology*, **235**, 27–37.

Magnetostratigraphy, fence diagrams and basin analysis

MAODU YAN[1,2]*, ROB VAN DER VOO[2], XIAO-MIN FANG[1,3] & CHUNHUI SONG[3]

[1]*Key Laboratory of Continental Collision and Plateau Uplift, Institute of Tibetan Plateau Research, Chinese Academy of Sciences, Beijing 100085, China*

[2]*Department of Earth and Environmental Sciences, University of Michigan, 1100 N University Ave, Ann Arbor, MI 48109–1005, USA*

[3]*National Laboratory of Western China's Environmental Systems, Ministry of Education of China and College of Resources and Environment, Lanzhou University, Gansu 730000, China*

**Corresponding author (e-mail: maoduyan@itpcas.ac.cn)*

Abstract: Correlation of lithostratigraphic sections is widely used to examine the nature of lateral facies changes within or between basins. It can provide significant clues for regional environmental and palaeogeographic reconstructions. There are problems associated with lithostratigraphic correlation; diachronous deposition of similar lithological units may not be recognized. We report here an attempt to determine lateral facies changes in coeval sedimentary sections, through three-dimensional magnetostratigraphic correlations in the Guide Basin, an intramontane basin in the northeastern part of the Tibetan Plateau. The method is successful for correlating lateral facies and helps to identify sediment sources in the basin.

As products of nearby relief, sediments in tectonically active basins usually reflect the related controls of erosional, depositional and structural processes. Studies of sedimentary basins can provide information on the nature of depositional systems, patterns of erosional unroofing and kinematic histories of fold–thrust systems (Chen *et al.* 1995; Sobel & Dumitru 1997; DeCelles *et al.* 1998; Horton *et al.* 2002).

Correlation of lithostratigraphic sections can reveal the nature of lateral facies changes in a basin, provide significant clues for environmental and palaeogeographic reconstructions, and may help to further understanding of a basin's evolution (Miall 2000). Diachroneity of basin infilling by similar lithological units may be difficult to identify when correlating rock units using lithostratigraphy alone. True time correlations are those based on widespread markers or other event beds that have chronostratigraphic value, which are not always available. When widespread markers and other event beds are absent, there is a need for other age constraints in order to arrive at useful lithostratigraphic correlations. Magnetostratigraphy, coupled with bioconstraints, has been proven successful in constraining the ages of the sediments elsewhere in and around the NE Tibetan Plateau (Li *et al.* 1997; Fang *et al.* 2003, 2007; Dai *et al.* 2006; Lu & Xiong 2009).

In this contribution, we present a hitherto underused method using three-dimensional magnetostratigraphy with the goal of adding chronological control to lateral lithofacies correlation of late Neogene sedimentary fill of the Guide Basin, northeastern Tibetan Plateau. This approach should improve our understanding of the evolution of the basin during a period of very active tectonism (Tapponnier *et al.* 1990; Burchfiel *et al.* 1991; Sobel *et al.* 2003). The method is successful in constraining the ages of lateral lithostratigraphic facies variations and allows us to reconstruct the sources of the sediments in those parts of the basin that were studied with magnetostratigraphy.

Geological setting

The Guide Basin is a sub-basin of the Gonghe–Guide Basin, which is a typical wrench-fault-bounded intramontane depression in the NE Tibetan Plateau. It is located between the sinistral Qilian–Haiyuan Fault to the north, and the sinistral Kunlun Fault to the south; on the west and east sides it is bounded by the dextral-transpressional Wenquan Fault and the dextral-transpressional Haiyan Fault, respectively (Fig. 1). The Guide Basin is bounded by the Qinghainan Shan (mountain range) to the NNW, the Laji Shan to the NNE, the Baji Shan to the south, and is separated from the Gonghe Basin to the west by the Waligong Shan. To the east, the Yamazari Shan is located between the Guide and Xunhua basins. It is truncated by the Longyang Gorge Fault and by at least six north-verging wedge-like thrust faults in the northeastern part of the basin (Fang *et al.* 2005*a*; Yan *et al.* 2006, F1–6 in Fig. 1b).

From: JOVANE, L., HERRERO-BERVERA, E., HINNOV, L. A. & HOUSEN, B. A. (eds) 2013. *Magnetic Methods and the Timing of Geological Processes*. Geological Society, London, Special Publications, **373**, 133–147.
First published online August 17, 2012, http://dx.doi.org/10.1144/SP373.3 © The Geological Society of London 2013.
Publishing disclaimer: www.geolsoc.org.uk/pub_ethics

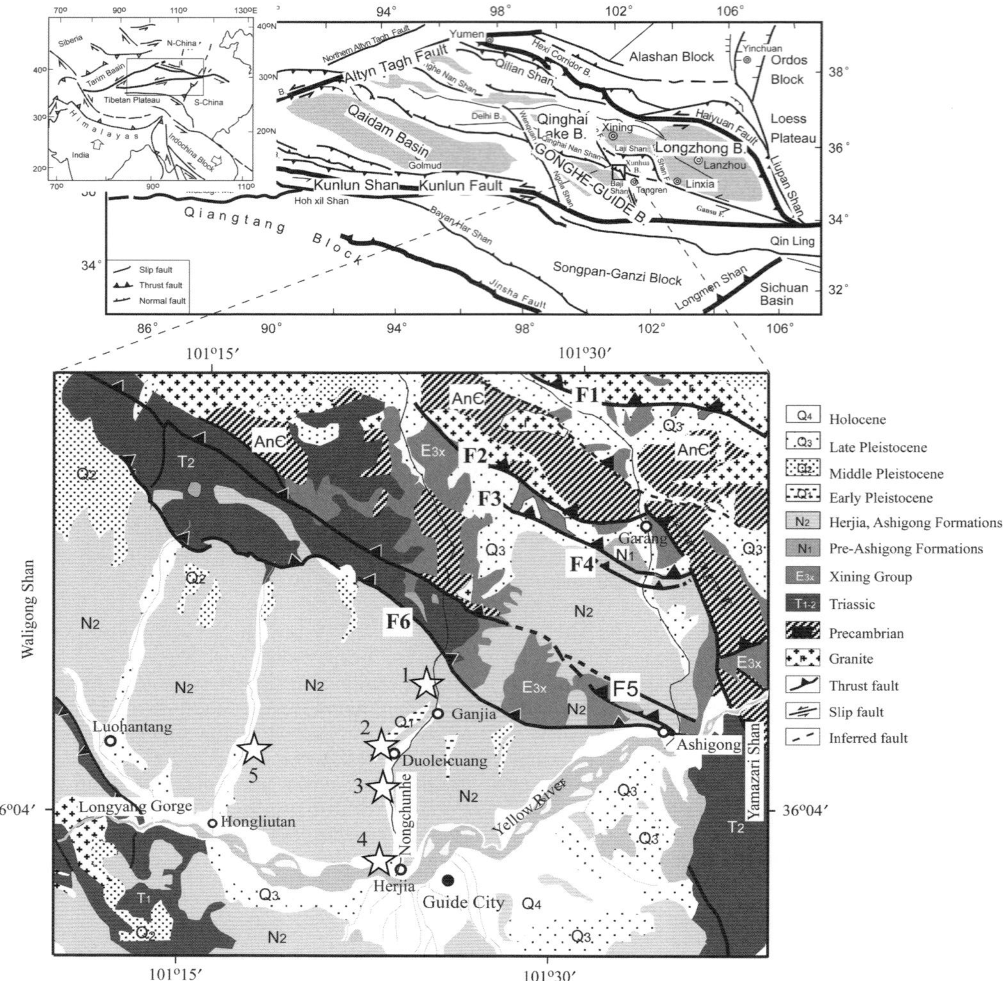

Fig. 1. Geological map of most of the Guide Basin, showing the locations of our sampling localities (stars) in Neogene strata (modified from Fang *et al.* 2005*a, b*). Magnetostratigraphic results from localities 1–4 along the Nongchunhe River have been described previously (Fang *et al.* 2005*a*). Locality 5 in the Hongliutan Valley is discussed in this study. The top part of the figure shows a tectonic map of central Asia compiled from Qinghai Geology Bureau (1989), Meyer *et al.* (1998), Song *et al.* (2001) and Yin and Harrison (2000). The NE Tibetan Plateau contains a small rectangle, which shows our study area located near 101.5°E and 36°N.

Detailed bio- and magnetostratigraphic studies have provided age constraints on the Neogene sedimentary infilling in the basin, above less well dated (likely Oligocene) redbeds (Parés *et al.* 2003; Fang *et al.* 2005*a*). Fang *et al.* (2005*a*) divided the sediments in the basin into eight units. Unit 8 (Oligocene?) consists of tan to orange-red sandy conglomerates to sandy mudstones, and is overlain by Miocene alluvial fan conglomeratic deposits of unit 7 (*c.* 19 to *c.* 20.8 Ma) and by braided river deposits of unit 6 (<16–*c.* 19 Ma). These are in turn overlain by a distinct sequence (unit 5) of alternating multi-coloured limestone, mudstone and siltstone, deposited between *c.* 7.8 and >11.5 Ma (Fig. 2). Unit 4 is a Mio-Pliocene fluvial succession of sediments deposited in shallow lake, braided river, fluvial over-bank flood and deltaic environments between 3.6 and *c.* 7.8 Ma; these sediments comprise cross-bedded sandstones and pebbly conglomerates along the margin of the basin, and much finer-grained deposits in the basin centre (i.e. alternating mudstones, siltstones, sandstones

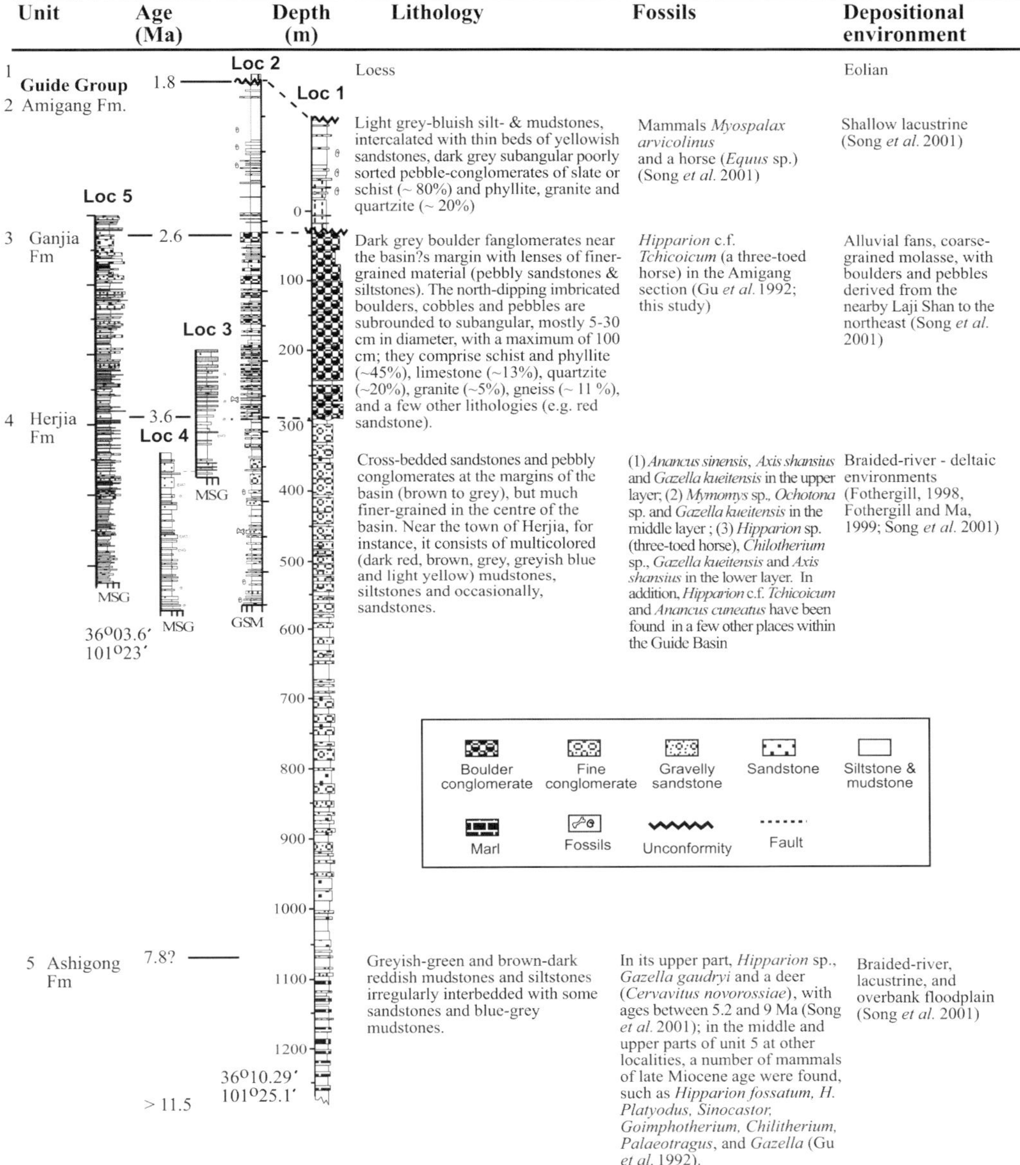

Fig. 2. Unit numbers, formation names and ages, lithological columns and descriptions of the lithologies, fossils and depositional environments of the five localities (Loc 1–5). Modified from Fang *et al.* (2005*a*).

and intercalations of some limestones; Figs 2 & 3b–e). These are overlain by thick (>263 m) massive conglomerates that make up unit 3, a succession of mostly alluvial fan deposits that accumulated between 2.6 and 3.6 Ma (Upper Pliocene, i.e. Piacenzian; Fig. 3a, b, f). Unit 2 consists mostly of 1.8–2.6 Ma shallow lacustrine limestones, mudstones and siltstones, intercalated with thin sandstones and poorly sorted conglomerates (Fig. 3f). Thin Pleistocene loess overlying Cenozoic and older formations and Pleistocene river terraces comprise unit 1.

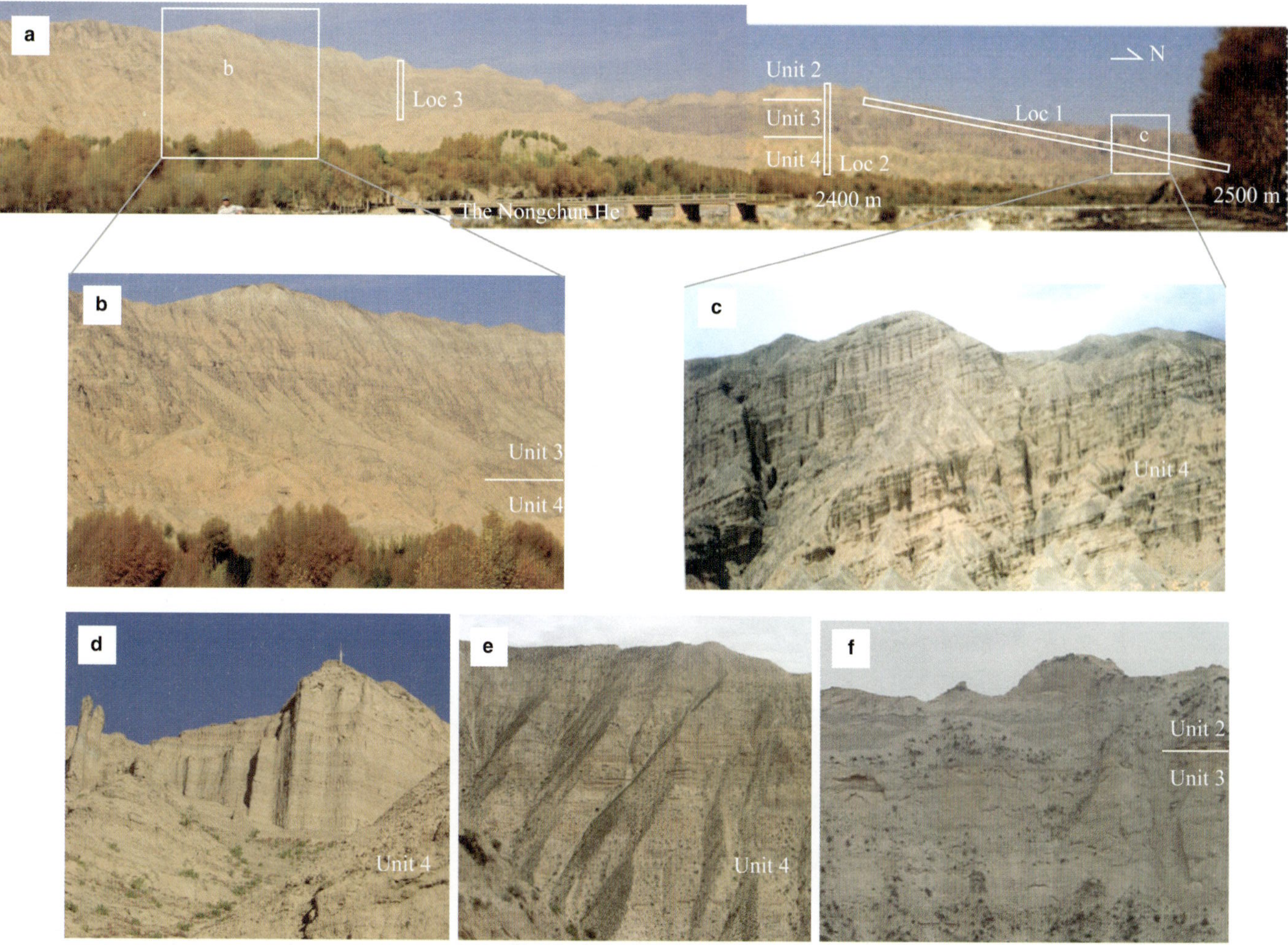
a
N
Loc 3
Loc 1
Unit 2
Unit 3
Unit 4
Loc 2
The Nongchun He
2400 m
2500 m
b
Unit 3
Unit 4
c
Unit 4
d
Unit 4
e
Unit 4
f
Unit 2
Unit 3

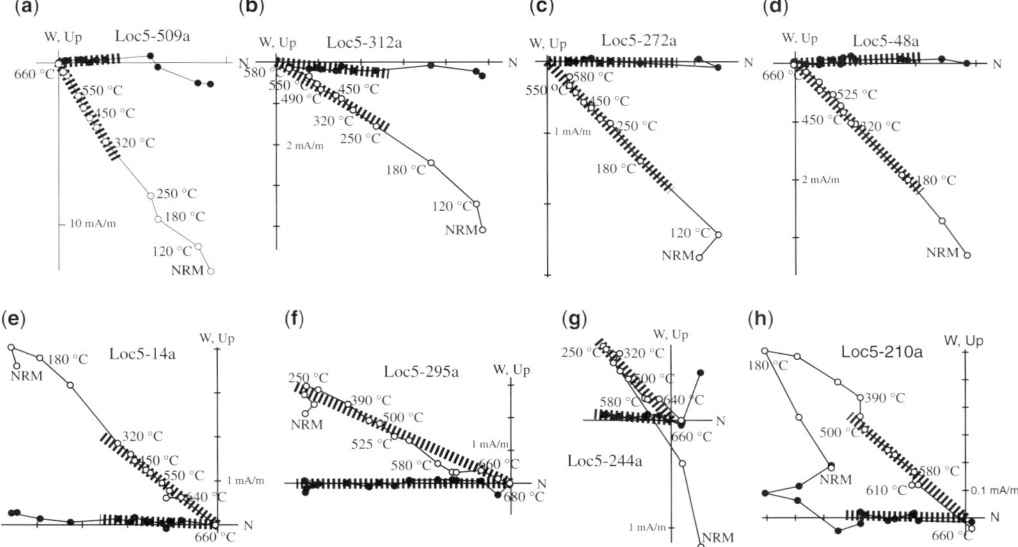

Fig. 4. Representative orthogonal thermal demagnetization diagrams for samples from locality 5. Open symbols represent vertical projections; solid symbols represent horizontal projections; highlighted trajectories represent those used to determine characteristic remanent magnetization (ChRM) directions in principal component analysis; intensities are given in mA m^{-1}; the numbers are the corresponding depths in the section. NRM, Natural remanent magnetization.

Sample-collecting and laboratory methods

Sediments older than Late Miocene (units 5–8) are mostly exposed around the margin of the basin, and their ages are not well established, whereas younger deposits (units 2–4) are widely distributed in the central and west part of the basin, and their ages have become well constrained (Fig. 2; Song *et al.* 2001; Parés *et al.* 2003; Yan *et al.* 2004; Fang *et al.* 2005*a*). In order to date the older deposits, we first measured and sampled four outcrop sections (1–4; Fang *et al.* 2005*a*) in the central part of the basin in the Nongchunhe Valley (Fig. 1b), subsequently expanded to the third dimension in a later field season by sampling at locality 5 in a valley further to the west. Oriented palaeomagnetic samples were taken at stratigraphic intervals of *c.* 2–5 m within these detailed stratigraphic sections. At a given site, typically three independently oriented cubic samples (*c.* 8 cm^3) were collected.

Results from localities (i.e. sections) 1–4 have been previously described (Fang *et al.* 2005*a*); results from locality 5 are presented herein.

Samples were thermally demagnetized (Fig. 4) and analysed by principal component analysis (Kirschvink 1980). The characteristic remanent magnetization (ChRM) directions were generally unambiguously identified, as can be seen by examining the highlighted trajectories of the demagnetization diagrams (Fig. 4). The ChRM directions were used to calculate virtual geomagnetic poles (VGP) for magnetostratigraphic analysis. Specimens excluded in our magnetostratigraphic analysis were rejected on the basis of three criteria (Yan *et al.* 2005, 2006): (1) ChRM directions could not be determined because of ambiguous or noisy orthogonal demagnetization diagrams; (2) ChRM directions revealed a maximum angular deviation angle greater than 15°; or (3) samples revealed magnetizations with VGP latitude values less than 30°.

Fig. 3. Field photos in a cross section along the Nongchunhe Valley. (**a**) The profile along the valley, showing lithostratigraphic units 2–4 and magnetostratigraphic sections at localities 1–3 (from Fang *et al.* 2005*a*). The composite photo shows the transition of coarse-grained conglomerates at the northern basin margin (to the right) to fine-grained mudstones, siltstones and sandstones at the basin centre at the left side of the picture. (**b**) Detail near locality 3. (**c**) Detail near locality 1. (**d, e**) Detail of unit 4. (**f**) Fine-grained lacustrine siltstones and mudstones of unit 2, on top of the conglomerates of unit 3 near locality 5.

Magnetostratigraphic age control

Localities 1–4

Detailed magnetostratigraphic analysis at localities 1–4 was carried out previously (Fang *et al.* 2005*a*). Unit 4 has been dated as late Tortonian through late Zanclean (late Miocene–Pliocene, *c.* 7.8–3.6 Ma; Fig. 2). The younger units 3 and 2 correspond to the Gauss geomagnetic interval (3.6–2.6 Ma) and the lower Matuyama (2.6–1.8 Ma), respectively (Song *et al.* 2003; Fang *et al.* 2005*a*; Fig. 2). These magnetostratigraphic ages are well supported by Pliocene and late Miocene fossils (Fang *et al.* 2005*a*). The observed polarity zones, their ages and related depths are listed in Table 1.

The localities are located along the Nongchunhe Valley, north of the town of Guide, in the north central part of the basin (Fig. 1). Most strata along the Nongchunhe Valley dip gently to the SSW; some marker layers can be traced laterally from north to south (Fig. 3a). Coarsening-upward and fining-southward patterns are observed among the various measured sections, and record lateral migration of depositional facies belts (Fig. 3a). Upward coarsening is most apparent in the vertical succession of silty–sandy conglomerates of unit 4, which grade into thick grey angular to subangular conglomeratic layers of unit 3 at locality 1 in the northernmost part of the valley (Fig. 3a, c). Laterally, silty–sandy conglomerates of unit 4 at locality 1 give way southward to cycles of silty sandstones and conglomerates at locality 2, and to mostly limestones, siltstones and mudstones, intercalated with some sandy pebbles at locality 4. Clearly, unit 4 consists of much finer-grained siltstones and mudstones at locality 4 in the southern part of the Nongchunhe Valley (Fig. 3d) than in the northern stretch (Fig. 3c; Song *et al.* 2001; Fang *et al.* 2005*a*). Similarly, the thick conglomerates of unit 3 in the north grade laterally (southward) into interlayers of sandstones, siltstones and finer (pebbly) conglomerates at locality 2 in the middle part of the valley. Lastly, above unit 3, deposits of bluish siltstones and mudstones, with some thin layers of pebbly conglomerates, represent unit 2. In the Nongchunhe Valley, unit 2 is only present around locality 2 (Fig. 3a; Fang *et al.* 2005*a*).

Locality 5

Locality 5 is located in the Hongliutan Valley, west of the Nongchunhe Valley (36°05.62′ N, 101°19.84′ E, see Fig. 1). The total exposed thickness used for the magnetostratigraphic section is 526 m. The lower 342 m consists of laminated bluish, greyish, light brownish siltstones and sandstones, irregularly interbedded with some conglomerates (Fig. 3e), overlain by 154 m of alternating greyish to light brownish sandstones, greyish conglomerates and siltstones. The top 30 m of the section comprises light bluish siltstones irregularly interbedded with sandstones and pebbly conglomerates; these are capped with a well-cemented pebbly sandstone (Fig. 3f).

A total of 192 samples (out of 230) yielded ChRM directions and the excellent dual-polarity record is shown in equal area projection on the left of Figure 5a. These ChRM directions demonstrate a positive bootstrap reversal test (Tauxe 1998; Fig. 5b), with classification of A (McFadden & McElhinny 1990). Virtual geomagnetic poles were calculated, and the present-day latitudes of these pole locations have been plotted as VGP-lat in Figure 6 as with other similar magnetostratigraphic studies of Cenozoic strata in northern Tibet, post-depositional processes have reduced the inclination of these strata; however, this does not influence the conclusions derived from the polarity determinations and age assignments (e.g. Yan *et al.* 2005).

A total of eight normal and eight reversed polarity intervals are recorded in the section (N1–8, R1–8 in Fig. 6). With constraints of late Neogene in age for the sediments in the Nongchunhe Valley (e.g. localities 1–4; Fang *et al.* 2005*a*), the observed polarities can be correlated without any ambiguity to the Geomagnetic Polarity Time Scale (GPTS) of Gradstein *et al.* (2004). R1 is correlated to the Lower Matuyama reversed polarity interval, N1–3 to the Gauss interval (Chron 2An), R4–R8 to the Gilbert interval with predominantly reversed polarities (Chrons 2Ar–3r), and N8 partly to Chron 3An. The resultant magnetostratigraphic ages for the section range from *c.* 6 to < 2.5 Ma.

Analogous to the stratigraphic divisions in the Nongchunhe Valley, lithological units were defined for locality 5 (see Fig. 6, units 2–4) and these have associated ages that are in surprisingly good agreement with the ages at localities 1–4. From lower to upper, these are >6 Ma–3.6 Ma for unit 4, 3.6–2.6 Ma for unit 3, and 2.6– < 2.5 Ma for unit 2 (Fig. 6).

Three-dimensional magnetostratigraphic analyses

The ages of each polarity zone and their related thicknesses in each section (Table 1) allow us to determinate sedimentation rates accurately and correlate lateral facies across the basin. As seen in Figure 7, sediments within the central part of the Guide Basin comprise an upward-coarsening sequence (units 4 and 3) that is overlain by an abrupt change to finer-grained deposits of unit 2.

The thicknesses of the sediments within each polarity zone also show a trend in a latitudinal (north–south) direction along the Nongchunhe Valley; equivalent sedimentary packages show

Table 1. *Summary of palaeomagnetic depths (or heights depending on field-based reference levels; in metres) for the polarity boundaries (defined as age, in Ma) for each of the five sections, with their thickness (Th) and sediment (Sed) accumulation rate*

Age (Ma)	Locality 1			Locality 2			Locality 3			Locality 4			Locality 5		
	Depth	Th (m)	Sed Rate	Depth	Th (m)	Sed Rate	Height	Th (m)	Sed Rate	Height	Th (m)	Sed Rate	Depth	Th (m)	Sed Rate
~2.581	37.5			217									30		
		66.5	>14.75		**90**	19.96								**75**	16.63
3.032	104			307?									105		
		49	58.33		**30**	35.71								**14**	16.67
3.116	153			337									119		
		47	55.29		**56**	65.88								**16.5**	19.41
3.201	200			393?			153.5						135.5		
		43.5	33.72		**40**	31.01		15.5	18.2					**12.5**	9.69
3.33	243.5			433			133.5						148		
		53.5	20.11		**46.5**	17.48		58	21.8					**36**	13.53
3.596	297			479.5									184		
							75.5						*"westward thinning" above*		
		111	18.78	*"eastward thinning" below*				70	11.84						
					65	11								**81.5**	13.79
4.187	408			544.5			5.5			205.5			265.5		
		14	12.39		**7**	6.19					7	6.19		**14.5**	12.83
4.3	422			551.5						198.5			280		
		40	20.73		**14.5**	7.51					14.5	7.51		**21**	10.88
4.493	462			566						184			301		
		25.5	18.48		**18**	13.04					18	13.04		**22.5**	16.3
4.631	487.5			584						166			323.5		
		28	16.67		**23**	13.69					23	13.69		**36.5**	21.73
4.799	515.5			607						143			360		
		15	15.46		**4**	4.12					4	4.13		**11**	11.34
4.896	530.5			611						139			371		
		21	20.79		**12**	11.88					12	11.88		**10**	9.9
4.997	551.5			623						127			381		
		32.5	13.66		**33**	13.87					33	13.87		**29.5**	12.39
5.235	584			656						94			410.5		
		133	16.67		**45**	5.64					45	5.64		**103.5**	12.97
6.033	717			701						49			514		

The ages are based on comparisons with the geomagnetic polarity time scale of Gradstein *et al.* (2004). The two depth levels with question marks indicate the existence of uncertainties owing to short-fluctuation transitional polarities; for these levels, the thickness listed is the average of the two juxtaposed depths. Sediment accumulation rates (Sed. rate, in cm/ka, in bold) are calculated from thickness divided by duration of each polarity interval. Th, thickness.

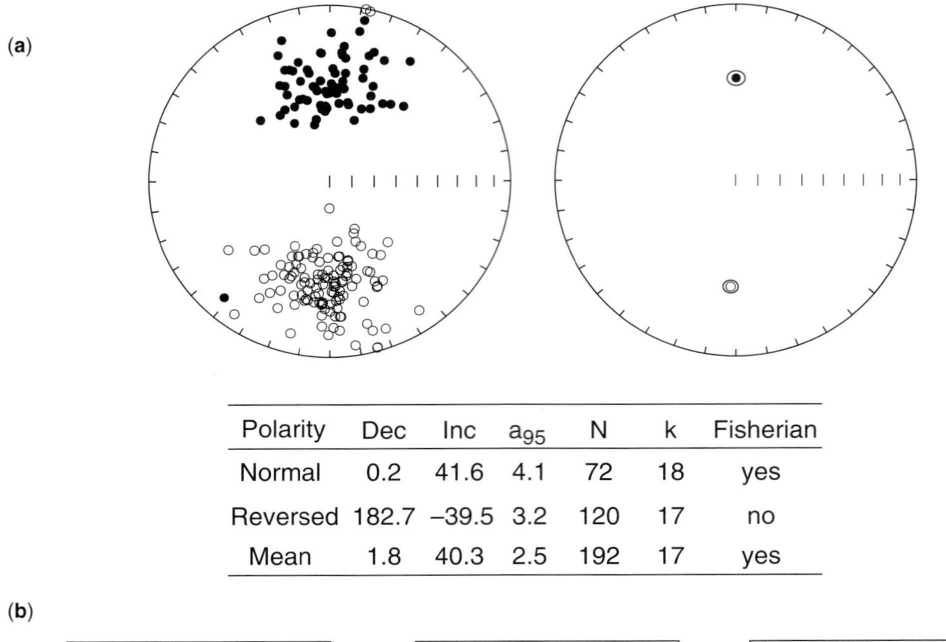

Polarity	Dec	Inc	a_{95}	N	k	Fisherian
Normal	0.2	41.6	4.1	72	18	yes
Reversed	182.7	−39.5	3.2	120	17	no
Mean	1.8	40.3	2.5	192	17	yes

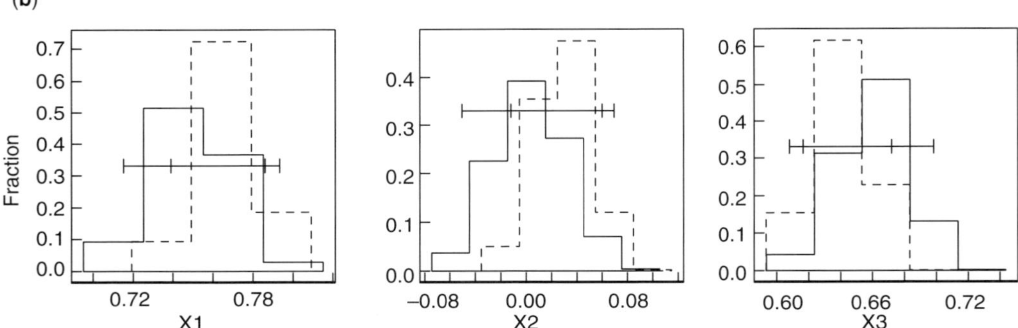

Fig. 5. Plot of ChRMs and the bootstrap reversal test for the locality 5. (**a**) Equal-area projections of ChRM directions (left) and their mean directions (with oval of 95% confidence, top right) for locality 5. Mean directions have been determined using the bootstrap method of Tauxe (1998). Downward directions are plotted as solid circles; upward directions as open circles; α_{95} and k are the statistical parameters associated with the mean directions, N is the number of directions. The column 'Fisherian' lists whether the ChRMs are circularly distributed around the mean: 'no' indicates an elliptical distribution, whereas 'yes' indicates a circular distribution in a statistically significant manner. (**b**) Bootstrap reversal test diagrams, following Tauxe (1998) for the results from locality 5. Reversed polarity directions have been inverted to their antipodes to test for a common mean for the normal and reversed magnetization directions. Confidence intervals for all components overlap, indicating a positive reversal test.

southward (downstream) thinning from localities 1–4 (Fig. 7 and Table 1). Sedimentary grain sizes follow this same trend and become relatively finer southward, with sediments changing from coarse conglomerates in the north to fine-grained sandstones and siltstones in the south (Fig. 7). Additionally, the present-day depositional surfaces of unit 4 in the Nongchunhe Valley indicate a southward decrease

Fig. 6. Magnetostratigraphic results for units 2–4 in locality 5 (see location in Fig. 1). Single sites indicating a different polarity from those of sites derived from layers above and below (e.g. sites at 40 or 77 m of apparently reversed polarity within N1) have been included in the virtual geomagnetic pole (VGP) latitude columns and have been drawn in grey in the column with 'observed polarities'. However, they have not been numbered in this column. Reference geomagnetic polarity time scale (GPTS) is from Gradstein *et al.* (2004). VGP-Lat (latitude) is explained in the text.

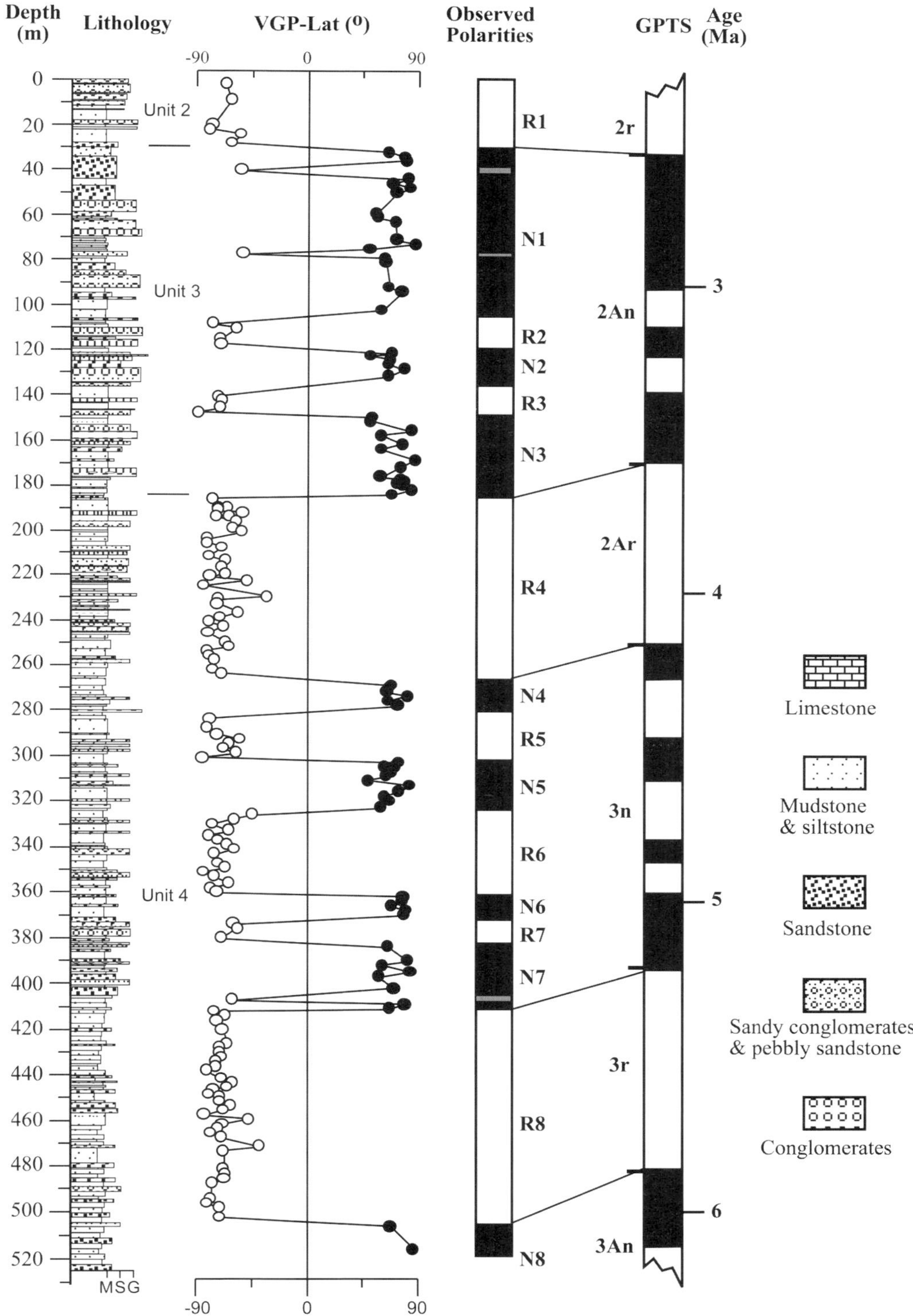

Depth (m)
Lithology
VGP-Lat (°)
Observed Polarities
GPTS
Age (Ma)
-90
0
90
0
20
40
60
80
100
120
140
160
180
200
220
240
260
280
300
320
340
360
380
400
420
440
460
480
500
520
Unit 2
Unit 3
Unit 4
MSG
R1
N1
R2
N2
R3
N3
R4
N4
R5
N5
R6
N6
R7
N7
R8
N8
2r
2An
2Ar
3n
3r
3An
3
4
5
6
Limestone
Mudstone & siltstone
Sandstone
Sandy conglomerates & pebbly sandstone
Conglomerates

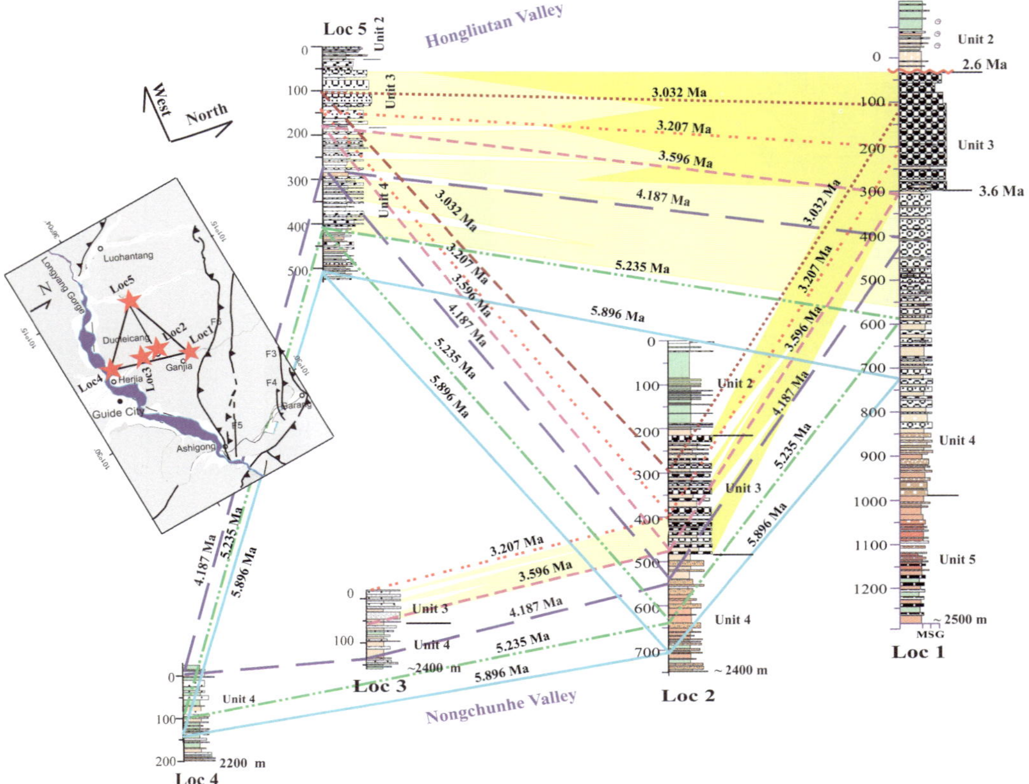

Fig. 7. Chronostratigraphic correlations, made possible by magnetostratigraphy and portrayed by fence diagrams. The lithologies of the five sections in the Guide Basin (see inset map, note that north is to the left) are shown together with unit designations and temporal tie lines (in Ma). The darker yellow colour represents massive conglomerates, whereas the lighter yellow colour represents pebble-conglomerates and coarse sandstones, in contrast to the finer-grained lithologies elsewhere (uncoloured). The ages shown are obtained from magnetostratigraphy and the global polarity time scale of Gradstein *et al.* (2004). Loc, Locality.

in elevation, for example >2500 m in locality 1, *c.* 2400 m in locality 2 and *c.* 2200 m in locality 4, as labelled at the bottom of each section in Figure 7.

The thicknesses of equivalent sedimentary packages within each polarity zone show westward thickening in a longitudinal (east–west) direction before 3.6 Ma, followed by later thinning during the Gauss interval (2.6–3.6 Ma), as can be seen in Table 1 by comparing localities 2 and 5 between the Nongchunhe and Hongliutan valleys.

Discussion

The observed sedimentary facies at locality 5 (Table 2) include a sequence of laminated siltstones and sandstones (unit 4), cycles of sandstones, conglomerates and siltstones (unit 3), and layers of laminated bluish siltstones and mudstones (unit 2).

If we were to base our correlation solely on sedimentary facies analysis, and with neither magnetostratigraphic nor faunal constraints or trace markers, then units 4 and 3 at locality 5 could perhaps have been correlated to units 4 and 3 in the Nongchunhe Valley at localities 2–4, but would more likely have been correlated to units 5 and 4–3 at locality 1. This is because units 4–3 at locality 1 are predominantly conglomeratic. However, the evidence of trace markers and bio-magnetostratigraphy shows that the latter correlation would have been incorrect. Thus, we observe that correlations based solely on lithostratigraphy in the Guide Basin would have been possible but probably partially incorrect. This example shows that our magnetostratigraphic study is useful in establishing temporal constraints on basin-wide lateral facies variations.

Similarly, Late Neogene lithostratigraphic correlations among several basins on the northeastern

Table 2. *Descriptions of the lithologies as used to define the units (2–5) in each of the five localities*

Unit	Locality 1	Locality 2	Locality 3	Locality 4	Locality 5
Unit 2		Mostly limestones, siltstones and mudstones, intercalated with some sandy pebbles			Light blue siltstones irregularly interbedded with sandstones and pebbly conglomerates, and capped with well-cemented pebbly sandstones
Unit 3	Thick layers of grey angular to subangular conglomerates	Interlayers of sandstones, siltstones, and conglomerates	Interlayers of siltstones, sandstones and conglomerates, but finer grained than locality 2		Alternations of siltstones, grey to light brown sandstones and greyish conglomerates
Unit 4	Silty–sandy conglomerates	Interlayers of mudstones, siltstones and sandy pebbles	Interlayers of mudstones, siltstones and sandy pebbles	Layers of mostly limestones, siltstones and mudstones, intercalated with some sandy pebbles	Laminated layers of bluish, grey, light brown siltstones and sandstones, irregularly interbedded with some conglomerates
Unit 5	Cycles of multi-coloured limestones, mudstones and siltstones				

Tibetan Plateau may continue to be unreliable unless chronostratigraphic data become more abundant (e.g. Fig. 8). These basins are the Guide Basin (Fang *et al.* 2005*a*, this study), the Linxia Basin (Li *et al.* 1997; Fang *et al.* 2003) and the Jiuquan Basin (Fang *et al.* 2005*b*; its location near Yumen can be seen in Figure 1(top) in the northerly Hexi Corridor). During the same interval as that of units 2–4 in Guide Basin, the late Neogene facies in the Linxia Basin coarsened from dominantly mudstones and sandstones (Liushu and Hewangjia Formations) to a thick conglomeratic package of molasse-type deposits (the Jishi Formation), which in turn is overlain by grey-green mudstones (Li *et al.* 1997; Fang *et al.* 2003). The similarities with the basin infill in Guide are rather evident (Fig. 8), as noted earlier by Fang *et al.* (2005*a*). Equally interesting is the similarity between Guide's stratigraphy and that of the sediments in the Jiuquan Basin (Fig. 8), which grade upward from cycles of poorly sorted coarse sandstones–sandy conglomerates and siltstones to a thick unit of poorly sorted conglomerates (Fang *et al.* 2005*b*). The fact that predominant molasse-type conglomerates were deposited at more or less the same time (3.6–2.6 Ma, Fig. 8) in three far-apart locations is important evidence for Pliocene tectonism in the area, albeit outside the scope of this contribution. Similar trends south of the Himalayas have been suggested in earlier studies that employed similar lines of reasoning (e.g. Keller *et al.* 1977; Barndt *et al.* 1978; Johnson *et al.* 1979).

The southward thinning of equivalent sedimentary packages and higher sediment accumulation rates in locality 1 than those in localities 3–5 (Fig. 9), may reflect either greater local subsidence around locality 1 near fault F6 (see Fig. 1), or it could indicate that the depositional surface was higher in locality 1 with a southward progradation of alluvial fans, or perhaps some combination of these two. The southward (downstream) fining trend in sedimentary grain sizes (Figs 3 & 7) demonstrates that sediments were derived from a northerly source, such as the nearby Qinghainan Shan and Laji Shan. Large-scale bedding dips are gently to the SSW, so that today's elevation of the depositional surface of unit 4 slopes from above 2500 m at locality 1 to above 2400 m at locality 2 and then to above 2200 m at locality 4 (Fig. 7). The southward fining

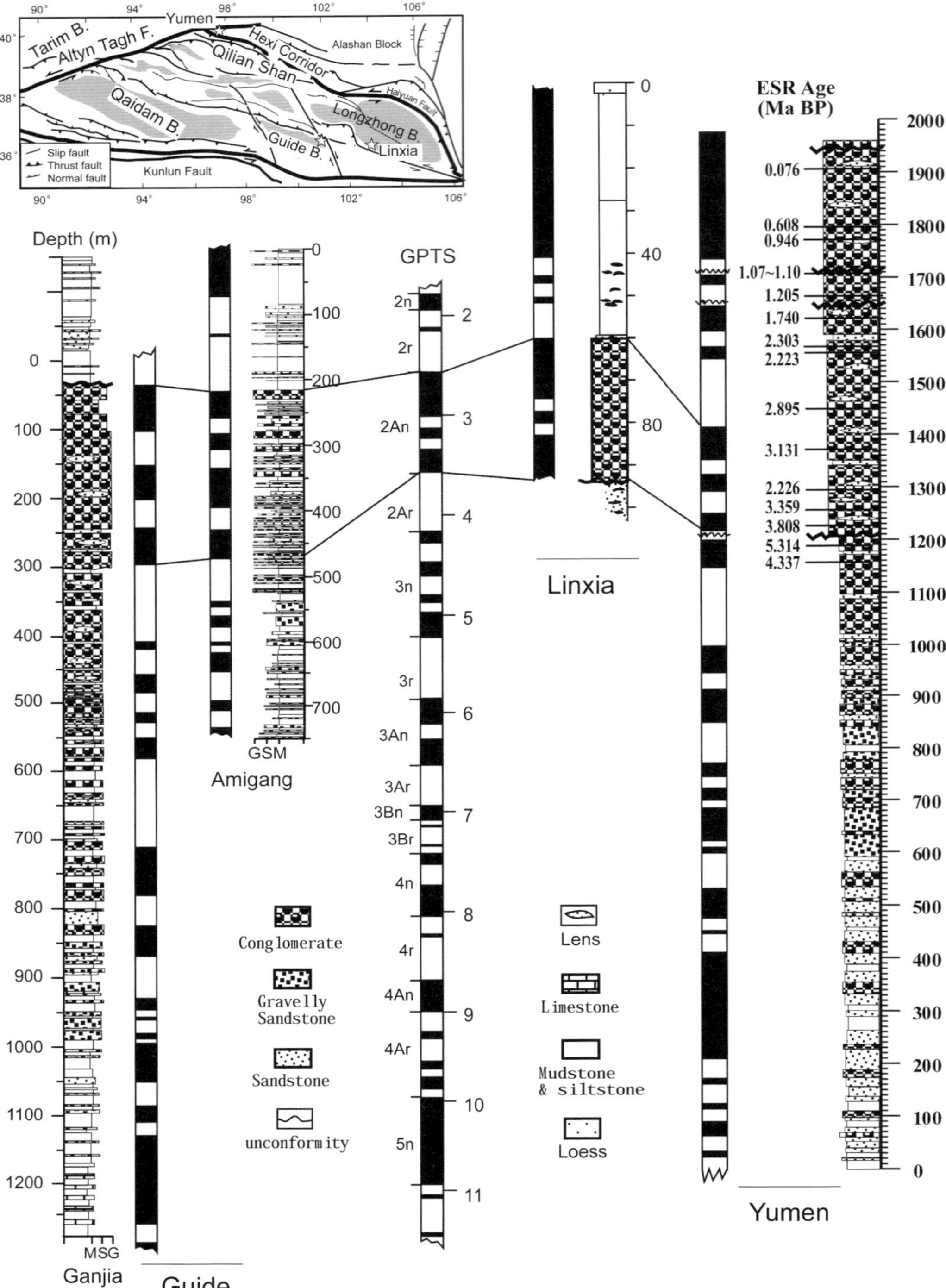

Fig. 8. Litho- and magneto-stratigraphic correlation of the Guide Basin (Ganjia and Amigang sections), Longzhong Basin near Linxia, and Jiuquan Basin near Yumen, with insert of simplified tectonic map of NE Tibetan Plateau on the left upper corner, showing the three locations as stars. The comparison reveals that a molasse-type massive

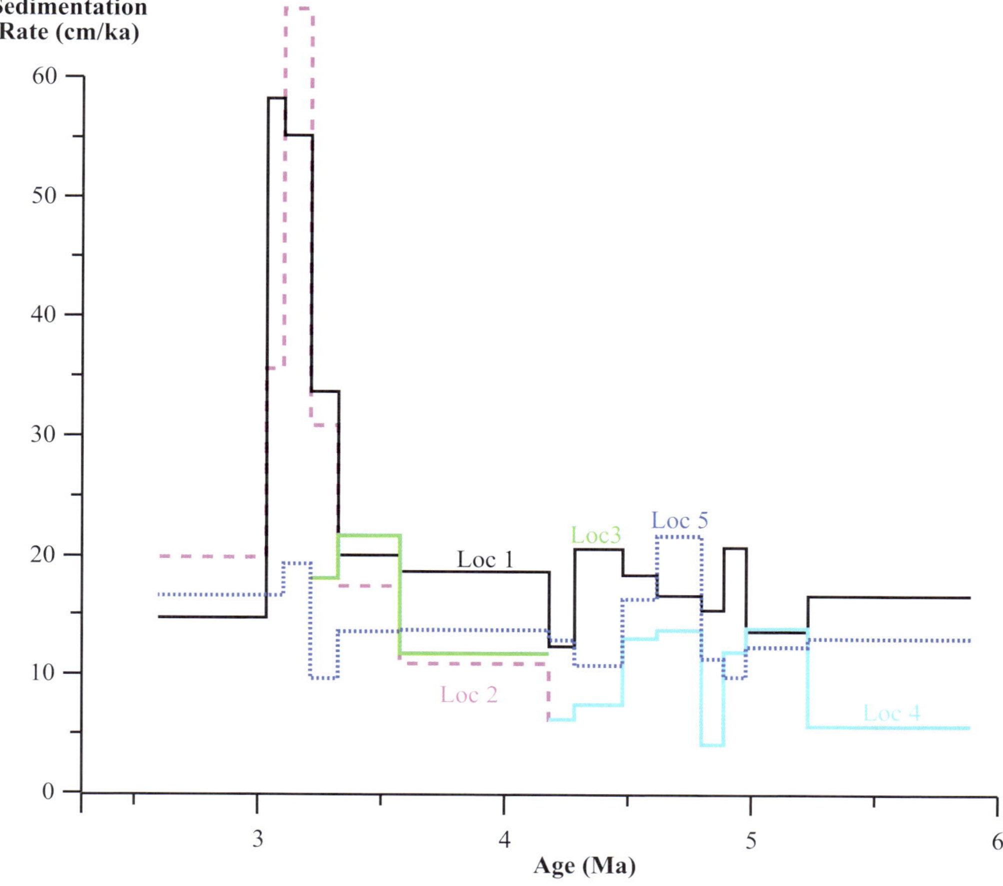

Fig. 9. Sediment accumulation rate v. age for the five sections (data from Table 1). Sediment accumulation rates for all localities show similar variation patterns that have relatively low values during the *c.* 5–3.6 Ma interval. Localities 1 and 2 have extremely high rates in the 2.6–3.6 Ma interval, whereas locality 5 maintained approximately the same sediment accumulation rate throughout the whole latest Miocene-to-Pleistocene interval.

and dip is consistent with southward thinning of units towards distal parts of his sequence. The transition at locality 1 of the depositional environment from shallow lacustrine (unit 5) to braided river (unit 4) and then alluvial fan environment (unit 3, with its massive conglomerates) is coupled with southward fining in grain size along the Nongchunhe River trajectory (Fig. 7). This suggests Pliocene uplift of the nearby mountains, especially between 3.6 and 2.6 Ma.

Figure 7 also reveals that equivalent sedimentary packages thin to the ESE before 3.6 Ma and westward after 3.6 Ma (Table 1 and Fig. 7). This gradient change in a longitudinal (east–west) sense in the basin may imply either a relocation of its depositional centre, or may represent different sedimentary sources supplying the two valleys, if the uplift in provenance areas was diachronous. Gu *et al.* (1992) suggested that the Duoleicuang–Herjia region (see locations in Fig. 1) became the depositional centre of the Guide Basin during the later Pliocene, in support of the latter explanation.

Field observations indicate that the sediments of unit 4 (> 3.6 Ma) not just thin but also fine towards

Fig. 8. *(Continued)* conglomerate unit was deposited synchronously from *c.* 3.6 to 2.6 Ma in each of the three basins. The litho-magnetostratigraphy of the Linxia Basin is from Li *et al.* (1997) and Fang *et al.* (2003), and that of the Yumen Basin is from Fang *et al.* (2005*b*).

the east and SE (from locality 5 in the direction of localities 2 and 4), whereas afterwards unit 3 thins to the west (compare localities 2, 3 and 5). The Nongchunhe Valley is close to thrust faults F5 and F6, whereas the Hongliutan Valley is located closer to the thrust fault that produced the uplift of the Longyang Gorge area upstream of the Yellow River (see locations in Fig. 1b). The time of the Longyang Gorge thrust movement is constrained by the age of the footwall sediments (unit 4) as early Pliocene (about 4.5–5 Ma); this age thus appears to be older than the late Pliocene movements along F5 and F6, in agreement with the above.

Because the Guide Basin was internally drained until 1.8 Ma when the Yellow River eroded into the basin (Fang *et al.* 2005*a*), sediment derived from the surrounding highlands was entirely trapped within the basin system. If so, the sediment-accumulation rates are a direct proxy for rates of denudation in the source areas (Horton *et al.* 2002). Although the influence of climate change cannot be ruled out (Zhang *et al.* 2001), increasing sedimentation may reflect intensified tectonic activity in the sedimentary provenance areas. The sediment accumulation rates in the Nongchunhe are different from that of the Hongliutan valleys (Fig. 9). This difference in sediment accumulation rates might have been a result of two different main sedimentary sources to the two valleys.

All these lines of evidence indicate that the main sedimentary source for units 4 and 3 in locality 5 (Hongliutan Valley) was from the Longyang Gorge area and the nearby Waligong Shan in the west, whereas for localities 1–4 along the Nongchunhe Valley the sources may have been in the Qinghainan Shan and/or the Laji Shan to the north and NE.

Conclusions

The results of our new magnetostratigraphy at locality 5 can easily and confidently be correlated to the GPTS (Gradstein *et al.* 2004), and provides age constraints on the sedimentary infilling along the Hongliutan Valley. The ages agree well with previous dates determined for the Nongchunhe Valley (Fang *et al.* 2005*a*).

Combining previous and new results, a three-dimensional analysis is performed using the spatial and temporal distribution of sedimentary packages in the basin. The grain sizes indicate trends of coarsening/fining in the vertical and north–south horizontal direction, whereas the thicknesses represented by polarity intervals show trends of horizontal thinning or thickening of equivalent sedimentary packages.

Combined with field studies and analysis of deposition rates, the trends can be taken to suggest

different sedimentary provenances for locality 5 in the Hongliutan Valley and localities 1–4 along the Nongchunhe Valley before and after 3.6 Ma. Thus, we find that the sediment accumulation rate during the early Pliocene at locality 5 may be related to the thrust faulting at *c.* 4.5 Ma of the nearby Longyang Gorge Fault to the SW, whereas the extremely high sediment accumulation rates of massive unit 3 conglomerates in localities 1 and 2 during the late Pliocene (3.6–2.6 Ma) can be related to movements on thrust faults F5 and F6 and significant denudation of the Laji Shan to the north.

In summary, our introduction of a method to study basin infill in three dimensions by magnetostratigraphy has been successful in allowing us to correlate lateral facies within the Guide Basin and suggests that similar success may be attainable on a larger scale in comparisons between the various basins of the NE Tibetan Plateau.

We thank Professor B. Wilkinson for sharing many valuable insights and suggestions, and help in improving English at an earlier stage of the manuscript. Y. Miao, W. Yang and L.n Xiong are thanked for generous help with the fieldwork. Two anonymous reviewers are thanked for making many valuable suggestions for improvement. The work was supported by the (973) National Basic Research Program of China (2011CB403000), the Chinese Academy of Science 'Hundred Talent Project' (to M.Y.), and the National Science Foundation of China (NSFC no. 41021001, 40920114001, KZCX-YW-Q09-04). Funding by the US National Science Foundation, Division of Earth Sciences, in grant EAR 0207257 to R. Van der Voo and J. M. Parés, is gratefully acknowledged.

References

BARNDT, J., JOHNSON, N. M. *ET AL.* 1978. The magnetic polarity stratigraphy and age of Siwalik Group near Dhok Pathan Village, Potwar Plateau, Pakistan. *Earth and Planetary Science Letters*, **41**, 355–364.

BURCHFIEL, B. C., ZHANG, P. *ET AL.* 1991. Geology of the Haiyuan fault zone, Ningxia-hui autonomous region, China, and its relation to the evolution of the northeastern margin of the Tibetan Plateau. *Tectonics*, **10**, 1091–1110.

CHEN, S. F., WILSON, C. J. L. & WORLEY, B. A. 1995. Tectonic transition from the Songpan-Garze fold belt to the Sichuan Basin, south-western China. *Basin Research*, **7**, 235–253.

DAI, S., FANG, X. M. *ET AL.* 2006. Magnetostratigraphy of Cenozoic sediments from the Xining Basin: tectonic implications for the northeastern Tibetan Plateau. *Journal of Geophysical Research*, **111**, B11102, http://dx.doi.org/10.1029/2005JB004187

DECELLES, P. G., GEHRELS, G. E., QUADE, J., OJHA, T. P., KAPP, P. A. & UPRETI, B. N. 1998. Neogene foreland basin deposits, erosional unroofing, and the kinematic history of the Himalayan fold–thrust belt, western Nepal. *Geological Society of America Bulletin*, **110**, 2–21.

FANG, X. M., GARZIONE, C., VAN DER VOO, R., LI, J. J. & FAN, M. 2003. Initial flexural subsidence by 29 Ma on the NE edge of Tibet from the magnetostratigraphy of Linxia Basin, China. *Earth and Planetary Science Letters* **10**, 545–560.

FANG, X. M., YAN, M. D. *ET AL*. 2005*a*. Late Cenozoic deformation and uplift of the NE Tibetan Plateau: Evidence from high-resolution magnetostratigraphy of the Guide Basin, Qinghai Province, China. *Geological Society of America Bulletin*, **117**, 1208–1225.

FANG, X. M., ZHAO, Z. J. *ET AL*. 2005*b*. Magnetostratigraphy of the late Cenozoic Laojunmiao anticline in the northern Qilian Mountains and its implications for the northern Tibetan Plateau uplift. *Science China (D)*, **48**, 1040–1051.

FANG, X. M., ZHANG, W. L. *ET AL*. 2007. High-resolution magnetostratigraphy of the Neogene Huaitoutala section in the eastern Qaidam Basin on the NE Tibetan Plateau, Qinghai Province, China and its implication on tectonic uplift of the NE Tibetan Plateau. *Earth and Planetary Science Letters*, **258**, 293–306.

GRADSTEIN, F., OGG, J. & SMITH, A. 2004. *A Geological Timescale*. Cambridge University Press, Cambridge, 344–383.

GU, Z., BAI, S., ZHANG, X., MA, Y., WANG, S. & LI, B. 1992. Neogene subdivision and correlation of sediments within the Guide and Hualong basins of Qinghai province. *Journal of Stratigraphy*, **16**, 96–104 (in Chinese).

HORTON, B. K., YIN, A., SPURLIN, M. S., ZHOU, J. & WANG, J. 2002. Paleocene–Eocene syncontractional sedimentation in narrow, lacustrine-dominated basins of east-central Tibet. *Geological Society of America Bulletin*, **114**, 771–786.

JOHNSON, G. D., JOHNSON, N. M., OPDYKE, N. D. & TAHIRKELI, R. A. K. 1979. Magnetic reversal stratigraphy and sedimentatary tectonic history of the Upper Siwalik Group, eastern Salt Range and southwestern Kashmir. *In*: FARAH, A. & DE JONG, K. A. (eds) *Geodynamics of Pakistan*. Geological Survey of Pakistan, Quetta, 149–166.

KELLER, H. M., TAHIRKHELI, R. A. K., MIRZA, M. A., JOHNSON, G. D. & JOHNSON, N. M. 1977. Magnetic polarity stratigraphy of the Upper Siwalik deposits, Pabbi Hills, Pakistan. *Earth and Planetary Science Letters*, **36**, 187–201.

KIRSCHVINK, J. L. 1980. The least-square line and plane and the analysis of paleomagnetic data. *Geophysical Journal of the Royal Astronomical Society*, **62**, 699–718.

LI, J.-J., FANG, X.-M. *ET AL*. 1997. Late Cenozoic magnetostratigraphy (11–0 Ma) of the Wangjiashan section in the Longzhong Basin, Western China: evidence for rapid mid-Pliocene uplift of the Tibetan Plateau. *Geologie en Mijnbouw*, **76**, 121–134.

LU, H. & XIONG, S. 2009. Magnetostratigraphy of the Dahonggou section, northern Qaidam Basin and its bearing on Cenozoic tectonic evolution of the Qilian Shan and Altyn Tagh Fault. *Earth and Planetary Science Letters*, **288**, 539–550.

MCFADDEN, P. L. & MCELHINNY, M. W. 1990. Classification of the reversal test in paleomagnetism. *Geophysical Journal International*, **103**, 725–729.

MEYER, B., TAPPONNIER, P. *ET AL*. 1998. Crustal thickening in Gansu-Qinghai, lithospheric mantle subduction, and oblique, strike-slip controlled growth of the Tibet plateau. *Geophysical Journal International*, **135**, 1–47.

MIALL, A. D. 2000. *Principles of Sedimentary Basin Analysis*, 3rd edn. Springer, New York.

PARÉS, J. M., VAN DER VOO, R., DOWNS, W. R., YAN, M. D. & FANG, X. M. 2003. Northeastward growth and uplift of the Tibetan Plateau: Magnetostratigraphic insights from the Guide Basin. *Journal of Geophysical Research*, **108**, 2017, http://dx.doi.org/10.1029/2001JB001349

QINGHAI GEOLOGY BUREAU. 1989. *Regional Geology of Qinghai Province*. Geology Press, Beijing, 215–217 (in Chinese).

SOBEL, E. R. & DUMITRU, T. A. 1997. Thrusting and exhumation around the margins of the western Tarim Basin during the India-Asia collision. *Journal of Geophysical Research*, **102**, 5043–5063.

SOBEL, E. R., HILLEY, G. E. & STRECKER, M. R. 2003. Formation of internally drained contractional basins by aridity-limited bedrock incision. *Journal of Geophysical Research*, **108**, 2344, http://dx.doi.org/10.1029/2002JB001883

SONG, C. H., FANG, X. M., GAO, J. P., SUN, D. & FAN, M. J. 2001. Cenozoic tectonic uplift and sedimentary evolution of the Guide Basin in the northeast margin of the Tibetan Plateau. *Sedimentology*, **19**, 498–506 (in Chinese with English abstract).

SONG, C., FANG, X. M., GAO, J. P., NIE, J., YAN, M. D., XU, X. & SUN, D. 2003. Magnetostratigraphy of Late Cenozoic fossil mammals in the northeastern margin of the Tibetan Plateau. *Chinese Science Bulletin*, **48**, 188–193.

TAPPONNIER, P., MEYER, B. *ET AL*. 1990. Active thrusting and folding in the Qilian Shan and decoupling between upper crust and mantle in northeastern Tibet. *Earth and Planetary Science Letters*, **97**, 382–403.

TAUXE, L. 1998. *Paleomagnetic Principles and Practice*. Kluwer Academic, Dordrecht.

YAN, M. D., VAN DER VOO, R., FANG, X. M., PARÉS, J. M., REA, D. K. & SONG, C. 2004. Neogene magnetostratigraphy of the Guide Basin: Mid-Miocene clockwise rotation and implications for uplift of the NE Tibetan Plateau. *Geological Society of America Abstracts with Programs*, abstract 18–13, **36**, 50.

YAN, M. D., VAN DER VOO, R., TAUXE, L., FANG, X. M. & PARÉS, J. M. 2005. Shallow bias in Neogene paleomagnetic directions from the Guide Basin, NE Tibet, caused by inclination error. *Geophysical Journal International*, **163**, 944–948.

YAN, M. D., VAN DER VOO, R., FANG, X. M., PARÉS, J. M. & REA, D. K. 2006. Paleomagnetic evidence for a mid-Miocene clockwise rotation of about 25° of the Guide Basin area in NE Tibet. *Earth and Planetary Science Letters*, **241**, 234–247.

YIN, A. & HARRISON, M. T. 2000. Geologic evolution of the Himalayan–Tibetan orogen. *Annual Review of Earth and Planetary Sciences*, **28**, 211–280.

ZHANG, P., MOLNAR, P. & DOWNS, W. R. 2001. Increased sediment accumulation rates and grain sizes 2–4 Myr ago due to the influence of climate change on erosion rates. *Nature*, **410**, 891–897.

Oligocene slow and Miocene–Quaternary rapid deformation and uplift of the Yumu Shan and North Qilian Shan: evidence from high-resolution magnetostratigraphy and tectonosedimentology

XIAOMIN FANG[1,2]*, DONGLIANG LIU[1,3], CHUNHUI SONG[2], SHUANG DAI[2] & QINGQUAN MENG[2]

[1]*Key Laboratory of Continental Collision and Plateau Uplift & Institute of Tibetan Plateau Research, Chinese Academy of Sciences, Shuangqing Road 18, Beijing 100085, China*

[2]*Key Laboratory of Western China's Environmental Systems (Ministry of Education of China) and College of Resources and Environment, Lanzhou University, Gansu 730000, China*

[3]*Key Laboratory of Continental Dynamics of the Ministry of Land and Resources, Institute of Geology, Chinese Academy of Geological Sciences, Beijing 100037, China*

**Corresponding author (e-mail: fangxm@itpcas.ac.cn)*

Abstract: Most existing tectonic models suggest Pliocene–Quaternary deformation and uplift of the NE Tibetan Plateau in response to the collision of India with Asia. Within the NE Tibetan Plateau, growth of the terranes was suggested to progress northeastward with the Yumu Shan (mountain) at the northeasternmost corner of the Qilian Shan (mountains) being uplifted only since about 1 Ma ago. Here we present a detailed palaeomagnetic dating and tectonosedimentological measurement of Cenozoic sediments in the eastern Jiuquan Basin related to the deformation and uplift of the North Qilian Shan and Yumu Shan. The results show that the eastern Jiuquan Basin is a Cenozoic foreland basin and received sediments at about 27.8 Ma at the latest. Eight subsequent tectonic events at about 27.8, 24.6, 13.7–13, 9.8–9.6, 5.1–3.6, 2.8–2.6, 0.8 and 0.1 Ma demonstrate the development of the foreland basin in response to Oligocene–Quaternary uplift of the North Qilian Shan and subsequent propagation of thrust–fold system owing to collision of India with Asia. The Yumu Shan is the late phase of deformation front in the thrust–fold system and commenced rapid uplift at about 9.8–9.6 Ma at the latest. A rigid block-floating model is proposed to interpret the mechanism of this deformation and uplift history.

The NE Tibetan Plateau is the furthest topographic and deformation edge of the Tibetan Plateau and is characterized by the overall features of WNW-trending ranges (Qilian Shan) and basins (Hexi to Qaidam) with high topographic reliefs (*c.* 4000–5000 m) and no Cenozoic volcanism, all truncated by the *c.* 1500 m long ENE-trending lithospheric strike-slip fault, the Altun fault (Fig. 1). This contrasts with the main Tibetan Plateau between the Himalayas and the Kunlun Shan, where a vast flat surface with widespread north–south-trending normal faults and grabens and Cenozoic volcanism occurs (Molnar & Tapponnier 1975; Fielding *et al.* 1994).

How and when these tectonomorphological features formed is less directly dealt with by many tectonic models of Tibet formation, because most working models are based on observations and data mainly from the southern edge (Himalayas–Gangdese) and in part from southern and central Tibet (e.g. Ni & Barazangi 1984; England & Houseman 1986, 1989; Dewey *et al.* 1988; Harrison *et al.* 1992;

Molnar *et al.* 1993); few data have come from the remote northeastern edge of Tibet. So far only two groups of end models based on recognition of the nature of the lithosphere have provided some information on the evolution of the NE Tibetan Plateau.

The first group regards the Asian lithosphere as a thin viscous sheet, which was continuously, broadly and homogeneously thickened on India–Asia collision. Convection removal of the thickened Asian lithospheric mantle root under Tibet led to rapid and uniform isostatic rebound of the vast Tibetan Plateau to an elevation higher than the present level (*c.* 5000 m). Subsequent gravity collapse gave rise to widespread east–west extensions (manifested as north–south directed normal faults and grabens) and volcanism (England & Houseman 1986, 1988, 1989) and associated Asian monsoon onset (Molnar *et al.* 1993). Dating these events has constrained this wholesale uplift, which occurred in the Late (Harrison *et al.* 1992; Molnar *et al.* 1993) or Middle Miocene (Turner *et al.* 1993;

From: JOVANE, L., HERRERO-BERVERA, E., HINNOV, L. A. & HOUSEN, B. A. (eds) 2013. *Magnetic Methods and the Timing of Geological Processes*. Geological Society, London, Special Publications, **373**, 149–171. First published online November 15, 2012, http://dx.doi.org/10.1144/SP373.5 © The Geological Society of London 2013. Publishing disclaimer: www.geolsoc.org.uk/pub_ethics

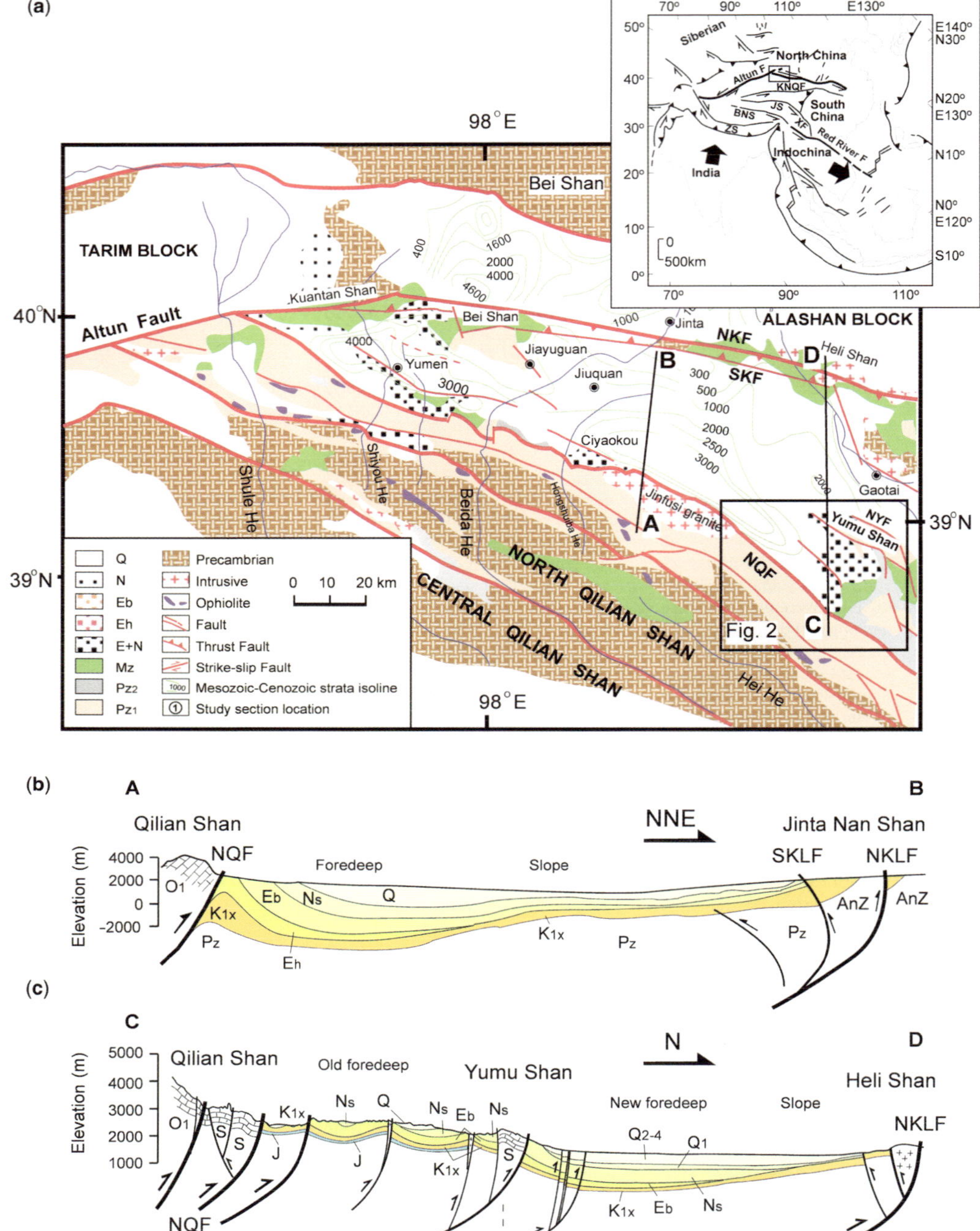
(a)
98°E
Bei Shan
TARIM BLOCK
Kuantan Shan
40°N
Altun Fault
Bei Shan
ALASHAN BLOCK
Jinta
NKF
Jiayuguan
SKF
Yumen
Jiuquan
Heli Shan
B
D
Ciyaokou
Gaotai
Shiyou He
Beida He
Hongshuiba He
Jinfusi granite
39°N
NYF
Yumu Shan
A
NQF
Fig. 2
C
39°N
NORTH QILIAN SHAN
CENTRAL QILIAN SHAN
Hei He
98°E
Q
Precambrian
N
Intrusive
Eb
Ophiolite
0   10   20 km
Eh
Fault
E+N
Thrust Fault
Mz
Strike-slip Fault
Pz2
1000 Mesozoic-Cenozoic strata isoline
Pz1
① Study section location

Siberian
North China
Altun F
KNQF
South China
BNS
JS
XF
Red River F
ZS
Indochina
India
70° 90° 110° E130°
E140°
N30°
50°
40°
N20°
E130°
30°
N10°
20°
N0°
10° E120°
0 S10°
500km
0°
70° 90° 110°

(b)
A
B
Qilian Shan
NNE
Jinta Nan Shan
NQF
SKLF NKLF
4000 Foredeep Slope
Elevation (m)
2000 O1 Eb Ns Q
0 K1x K1x AnZ AnZ
-2000 Pz Pz Pz
Eh

(c)
C
N
D
Qilian Shan Old foredeep Yumu Shan Heli Shan
5000
4000 NKLF
Elevation (m) 3000 O1 K1x Ns Q Ns Eb Ns New foredeep Slope
2000 S S J Q2-4 Q1
1000 J J S K1x
K1x Eb Ns
NQF NYF GSF

Time NE
Yumu Shan thrust-fold NYF Maying drepression

Coleman & Hodges 1995) or earlier (Chung *et al.* 1998). This model focuses on interpretation of first-order features of flatness, grabens and volcanism of the vast area north of Himalayas and south of Kunlun Shan, regarding the NE Tibetan Plateau only as a northern boundary of the model. From this model, the isostatic rebound of Tibet would have caused strong coeval activations of the northeastern boundary edge.

This group of models is derived from the hypothesis that distributed and continuous creep deformation occurs only in the lower crust (Royden 1996). Such lower crust flow moves progressively northwards and eastwards with the India–Asia collision and a later phase of deformation and uplift of the north and east margins of Tibet would be predicted (Royden *et al.* 1997, 2008).

The second group of end models regards the Asian lithosphere as rigid blocks, which were progressively broken along some former suture zones in Tibet and then squeezed out eastwards along some newly formed large lithospheric slip faults at the Tibet margin in response to the India–Asia collision, leading to northeastward oblique stepwise rise and growth of the Tibetan Plateau (Meyer *et al.* 1998; Tapponnier *et al.* 2001). South Tibet was raised with great eastward extrusion along the Red River fault in the Eocene; central-north Tibet slipped eastwards along Jinshajiang suture–Xianshuihe fault and Kunlun fault, and uplifted in the Oligocene–Miocene; and NE Tibet was extruded and grew northeastwards along the Altun fault in the Plio-Quaternary (Meyer *et al.* 1998; Tapponnier *et al.* 2001; see Fig. 1 insert for locations). Within NE Tibet, as the Altun fault propagates northeastwards, the upper crust of the NE Tibet is progressively decoupled from the Asian lower crust and mantle dipping and moving southwards beneath the Kunlun Shan (Burchfiel *et al.* 1989; Tapponnier *et al.* 1990), causing northeastward stepwise rise of the south, central and north Qilian Shan in the Plio-Quaternary and the northeastmost Yumu Shan at *c.* 1 Ma (see Fig. 15c in Tapponnier *et al.* 1990, 2001; Metiver *et al.* 1998; Meyer *et al.* 1998; Fig. 1).

Therefore, these models outlined above can be used to test the timing and processes that formed the present macrofeature of the NE Tibetan Plateau, and will provide help in understanding the dynamic mechanism of the Tibetan Plateau formation and continental deformation. The timing of deformation events holds the key.

For late-phase deformation and uplift of NE Tibet predicted from the models above, fission track analysis shows that the Qilian Shan experienced a rapid cooling in the Miocene (George *et al.* 2001; Jolivet *et al.* 2001), and some preliminary palaeomagnetic work shows that the Danghe Nan Shan (South Qilian Shan) and western North Qilian Shan may have uplifted in the Eocene or Oligocene (Yin *et al.* 2002; Dai *et al.* 2005). Recent U–Th/He dating of rocks in the central East Kunlun Shan (mountains) and West Qinling (mountains) and basin sedimentological analysis indicate an Eocene to Oligocene deformation and uplift of the NE Tibetan Plateau (Clark *et al.* 2010; Zhang *et al.* 2010; Fang *et al.* 2003). Our previous high-resolution palaeomagnetic dating of the Laojunmiao section in Yumen in the western Jiuquan Basin provided the first detailed time constraint for the late Cenozoic stratigraphy and demonstrated that the western North Qilian Shan was rapidly uplifted at latest about 8 Ma (Fang *et al.* 2005*b*). These studies show the lack of consensus and call for more detailed work to identify the precise timing of the Cenozoic deformation and uplift history of the Qilian Shan. Here we present a detailed palaeomagnetic dating and tectonosedimentological analysis of Cenozoic sediments in the margins of the Yumu Shan to test if the mountain was the most recently (*c.* 1 Ma) uplifted part of the NE Tibetan Plateau as predicted by the commonly accepted model of Tapponnier's group (Tapponnier *et al.* 1990, 2001; Meyer *et al.* 1998; Metiver *et al.* 1998).

Geological setting and stratigraphy

The NE Tibetan Plateau is a terrane delineated by the major sinistral Kunlun–North Qinling strike-slip fault in the south, the Altun fault to the west and the North Qilian–Haiyuan–Liupan Shan fault in the north and east (Fig. 1). From SW to NE, it consists of Qaidam Basin, Qilian Shan, Longzhong Basin and Liupan Shan. The Qilian Shan is

Fig. 1. Geological and location map of the Qilian Shan and Hexi Corridor Basin (**a**). Note that the basin-cross sections (**b** and **c**) show the nature of a foreland basin owing to compressive flexure and the northward migration of the foredeep owing to uplift of the Yumu Shan (c; data were provided by Yumen Oil Field Co.). NQF, North Qilian fault; NCQF, northern marginal fault of Central Qilian Shan; NKLF, northern marginal fault of Kuantan Shan–Longshou Shan; SKLF, southern marginal fault of Kuantan Shan–Longshou Shan; NYF, northern Yumu Shan fault; KNQF, Kunlun–North Qinling fault; XF, Xianshuihe fault; JS, Jinshajiang suture; BS, Bangong suture; ZS, Zangbo suture. O_1, Early Ordovician; S, Silurian; J, Jurassic; K_{1x}, Early Cretaceous Xiagou Formation; E_b, Palaeogene Baiyanghe Formation; N_s, Neogene Shulehe Formation; Q_1, Early Quaternary; Q_{2-4}, Late Quaternary.

of interest in this study, and consists of south, central and north Qilian Shan. The Yumu Shan is the northeasternmost corner of the North Qilian Shan (Fig. 1).

The Qilian Shan orogenic belt formed in the Caledonian through collisions of Proterozoic south and central Qilian terranes with North China block and Eastern Kunlun–Qaidam terrane, mostly in the Silurian (see Fig. 1 inset for locations). Subsequent tectonic movements in response to collisions of older Tibetan terranes (Qiangtang and Lhasa) in the Tethys Sea with Palaeo-Asia (here southern margin of Eastern Kunlun–Qaidam terrane) resulted in relatively minor modification of the existing structural setting. The sea receded completely from the North Qilian Shan from the Middle Permian (Gansu Geologic Bureau 1989; Feng & Wu 1992; Yin & Harrison 2000; Fig. 2). The purpose of this paper is to examine the last rejuvenation (uplift) of the Qilian Shan caused by remote response to the collision of the Indian block with Palaeo-Asia (here the southern margin of the Lhasa terrane) in the Neotethys Sea.

The southern and central Qilian terranes and Alashan Block consist mainly of very thick (>15 km) basement rocks of Proterozoic gneisses, schists, slates, phyllite, dolomite and limestones, intercalated with some quartzite, chert, shales, migmatite, tuff and basalt. Early Palaeozoic and Silurian–Devonian stratigraphy were absent from these terranes, with only a small amount of Middle Cambrian and Lower and Middle Ordovician marine limestone and clasts deposited in marginal areas. The cover rocks are mainly Carboniferous–Triassic neritic and paralic clast and carbonates, intermontane Jurassic coal-bearing sediments and Cretaceous–Quaternary clast and molasse (Gansu Geologic Bureau 1989; Feng & Wu 1992; Fig. 2).

The North Qilian Shan is composed mainly of Cambrian–Devonian stratigraphy and Caledonian granite and diorite, which form the core of the mountain. Carboniferous to Quaternary stratigraphy is restricted mostly along the northern margin of the core. The Cambrian–Ordovician rocks form the base of the North Qilian Shan and are mainly basic–intermediate volcanic and pyroclastic rock, chert, phyllite and slate, recording the Qilian ocean history (Fig. 2). The Silurian consists of low-grade metamorphosed continental flysch consisting of purple sandstone and phyllite, representing the closure of the Qilian Ocean. The Devonian is a molasse sequence of post-orogenic purple conglomerates and sandstones. The Carboniferous and Lower Permian are neritic and paralic grey, green sandstone, carbonaceous shale and limestone intercalated with some coal beds. The Upper Permian consists of continental purple conglomerate and sandstone, recording the complete emergence of the region from the sea (Fig. 2). The Triassic to Cretaceous are intermontane multi-coloured (green, yellow-green, blue,

grey, purple-grey, purple, brown) sandstone and conglomerate, intercalated with carbonaceous shale, coal and mudstone (Gansu Geologic Bureau 1989; Feng & Wu 1992; Fig. 2).

The Yumu Shan is an independent massif located to the north of the North Qilian Shan uplifted on the North Yumu Shan progradation fault (NYF; Tapponnier *et al.* 1990; Fig. 1c). It comprises mainly Silurian sedimentary rocks and small proportions of other Palaeozoic and Mesozoic rocks having the same lithology as those in the North Qilian Shan (Gansu Geologic Bureau 1971; Figs 3 & 4).

To the north of the Qilian lies the Alashan Block and its marginal mountains of Kuantan Shan–Bei Shan–Longshou Shan, consisting of Proterozoic rocks similar to those in the Qilian Shan (Gansu Geologic Bureau 1989; Feng & Wu 1992; Figs 1 & 2).

Between the Qilian Shan and the Alashan Block is the long and narrow Hexi Corridor Basin, a Cenozoic foreland basin developed atop a Jurassic foreland basin (Wang & Coward 1993) with its base as a passive continental margin of the Alashan Block (Feng & Wu 1992; Fig. 2), or developed on Early Cretaceous extensive basins in the front of these mountains (Gansu Geologic Bureau 1989; He *et al.* 2004). This foreland basin is divided from NW to SE into Jiuquan, Zhangye and Wuwei sub-basins by several NNW dextral faults and their related uplands in the Yumu Shan and the Dahuang Shan (EGPGYO 1989; Fang *et al.* 2005*b*). The studied sections are located in the east end of the Jiuquan Basin surrounded by the North Qilian Shan, Altun fault, Kuantan Shan and Yumu Shan to the south, west, north and east, respectively (Figs 1 & 3).

Thick Cenozoic stratigraphy is deposited in the Hexi Corridor Basin and its distribution is strictly controlled by faults. Cenozoic stratigraphy is exposed only north of the Northern Qilian fault and south of the southern marginal fault of Kuantan Shan–Longshou Shan. Stratigraphy thickness is over 3000 m in proximity to the North Qilian Shan and gradually thins northwards to <300 m in the southern foot of the Longshou Shan, presenting a typical foreland basin clastic wedge (Fig. 1a, b). Within the basin, several propagation faults control some new foredeep formation and sedimentary deposition, which caused a second depocentre in the new foredeep. This is quite obvious in the western part of the Jiuquan Basin (not shown by cross section in our Fig. 1 owing to space limit, but can be seen on Fig. 1 in Fang *et al.* 2005*b*). North of the Yumu Shan in the eastern end of the Jiuquan Basin, the northern Yumu Shan fault clearly controls a new foredeep and deposition of the Neogene Shulehe Formation and Quaternary sediments to the north of the fault (Fig. 1c). Isobath lines obtained from boreholes and seismostratigraphy of PetroChina demonstrate clearly a second depocentre of the

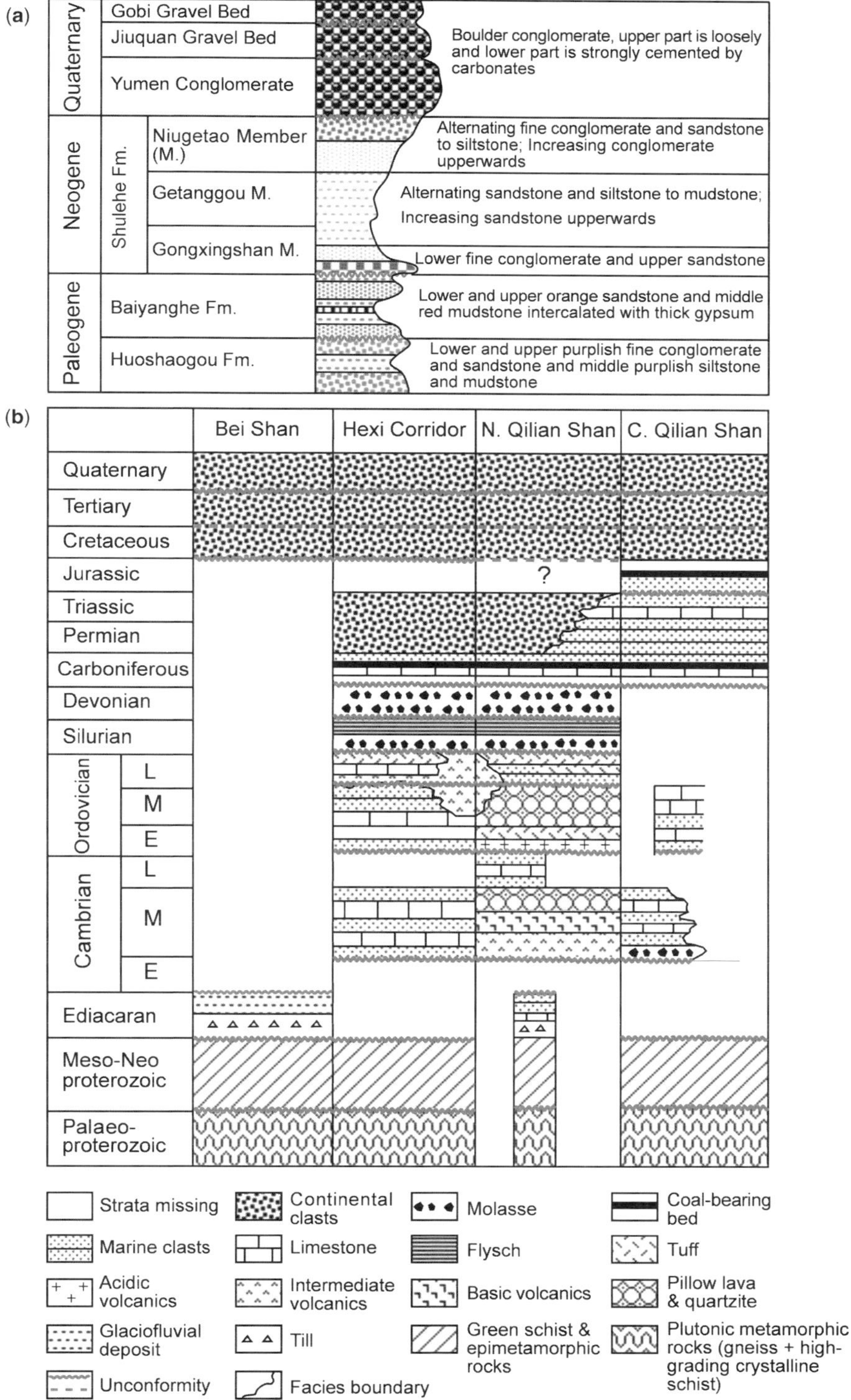

Fig. 2. Tectonostratigraphic evolution of the studied region.

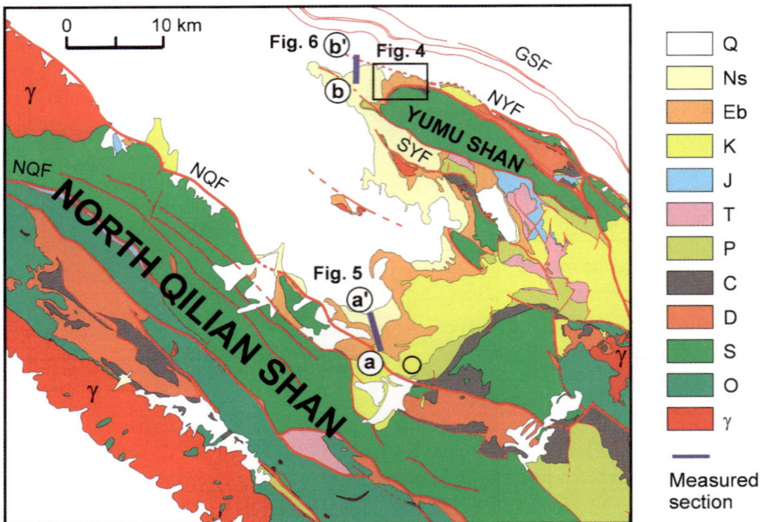

Fig. 3. Geological map of the Yumu Shan and North Qilian Shan showing domination of Palaeozoic rocks and Caledonian granite and locations of two studied sections. Note that the distribution of the Oligocene stratigraphy of the Baiyanghe Formation extends into the North Qilian Shan. NQF, Northern Qilian fault; NYF, northern Yumu Shan fault; SYF, southern Yumu Shan fault; GSF, Gaotai Station fault. T, Triassic; P, Permian; C, Carboniferous; D, Devonian; γ, Caledonian granites. Other stratigraphic units are same as in Figure 1.

Cenozoic stratigraphy appearing in the front of the northern Yumu Shan fault (Fig. 1a).

The Cenozoic stratigraphy consists upwards of Palaeogene Huoshaogou Formation (Eh) and Baiyanghe Formation (Eb) of alluvial and fluviolacustrine red beds of fine conglomerate to mudstone intercalated with some playa gypsum beds, Neogene Shulehe Formation (Ns) of alternated fluviolacustrine grey and brown fine conglomerate, sandstone and siltstone, and Quaternary Yumen Conglomerate Bed (Yumen Formation, Q_1), Jiuquan Gravel Bed (Jiuquan Formation, Q_2) and Gobi Gravel Bed (Gobi Formation, Q_{3-4}; Figs 2 & 3). The Huoshaogou Formation is only distributed in the proximity to the North Qilian Shan and Kuantan Shan–Longshou Shan (Fig. 1; Gansu Geologic Bureau 1989; EGPGYO 1989; Dai *et al.* 2005; Fang *et al.* 2005*b*). The studied area only has the Cenozoic stratigraphic sequence from the Baiyanghe Formation, which is superimposed unconformably on the Lower Cretaceous rocks (Gansu Geologic Bureau 1971). This sequence is completely exposed in folds in the fronts of the North Qilian Shan and Yumu Shan (Figs 1 & 3).

Studied sections

Two sections were chosen for detailed measurements and sampling. One is located on the southern limb of the Sunan syncline between the North Qilian Shan and the Yumu Shan, called the Sunan section (3853'54.9"N, 99°35'39.6"E), the other on the outer part of the northern limb of the Yumu Shan anticline, called the Upper Yumu Shan section (39°15'19.43"N, 99°29'46.13"E; Figs 1c & 3).

The Sunan section is 916 m thick and exposes the Baiyanhe Formation and the Shulehe Formation (Fig. 5). An angular unconformity (U1) exists between the Baiyanhe Formation and the underlying Cretaceous Ximinpu Group. The Baiyanhe Formation (from 0 to 272 m) consists predominantly of fine-grained distinct red-orange, brownish red and purple mudstones and sandstones, intercalated with thin fine-grained grey conglomerate layers. It contains a characteristic thick sandy gypsum bed near the bottom (Fig. 5). The Shulehe Formation (from 272–916 m) is an upward coarsening sequence with the lower part consisting of alternating layers of grey conglomerate and yellow-brown mudstone and siltstone, and the upper part of predominant thick grey conglomerate layers, intercalated with thin brownish yellow sandstone and siltstone lenses. An angular unconformity exists between the Baiyanghe and Shulehe Formations (U2). Growth strata exist at the bottom of the Shulehe Formation, with strata dip shallowing from 25 to 18° between 272 and 343 m (Fig. 5).

The Upper Yumu Shan section contains sediments from the Beiyanghe Formation to the Quaternary (Figs 3 & 4), but it is complicated by

Fig. 4. Southeastern view of the western Yumu Shan and the Cenozoic stratigraphy in the Lower Yumu Shan section. Note that the Yumu Shan thrusts over the Cenozoic stratigraphy along the North Yumu Shan fault (NYF). See Figure 2 for location.

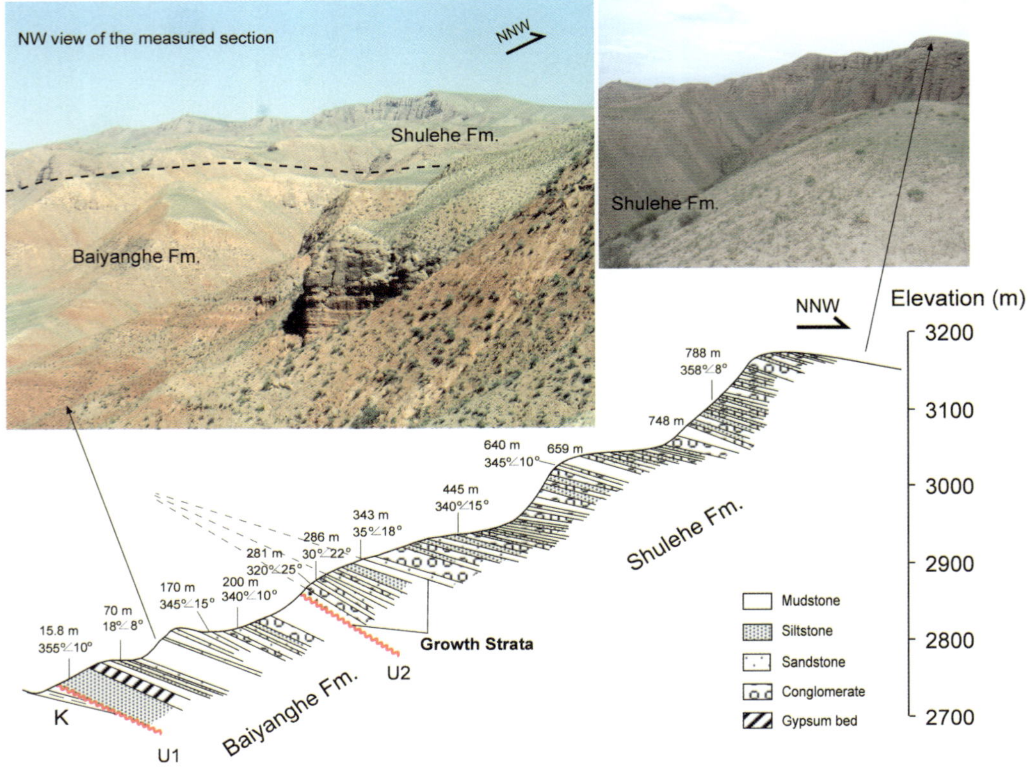

Fig. 5. The measured Sunan section along the southern limb of the Sunan syncline between the North Qilian Shan and the Yumu Shan. See Figures 1*c* and 2 for location.

many faulted and buried segments owing to close proximity to the Yumu Shan, and thus is not suitable for main palaeomagnetic dating. Only the uppermost part of this section was studied to supplement the Sunan section. The measured Upper Yumu Shan section is 400 m thick. Its bottom part consists of 69 m of the uppermost Shulehe Formation, characterized by interbedded thin yellow–grey fine conglomerate and sandstone. Its middle and upper parts are all thick layers of Quaternary greyconglomerates intercalated with some thin yellowish sandstone and siltstone lenses, with gravel diameter coarsening upwards. Four unconformities (U3–U6) occur at thicknesses 69, 148, 296 and 400 m, dividing the Quaternary conglomerates respectively into the Lower and Upper Yumen Conglomerate Beds, the Jiuquan Gravel Bed and the Gobi Gravel Bed (unmeasured; Fig. 6).

Sampling and laboratory measurements

Within the two measured sections, orientated block samples of about $10 \times 10 \times 8$ cm in size were taken at intervals of 1–2 m in the mudstone and sandstone layers, and *c.* 3–4 m in the conglomerate layers, depending on the occurrence of sandstone and siltstone lenses. Each oriented sample was cut into three cubic specimens of $2 \times 2 \times 2$ cm in size in the laboratory. A total of 630 block samples and 1890 specimens were obtained.

According to previous studies in this region, thermal demagnetization provides better results than alternating field demagnetization, as the magnetization carrier is mostly hematite (Dai *et al.* 2005; Fang *et al.* 2005*b*). We only used thermal demagnetization analysis in this study. The samples were measured on a 2 G magnetometer in a magnetically sheltered room in the Paleomagnetism Laboratory of the Institute of Geology and Geophysics, Chinese Academy of Sciences.

Eighteen systematic stepwise thermal demagnetizations were carried out for pilot samples from different lithologies and layers, from room temperature to 680 °C with intervals of 10–50 °C. Representative thermal demagnetization diagrams show that most samples present a similar demagnetization behaviour. The low temperature component can be

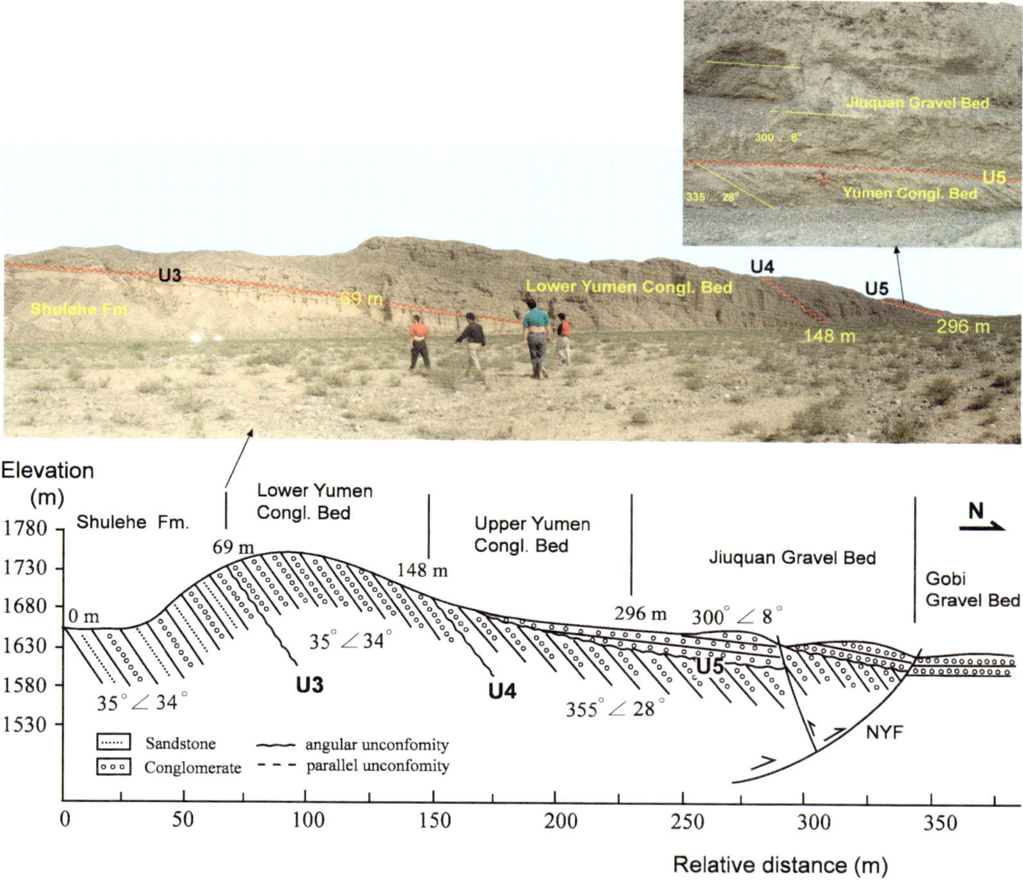

Fig. 6. The measured Upper Yumu Shan section showing the occurrences of unconformities in the Plio-Quaternary stratigraphy.

readily removed around 220 °C, and the characteristic remanent magnetization (ChRM) can be clearly isolated above 300 °C. An obviously rapid decay of the remanent magnetization occurs at 650–680 °C, with a small drop at around 580 °C, indicating that hematite is the major ChRM-carrier and magnetite is the second one (Fig. 7). Based on the pilot samples, the rest of the samples were measured with 12–14 steps from 300 to 680 °C. The ChRM directions were obtained by principal component analysis, and the virtual geomagnetic poles (VGPs) were then calculated. Specimens not included in the magnetostratigraphic analysis were rejected based on three criteria: (1) ChRM directions could not be determined because of ambiguous or noisy orthogonal demagnetization diagrams; (2) ChRM directions revealed maximum angular deviation angles >15°; and (3) Specimens yielded magnetizations with VGP latitude values <30°.

All the accepted ChRM directions of the Sunan section were used for Fisherian statistics and the reversal test. The difference between the normal and reversed mean magnetic declinations was almost 180° (353.5 v. 172.4°), and that of the normal and reversed mean inclination was around 90° (35 v. −34.4°; Fig. 8a). The statistical bootstrap technique (Tauxe 1998) was used to examine possible non-Fisherian distributions of ChRM vectors, and to characterize the associated uncertainties for both normal and reversed ChRM directions (Fig. 8c). The reversed-polarity directions were inverted to their antipodes to test for a common mean for the normal and reversed magnetization directions. The confidence intervals for all components overlap, indicating a positive reversal test (Tauxe 1998; Fig. 8c). The results in Figure 7a, c indicate that the obtained ChRMs are most likely the primary remanences.

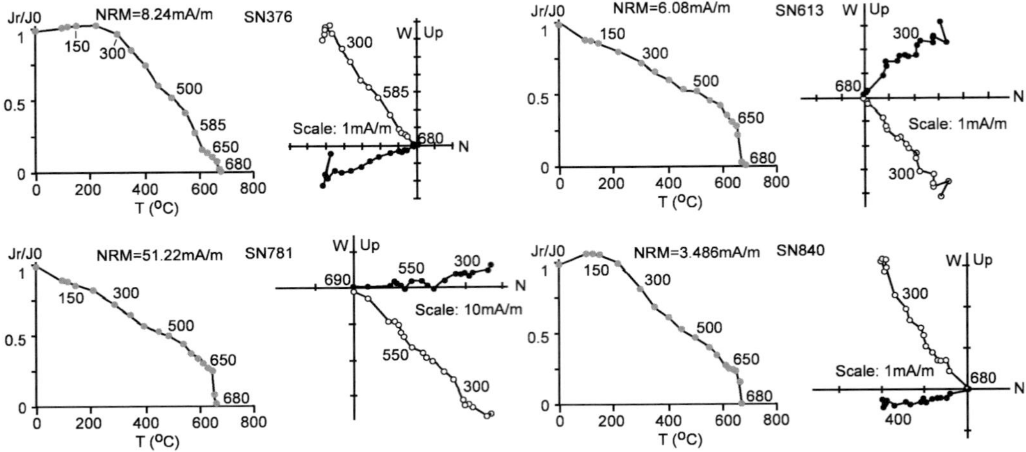

Fig. 7. Orthogonal presentation of thermal demagnetization of some representative samples from the Sunan section. Solid and open circles represent horizontal and vertical projections, respectively.

For the fold test, 100 representative high-quality (maximum angular deviation <5) site-mean ChRMs from different parts of the section were averaged again to fall into 11 grouped sites according to stratigraphic dip and were used for a calculation by the method of McElhinny (1964), which indicates a positive fold test with the tilt-dependent dispersed ChRM directions tending to cluster together around their antipolar means (Fig. 9).

Magnetostratigraphy

The VGP results were plotted as a function of thickness for the Sunan section. We regarded single VGP direction as the unstable or unreliable direction, and thus it was not included in the magnetic zone for interpretation. A total of 22 normal (N1–N22) and 22 reversed (R1–R22) polarity zones are clearly observed in the section (Fig. 10). Palaeontological findings in the Jiuquan Basin were used to control the magnetostratigraphic interpretation.

Many fossil mammals have been found in the upper and middle parts of the Baiyanghe Formation in sites near our section and throughout the Jiuquan Basin (Fig. 10). These include *Tataromys grangeri, T. Sigmoden, Leptotataromys minor, Parasminthus* cf. *Asiae-centralis, P. tangingili, P. parvulus, Eucricetodon asiaticus, Desmatolagus* sp., *Sinolagomys* sp. and *Amphechinus* sp. All of these fossils are major components of the Chinese Taben buluk fauna, which lived during late Oligocene time (Wang 1965; Exploration and Exploitation Department of the Yumen Oilfield, 1990, 'The Tertiary in the Jiuquan Basin', unpublished). Furthermore, fossils of stoneworts (*Tectochaea, Kosmogyra*

ovalis and *Charites huangi* sp.), ostracods and gastropods (*Metacypris* sp., *Ostrea* sp., *Nyocypris* sp., *Chara* sp., *Plannorbis* sp., *Hydrobia* sp. and *Ancykes* sp.) were found in the lower and middle parts of the Shulehe Formation, suggesting a Miocene age (Wang 1965; Exploration and Exploitation Department of the Yumen Oilfield, 1990, 'The Tertiary in the Jiuquan Basin', unpublished).

Based on the above age constraints, the observed chrons in the Baiyanghe Formation can be correlated with the GPTS chrons between 6Cr and 9n (Cande & Kent 1995), with the prominent two long normal polarity zones N21 and N20 correlating with the characteristic long normal chrons 9n and 8n, and N17 with 7n (Fig. 10). The 16 observed normal (N1–N16) and 17 reversed chrons (R1–R17a) in the Shulehe Formation match well with the GPTS chron interval between 4Ar and 5ABr. The distinct long, mostly normal zone interval (N2–N6) is correlated with striking long normal chron of 5n, with short reversed zones R3–R6 being regarded as analogues of several cryptochrons in chron 5n (Cande & Kent 1995; Fig. 10). These cryptochrons are frequently recorded in chron 5n in the NE Tibetan Plateau (Li *et al.* 1997; Fang *et al.* 2005*a, b*), nearby regions (Charreau *et al.* 2009) and other parts of the world (Cande & Kent 1995; Garcés *et al.* 1996). The normal zones N11–N16 can be correlated with Chrons 5An–5ABn. We correlated the mostly reversed zone interval (R7–R11) with the long striking reversed chron of 5r, and the two short normal zones N7 and N8 with the short normal chrons of 5r.1n and 5r.2n, and regarded other two short normal zones N9 and N10 as false signals probably owing to non-fresh samples in the section (Fig. 10).

Polarity	Dec.	Inc.	α_{95}	N	K	Fisherian
N	353.5	35.0	3.5	266	7	no
R	172.4	-34.4	4.1	229	6	yes

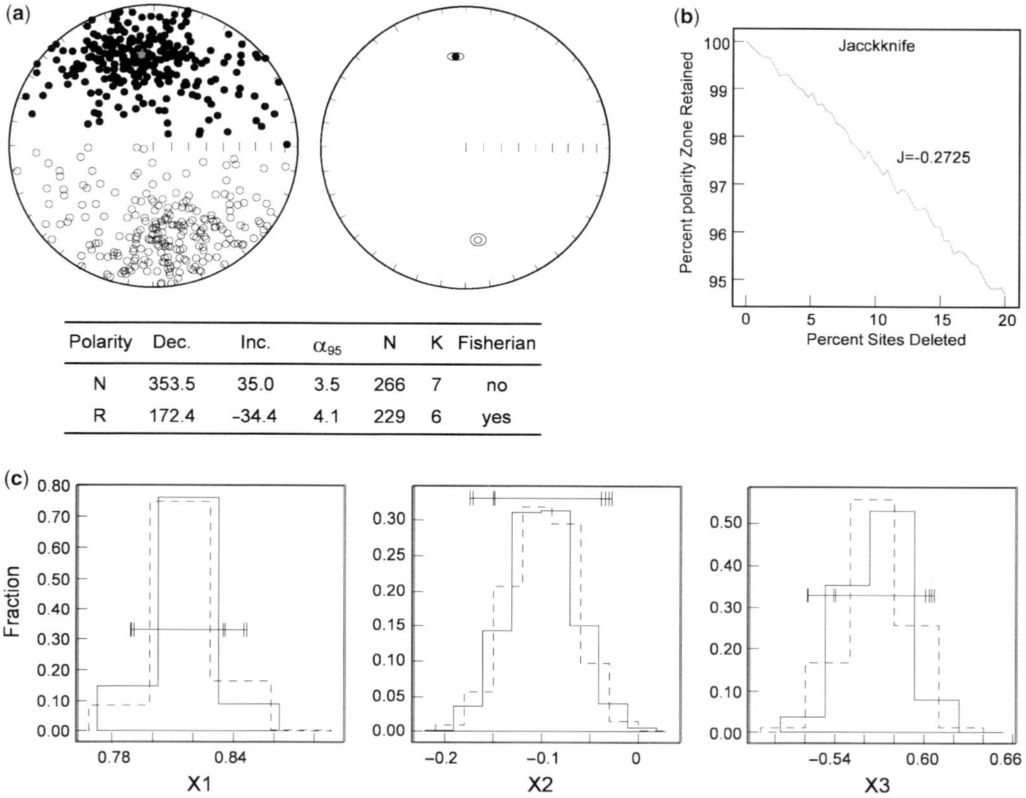

Fig. 8. (**a**) Equal-area projections of the obtained ChRM directions and mean directions (with oval of 95% confidence and their Fisherian statistics in the table below) for the Sunan section determined with the bootstrap method (Tauxe 1998). Downward (upward) directions are shown as solid (open) circles. (**b**) Magnetostratigraphic jack-knife analysis (Tauxe & Gallet 1991) for the Sunan section. The plot indicates the relationship between average percentage of polarity zones retained and the percentage of sampling sites deleted, where the slope *J* is directly related to the robustness of the results. The obtained slopes *J* have values of −0.2725 in the study section, which predicts that the section has recovered more than 95% of the true number of polarity intervals. (**c**) Bootstrap reversal test diagram for the Sunan section. Reversed polarity directions have been inverted to their antipodes to test for a common mean for the normal and reversed magnetization directions. The confidence intervals for all components overlap, indicating a positive reversal test.

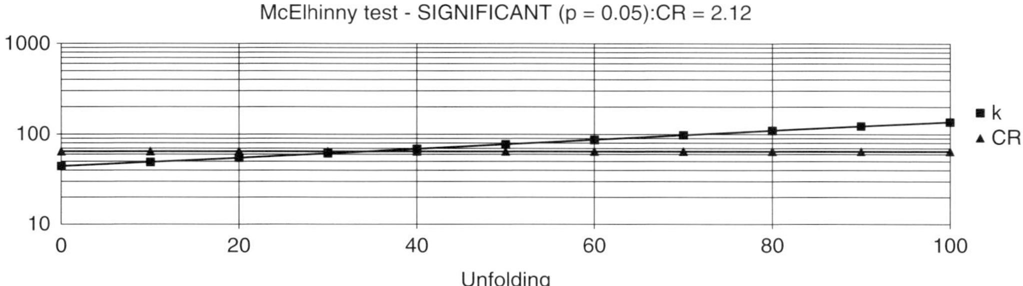

Fig. 9. Fold test of high-quality samples Fisher-averaged for 11 sites along different heights (and thus dips) of the Sunan section.

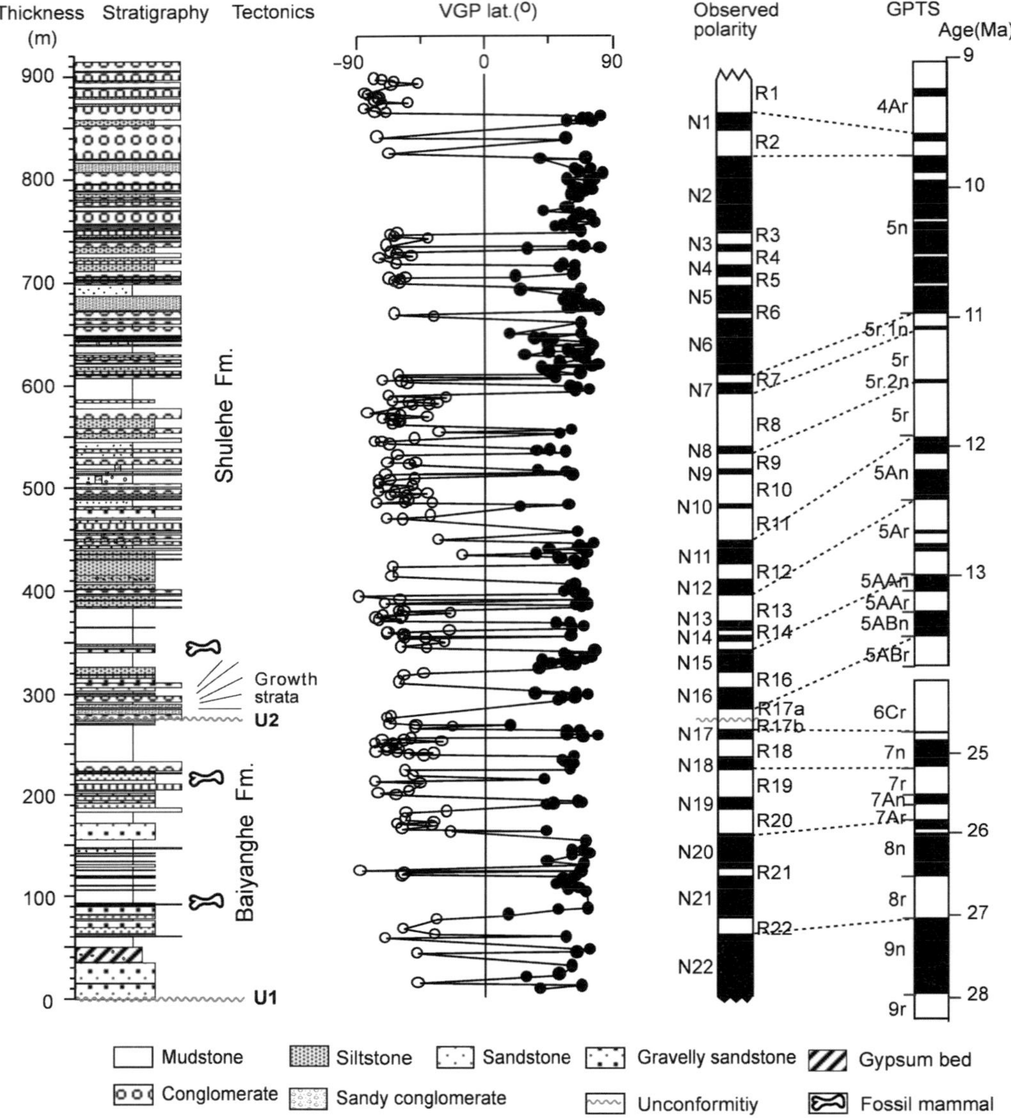

Fig. 10. Magnetostratigraphy of the Sunan section and its correlation with the geomagnetic polarity time scale (GPTS) of Cande and Kent (1995). VGP, Virtual geomagnetic pole. Fossil levels were roughly equivalent to those in nearby sections.

We performed a robust test for the obtained magnetostratigraphy. The jack-knife parameter (J) obtained for the observed polarity zones of this section has a value of -0.2725, which falls within the range of 0 to -0.5 recommended for a robust magnetostratigraphic data set by Tauxe & Gallet (1991). This result indicates that sampling of this section has recovered more than 95% of the true number of polarity intervals (Fig. 8b).

Palaeomagnetic directions and polarity zones of the Upper Yumu Shan section are plotted in Figure 9 for a whole view of the magnetostratigraphic sequence in the Yumu Shan area. Reasons for its interpretation and correlation with the GPTS were described in an earlier publication (Liu *et al.* 2010).

Based on these new data, the age of the Sunan section was constrained between about 27.8 and 9.33 Ma, with the Baiyanghe and Shulehe Formations forming at about 27.8–24.63 and >13.69–9.33 Ma, respectively. The age of the Upper Yumu Shan section was dated between about 6.1 and 0.1 Ma, with the lower and Upper Yumen Conglomerate Beds being constrained at about 3.6–2.8 and 2.6–0.9 Ma, and the Jiuquan Gravel Bed at about 0.8–0.1 Ma (Liu *et al.* 2010; Fig. 11).

These age determinations are in good agreement with those we obtained for the Laojunmiao section at Yumen in the western Jiuquan Basin (see Fig. 1a for location), where the Shulehe Formation and the Yumen and Jiuquan Conglomerates were dated at >13–4.9, 3.66–0.93 and 0.84–0.14 Ma, respectively (Fang *et al.* 2005b). They also generally match previous preliminary magnetostratigraphic determinations of the Baiyanghe Formation at about 31.6–24.6 Ma and the Shulehe Formation at about 22.5–4.2 Ma in the Jiuquan Basin (Huang *et al.* 1993).

Sedimentology and palaeocurrents

The Sunan section

Field investigations and quantification of lithology (Fig. 12) show that the Baiyanghe Formation consists dominantly of fine-grained distinct red-orange, brown-red and purple sandstones and mudstones, intercalated with some thin fine-grained grey conglomerate layers below sandstones. In the lower part, there is a characteristic thick sandy gypsum bed (Figs 5 & 13c). The sandstones are moderately to well sorted, massive or weakly laminated and occasionally cross-bedded. The siltstone and mudstones are usually massive with rare mud cracks. Both sandstones and siltstones–mudstones contain some carbonate concretions, with some sandstones cemented by carbonates. We interpreted these as flood plain–flood basin deposits with some playa deposits formed in an arid environment (Reading 1978; Fig. 13c).

The Shulehe Formation is an upward-coarsening sequence that can be divided into three lithological units. The lower unit (272–383 m) is an up-fining sequence with a fine conglomerate and sandstone bed complex at the bottom (272–315 m; corresponding to growth strata), and mostly brown mudstones in the upper section (315–383 m; Figs 5, 12 & 13c). The conglomerates develop clear imbricate structure and parallel bedding, and the sandstones are mostly massive and occasionally

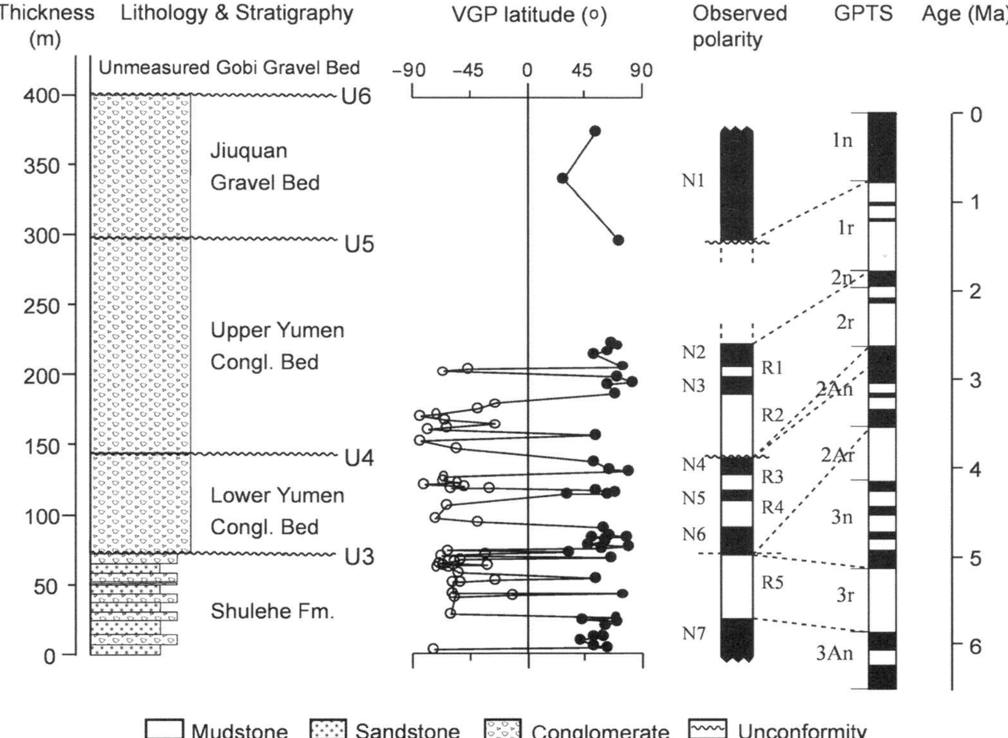

Fig. 11. Magnetostratigraphy of the Upper Yumu Shan section and its correlation with the GPTS of Cande & Kent (1995).

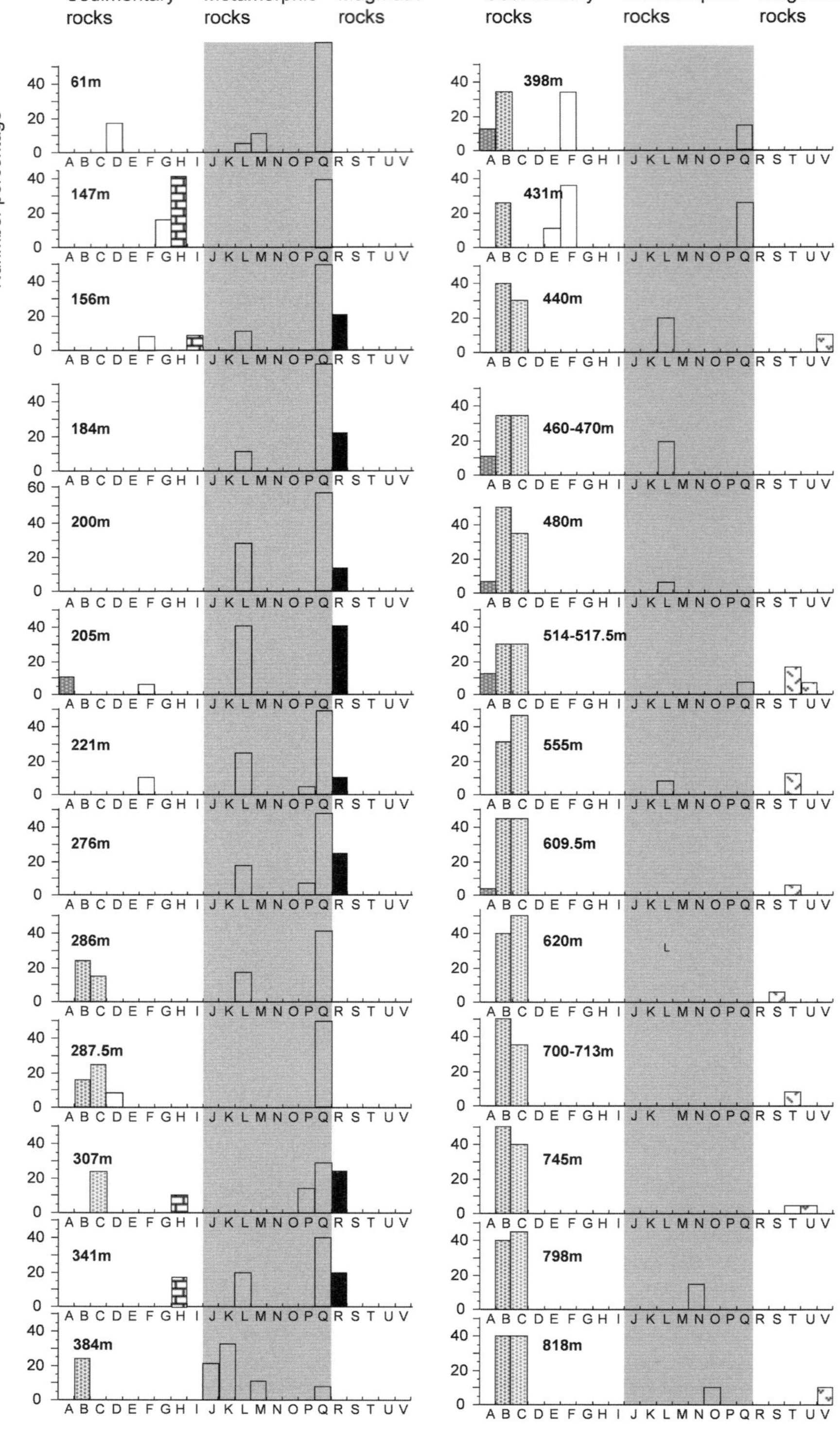
Sedimentary rocks
Metamorphic rocks
Magmatic rocks
Nunmber percentage
61m
147m
156m
184m
200m
205m
221m
276m
286m
287.5m
307m
341m
384m
398m
431m
440m
460-470m
480m
514-517.5m
555m
609.5m
620m
700-713m
745m
798m
818m
A B C D E F G H I J K L M N O P Q R S T U V

oblique- (cross-)bedded. The siltstones and mudstones are mostly massive with rare weak, horizontal laminations. We interpreted these features as channel and overbank deposits of moderately meandering rivers (Reading 1978; Fig. 13c).

The middle unit of the Shulehe Formation (383–823 m) consists of an alternating grey conglomerates–sandstones and yellowish brown siltstones–mudstones (Figs 12 & 13c). The conglomerates occur in large lenses and have imbricate structure, parallel and trough cross-beddings. The sandstones also occur in large lenses and have massive, parallel and occasionally cross-beddings. The siltstones and mudstones are generally massive. We interpreted these as braided river deposits (Reading 1978; Fig. 13c).

The third unit (823–916 m) consists predominantly of thick grey conglomerate layers, intercalated with thin brownish yellow sandstone and siltstone lens (Figs 12 & 13c). The conglomerates occur as thick layers and develop massive bedding with some boulders and weakly imbricate structure. We regarded these as debris flow and alluvial fan deposits (Reading 1978; Fig. 13c).

Variation of the sedimentation rate in the Sunan section lends support to the facies interpretation above (Fig. 13c, f). The sedimentation rate during deposition of the Baiyanghe Formation was the slowest (average 87.1 m/Ma) in the Sunan section (Fig. 13f). It matches the interpretation of a flood plain–playa depositional system with a deficiency of detrital supply (Reading 1978). The lower, middle and upper units of the Shulehe Formation have average sedimentation rates of about 111, 157 and 250 m/Ma, respectively (Fig. 13f). They agree well with the interpreted corresponding facies of a moderately meanding or braided river and alluvial fan (Fig. 13c) that require an increasing supply of detrital sediments (Reading 1978).

The Upper Yumu Shan section

The Plio-Pleistocene conglomerate sequence in the Upper Yumu Shan section comprises conglomerate beds with a few small sandstone and siltstone lenses. These conglomerates develop weakly imbricate or massive structures, with many boulders. They are interpreted as debris flows and alluvial fan deposits (Reading 1978; Fig. 13c).

The gravel composition and palaeocurrent data provide information on mountain uplift and erosion.

The main components of the gravels along the Sunan section are purple, green and yellow sandstones, granite and metamorphic rocks of quartzite and gneiss, with minor components of marl, mudstone, conglomerate, schist, mylonite, granulite, basalt and andesite in some beds (Fig. 12). Three distinct changes in gravel composition were found in the Sunan section. The measurements refer to heights within the measured section. Granite appears only in the rocks below the height 341–384 m (equivalent to *c.* 13 Ma), distinct purple, green and yellow sandstones generally occur above 276–286 m (*c.* 13.5 Ma), and metamorphic rocks decrease upwards and do not occur above the 559–609.5 m layer (*c.* 11.5–12 Ma; Fig. 12).

Palaeocurrent directions in the Sunan section trend weakly northward in the Baiyanghe Formation, persist northward in the Lower and Middle Shulehe Formation, and are stable southward from thickness 820 m (*c.* 9.8 Ma) in the Upper Shulehe direction (Fig. 13e). Stable northward palaeocurrents were also observed for the Upper Yumu Shan section (Fig. 13e).

Discussion

Tectonic events and deformation: uplift history

Based on the magnetostratigraphy and field tectonosedimentary investigations outlined above, we suggest that the North Qilian Shan and its associated Jiuquan foreland basin began to form from the Oligocene at the latest and the Yumu Shan uplift started in the late Miocene owing to a basinward transfer of deformation by propagation faults. This deformation and uplift are recorded as eight episodic tectonic events (Figs 13 & 14).

The occurrence of unconformity U1 at the base of the Sunan section and the deposition of the Beiyanghe Formation with dominantly northward current and gravels dominated by granite, gneiss and quartzite (Figs 5, 10 & 13) indicate that the eastern Jiuquan Basin began to subside and receive sediments from the North Qilian Shan beginning no later than 27.8 Ma in the Oligocene (Fig. 12a). The gravels of metamorphic rocks of gneiss and quartzite are typical components of Proterozoic rocks dominating Central Qilian Shan and southern part of the North Qilian Shan and do not appear in the Yumu Shan (Fig. 1a), suggesting an early uplift

Fig. 12. Variation of the gravel composition in the Sunan section. Gravel composition is expressed as a number percentage of an identified gravel to total counting gravels within an area of 1 × 1 m in the section. (**a**) Conglomerate; (**b**) purple sandstone; (**c**) greensandstone; (**d**) yellow sandstone; (**e**) grey sandstone; (**f**) other sandstones; (**g**) green-blue mudstones; (**h**) marl; (**i**) limestone; (**j**) leptynite; (**k**) schist; (**l**) gneiss; (**m**) jasper rock; (**n**) phyllonite; (**o**) chlorite schist; (**p**) mylonite; (**q**) siliceous rocks; (**r**) granite; (**s**) granodiorite; (**t**) diorite; (**u**) andesite; U, basalt.

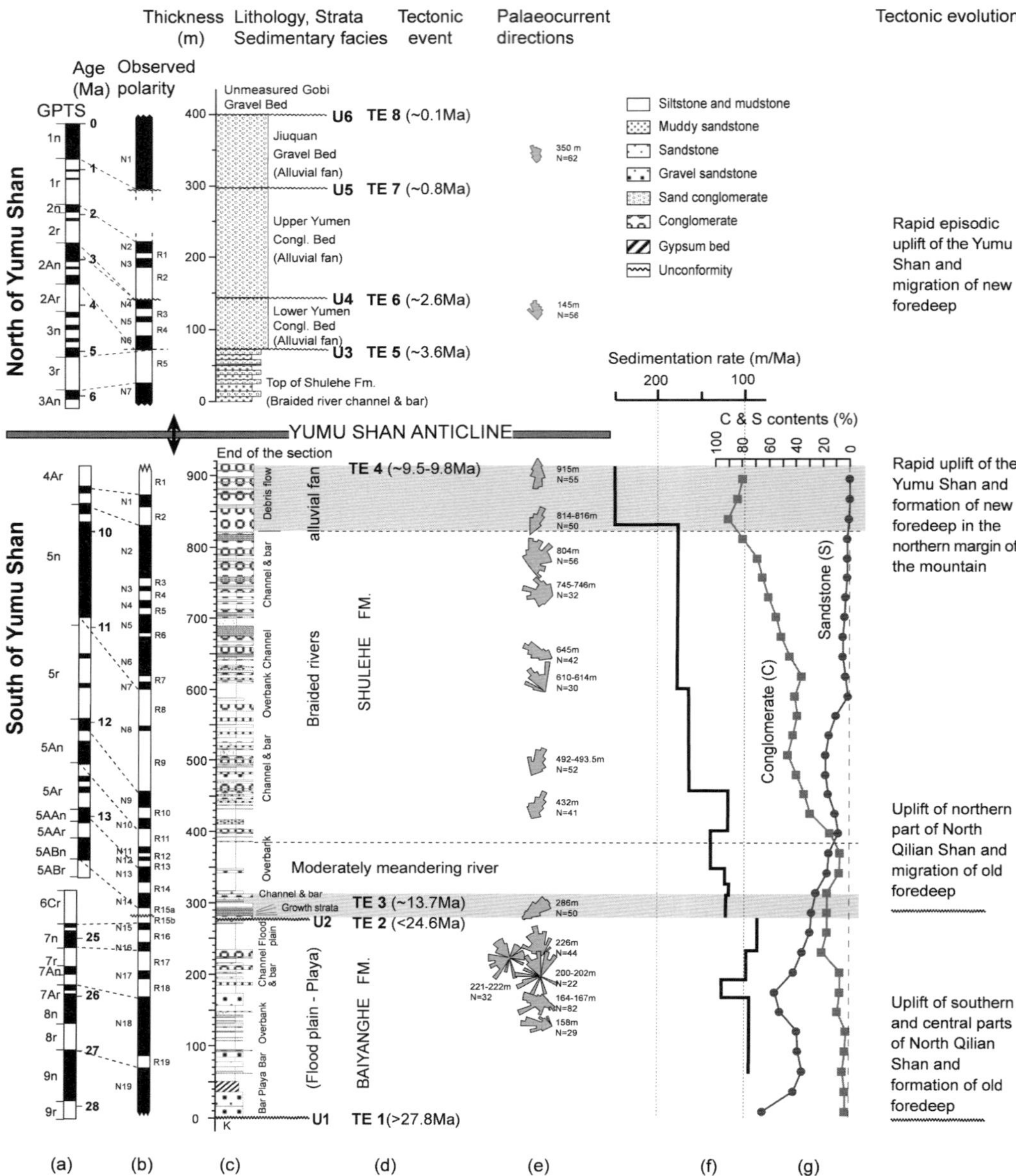

Fig. 13. Tectonosedimentary evolution of the Sunan and Upper Yumu Shan sections. River palaeocurrent directions (**e**) were obtained by statistical measurements of directions of gravel ab-faces. Sedimentation rate (**f**) was based on interpretation of magnetostratigraphy of the Sunan section in Figure 8. Percentage conglomerate and sandstone (**g**) were calculated from each 100 m stratigraphic interval using 25 m moving-window increments based on category of three statistic units: conglomerate, sandstone and siltstone–mudstone. This means that the rest is the percentage of siltstone–mudstone in the section, which is not present in the diagram.

and erosion of the North Qilian Shan. The fine sediments, playa–floodplain sedimentary environments, the weakly north-trending currents and the low sedimentation rate (average 87.1 m/Ma) during deposition of the Baiyanghe Formation in the Sunan section (Fig. 13) indicate slow subsidence of the Jiuquan Basin, a lowland with broad flat surface in the basin, and a deficiency of detrital supply

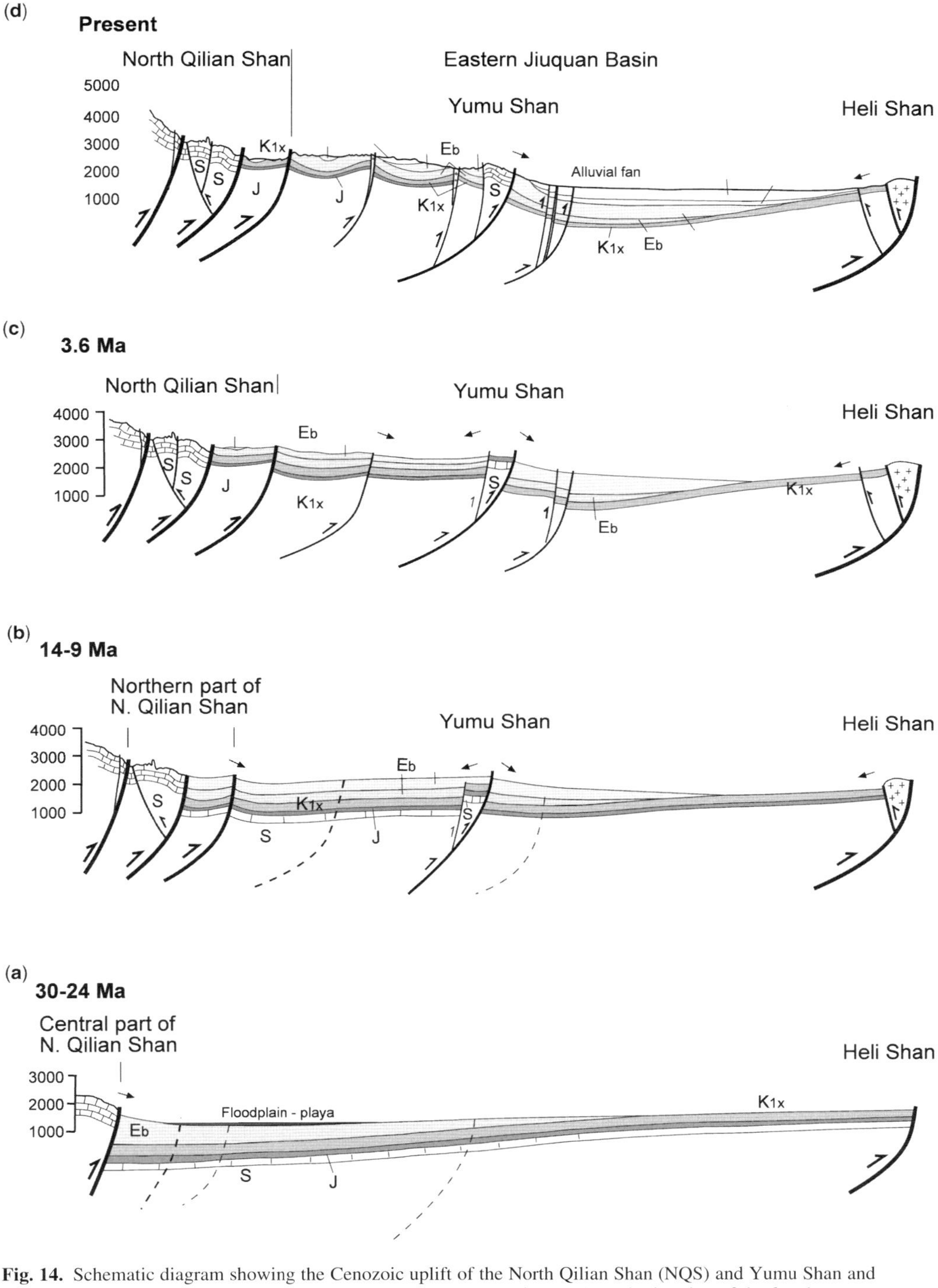

Fig. 14. Schematic diagram showing the Cenozoic uplift of the North Qilian Shan (NQS) and Yumu Shan and formation of the eastern Jiuquan foreland basin. Note the progressive northward migration of the foredeep and wedge-top of the Jiuquan foreland basin through progradation faults as responses in turn to uplifts of southern and central NQS, northern NQS and Yumu Shan. Marks for stratigraphic units and faults are same as those in Figure 1.

in the source area. By facies association, there must be coeval alluvial fan and braided river deposits further south in the Sunan section during deposition of the Baiyanghe Formation In fact, the Baiyanghe Formation is distributed in the northern part of the North Qilian Shan about 15–20 km south of the Sunan section, or further south in the western North Qilian Shan (Figs 1 & 3). Sedimentary facies and basin analyses revealed that these sediments do not reflect intermontane basin deposition but are connected with those in the Jiuquan Basin (Gansu Geologic Bureau 1989; EGPGYO 1989).

These data collectively suggest slow uplift of the southern and central parts of the North Qilian Shan (Fig. 13a). Southward thickening of the Baiyanghe Formation interpreted from a PetroChina borehole and seismostratigraphy (Fig. 1) suggests a flexural origin for the basin. The collision of India with Asia probably caused the rapid movement of Altun slip fault in Oligocene times (Yin *et al.* 2002) with subsequent compression and uplift from the North Qilian Shan (EGPGYO 1989) (Fig. 14a).

Deposition of the Baiyanghe Formation terminated at about 24.63 Ma, followed by a big gap in the unconformity U2 (24.63–13.7 Ma; Figs 10 & 13a–c). This suggests a second deformation and uplift of the basin and the Qilian Shan, most likely just at the end of the Baiyanghe Formation, because this unconformity is widely distributed in the Jiuquan Basin, and in the western Jiuquan Basin, the Shulehe Formation above the U2 is much thicker and has lower successions not occurring in the Sunan section. Its base age was estimated at the early Miocene by extrapolation of palaeomagnetically dated upper part of the Shulehe Formation (Fang *et al.* 2005*b*). This can be confirmed in the northern part of the eastern Jiuquan Basin just to the north of the Yumu Shan, where borehole and seismostratigraphy show that very thick Shulehe Formation (over twice as thick as that in the Sunan section and over three times thicker than the underlying Baiyanghe Formation) was deposited in the basin (Fig. 1c), suggesting a strong persistent flexing and subsiding of the Jiuquan Basin beginning in the early Miocene. We argue that this broad area of strong deformation of the Jiuquan Basin was caused by the intense uplift of the Qilian Shan probably in response to the growth of the Himalayas in the early Miocene (Yin & Harrison 2000; Tapponnier *et al.* 2001).

The bottom of the Shulehe Formation above the U2 unconformity starts with clear growth strata that occurred at about 13.7–13 Ma, accompanied by an increase of sedimentation rate over 115–150 m/Ma, a substantial conglomerate deposition, the first and important appearance of characteristic gravels of purple, green and yellow sandstones and the end or marked reduction of gravels of granite and

metamorphic rocks, clear northward currents and a change in sedimentary environment from the previous playa–floodplain to a braided river (Figs 5 & 13). These observations collectively suggest another intense tectonic deformation of the basin through progradational faults and uplift of the northern part of the North Qilian Shan at latest at about 13.7–13 Ma (Fig. 14b), because the northern part of the North Qilian Shan consists of only Palaeozoic and Mesozoic rocks containing the characteristic gravels of purple, green and yellow sandstones (mostly from stratigraphy S, J_3 and K_1; Figs 1 & 3). In comparison with the much thicker Shulehe Formation in the northern part of the eastern Jiuquan Basin to the north of the Yumu Shan (Fig. 1c), we infer that the lower successions of the Shulehe Formation in the Sunan section were eroded away in this tectonic event. This site was probably subjected to rapid upheaval caused by the propagation fault. The subsequent deposition of sedimentary rocks with growth strata indicates a continuous but slow uplift of this site by the propagation fault. Taking into account that these sedimentary rocks in this rising site have a much higher sedimentation rate than previously deposited rocks (Fig. 13f) and much thicker equivalent stratigraphy of the Shulehe Formation in the north of the Yumu Shan (Fig. 1c), we argue that the sedimentation rate increase in this site could not be related to the approach of the deformation front but must point to a much higher sedimentation rate increase and uplift from the North Qilian Shan.

From 820 m (*c.* 9.8 Ma), the former braided river system was replaced by debris flows and alluvial fan deposits, accompanied by a rapid increase in sedimentation rate (250 m/Ma), persistent occurrence of coarse conglomerates and change in former northward currents to stable southward currents. The Sunan section ends at 916 m (*c.* 9.6 Ma). All of this suggests that rapid uplift of the Yumu Shan, to the north of the Sunan section, began to occur at about 9.8–9.6 Ma, adding new sediments to the studied site (thus contributing to further increase of sedimentation rate), suggesting that the deformation was propagated into the inner basin through initiation of the NYF and its induced back-thrust fault (South Yumu Shan fault) from the North Qilian Shan (Figs 1, 3, 4 & 14b). Soon thereafter, the section was ended, suggesting that deformation and uplift of the North Qilian Shan and Yumu Shan accelerated, causing the studied site to be strongly faulted and folded. Thus, sedimentation in this region ceased and erosion commenced. This rise is about 8 million years earlier than the previous hypothesis that the Yumu Shan terrane was raised by the NYF at *c.* 1 Ma (see Fig. 15c in Meyer *et al.* 1998; Tapponnier *et al.* 1990, 2001). Rapid progradation of the NYF caused the previous (old) foredeep area to the

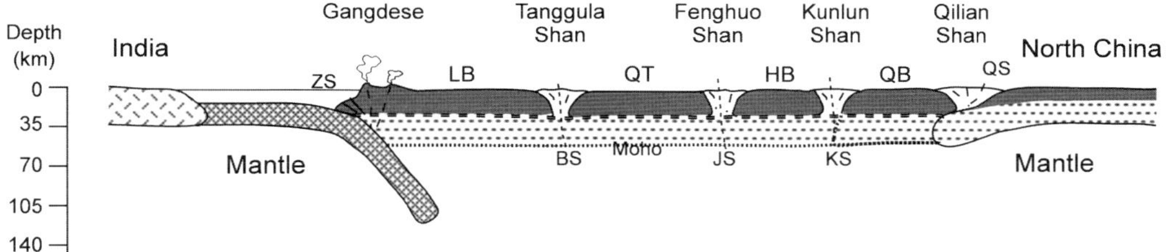

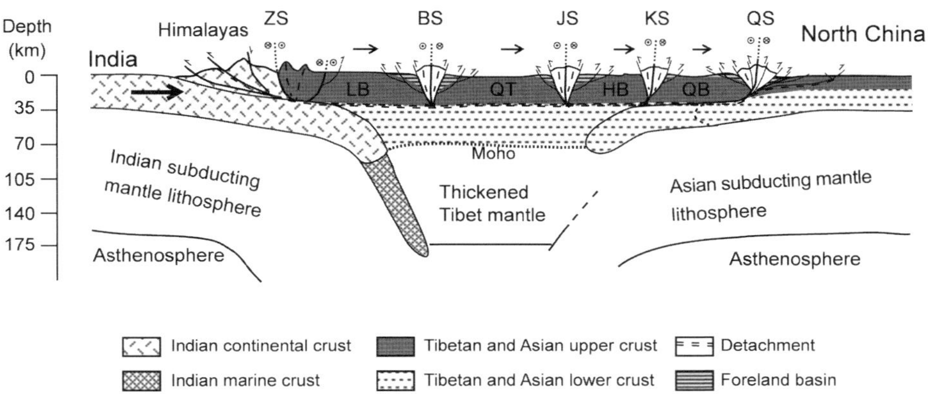

Fig. 15. Schematic diagram showing the process of rigid upper crust blocks detached from and floating on viscous lower crust and mantle of Tibet and synchronous differential deformation uplift of the Tibetan Plateau. Positions and boundaries of crusts, mantle lithospheres of India, Asia and Tibet (part of Asia) are referred to the literature (Tapponnier *et al.* 2001; Kind, *et al.* 2002; Kumar *et al.* 2006; Zhao *et al.* 2010). Arrows indicate moving direction and relative velocity. ZS, Zangbo suture; BS, Bangong suture; JS, Jinshan suture; KS, Kunlun suture; QS, Qilian suture; LB, Lhasa block; QB, Qiangtang block; HB, Hoxil block; QB, Qaidam block.

south of the Yumu Shan to be heaved up as wedge-top area (DeCelles *et al.* 1998) for erosion and new source rock supply and the previous slope area to the north of the Yumu Shan to be strongly flexed as a new foredeep area (Figs 1 & 14b).

The occurrences of the unconformities U3–U6 at about 3.6, 2.6, 0.8 and 0.1 Ma in the Upper Yumu Shan section (Figs 5, 10 & 13a–c), accompanied by persistent debris flow deposition, indicate the continuing episodic thrusting from the NYF and rapid uplifts of the Yumu Shan, resulting in very thick (over 2000 m) Plio-Quaternary sediments in the new foredeep (Fig. 14c, d).

Dynamic mechanism of deformation: uplift of NE Tibet: a proposed 'upper crust-floating model'

The Oligocene deformation, uplift of the North Qilian Shan and flexing of the Hexi foreland basin

revealed from detailed magnetostratigraphy and basin analysis are generally known from previous studies of the Kunlun–Qinling slip fault and Altun slip fault (e.g. Jolivet *et al.* 2001; Yin *et al.* 2002; Fang *et al.* 2003; Clark *et al.* 2010; Zhang *et al.* 2010) which may have been activated in the Eocene–Oligocene, and compression may have reached the North Qilian and east end of the Qilian Shan roughly at the same time (Fang *et al.* 2003; Dai *et al.* 2005). This challenges the current working models that are based either on the viscous or rigid lithosphere assumption. Both of these models regard the NE Tibetan Plateau, especially the northernmost part, the North Qilian Shan, as the final phase (Plio-Quaternary) of deformation and uplift of Tibet (Tapponnier *et al.* 1990, 2001; Meyer *et al.* 1998; England & Houseman 1988, 1989; Molnar *et al.* 1993; Royden *et al.* 1997, 2008).

The roughly synchronous (Eocene–Oligocene) response of deformation of the furthest north edge of the Tibetan Plateau to the main collision of

India with Asia at *c.* 50–40 Ma (Molnar & Tapponnier 1975; Patriat & Achache 1984; Rowley 1996) suggests that the collision stress reached the northern edge of the Tibetan Plateau rapidly. This requires the block to have a rigid nature. Mechanically, for a vast block like Tibet, *c.* 3500 × 1500 km (*c.* 2 500 000 km^2; Fielding *et al.* 1994), if its whole lithosphere acts as a rigid block, it is impossible to pass push-stress from the southern edge to the northern one; instead it will break the block immediately north of the push zone (or along some nearby weak zone like a former suture, as indicated by Tapponnier *et al.* 2001), owing to the great resistance of the rigid block. Further push will accumulate stress and migrate northward to make a new break. This process will follow the pathway to cause a stepwise rise and growth of Tibet, such as that proposed by Tapponnier's group (Tapponnier *et al.* 1990, 2001; Meyer *et al.* 1998).

In such a large rigid block (upper crust) floating on a viscous lower crust with a detachment, a push from the southern edge of a rigid block will easily and immediately pass on to the northern edge, just like pushing on a large wood board floating on water. Since the Tibetan Plateau consists of several Tethys blocks separated by narrow ranges and sutures (Fig. 1 inset), we regard the Tibet upper crust as rigid blocks separated by narrow 'soft' ranges and sutures, and assume that these 'range-conjuncted' upper crust blocks were detached to a considerable extent (if not entirely) and floated on their viscous lower crusts and mantles at each collision of the Tethys blocks with Palaeo-Asia before the India–Asia collision. Thus, when India collided with Asia, the Tibet upper crust would rapidly float northwards, passing compressive stress and deformation northwards immediately into inner and NE Tibet through rigid blocks, activating former sutures and deep faults, uplifting inter-block ranges and forming new thrust–fold systems, probably connecting with former (or new) detachments and foreland basins along margins of ranges, while the Tibet lower crust and mantle would be subjected to continuous thickening differentially from both the southern and northern sides (larger in the south and smaller in the north), thus probably also creeping northwards, because the southern and northern margins of Tibet experience positive and passive underthrusting and compression, respectively, and the buffering (thus attenuated deformation in magnitude) of 'soft ranges' between rigid blocks (Fig. 15b). In consequence, the Altun fault would have formed shortly after the India–Asia collision initially at *c.* 55 Ma and mainly at *c.* 50–40 Ma (Molnar & Tapponnier 1975; Patriat & Achache 1984; Rowley 1996), and the former sutures and deep faults bounding the Qilian Shan (Molnar & Tapponnier 1975; Gansu Geologic Bureau 1989;

Feng & Wu 1992) would have been activated coevally and detached from the North China lower crust and mantle as the latter passively moved southwards and plunged beneath the Kunlun Shan, causing synchronous Eocene–Oligocene deformation, uplift of the Qilian Shan and formation of flexural foreland basins along its rims (Fig. 15b). When there was no further accommodation for deformation of 'soft ranges' south of the Qilian Shan, compressive stress would have transferred to the NE edge, leading to a later-phase (Miocene–Quaternary) of rapid deformation and uplift of the North Qilian Shan and Yumu Shan (Fig. 15b).

A complete discussion and summary of evidence to support this model is beyond the scope of the paper. Here we outline evidence which supports the model.

Theoretical calculation and modelling have shown that the lower crust may act as a viscous thin layer that might flow out mostly southwards, northwards and eastwards along channels (Royden 1996; Royden *et al.* 1997, 2008). Exposed leucogranites and the metamorphic sequence in the Himalayas have been recognized to indicate lower crust partial melting and channel flow (Le Fort *et al.* 1987; Yin & Harrison 2000; Beaumont *et al.* 2001; Grujic *et al.* 2002).

Recent magnetotelluric data confirm that such a channel flow existed along at least 1000 km of the southern margin of the Tibetan plateau (Unsworth *et al.* 2005), and also in a vast area from central-east Tibet to SW China at a depth of 20–40 km (Bai *et al.* 2010). This is concordant with the notion that a lower-velocity (thus weaker) zone of P-wave in middle–lower crust is commonly observed beneath Tibet (Kind *et al.* 2002; Kumar *et al.* 2006; Zhao *et al.* 2010). Seismic tomography clearly shows the existence of detachment surfaces between upper and middle–lower crusts beneath the Qilian Shan, Kunlun Shan and other ranges of Tibet and Himalayas (Gao *et al.* 1995; Tapponnier *et al.* 2001; Kumar *et al.* 2006; Zhao *et al.* 2010).

Recent recognition and dating of the slip-related and flexural-depressed basins along the margins of ranges on the Tibetan Plateau demonstrate that they formed at roughly the same time, at *c.* 50–40 Ma (e.g. Decelles *et al.* 1998; Horton *et al.* 2002; Fang *et al.* 2003; Dai *et al.* 2005; Wang *et al.* 2008), corroborating the Eocene rapid rise and exhumation events detected in related ranges from the south to the NE Tibetan Plateau by a variety of thermochronology methods (Jolivet *et al.* 2001; Wang *et al.* 2008; Clark *et al.* 2010).

Modern GPS observation indicates clearly that the whole Tibetan Plateau is primarily under simultaneous but differential shortening in a north–south direction, with the shortening rate decreasing from *c.* 15–20 mm a^{-1} in the Himalayas, through *c.* 12–9 mm a^{-1} in the central and north Tibet, to *c.* 6 mm a^{-1} in northeastern Tibet (Zhang *et al.* 2004). This provides a robust support for our model that Tibetan rigid upper crust blocks buffered by inter-block soft ranges and sutures float and move northeastwards on a viscous lower crust (Fig. 15).

Conclusions

(1) High-resolution magnetostratigraphy constrains the age range of the Sunan section between the North Qilian Shan and the Yumu Shan at about 27.8–9.6 Ma, and the Upper Yumu Shan section to the north of the Yumu Shan at about 6.1 and 0.1 Ma, with stratigraphic units of the Baiyanghe Formation at about 27.8–24.63 Ma, the Shulehe Formation at about 13.69–3.6 Ma, and the Yumen, Jiuquan and Gobi Conglomerate (Gravel) Beds at about 3.6–0.9, 0.8–0.1 and 0.1 Ma, respectively.

(2) Tectonosedimentological studies and basin analysis suggest that eight tectonic events occurred at about 27.8, 24.6, 13.7, 9.8–9.5, 3.6, 2.6, 0.8 and 0.1 Ma, recording an early (Oligocene) slow and later (Miocene–Quaternary) episodic rapid uplift of the North Qilian Shan and formation of the Jiuquan foreland basin. The Yumu Shan at the northeasternmost corner of the Qilian Shan began to uplift rapidly at latest at about 9.8–9.6 Ma, rather than the previously thought much later (*c.* 1 Ma) rise, thus challenging the commonly accepted tectonic model of the oblique stepwise rise of the Tibetan Plateau and northeastward growth of the Qilian Shan and Yumu Shan (Tapponnier *et al.* 1990, 2001). Rigid blocks floating on viscous lower crust, called the 'block-floating model', is proposed to interpret this early and episodic response to Indian–Asian collision.

This work was supported by National Natural Science Foundation of China grants (41021001, 40920114001) and the (973) National Basic Research Program of China (grant no. 2011CB403000). We thank L.L.X. Zhang, W. Ma, Y. Tang, Y. Liu, Q. Xu, S. Hu and Z. Zhang for field-work assistance. We also thank Y. Liu, J. Wang and Y. Chen for laboratory help. Special thanks are due to Professor R. Zhu for his continual laboratory support, Professors Z. Junmeng and H. Jiankun for their constructive discussions, Professor Y. An for critical comments on an earlier version of the manuscript, and Yumen Oil Field Com., PetroChina for providing and permitting use of their seismic data. We finally thank Professor R. Burmester and an anonymous reviewer for their constructive comments which helped to improve the manuscript greatly.

References

BAI, D. H., UNSWORTH, M. J. *ET AL.* 2010. Crustal deformation of the eastern Tibetan plateau revealed by magnetotelluric imaging. *Nature Geosciences*, http://dx.doi.org/10.1038/NGEO830

BEAUMONT, C., JAMIESON, R. A., NGUYEN, M. H. & LEE, B. 2001. Himalayan tectonics explained by extrusion of a low-viscosity crustal channel coupled to focused surface denudation. *Nature*, **414**, 738–742.

BURCHFIEL, B. C., DENG, Q. *ET AL.* 1989. Intracrustal detachment within zones of continental deformation. *Geology*, **17**, 448–452.

CANDE, S. C. & KENT, D. V. 1995. Revised calibration of the geomagnetic polarity timescale for the late Cretaceous and Cenozoic. *Journal of Geophysical Research*, **100**, 6093–6095.

CHARREAU, J., CHEN, Y. *ET AL.* 2009. Neogene uplift of theTian Shan Mountains observed in the magnetic record of the Jingou River section (northwest China). *Tectonics*, **28**, http://dx.doi.org/10.1029/2007TC002137

CHUNG, S. L., LO, C. C. *ET AL.* 1998. Diachronous uplift of the Tibetan plateau starting 40 Myr ago. *Nature*, **394**, 769–73.

CLARK, M. K., FARLEY, K. A., ZHENG, D. W., WANG, Z. C. & DUVALL, A. R. 2010. Early Cenozoic faulting of the northern Tibetan Plateau margin from apatite (U–Th)/He ages. *Earth and Planetary Science Letters*, **296**, 78–88.

COLEMAN, M. E. & HODGES, K. V. 1995. Evidence for Tibetan Plateau uplift before 14 Myr ago from a new minimum estimate for east–west extension. *Nature*, **374**, 49–52.

DAI, S. A., FANG, X. M., SONG, C. H., GAO, J. P., GAO, D. L. & LI, J. J. 2005. Early tectonic uplift of the northern Tibetan Plateau. *Chinese Science Bulletin*, **50**, 1642–1652.

DECELLES, P. G., GEHRELS, G. E., QUADE, J. & OJHA, T. P. 1998. Eocene–early Miocene foreland basin development and the histroy of Himalayas thrusting, western and central Nepal. *Tectonics*, **17**, 741–765.

DEWEY, J. F., SHACKLETON, R. M., CHANG, C. & YIYIN, S. 1988. The tectonic evolution of the Tibetan Plateau. *Philosophical Transactions of the Royal Society of London, A*, **327**, 379–413.

EGPGYO (EDITORAL GROUP OF PETROLEUM GEOLOGY OF YUMEN OILFIELD), 1989. *Petroleum Geology of China. Yumen Oilfield*, Vol. **13**. Petroleum Industry Publishing, Beijing, 1–441.

ENGLAND, P. C. & HOUSEMAN, G. A. 1986. Finite strain calculations of continental deformation, 2. *comparison with the India–Asia collision zone. Journal of Geophysical Research*, **91**, 3664–3676.

ENGLAND, P. C. & HOUSEMAN, G. A. 1988. The mechanics of the Tibetan Plateau [and discussion]. *Philosophical Transactions of the Royal Society of London A*, **326**, 301–320.

ENGLAND, P. C. & HOUSEMAN, G. A. 1989. Extension during continental convergence, with application to the Tibetan Plateau. *Journal of Geophysical Research*, **94**, 17 561–17 579.

FANG, X. M., GARZIONE, C., VAN DER VOO, R., LI, J. J. & FAN, M. J. 2003. Flexural subsidence by 29 Ma on the

NE edge of Tibet from the magnetostratigraphy of Linxia Basin, China. *Earth and Planetary Science Letters*, **210**, 545–560.

FANG, X. M., YAN, M. D. *ET AL.* 2005*a*. Late Cenozoic deformation and uplift of the NE Tibetan plateau: evidence from high-resolution magneto stratigraphy of the Guide Basin, Qinghai Province, China. *GSA Bulletin*, **117**, 1208–1225.

FANG, X. M., ZHAO, Z. J. *ET AL.* 2005*b*. Magnetostratigraphy of the late Cenozoic Laojunmiao anticline in the northern Qilian Mountains and its implications for the northern Tibetan Plateau uplift. *Science in China Series D: Earth Science*, **48**, 1040–1051.

FENG, Y. M. & WU, H. Q. 1992. Tectonic evolution of North Qilian Mountains and its neighbourhood since Paleozoic. *Northwest Geoscience*, **13**, 61–74 (in Chinese with English abstract).

FIELDING, E., ISACKS, B., BARAZANGI, M. & DUNCAN, C. C. 1994. How flat is Tibet? *Geology*, **22**, 163–167.

GANSU GEOLOGIC BUREAU. 1971. *Geologic Map of Sunan, 1:200 000 (with Guide Book)*. Gansu Publishing House, Lanzhou (in Chinese).

GANSU GEOLOGIC BUREAU. 1989. *Regional Geology Evolution of Gansu Province*. Geological Publishing House, Beijing (in Chinese).

GAO, R., CHENG, X. Z. & DING, Q. 1995. Preliminary geodynamic model of Golmud-Ejin Qi geoscience transect. *Journal of Geophysics*, **38**(suppl. I), 3–14 (in Chinese with English abstract).

GARCÉS, M., AGUST, I. J., CABRERA, L. & PARÉS, J. M. 1996. Magnetostratigraphy of the Vallesian (late Miocene) in the Vallès-Penedès basin (NE Spain). *Earth and Planetary Science Letters*, **142**, 381–396.

GEORGE, A. D., MARSHALLSEA, S. J., WYRWOLL, K. H., CHEN, J. & LU, Y. C. 2001. Miocene cooling in the Northern Qilian Shan, northeastern margin of the Tibetan Plateau, revealed by apatite fission-track and vitrinite reflectance analysis. *Geology*, **29**, 939–942.

GRUJIC, D., HOLLISTER, L. S. & PARRISH, R. R. 2002. Himalayan metamorphic sequence as an orogenic channel: insight from Bhutan. *Earth and Planetary Science Letters*, **198**, 177–191.

HARRISION, T. M., COPELAND, P., KIDD, W. S. F. & YIN, A. 1992. Raising Tibet. *Science*, **255**, 1663–1670.

HE, G. Y., YANG, S. F., CHEN, H. L., XIAO, A. C. & CHENG, X. G. 2004. New ideas about the Early Cretaceous basins in western Gansu Corridor and nearby regions. *Acta Petrologica Sinica*, **25**, 18–22 (in Chinese with English Abstract).

HORTON, B. K., YIN, A., SPURLIN, M. S., ZHOU, J. & WANG, J. 2002. Paleocene–Eocene syncontractional sedimentation in narrow, lacustrine-dominated basins of east-central Tibet. *Geologic Society of American Bulletin*, **114**, 771–786.

HUANG, H. F., PENG, Z. L., LU, W. & ZHENG, J. J. 1993. Paleomagnetic division and comparison of the tertiary system in Jiuxi and Jiudong basins. *Acta Geologica Gansu*, **2**, 6–16 (in Chinese with English Abstract).

JOLIVET, M., BRUNEL, M. *ET AL.* 2001. Mesozoic and Cenozoic tectonics of the northern edge of the Tibetan plateau: fission-track constraints. *Tectonophysics*, **343**, 111–134.

KIND, R., YUAN, X. *ET AL.* 2002. Seismic images of crust and upper mantle beneath Tibet: evidence for Eurasian plate subduction. *Science*, **298**, 1219–1221.

KUMAR, P., YUAN, X., KIND, R. & NI, J. (2006) Imaging the colliding Indian and Asian lithospheric plates beneath Tibet. *Journal of Geophysical Research*, **111**, http://dx.doi.org/10.1029/2005JB003930

LE FORT, P., CUNEY, M. *ET AL.* 1987. Crustal generation of the Himalayan leucogranites. *Tectonophysics*, **134**, 39–57.

LI, J. J., FANG, X. M. *ET AL.* 1997. Late Cenozoic magnetostratigraphy (11–0 Ma) of the Dongshanding and Wangjiashan sections in the Longzhong Basin, western China. *Geologie en Mijnbouw*, **76**, 121–134.

LIU, D. L., FANG, X. M. *ET AL.* 2010. Stratigraphic and paleomagnetic evidence of mid-Pleistocene rapid deformation and uplift of the NE Tibetan Plateau. *Tectonophysics*, **486**, 108–119.

MCELHINNY, M. W. 1964. Statistical significance of the Fold Test in paleomagnetism. *Geophysical Journal of the Royal Astronomical Society*, **8**, 33–40.

METIVER, F., GAUDEMER, Y., TAPPONNIER, P. & MEYER, B. 1998. Northeastward growth of the Tibet plateau deduced from balanced reconstruction of two depositional areas: the Qaidam and Hexi Corridor basins, China. *Tectonics*, **17**, 823–842.

MEYER, B., TAPPONNIER, P. *ET AL.* 1998. Crustal thickening in Gansu–Qinghai, lithospheric mantle subduction, and oblique, strike-slip controlled growth of the Tibet plateau. *Geophysics Journal International*, **135**, 1–47.

MOLNAR, P. & TAPPONNIER, P. 1975. Cenozoic tectonics of Asia – effects of a continental collision. *Science*, **189**, 419–426.

MOLNAR, P., ENGLAND, P. C. & MARTINOD, J. 1993. Mantle dynamics, uplift of the Tibetan Plateau, and the Indian Monsoon. *Review of Geophysics*, **31**, 357–396.

NI, J. & BARAZANGI, M. 1984. Seismotectonics of the Himalayan collision zone: geometry of the underthrusting Indian Plate beneath the Himalaya. *Journal of Geophysical Research*, **89**, 1147–1163.

PATRIAT, P. & ACHACHE, J. 1984. India–Eurasia collision chronology has implications for crustal shortening and driving mechanism of plates. *Nature*, **311**, 615–621.

READING, H. G. 1978. *Sedimentary Environments and Facies*. Blackwell Science, Oxford, 1–557.

ROWLEY, D. B. 1996. Age of initiation of collision between Indian and Asia: a review of stratigraphic data. *Earth and Planetary Science Letters*, **145**, 1–13.

ROYDEN, L. H. 1996. Coupling and decoupling of crust and mantle in convergent orogens: implications for strain partitioning in the crust. *Journal of Geophysical Research*, **101**, 17 679–17 705.

ROYDEN, L. H., BURCHFIEL, B. C. *ET AL.* 1997. Surface deformation and lower crustal flow in eastern Tibet. *Science*, **276**, 788–790.

ROYDEN, L. H., BURCHFIEL, B. C. & VAN DER HILST, R. D. 2008. The geological evolution of the Tibetan Plateau. *Science*, **321**, 1054–1058.

TAPPONNIER, P., MEYER, B. *ET AL.* 1990. Active thrusting and folding in the Qilian Shan, and decoupling between upper crust and mantle in northeastern Tibet. *Earth and Planetary Science Letters*, **97**, 382–403.

TAPPONNIER, P., XU, Z. Q. *ET AL.* 2001. Geology – oblique stepwise rise and growth of the Tibet plateau. *Science,* **294**, 1671–1677.

TAUXE, L. & GALLET, Y. 1991. A jackknife for magnetostratigraphy. *Geophysical Research Letters,* **18**, 1783–1786.

TAUXE, L. 1998. *Paleomagnetic Principles and Practice.* Kluwer Academic, Dordrecht, 121–170.

TURNER, S., HAWKESWORTH, C., LIU, J., ROGERS, N., KELLEY, S. & VAN CALSTEREN, P. 1993. Timing of Tibetan uplift constrained by analysis of volcanic rocks. *Nature,* **364**, 50–54.

UNSWORTH, M. J., JONES, A. G., WEI, W., MARQUIS, G., GOKARN, S. G., SPRATT, J. E. & THE INDEPTH-MT TEAM 2005. Crustal rheology of the Himalaya and Southern Tibet inferred from magnetotelluric data. *Nature,* **438**, 78–81.

WANG, S. 1965. Cenozoic stoneworts in the Jiuquan Basin, Gansu. *Acta Petrologica Sinica,* **13**, 463–509 (in Chinese with English abstract).

WANG, Q. M. & COWARD, M. P. 1993. The Jiuxi Basin, Hexi Corridor, NW China: Foreland structural features and hydrocarbon potential. *Journal of Petroleum Geology,* **16**, 169–182.

WANG, C. S. *ET AL.* 2008. Constraints on the early uplift history of the Tibetan Plateau. *Proceedings of the National Academy of Sciences USA,* **105**, 4987–4992.

YIN, A. & HARRISON, T. M. 2000. Geologic evolution of the Himalayan–Tibetan orogen. *Annual Review of Earth and Planetary Sciences,* **28**, 211–280.

YIN, A., RUMELHART, P. E. *ET AL.* 2002. Tectonic history of the Altyn Tagh fault system in northern Tibet inferred from Cenozoic sedimentation. *GSA Bulletin,* **114**, 1257–1295.

ZHANG, P. Z. *ET AL.* 2004. Continuous deformation of the Tibetan Plateau from global positioning system data. *Geology,* **32**, 809–812.

ZHANG, J., CUNNINGHAM, W. D. & CHENG, H. Y. 2010. Sedimentary characteristics of Cenozoic strata in central-southern Ningxia, NW China: Implications for the evolution of the NE Qinghai-Tibetan Plateau. *Journal of Asian Earth Sciences,* **39**, 740–759.

ZHAO, J. M. *ET AL.* 2010. The boundary between the Indian and Asian tectonic plates below Tibet. *Proceedings of the National Academy of Sciences USA,* **107**, 11 229–11 233.

Neogene rotations in the Jiuquan Basin, Hexi Corridor, China

MAODU YAN[1]*, XIAOMIN FANG[1,2], ROB VAN DER VOO[3], CHUNHUI SONG[2] & JIJUN LI[2]

[1]*Key Laboratory of Continental Collision and Plateau Uplift, Institute of Tibetan Plateau Research, Chinese Academy of Sciences, Beijing 100085, China*

[2]*Key Laboratory of Western China's Environmental Systems (Ministry of Education of China) and College of Resources and Environment, Lanzhou University, Gansu 730000, China*

[3]*Department of Earth and Environmental Sciences, University of Michigan, Ann Arbor, MI 48109-1063, USA*

**Corresponding author (e-mail: maoduyan@itpcas.ac.cn)*

Abstract: Vertical-axis rotations of blocks in/around the Tibetan Plateau can be attributed to the India–Asia collision. Study of the vertical-axis rotations of these blocks will increase our understanding of the mechanisms and kinematics of continent–continent collisions. We report here a new palaeomagnetic study of rotations using data from four localities (five magnetostratigraphy sections) in the Jiuquan Basin. Our study indicates that the mean declinations of each section are different from each other, similar to what has been observed in the other localities in the NE Tibetan Plateau. However, using the mean directions of every 100 m of section, we observe that the four localities have similar sequential patterns of rotations during the last 13 Ma: significant continuous counterclockwise before *c.* 8.0 Ma, insignificant rotations between 8.0–4.0 Ma, and slight clockwise rotation after 4.0 Ma. This indicates that, rather than being a record of spatially varying declinations, it is a temporal variation in the occurrence of regional rotations. Combined with other geological evidence, the rotation patterns may suggest two major tectonic activity phases of the northeastern Tibetan Plateau during the last 13 Ma: an eastward extrusion and strike-slip dominant phase before 8.0 Ma, a significant shortening and a rapid uplift dominant phase after 8.0 Ma.

Supplementary material: Magnetostratigraphic results of the Hongshuiba and Wenshushan sections are available at: http://www.geolsoc.org.uk/SUP18540.

The Tibetan Plateau, resulting from India–Asia collision since *c.* 55 Ma (e.g. Molnar & Tapponnier 1975; Dewey *et al.* 1988; Yin & Harrison 2000; Tapponnier *et al.* 2001), is an ideal natural laboratory for studying tectonic and geodynamic processes involved in intracontinental collision and mountain building. Knowledge of the evolution and deformation of this orogen will increase our understanding of the mechanisms and kinematics of continent–continent collisions. Because continental crust typically responds heterogeneously to convergence, vertical-axis rotations of crustal elements are often a consequence of collision. Many efforts have been made to understand vertical-axis rotations of the Tibetan crustal blocks and basins, using velocity field and strain distribution through Global Positioning System (GPS) measurements (e.g. Chen *et al.* 2000; Shen *et al.* 2001; Zhang *et al.* 2004), seismic moment tensor analyses of earthquakes (e.g. Molnar & Lyon-Caen 1989; Holt & Haines 1993; Holt *et al.* 1995) and neotectonic studies of Quaternary slip rates on major faults (e.g. van der Woerd *et al.* 1998; Meyer *et al.* 1998). The observed GPS and seismic moment tensors suggest distributed deformation in the eastern part of the Tibetan Plateau (e.g. Molnar & Lyon-Caen 1989; Shen *et al.* 2001; Zhang *et al.* 2004). However, the measured Quaternary deformation rates can be fitted by numerical models assuming either distributed or localized deformation (Avouac & Tapponnier 1993; Peltzer & Saucier 1996; England & Molnar 1997; Holt 2000). Whether the plateau's lithosphere has been deformed as a viscous continuum (e.g. England & Houseman 1989; Molnar *et al.* 1993; Zhang *et al.* 2004) or as quasi-rigid blocks (e.g. Avouac & Tapponnier 1993; Meyer *et al.* 1998; Tapponnier *et al.* 2001) is still under debate. The lack of resolution concerning these two contrasting tectonic models is largely due to insufficient kinematic constraints. Most of the evidence is time-limited, and is mainly constrained by GPS measurements and seismic moment tensor analysis for the last few decades or, at best, based on slip rates during the Quaternary.

To better understand the deformation of the Tibetan Plateau, more and more attention has been paid to the NE Tibetan Plateau (e.g. Yin *et al.* 2002; Fang *et al.* 2003; Parés *et al.* 2003; Lease *et al.* 2007; Clark *et al.* 2010; Molnar *et al.* 2010; Zheng *et al.* 2010). The ongoing sequence of uplift

From: JOVANE, L., HERRERO-BERVERA, E., HINNOV, L. A. & HOUSEN, B. A. (eds) 2013. *Magnetic Methods and the Timing of Geological Processes*. Geological Society, London, Special Publications, **373**, 173–189. First published online November 16, 2012, http://dx.doi.org/10.1144/SP373.6 © The Geological Society of London 2013. Publishing disclaimer: www.geolsoc.org.uk/pub_ethics

and deformation phases of the northeastern Tibetan Plateau has been suggested to be a young, continuous outgrowth of the Tibetan Plateau (e.g. Burchfiel *et al.* 1991; Meyer *et al.* 1998; Tapponnier *et al.* 2001). Knowledge of vertical-axis rotations of this part of the plateau would provide a more comprehensive understanding of the deformation mechanisms and kinematics of the region.

Palaeomagnetism can quantify vertical axis rotations experienced since the rocks became magnetized, which in turn is useful in defining how the deformation was accommodated. Many palaeomagnetic studies have been carried out in the northeastern Tibetan Plateau (e.g. Frost *et al.* 1995; Halim *et al.* 1998; Rumelhart *et al.* 1999; Chen *et al.* 2002*a*, *b*; Dupont-Nivet *et al.* 2002, 2004; Fang *et al.* 2003, 2005*a*, 2007; Dai *et al.* 2006; Yan *et al.* 2006). However, even with these major efforts, so far no consensus has been approached on when and how the region has been deformed. Recent palaeomagnetic data within the Qaidam–Altyn Tagh–Qilian region (Table 1) are contradictory, in that they are being used as arguments for significant clockwise and counterclockwise rotations (Frost *et al.* 1995; Halim *et al.* 1998; Gilder *et al.* 2001; Chen *et al.* 2002*a*, *b*), as well as insignificant rotations (Rumelhart *et al.* 1999; Dupont-Nivet *et al.* 2002, 2003) of the region, without extensive knowledge of the causes of the contradictory results. In addition, palaeomagnetic studies in the Gonghe–Guide and Longzhong basins are marred by a lack of conclusive recognition as to when the significant clockwise rotations of the region have occurred (Dupont-Nivet *et al.* 2004; Yan *et al.* 2006).

The Jiuquan Basin, which is adjacent to the Qaidam–Altyn Tagh–Qilian region, is a foreland basin of the northern Qilian Shan (Mountains) at the northern margin of the Tibetan Plateau. The deposits of sediments in the basin are sensitive to the northward indentation of India into Eurasia, and study of the evolution history of the basin could provide comprehensive knowledge of the uplift and deformation of the Tibetan Plateau (Zhao *et al.* 2001; Fang *et al.* 2005*b*). We report here on a palaeomagnetic rotation study of five magnetostratigraphic sections in the Jiuquan Basin in the west Hexi Corridor. Detailed magnetostratigraphic results are used to provide age constraints on the Neogene deformation of the basin.

Geological setting and sample collection

The Jiuquan Basin is centred at *c.* 39.6°N, 97.5°E in the NW corner of the Hexi Corridor (Fig. 1). It is bounded by the transpressional north Qilian Fault (NQF) to the south, the sinistral Altyn Tagh Fault (ATF) to the west, and the dextral Longshou–Gaotai Fault (LGF) to the east (Fig. 1). The basin is further divided into two sub-basins (the Jiuxi (west) and Jiudong (east) sub-basins) by the detxtral (transpressional) Jiayuguan–Wenshushan fault (JWF in Fig. 1).

The palaeomagnetic samples for the rotation study of the basin are from five magnetostratigraphic sections at four localities: the Laojunmiao (LJM) section along the LJM anticline in the west of the basin, two sections at the Hongshuiba (HSB-A and HSB-B) locality along the HSB anticline, the Wenshushan (WSS) section along the WSS anticline in the middle of the basin, and the Yumushan (YMS) section along the YMS anticline in the east of the basin (Fig. 1). Palaeomagnetic samples were mostly taken from fine-grained sediments, such as mudstones, siltstones and fine-grained sandstones.

The LJM section is located at *c.* 39.5°N, 97.6°E, SW of the Yumen City (Fig. 1). The Shiyouhe ('he' means 'river') drains the section, and the Late Cenozoic sediments are well exposed. The magnetostratigraphy section was taken along the north limb of the Laojunmiao anticline, which has a gently dipping south limb and a steep north limb, where the strata are almost vertical near the core of the anticline. The total thickness of the section is 1960 m.

The HSB locality at *c.* 39.4°N, 98.3°E is located to the east of the LJM section (Fig. 1). The HSB sections were sampled along the south limb of the Hongshuiba anticline. The strata in the sections dip *c.* 52° southwestward. The locality consists of a lower and an upper section, which overlap slightly. The total thickness is 420 m for the lower section and 800 m for the upper section.

The Wenshushan anticline is located *c.* 10 km SW of Jiuquan City, east of the LJM section and NW of the HSB sections (Fig. 1). The WSS section was taken along the north limb of the WSS anticline, where dip directions are N50–60°E, with dips *c.* 40°. The total thickness of the section is 1124 m.

The Yumushan section lies at the eastern corner of the Jiuquan Basin, west of the dextral Longshou–Gaotai Fault. The Yumushan section was taken at the north limb of the YMS anticline, where dip directions are *c.* N35°E, with dips *c.* 34°. The total thickness of the section is 900 m.

As palaeomagnetic sampling was carried out for magnetostratigraphic purposes, palaeomagnetic sites are defined as particular horizons (*c.* 2–6 m apart vertically) within carefully measured stratigraphic sections. At a given site, typically three independently oriented cubic samples ($2 \times 2 \times 2$ cm^3) were collected. This resulted in a collection of more than 4400 specimens (522×3 for the LJM section, 155×3 and 116×3 for the two HSB sections, 432×3 for the WSS section, and 261×3 for the YMS section).

Table 1. *Summary of reported palaeomagnetic directions of the NE Tibetan Plateau*

Age	Site					Palaeopole (VGP)					Reference pole				Expected		Rotation/ Eurasia	References or GPDP
	Latitude (°N)	Longitude (°E)	Declination (deg)	Inclination (deg)	α_{95}	Latitude (°N)	Longitude (°E)	DP	DM	Site no.	Latitude (°N)	Longitude (°E)	α_{95}	Age	Declination (deg)	Inclination (deg)		
1.8–2.6	36.0	101.4	357.1	40.4	6.1	76.8	293.1	4.5	7.4	67	86.7	178.7	3.0	2.1	4.0	56.1	−7.9 ± 9.6	Yan *et al.* (2006)
2.6–3.6	36.1	101.4	358.7	42.2	3.9	78.2	287.2	2.9	4.8	166	86.3	172.0	2.6	3.1	4.4	56.6	−5.7 ± 7.1	Yan *et al.* (2006)
N2	38.4	90.9	359.2	46.9	3.2	79.7	274.9	2.7	4.1	19	86.7	178.7	3.0	2.1	4.2	57.8	−5 ± 7.3	Chen *et al.* (2002a)
N2	38.7	91.1	1.3	54.5	4.4	86.3	254.5	4.4	6.2	10	86.7	178.7	3.0	2.1	4.2	58.1	2.9 ± 9.5	Chen *et al.* (2002a)
N2	37.4	95.3	0.4	48.6	3.6	82.8	272.8	3.1	4.7	30	86.7	178.7	3.0	2.1	4.7	57.1	−4.3 ± 7.7	Dupont-Nivet *et al.* (2002)
3.6–7.8	36.1	101.4	3.3	42.7	1.5	78.3	266.5	1.1	1.9	635	86.3	172.0	2.6	3.1	4.4	56.6	−1.1 ± 5.1	Yan *et al.* (2006)
0–13.0	39.5	97.4	0.0	45.5	2.0	77.5	277.4	1.6	2.5	19	86.3	172.0	2.6	3.1	4.7	59.7	−4.7 ± 2.9	This study
6.7–11	39.4	98.1	321.5	42.7	4.2	54.5	354.8	3.2	5.2	5	85.0	155.7	3.1	11.9	5.7	61.1	−44.2 ± 5.7	This study
7.5–11.5	36.1	101.4	4.4	44.6	3.0	72.5	267.7	3.7	5.5	103	85.0	155.7	3.1	11.9	5.2	58.2	−0.8 ± 8.7	Yan *et al.* (2006)
N1	38.0	86.5	358.4	39.6	6.7	74.4	272.0	4.8	8.0	28 m	84.2	154.9	3.2	14.8	7.0	59.1	−8.6 ± 10.7	Rumelhart *et al.* (1999)
N1	36.1	89.3	20.7	47	10.2	70.8	197.6	8.5	13.2	8	84.2	154.9	3.2	14.8	6.7	57.6	14 ± 16.1	Chen *et al.* (2002a)
N1	38.2	88.7	2.1	37.6	20.7	72.8	262.1	14.4	24.4	3	84.2	154.9	3.2	14.8	7.0	59.5	−4.9 ± 26.1	Chen *et al.* (2002a)
N1	38.4	90.9	354.1	44.4	5.8	76.7	294.6	4.6	7.3	4	84.2	154.9	3.2	14.8	6.9	59.9	−12.8 ± 10.3	Chen *et al.* (2002a)
N1	38.8	95.5	5.7	46.7	8.9	78.2	250.2	7.4	11.5	12	84.2	154.9	3.2	14.8	6.6	60.6	−0.9 ± 14.5	Dupont-Nivet *et al.* (2003)
N1	39.0	95.7	38.5	28.3	14.4	48.7	210.2	8.7	15.8	5	84.2	154.9	3.2	14.8	6.7	60.8	31.8 ± 17.6	Dupont-Nivet *et al.* (2003)
N1	39.7	97.7	358.4	41.3	12.4	74.0	283.0	9.2	15.1	12	84.2	154.9	3.2	14.8	6.6	61.5	−8.2 ± 17.8	Dupont-Nivet *et al.* (2003)
N1	38.7	98.0	11.6	30.0	8.8	65.2	250.6	5.4	9.8	10	84.2	154.9	3.2	14.8	6.5	60.7	5.1 ± 12.1	Dupont-Nivet *et al.* (2003)
N1	34.7	100.7	19.8	33.5	6.0	66.0	228.6	3.9	6.8	9	84.2	154.9	3.2	14.8	6.0	57.3	13.8 ± 9.3	Congé *et al.* (1999)
N1	36.7	101.9	10.3	39.2	8.4	73.0	247.4	6.0	10.0	7	84.2	154.9	3.2	14.8	6.0	59.2	4.3 ± 12.5	Dupont-Nivet *et al.* (2003)
8.5–24	39.4	98.1	327.2	43.2	3.9	59.0	349.9	3.0	4.8	13	84.2	154.9	3.2	14.8	6.6	61.5	−39.4 ± 5.4	Yan (2006)
17.3–19	36.1	101.6	31.1	35.2	9.8	58.1	214.6	6.5	11.3	130 m	81.4	149.7	4.5	19.6	8.6	60.6	22.5 ± 15.1	Yan *et al.* (2006)
E3-N1	39.5	94.7	349	46.4	13.0	75.1	315.7	10.7	16.7	4	81.4	149.7	4.5	19.6	9.8	62.7	−20.8 ± 21.2	Chen *et al.* (2002a)

(Continued)

Table 1. *Continued*

Age	Site					Palaeopole (VGP)					Reference pole				Expected		Rotation/ Eurasia	References or GPDP
	Latitude (°N)	Longitude (°E)	Declination (deg)	Inclination (deg)	α_{95}	Latitude (°N)	Longitude (°E)	DP	DM	Site no.	Latitude (°N)	Longitude (°E)	α_{95}	Age	Declination (deg)	Inclination (deg)		
19–20.8	36.1	101.6	43.9	22.5	10.4	43.6	212.0	5.8	11.0	136 m	81.4	149.7	4.5	19.6	8.6	60.6	35.3 ± 12.4	Yan *et al.* (2006)
24.6–29	36.2	101.6	35.7	34.0	6.8	54.1	212.1	4.4	7.8	275 m	83.8	153.2	5.3	26.0	7.0	58.2	28.7 ± 13.1	Yan *et al.* (2006)
E3	38.9	91.4	5	49.7	5.4	80.6	243.9	4.8	7.2	–	82.8	158.1	3.8	29.5	8.8	60.5	−3.8 ± 11.4	Dupont-Nivet *et al.* (2003)
E3	38.4	94.3	26.2	40	10.5	62.7	211.7	7.6	12.6	7	82.8	158.1	3.8	29.5	8.6	60.3	17.6 ± 15.7	Chen *et al.* (2002a)
E3	39.2	94.3	22.5	38.2	9.7	23.9	220.4	6.8	11.5	6	82.8	158.1	3.8	29.5	8.7	61.0	13.8 ± 14.6	Chen *et al.* (2002a)
E3	39.5	94.8	344.7	39.7	6.6	68.6	316.8	4.8	7.9	21	82.8	158.1	3.8	29.5	8.7	61.3	−24 ± 11.7	Rumelhart *et al.* (1999)
E3	40.0	97.7	8.6	33	6.0	66.8	256.5	3.9	6.8	12	82.8	158.1	3.8	29.5	8.6	62.0	0 ± 10.8	Dupont-Nivet *et al.* (2003)
E3	36.7	101.9	12.8	40.2	5.3	72.4	239.5	3.9	6.4	17	82.8	158.1	3.8	29.5	7.9	59.6	4.9 ± 10.2	Dupont-Nivet *et al.* (2004)
E3	38.7	92.8	8.0	43.6	5.1	75.1	243.5	4.0	6.4	16	82.8	158.1	3.8	29.5	8.8	60.4	−0.8 ± 10.4	Dupont-Nivet *et al.* (2002a)
E3	37.5	95.2	11.0	37.3	11.5	70.8	242.3	7.9	13.5	6	82.8	158.1	3.8	29.5	8.4	59.6	2.6 ± 16.3	Dupont-Nivet *et al.* (2002a)
E2	36.6	101.9	26.8	33.7	11.9	60.3	222.3	12.3	11.5	9	81.3	162.4	3.3	40.0	10.0	59.6	16.8 ± 15.7	Dupont-Nivet *et al.* 2003
E2	36.5	102.0	29.3	40.8	13.2	61.6	211.3	13.6	14.0	5	81.3	162.4	3.3	40.0	9.9	59.6	19.4 ± 18.6	Congé *et al.* (1999)
J3–E3	37.5	86.2	8.3	28.6	5.5	66.6	245.7	5.8	5.0	41	82.2	202.1	5.2	90.0	8.4	53.2	−2.0 ± 13.5	Dupont-Nivet *et al.* (2002a)
K1–2	38.4	90.7	20.7	36.7	6.7	64.6	220.1	7.2	6.5	5	78.2	189.4	2.4	119.1	14.4	55.2	6.3 ± 9.4	Chen *et al.* (2002a)
K1–2	39.9	97.7	18.9	61.7	5.7	75.5	169.7	6.8	8.8	9	78.2	189.4	2.4	119.1	15.2	57.9	3.7 ± 12.8	Chen *et al.* (2002b)
K1–2	39.0	99.6	41.9	39	5.1	48.7	199.7	5.6	5.1	21	78.2	189.4	2.4	119.1	15.1	57.1	26.8 ± 7.9	Frost *et al.* (1995)
K1–2	39.1	100.5	6.8	53.4	3.8	82.5	231.7	4.2	4.9	36	78.2	189.4	2.4	119.1	15.1	57.5	−8.3 ± 7.8	Dupont-Nivet (2003)
K1–2	36.5	101.2	35.8	37.8	6.3	55.4	207.4	6.5	6.4	1	78.2	189.4	2.4	119.1	14.6	55.4	21.2 ± 9.0	Halim *et al.* (1998)
K1–2	37.5	101.5	225	−39.6	4.4	48.7	199.7	4.1	4.1	21	78.2	189.4	2.4	119.1	14.8	56.4	30.2 ± 7.2	GPDP
J3–E3	37.5	90.8	40.4	34.8	7.5	68.8	118.9	8.1	7.1	9	75.0	159.9	6.6	151.6	19.1	61.2	21.3 ± 16.5	Chen *et al.* (2002a)

DP, DM, confidence limits of poles calculated from tilt corrected site mean directions.

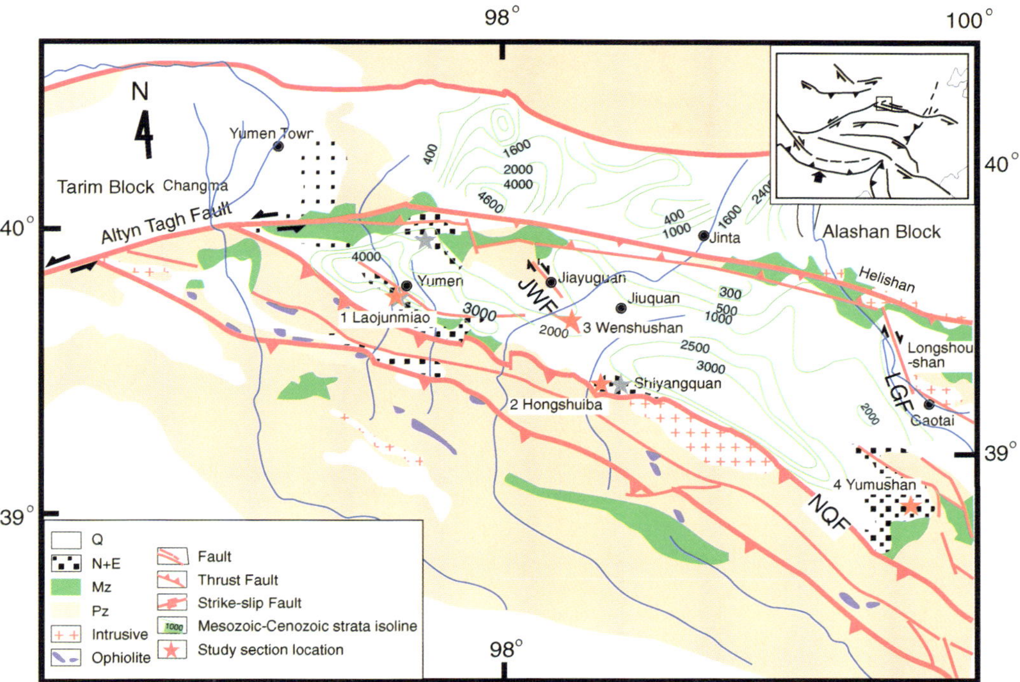

Fig. 1. Simplified geologic map of the Jiuquan Basin (modified from Qinghai Geology Bureau 1989). The red stars indicate the locations of our four localities (in bold) of Laojunmiao (LJM), Hongshuiba (HSB), Wenshushan (WSS) and Yumushan (YMS), labelled as 1–4, respectively. The grey stars are the four sites studied by others; NQF is the North Qilian Fault. Inset is a simplified tectonic map of the Tibetan Plateau showing the location of the Jiuquan Basin.

Laboratory methods

Remanent directions were measured with a 2 G cryogenic magnetometer at the Beijing Institute of Geology and Geophysics of Chinese Academy of Science, and at the University of Michigan. Only thermal demagnetization was used and proceeded with 10–16 measurement steps up to 680 °C or until the intensity was near the noise level of the cryogenic magnetometer. Characteristic remanent magnetization (ChRM) directions were calculated from at least four successive temperature steps by principal component analysis (Kirschvink 1980), guided by visual inspection of the orthogonal demagnetization diagrams such as those shown in Figure 2a, b. ChRM directions that could not be determined because of ambiguous or noisy orthogonal demagnetization diagrams were discarded (Fig. 2c). Samples yielding maximum angular deviation angles larger than 15° or virtual geometric pole (VGP) latitudes less than 45° were also rejected. VGP latitudes for samples from the LJM, WSS and YMS sections were determined by calculating mean directions and their corresponding pole positions, as their mean declination is close to due north (Table 1),

whereas for results from the HSB-A and -B sections, VGP latitudes are relative to those for the mean declinations of *c.* 324.5 and *c.* 336.5°, respectively. Accordingly, the declinations were restored to 35.5 or 23.5° in a clockwise sense so that they then coincided with due north. To analyse any sequential rotations in detail, ChRM directions of every 100 m of section were analysed as a group, so that a mean for each group could be calculated (inverting all negative polarity directions). Groups with fewer than four accepted samples (e.g. 1500–1600 and 1700–1800 m in the LJM Section) or with α_{95} larger than 20° (1500–1600 m in the LJM Section) were excluded from further analysis.

Palaeomagnetic results

Representative thermal demagnetization diagrams of samples from four localities are shown in Figure 2. Most samples show an uncomplicated demagnetization behaviour: after removal of a low unblocking temperature component of 150–350 °C, a ChRM is isolated and decays nearly linearly to the origin (Fig. 2a, b). Maximum unblocking

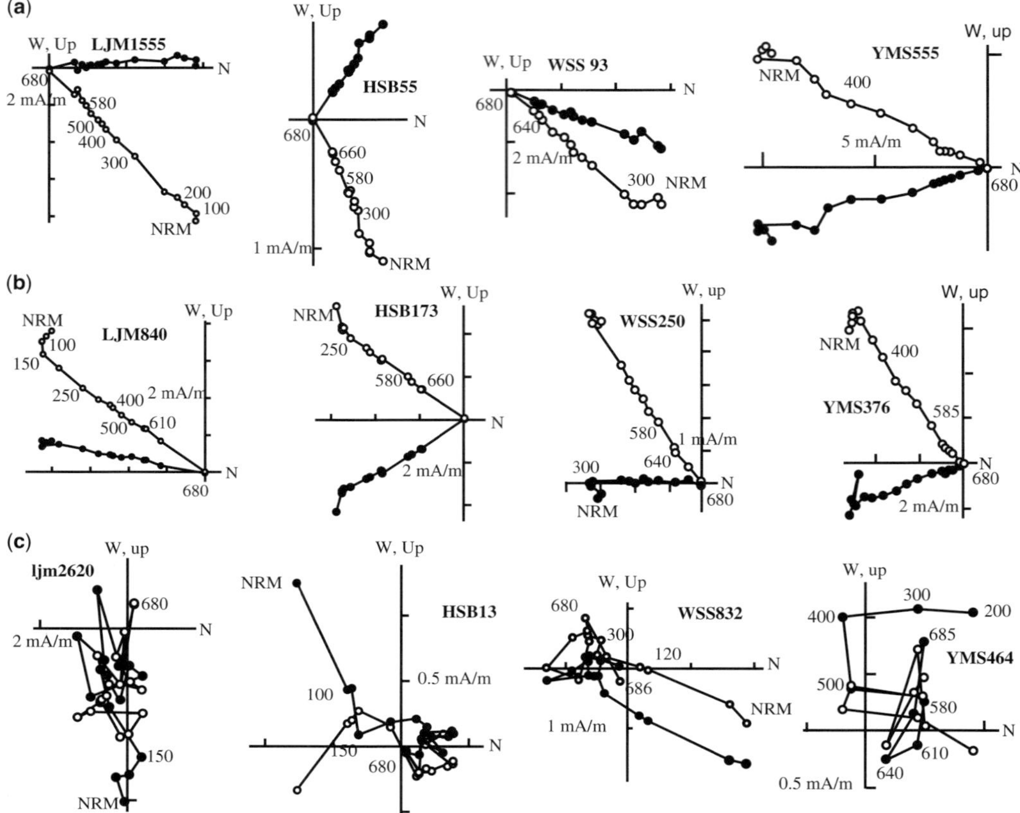

Fig. 2. Representative orthogonal demagnetization diagrams showing thermal (degrees Celsius) demagnetization results from the Laojunmiao, Hongshuiba, Wenshushan and Yumushan sections. Tilt-corrected representative demagnetization diagrams of the four localities for good normal polarity samples (**a**), good reversed polarity samples (**b**) and discarded samples (**c**). Solid squares represent horizontal projections, and open circles represent vertical projections.

temperatures of up to 680 °C and rock magnetic analysis (Fig. 3) indicate that magnetite and hematite are both likely carriers of the magnetization. The final mean direction for each site was obtained by Fisher averaging of the ChRM directions from the samples for that site (Fig. 3a).

The Laojunmiao section

Magnetostratigraphic results for the Laojunmiao section were previously reported in a Chinese journal (Fang *et al.* 2005*b*, reproduced in Fig. 4). Combined with a focus on lithostratigraphy, that study identified detailed depositional environments and provided age constraints (13–0 Ma) on the sedimentary infilling of the basin. The combined age, lithological and magnetic data suggest an increase in significant tectonic activity in the region since

the late Miocene (*c.* 8 Ma), especially after 3.6 Ma (Fang *et al.* 2005*b*).

Using the criteria mentioned above, a total of 370 (71%) ChRM directions were used for our palaeomagnetic rotation determination of LJM. These directions pass a tilt test (Fig. 3b; Fang *et al.* 2005*b*), and have a positive reversal test (Table 2), with the classification A (McFadden & McElhinny 1990; Tauxe, 1998). The mean direction for the section is $D_s = 3.0°, I_s = +45.5°, \alpha_{95} = 2.0°$ (Fig. 3a); a mean direction for every 100 m group was calculated and is shown in Table 2.

The Hongshuiba sections

Detailed magnetostratigraphic analysis of the two sections at the HSB locality reveals seven normal and seven reversed-polarity intervals for its lower

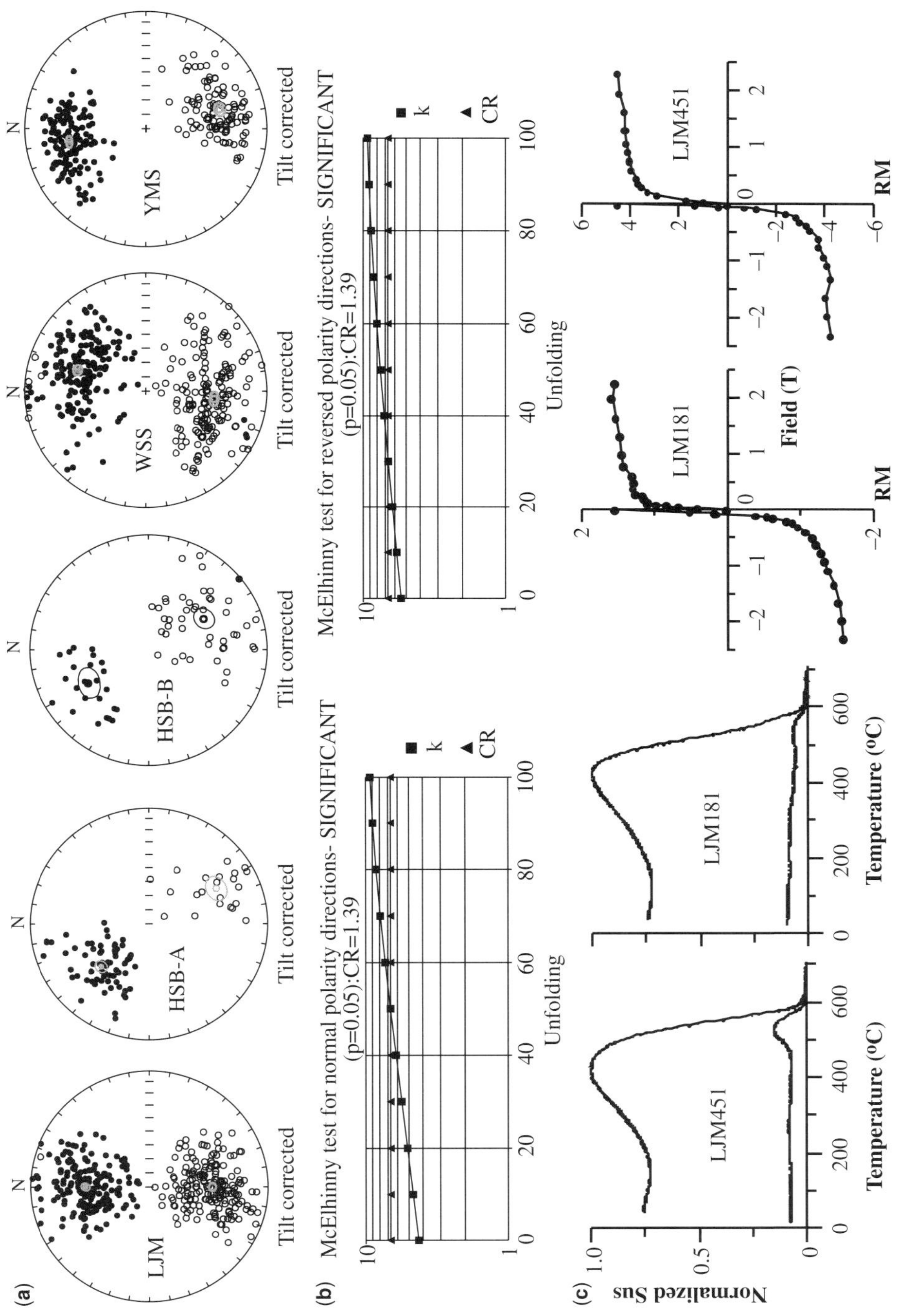

Fig. 3. (**a**) Equal-area projections of the tilt-corrected characteristic remanent magnetization directions (site-means) and their means (with oval of 95% confidence) for the Laojunmiao, Hongshuiba, Wenshushan and Yumushan sections. Mean directions are in grey; downward (upward) directions are shown as solid (open) circles. (**b**) McElhinny's tilt test for the Laojunmiao anticline. Both normal and reversed ChRM directions have a positive tilt test. (**c**) Representative rock magnetic diagrams (susceptibility *v.* temperature, isothermal remanent magnetization acquisition with reverse magnetic field). The results indicate that magnetite and hematite are the two major magnetization carriers of the samples.

part (section A), and six normal and seven reversed intervals for its upper part (section B). Based on pollen ages (Sung 1958) and previous magnetostratigraphic studies for the Cenozoic red beds in the region (Huang *et al.* 1993), as well as the magnetostratigraphic results of the Laojunmiao section, these intervals are most reasonably correlated with the Geomagnetic Polarity Time Scale (GPTS; Gradstein *et al.* 2004) with magnetostratigraphic ages of *c.* 11–6.7 Ma for the lower section and 7.2 to <2.58 Ma for the slightly overlapping upper section.

A total of 89 (57%) and 73 (63%) ChRM directions were used from the two sections for determination of palaeomagnetic rotations. The directions pass the reversal test (McFadden & McElhinny 1990), with a classification of C for the lower section and B for the upper section. The mean directions are $D_s = 324.5°$, $I_s = +42.7°$, $\alpha_{95} = 4.2°$ for the lower section, and $D_s = 336.5°$, $I_s = +45.8°$, $\alpha_{95} = 6.5°$ for the upper section (Fig. 3a). The mean of every 100 m interval was also calculated (Table 2).

The Wenshushan section

Detailed magnetostratigraphic analysis of the WSS Section reveals 20 normal and 20 reversed-polarity intervals. Constrained by pollen ages (Sung 1958) in the region and the magnetostratigraphic ages of the Laojunmiao section, these intervals are best correlated to the chrons between 5r and 2r of the GPTS (Gradstein *et al.* 2004), with magnetostratigraphic ages of *c.* 11 to <2.58 Ma.

A total of 347 (80%) ChRM directions were used for this study. Directions had a positive reversal test (McFadden & McElhinny 1990), with classification of B. The mean direction for the section is $D_s = 15.6°$, $I_s = +41.7°$, $\alpha_{95} = 2.6°$ (Fig. 3a), and the means of every 100 m section were calculated (Table 2).

The Yumushan section

Magnetostratigraphic results for the Yumushan section are presented in Fang *et al.* (2012), with magnetostratigraphically determined ages of *c.* 27.5–9.5 Ma. Our study only focuses on any late Neogene rotation of the basin; the results with Oligocene ages are not discussed.

A total of 230 (88%) ChRM directions of the Yumushan Section are used for our analysis. The directions have a positive reversal test, with a classification of A (McFadden & McElhinny 1990; Tauxe 1998; Fang *et al.* 2012). The mean direction for the section is $D_s = -9.2°$, $I_s = +36.4°$, $\alpha_{95} = 2.3°$ (Fig. 3a); mean directions of every 100 m section are also calculated (Table 2).

Discussion and conclusions

Inclination shallowing

The average inclinations show no significant temporal trend in the Laojunmiao, Hongshuiba, Wenshushan or Yumushan sections, with means of 45.5, 42.7/45.8, 41.7 and 36.4°, respectively (Tables 2 & 3). These values are 15–30° lower than those expected from the Late Neogene Eurasia reference prediction for the area (Table 3). This is entirely similar to what has been observed in other studies of Central Asian Neogene sediments (e.g. Cogné *et al.* 1999; Gilder *et al.* 2001; Chen *et al.* 2002b; Dupont-Nivet *et al.* 2003; Fang *et al.* 2003; Yan *et al.* 2005; Tauxe *et al.* 2008; Tan *et al.* 2010). As many of these studies in the region have suggested that syn- or post-depositional processes such as compaction are the most likely cause of the shallowing (Tauxe & Kent 2004; Tauxe 2005; Yan *et al.* 2005), we similarly attribute the too shallow inclinations observed in our study to flattening during syn- and/or post-depositional processes.

Palaeomagnetic rotations

Previously observed declination patterns in the Qaidam–Qilian thrust belt region resulted in conflicting interpretations based on then-existing palaeomagnetic results (Table 1). Frost *et al.* (1995) suggested significant post-Cretaceous clockwise rotations of the Hexi Corridor, followed later by publications by Halim *et al.* (1998) and Chen *et al.* (2002a, b), who proposed significant Late Oligocene–Late Miocene clockwise rotations of the NE Tibetan Plateau. Rumelhart *et al.* (1999) and Dupont-Nivet *et al.* (2002, 2003) argued for insignificant rotation of the Qaidam Basin and the entire NE Tibetan Plateau since the Eocene, whereas Gilder *et al.* (2001) suggested significant Early–Middle Miocene counterclockwise rotations of the Subei Basin in the Qilian thrust belt. The discordant declination pattern in the region may have been the result of

---

Fig. 4. Magnetostratigraphic results v. lithostratigraphic position in the Laojunmiao section (reproduced from Fang *et al.* 2005b). The ages listed under depositional environment were estimated from magnetostratigraphic correlations (Fang *et al.* 2005b); electron spin resonance (ESR) ages are from Shi *et al.* (2001). GPTS is the reference geomagnetic polarity time scale from Gradstein *et al.* (2004); VGP-Lat is virtual geomagnetic pole latitude.

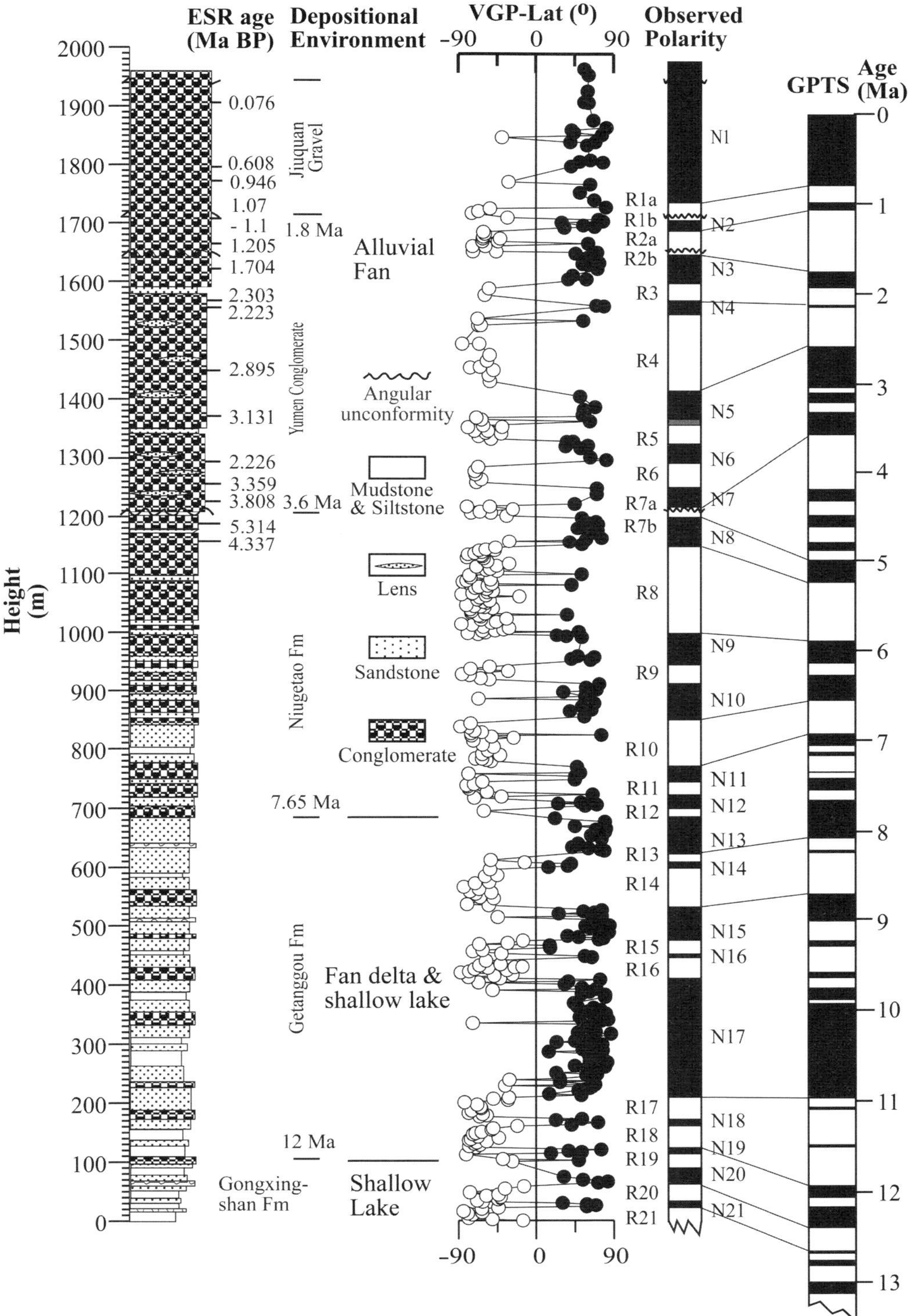
ESR age
(Ma BP)
Depositional
Environment
VGP-Lat (⁰)
Observed
Polarity
GPTS
Age
(Ma)
Height
(m)
2000
1900
1800
1700
1600
1500
1400
1300
1200
1100
1000
900
800
700
600
500
400
300
200
100
0
0.076
0.608
0.946
1.07
- 1.1
1.205
1.704
2.303
2.223
2.895
3.131
2.226
3.359
3.808
5.314
4.337
Jiuquan Gravel
Yumen Conglomerate
Niugetao Fm
Getanggou Fm
Gongxing-
shan Fm
1.8 Ma
3.6 Ma
7.65 Ma
12 Ma
Alluvial
Fan
Angular
unconformity
Mudstone
& Siltstone
Lens
Sandstone
Conglomerate
Fan delta &
shallow lake
Shallow
Lake
-90   0   90
-90   0   90
N1
R1a
R1b
R2a
R2b
R3
R4
R5
R6
R7a
R7b
R8
R9
R10
R11
R12
R13
R14
R15
R16
R17
R18
R19
R20
R21
N2
N3
N4
N5
N6
N7
N8
N9
N10
N11
N12
N13
N14
N15
N16
N17
N18
N19
N20
N21
0
1
2
3
4
5
6
7
8
9
10
11
12
13

Table 2. *Summary of palaeomagnetic directions from the Laojunmiao, Hongshuiba, Wenshushan and Yumushan sections*

Depth interval (m)	Age interval (Ma)	Age error (Ma)	Declination (deg)	Inclination (deg)	α_{95}	n/N	k	Fisherian
Laojunmiao								
1800–1900	0.27	0.18	13.1	32.1	10.5	9/13	25	Yes
1700–1800	*0.68*	*0.32*	*−4.2*	*33.1*	*14.2*	*2/8*	*–*	*–*
1600–1700	1.38	0.39	12.6	36.7	13.2	11/20	13	Yes
1500–1600	*2.04*	*0.27*	*−0.6*	*42.3*	*30.2*	*3/9*	*18*	*No*
1400–1500	2.5	0.19	7.6	30.1	11.1	7/11	22	Yes
1300–1400	3	0.22	10.0	34.1	11.4	11/17	17	No
1200–1300	3.4	0.2	10.8	40.1	8.6	12/18	26	Yes
1100–1200	5.2	0.3	−0.4	40.1	8.6	19/28	26	Yes
1000–1100	5.65	0.15	3.9	42.7	7.9	28/34	13	No
900–1000	6.05	0.25	−1.0	40.4	8.2	24/32	14	Yes
800–900	6.57	0.28	4.8	40.1	7.1	29/36	15	Yes
700–800	7.14	0.32	2.8	39.8	9.2	25/38	11	Yes
600–700	7.85	0.41	4.6	50.1	7.1	24/35	18	Yes
500–600	8.56	0.3	−1.9	61.1	6.3	26/36	21	Yes
400–500	9.37	0.51	5.3	60.0	5.2	30/43	26	Yes
300–400	10.16	0.28	−5.0	53.2	7.8	25/33	15	No
200–300	10.71	0.28	−6.5	42.9	7.4	27/37	15	Yes
100–200	11.56	0.57	10.4	45.8	6.6	27/35	19	Yes
0–100	12.55	0.42	−1.0	46.0	7.8	29/38	13	Yes
Mean (normal)			3.1	45.4	2.9	186/271	14	No
Mean (reversed)			183.0	−45.6	2.8	184/251	14	Yes
Mean			**3.0**	**45.5**	**2.0**	**370/522**	**14**	**No**
Hongshuiba								
Section A								
0–100	7.08	0.35	−31.3	42.3	16.0	10/16	10	Yes
100–200	7.93	0.5	−24.7	31.6	9.1	20/37	14	Yes
200–300	9.21	0.79	−36.1	41.8	7.7	27/46	14	Yes
300–400	10.4	0.4	−46.2	50.4	5.3	27/49	28	Yes
400–420	10.87	0.07	−39.4	44.5	16.0	5/7	24	Yes
Mean (normal)			−40.3	44.8	4.5	67/101	16	Yes
Mean (reversed)			154.7	−35.7	9.3	22/54	12	Yes
Mean			**−35.5**	**42.7**	**4.2**	**89/155**	**14**	**No**
Section B								
800–637	2.4	0.2	−22.7	46.9	11.2	21/29	9	Yes
635–525	2.85	0.27	−22.9	42.5	13.9	16/20	10	Yes
515–415	*3.22*	*0.11*	*−21.8*	*27.9*	*33.7*	*3/9*	*14*	*Yes*
415–315	*3.46*	*0.13*	*−38.4*	*41.9*	*24.4*	*6/10*	*8*	*Yes*
315–215	*6.01*	*0.12*	*−15.8*	*43.8*	*22.2*	*5/14*	*13*	*Yes*
200–100	*6.5*	*0.36*	*−24.9*	*59.5*	*21.2*	*14/17*	*4*	*No*
100–0	7.17	0.31	−27.3	43.3	18.1	8/17	10	Yes
Mean (normal)			−17.3	46.8	11.3	27/50	7	No
Mean (reversed)			150.0	45.0	7.8	46/67	8	Yes
Mean			**−23.5**	**45.8**	**6.5**	**73/116**	**8**	**Yes**
Wenshushan								
1064–1124	<2.58		14.7	40	17.7	11/11	8	Yes
1000–1064	2.81	0.23	20.8	36.4	13.6	22/29	6	Yes
900–1000	3.28	0.19	21.4	38.2	9	30/49	10	No
800–900	3.86	0.38	27.7	45.9	8.4	25/37	13	Yes
700–800	4.78	0.54	17.6	45.6	7.6	38/48	10	Yes
600–700	5.79	0.47	21.1	44.9	6.5	49/60	11	Yes
500–600	6.67	0.38	19.8	43.9	6.6	42/46	12	No
400–500	7.57	0.5	19.3	36.7	9.5	37/39	7	Yes

(Continued)

Table 2. *Continued*

Depth interval (m)	Age interval (Ma)	Age error (Ma)	Declination (deg)	Inclination (deg)	α_{95}	n/N	k	Fisherian
300–400	8.56	0.49	10.4	35	11.5	22/27	8	Yes
200–300	9.39	0.34	4.8	36.7	10.6	25/32	8	No
100–200	10.06	0.36	−1.8	45.8	5.8	23/27	29	Yes
0–100	10.87	0.43	4.1	40.6	11.1	23/27	8	Yes
Mean (normal)			20.3	40.7	3.4	187/227	10	Yes
Mean (reversed)			190	−42.7	4	160/205	9	No
Mean			**15.6**	**41.7**	**2.6**	**347/432**	**9**	**No**
Yumushan								
800–900	9.61	0.3	3.1	43.2	4.1	27/31	44	Yes
700–800	10.19	0.27	−10.2	33.9	4.3	44/49	26	No
600–700	10.74	0.28	−10	35	5.1	37/49	22	No
500–600	11.41	0.39	−13.1	40.1	6.3	42/45	13	Yes
400–500	12.09	0.29	−7.4	35.7	7.2	29/35	15	No
300–400	12.9	0.52	−7.7	32.5	5.3	51/52	15	Yes
Mean (normal)			−7.1	35.8	3	130/146	18	No
Mean (reversed)			168	−37.5	3.6	100/115	17	Yes
Mean			**−9.2**	**36.4**	**2.3**	**230/261**	**17**	**No**
Oligocene								
200–280			1.7	40.2	7.6	37	11	Yes
100–199			−9.2	34.9	7.6	23	17	Yes
0–88			5.8	39.4	15.6	13	8	Yes

Expected directions are calculated from the reference poles for Eurasia (Besse & Courtillot 2002). N is the number of sites (stratigraphic levels) included in the mean directions; α_{95} and k are the statistical parameters associated with the means. Fisherian statistics 'yes' represents a circular distribution of the observed palaeomagnetic directions, whereas 'no' represents an elliptical distribution. Data discarded for rotation analysis are in italics, owing to $n < 5$ or $\alpha_{95} > 20°$.

local deformation, and that no significant rotation has occurred since the Eocene for the area as a whole (Dupont-Nivet *et al.* 2002, 2003). However, several of the palaeomagnetic results lack well-determined age constraints, and studied intervals may represent long and uncertain durations (Table 1). Owing to the possibility that the reported rotations are averages of alternating clockwise and counter-clockwise declinations during long durations, it is not clear whether the above-reported palaeomagnetic results are representative of rotations in one locality only or must be seen as valid for a much larger region.

The obtained means of Neogene declinations of the Laojunmiao, Hongshuiba, Wenshushan and Yumushan sections in the Jiuquan Basin are +3.0, −35.5/ −23.5, +15.6 and −9.2°, respectively. These declinations deviate from the expected direction by the following angles (negative is counterclockwise) −1.7, −41.3/ − 28.3, +10.9 and −14.7°, respectively. These values have been calculated with respect to reference directions extrapolated from Eurasia's Apparent Polar Wander Path (APWP) (Besse & Courtillot 2002). Assuming that these deviations represent vertical-axis rotations

within the Hexi Corridor, these values indicate (i) insignificant rotations of the Laojunmiao Section, (ii) significant counterclockwise rotations of the Hongshuiba Sections and the Yumushan Section and (iii) clockwise rotations of the Wenshushan Section. Given that the five sections are located in different parts of the Jiuquan Basin (e.g. west, middle and east; see Fig. 1), their means could be taken to imply that, during the late Neogene, different parts of the basin underwent rather different magnitudes and senses of rotation, amounting to an incoherent pattern of deformation. This observation would resemble the rotation pattern inferred for the nearby Qaidam Basin.

There is an alternative way of interpreting these data, which involves comparing the declinations of shorter intervals, and attempting to see whether temporal variations show similarities between sections. Indeed, the variations (trends) of declinations as a function of time are similar throughout the last 13 Ma for each section (Fig. 5), although the magnitude of rotations for the Hongshuiba Sections is different from those of the three other localities: (1) increased declination through time during *c.* 11.0–8.0 Ma; (2) no significant variations in

M. YAN *ET AL.*

Table 3. *Palaeomagnetic directions of the Jiuquan Basin compared with expected Eurasia reference directions in the basin*

Site no.	Locality	Age (Ma)	Location		Observed direction				Reference pole			Rotation (deg)		Flattening (deg)		Reference
			Latitude (°N)	Longitude (°E)	Declination (deg)	Inclination (deg)	α_{95}	N	Latitude (°N)	Longitude (°E)	α_{95}	R	$\pm\Delta R$	F	$\pm\Delta F$	
1	Laojunmiao	0–13	39.7	97.7	3.0	45.5	2.0	19	86.3	172.0	2.6	−1.7	±4.7	14.4	±3.3	This study
2	Shiyougou	N1	39.69	97.66	−1.6	41.3	12.4	12	84.2	155.3	1.9	−8.3	±13.4	20.2	±10.0	Dupont-Nivet et al. (2003)
3	Yaoquanzi	E3	39.7	97.68	8.6	33.0	6.0	12	82.4	171.7	3.7	−1.2	±7.0	27.7	±5.4	Dupont-Nivet et al. (2003)
4	Yumen Town	K1	39.9	97.7	18.9	61.7	5.7	9	79.8	188.0	2.5	5.7	±10.0	−3.3	±4.9	Chen et al. (2002b)
5	Hongshuiba Lower	6.7–11	39.4	98.3	−35.5	42.7	4.2	5	85.0	155.7	3.1	−41.3	±6.2	18.6	±7.0	This study
5	Hongshuiba Upper	7.2–<2.5	39.4	98.3	−23.5	45.8	6.5	73	86.3	172.0	2.6	−28.3	±10.6	13.7	±5.3	This study
6	Shiyangjuan	8.5–24	39.4	98.4	−32.8	43.2	3.9	13	84.2	154.9	3.2	−36.5	±5.6	18.5	±5.0	Yan (2006)
7	Wenshushan	<2.5–11	39.6	98.2	15.6	41.7	2.6	12	86.3	172.0	2.6	10.9	±4.9	18.0	±3.7	This study
8	Yumushan	13.5–9.5	38.9	99.6	−9.2	36.4	2.3	6	85.0	155.7	3.1	−14.7	±4.8	24.2	±3.9	This study

α_{95} is the cone of 95% confidence associated with the mean; N is the number of sites used for a mean direction. Positive (negative) R represents clockwise (counterclockwise) rotation compared with the expected direction. $\Delta R/\Delta F$, which is calculated by using the method of Demarest (1983), is the relative declination/inclination deviation observed in the Jiuquan Basin with respect to the expected reference directions extrapolated from Eurasia's APWP (Besse & Courtillot 2002).

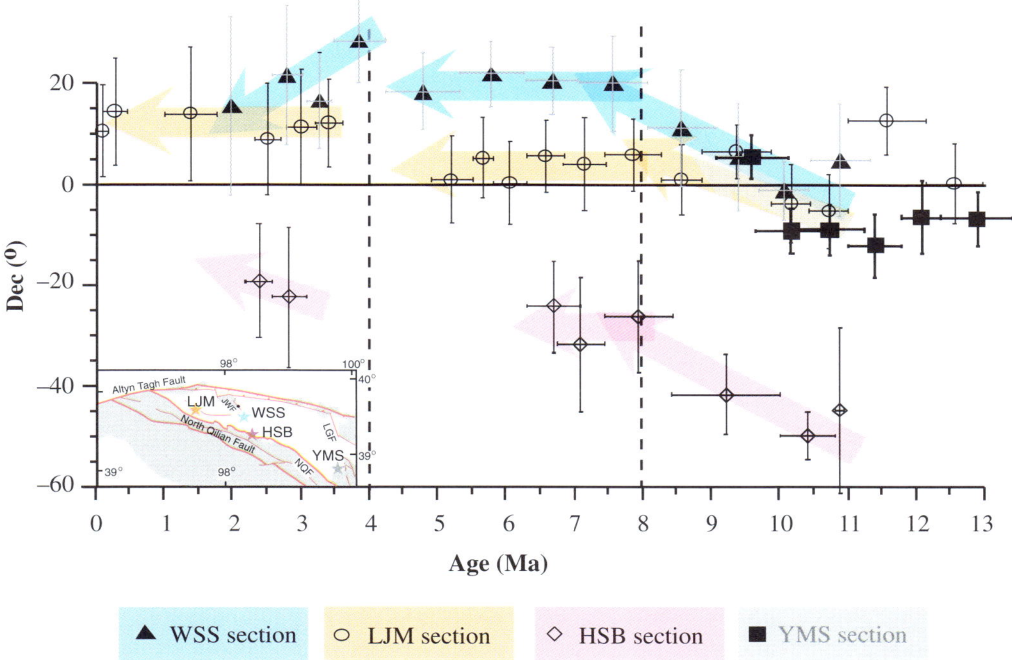

Fig. 5. Plot of mean palaeomagnetic declination v. age for the Laojunmiao, Hongshuiba, Wenshushan and Yumushan sections in the Jiuquan Basin. In general, positive declination implies subsequent clockwise rotation, whereas a negative one suggests counterclockwise rotation. Steadily changing declinations as a function of time imply ongoing rotation, clockwise (counterclockwise) when they are decreasing (increasing). Average declinations were calculated for every 100 m of section. Ages were obtained from magnetostratigraphic correlations with the Geomagnetic Polarity Timescale (Gradstein & Ogg 2004; Fig. 4). Open circles represent the Laojunmiao section, open diamonds the Hongshuiba section, solid triangles the Wenshushan section, and solid squares the Yumushan section. Vertical lines indicate the error limits of declination and horizontal lines show age uncertainties. Dashed vertical lines are the boundaries of the three stages described in the text. Declination means with number of samples less than five and/or α_{95} greater than 20° were excluded from further palaeomagnetic rotation study. Coloured arrows indicate the declination variation trend directions of each section.

declination during c. 8.0–4.0 Ma, although the data for this interval are a little sparse; and (3) no significant change of direction after c. 4.0 Ma, although there was a higher declination value than during c. 8.0–4.0 Ma. This observation leads to the conclusion that the Jiuquan Basin experienced significant (but successively counterclockwise and clockwise) rotations in a similar pattern across the basin during the last 13 Ma. This pattern is summarized as follows: (1) significant and continuous counterclockwise rotations during c. 11.0–8.0 Ma; (2) insignificant rotations during c. 8.0–4.0 Ma, representing a relatively stable stage; and (3) a small clockwise rotation during the last 4 million years.

We thus argue that the reason why the mean of each section appears to indicate discordant palaeomagnetic rotations is mostly (but not entirely) due to the averaging of different rotation angles and senses during different time intervals. Although the magnitude of rotations for the Laojunmiao,

Wenshushan and Yumushan sections are similar, the magnitude of the rotation observed for the Hongshuiba Sections (as well as the 8.3–24 Ma Shiyangquan Section (Yan 2006)) is different from those of the three other localities (Fig. 5). The Hongshuiba Section is just next to the North Qilian Fault (as is the case also for the Shiyangquan Section), where the strike of the NQF changes from WNW–ESE to west–east. This transition may have caused an increase in the magnitude of counterclockwise rotations of the HSB sections. Hence, it is likely that local deformations did occur in some parts of the Jiuquan Basin, but that the basin as a whole also underwent deformation (rotation) as a quasi-rigid block.

Based on the variation patterns from the 100 m average palaeomagnetic declinations (Fig. 5), we now discuss the deformation history of the Jiuquan Basin during the last 13 Ma. Figure 6 presents a simple model to clarify how our findings relate to

(a) Strike-slipping, eastward extrusion and rotations

Eastward extrusion of the northeast Tibetan Plateau causes counterclockwise rotation of the Jiuquan Basin

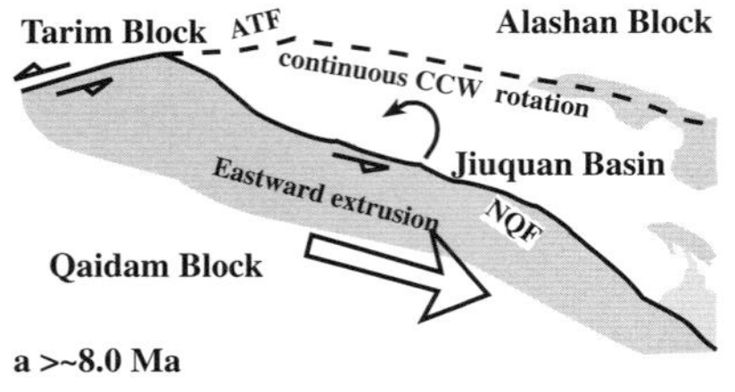

(b) Shortening and rapid uplift

North-southward shortening and rapid uplift of the north Qilian Shan

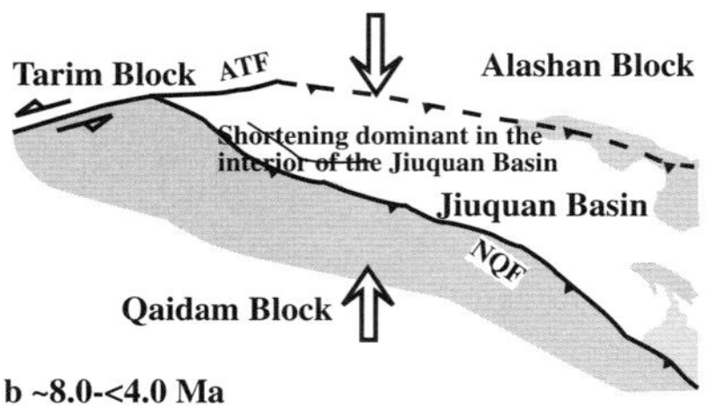

Shortening and uplift are dominant, and northeast propagation of the ATF to the Jiuquan Basin, with right lateral slip of the JWF and LGF, caused slight clockwise rotation of the Jiuquan Basin

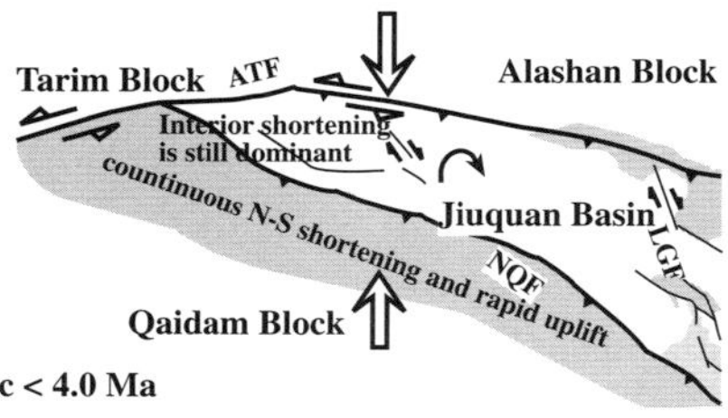

Fig. 6 Two-stage model of the tectonic evolution of the Jiuquan Basin. (**a**) *c.* 13.0–8.0 Ma, a period dominated by strike-slip displacements, eastward extrusion and rotation. During this period, the Jiuquan Basin experienced counterclockwise rotations. (**b**) *c.* 8.0–0 Ma, a period dominated by shortening and rapid uplift. No significant rotation occurred during the period of 8–4 Ma, followed by small clockwise rotations of the basin after 4.0 Ma.

the kinematic evolution of the Jiuquan Basin in the later Miocene, Pliocene and Pleistocene.

The first major phase occurred before *c.* 8.0 Ma, a period dominated by strike-slip movements, eastward extrusion and counterclockwise rotations of the Jiuquan Basin. During this time, the NE Tibetan Plateau was extruded eastward, accompanied by left-lateral slip faulting of the Altyn Tagh, Qilian–Haiyuan and Kunlun faults. This eastward extrusion of the plateau with a relatively stationary foreland basin of the Hexi Corridor caused significant counterclockwise rotations of the basin before *c.* 8.0 Ma, by eastward dragging of the Jiuquan Basin. Additional strike-slip related tectonic activity in the northeastern Tibetan Plateau occurred before 8.0 Ma as well, as seen in (1) significant clockwise rotations of the Guide Basin before *c.* 17–11 Ma (Yan *et al.* 2006), (2) outgrowth of the north Qilian Shan at 10 Ma (Zheng *et al.* 2010), (3) right-lateral slip of the Wenquan Fault around the middle Miocene (Wang & Burchfiel 2004), (4) uplift of the Liupan Mountains in the NE corner of the Tibetan Plateau around 8 Ma (Zheng *et al.* 2006), and (5) volcanic eruptions in the Xihe-Lixian, Nanyang and Niudingshan basins before 14–7 Ma (Yu *et al.* 2001; Wang & Li 2003; Horton *et al.* 2004).

The second major phase took place in the interval *c.* 8.0 Ma–present, a period dominated by significant shortening and rapid uplift. Based on thermochronology (Zheng *et al.* 2006, 2010) and basin analysis of the Jiuquan Basin (Song *et al.* 2001; Fang *et al.* 2003, 2005*a*, 2007), the northeastern Tibetan Plateau and north Qilianshan started to be uplifted rapidly after 8.0 Ma. Multiple processes of deformation and uplift followed, increasing in magnitude and frequency, such as significant uplift around 3.6 Ma (marked by huge molasse deposits elsewhere in the NE Tibetan Plateau; Li 1995; Fang *et al.* 2003, 2005*a*; Parés *et al.* 2003; Yan *et al.* 2012), and the appearance of the Yellow River around 1.8 Ma (Li 1995; Fang *et al.* 2005*a*), followed by multiple headward incision events (e.g. at *c.* 1.5, 1.2, 0.9, 0.6 and 0.14 Ma; Li 1995). Molasse deposition started around 8.0 Ma in the Jiuquan Basin in response to the second major phase, with increases in size and amount of conglomerates through time, especially after 3.6 Ma. Six angular unconformities and disconformities are observed with ages <8 Ma (Fig. 4; Fang *et al.* 2005*b*). Because the basin mainly underwent north–south shortening, no significant rotation occurred during the *c.* 8.0–4.0 Ma interval (Fig. 6b, c). Then, after 4.0 Ma, the Altyn Tagh Fault propagated into the Jiuquan Basin, and the Jiuyuguna–Wenshushan and Yumushan faults underwent right-lateral slip movements, together causing minor clockwise rotations of the basin sometime after 4.0 Ma (Fig. 6c). A change in strike direction along the North Qilian Fault from WNW to ENE took place near the Hongshuiba and Shuyangquan sections, and occurred sometime after 2.5 Ma, causing significant counterclockwise rotation of these three sections (Figs 1 & 5).

Conclusions

Detailed magnetostratigraphic studies in the Jiuquan Basin provide well-dated age constraints on the late Neogene sedimentary infilling of the basin. The magnetostratigraphy reveals ages as follows for the sampled sections: *c*. 12–0 Ma for the Laojunmiao Section, *c*. 11 to <2.5 for the Hongshuiba (HSB) Section, 11 to <2.5 Ma for the Wenshushan Section, and *c*. 13–9 Ma for the Yumushan Section. These ages allow us to analyse the temporal patterns of basin-wide as well as local rotations in great detail.

Mean declinations of the five sections are: 3.0, −35.5/−23.5, 15.6 and −9.2°, for the LJM, HSB upper/lower, WSS and YMS sections, respectively. The declinations represent cumulative rotations as calculated by comparison to a reference APWP (Besse & Courtillot 2002), as follows: −1.7, −41.3/−28.3, +10.9 and −14.7°, where positive is clockwise. Although the overall declination means of the five sections are discordant, mean declinations from 100 m average section intervals show similar patterns of rotations throughout the five sections during the last 13 Ma: (1) significant and continuous counterclockwise rotations before *c*. 8.0 Ma; (2) insignificant rotations during *c*. 8.0–4.0 Ma (relatively stable stage in rotations); and (3) slight clockwise rotations sometime after 4.0 Ma. These data reveal that the Jiuquan Basin experienced multiple changes in the sense and magnitude of rotation during the last 13 Ma. Calculating a mean declination for an entire section averages potential clockwise and counterclockwise declination deviations and thereby masks the details of the evolution of the basin. Thus, an important outcome of this contribution is that taking spatial declination patterns and averages should be replaced by analysing temporal rotation patterns and averages, when well-dated and continuous sections are available for analysis.

The work was supported by the (973) National Basic Research Program of China (2011CB403000), the Chinese Academy of Science 'Hundred Talent Project' (to M.Y.), and the National Science Foundation of China (NSFC no. 41021001, 40920114001), the Chinese Academy of Science 'Hundred Talent Project' (to M.Y.), the Knowledge Innovation Project (KZCX-YW-Q09-04) and the (973) National Basic Research Program of China (2011CB403000). The reviewers, A. Weil and S. Titus are thanked for making many valuable suggestions for improvement.

References

AVOUAC, J. P. & TAPPONNIER, P. 1993. Kinematic model of active deformation in central Asia. *Geophysical Research Letters*, **20**, 895–898.

BESSE, J. & COURTILLOT, V. 2002. Apparent and true polar wander and the geometry of the geomagnetic field over the last 200 Myr. *Journal of Geophysical Research*, **107**, 2300, http://dx.doi.org/10.1029/2000JB000050

BURCHFIEL, B. C., ZHANG, P. Z. *ET AL.* 1991. Geology of the Haiyuan fault zone, Ningxia-hui autonomous region, China, and its relation to the evolution of the northeastern margin of the Tibetan Plateau. *Tectonics*, **10**, 1091–1110.

CHEN, Z., BURCHFIEL, B. C. *ET AL.* 2000. Global Positioning System measurements from eastern Tibet and their implications for India/Eurasia intercontinental deformation. *Journal of Geophysical Research*, **105**, 16 215–16 227.

CHEN, Y., GILDER, S., HALIM, N., COGNÉ, J. P. & COURTILLOT, V. 2002*a*. New paleomagnetic constraints on central Asian kinematics: displacement along the Altyn Tagh fault and rotation of the Qaidam Basin. *Tectonics*, **21**, 1042, http://dx.doi.org/10.1029/2001 TC901030

CHEN, Y., WU, H., COURTILLOT, V. & GILDER, S. 2002*b*. Large N–S convergence at the northern edge of the Tibetan Plateau? New Early Cretaceous paleomagnetic data from the Hexi Corridor, NW China *Earth and Planetary Science Letters*, **201**, 293–307.

CLARK, M. K., FARLEY, K. A., ZHENG, D. W., WANG, Z. C. & DUVALL, A. R. 2010. Early Cenozoic faulting of the northern Tibetan Plateau margin from apatite (U–Th)/He ages. *Earth and Planetary Science Letters*, **29**, 78–88.

COGNÉ, J. P., HALIM, N., CHEN, Y. & COURTILLOT, V. 1999. Resolving the problem of shallow magnetizations of Tertiary age in Asia: insights from paleomagnetic data from the Qiangtang, Kunlun, and Qaidam blocks (Tibet, China), and a new hypothesis. *Journal of Geophysical Research*, **104**, 17 715–17 734.

DAI, S., FANG, X. M. *ET AL.* 2006. Magnetostratigraphy of Cenozoic sediments from the Xining Basin: tectonic implications for the northeastern Tibetan Plateau. *Journal of Geophysical Research*, **111**, http://dx.doi.org/10.129/2005JB004187

DEMAREST, H. H. 1983. Error analysis for the determination of tectonic rotation from paleomagnetic data. *Journal of Geophysical Research*, **88**, 4321–4328.

DEWEY, J. F., SHACKLETON, R. M., CHANG, C. & SUN, Y. 1988. The tectonic evolution of the Tibetan Plateau. *Philosophical Transactions of the Royal Society of London*, **327**, 379–413.

DUPONT-NIVET, G., BUTLER, R. F., YIN, A. & CHEN, X. H. 2002. Paleomagnetism indicates no Neogene rotation of the Qaidam basin in north Tibet during Indo-Asian collision. *Geology*, **30**, 263–266.

DUPONT-NIVET, G., BUTLER, R. F., YIN, A. & CHEN, X. H. 2003. Paleomagnetism indicates no Neogene rotation of the Northeastern Tibetan Plateau. *Journal of Geophysical Research*, **108**, 2386, http://dx.doi.org/10.1029/2003JB002399

DUPONT-NIVET, G., HORTON, B. K., BUTLER, R. F., WANG, J., ZHOU, J. & WAANDERS, G. L. 2004. Paleogene clockwise tectonic rotation of the Xining-Lanzhou region, northeastern Tibetan Plateau. *Journal of Geophysical Research*, **109**, B04401, http://dx.doi.org/10.1029/2003JB002620

ENGLAND, P. C. & HOUSEMAN, G. A. 1989. Extension during continental convergence, with application to the Tibetan plateau. *Journal of Geophysical Research*, **B94**, 17 561–17 579.

ENGLAND, P. & MOLNAR, P. 1997. The field of crustal velocity in Asia calculated from Quaternary rates of slip on faults. *Geophysical Journal International*, **130**, 551–582.

FANG, X. M., GARZIONE, C., VAN DER VOO, R., LI, J. J. & FAN, M. 2003. Initial Flexural Subsidence by 29 Ma on the NE Edge of Tibet from the Magnetostratigraphy of Linxia Basin, China. *Earth and Planetary Science Letters*, **210**, 545–560.

FANG, X. M., YAN, M. D. *ET AL.* 2005*a*. Late Cenozoic deformation and uplift of the NE Tibetan plateau: evidence from high resolution magnetostratigraphy of the Guide Basin, Qinghai Province, China. *Geological Society of America Bulletin*, **107**, 1208–1225.

FANG, X. M., ZHAO, Z. J. *ET AL.* 2005*b*. Magnetostratigraphy of the late Cenozoic Laojunmiao anticline in the northern Qilian Mountains and its implications for the northern Tibetan Plateau uplift. *Science China (D)*, **48**, 1040–1051.

FANG, X. M., ZHANG, W. L. *ET AL.* 2007. High-resolution magnetostratigraphy of the Neogene Huaitoutala section in the eastern Qaidam Basin on the NE Tibetan Plateau, Qinghai Province, China and its implication on tectonic uplift of the NE Tibetan Plateau. *Earth and Planetary Science Letters*, **258**, 293–306.

FANG, X. M., LIU, D. L., SONG, C. H., DAI, S. & MENG, Q. Q. 2012. Oligocene slow and Miocene–Quaternary rapid deformation and uplift of the Yumu Shan and North Qilian Shan: evidence from high-resolution magnetostratigraphy and tectonosedimentology. *In*: JOVANE, L., HERRERO-BERVERA, E., HINNOV, L. A. & HOUSEN, B. A. (eds) *Magnetic Methods and the Timing of Geological Processes*. Geological Society, London, Special Publications, **373**, http://dx.doi.org/10.1144/SP373.5.

FROST, G. M., COE, R. S. *ET AL.* 1995. Preliminary early Cretaceous paleomagnetic results from the Gansu Corridor, China. *Earth and Planetary Science Letters*, **129**, 217–232.

GILDER, S., CHEN, Y. & SEN, S. 2001. Oligo-Miocene magnetostratigraphy and environmental magnetism of the Xishuigou section, Subei (Gansu Province, western China): further implication on the shallow inclination of central Asia. *Journal of Geophysical Research*, **106**, 30 505–30 521.

GRADSTEIN, F., OGG, J. & SMITH, A. 2004. *A Geological Timescale*. Cambridge University Press, Cambridge, 344–383.

HALIM, N., COGNÉ, J. P. *ET AL.* 1998. New Cretaceous and early Tertiary paleomagnetic results from Xining-Lanzhou basin, Kunlun and Qiangtang blocks, China: implications on the geodynamic evolution of Asia. *Journal of Geophysical Research*, **103**, 21 025–21 045.

HOLT, W. E. 2000. Correlated crust and mantle strain fields in Tibet. *Geology*, **28**, 67–70.

HOLT, W. E. & HAINES, A. J. 1993. Velocity fields in deforming Asia from the inversion of earthquake-released strains. *Tectonics*, **12**, 1–20.

HOLT, W. E., LI, M. & HAINES, A. J. 1995. Earthquake strain rates and instantaneous relative motions within central and eastern Asia. *Geophysical Journal International* **122**, 569–593.

HORTON, B. K., DUPONT-NIVET, G., ZHOU, J., WAANDERS, G. L., BUTLER, R. F. & WANG, J. 2004. Mesozoic–Cenozoic evolution of the Xining–Minhe and Dangchang basins, northeastern Tibetan plateau: magnetostratigraphic and biostratigraphic results. *Journal of Geophysical Research*, **109**, 1–15.

HUANG, H. F., PENG, Z. L., LU, W. & ZHENG, J. J. 1993. Paleomagnetic devision and comparison of the Tertiary system in Jiuxi and Jiudong basins. *Acta Geologica Gansu*, **2**, 6–16 (in Chinese).

KIRSCHVINK, J. L. 1980. The least-square line and plane and the analysis of paleomagnetic data. *Geophysical Journal of the Royal Astronomical Society*, **62**, 699–718.

LEASE, R. O., BURBANK, D. W., GEHRELS, G. E., WANG, Z. C. & YUAN, D. Y. 2007. Signatures of mountain building: detrital zircon U/Pb ages from northeastern Tibet. *Geology*, **35**, 239–242.

LI, J. 1995. *Uplift of Qinghai–Xizang (Tibet) Plateau and Global Change*. University Press, Lanzhou, 207.

MCFADDEN, P. L. & MCELHINNY, M. W. 1990. Classification of the reversal test in paleomagnetism. *Geophysical Journal International* **103**, 725–729.

MEYER, B., TAPPONNIER, P. *ET AL.* 1998. Crustal thickening in the Gansu- Qinghai, lithospheric mantle, oblique and strike-slip controlled growth of the Tibetan Plateau. *Geophysical Journal International* **135**, 1–47.

MOLNAR, P. & LYON-CAEN, H. 1989. Fault plane solutions of earthquakes and active tectonics of the Tibetan Plateau and its margins. *Geophysical Journal International* **99**, 123–153.

MOLNAR, P. & TAPPONNIER, P. 1975. Cenozoic tectonics of Asia: effects of a continental collision. *Science*, **189**, 419–426.

MOLNAR, P., ENGLAND, P. & MARTINOD, J. 1993. Mantle dynamics, uplift of the Tibetan Plateau, and the Indian monsoon. *Reviews in Geophysics*, **31**, 357–396.

MOLNAR, P., BOOS, W. R. & BATTISTI, D. S. 2010. Orographic controls on climate and Paleoclimate of Asia: thermal and mechanical roles for the Tibetan Plateau. *Annual Review of Earth and Planetary Sciences*, **38**, 77–102.

PARÉS, J. M., VAN DER VOO, R., DOWNS, W. R., YAN, M. D. & FANG, X. M. 2003. Northeastward growth and uplift of the Tibetan Plateau: magnetostratigraphic insights from the Guide Basin. *Journal of Geophysical Research*, **108**, 1–11.

PELTZER, G. & SAUCIER, F. 1996. Present-day kinematics of Asia derived from geologic fault rates. *Journal of Geophysical Research*, **101**, 27 943–27 956.

QINGHAI GEOLOGY BUREAU. 1989. *Regional Geology of Qinghai Province*. Geology Press, Beijing, 215–217 (in Chinese).

RUMELHART, P. E., YIN, A., COWGILL, E., BUTLER, R., ZHANG, Q. & WANG, X. 1999. Cenozoic vertical-axis rotation of the Altyn Tagh fault system. *Geology*, **27**, 819–822.

SHEN, Z. K., WANG, M. *ET AL.* 2001. Crustal deformation along the Altyn Tagh Fault system, western China, from GPS. *Journal of Geophysical Research*, **106**, 30 607–30 621.

SHI, Z., YE, Y., ZHAO, Z., FANG, X. M. & LI, J. 2001. ESR dating of the Late Cenozoic Jiuxi Basin and Tibetan uplift. *Science of China*, **31**, 163–168.

SONG, C. H., FANG, X. M., LI, J. J., GAO, J. P. & FAN, M. J. 2001. Tectonic uplift and sedimentary evolution of the Jiuxi Basin in the northern margin of the Tibetan

Plateau since 13 Ma BP. *Science China*, **44**(Suppl.), 192–202.

SUNG, T. C. 1958. Tertiary spore and pollen complexes from the red beds of Chiuchuan, Kansu and their geological and botanical significance. *Acta Palaeontologica Sinica*, **6**, 163–170.

TAN, X. D., GILDER, S. *ET AL.* 2010. New paleomagnetic results from the Lhasa block: revised estimation of latitudinal shortening across Tibet and implications for dating the India–Asia collision. *Earth and Planetary Science Letters*, **293**, 396–404.

TAPPONNIER, P., XU, Z. Q. *ET AL.* 2001. Oblique stepwise rise and growth of the Tibetan Plateau. *Science*, **294**, 1671–1677.

TAUXE, L. 1998. *Paleomagnetic Principles and Practice.* Kluwer Academic, Dordrecht.

TAUXE, L. 2005. Inclination flattening and the geocentric axial dipole hypothesis. *Earth and Planetary Science Letters*, **233**, 247–261.

TAUXE, L. & KENT, D. V. 2004. A simplified statistical model for the geomagnetic field and the detection of shallow bias in paleomagnetic inclinations: was the ancient magnetic field dipolar? *In*: CHANNELL, J. E. T., KENT, D. V., LOWRIE, W. & MEERT, J. G. (eds) *Timescales of the Paleomagnetic Field.* Geophysical Monograph **145**. American Geophysical Union, Washington, DC, 101–117.

TAUXE, L., KODAMA, K. P. & KENT, D. V. 2008. Testing corrections for paleomagnetic inclination error in sedimentary rocks: a comparative approach. *Physics of the Earth and Planetary Interiors*, **169**, 152–165.

VAN DER WOERD, J., RYERSON, F. J. *ET AL.* 1998. Holocene leftslip rate determined by cosmogenic surface dating on the Xidatan segment of the Kunlun fault (Qinghai, China). *Geology*, **26**, 695–698.

WANG, E. & BURCHFIEL, B. C. 2004. Late Cenozoic right-lateral movement along the Wenquan Fault and associated deformation: implications for the kinematic history of the Qaidam Basin, Northeastern Tibetan Plateau. *International Geology Review*, **46**, 861–879.

WANG, J. & LI, J. 2003. Geochemical features and its geological significance of the Cenozoic kamafugite in West Qinling near Xihe–Lixian. *Journal of Petrology and Mineralogy*, **22**, 12–19.

YAN, M. D., VAN DER VOO, R., TAUXE, L., FANG, X. M. & PARÉS, J. M. 2005. Shallow bias in Neogene paleomagnetic directions from the Guide Basin, NE Tibet, caused by inclination error. *Geophysical Journal International* **163**, 944–948.

YAN, M. D. 2006. *Paleomagnetic insights into the Neogene evolution of the Guide and Jiuxi Basins, NE Tibetan Plateau.* PhD thesis, University of Michigan.

YAN, M. D., VAN DER VOO, R., FANG, X. M., PARES, J. M. & REA, D. K. 2006. Paleomagnetic evidence for a mid-Miocene clockwise rotation of about 25° of the Guide Basin area in NE Tibet, Earth and Planet. *Science Letters*, **241**, 234–247.

YAN, M. D., VAN DER VOO, R., FANG, X. M. & SONG, C. H. 2012. Magnetostratigraphy, fence diagrams, and basin analysis. *In*: JOVANE, L., HERRERO-BERVERA, E., HINNOV, L. A. & HOUSEN, B. A. (eds) *Magnetic Methods and the Timing of Geological Processes.* Geological Society, London, Special Publications, **373**, http://dx.doi.org/10.1144/SP373.3.

YIN, A. & HARRISON, M. T. 2000. Geologic evolution of the Himalayan–Tibetan orogen. *Annual Reviews, Earth and Planetary Science*, **28**, 211–280.

YIN, A., RUMELHART, P. E. *ET AL.* 2002. Tectonic history of the Altyn Tagh fault system in northern Tibet inferred from Cenozoic sedimentation. *Geological Society of America Bulletin*, **114**, 1257–1295.

YU, X. H., ZHAO, Z. D., ZHOU, S., MO, X. X., ZHU, D. Q. & WANG, Y. L. 2001. 40Ar/39Ar dating for Cenozoic kamafugite from western Qinling in Gansu Province. *Chinese Science Bulletin*, **51**, 1621–1627.

ZHANG, P. Z., SHENG, K. *ET AL.* 2004. Continuous deformation of the Tibetan Plateau from global positioning system data. *Geology*, **32**, 809–812.

ZHAO, Z. J., FANG, X. M., LI, J. J., PAN, B. T., YAN, M. D. & SHI, Z. T. 2001. Paleomagnetic dating of the Jiuquan Gravel in the Hexi Corridor: implication on mid-Pleistocene uplift of the Qinghai–Tibetan Plateau. *Chinese Science Bulletin*, **46**, 2001–2005.

ZHENG, D. W., ZHANG, P. Z. *ET AL.* 2006. Rapid exhumation at *c.* 8 Ma on the Liupan Shan thrust fault from apatite fission-track thermochronology: implications for growth of the northeastern Tibetan plateau margin. *Earth and Planetary Science Letters*, **248**, 183–193.

ZHENG, D. W., CLARK, M. K., ZHANG, P. Z., ZHENG, W. J. & FARLEY, K. A. 2010. Erosion, fault initiation and topographic growth of the North Qilian Shan (northern Tibetan Plateau). *Geosphere*, **6**, 937–941.

Magnetostratigraphic results from sedimentary rocks of IODP's Nankai Trough Seismogenic Zone Experiment (NanTroSEIZE) Expedition 322

XIXI ZHAO[1,7]*, HIROKUNI ODA[2], HUAICHUN WU[3], TOMOHIRO YAMAMOTO[4], YUHJI YAMAMOTO[5], YUZURU YAMAMOTO[6], TAKESHI NAKAJIMA[2], YUJIN KITAMURA[6] & TOSHIYA KANAMATSU[6]

[1]*Earth and Planetary Sciences Department, University of California Santa Cruz, 1156 High Street, Santa Cruz, CA 95064, USA*

[2]*Geological Survey of Japan, National Institute of Advanced Industrial Science and Technology, Central 7, 1-1-1 Higashi, Tsukuba 305-8567, Japan*

[3]*School of Marine Science, China University of Geosciences, Xueyuan Road 29, Haidian District, Beijing 100083, People's Republic of China*

[4]*Department of Environmental Systems Science, Doshisha University, 1-3 Tatara Miyakodani, Kyotanabe City 610-0394, Japan*

[5]*Center for Advanced Marine Core Research, Kochi University, B200 Monobe, Nankoku, Kochi 783-8502, Japan*

[6]*Japan Agency for Marine-Earth Science and Technology, 3173-25 Showa-machi, Kanazawa-ku, Yokohama 236-0001, Japan*

[7]*Institute of Geophysics, University of Sao Paulo, Rua do Matao, 1226. CEP 05508-900, Sao Paulo, Brazil*

**Corresponding author (e-mail: xzhao@ucsc.edu)*

Abstract: We conducted a palaeomagnetic study on the Cenozoic sedimentary sequences of the Nankai Trough, recovered by the Integrated Ocean Drilling Program Expedition 322 in SE Japan. Sedimentary sections of Late Miocene age from the two subduction input sites (sites C0011 and C0012) recorded a pattern of magnetic polarity reversals that correlates well with the known magnetic polarity time scale. The polarity of characteristic remanent magnetization could be identified throughout the majority of the recovered cores of the two sites, following removal of a low-stability drilling-induced remanence. Most of the observed magnetostratigraphy from the characteristic directions is in good agreement with that to be expected from the stratigraphic position of the sequence deduced from the biostratigraphic data. Palaeomagnetic data from both shipboard and shore-based studies indicate changes in the rate of sedimentation from 9.5 to 2.7 cm/kyr at about 11 Ma, suggesting that some fundamental palaeoenvironmental change in the Shikoku Basin and/or significant tectonic event may have occurred in Late Miocene.

Most earthquakes are generated in the seismogenic zone, a relatively small portion of the subduction zone where elastic strain accumulates and is eventually released to generate damaging earthquakes and tsunamis (e.g. Lay *et al.* 2005). Despite the current quantitative knowledge of plate motions monitored by an array of geodetic measurements and the Global Positioning System, we still have little understanding about the sudden release of long-term accumulation of strain in the seismogenic zone. Understanding the processes that govern the strength, nature and distribution of slip along these plate boundary systems is a fundamental and societally relevant goal of modern earth science. To this end, a multistage, multiexpedition, complex drilling project known as the Nankai Trough Seismogenic Zone Experiment (NanTroSEIZE) of the Integrated Ocean Drilling Program (IODP) is an outstanding example of such efforts. The fundamental goal of NanTroSEIZE is to sample and instrument the

From: JOVANE, L., HERRERO-BERVERA, E., HINNOV, L. A. & HOUSEN, B. A. (eds) 2013. *Magnetic Methods and the Timing of Geological Processes*. Geological Society, London, Special Publications, **373**, 191–243.
First published online March 25, 2013, http://dx.doi.org/10.1144/SP373.14 © The Geological Society of London 2013.
Publishing disclaimer: www.geolsoc.org.uk/pub_ethics

plate boundary system at several locations offshore the Kii Peninsula of SW Japan, where violent, large-scale earthquakes have occurred repeatedly throughout history (Ando 1975; Tobin & Kinoshita 2006*a*, *b*; Fig. 1). To accomplish this ambitious goal, several key components of the plate-boundary system have been investigated starting with the pre-subduction inputs of sediment and oceanic basement, moving landward into the shallow plate interface and finally drilling to depths where earthquakes occur.

IODP Expedition 322 (Exp. 322) is one part of the NanTroSEIZE complex drilling project to investigate the characteristics of incoming sedimentary strata and igneous basement entering the subduction front of Nankai Trough offshore the Kii Peninsula of SW Japan (Fig. 1). Coring was conducted at two sites in the Shikoku Basin on the subducting

Philippine Sea plate. The resulting data, which include logging while drilling during IODP Expedition 319, provide a wealth of new information on pre-subduction equivalents of the seismogenic zone. When viewed together with subsequent drilling during IODP Expedition 333, the cored material documents the evolution of the Shikoku Basin and provides a reference to study changes in geological properties down the subduction zone to seismogenic depth (Underwood *et al.* 2009; Henry *et al.* 2012).

Magnetostratigraphy, biostratigraphy and radiometric dating are three principal techniques for chronostratigraphic analysis of the recovered sequences. During Exp. 322, palaeomagnetic measurements were taken on discrete samples from split-core archive sections. In order to isolate the characteristic remanent magnetization (ChRM), core samples were subjected to alternating field

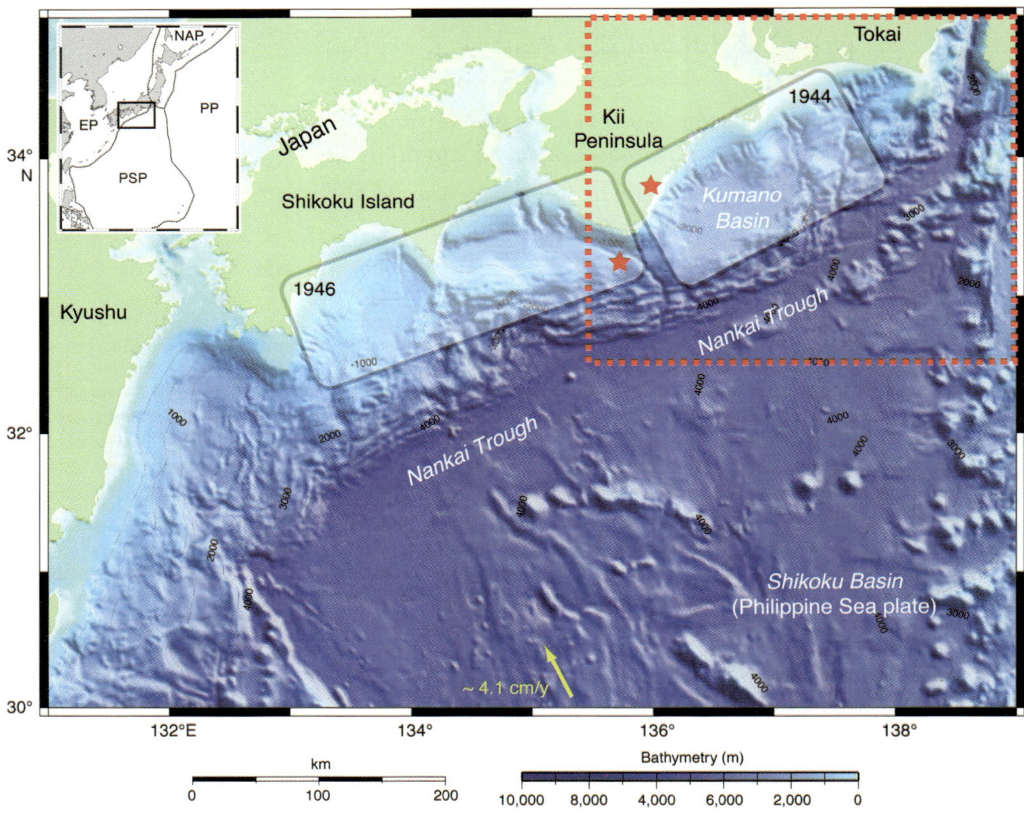

Fig. 1. Nankai Trough off SW Japan, locus of Philippine Sea plate (PSP) subduction beneath Honshu and Shikoku islands. Yellow arrow, convergence direction between PSP and Japan (Seno *et al.* 1993). Rupture zones of 1944 and 1946 earthquakes are also shown. Stars, epicentre locations for earthquake nucleation. Red dashed line, Kumano Basin drilling area including Exp. 322 sites in Figure 2. Inset shows location of Nankai Trough. Previous transects of Nankai Trough were positioned off Ashizuri and Muroto peninsulas of Shikoku. EP, Eurasian plate; PP, Pacific plate; NAP, North American plate (after Underwood *et al.* 2009).

(AF) and thermal demagnetization. Preliminary shipboard palaeomagnetic results have demonstrated that sediments and basement rocks from the drill sites contain both reverse and normally magnetized intervals, allowing temporal controls on site geological history to be determined (Underwood *et al.* 2009). However, the ChRM was not resolved onboard in some cases owing to the failure of the ship's cryogenic magnetometer from the beginning of the expedition and the limited sensitivity of the available spinner magnetometer.

Here we present new results of a study based on the palaeo- and rock-magnetism, magnetostratigraphy, fission track dating and biostratigraphy of samples from Exp. 322 drill sites. We first summarize and discuss the palaeomagnetic data and polarity patterns of the sediments recovered from Sites C0011 and C0012. We subsequently consolidate and expand the preliminary interpretations of the polarity sequence in conjunction with the Exp. 333 results (Henry *et al.* 2012), and present an updated magnetostratigraphy for these two sites. We then use the magnetostratigraphy to estimate the sedimentation rates and the times at which significant changes in these rates occurred.

Site setting and background information

The Nankai Trough is located in the Pacific Ocean, *c.* 100 km SE off the coast of Japan (Fig. 1) and is one of the world's best studied seismogenic subduction zones, where large megathrust earthquakes have repeatedly occurred every 90–150 years.

Tectonically, the Nankai Trough is a sediment-dominated subduction zone formed by subduction of the Philippine Sea plate beneath the Eurasian plate at rates of 4.0–6.0 cm a^{-1} (Seno *et al.* 1993; Miyazaki & Heki 2001). The subducting lithosphere of the Shikoku Basin was formed by backarc spreading during a time period of approximately 15–25 Ma (Okino *et al.* 1994). The current convergence direction is approximately normal to the trench with an azimuth of 300–315°N down an interface dipping 3–7° (Kodaira *et al.* 2000), and sediments of the Shikoku Basin are actively accreting at the deformation front (Kimura *et al.* 2007; Tobin *et al.* 2009). Land-based geodetic studies suggest that the plate boundary thrust here is strongly locked (Miyazaki & Heki 2001), implying significant interseismic strain accumulation on the megathrust (Miyazaki & Heki 2001).

Drilling by Exp. 322 was conducted at two input sites (before the involvement of subduction processes) in the Shikoku Basin (Fig. 2). Site C0011 is located on the NW flank of a prominent bathymetric high (Kashinosaki Knoll) on the subducting Philippine Sea plate, whereas Site C0012 is located near the crest of the knoll (Fig. 2). At Site C0011 (Hole C0011B), coring started at 340 m core depth below seafloor (CSF) and continued until the drill bit failed at 881 m CSF. Core recovery was 68.1%. Five lithologic units of Middle to Late Miocene sediments were identified in Hole C0011B (Fig. 3). Fortunately, Exp. 333 reoccupied Site C0011 and drilled the top part of 380 m, expanding the recovered sediments from Miocene into the Pliocene and Quaternary (Expedition 333 Scientists 2011). At Site C0012, six sedimentary lithologic units were recognized on the basis of sediment composition, sediment texture and sedimentary structures (Fig. 4). The sediment–basalt interface was penetrated at 537.81 m CSF and 38.2 m of basement was cored with 18% recovery. The recovered sediments range in age from Early to Late Miocene, and the age of basal sediment (reddish-brown pelagic claystone) is >18.9 Ma (Underwood *et al.* 2009; Saito *et al.* 2010).

At Sites C0011 and C0012, the deposition of both volcanic ash (by air fall) and sand/silt sediments (by turbidity currents) ceased during the Middle–Late Miocene (*c.* 9.1 to *c.* 12.8 Ma). The composite stratigraphic succession at Sites C0011 and C0012 captures all-important lithologies of the Shikoku Basin and the underlying basement with a good correlation of unit boundary ages between the two sites (Saito *et al.* 2010).

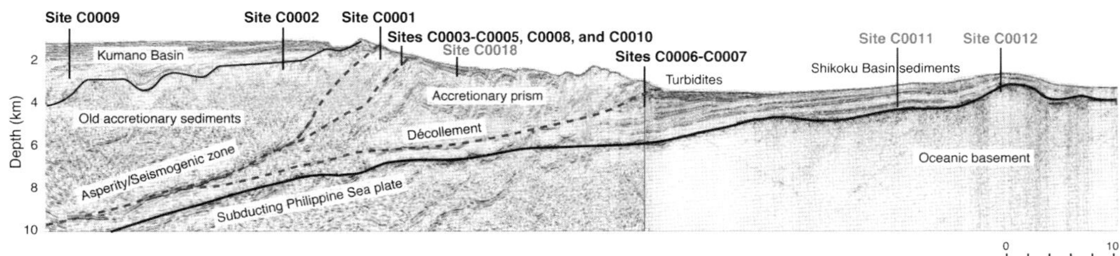

Fig. 2. Locations of the two drilling sites occupied during IODP Expedition 322 (after Expedition 333 Scientists 2011). Projected positions of Stages 1 and 2 drilling sites of the NanTroSEIZE programme are also shown.

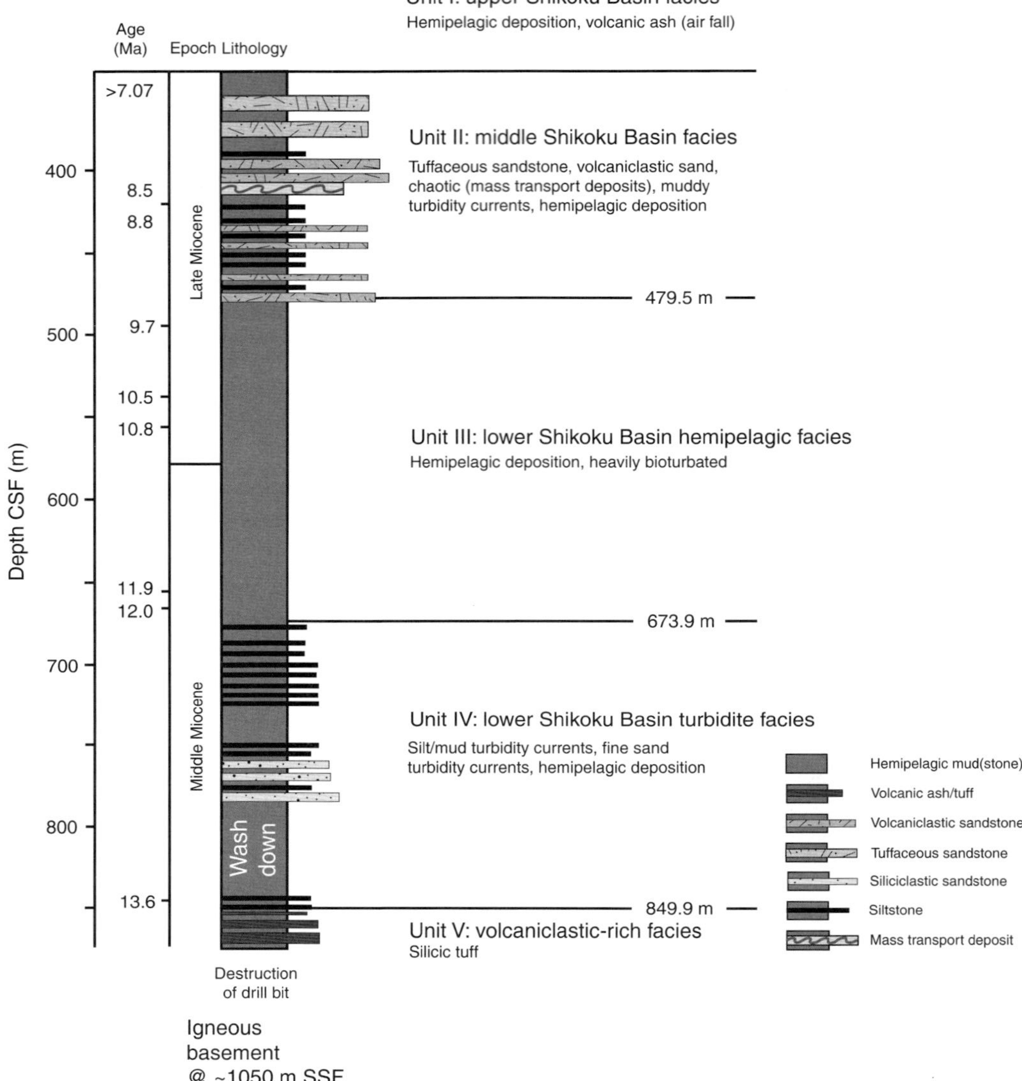

Fig. 3. Lithologic units for Hole C0011B. Depositional ages are from nannofossil datums (after Underwood *et al.* 2009).

Laboratory and analytical methods

Palaeomagnetic sampling

During Exp. 322, a total of 1459 discrete palaeomagnetic samples were taken for shipboard and shore-based magnetostratigraphic and rock magnetic studies. In the case of unconsolidated sediments, palaeomagnetic samples were taken by pushing non-magnetic cubes into the split working halves. In order to reduce the deformation/disturbance of the sediment, the core was carefully cut using a thin stainless-steel spatula before pressing the plastic sampling boxes into the sediment. In lithified sediments, 2.2 cm-long cylindrical samples of 1 inch diameter were drilled from the core sections using a water-cooled nonmagnetic drill bit attached to the standard drill press, or 2.2 × 2.2 × 2.2 cm cube-shaped samples were cut by diamond saw. These cylindrical samples were drilled

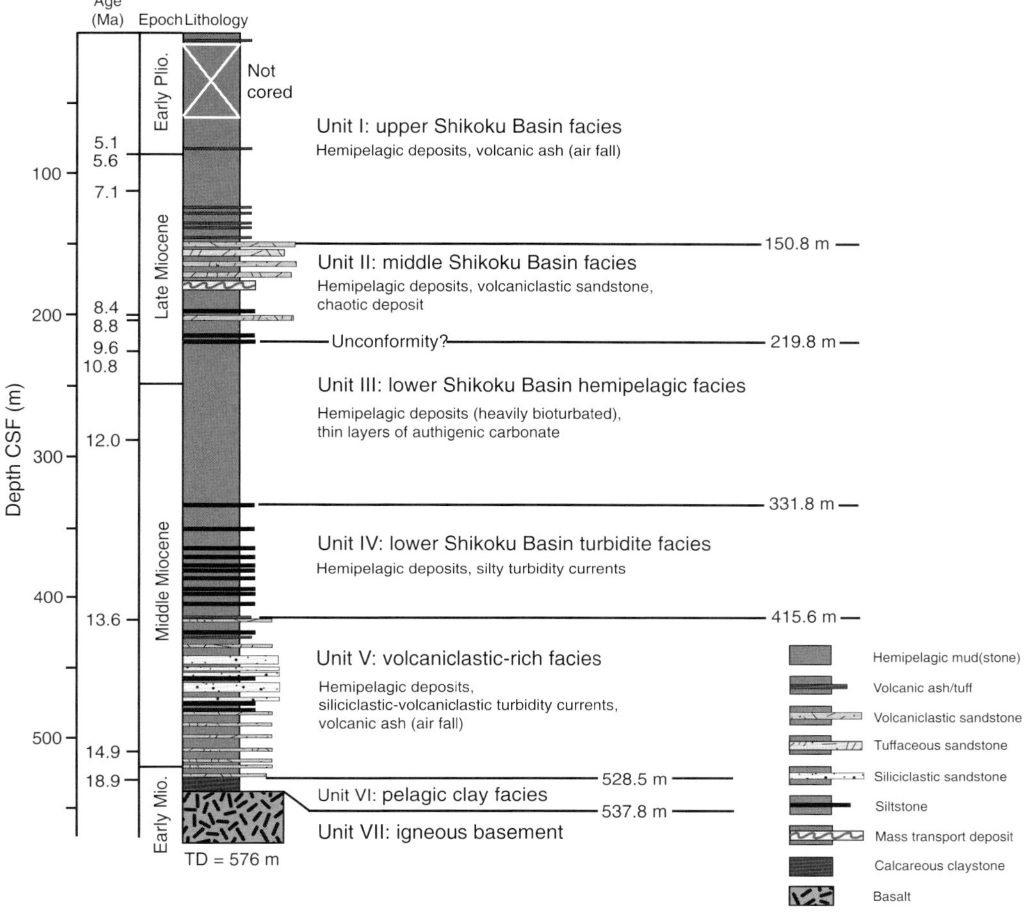

Fig. 4. Lithologic units for Hole C0012A. Depositional ages are from nannofossil datums (after Underwood *et al.* 2009).

from the core sections that contained long pieces securing the up–down directions, generally taken from the least deformed parts inspected by whole round X-ray CT images taken prior to the splitting of the cores. In all cases the uphole direction was carefully recorded on the sample by means of an orientation arrow before removal from the core section, and only sediments showing no visible signs of deformation were sampled. All samples were kept in a relatively cold temperature and low magnetic field environment (in magnetically shielded spaces) to inhibit water loss and prevent viscous remanence acquisition.

Magnetic measurement procedure

The palaeomagnetic and rock magnetic data presented in this paper are from measurements performed both on *D/V Chikyu* and in the palaeomagnetism laboratories at University of California at Santa Cruz (UCSC), Geological Survey of Japan and at the China University of Geosciences (Beijing) (CUGB). One of the major experimental requirements in palaeomagnetic research is to isolate the characteristic remanent magnetization by selective removal of secondary magnetization. To investigate the nature of the remanent magnetization of the rocks from Sites C0011 and C0012, discrete samples were AF or thermally demagnetized during shipboard and shore-based studies to identify the magnetic components and to evaluate the directional stability and coercivity/unblocking temperature spectra of each sample.

For rock magnetic characterization, samples were subjected to several magnetic measurements. Curie temperatures were determined by measurement of low-field magnetic susceptibility v. temperature (using the Kappabridge susceptometer at UCSC and CUGB). To avoid oxidation that could lead to chemical alteration, we conducted thermomagnetic analyses in an inert argon atmosphere. We used a graphic method (Grommé *et al.* 1969) to determine the Curie temperature that uses the intersection of two tangents to the thermomagnetic curve that bounds the Curie temperature. This method is most straightforward to do by hand even though it tends to underestimate Curie temperatures compared with the two other methods presented by Moskowitz (1981) and Tauxe (1998).

Isothermal remanent magnetization acquisition and hysteresis loop parameter measurements were performed on a Princeton Measurements Corporation Alternating Gradient Field Magnetometer (Princeton MicroMag 2900) at UCSC and CUGB using 50–100 mg rock chips. Hysteresis loop parameters are useful in characterizing the intrinsic magnetic behaviour of rocks. Thus, they are helpful in studying the origin of remanence. Saturation magnetization (J_s) is the largest magnetization a sample can have and is thus a measure of the total amount of magnetic mineral in the sample. The saturation remanence is the remanent magnetization when the applied field is removed at zero. The coercivity (H_c) and remanent coercivity (H_{cr}) are measures of magnetic stability. The two ratios, J_r/J_s and H_{cr}/H_c, are commonly used as indicators of domain states and, indirectly, grain size (e.g. Dunlop & Özdemir 1997). For magnetite, high values of J_r/J_s (>0.5) indicate small (c. 0.05–0.08 μm) single-domain grains, and low values (<0.1) are characteristic of large (>10–20 μm) multidomain grains. The intermediate regions are usually referred to as pseudosingle domain. Magnetic properties in this pseudosingle domain range have been shown to change gradually from single-domain-like at the smaller sizes to multidomain-like at the larger sizes. H_{cr}/H_c is a much less reliable parameter, but conventionally single-domain grains have a value close to 1.1, and multidomain grains should have values >3–4 (Day *et al.* 1977; Dunlop 2002).

Low-temperature measurements were made on eight representative samples to help characterize the magnetic minerals. These measurements were designed to determine critical temperatures of a magnetic substance (such as Verwey transition for magnetite/titanomagnetite) and were made from 6 K to room temperature on 100–300 mg subsamples in a Quantum Design magnetic property measurement system at the Kochi Core Center and Geological Survey of Japan. Samples were given a saturation isothermal remanent magnetization in a steady magnetic field of 2.5 T at room temperature (300 K), cooled down to 6 K and then heated to 300 K in a zero field (all samples); the remanence

was measured at 2 K intervals (five samples). Subsequently, the sample was cooled down to 6 K in a zero field, given a saturation isothermal remanent magnetization in a field of 2.5 T, then the remanence was measured while warming it to 300 K in a zero field at 2 K intervals (all samples). Secondly, the sample was cooled down to 6 K in a field of 2.5 T, and then the remanence was measured in a zero field at 2 K steps (five samples). Finally, the sample was cooled down to 6 K in a zero field, and then the remanence was measured in a field of 500 mT at 2 K steps (four samples).

Data analysis

Changes in the intensity and direction of remanent magnetization vectors during demagnetization experiments were analysed using orthogonal vector end-point projections (Zijderveld 1967). For samples that gave reliable demagnetization results, magnetic components were determined by fitting least-squares lines to segments of the vector demagnetization plots that were linear in three-dimensional space (Principal Component Analysis - PCA; Kirschvink 1980). For some samples with overlapping magnetic components, the methods of combined remagnetization planes and least-square-best-fit-line data of McFadden & McElhinny (1988) were required. Data were processed using software PMGSC developed by Enkin (1994), PalaeoMag by Craig Jones (http://cires.colorado.edu/people/jones.craig/CHJ_PMag_overview.html) and maximum likelihood solution for inclination-only data by Arason & Levi (2010).

Because of the rotary technique used for drilling Exp. 322 cores, relative rotation frequently occurs between different segments of sediment within the core. This may cause apparent changes in the declination of stable remanent magnetization. Consequently, the magnetic polarity has been assigned on the basis of the inclination of the stable remanent magnetization alone. Since all sites are situated at moderate latitudes in the Northern Hemisphere, positive (downward-directed) inclinations are taken to signify a normal polarity, and negative (upward-directed) inclination signifies reversed polarity. In this study, we adhere to the chronostratigraphic nomenclature and geochronology of Cande & Kent (1995) as the standard Geomagnetic Polarity Time Scale (GPTS). We correlate polarity zones with the GPTS in the manner that appears most consistent with both magnetic and biostratigraphic data. In naming the various polarity intervals, we use the familiar proper names for the Pliocene Pleistocene magnetic chrons (Brunhes, Matuyama, Gauss and Gilbert) and subchrons (e.g. Jaramillo and Olduvai).

In order to obtain reliable estimates of the characteristic remanent magnetization from the demagnetization data, we devised several criteria to avoid overprinted and magnetically unstable samples. Our criteria included (1) rejecting any measurements with inclination steeper than $\pm 70°$, a very likely sign that the drilling-induced remagnetization is still present; (2) accepting samples with principal component maximum angular deviation angles $<20°$; and (3) rejecting samples with inclinations less than $\pm 10°$, whose magnetization was probably affected by coring disturbance, inclination shallowing, palaeosecular variation, tilt of drill holes and/or tectonic rotations; and (4) using the highest coercivity component for polarity determination. We generally accept data if the higher coercivity component of a sample shows linear trend to the origin with at least three data points. This sample then is interpreted to have primary ChRM of positive or negative polarity according to the inclination. If the higher coercivity data points move steadily towards the negative inclination regions to form a great circle, we assign the sample to have primary ChRM of negative polarity. Finally, results from samples exhibiting no stable directional endpoint were discarded on the grounds that they were either unstable or heavily contaminated by drilling induced remanence that could not be satisfactorily resolved by available laboratory techniques.

Palaeomagnetic results

Palaeomagnetic results from shipboard measurements

During our shipboard palaeomagnetic studies for Exp. 322, we characterized the stable components of natural remanent magnetization (NRM) through thermal and AF demagnetization. Details are illustrated and described in Underwood *et al.* (2009) and Expedition 322 Scientists (2010*a*, *b*). Here we only highlight the most relevant features in terms of demagnetization behaviour and magnetic polarity.

The most common property during demagnetization experiments on Sites C0011 and C0012 samples is a pervasive remagnetization (overprint) imparted by the coring process, which is characterized by NRM inclinations that are strongly biased towards vertical (towards $+90°$) in all cores. In most cases, this steep downward component of magnetization can be removed after $10-40$ mT or $200 °C$ demagnetization. Figure 5 shows demagnetization behaviour for several representative samples. The behaviour demonstrates the removal of the nearly vertical downward overprint component after AF or thermal demagnetizations and the isolation of a stable ChRM that univectorially decays towards the origin of the vector plots (Zijderveld 1967). Although the amount and coercivity of

overprinting varied, most of it seems to be removed with $20-75$ mT AF demagnetization for the majority of samples, allowing us to isolate the ChRM direction using the principal component analysis method on higher field demagnetization steps (Tables 1 & 2).

The inclination values of ChRM are moderate downward or upward, roughly consistent with the expected inclination for the drill sites (expected inclination at drill sites $+52.2°$ for normal polarity or $-52.2°$ for reversed polarity). A number of samples, however, have steeper inclinations that do not resemble the time-averaged geomagnetic field. The steeper inclinations may suggest that the drilling-induced remagnetization was not completely removed from the recovered sediments and therefore that these sediments do not necessarily have an inclination corresponding to that expected from a geocentric axial dipole.

The response of ChRM to AF and thermal demagnetization suggests that ChRM in most mudstone samples might be carried by fine (i.e. single-domain to pseudosingle-domain), low-Ti titanomagnetite grains as revealed by moderate unblocking coercivity (i.e. resistance to AF demagnetization) and near $580 °C$ unblocking temperatures. Some samples from recovered lithologic units, however, have very high coercivity. For example, several muddy turbidite samples from Lithostratigraphic Unit IV exhibit very high coercivity (>180 mT) during AF demagnetization, suggesting that hematite might exist in these samples as well. There are also other lithologies that are dominated by softer magnetization, with unblocking coercivity about 20 mT and carried by coarser multidomain grains. Nevertheless, the ChRM components extracted from the sedimentary cores in Nankai Trough have recorded a stable component of magnetization with both normal and reversed inclinations and might be the primary magnetization acquired when sediments were deposited (Underwood *et al.* 2009; Expedition 322 Scientists 2010*a*, *b*).

Demagnetization results of discrete samples from shore-based studies

Discrete palaeomagnetic samples were kept in a low-field environment (field-free room) to prevent viscous remanence acquisition; the NRM intensity or direction of the minicore samples did not change significantly ($<5\%$) after zero-field storage for 2 weeks. Samples were then stepwise AF or thermally demagnetized to extract the primary component of magnetizations acquired at the time of deposition. Whenever possible, demagnetization was continued until the stable component of magnetization had been achieved.

X. ZHAO *ET AL.*

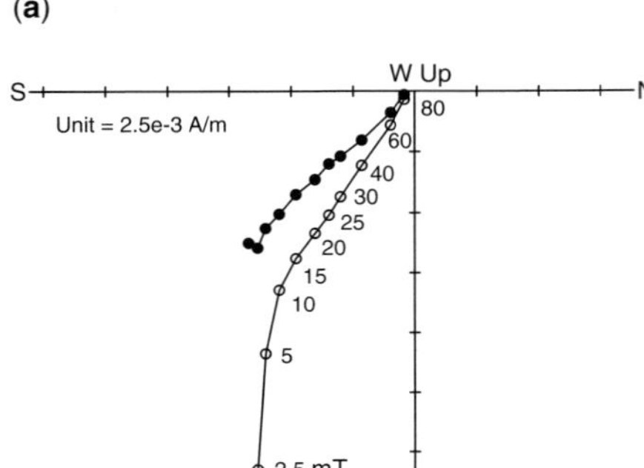

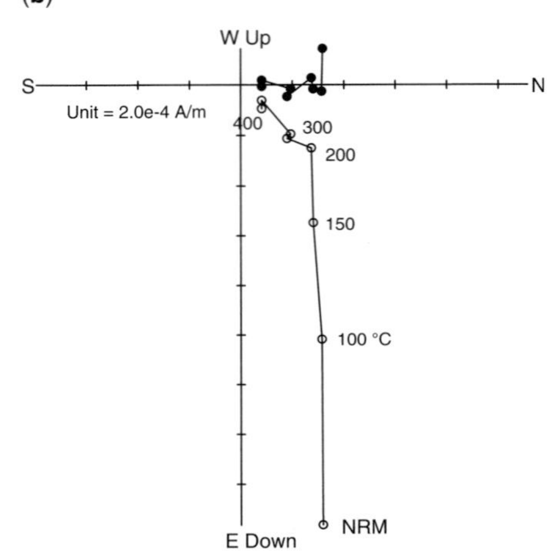

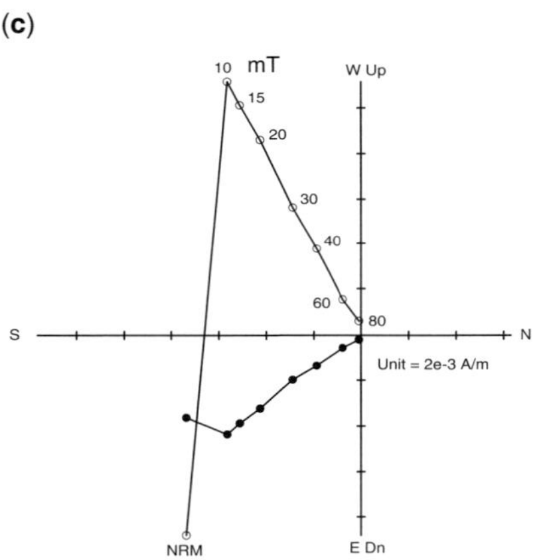

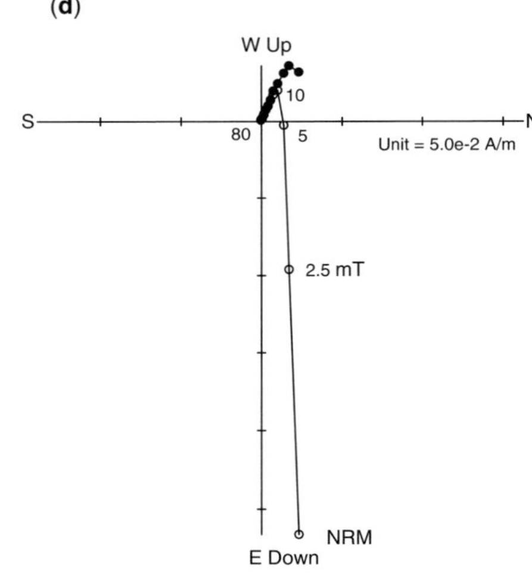

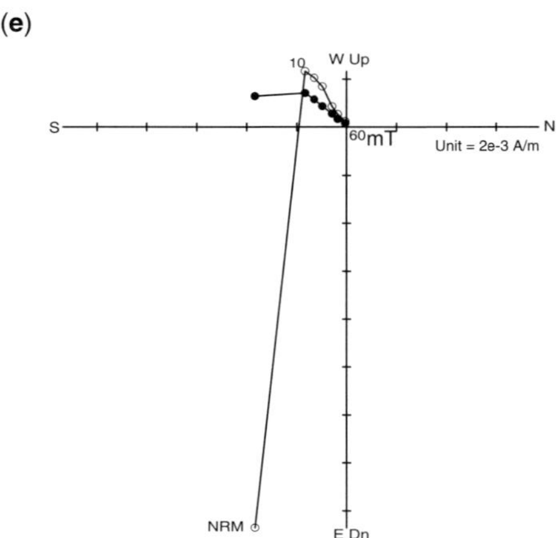

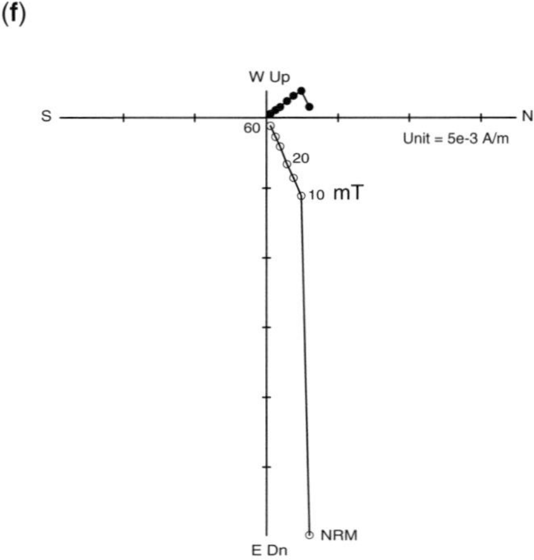

Figure 6 illustrates the demagnetization behaviour of several samples from different lithologic units. These samples demonstrate the removal of the vertically directed drilling-induced magnetization component and the isolation of the ChRM. The high stability shown by the normally magnetized Sample 322-C0011B-3R-6W, 71–73 cm (Fig. 6a) and reverse-magnetized Sample 322-C0012A-17R-2W, 6–8 cm (Fig. 6e) is common in a high proportion of the samples from Sites C0011 and C0012. These samples maintain an inclination close to the theoretically predicted value for the latitude of these sites, indicating that they may represent the primary ChRM. Moreover, the mean normal and reversed directions for Units I–III of Site C0012 are not significantly different from antipodal. Application of the reversal test (Arason & Levi 2010) on the mean normal and reversed inclinations of Units I–III of Site C0012 yields positive results (Table 3): the two polarity inclinations are not statistically different from each other at the 95% confidence level (McFadden & McElhinny 1990).

Certain lithology is more susceptible to the vertically directed drilling-induced magnetization overprint. As shown in Figure 7, the drilling-induced component persisted at least to 40 mT AF demagnetization on the silty sandstone Samples 322-C0011B-43R-1W, 15–17 cm and 322-C0011B-44R-1W, 65–67 cm. For these samples, we used the method of combined remagnetization planes and least-square-best-fit-line data of McFadden & McElhinny (1988) to reveal the ChRM. Figure 7 demonstrates that ChRM components of the two samples are evidenced in their demagnetization planes. For both samples, the characteristic component with upward inclinations is revealed after the removal of a normal field component. We used the same criteria by which we accepted the shipboard data. When compared with shipboard data, our results show that there is a discrepancy in the mean inclinations. Steeper inclination values observed in the shipboard measurements may be due to an overprint that has not been completely removed. It is worth pointing out that, although the shipboard demagnetization of sediments revealed initial polarity patterns, it was only the subsequent shore-based detailed AF demagnetization that were able to resolve a higher-coercivity component, which improved the quality of the magnetostratigraphy.

Rock magnetic characterization

Curie temperature determination and isothermal remanent magnetization acquisition

The Curie temperature is a sensitive indicator of composition of the magnetic mineralogy in a given rock sample. Representative Curie temperature determinations of samples from Exp. 322 sites are shown in Fig. 8. According to Curie temperatures, three different groups can be recognized. Group 1 is characterized by a single ferromagnetic phase with Curie temperatures between 450° and 500 °C, compatible with that of typical Ti-rich titanomagnetite or low-temperature oxidized titanomaghemites. The cooling and heating curves are reasonably reversible (e.g. Sample 322-C0011B-5R-1W, 43–45 cm; Sample 322-C0012A-11R-6W, 42–44 cm; Fig. 8a, f, respectively). Group 2 also has one magnetic phase. The irreversible thermomagnetic curve of this type displays the magnetic phase with Curie temperature around 570 °C on heating and with much higher intensity on cooling (e.g. Sample 322-C0012A-13R-4W, 40–42 cm; Fig. 8c, g–i). Group 3 has multiple magnetic phases and also irreversible thermomagnetic curves. The thermomagnetic curves display one magnetic phase with Curie temperature around 300–390 °C on heating, most likely titanomaghemite. The second high Curie temperature phase is observed around 580–600 °C (e.g. Sample 322-C0011B-13R-1W, 34–36 cm; Fig. 8d, e, j). The large difference between heating and cooling of the sample suggests that the magnetic mineral is oxidized or more generally altered.

Isothermal remanent magnetization (IRM) acquisition results incorporate the Curie temperature measurements mentioned above and reflect the difference in the Expedition 322 lithologies we considered in this study. The IRM acquisition curves of a claystone from Lithostratigraphic Unit IIb (Sample 322-C0011B-5R-4W, 81–83 cm; Fig. 9a), a silt sand from Lithostratigraphic Unit IV (Sample 332-C0011B-55R-3W, 62–64 cm;

Fig. 5. Representative vector end point diagrams (Zijderveld 1967) showing the shipboard results of AF and thermal demagnetization for sedimentary rock samples from Exp. 322 drill sites with well defined reversed and normal polarity ChRM magnetization. (**a**) AF demagnetization on Sample 322-C0011B-4R-1, 72–74 cm; (**b**) thermal demagnetization on Sample 322-C0011B-15R-4, 96–98 cm; (**c**) Sample 322-C0012A-11R-1, 82–84 cm; (**d**) Sample 322-C0011B-7R-1, 118–120 cm; (**e**) Sample 322-C0012A -24R-4, 4–6 cm; and (**f**) Sample 322-C0012A-23R-4, 24–26 cm. Samples in (a), (b) and (f) are normally magnetized; and samples in (c), (d) and (e) are reverse magnetized. Black and open circles, projection of the magnetization vector end point on the horizontal and vertical planes, respectively. NRM, natural remanent magnetization.

Table 1. *Summary of palaeomagnetic data of sediments for Hole C0011B*

Hole, core, section	Top (cm)	Bottom (cm)	Sample centre depth (cm)	Lithological unit	Lithology	AFD/ThD	Linear fitting (preliminary)					Polarity
							Min	Max	Dec	Inc	MAD	
C0011B-1R-1	0.34	0.36	340.35	Unit IIa	Mudstone	AFD	30	70	195.6	69.3	0.9	N
C0011B-1R-1	0.36	0.38	340.37	Unit IIa	Mudstone	AFD	5	100	19.8	58.8	1.5	N
C0011B-1R-1	0.81	0.83	340.82	Unit IIa	Mudstone	AFD	15	60	−123.6	39.7	5.5	N
C0011B-1R-2	1.06	1.09	341.95	Unit IIa	Mudstone	AFD	20	80	−153.5	68	3.3	N
C0011B-1R-CC	0.3	0.32	342.32	Unit IIa	Mudstone	AFD	15	40	144	−48.2	1.4	R
C0011B-3R-1	0.27	0.3	359.27	Unit IIa	Mudstone	AFD	10	80	139.5	31.1	1.7	N
C0011B-3R-3	0.15	0.17	360.28	Unit IIa	Mudstone	AFD	10	80	46	29.3	3.2	N
C0011B-3R-3	0.5	0.52	360.61	Unit IIa	Volcaniclastic sand	AFD	10	30	−46.9	62.1	8.8	N
C0011B-3R-3	1.18	1.2	361.26	Unit IIa	Mudstone	AFD	20	80	339.5	36.5	0.9	N
C0011B-3R-4	0.51	0.53	361.96	Unit IIa	Mudstone	AFD	20	70	−51.6	50.1	2.2	N
C0011B-3R-4	0.66	0.68	362.09	Unit IIa	Mudstone	AFD	20	80	286	61	1	N
C0011B-3R-4	0.75	0.77	362.17	Unit IIa	Mudstone	AFD	15	100	94.8	56.1	1.7	N
C0011B-3R-4	1.05	1.07	362.46	Unit IIa	Mudstone	AFD	20	60	297.4	51.4	1.3	N
C0011B-3R-5	0.41	0.43	363.22	Unit IIa	Mudstone	AFD	30	80	−149.4	43.6	2.9	N
C0011B-3R-5	0.48	0.5	363.27	Unit IIa	Mudstone	AFD	15	80	21.2	54.9	3	N
C0011B-3R-5	0.88	0.9	363.65	Unit IIa	Mudstone	AFD	20	80	13.9	54.6	2.7	N
C0011B-3R-5	0.98	1	363.75	Unit IIa	Mudstone	AFD	15	80	−174.9	45.7	2.3	N
C0011B-3R-5	1.24	1.26	364.05	Unit IIa	Mudstone	AFD	30	70	−1	41.2	2.9	N
C0011B-3R-5	1.34	1.36	364.09	Unit IIa	Mudstone	ThD	25	580	96.8	52.7	2.9	N
C0011B-3R-6	0.14	0.16	364.29	Unit IIa	Mudstone	AFD	40	80	296.7	60.7	4.5	N
C0011B-3R-6	0.34	0.36	364.48	Unit IIa	Mudstone	AFD	10	80	99.2	50.6	1.8	N
C0011B-3R-6	0.47	0.49	364.62	Unit IIa	Mudstone	AFD	20	80	−56.1	48.9	1.7	N
C0011B-3R-6	0.69	0.71	364.81	Unit IIa	Mudstone	AFD	20	80	286.8	53.6	1.6	N
C0011B-4R-1	0.15	0.17	365.16	Unit IIa	Mudstone	AFD	15	80	23.1	38.8	3.5	N
C0011B-4R-1	0.45	0.47	365.46	Unit IIa	Mudstone	AFD	20	80	−171.3	52.2	1.9	N
C0011B-4R-1	0.72	0.74	365.73	Unit IIa	Mudstone	AFD	10	80	137.1	47.9	1.5	N
C0011B-4R-1	0.91	0.93	365.92	Unit IIa	Mudstone	AFD	20	80	308.1	48.2	2.1	N
C0011B-4R-1	1.25	1.27	366.26	Unit IIa	Mudstone	AFD	20	60	80.8	35.6	4.7	N
C0011B-4R-2	0.04	0.06	366.46	Unit IIa	Mudstone	AFD	20	80	229.7	38	3.2	N
C0011B-4R-2	0.2	0.22	366.62	Unit IIa	Mudstone	AFD	15	80	45.8	37.8	6	N
C0011B-4R-4	0.03	0.05	367.28	Unit IIa	Mudstone	AFD	30	70	95.5	42.7	11.4	N
C0011B-4R-4	0.19	0.21	367.44	Unit IIa	Tuffaceous sand	AFD	20	60	−70.4	43.4	9.7	N
C0011B-5R-1	0.43	0.45	374.94	Unit IIa	Tuffaceous sand	AFD	10	60	−79	64.8	11.7	N
C0011B-5R-3	1.17	1.19	376.82	Unit IIa	Tuffaceous sand	AFD	10	60	9.5	52.1	15.8	N
C0011B-5R-4	0.72	0.74	377.69	Unit IIa	Mudstone	AFD	20	80	−13.2	59.6	6.6	N

C0011B-5R-4	0.74	0.76	377.71	Unit IIa	Mudstone	AFD	15	60	158.2	54.7	3.5	N
C0011B-5R-4	0.81	0.83	377.78	Unit IIb	Mudstone	AFD	40	70	335.7	52.4	4.4	N
C0011B-6R-1	0.07	0.09	384.08	Unit IIb	Mudstone	AFD	20	80	−44.5	64.6	4.7	N
C0011B-6R-1	0.34	0.37	384.34	Unit IIb	Mudstone	AFD	15	30	122.8	61.6	2.3	N
C0011B-6R-1	0.39	0.41	384.39	Unit IIb	Mudstone	AFD	–	–	334.2	25.1		N
C0011B-6R-1	0.78	0.8	384.77	Unit IIb	Mudstone	AFD	20	60	252.5	62.8	6.4	N
C0011B-6R-1	1.2	1.22	385.18	Unit IIb	Mudstone	AFD	10	80	116.5	48.8	3.1	N
C0011B-6R-2	0.2	0.23	385.58	Unit IIb	Mudstone	AFD	30	60	26.8	64.7	1.9	N
C0011B-6R-2	0.39	0.41	385.77	Unit IIb	Mudstone	AFD	20	50	176.2	66.6	1.9	N
C0011B-6R-2	0.79	0.81	386.16	Unit IIb	Mudstone	AFD	20	80	239.7	46.1	7	N
C0011B-6R-2	0.91	0.93	386.27	Unit IIb	Mudstone	AFD	10	60	−72.3	53.7	1.3	N
C0011B-6R-2	1.26	1.28	386.65	Unit IIb	Mudstone	AFD	20	80	97.7	63.6	5.2	N
C0011B-6R-3	0.16	0.19	386.92	Unit IIb	Mudstone	AFD	15	40	−157.6	52.5	15.4	N
C0011B-6R-3	0.7	0.72	387.44	Unit IIb	Mudstone	AFD	15	40	172.2	56.4	7.9	N
C0011B-6R-3	1.38	1.4	388.14	Unit IIb	Mudstone	AFD	25	80	−42.5	37.9	4.4	N
C0011B-6R-4	0.06	0.08	388.20	Unit IIb	Mudstone	AFD	20	80	132.9	33.9	10	N
C0011B-6R-6	0.15	0.17	388.98	Unit IIb	Mudstone	AFD	20	80	21.7	62.1	2.8	N
C0011B-6R-6	0.36	0.38	389.19	Unit IIb	Mudstone	AFD	15	80	−155.1	47.8	2.2	N
C0011B-6R-6	0.39	0.41	389.21	Unit IIb	Mudstone	AFD	30	70	21.9	54.3	1.7	N
C0011B-6R-6	1.17	1.19	389.98	Unit IIb	Mudstone	AFD	30	60	138.1	56.3	6.2	N
C0011B-6R-6	1.27	1.29	390.07	Unit IIb	Mudstone	AFD	20	60	334.9	64.8	10.4	N
C0011B-6R-7	0.16	0.18	390.38	Unit IIb	Mudstone	AFD	20	80	−54.8	59.8	2.4	N
C0011B-6R-7	0.21	0.23	390.42	Unit IIb	Mudstone	AFD	10	60	105.3	57.6	1.9	N
C0011B-6R-7	0.42	0.44	390.63	Unit IIb	Mudstone	AFD	40	70	293.1	58.7	5.2	N
C0011B-6R-7	0.87	0.89	391.06	Unit IIb	Mudstone	AFD	10	40	104.1	57.6	6.6	N
C0011B-6R-7	1.15	1.17	391.34	Unit IIb	Mudstone	AFD	20	60	287.7	69.9	2.2	N
C0011B-6R-8	0.2	0.22	391.8	Unit IIb	Mudstone	AFD	20	80	−97	47.1	4.3	N
C0011B-6R-8	0.36	0.38	391.95	Unit IIb	Mudstone	AFD	15	60	74.2	51.8	3.9	N
C0011B-6R-8	0.81	0.83	392.39	Unit IIb	Mudstone	AFD	40	70	5.3	41.7	3.1	N
C0011B-6R-8	1.27	1.29	392.84	Unit IIb	Mudstone	AFD	15	60	77.2	47.4	1.5	N
C0011B-6R-9	0.18	0.2	393.15	Unit IIb	Mudstone	AFD	15	60	88.2	59.5	2.6	N
C0011B-7R-1	0.14	0.16	393.65	Unit IIb	Mudstone	AFD	25	80	19.7	52.8	9.4	N
C0011B-7R-1	0.18	0.2	393.69	Unit IIb	Volcaniclastic sand	AFD	30	80	−143.5	39	1.6	N
C0011B-7R-1	0.72	0.74	394.23	Unit IIb	Mudstone	AFD	40	90	160.1	−37.3	1.2	N
C0011B-7R-1	1.03	1.05	394.54	Unit IIb	Mudstone	AFD	20	100	135.9	−31.4	3	N
C0011B-7R-1	1.18	1.2	394.69	Unit IIb	Volcaniclastic sand	AFD	20	80	−65.9	−40.5	1.8	R
C0011B-7R-4	0.37	0.39	395.88	Unit IIb	Mudstone	AFD	15	80	−114.8	−47.9	3.5	R
C0011B-7R-4	0.87	0.89	396.38	Unit IIb	Mudstone	AFD	30	50	288.7	−27.9	5.1	R

(*Continued*)

Table 1. *Continued*

Hole, core, section	Top (cm)	Bottom (cm)	Sample centre depth (cm)	Lithological unit	Lithology	AFD/ThD	Linear fitting (preliminary)					Polarity
							Min	Max	Dec	Inc	MAD	
C0011B-7R-4	1.19	1.21	396.7	Unit IIb	Mudstone	AFD	40	90	236	−18.8	9.2	R
C0011B-7R-5	0.31	0.33	397.16	Unit IIb	Mudstone	AFD	40	60	195	−38.8	5.3	R
C0011B-7R-5	0.65	0.67	397.5	Unit IIb	Mudstone	AFD	40	70	−27.5	−13.1	13.7	R
C0011B-7R-6	0.11	0.13	398.38	Unit IIb	Volcaniclastic sand	AFD	40	80	−43.5	−43	3.8	R
C0011B-7R-6	0.17	0.19	398.44	Unit IIb	Mudstone	AFD	50	80	34.9	−29.7	13	R
C0011B-7R-6	0.6	0.62	398.87	Unit IIb	Mudstone	AFD	20	70	305.8	45.5	2.3	N
C0011B-7R-6	0.79	0.81	399.06	Unit IIb	Mudstone	AFD	20	80	115.2	30.6	3	N
C0011B-7R-7	0.32	0.34	400.00	Unit IIb	Mudstone	AFD	20	60	36.5	20.6	3.1	N
C0011B-7R-7	0.4	0.42	400.08	Unit IIb	Mudstone	AFD	20	80	245.9	19.7	3.9	N
C0011B-7R-7	0.97	0.99	400.65	Unit IIb	Volcaniclastic sand	AFD	30	70	202	22.4	8.3	N
C0011B-8R-5	0.17	0.19	407.59	Unit IIb	Mudstone	AFD	40	80	−120.9	−18.9	18	R
C0011B-8R-5	0.21	0.23	407.63	Unit IIb	Mudstone	AFD	25	60	49.7	−21.8	6.2	R
C0011B-8R-5	0.99	1.01	408.41	Unit IIb	Mudstone	AFD	10	80	59.6	−41.8	6.2	R
C0011B-8R-7	0.47	0.49	410.71	Unit IIb	Mudstone	AFD	40	80	113	−20.6	8.4	R
C0011B-8R-7	0.57	0.59	410.81	Unit IIb	Mudstone	AFD	30	60	118.5	−27.9	4.1	R
C0011B-9R-3	0.23	0.25	414.28	Unit IIb	Mudstone	AFD	20	80	−85.2	−60.4	3.9	R
C0011B-9R-3	0.58	0.6	414.63	Unit IIb	Mudstone	AFD	20	80	−93	−62.6	9.6	R
C0011B-9R-3	0.78	0.8	414.83	Unit IIb	Mudstone	AFD	30	50	325.7	−60.9	18.6	R
C0011B-9R-3	0.96	0.98	415.01	Unit IIb	Mudstone	AFD	30	60	153	−60.5	7.9	R
C0011B-9R-5	0.6	0.62	417.47	Unit IIb	Mudstone	AFD	15	80	68.9	47.6	7.5	N
C0011B-10R-1	0.62	0.64	422.63	Unit IIb	Mudstone	AFD	40	80	331.3	50.2	8.2	N
C0011B-10R-1	1.04	1.06	423.05	Unit IIb	Mudstone	AFD	10	60	−60.9	65.1	15.7	N
C0011B-10R-1	1.24	1.26	423.25	Unit IIb	Mudstone	AFD	30	60	116.2	24.7	13.4	N
C0011B-10R-2	0.02	0.04	423.44	Unit IIb	Mudstone	AFD	30	60	1.9	−10.8	11.9	R
C0011B-10R-2	0.16	0.18	423.58	Unit IIb	Mudstone	AFD	20	100	114.9	−20.5	14.2	R
C0011B-10R-2	0.25	0.27	423.67	Unit IIb	Mudstone	AFD	30	80	−54.3	−21.1	11.8	R
C0011B-10R-2	0.55	0.57	423.97	Unit IIb	Mudstone	AFD	20	100	114.9	−20.5	14.2	R
C0011B-10R-2	1.05	1.07	424.47	Unit IIb	Mudstone	AFD	20	80	77.3	45	17.4	N
C0011B-10R-2	1.28	1.3	424.7	Unit IIb	Mudstone	AFD	30	100	245.9	−15	9.7	R
C0011B-11R-1	0.41	0.43	431.92	Unit IIb	Mudstone	AFD	25	60	61.9	−40.9	12.7	R
C0011B-11R-1	0.77	0.79	432.28	Unit IIb	Mudstone	AFD	30	50	226.4	−17.9	18.1	R
C0011B-11R-1	1.31	1.33	432.82	Unit IIb	Mudstone	AFD	30	60	179.7	57.4	10.9	N
C0011B-11R-2	0.56	0.58	433.47	Unit IIb	Mudstone	AFD	40	80	72.3	48.5	6.6	N
C0011B-11R-2	0.99	1.01	433.90	Unit IIb	Mudstone	AFD	25	60	−4.3	−16.3	8	R
C0011B-11R-2	1.34	1.36	434.25	Unit IIb	Mudstone	AFD	30	150	28.9	−30.2	3.6	R

C0011B-11R-3	0.15	0.17	434.47	Unit IIb	Mudstone	AFD	20	120	176.9	−42.7	5.1	R
C0011B-11R-5	0.17	0.19	435.46	Unit IIb	Silty Sand	AFD	15	140	157.5	−46.5	9.6	R
C0011B-11R-5	0.28	0.3	435.57	Unit IIb	Mudstone	AFD	30	80	337.7	−38.2	3.9	R
C0011B-11R-5	1.16	1.18	436.45	Unit IIb	Mudstone	AFD	50	80	−56.6	−15.7	12.1	R
C0011B-11R-5	1.26	1.28	436.55	Unit IIb	Mudstone	AFD	30	100	316.9	−15.1	5.8	R
C0011B-11R-6	0.24	0.26	436.94	Unit IIb	Mudstone	AFD	20	80	110.5	−41.1	13.1	R
C0011B-11R-6	0.42	0.44	437.12	Unit IIb	Mudstone	AFD	20	60	291.8	−14.8	12.4	R
C0011B-11R-6	0.58	0.6	437.28	Unit IIb	Mudstone	AFD	40	70	288.2	−23.4	9.6	R
C0011B-11R-6	0.87	0.89	437.57	Unit IIb	Mudstone	AFD	25	60	−70.3	−49	12.2	R
C0011B-11R-6	1.12	1.14	437.82	Unit IIb	Mudstone	AFD	20	60	156.7	67.5	11.7	N
C0011B-11R-6	1.35	1.37	438.05	Unit IIb	Mudstone	AFD	30	60	145.8	33.8	7.8	N
C0011B-11R-7	0.37	0.39	438.48	Unit IIb	Mudstone	AFD	25	50	130.1	23.2	15.4	N
C0011B-11R-7	0.44	0.46	438.55	Unit IIb	Mudstone	AFD	60	80	259.4	30	12.1	N
C0011B-11R-7	0.79	0.81	438.9	Unit IIb	Mudstone	AFD	40	80	141.8	−21.4	12.3	R
C0011B-11R-8	0.12	0.14	439.64	Unit IIb	Mudstone	AFD	25	60	−61.8	−22.8	7.5	R
C0011B-11R-8	0.52	0.54	440.04	Unit IIb	Mudstone	AFD	25	60	108.8	−40.4	9.4	R
C0011B-11R-8	0.83	0.85	440.35	Unit IIb	Mudstone	AFD	20	60	84.5	−46.4	12.9	R
C0011B-12R-2	0.3	0.32	442.72	Unit IIb	Mudstone	AFD	20	80	17.8	−16.9	4.7	R
C0011B-12R-2	1.11	1.13	443.53	Unit IIb	Mudstone	AFD	50	90	189.5	−12	4.5	R
C0011B-12R-2	1.23	1.25	443.65	Unit IIb	Mudstone	AFD	30	60	−65.1	−38.3	14.1	R
C0011B-12R-3	0.04	0.06	443.88	Unit IIb	Mudstone	AFD	40	60	198	25.2	0	N
C0011B-12R-3	0.45	0.47	444.29	Unit IIb	Mudstone	AFD	25	60	175.2	−63.4	15.2	R
C0011B-12R-3	0.94	0.96	444.78	Unit IIb	Mudstone	AFD	30	70	158.3	−14	4.5	R
C0011B-12R-4	0.33	0.35	445.59	Unit IIb	Mudstone	AFD	20	40	159.3	−29.8	11.1	R
C0011B-12R-4	1.15	1.17	446.41	Unit IIb	Mudstone	AFD	40	70	313	38.1	2.1	N
C0011B-12R-7	0.28	0.3	447.62	Unit IIb	Mudstone	AFD	30	50	288.8	−20.9	12	R
C0011B-13R-1	0.25	0.27	450.76	Unit IIb	Mudstone	AFD	20	60	0.1	56.7	7.9	N
C0011B-13R-1	0.34	0.36	450.85	Unit IIb	Mudstone	AFD	10	40	−0.2	57.6	12.1	N
C0011B-13R-1	0.68	0.7	451.19	Unit IIb	Mudstone	AFD	30	80	−17.7	62.9	5.4	N
C0011B-13R-1	1.17	1.19	451.68	Unit IIb	Mudstone	AFD	20	60	179	66.1	8.6	R
C0011B-13R-2	0.15	0.17	452.07	Unit IIb	Mudstone	AFD	10	20	172.7	66.4	6	R
C0011B-13R-2	0.41	0.43	452.33	Unit IIb	Mudstone	AFD	30	60	34.4	−37.7	8.5	R
C0011B-13R-2	0.62	0.64	452.54	Unit IIb	Mudstone	AFD	20	60	6.5	−14.4	13.7	R
C0011B-13R-2	1.2	1.22	453.12	Unit IIb	Mudstone	AFD	30	60	−10.1	38.1	17.4	N
C0011B-13R-4	0.89	0.91	454.58	Unit IIb	Mudstone	AFD	20	70	−7.2	68.5	16.6	N
C0011B-13R-4	1.24	1.26	454.93	Unit IIb	Mudstone	AFD	30	70	312	67.8	8.4	N
C0011B-13R-5	0.44	0.46	455.54	Unit IIb	Mudstone	AFD	15	60	−36.7	54.8	6.2	N
C0011B-13R-5	1.23	1.25	456.33	Unit IIb	Mudstone	AFD	15	80	107.1	45.4	7.9	N
C0011B-13R-5	1.35	1.37	456.45	Unit IIb	Mudstone	AFD	30	70	303	51	4.9	N
C0011B-13R-6	0.04	0.06	456.55	Unit IIb	Mudstone	AFD	40	60	313.4	17.9	9.2	N

(Continued)

Table 1. *Continued*

Hole, core, section	Top (cm)	Bottom (cm)	Sample centre depth (cm)	Lithological unit	Lithology	AFD/ThD	Linear fitting (preliminary)					Polarity
							Min	Max	Dec	Inc	MAD	
C0011B-13R-6	0.59	0.61	457.10	Unit IIb	Mudstone	AFD	10	80	82.4	68.6	8.7	N
C0011B-13R-6	1.05	1.07	457.56	Unit IIb	Mudstone	AFD	10	40	−26.8	58.8	7.8	N
C0011B-13R-6	1.22	1.24	457.73	Unit IIb	Mudstone	AFD	30	70	194.8	54	2.3	N
C0011B-13R-7	0.13	0.15	458.05	Unit IIb	Mudstone	AFD	30	80	−13	37.3	4.5	N
C0011B-13R-7	0.19	0.21	458.11	Unit IIb	Mudstone	AFD	40	70	223.4	46.1	4.4	N
C0011B-13R-7	0.47	0.49	458.39	Unit IIb	Mudstone	AFD	20	80	−149.3	58.9	5.6	N
C0011B-14R-1	0.56	0.58	460.57	Unit IIb	Mudstone	AFD	30	60	101.7	59.6	6.5	N
C0011B-14R-1	0.79	0.81	460.8	Unit IIb	Mudstone	AFD	20	70	99.2	41.7	3	N
C0011B-14R-1	1.05	1.07	461.06	Unit IIb	Mudstone	AFD	20	80	93.6	52.9	5.3	N
C0011B-14R-1	1.22	1.24	461.23	Unit IIb	Mudstone	AFD	10	80	−88.3	56.8	6.3	N
C0011B-14R-2	0.23	0.25	461.65	Unit IIb	Mudstone	AFD	40	80	228.2	58.4	10	N
C0011B-14R-2	0.64	0.66	462.06	Unit IIb	Mudstone	AFD	15	80	56.6	39.6	6.7	N
C0011B-14R-2	0.79	0.81	462.21	Unit IIb	Mudstone	AFD	40	90	240.8	32.5	3.3	N
C0011B-14R-2	1	1.02	462.42	Unit IIb	Mudstone	AFD	30	80	−81.7	66.3	14.5	N
C0011B-14R-2	1.29	1.31	462.71	Unit IIb	Mudstone	AFD	15	80	22	36.3	7.5	N
C0011B-14R-3	0.04	0.06	462.88	Unit IIb	Mudstone	AFD	25	100	−49.9	26.2	5.8	N
C0011B-14R-3	0.17	0.19	463.01	Unit IIb	Mudstone	AFD	30	90	131.3	30.7	7	N
C0011B-14R-5	0.04	0.06	463.67	Unit IIb	Mudstone	AFD	30	80	133.3	53	10.9	N
C0011B-14R-5	1.24	1.26	464.87	Unit IIb	Mudstone	AFD	10	60	122.9	45.6	8.2	N
C0011B-14R-5	1.38	1.4	465.01	Unit IIb	Mudstone	AFD	15	80	−27.6	52.4	4.6	N
C0011B-14R-6	0.35	0.37	465.40	Unit IIb	Mudstone	AFD	40	80	172.6	61.2	5.7	N
C0011B-14R-6	0.53	0.55	465.58	Unit IIb	Mudstone	AFD	40	60	282.5	55.8	6.1	N
C0011B-14R-6	0.67	0.69	465.72	Unit IIb	Mudstone	AFD	15	80	−76.3	29	3.4	N
C0011B-15R-1	0.41	0.43	469.92	Unit IIb	Mudstone	AFD	15	80	−89.4	48.4	3.1	N
C0011B-15R-1	0.46	0.48	469.96	Unit IIb	Mudstone	AFD	15	60	90.6	27.5	3.2	N
C0011B-15R-1	0.77	0.79	470.26	Unit IIb	Mudstone	AFD	30	80	263.4	22.8	5.3	N
C0011B-15R-1	0.98	1	470.47	Unit IIb	Mudstone	AFD	25	60	57.7	51.9	9	N
C0011B-15R-2	0.04	0.06	470.92	Unit IIb	Mudstone	AFD	15	60	−83	59.7	5.2	N
C0011B-15R-2	0.07	0.09	470.95	Unit IIb	Mudstone	AFD	30	80	93	60.4	4.6	N
C0011B-15R-2	0.48	0.5	471.36	Unit IIb	Mudstone	AFD	15	80	−175.8	68.3	8.8	N
C0011B-15R-2	0.87	0.89	471.73	Unit IIb	Mudstone	AFD	30	70	220.3	38	12.5	N
C0011B-15R-2	0.98	1	471.84	Unit IIb	Mudstone	AFD	20	80	−19.6	47.8	7.4	N
C0011B-15R-4	0.11	0.13	472.60	Unit IIb	Mudstone	AFD	20	60	−17.4	36.7	4.2	N
C0011B-15R-4	0.47	0.49	472.96	Unit IIb	Mudstone	AFD	20	80	169.9	53	8.2	N
C0011B-15R-4	0.99	1.01	473.46	Unit IIb	Mudstone	AFD	20	60	8.5	17.8	13.1	N
C0011B-15R-4	1.29	1.31	473.75	Unit IIb	Mudstone	AFD	20	70	172.1	49.1	5.1	N
C0011B-15R-5	0.03	0.05	473.90	Unit IIb	Mudstone	AFD	15	60	25.6	47.6	14.1	N

C0011B-15R-5	0.71	0.73	474.58	Unit IIb	Mudstone	AFD	20	80	−28.3	44.1	6.3	N
C0011B-15R-5	1.07	1.09	474.91	Unit IIb	Mudstone	AFD	20	70	296.8	56.5	4.4	N
C0011B-15R-6	0.05	0.07	475.31	Unit IIb	Mudstone	AFD	10	100	225.7	36.6	1.6	N
C0011B-15R-6	0.09	0.11	475.34	Unit IIb	Mudstone	AFD	10	80	37.7	40.1	6	N
C0011B-15R-6	0.3	0.32	475.56	Unit IIb	Mudstone	AFD	20	80	−56.6	69.8	6.3	N
C0011B-15R-6	0.47	0.49	475.72	Unit IIb	Mudstone	AFD	30	70	289.6	66.2	6.5	N
C0011B-15R-6	0.5	0.52	475.75	Unit IIb	Mudstone	AFD	30	80	122.3	43.1	5.9	N
C0011B-16R-1	0.35	0.37	477.16	Unit IIb	Tuffaceous sand	AFD	20	40	173.8	66.8	10.9	N
C0011B-19R-1	0.43	0.45	491.33	Unit III	Mudstone	AFD	15	80	26.8	−41.3	1.8	R
C0011B-19R-1	0.67	0.69	491.58	Unit III	Mudstone	AFD	25	80	−166.5	−38.2	2	R
C0011B-19R-1	1.09	1.11	491.97	Unit III	Mudstone	AFD	20	30	12.1	−44.2	14.3	R
C0011B-19R-1	1.19	1.21	492.07	Unit III	Mudstone	AFD	30	70	216.2	−37.3	14.5	R
C0011B-19R-2	0.14	0.16	492.40	Unit III	Mudstone	AFD	20	40	−112.9	63.4	7.8	N
C0011B-19R-2	0.19	0.21	492.45	Unit III	Mudstone	AFD	30	50	289.3	59.2	13	N
C0011B-19R-4	0.8	0.82	493.77	Unit III	Mudstone	AFD	25	80	34.6	63.7	7.1	N
C0011B-19R-4	1.08	1.1	494.02	Unit III	Mudstone	AFD	30	100	−150.7	31.3	6.8	N
C0011B-19R-5	0.37	0.39	494.67	Unit III	Mudstone	AFD	20	40	45.4	51.3	5.1	N
C0011B-19R-5	1.09	1.11	495.36	Unit III	Mudstone	AFD	20	40	−86.1	−40	10.6	R
C0011B-19R-5	1.18	1.2	495.45	Unit III	Mudstone	AFD	20	60	34.4	−20.1	6.4	R
C0011B-19R-6	0.3	0.32	495.92	Unit III	Mudstone	AFD	20	60	−25	−34.8	1.6	R
C0011B-19R-6	0.89	0.91	496.52	Unit III	Mudstone	AFD	25	80	−164.6	−40	6.7	R
C0011B-19R-6	1	1.02	496.60	Unit III	Mudstone	AFD	20	60	−24.3	−25.8	3.2	R
C0011B-19R-6	1.35	1.37	496.94	Unit III	Mudstone	AFD	20	60	145.6	−24.4	6.3	R
C0011B-19R-7	0.1	0.12	497.11	Unit III	Mudstone	AFD	40	90	161.6	−40.6	6.3	R
C0011B-19R-7	0.15	0.17	497.16	Unit III	Mudstone	AFD	20	60	−12	−14	3.5	R
C0011B-19R-7	0.41	0.43	497.42	Unit III	Mudstone	AFD	20	80	16.2	64.2	9.9	N
C0011B-21R-3	0.04	0.06	502.31	Unit III	Mudstone	AFD	15	30	166.3	53	11.7	N
C0011B-21R-3	0.62	0.64	502.89	Unit III	Mudstone	AFD	30	50	341.3	51.4	8.4	N
C0011B-21R-3	1.14	1.16	503.41	Unit III	Mudstone	AFD	20	60	163.6	64.2	9	N
C0011B-21R-4	1.12	1.14	504.80	Unit III	Mudstone	AFD	20	40	63.6	52.7	2.6	N
C0011B-21R-5	0.22	0.24	505.31	Unit III	Mudstone	AFD	30	60	97.9	69.9	7.2	N
C0011B-21R-5	0.43	0.45	505.52	Unit III	Mudstone	AFD	20	80	−10.8	57.4	19.9	N
C0011B-21R-5	0.72	0.74	505.81	Unit III	Mudstone	AFD	30	70	173.9	15.9	8.8	N
C0011B-21R-5	1.06	1.08	506.15	Unit III	Mudstone	AFD	20	60	48.2	−31.8	9.1	R
C0011B-21R-6	0.03	0.05	506.53	Unit III	Mudstone	AFD	20	80	136	−48.1	3.2	R
C0011B-21R-6	0.17	0.19	506.67	Unit III	Mudstone	AFD	30	80	124.2	−41.6	0.9	R
C0011B-21R-6	0.75	0.77	507.25	Unit III	Mudstone	AFD	20	80	99.4	−51.6	3	R
C0011B-21R-6	1	1.02	507.50	Unit III	Mudstone	AFD	15	80	−7.7	−45.6	2.7	R
C0011B-21R-7	0.1	0.12	508.01	Unit III	Mudstone	AFD	20	60	103.5	−40.9	3.4	R

(*Continued*)

Table 1. *Continued*

Hole, core, section	Top (cm)	Bottom (cm)	Sample centre depth (cm)	Lithological unit	Lithology	AFD/ThD	Linear fitting (preliminary)					Polarity
							Min	Max	Dec	Inc	MAD	
C0011B-21R-7	0.81	0.83	508.72	Unit III	Mudstone	AFD	15	40	−67	45.9	6.7	N
C0011B-21R-7	1.31	1.33	509.22	Unit III	Mudstone	AFD	30	70	77.2	53.4	4.1	N
C0011B-21R-8	0.04	0.06	509.36	Unit III	Mudstone	AFD	10	40	−27.1	69.6	4	N
C0011B-21R-8	0.44	0.46	509.76	Unit III	Mudstone	AFD	40	90	157	55.3	12.2	N
C0011B-21R-8	0.71	0.73	510.03	Unit III	Mudstone	AFD	10	40	−11.6	68.8	7.9	N
C0011B-23R-1	0.42	0.44	520.83	Unit III	Mudstone	AFD	20	40	−2.7	67.9	1.2	N
C0011B-23R-1	0.63	0.65	521.04	Unit III	Mudstone	AFD	30	90	147.5	64.3	3.4	N
C0011B-23R-3	0.01	0.03	521.62	Unit III	Mudstone	AFD	30	70	179.2	67.2	7.3	N
C0011B-23R-3	0.27	0.29	521.88	Unit III	Mudstone	AFD	20	40	15.5	68.6	4.3	N
C0011B-23R-3	0.8	0.82	522.41	Unit III	Mudstone	AFD	20	40	−37.2	55.8	5.6	N
C0011B-23R-4	0.14	0.16	523.15	Unit III	Mudstone	AFD	30	70	128.4	54.7	4.8	N
C0011B-23R-4	0.34	0.36	523.35	Unit III	Mudstone	AFD	15	30	−43.3	60.3	3.4	N
C0011B-23R-4	0.94	0.96	523.95	Unit III	Mudstone	AFD	15	40	−31.6	59.4	1.5	N
C0011B-23R-4	1.34	1.36	524.35	Unit III	Mudstone	AFD	30	60	136.2	61.2	5	N
C0011B-23R-5	0.15	0.17	524.59	Unit III	Mudstone	AFD	30	70	145.4	67.2	11.6	N
C0011B-23R-5	0.37	0.39	524.81	Unit III	Mudstone	AFD	15	30	−104.7	57.5	0.8	N
C0011B-23R-5	0.67	0.69	525.1	Unit III	Mudstone	AFD	20	80	100.3	63.8	2.3	N
C0011B-23R-5	0.96	0.98	525.40	Unit III	Mudstone	AFD	15	30	−112.5	63.4	7.7	N
C0011B-23R-6	0.05	0.07	525.93	Unit III	Mudstone	AFD	10	30	−67.2	65.2	7.6	N
C0011B-23R-7	0.12	0.14	526.46	Unit III	Mudstone	AFD	20	80	139.8	62.2	4.7	N
C0011B-23R-7	0.3	0.32	526.64	Unit III	Mudstone	AFD	20	40	−53.3	57	1.8	N
C0011B-23R-7	0.98	1	527.32	Unit III	Mudstone	AFD	15	30	−61.3	62.5	0.7	N
C0011B-23R-7	1.28	1.3	527.62	Unit III	Mudstone	AFD	40	70	132.7	67.9	6.9	N
C0011B-23R-8	0.1	0.12	527.85	Unit III	Mudstone	AFD	20	80	61.3	60.3	8.8	N
C0011B-23R-8	1.04	1.06	528.79	Unit III	Mudstone	AFD	20	40	−66.7	53.2	2.2	N
C0011B-23R-9	0.13	0.15	529.29	Unit III	Mudstone	AFD	20	40	−68	57.7	0.7	N
C0011B-24R-2	0.4	0.42	530.88	Unit III	Mudstone	AFD	20	80	−86.4	68.4	4	N
C0011B-24R-2	0.56	0.58	531.04	Unit III	Mudstone	AFD	20	40	103.6	62.3	1.8	N
C0011B-24R-2	0.9	0.92	531.38	Unit III	Mudstone	AFD	30	90	268.2	67.6	5.2	N
C0011B-24R-2	1.09	1.11	531.57	Unit III	Mudstone	AFD	15	40	97.3	63.8	1.3	N
C0011B-24R-2	1.26	1.28	531.74	Unit III	Mudstone	AFD	20	70	253.1	68.8	3.9	N
C0011B-24R-3	0.39	0.41	532.28	Unit III	Mudstone	AFD	20	40	3.3	39.3	5.2	N
C0011B-24R-3	1.08	1.1	532.97	Unit III	Mudstone	AFD	15	40	71.4	52.6	1.5	N
C0011B-24R-3	1.23	1.25	533.12	Unit III	Mudstone	AFD	20	80	249.6	55.3	3.6	N
C0011B-24R-4	0.11	0.13	533.42	Unit III	Mudstone	AFD	25	80	−160.4	66.2	3.6	N
C0011B-24R-4	0.46	0.48	533.77	Unit III	Mudstone	AFD	20	40	29.3	42.6	1.2	N
C0011B-24R-4	0.98	1	534.29	Unit III	Mudstone	AFD	20	40	8.5	65.2	1.7	N

C0011B-24R-4	1.2	1.22	534.51	Unit III	Mudstone	AFD	20	70	195.5	56.6	2.8	N
C0011B-24R-5	0.06	0.08	534.77	Unit III	Mudstone	AFD	15	40	7.5	69.1	4.8	N
C0011B-25R-1	0.25	0.27	539.66	Unit III	Mudstone	AFD	15	40	155.2	57.6	4.4	N
C0011B-25R-3	0.25	0.27	540.75	Unit III	Mudstone	AFD	15	40	−51.7	60.7	3.1	N
C0011B-25R-3	1.33	1.35	541.83	Unit III	Mudstone	AFD	20	60	−155.7	55.5	2.8	N
C0011B-25R-4	0.07	0.09	541.97	Unit III	Mudstone	AFD	30	80	311.9	55.8	2.7	N
C0011B-25R-4	0.77	0.79	542.67	Unit III	Mudstone	AFD	30	60	116.7	49.6	2.8	N
C0011B-25R-4	1.28	1.3	543.18	Unit III	Mudstone	AFD	30	60	175.3	55	0.6	N
C0011B-26R-1	0.29	0.31	549.30	Unit III	Mudstone	AFD	20	60	−128.8	56.6	3.5	N
C0011B-26R-1	0.35	0.37	549.36	Unit III	Mudstone	AFD	30	80	52.3	54.2	11.6	N
C0011B-26R-1	0.62	0.64	549.63	Unit III	Mudstone	AFD	20	40	44.6	63.3	2.4	N
C0011B-26R-1	0.77	0.79	549.78	Unit III	Mudstone	AFD	20	40	−125.5	49.6	1.6	N
C0011B-26R-1	0.87	0.89	549.88	Unit III	Mudstone	AFD	20	70	46.8	61.2	2.8	N
C0011B-26R-3	0.32	0.34	551.35	Unit III	Mudstone	AFD	15	60	−162.3	63.7	2.2	N
C0011B-26R-3	0.36	0.38	551.39	Unit III	Mudstone	AFD	30	80	29.5	50.7	1.7	N
C0011B-26R-3	0.69	0.71	551.72	Unit III	Mudstone	AFD	20	80	31.6	60.8	2.5	N
C0011B-26R-3	0.92	0.94	551.95	Unit III	Mudstone	AFD	20	40	−161.7	58.1	0.9	N
C0011B-26R-3	1.09	1.11	552.12	Unit III	Mudstone	AFD	30	70	297.7	64.6	15.4	N
C0011B-26R-4	0.11	0.13	552.4	Unit III	Mudstone	AFD	30	60	214.6	61.9	0.9	N
C0011B-26R-4	0.2	0.22	552.49	Unit III	Mudstone	AFD	15	40	81.1	61.7	4.3	N
C0011B-26R-4	1.05	1.07	553.34	Unit III	Mudstone	AFD	15	40	16.6	58.2	7.9	N
C0011B-26R-4	1.26	1.28	553.55	Unit III	Mudstone	AFD	20	60	219.4	67	3.5	N
C0011B-26R-5	0.2	0.22	553.89	Unit III	Mudstone	AFD	20	60	11.6	54.3	1.9	N
C0011B-26R-5	0.52	0.54	554.21	Unit III	Mudstone	AFD	40	80	180.6	50.3	5	N
C0011B-26R-5	0.87	0.89	554.56	Unit III	Mudstone	AFD	15	40	8.5	53.3	1.5	N
C0011B-26R-5	0.87	0.89	554.56	Unit III	Mudstone	AFD	20	70	178.3	65.2	3.8	N
C0011B-26R-6	0.17	0.19	555.27	Unit III	Mudstone	AFD	20	40	−30.9	63.8	0.2	N
C0011B-26R-6	0.9	0.92	556.00	Unit III	Mudstone	AFD	15	40	−20.4	60.1	5.5	N
C0011B-26R-6	1.27	1.29	556.37	Unit III	Mudstone	AFD	20	80	144	67.1	4.4	N
C0011B-26R-7	0.09	0.11	556.62	Unit III	Mudstone	AFD	15	40	−27	59.7	7.7	N
C0011B-26R-7	0.27	0.29	556.8	Unit III	Mudstone	AFD	20	50	147.8	60.3	5.8	N
C0011B-26R-7	0.7	0.72	557.23	Unit III	Mudstone	AFD	20	80	159.3	67.7	3.2	N
C0011B-26R-7	0.8	0.82	557.33	Unit III	Mudstone	AFD	15	30	−34.7	54.7	2.7	N
C0011B-27R-1	0.3	0.32	558.81	Unit III	Mudstone	AFD	15	40	−19.2	69.3	6.1	N
C0011B-27R-1	0.46	0.48	558.97	Unit III	Mudstone	AFD	35	80	144.5	48	9.4	N
C0011B-27R-1	0.65	0.67	559.16	Unit III	Mudstone	AFD	20	60	192.1	67.2	3.5	N
C0011B-27R-3	0.03	0.05	559.96	Unit III	Mudstone	AFD	20	60	76.6	55.6	4.2	N
C0011B-27R-3	0.2	0.22	560.13	Unit III	Mudstone	AFD	15	80	63.8	63.8	4.3	N
C0011B-27R-3	0.38	0.4	560.31	Unit III	Mudstone	AFD	20	40	−105.5	53.8	4.8	N
C0011B-27R-3	0.54	0.56	560.47	Unit III	Mudstone	AFD	30	80	65.6	58.1	5	N

(Continued)

Table 1. *Continued*

Hole, core, section	Top (cm)	Bottom (cm)	Sample centre depth (cm)	Lithological unit	Lithology	AFD/ThD	Linear fitting (preliminary)					Polarity
							Min	Max	Dec	Inc	MAD	
C0011B-27R-3	1.28	1.3	561.21	Unit III	Mudstone	AFD	20	40	−153.5	50	2.5	N
C0011B-27R-4	0.31	0.33	561.62	Unit III	Mudstone	AFD	20	40	−85.7	63.2	1.1	N
C0011B-27R-4	0.95	0.97	562.26	Unit III	Mudstone	AFD	10	40	−92.6	69.4	6	N
C0011B-27R-5	0.1	0.12	562.82	Unit III	Mudstone	ThD	500	580	69.7	62.6	9.3	N
C0011B-27R-5	0.21	0.23	562.93	Unit III	Mudstone	AFD	15	40	−111.3	66	1.4	N
C0011B-27R-5	0.72	0.74	563.44	Unit III	Mudstone	AFD	15	40	−134.5	59.1	4.7	N
C0011B-27R-5	1.21	1.23	563.93	Unit III	Mudstone	AFD	30	80	146.5	50.4	4.3	N
C0011B-27R-6	0.08	0.1	564.22	Unit III	Mudstone	AFD	30	60	342	66.3	13.1	N
C0011B-27R-6	0.23	0.25	564.37	Unit III	Mudstone	AFD	15	40	127.9	68.3	18.2	N
C0011B-28R-1	0.86	0.88	568.87	Unit III	Mudstone	AFD	20	40	156.4	33	7.3	N
C0011B-28R-3	0.15	0.17	570.15	Unit III	Mudstone	AFD	30	60	−162	14.2	8.7	N
C0011B-30R-1	0.63	0.65	581.04	Unit III	Mudstone	AFD	30	50	73.1	46.1	13.3	N
C0011B-30R-1	1.26	1.28	581.67	Unit III	Mudstone	AFD	40	80	3.4	−24.5	17.1	R
C0011B-30R-3	0.02	0.04	582.34	Unit III	Mudstone	AFD	30	50	2.2	−17.2	13.1	R
C0011B-30R-4	0.08	0.1	583.81	Unit III	Mudstone	AFD	40	60	48.7	−59.9	9.8	R
C0011B-30R-4	0.18	0.2	583.91	Unit III	Mudstone	AFD	30	50	75.2	56.4	14.6	N
C0011B-30R-4	0.35	0.37	584.08	Unit III	Mudstone	AFD	15	40	−90.1	60.7	13.1	N
C0011B-31R-1	0.36	0.38	586.87	Unit III	Mudstone	AFD	20	60	112.4	−40.3	9.3	R
C0011B-31R-1	1.26	1.28	587.77	Unit III	Mudstone	AFD	30	70	71.7	53.9	10.7	N
C0011B-31R-3	1.27	1.29	589.65	Unit III	Mudstone	AFD	20	50	177.9	68.3	12.1	N
C0011B-31R-5	0.38	0.4	590.53	Unit III	Mudstone	AFD	30	80	−19.5	−25	12.6	R
C0011B-31R-5	0.81	0.83	590.96	Unit III	Mudstone	AFD	30	50	344.3	−27	10.7	R
C0011B-32R-1	0.16	0.18	596.17	Unit III	Mudstone	AFD	30	80	69.4	59.9	10.6	N
C0011B-32R-1	0.32	0.34	596.33	Unit III	Mudstone	AFD	15	40	19.9	66.1	13.7	N
C0011B-32R-1	0.86	0.88	596.87	Unit III	Mudstone	AFD	20	40	−51	55.9	3.8	N
C0011B-32R-1	1.2	1.22	597.21	Unit III	Mudstone	AFD	20	30	103.5	58.1	10.9	N
C0011B-32R-2	1.13	1.15	598.57	Unit III	Mudstone	AFD	20	40	26.6	67	6.5	N
C0011B-35R-1	0.42	0.44	624.91	Unit III	Mudstone	AFD	40	60	151.8	46.4	5.3	N
C0011B-35R-1	0.93	0.95	625.39	Unit III	Mudstone	AFD	10	40	−25.4	51.4	12.2	N
C0011B-35R-5	0.97	0.99	629.63	Unit III	Mudstone	AFD	20	80	−65.2	66.9	9.1	N

C0011B-35R-5	1.07	1.09	629.67	Unit III	Mudstone	AFD	30	70	273.5	64.4	4.2	N
C0011B-35R-6	0.17	0.19	630.15	Unit III	Mudstone	AFD	15	40	−82.3	63.7	4.8	N
C0011B-35R-6	0.64	0.66	630.59	Unit III	Mudstone	AFD	20	60	−58.7	67.7	14.6	N
C0011B-35R-8	1.02	1.04	633.63	Unit III	Mudstone	AFD	20	40	−168.1	60.5	7.9	N
C0011B-36R-3	0.11	0.13	635.33	Unit III	Mudstone	AFD	60	80	61	−46.3	5.7	R
C0011B-36R-3	0.25	0.27	635.47	Unit III	Mudstone	AFD	20	40	−109.5	−41.7	2.3	R
C0011B-37R-6	0.03	0.05	649.21	Unit III	Mudstone	AFD	15	30	84.1	49	7.8	N
C0011B-37R-6	0.8	0.82	649.93	Unit III	Mudstone	AFD	20	40	36.3	42.2	7.6	N
C0011B-37R-6	1	1.02	650.18	Unit III	Mudstone	AFD	30	80	−47.5	51.6	18.5	N
C0011B-37R-7	0.35	0.5	650.92	Unit III	Mudstone	AFD	6.5	20	32.1	68.3	6.2	N
C0011B-37R-7	1.38	1.4	651.89	Unit III	Mudstone	AFD	15	80	40.8	61.4	12.5	N
C0011B-38R-1	1.23	1.25	654.24	Unit III	Mudstone	AFD	20	40	37.2	61.9	5	N
C0011B-38R-3	0.93	0.95	655.87	Unit III	Mudstone	AFD	20	40	−157.8	35.1	5.2	N
C0011B-38R-3	1.12	1.14	656.06	Unit III	Mudstone	AFD	30	60	28	65.9	5.9	N
C0011B-38R-4	0.05	0.07	656.4	Unit III	Mudstone	AFD	15	80	27.3	69.1	12.5	N
C0011B-38R-4	0.56	0.58	656.91	Unit III	Mudstone	AFD	30	50	356	63.7	3	N
C0011B-38R-4	1.06	1.08	657.41	Unit III	Mudstone	AFD	15	40	124.4	60.2	5.6	N
C0011B-38R-5	0.46	0.48	658.23	Unit III	Mudstone	AFD	20	40	−173.7	64.8	3.9	N
C0011B-38R-5	0.87	0.89	658.64	Unit III	Mudstone	AFD	20	40	53.7	57.3	3.7	N
C0011B-38R-5	1.35	1.37	659.12	Unit III	Mudstone	AFD	20	80	−115.2	53.5	10.2	N
C0011B-38R-6	0.83	0.85	660.01	Unit III	Mudstone	AFD	30	80	−150.3	68.3	17.8	N
C0011B-38R-7	0.4	0.42	660.98	Unit III	Mudstone	AFD	20	80	105.9	66.9	12.2	N
C0011B-38R-7	0.47	0.49	661.05	Unit III	Mudstone	AFD	0	50	105	66.6	2.5	N
C0011B-38R-7	0.97	0.99	661.55	Unit III	Mudstone	AFD	10	60	356.6	64.4	3.9	N
C0011B-38R-8	0.08	0.1	661.67	Unit III	Mudstone	AFD	25	80	−96.3	65.9	16.4	N
C0011B-39R-2	1.03	1.05	664.95	Unit III	Mudstone	AFD	30	60	285.1	53.1	8.3	N
C0011B-39R-3	0.8	0.82	666.13	Unit III	Mudstone	AFD	15	30	60.2	48.1	2.5	N
C0011B-40R-1	1.25	1.27	673.26	Unit III	Mudstone	AFD	30	70	23.9	68.3	2.8	N
C0011B-40R-2	0.36	0.38	673.77	Unit III	Mudstone	AFD	30	80	40.8	50.5	11.7	N
C0011B-40R-2	1.36	1.38	674.77	Unit IV	Mudstone	AFD	30	50	57	37	14.7	N
C0011B-42R-3	0.03	0.05	682.45	Unit IV	Mudstone	AFD	20	90	124.4	69.2	12.5	N
C0011B-42R-3	0.51	0.53	682.93	Unit IV	Mudstone	AFD	40	60	354.3	53.8	10	N
C0011B-43R-1	0.16	0.18	685.08	Unit IV	Mudstone	AFD	40	50	187.9	68.6	2.5	N
C0011B-43R-2	0.45	0.47	686.09	Unit IV	Mudstone	AFD	20	80	122.5	64.6	6	N

(Continued)

Table 1. *Continued*

Hole, core, section	Top (cm)	Bottom (cm)	Sample centre depth (cm)	Lithological unit	Lithology	AFD/ThD	Linear fitting (preliminary)					Polarity
							Min	Max	Dec	Inc	MAD	
C0011B-43R-4	0.27	0.29	686.24	Unit IV	Mudstone	AFD	15	30	−132.8	55.1	7.1	N
C0011B-43R-4	0.31	0.33	686.26	Unit IV	Mudstone	AFD	30	70	22.1	64.3	7.7	N
C0011B-43R-4	0.72	0.74	686.45	Unit IV	Mudstone	AFD	20	60	267.4	67.5	3.8	N
C0011B-43R-4	1.07	1.09	687.19	Unit IV	Mudstone	AFD	20	80	−72.7	63	6.3	N
C0011B-44R-1	0.62	0.64	689.63	Unit IV	Mudstone	AFD	30	60	233.8	69.5	5.4	N
C0011B-44R-1	1.03	1.05	690.04	Unit IV	Mudstone	AFD	30	70	7.3	68.2	2.5	N
C0011B-44R-1	1.15	1.17	690.16	Unit IV	Mudstone	AFD	80	120	−130.4	46.4	5.5	N
C0011B-44R-4	0.04	0.06	691.74	Unit IV	Mudstone	AFD	30	50	266.8	67.2	3.7	N
C0011B-44R-4	1.1	1.12	692.8	Unit IV	Mudstone	AFD	20	50	165.8	65.6	3.2	N
C0011B-44R-5	0.66	0.68	693.77	Unit IV	Mudstone	AFD	20	80	111.3	66.6	12.4	N
C0011B-44R-5	1.33	1.35	694.44	Unit IV	Mudstone	AFD	30	60	55	55.1	5.1	N
C0011B-44R-6	1.13	1.15	695.66	Unit IV	Mudstone	AFD	30	50	93.2	58.8	10.3	N
C0011B-44R-7	0.61	0.63	696.56	Unit IV	Mudstone	AFD	20	60	−5.4	65.6	9	N
C0011B-44R-7	0.82	0.85	696.78	Unit IV	Mudstone	AFD	40	80	290.9	42.7	6.1	N
C0011B-44R-8	0.11	0.13	697.09	Unit IV	Mudstone	AFD	20	70	359.6	67.9	4.3	N
C0011B-44R-8	0.36	0.38	697.34	Unit IV	Mudstone	AFD	40	100	−132.8	−10.9	8	R
C0011B-45R-1	0.45	0.47	698.96	Unit IV	Mudstone	AFD	20	70	85.1	60.1	3.4	N
C0011B-45R-1	0.7	0.72	699.21	Unit IV	Mudstone	AFD	20	80	98.2	63.1	9.7	N
C0011B-45R-4	0.11	0.13	700.93	Unit IV	Mudstone	AFD	140	160	−143.5	−10.4	8.6	R
C0011B-45R-4	0.89	0.91	701.71	Unit IV	Mudstone	AFD	30	70	255.5	−50.1	7.9	R
C0011B-45R-4	1.05	1.07	701.87	Unit IV	Mudstone	AFD	40	60	91.7	45.9	6.6	R
C0011B-45R-6	0.31	0.33	703.98	Unit IV	Mudstone	AFD	80	140	−15.7	−39.8	4.5	R
C0011B-46R-1	1.17	1.19	709.18	Unit IV	Mudstone	AFD	80	160	−137.4	−26.1	15.4	R
C0011B-47R-5	0.38	0.4	716.41	Unit IV	Mudstone	AFD	100	160	62.6	−35.9	13.4	R
C0011B-47R-5	0.88	0.9	716.91	Unit IV	Mudstone	AFD	100	160	−137.8	−15.6	4.7	R
C0011B-48R-1	0.65	0.67	721.66	Unit IV	Mudstone	AFD	30	70	140.2	47.8	4.1	N
C0011B-48R-2	0.32	0.34	722.33	Unit IV	Mudstone	AFD	30	70	310.2	67.6	2.4	N
C0011B-48R-2	0.41	0.43	722.42	Unit IV	Mudstone	AFD	80	160	−154.5	43.3	2.3	N
C0011B-49R-5	0.63	0.65	734.2	Unit IV	Mudstone	AFD	20	45	132.6	40.4	3.9	N
C0011B-51R-8	0.18	0.2	755.18	Unit IV	Mudstone	AFD	20	40	−43.1	44.9	7.1	N

Sample	Min	Max	Depth	Unit	Lithology	Method	Min	Max	Dec	Inc	MAD	Pol
C0011B-51R-8	0.58	0.6	755.55	Unit IV	Mudstone	AFD	30	50	93.3	64.6	2.2	N
C0011B-51R-8	0.78	0.8	755.74	Unit IV	Mudstone	AFD	20	60	−93.9	50.5	7.6	N
C0011B-51R-8	0.86	0.88	755.81	Unit IV	Mudstone	AFD	30	50	99.2	48.5	2	N
C0011B-51R-9	0.3	0.32	756.23	Unit IV	Mudstone	AFD	30	70	95.1	48	4	N
C0011B-51R-9	0.68	0.7	756.63	Unit IV	Mudstone	AFD	20	80	64.4	55.3	8.5	N
C0011B-52R-5	0.2	0.22	762.43	Unit IV	Mudstone	AFD	80	160	−146.6	−12	13.5	R
C0011B-54R-4	0.03	0.05	772.87	Unit IV	Mudstone	AFD	30	70	156.2	63.6	2.8	N
C0011B-54R-4	0.28	0.3	773.12	Unit IV	Mudstone	AFD	20	80	171.4	64.9	6.7	N
C0011B-54R-5	0.01	0.03	774.26	Unit IV	Volcaniclastic sand	AFD	100	180	145.6	−9.9	8.9	R
C0011B-55R-1	0.16	0.18	779.17	Unit IV	Mudstone	AFD	30	80	74.6	68.9	4.4	N
C0011B-55R-3	0.15	0.17	780.26	Unit IV	Silty Sand	AFD	30	60	47.8	65.8	8.7	N
C0011B-56R-1	0.04	0.06	844.05	Unit IV	Mudstone	AFD	20	60	47.1	61.4	4.4	N
C0011B-56R-1	0.44	0.46	844.42	Unit IV	Mudstone	AFD	30	60	46.1	49.7	1.1	N
C0011B-57R-5	1.23	1.41	852.37	Unit V	Mudstone	AFD	20	60	128.8	60	6.2	N
C0011B-58R-1	0.04	0.06	855.95	Unit V	Fine grained tuff	AFD	30	60	42.7	−54.1	3.3	R
C0011B-58R-1	0.16	0.18	856.06	Unit V	Fine grained tuff	AFD	30	60	30.9	−47.9	6.8	R
C0011B-58R-1	0.19	0.21	856.09	Unit V	Mudstone	AFD	40	100	207	−36.9	3.1	R
C0011B-58R-1	0.47	0.49	856.36	Unit V	Mudstone	AFD	30	70	17.6	67.1	1.3	N
C0011B-58R-4	1.38	1.4	860.28	Unit V	Mudstone	AFD	40	80	−47.7	38.4	16.7	N
C0011B-58R-6	0.11	0.13	861.56	Unit V	Mudstone	AFD	30	80	46	63.6	2.7	N
C0011B-58R-7	0.15	0.17	862.88	Unit V	Tuffaceous silty clay	AFD	30	80	−19.9	−19.8	8.9	R
C0011B-59R-1	0.43	0.45	865.84	Unit V	volcaniclastic clayey silt	AFD	30	60	−143.8	−12.2	3.1	R
C0011B-59R-1	0.53	0.55	865.94	Unit V	Tuffaceous sandy silt	AFD	30	60	−168.5	−38.3	1	R
C0011B-59R-1	0.69	0.71	866.10	Unit V	Tuffaceous sandy silt	AFD	40	80	−173.2	−35.4	5.6	R
C0011B-59R-2	0.17	0.19	866.93	Unit V	Mudstone	AFD	40	80	148	−12.5	8.5	R
C0011B-59R-4	0.72	0.74	868.72	Unit V	Tuffaceous sandy silt	AFD	100	180	−137.1	−31.7	3.2	R

Note: AFD/ThD stand for alternating field and thermal demagnetization experiment, respectively; Min and Max, minimum and maximum demagnetization fields and temperatures to define the ChRM; Dec and Inc, declination and inclination; MAD, maximum angular derivation; N, normal; R, reversed.

Table 2. *Summary of palaeomagnetic data of sediments for Hole C0012A*

Sample			Sample centre depth (cm)	Lithological Unit	Lithology	AFD/ThD	Linear fitting (preliminary)					Polarity
Hole, core, section	Top (cm)	Bottom (cm)					Min	Max	Dec	Inc	MAD	
C0012A-2R-4	39	41	64.25	Unit I	Mudstone	AFD	20	80	93.8	−62.1	13.4	R
C0012A-4R-2	17	19	80.19	Unit I	Mudstone	AFD	20	60	−42.1	34.9	2.3	N
C0012A-5R-1	30	32	88.81	Unit I	Mudstone	AFD	40	80	302.7	22.7	1.2	N
C0012A-5R-1	81	83	89.32	Unit I	Mudstone	AFD	20	80	7.2	65.2	1	N
C0012A-5R-3	84	86	90.74	Unit I	Mudstone	AFD	20	60	156.2	49.4	4.7	N
C0012A-5R-3	94	96	90.84	Unit I	Mudstone	AFD	30	80	308.4	65.3	4.6	N
C0012A-5R-3	108	110	90.98	Unit I	Mudstone	AFD	20	100	83.5	44.1	0.6	N
C0012A-5R-4	58	60	91.88	Unit I	Mudstone	AFD	20	80	219.4	−67.5	4.1	R
C0012A-5R-4	85	87	92.15	Unit I	Mudstone	AFD	20	60	140.8	−61	1.7	R
C0012A-5R-4	91	93	92.21	Unit I	Mudstone	AFD	10	100	342.6	−56.6	3.3	R
C0012A-6R-1	32	34	98.33	Unit I	Mudstone	AFD	20	60	−66.2	−30	4.2	R
C0012A-6R-1	89	91	98.90	Unit I	Mudstone	AFD	–	–	321.8	−43.1	0	R
C0012A-6R-2	37	39	99.79	Unit I	Mudstone	AFD	20	60	−165.3	−51.3	1.3	R
C0012A-6R-2	55	57	99.97	Unit I	Mudstone	AFD	20	80	302.3	−58.9	1.5	R
C0012A-6R-2	62	64	100.04	Unit I	Mudstone	AFD	15	40	67.5	−60.2	2	R
C0012A-6R-2	80	82	100.22	Unit I	Mudstone	AFD	10	100	153.3	−52.7	3.1	R
C0012A-6R-4	15	17	100.84	Unit I	Mudstone	AFD	10	100	305.1	−56.2	1.2	R
C0012A-6R-4	24	26	100.93	Unit I	Mudstone	AFD	15	60	100.2	−51.5	1.6	R
C0012A-6R-4	53	55	101.22	Unit I	Mudstone	AFD	15	60	14.9	−42.7	1.3	R
C0012A-6R-4	58	60	101.27	Unit I	Mudstone	AFD	20	80	−171.9	−53	1.1	R
C0012A-7R-5	63	65	107.40	Unit I	Mudstone	AFD	20	60	−123.7	−57	2.1	R
C0012A-7R-CC	4	6	107.91	Unit I	Mudstone	AFD	15	60	−159.8	−49.6	1.4	R
C0012A-8R-2	36	38	113.28	Unit I	Mudstone	AFD	20	80	−11.6	63.4	6.1	N
C0012A-8R-2	59	61	113.51	Unit I	Mudstone	AFD	30	80	303.8	52.5	2.4	N
C0012A-8R-2	69	71	113.61	Unit I	Mudstone	AFD	20	60	100.1	51	8.5	N
C0012A-8R-2	97	99	113.89	Unit I	Mudstone	AFD	20	80	96.4	67.9	1.9	N
C0012A-8R-2	120	122	114.12	Unit I	Mudstone	AFD	20	60	−131.6	63.7	1.7	N
C0012A-8R-3	4	6	114.37	Unit I	Mudstone	AFD	30	80	48.9	56	1.6	N
C0012A-8R-3	18	20	114.51	Unit I	Mudstone	AFD	15	30	139.2	56.9	0.4	N
C0012A-8R-3	53	55	114.86	Unit I	Mudstone	AFD	20	60	−5.7	51.2	3.6	N
C0012A-8R-3	106	108	115.39	Unit I	Mudstone	AFD	20	60	152.5	53.8	5.6	N
C0012A-8R-4	5	7	115.79	Unit I	Mudstone	AFD	50	80	208.5	64	7.8	N
C0012A-8R-4	43	45	116.17	Unit I	Mudstone	AFD	15	40	137.3	−49.4	4.3	R
C0012A-8R-4	56	58	116.29	Unit I	Mudstone	AFD	15	80	−137.8	−61.1	4.8	R
C0012A-8R-4	84	86	116.58	Unit I	Mudstone	AFD	20	60	326	64.9	1.9	N

C0012A-8R-6	7	9	116.94	Unit I	Mudstone	AFD	20	80	−59.2	65.5	1.4	N
C0012A-8R-6	17	19	117.04	Unit I	Mudstone	AFD	30	80	107.4	61.4	2	N
C0012A-8R-6	81	83	117.68	Unit I	Mudstone	AFD	30	80	163.2	56.7	4.7	N
C0012A-8R-6	100	102	117.87	Unit I	Mudstone	AFD	30	60	−35.5	69.1	5.9	N
C0012A-9R-1	135	137	122.36	Unit I	Mudstone	AFD	15	60	151.4	−36.8	4.6	R
C0012A-9R-2	21	23	122.62	Unit I	Mudstone	AFD	15	60	13.7	−37.5	6.8	R
C0012A-9R-2	48	50	122.89	Unit I	Mudstone	AFD	20	80	100.6	−55.9	4.1	R
C0012A-9R-2	69	71	123.10	Unit I	Mudstone	AFD	10	80	328.2	−39.7	4	R
C0012A-9R-2	96	98	123.37	Unit I	Mudstone	AFD	40	80	283.7	−45.6	7.3	R
C0012A-9R-2	106	108	123.47	Unit I	Mudstone	AFD	15	60	90.7	−44.5	4.5	R
C0012A-9R-3	10	12	123.73	Unit I	Mudstone	AFD	20	80	252.6	−39.6	5.5	R
C0012A-9R-3	21	23	123.84	Unit I	Mudstone	AFD	20	80	60.1	−59.8	4.5	R
C0012A-9R-5	4	6	124.32	Unit I	Mudstone	AFD	20	60	−52.7	−61.1	5.2	R
C0012A-9R-5	44	46	124.72	Unit I	Mudstone	AFD	15	100	71.8	−50.4	2.3	R
C0012A-9R-5	81	83	125.09	Unit I	Mudstone	AFD	20	80	29	−63	4.5	R
C0012A-9R-5	101	103	125.29	Unit I	Mudstone	AFD	15	60	−152.1	−56	3.8	R
C0012A-9R-5	131	133	125.59	Unit I	Mudstone	AFD	20	80	15.2	−53.9	2.2	R
C0012A-9R-6	6	8	125.75	Unit I	Mudstone	AFD	15	60	136.6	−62	2.6	R
C0012A-9R-6	12	14	125.81	Unit I	Mudstone	AFD	20	80	241.7	−55.8	4.1	R
C0012A-9R-6	33	35	126.02	Unit I	Mudstone	AFD	20	100	294.1	−56	1.7	R
C0012A-9R-6	79	81	126.48	Unit I	Mudstone	AFD	15	60	6.9	−61.8	3.9	R
C0012A-9R-6	121	123	126.90	Unit I	Mudstone	AFD	20	80	−152.8	−51.6	3.7	R
C0012A-9R-7	4	6	127.14	Unit I	Mudstone	AFD	20	80	−164.7	−46.5	2.1	R
C0012A-9R-7	11	13	127.21	Unit I	Mudstone	AFD	20	60	−7.2	−60.7	5.6	R
C0012A-9R-7	87	89	127.97	Unit I	Mudstone	AFD	20	60	94.7	−27	12.1	R
C0012A-9R-8	28	30	128.80	Unit I	Mudstone	AFD	20	60	109.6	58.6	4.7	N
C0012A-10R-1	11	13	130.62	Unit I	Mudstone	AFD	20	60	26.9	−45	2.7	R
C0012A-10R-1	28	30	130.79	Unit I	Mudstone	AFD	20	80	−179.8	−40.9	5.4	R
C0012A-10R-1	62	64	131.13	Unit I	Mudstone	AFD	20	80	334	−67.8	3.8	R
C0012A-10R-1	80	82	131.31	Unit I	Mudstone	AFD	20	90	324	−63.8	2.6	R
C0012A-10R-1	93	95	131.44	Unit I	Mudstone	AFD	20	60	125	−65.7	3.7	R
C0012A-10R-2	2	4	131.65	Unit I	Mudstone	AFD	15	60	158.7	−55.1	5	R
C0012A-10R-2	21	23	131.84	Unit I	Mudstone	AFD	20	80	326.5	−57.7	4	R
C0012A-10R-4	35	37	133.00	Unit I	Mudstone	AFD	20	90	334.8	53.9	1.5	N
C0012A-10R-4	50	52	133.14	Unit I	Mudstone	AFD	20	80	−17.5	65	3.1	N
C0012A-10R-4	65	67	133.30	Unit I	Mudstone	AFD	10	60	−171.8	61	4.4	N
C0012A-10R-4	97	99	133.62	Unit I	Mudstone	AFD	10	80	340.8	64.9	1.4	N
C0012A-10R-4	124	126	133.89	Unit I	Mudstone	AFD	10	60	71.8	74.6	3.8	N
C0012A-10R-5	7	9	134.12	Unit I	Mudstone	AFD	20	80	−69.8	59.3	4.8	N
C0012A-10R-5	52	54	134.57	Unit I	Mudstone	AFD	15	60	52.7	46.7	4.9	N

(*Continued*)

Table 2. *Continued*

Sample			Sample centre depth (cm)	Lithological Unit	Lithology	AFD/ThD	Linear fitting (preliminary)					Polarity
Hole, core, section	Top (cm)	Bottom (cm)					Min	Max	Dec	Inc	MAD	
C0012A-10R-5	103	105	135.08	Unit I	Mudstone	AFD	10	80	154.2	65.9	3.2	N
C0012A-10R-5	130	132	135.35	Unit I	Mudstone	AFD	10	60	12.4	62.7	2.7	N
C0012A-10R-6	19	21	135.64	Unit I	Mudstone	AFD	10	60	−34.9	61.2	5.3	N
C0012A-10R-6	31	33	135.76	Unit I	Mudstone	AFD	30	90	112.6	60.9	2.2	N
C0012A-10R-6	72	74	136.17	Unit I	Mudstone	AFD	20	80	−108.3	60.7	1.6	N
C0012A-10R-7	21	23	137.07	Unit I	Mudstone	AFD	15	60	−129.4	61.1	1.6	N
C0012A-10R-7	56	58	137.42	Unit I	Mudstone	AFD	10	100	107.2	61.9	0.9	N
C0012A-10R-7	66	68	137.52	Unit I	Mudstone	AFD	20	80	18.8	50.3	1.5	N
C0012A-10R-7	77	79	137.63	Unit II	Mudstone	AFD	10	60	61.8	41.8	1.8	N
C0012A-10R-CC	4	6	137.75	Unit II	Mudstone	AFD	10	60	−164.8	50.8	2.5	N
C0012A-11R-1	14	16	140.15	Unit II	Mudstone	AFD	40	80	295.9	−46.1	2.6	R
C0012A-11R-1	36	38	140.37	Unit II	Mudstone	AFD	15	60	164.5	−51.8	3.9	R
C0012A-11R-1	47	49	140.48	Unit II	Mudstone	AFD	30	60	315.6	−51.8	1	R
C0012A-11R-1	82	84	140.83	Unit II	Mudstone	AFD	10	80	142.7	−57.1	1.2	R
C0012A-11R-1	111	113	141.12	Unit II	Mudstone	AFD	25	80	4.4	−43.2	12.2	R
C0012A-11R-2	21	23	141.46	Unit II	Mudstone	AFD	10	90	333.8	−57	2	R
C0012A-11R-2	30	32	141.55	Unit II	Mudstone	AFD	10	60	−131.1	−57.1	1.7	R
C0012A-11R-2	63	65	141.88	Unit II	Mudstone	AFD	15	80	44	−58.6	2.1	R
C0012A-11R-2	86	88	142.11	Unit II	Mudstone	AFD	15	60	172.2	−43.1	5.2	R
C0012A-11R-2	100	102	142.25	Unit II	Mudstone	AFD	20	80	227.3	55	4.4	R
C0012A-11R-3	53	55	143.19	Unit II	Mudstone	AFD	15	60	42.6	62.1	3.2	N
C0012A-11R-3	56	58	143.22	Unit II	Mudstone	AFD	30	80	243.4	62.7	2.6	N
C0012A-11R-3	92	94	143.58	Unit II	Mudstone	AFD	20	90	241.1	67.6	1.6	N
C0012A-11R-3	114	116	143.80	Unit II	Mudstone	AFD	20	60	34.8	52.7	4.1	N
C0012A-11R-3	123	125	143.89	Unit II	Mudstone	AFD	50	80	−124.3	33.9	5.8	N
C0012A-11R-5	8	10	144.39	Unit II	Mudstone	AFD	20	80	277.5	66.7	2.6	N
C0012A-11R-5	37	39	144.68	Unit II	Mudstone	AFD	10	60	101.8	46.8	1.5	N
C0012A-11R-5	53	55	144.84	Unit II	Mudstone	AFD	40	80	248.8	51.9	2.2	N
C0012A-11R-5	112	114	145.43	Unit II	Mudstone	AFD	20	60	71.3	59.4	4.3	N
C0012A-11R-6	14	16	145.66	Unit II	Mudstone	AFD	20	80	226.1	49	5.7	N
C0012A-11R-6	42	44	145.94	Unit II	Mudstone	AFD	20	80	217.3	48.4	2.3	N
C0012A-11R-6	59	61	146.11	Unit II	Mudstone	AFD	15	60	83.1	31.8	2.5	N
C0012A-11R-6	100	102	146.52	Unit II	Mudstone	AFD	10	60	17.4	65.7	1.5	N
C0012A-11R-7	8	10	146.99	Unit II	Mudstone	AFD	40	90	173.3	55.5	2.1	N
C0012A-11R-7	33	35	147.24	Unit II	Mudstone	AFD	15	60	−71.8	68.6	1.6	N
C0012A-11R-7	39	41	147.30	Unit II	Mudstone	AFD	20	80	99.9	63.2	1.8	N

C0012A-11R-7	68	70	147.59	Unit II	Mudstone	AFD	15	80	−27.4	57.1	3.6	N
C0012A-11R-8	11	13	148.20	Unit II	Mudstone	AFD	20	80	18.1	59.7	6.6	N
C0012A-11R-8	33	35	148.42	Unit II	Mudstone	AFD	15	60	149.2	53	1.9	N
C0012A-12R-1	56	58	150.07	Unit II	Mudstone	AFD	20	60	156.3	14.3	2.4	R
C0012A-12R-1	74	76	150.25	Unit II	Mudstone	AFD	20	80	227.7	41.9	2.9	N
C0012A-12R-1	79	81	150.30	Unit II	Mudstone	AFD	10	60	48.5	37.6	1.9	N
C0012A-13R-1	43	45	159.44	Unit II	Mudstone	AFD	15	60	−28	14	2.6	N
C0012A-13R-1	46	48	159.47	Unit II	Mudstone	AFD	20	90	287.6	−31.9	3.8	R
C0012A-13R-4	18	20	161.73	Unit II	Mudstone	AFD	15	60	−55.9	53.8	2	N
C0012A-13R-4	22	24	161.77	Unit II	Mudstone	AFD	20	50	65.8	28.7	8.1	N
C0012A-13R-4	70	72	162.25	Unit II	Mudstone	AFD	15	60	110.4	61.4	2.1	N
C0012A-13R-4	83	85	162.38	Unit II	Mudstone	AFD	20	60	265	65.1	3.8	N
C0012A-13R-5	8	10	162.58	Unit II	Mudstone	AFD	35	90	210.8	63.1	3	N
C0012A-13R-5	18	20	162.68	Unit II	Mudstone	AFD	10	60	44.7	64.7	1.2	N
C0012A-14R-1	35	37	168.86	Unit II	Mudstone	AFD	10	60	61.8	59.9	1.5	N
C0012A-14R-1	88	90	169.39	Unit II	Mudstone	AFD	10	60	−178.6	49.3	3.1	N
C0012A-14R-3	43	45	170.42	Unit II	Volcanic silt	AFD	10	60	−74.4	−59.3	1.8	R
C0012A-14R-3	108	110	171.07	Unit II	Volcanic silt	AFD	10	60	31	−47.2	1.5	R
C0012A-14R-4	7	9	171.38	Unit II	Sand	AFD	15	60	−112.1	−38.1	2.4	R
C0012A-15R-1	38	40	178.39	Unit II	Mudstone	AFD	20	60	35.2	59	3	N
C0012A-15R-1	66	68	178.67	Unit II	Mudstone	AFD	30	60	−119.8	−49.6	14.8	R
C0012A-15R-2	1	3	179.43	Unit II	Mudstone	AFD	15	60	129.9	−23.8	8.4	R
C0012A-15R-2	39	41	179.81	Unit II	Mudstone	AFD	20	80	19.8	−31.3	4.9	R
C0012A-15R-2	39	41	179.81	Unit II	Mudstone	AFD	20	80	23.3	−35.3	5.4	R
C0012A-15R-2	39	41	179.81	Unit II	Mudstone	AFD	20	70	23.9	−26.8	4.1	R
C0012A-15R-2	44	46	179.86	Unit II	Mudstone	AFD	20	100	18.3	−40.1	4.5	R
C0012A-15R-2	49	51	179.91	Unit II	Mudstone	AFD	15	60	−158.6	−44.8	1.4	R
C0012A-16R-1	34	36	187.85	Unit II	Mudstone	AFD	−	−	67.7	−30	0	R
C0012A-16R-1	43	45	187.94	Unit II	Mudstone	AFD	15	60	−12.7	66	4.9	R
C0012A-16R-2	11	13	188.54	Unit II	Mudstone	AFD	20	80	−70	−62.3	2.8	R
C0012A-16R-2	17	19	188.60	Unit II	Mudstone	AFD	15	60	−147.6	−64.7	5.9	R
C0012A-16R-2	54	56	188.97	Unit II	Mudstone	AFD	10	60	−150.1	−61.8	3.8	R
C0012A-16R-4	13	15	189.45	Unit II	Mudstone	AFD	10	60	104.2	−31.8	5.9	R
C0012A-16R-5	6	8	190.80	Unit II	Mudstone	AFD	10	40	45.8	64.5	6.1	N
C0012A-16R-5	44	46	191.18	Unit II	Mudstone	AFD	10	60	−170.2	54.2	5.2	N
C0012A-17R-1	32	34	197.33	Unit II	Mudstone	AFD	20	60	−1.8	−35.4	3.7	R
C0012A-17R-1	38	40	197.39	Unit II	Mudstone	AFD	20	80	208.7	−62.6	5.8	R
C0012A-17R-1	105	107	198.06	Unit II	Mudstone	AFD	15	60	−111	−58.8	2.7	R
C0012A-17R-2	6	8	198.27	Unit II	Mudstone	AFD	20	80	13.2	−61.2	2.4	R
C0012A-17R-2	13	15	198.34	Unit II	Mudstone	AFD	20	60	−156.3	−65.3	2.6	R

(*Continued*)

Table 2. *Continued*

Sample			Sample centre depth (cm)	Lithological Unit	Lithology	AFD/ThD	Linear fitting (preliminary)					Polarity
Hole, core, section	Top (cm)	Bottom (cm)					Min	Max	Dec	Inc	MAD	
C0012A-17R-2	42	44	198.63	Unit II	Mudstone	AFD	20	80	63.6	−47.9	4.3	R
C0012A-17R-4	34	36	199.45	Unit II	Mudstone	AFD	30	80	30	−12.3	6.1	R
C0012A-17R-4	54	56	199.66	Unit II	Mudstone	AFD	20	60	249	−62.2	7.2	R
C0012A-18R-2	18	20	208.09	Unit II	Mudstone	AFD	15	60	−122.3	−63.2	14.1	R
C0012A-18R-CC	12	14	208.64	Unit II	Mudstone	AFD	40	60	−142.4	−29.8	0	R
C0012A-19R-1	14	16	215.85	Unit II	Mudstone	AFD	10	60	−154.3	65.7	7.1	R
C0012A-19R-1	50	52	216.21	Unit II	Mudstone	AFD	40	70	133.9	20.8	11.1	R
C0012A-19R-1	73	75	216.44	Unit II	Mudstone	AFD	10	60	86.1	67.5	4.3	N
C0012A-19R-1	126	128	216.97	Unit II	Mudstone	AFD	20	70	109.3	59.5	3.3	N
C0012A-19R-3	3	5	217.62	Unit II	Mudstone	AFD	10	60	84.7	60.7	8.1	N
C0012A-19R-3	46	48	218.05	Unit II	Mudstone	AFD	25	80	−48.4	43	3.5	N
C0012A-19R-3	67	69	218.26	Unit II	Mudstone	AFD	20	70	257.2	56.1	1.3	N
C0012A-19R-3	90	92	218.49	Unit II	Mudstone	AFD	10	80	256.8	56.8	2.8	N
C0012A-19R-3	131	133	218.90	Unit II	Mudstone	AFD	10	60	94.7	60.7	2.6	N
C0012A-19R-4	66	68	219.65	Unit II	Mudstone	AFD	20	70	103.3	65.4	1.6	N
C0012A-19R-4	98	100	219.97	Unit II	Mudstone	AFD	20	80	303.2	51.9	2.2	N
C0012A-19R-4	126	128	220.25	Unit III	Mudstone	AFD	10	60	123.5	56.2	1.4	N
C0012A-19R-5	27	29	220.58	Unit III	Mudstone	AFD	15	60	24	55.6	1.3	N
C0012A-19R-CC	6	8	220.74	Unit III	Mudstone	AFD	20	80	214.3	56.3	2.6	N
C0012A-20R-1	13	15	225.34	Unit III	Mudstone	AFD	10	60	20.4	50.5	1.2	N
C0012A-20R-1	66	68	225.87	Unit III	Mudstone	AFD	20	80	180.6	60.4	1.9	N
C0012A-20R-1	90	92	226.11	Unit III	Mudstone	AFD	10	60	7.5	55.8	1.4	N
C0012A-20R-1	113	115	226.34	Unit III	Mudstone	AFD	20	80	33	45.5	3.6	N
C0012A-20R-1	128	130	226.49	Unit III	Mudstone	AFD	15	120	17.7	43.5	1.8	N
C0012A-20R-2	9	11	226.70	Unit III	Mudstone	AFD	10	60	−167.3	60	2	N
C0012A-20R-2	45	48	227.07	Unit III	Mudstone	AFD	12	100	53.6	52.6	1.7	N
C0012A-20R-2	56	58	227.17	Unit III	Mudstone	AFD	15	60	−136.2	50.4	1.8	N
C0012A-20R-2	67	69	227.28	Unit III	Mudstone	AFD	20	80	38.7	47.4	1.4	N
C0012A-20R-4	19	21	228.04	Unit III	Mudstone	AFD	10	60	−161.1	50.3	1.6	N
C0012A-20R-4	41	43	228.26	Unit III	Mudstone	AFD	20	50	249.6	47.3	4.4	N
C0012A-20R-4	57	59	228.42	Unit III	Mudstone	AFD	15	60	78.6	48.2	4.4	N
C0012A-20R-4	120	122	229.05	Unit III	Mudstone	AFD	15	100	315.5	45.9	3.8	N
C0012A-20R-4	134	136	229.19	Unit III	Mudstone	AFD	20	80	−38.7	46.3	4.1	N
C0012A-20R-4	134	136	229.19	Unit III	Mudstone	AFD	20	80	−36.5	45.8	4.8	N
C0012A-20R-4	134	136	229.19	Unit III	Mudstone	AFD	20	80	−33.9	47.7	2.5	N
C0012A-20R-5	4	6	229.29	Unit III	Mudstone	AFD	10	60	−151.7	39.4	1.3	N

C0012A-21R-1	37	39	235.08	Unit III	Mudstone	AFD	10	60	44.4	69.8	2.4	N
C0012A-21R-1	52	54	235.23	Unit III	Mudstone	AFD	20	70	−158.2	67.5	3.6	N
C0012A-21R-3	26	28	235.99	Unit III	Mudstone	AFD	10	30	30.2	44.8	8.6	N
C0012A-21R-3	111	113	236.84	Unit III	Mudstone	AFD	15	80	182.5	52.5	2.6	N
C0012A-21R-4	6	8	237.11	Unit III	Mudstone	AFD	30	70	270.7	51	9.6	N
C0012A-21R-4	29	31	237.34	Unit III	Mudstone	AFD	20	50	−91.5	63.9	2.7	N
C0012A-22R-1	40	42	244.61	Unit III	Mudstone	AFD	20	60	251.4	14.5	19.9	R
C0012A-22R-1	43	45	244.64	Unit III	Mudstone	AFD	10	60	39.1	−22	11.7	R
C0012A-22R-1	59	61	244.80	Unit III	Mudstone	AFD	10	40	−70.1	54.9	12.6	N
C0012A-22R-1	74	76	244.95	Unit III	Mudstone	AFD	10	40	173.6	58.1	6.2	N
C0012A-22R-1	106	108	245.27	Unit III	Mudstone	AFD	20	70	32.1	38.1	2.7	N
C0012A-22R-3	24	26	246.05	Unit III	Mudstone	AFD	10	30	−72.3	50.8	5.3	N
C0012A-22R-3	82	84	246.63	Unit III	Mudstone	AFD	10	60	37.1	−60.4	1.1	R
C0012A-22R-3	89	91	246.70	Unit III	Mudstone	AFD	20	80	−162.5	−49.7	1.8	R
C0012A-23R-1	21	23	253.92	Unit III	Mudstone	AFD	15	60	−84.1	−36.9	3.5	R
C0012A-23R-1	34	36	254.05	Unit III	Mudstone	AFD	20	70	114.1	−37.9	3.5	R
C0012A-23R-1	80	82	254.51	Unit III	Mudstone	AFD	30	80	104	−28.3	8.3	R
C0012A-23R-1	100	102	254.71	Unit III	Mudstone	AFD	30	60	166.4	−58.7	12.4	R
C0012A-23R-1	130	132	255.01	Unit III	Mudstone	AFD	15	80	283.6	64.7	3.1	N
C0012A-23R-2	9	11	255.17	Unit III	Mudstone	AFD	10	60	−107.5	66.1	2.2	N
C0012A-23R-2	60	62	255.68	Unit III	Mudstone	AFD	10	60	161.3	65.8	3.5	N
C0012A-23R-2	83	85	255.91	Unit III	Mudstone	AFD	20	70	218.4	65.4	3.9	N
C0012A-23R-4	7	9	256.62	Unit III	Mudstone	AFD	20	70	155.8	59.8	5.1	N
C0012A-23R-4	24	26	256.79	Unit III	Mudstone	AFD	10	60	−38.1	61.6	0.9	N
C0012A-23R-4	52	54	257.07	Unit III	Mudstone	AFD	20	80	−175.1	63.6	2.4	N
C0012A-23R-4	89	91	257.44	Unit III	Mudstone	AFD	10	60	4	60.7	2.8	N
C0012A-23R-4	120	122	257.75	Unit III	Mudstone	AFD	15	80	221.4	65.9	3.6	N
C0012A-23R-5	57	59	258.42	Unit III	Mudstone	AFD	20	70	−118.2	66.3	7.4	N
C0012A-23R-5	97	99	258.82	Unit III	Mudstone	AFD	15	60	58.2	−61.7	11.3	R
C0012A-23R-6	3	5	259.12	Unit III	Mudstone	AFD	20	70	−115.5	−51.8	4.1	R
C0012A-23R-6	9	11	259.18	Unit III	Mudstone	AFD	15	40	33.5	−59.4	3.4	R
C0012A-23R-6	61	63	259.70	Unit III	Mudstone	AFD	30	70	185.9	−36.8	16.4	R
C0012A-23R-6	72	74	259.81	Unit III	Mudstone	AFD	20	60	0.2	−43.8	11.9	R
C0012A-24R-1	8	10	263.29	Unit III	Mudstone	AFD	10	60	158.4	−57.4	3	R
C0012A-24R-1	24	26	263.45	Unit III	Mudstone	AFD	30	70	349.4	44.9	12.4	R
C0012A-24R-3	3	5	264.32	Unit III	Mudstone	AFD	30	60	−78	−59.4	12.4	R
C0012A-24R-3	13	15	264.42	Unit III	Mudstone	AFD	25	80	90.3	−32.2	9.3	R
C0012A-24R-3	57	59	264.86	Unit III	Mudstone	AFD	20	50	56	−60.1	3.7	R
C0012A-24R-3	80	82	265.09	Unit III	Mudstone	AFD	15	60	−62.3	−52.2	4.5	R
C0012A-24R-3	105	107	265.34	Unit III	Mudstone	AFD	20	40	74.7	−44.8	5.5	R

(*Continued*)

Table 2. *Continued*

Sample			Sample centre depth (cm)	Lithological Unit	Lithology	AFD/ThD	Linear fitting (preliminary)					Polarity
Hole, core, section	Top (cm)	Bottom (cm)					Min	Max	Dec	Inc	MAD	
C0012A-24R-4	4	6	265.74	Unit III	Mudstone	AFD	15	60	−140.7	−50.5	3.6	R
C0012A-24R-4	25	27	265.95	Unit III	Mudstone	AFD	20	80	47.5	−49.5	2.3	R
C0012A-24R-4	43	45	266.13	Unit III	Mudstone	AFD	30	70	9.2	−62.7	4.6	R
C0012A-24R-4	61	63	266.31	Unit III	Mudstone	AFD	15	60	−157.2	−65.2	1.7	R
C0012A-24R-4	122	124	266.92	Unit III	Mudstone	AFD	15	80	47	−57.5	1.7	R
C0012A-24R-5	4	6	267.09	Unit III	Mudstone	AFD	30	70	86.8	−63.8	4.2	R
C0012A-24R-5	16	18	267.21	Unit III	Mudstone	AFD	10	60	−89.8	−56.6	1.9	R
C0012A-24R-5	29	31	267.34	Unit III	Mudstone	AFD	15	80	65.1	−57	3.8	R
C0012A-24R-5	43	45	267.48	Unit III	Mudstone	AFD	20	80	66.4	−46.7	5.7	R
C0012A-24R-5	127	129	268.32	Unit III	Mudstone	AFD	10	60	117.5	−62	2.9	R
C0012A-24R-6	25	27	268.71	Unit III	Mudstone	AFD	30	60	22	−33	11.8	R
C0012A-25R-1	65	67	273.36	Unit III	Mudstone	AFD	15	60	−139.2	−49.5	11.9	R
C0012A-25R-3	5	7	274.57	Unit III	Mudstone	AFD	30	50	248.6	−52.5	9.3	R
C0012A-25R-4	1	3	275.58	Unit III	Mudstone	AFD	30	50	61.3	−47.7	10.2	R
C0012A-25R-4	39	41	275.96	Unit III	Mudstone	AFD	30	60	78.3	−57.4	18.9	R
C0012A-25R-4	78	80	276.35	Unit III	Mudstone	AFD	20	60	154.3	−20.2	13.3	R
C0012A-25R-5	90	92	277.68	Unit III	Mudstone	AFD	20	50	269.3	−67.2	4.5	R
C0012A-25R-5	110	112	277.88	Unit III	Mudstone	AFD	20	40	139.6	−22.4	13.8	R
C0012A-25R-6	17	19	278.31	Unit III	Mudstone	AFD	15	40	67.4	−43	18.1	R
C0012A-25R-6	18	20	278.32	Unit III	Mudstone	AFD	25	80	−128.2	−22	13.1	R
C0012A-25R-7	3	5	278.95	Unit III	Mudstone	AFD	10	60	168.2	−57.4	9.3	R
C0012A-25R-7	11	13	279.03	Unit III	Mudstone	AFD	20	80	348	−53.8	12.3	R
C0012A-26R-1	10	12	282.31	Unit III	Mudstone	AFD	30	60	−29.9	−41.1	8.6	R
C0012A-27R-3	2	4	292.40	Unit III	Mudstone	AFD	10	30	93.7	48.4	11.2	R
C0012A-27R-3	57	59	292.95	Unit III	Mudstone	AFD	20	40	23	−65.6	14.9	R
C0012A-28R-1	57	59	301.78	Unit III	Mudstone	AFD	15	60	70.7	69.2	3.1	N
C0012A-28R-1	77	79	301.98	Unit III	Mudstone	AFD	15	40	−140.1	54.8	7.4	N
C0012A-28R-3	13	15	302.75	Unit III	Mudstone	AFD	20	70	311.2	62.7	5.9	N
C0012A-28R-3	25	27	302.87	Unit III	Mudstone	AFD	30	60	−78	42.7	1.5	N
C0012A-28R-3	33	35	302.95	Unit III	Mudstone	AFD	20	80	−6.2	64.9	3.3	N
C0012A-28R-3	95	97	303.57	Unit III	Mudstone	AFD	10	80	127.3	51.4	3.6	N
C0012A-28R-3	119	121	303.81	Unit III	Mudstone	AFD	15	60	116	68.1	4.4	N
C0012A-28R-4	2	4	304.05	Unit III	Mudstone	AFD	20	60	−140.8	71.2	9	N
C0012A-28R-4	35	37	304.38	Unit III	Mudstone	AFD	15	60	−41.2	61.5	12.2	N
C0012A-29R-1	130	132	312.01	Unit III	Mudstone	AFD	40	60	−95.4	−57	17.4	R
C0012A-31R-4	17	19	331.25	Unit III	Mudstone	AFD	20	60	−23.2	60.3	8.2	N

C0012A-31R-4	73	75	331.81	Unit III	Mudstone	AFD	20	60	122.7	63.6	11.3	N
C0012A-31R-4	97	99	332.05	Unit III	Mudstone	AFD	15	60	36.2	68.5	2.6	N
C0012A-31R-5	51	53	332.91	Unit III	Mudstone	AFD	40	80	69.8	52.4	11.1	N
C0012A-31R-5	84	86	333.24	Unit III	Mudstone	AFD	10	60	139.7	59	19.3	R
C0012A-32R-1	57	59	339.78	Unit III	Mudstone	AFD	30	80	274.6	−25.7	14.6	R
C0012A-32R-3	63	65	341.46	Unit III	Mudstone	AFD	–	–	322.6	−61.5	0	R
C0012A-32R-3	93	95	341.76	Unit III	Mudstone	AFD	30	80	330.6	−13.9	17.7	R
C0012A-32R-3	110	112	341.93	Unit IV	Mudstone	AFD	20	60	82.7	−13.9	17.5	R
C0012A-32R-4	22	24	342.32	Unit IV	Mudstone	AFD	30	60	253.9	10.7	16.8	R
C0012A-32R-4	58	60	342.68	Unit IV	Mudstone	AFD	40	60	−141.5	−62.5	0.2	R
C0012A-32R-4	73	75	342.83	Unit IV	Mudstone	AFD	–	–	95.8	−11.9	0	R
C0012A-33R-1	11	13	348.82	Unit IV	Mudstone	AFD	20	30	146	53.3	6.5	N
C0012A-33R-1	31	33	349.02	Unit IV	Mudstone	AFD	30	60	24.7	63.2	14	N
C0012A-33R-3	1	3	349.75	Unit IV	Mudstone	AFD	20	70	75	64.6	4.2	N
C0012A-33R-3	34	36	350.08	Unit IV	Mudstone	AFD	15	80	72.7	60.4	9	N
C0012A-33R-3	99	101	350.73	Unit IV	Mudstone	AFD	30	60	−138.3	57.5	7.8	N
C0012A-34R-1	52	54	357.53	Unit IV	Mudstone	AFD	15	60	51.4	38.4	6.7	N
C0012A-34R-2	9	11	358.44	Unit IV	Mudstone	AFD	20	70	43.2	52.4	4.9	N
C0012A-34R-2	42	44	358.77	Unit IV	Mudstone	AFD	15	80	−110.8	60.7	6.6	N
C0012A-34R-2	89	91	359.24	Unit IV	Mudstone	AFD	30	60	−92.5	59.9	19.8	N
C0012A-34R-2	104	106	359.39	Unit IV	Mudstone	AFD	30	70	186.4	65.1	6.7	N
C0012A-34R-4	24	26	360.23	Unit IV	Mudstone	AFD	15	60	152.3	56.7	6.1	N
C0012A-34R-4	33	35	360.32	Unit IV	Mudstone	AFD	10	80	56.8	58.4	3.5	N
C0012A-36R-3	38	40	376.92	Unit IV	Mudstone	AFD	20	40	294.6	55.8	16.5	R
C0012A-37R-1	14	16	385.65	Unit IV	Mudstone	AFD	15	60	−7.1	42.1	6	N
C0012A-37R-1	22	24	385.73	Unit IV	Mudstone	AFD	15	80	−68.9	58.8	6.2	N
C0012A-37R-1	35	37	385.86	Unit IV	Mudstone	AFD	15	120	299.9	62.2	2.3	N
C0012A-37R-1	52	54	386.03	Unit IV	Mudstone	AFD	15	60	−159.2	58.2	5.6	N
C0012A-38R-1	3	5	395.04	Unit IV	Mudstone	AFD	30	60	−47.1	67.6	10.7	N
C0012A-38R-1	77	79	395.78	Unit IV	Mudstone	AFD	30	50	340.4	48.7	8.4	N
C0012A-38R-2	19	21	396.31	Unit IV	Mudstone	AFD	15	30	79.8	38.6	9.5	N
C0012A-39R-1	107	109	405.58	Unit IV	Mudstone	AFD	20	60	−168	34.3	2	N
C0012A-39R-2	2	4	405.89	Unit IV	Mudstone	AFD	15	60	114.7	36	3.6	N
C0012A-39R-2	95	97	406.82	Unit IV	Mudstone	AFD	20	40	89.3	50.5	1.6	N
C0012A-39R-4	1	3	407.41	Unit IV	Mudstone	AFD	30	60	14.6	17.1	1.8	N
C0012A-39R-4	8	10	407.47	Unit IV	Mudstone	AFD	20	50	−158.6	66.4	5.7	N
C0012A-39R-4	28	30	407.68	Unit IV	Mudstone	AFD	20	40	180.5	58.1	8.8	N
C0012A-39R-4	36	38	407.76	Unit IV	Mudstone	AFD	10	40	182.5	61.1	7.5	N
C0012A-39R-4	44	46	407.84	Unit IV	Mudstone	AFD	15	50	−170.5	62.8	3.9	N
C0012A-39R-4	52	54	407.92	Unit IV	Mudstone	AFD	15	40	−29.6	53.7	19.7	N

(Continued)

Table 2. *Continued*

Sample			Sample centre depth (cm)	Lithological Unit	Lithology	AFD/ThD	Linear fitting (preliminary)					Polarity
Hole, core, section	Top (cm)	Bottom (cm)					Min	Max	Dec	Inc	MAD	
C0012A-40R-3	57	59	417.30	Unit IV	Mudstone	AFD	30	90	179.9	64	3.5	N
C0012A-40R-5	98	100	419.09	Unit V	Mudstone	AFD	15	40	13.9	67.2	4.8	N
C0012A-40R-6	3	5	419.35	Unit V	Mudstone	AFD	15	70	28.3	65.5	5.7	N
C0012A-41R-1	17	19	423.68	Unit V	Mudstone	AFD	20	60	97	37.5	5.9	N
C0012A-41R-1	56	58	424.07	Unit V	Mudstone	AFD	10	60	227	67.7	5.8	N
C0012A-41R-2	29	31	425.18	Unit V	Mudstone	AFD	20	60	269.5	59.3	4.9	N
C0012A-41R-2	48	50	425.37	Unit V	Mudstone	AFD	20	50	259.7	68.9	2.6	N
C0012A-41R-4	61	63	427.13	Unit V	Mudstone	AFD	15	60	−41.2	52.7	2.5	N
C0012A-41R-5	5	7	428.00	Unit V	Mudstone	AFD	20	60	−90.9	−20.9	17.5	R
C0012A-41R-5	19	21	428.14	Unit V	Mudstone	AFD	30	60	104.9	−35.4	7.6	R
C0012A-41R-5	53	55	428.48	Unit V	Mudstone	AFD	20	80	−59.1	−49.2	7.7	R
C0012A-41R-5	71	73	428.66	Unit V	Mudstone	AFD	20	60	126.5	−61.9	4.1	R
C0012A-41R-5	94	96	428.89	Unit V	Mudstone	AFD	30	70	47	−42.4	7.2	R
C0012A-41R-6	5	7	429.26	Unit V	Mudstone	AFD	20	60	266	−46.2	5.4	R
C0012A-41R-6	31	33	429.52	Unit V	Mudstone	AFD	20	60	159.8	−32	8.2	R
C0012A-42R-1	4	6	433.05	Unit V	Mudstone	AFD	40	70	346.3	−19.8	6.5	R
C0012A-42R-1	11	13	433.12	Unit V	Mudstone	AFD	30	60	110.2	−15.5	10	R
C0012A-42R-1	57	59	433.58	Unit V	Mudstone	AFD	20	60	79	−31.8	12.7	R
C0012A-42R-3	103	105	435.70	Unit V	Mudstone	AFD	30	60	168.3	69.6	12.2	N
C0012A-42R-4	2	4	435.98	Unit V	Mudstone	AFD	20	80	120.5	65.7	7.3	N
C0012A-43R-1	19	21	442.70	Unit V	Mudstone	AFD	15	80	−21.4	58.4	4.4	N
C0012A-43R-1	62	64	443.13	Unit V	Mudstone	AFD	20	70	178.5	65.7	4	N
C0012A-43R-1	134	136	443.85	Unit V	Mudstone	AFD	30	60	223.1	38.5	11.5	R
C0012A-43R-2	57	59	444.48	Unit V	Mudstone	AFD	15	80	0.5	62	6.4	N
C0012A-43R-2	73	75	444.64	Unit V	Mudstone	AFD	30	80	326.5	57	5.6	N
C0012A-43R-2	113	115	445.04	Unit V	Sand	AFD	20	60	178.6	54.7	15.9	N
C0012A-43R-2	135	137	445.26	Unit V	Mudstone	AFD	20	80	12.4	61.3	2.3	N
C0012A-43R-3	1	3	445.31	Unit V	Mudstone	AFD	25	80	48.5	56	4.7	N
C0012A-43R-3	20	22	445.50	Unit V	Mudstone	AFD	15	80	43	60	5.7	N

C0012A-43R-3	34	36	445.64	Unit V	Sand	AFD	15	60	−117.8	66.6	11.5	N
C0012A-43R-5	8	10	446.51	Unit V	Volcaniclastic Sand	AFD	20	60	9.1	69.4	3.6	N
C0012A-43R-5	30	32	446.73	Unit V	Mudstone	AFD	10	60	239.8	67.1	2.6	N
C0012A-43R-5	77	79	447.20	Unit V	Mudstone	AFD	15	80	−99.1	53.9	5.1	N
C0012A-44R-1	53	55	452.54	Unit V	Mudstone	AFD	15	40	−106.1	63.1	3.3	N
C0012A-44R-1	59	61	452.60	Unit V	Mudstone	AFD	25	80	76	63.7	3.5	N
C0012A-44R-3	10	12	453.81	Unit V	Mudstone	AFD	30	60	35.1	48.9	10.9	N
C0012A-44R-3	35	37	454.06	Unit V	Mudstone	AFD	20	80	326.3	60.3	2.9	N
C0012A-44R-3	61	63	454.32	Unit V	Mudstone	AFD	20	80	118.3	57.8	2.9	N
C0012A-44R-3	122	124	454.93	Unit V	Mudstone	AFD	20	70	311.7	42.6	1.8	N
C0012A-44R-3	127	129	454.98	Unit V	Mudstone	AFD	15	60	137.2	45.3	1.9	N
C0012A-45R-1	4	6	461.55	Unit V	Sand	AFD	20	60	72.8	48.8	6.1	N
C0012A-45R-1	50	52	462.01	Unit V	Mudstone	AFD	30	80	252.9	68.4	6.3	N
C0012A-45R-1	107	109	462.58	Unit V	Mudstone	AFD	15	80	69.3	67.7	4.5	N
C0012A-45R-2	77	79	463.59	Unit V	Mudstone	AFD	20	80	353.6	69.5	3.2	N
C0012A-45R-2	96	98	463.78	Unit V	Sand	AFD	20	60	−119.2	69	3.6	N
C0012A-45R-4	1	3	464.81	Unit V	Sand	AFD	15	40	−135.2	48.8	5.3	N
C0012A-45R-4	62	64	465.42	Unit V	Sand	AFD	20	60	−45.4	66.6	4.8	N
C0012A-45R-CC	2	4	465.78	Unit V	Mudstone	AFD	30	120	208.3	68.4	3.7	N
C0012A-46R-1	5	7	471.06	Unit V	Sand	AFD	20	60	38.9	65.9	10	N
C0012A-46R-1	29	31	471.30	Unit V	Mudstone	AFD	10	150	328.6	63.1	2.5	N
C0012A-46R-2	19	21	471.56	Unit V	Mudstone	AFD	20	70	−8.7	52.4	6.5	N
C0012A-46R-2	81	83	472.18	Unit V	Mudstone	AFD	30	90	211.8	68.4	6	N
C0012A-46R-3	71	73	472.65	Unit V	Mudstone	AFD	20	60	26.3	61.5	12.2	N
C0012A-46R-4	1	3	473.36	Unit V	Mudstone	AFD	20	90	79.9	60.6	3.4	N
C0012A-46R-4	10	12	473.45	Unit V	Sand	AFD	20	60	−104.8	55.2	6.8	N
C0012A-46R-CC	28	30	474.37	Unit V	Mudstone	AFD	30	70	−64.1	−55.8	15.5	R
C0012A-47R-3	21	23	481.34	Unit V	Mudstone	AFD	20	70	303.6	−67.9	11.8	R
C0012A-47R-3	39	41	481.52	Unit V	Mudstone	AFD	20	60	51	−67.7	3.1	R
C0012A-47R-3	46	48	481.59	Unit V	Mudstone	AFD	20	80	226.3	−44.6	9.1	R
C0012A-47R-CC	10	12	481.85	Unit V	Mudstone	AFD	40	70	−145.2	−59.1	5.1	R
C0012A-47R-CC	16	18	481.91	Unit V	Mudstone	AFD	15	60	157.1	−46.4	0.9	R
C0012A-48R-2	17	19	490.62	Unit V	Mudstone	AFD	20	60	−98.5	61	3.3	N

(*Continued*)

Table 2. *Continued*

Sample			Sample centre depth (cm)	Lithological Unit	Lithology	AFD/ThD	Linear fitting (preliminary)					Polarity
Hole, core, section	Top (cm)	Bottom (cm)					Min	Max	Dec	Inc	MAD	
C0012A-48R-2	41	43	490.94	Unit V	Mudstone	AFD	20	60	−71.7	51.5	0.8	N
C0012A-48R-2	75	77	491.28	Unit V	Mudstone	AFD	20	70	307.7	−16.3	10.1	R
C0012A-48R-CC	7	9	491.46	Unit V	Mudstone	AFD	10	80	288.8	−34	4.5	R
C0012A-49R-1	6	8	499.57	Unit V	Mudstone	AFD	10	60	−108.3	66.5	0.9	N
C0012A-49R-1	29	31	499.80	Unit V	Mudstone	AFD	25	70	−170.1	−37.8	4.9	R
C0012A-49R-3	29	31	500.62	Unit V	Mudstone	AFD	20	80	136.3	−48.5	8.2	R
C0012A-49R-3	45	47	500.78	Unit V	Mudstone	AFD	20	50	122	−14	12.3	R
C0012A-49R-3	60	62	500.93	Unit V	Mudstone	AFD	15	60	117.7	−45.6	1.7	R
C0012A-50R-3	8	10	509.80	Unit V	Mudstone	AFD	20	50	52.8	38.8	2.5	N
C0012A-50R-3	15	17	509.87	Unit V	Mudstone	AFD	20	70	37.9	39.4	3.6	N
C0012A-50R-3	21	23	509.93	Unit V	Mudstone	AFD	20	60	−138.2	21.6	1.2	N
C0012A-50R-3	27	29	509.99	Unit V	Mudstone	AFD	20	70	263.6	37.4	1.5	N
C0012A-50R-3	77	79	510.49	Unit V	Mudstone	AFD	20	60	129.1	32.7	0.7	N
C0012A-51R-1	5	7	518.56	Unit V	Mudstone	AFD	20	60	66	39	2.9	N
C0012A-52R-1	82	84	528.83	Unit V	Mudstone	AFD	40	80	29.5	57.8	4.5	N
C0012A-52R-3	2	4	529.57	Unit VI	Mudstone	AFD	20	60	22.7	51.5	4.7	N
C0012A-52R-3	27	29	529.82	Unit VI	Mudstone	AFD	30	80	189.1	60.5	7.2	N
C0012A-52R-3	55	57	530.10	Unit VI	Mudstone	AFD	40	130	70.7	67.5	7.5	N
C0012A-52R-3	61	63	530.16	Unit VI	Mudstone	AFD	20	50	−58.1	19.4	3.9	N
C0012A-52R-CC	4	6	530.28	Unit VI	Mudstone	AFD	30	60	135.9	65	6.1	N

Note: AFD/ThD stand for alternating field and thermal demagnetization experiment, respectively; Min and Max, minimum and maximum demagnetization fields and temperatures to define the ChRM; Dec and Inc, declination and inclination; MAD, maximum angular derivation; N, normal; R, reversed.

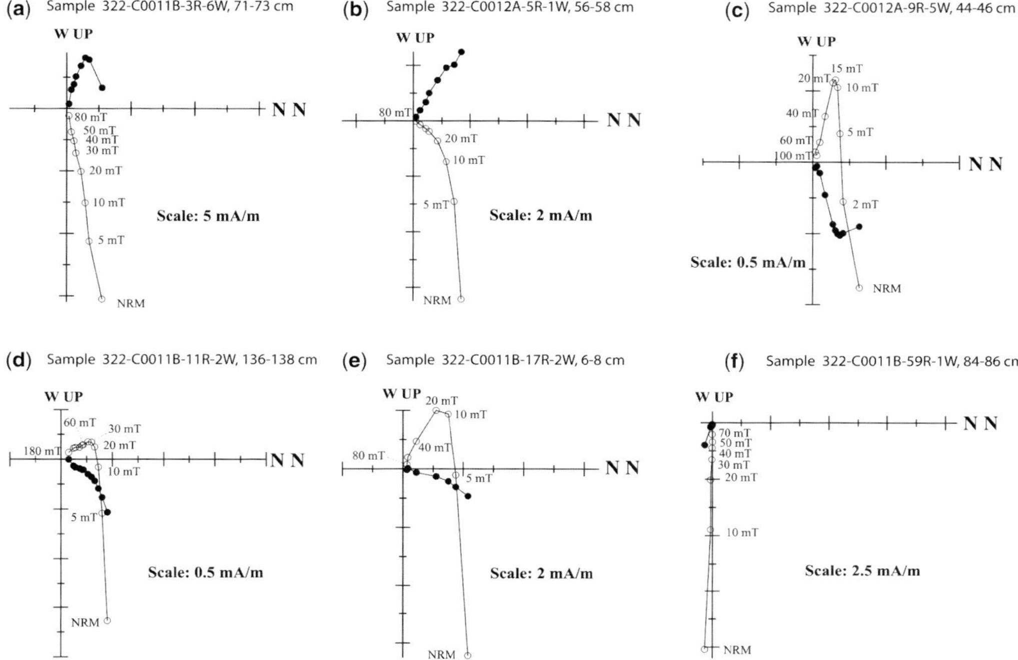

Fig. 6. Representative vector end point diagrams (Zijderveld 1967) showing the shore-based results of AF demagnetization for sedimentary rock samples from Exp. 322 drill sites with well-defined reverse and normal polarity ChRM magnetization, showing in all cases the removal of a normal component of magnetization due to drilling and isolation of a more stable component that is decaying towards the origin of the vector plot. Solid and open circles represent the projection of the magnetization vector endpoint on the horizontal and vertical planes, respectively. NRM, natural remanent magnetization.

Fig. 9b), and a hemipelagic mudstone from Lithostratigraphic Unit V (Sample 322-C0011B-59R-2W, 45–47 cm, Fig. 9c) display similar signatures, characterized by very rapid increase in IRM acquisition at low applied fields and reach saturation by 100 mT. This combination indicates magnetite

Table 3. *Results of reversal tests using inclination-only statistics on cores from Site C0012s*

Unit	Polarity	N	I	α_{95}	k	R	λ_o	λ_c	R-Test results
Site C0012A									
I	N	41	58.04	5.0	20.6	39.060	4.4	6.4	B class
	R	48	−53.7	3.7	31.3	46.500			
II	N	56	55.9	5.7	12.2	51.500	7.8	8.4	B class
	R	42	−48.1	6.5	12.3	38.680			
III	N	81	56.1	3.6	20.1	77.020	6.2	6.5	B class
	R	58	−49.9	6.5	14.5	54.060			
IV	N	53	55.8	6.0	11.5	48.480	26.4	15	Negative
	R	12	−29.4	13.9	10.8	10.980			
V	N	82	59.9	3.9	17.0	77.230	9.7	16.3	Negative
	R	23	−43.6	9.1	12.0	21.170			

Notes: N, normal; R, reversed; N, number of samples exhibiting the ChRM inclinations and used in statistical analysis; I, estimated mean inclination using the method of Arason & Levi (2010); α_{95}, radius of circle of 95% confidence about the mean inclination in degrees; k, Fisher (1953) precision parameter for the mean inclination; R, length of resultant vector; λ_o/λ_c, observed/critical (at the 95% confidence level) angles between the normal and reversed mean inclinations (McFadden & McElhinny 1990); if $\lambda_o > \lambda_c$, then the hypothesis of a common mean direction may be rejected at the 95% confidence level. Otherwise, the reversal test is considered to be positive.

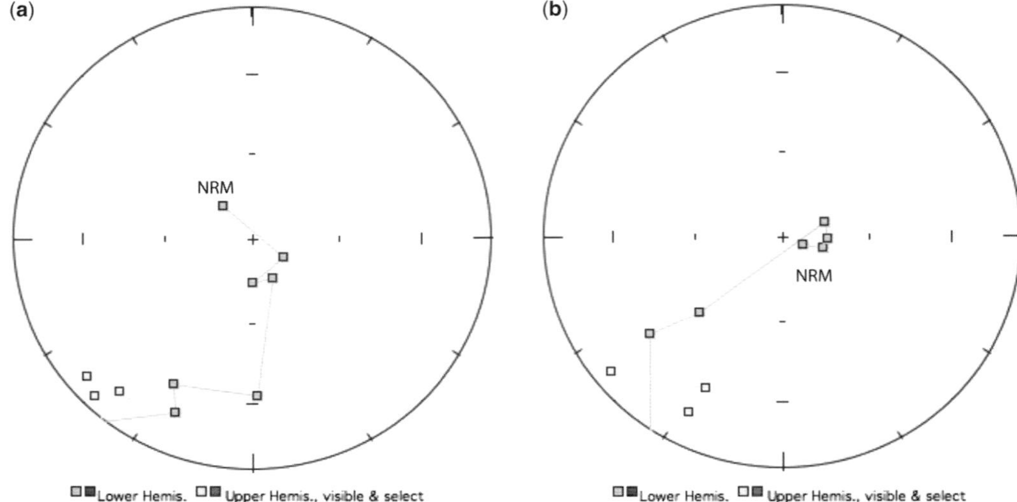

Fig. 7. Representative remagnetization circles (McFadden & McElhinny 1988) for Samples 322-C0011B-44R-1W, 65–67 cm (**a**) and 322-C0011B-43R-1W, 15–17 cm (**b**), showing remagnetization great circles with reversed polarity (negative inclination) ChRM.

behaviour. In mud Sample 332-C0011B-3R-5W, 134–136 cm of Unit IIa, on the other hand, the saturation IRM is not reached at 100 mT applied field, indicating that this sample may have higher coercivity mineral such as hematite as the magnetic carrier (Fig. 9d).

Hysteresis loop parameters

Hysteresis loops were measured on representative samples to estimate the domain structure of magnetic minerals. The samples analysed in this study indicate that rock samples from Exp. 322 sites show the predominance of the pseudosingle domain grain (e.g. Dunlop 2002), and some multidomain grains (Table 4). Several samples also show constricted hysteresis loops (or 'wasp-waisted' loops), indicating that either a combination of magnetic minerals with strongly contrasting coercivities (such as magnetite and hematite, Tauxe *et al.* 1996) or a mixture of superparamagnetic and single-domain grains of the same phase must coexist (Dunlop & Özdemir 1997). We plot examples of room-temperature hysteresis loops for representative samples from Sites C0011 and C0012 in Figures 10 and 11, respectively.

Low-temperature properties

Low-temperature measurements were made on representative samples to help characterize the magnetic minerals and understand their rock magnetic properties. As shown in Figure 12, the low-temperature curves of saturation isothermal remanent magnetization in both zero-field cooling and field cooling display a variety of features. These include an unblocking temperature in the vicinity of 40–50 K and a decrease in remanence in the 100–120 K range. The latter is most likely caused by the Verwey transition (Verwey *et al.* 1947) whereas the former could be caused by various reasons, such as the relaxation of superparamagnetic magnetite particles (Moskowitz *et al.* 1993), the unpinning of domain walls of coarse-grain titanomagnetite (Moskowitz *et al.* 1998) or the unblocking of magnetization of ilmenite grains (Dunlop & Özdemir 1997). Figure 12a, b shows cooling and warming curves for Samples 322-C0011B-7R-5W, 65–67 cm and 322-C0012A-13R-4, 22–24 cm, respectively. Remanence is lost at *c.* 120–130 K as the samples both cool and warm through the Verwey transition, suggesting the presence of magnetite and titanomagnetite in the samples. On the other hand, no obvious Verwey transition is observed for Samples 322-C0011B-32-R-4W, 43–45 cm and 322-C0011B-58R-4W, 138–140 cm during cooling to 6 K. Upon warming from 6 K, however, a tiny bend in remanence occurs near 120 K for both samples (Fig. 12c & d, Table 5). According to Moskowitz *et al.* (1998), titanomagnetites with high Ti replacement level ($x > 0.04$) could not exhibit the Verwey temperature. Alternatively, there may be a dominant shape anisotropy in the sample.

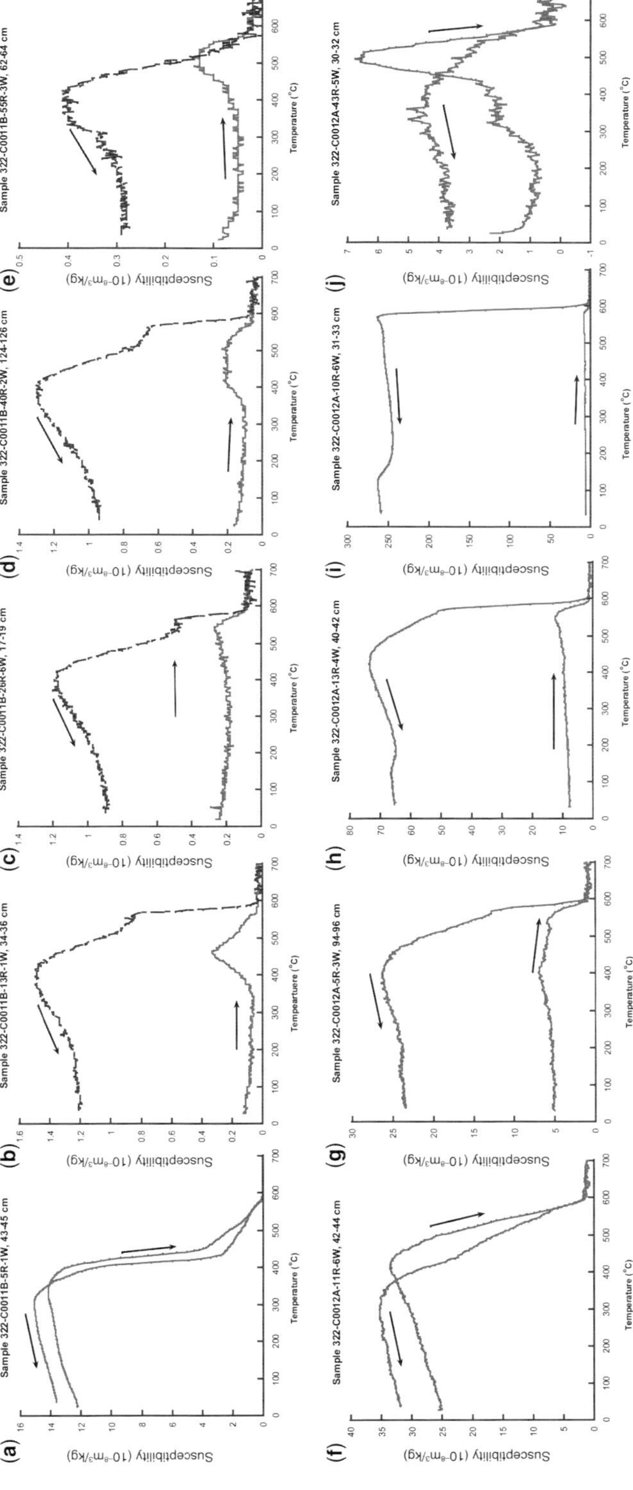

Fig. 8. Typical thermomagnetic curves for representative samples. (**a**) Sample 322-C0011B-5R-1W, 43–45 cm; (**b**) Sample 322-C0011B-13R-1W, 34–36 cm; (**c**) Sample 322-C0011B-26R-6W, 17–19 cm; (**d**) Sample 322-C0011B-40R-2W, 124–126 cm; (**e**) Sample 322-C0011B-55R-3W, 62–64 cm. (**f**) Sample 322-C0012A-11R-6W, 42–44 cm; (**g**) Sample 322-C0012A-5R-3W, 94–96 cm; (**h**) Sample 322-C0012A-13R-4W, 40–42 cm; (**i**) Sample 322-C0012A-10R-6W, 31–33 cm; and (**j**) Sample 322-C0012A-43R-5W, 30–32 cm. The directions of arrows indicate heating and cooling curves.

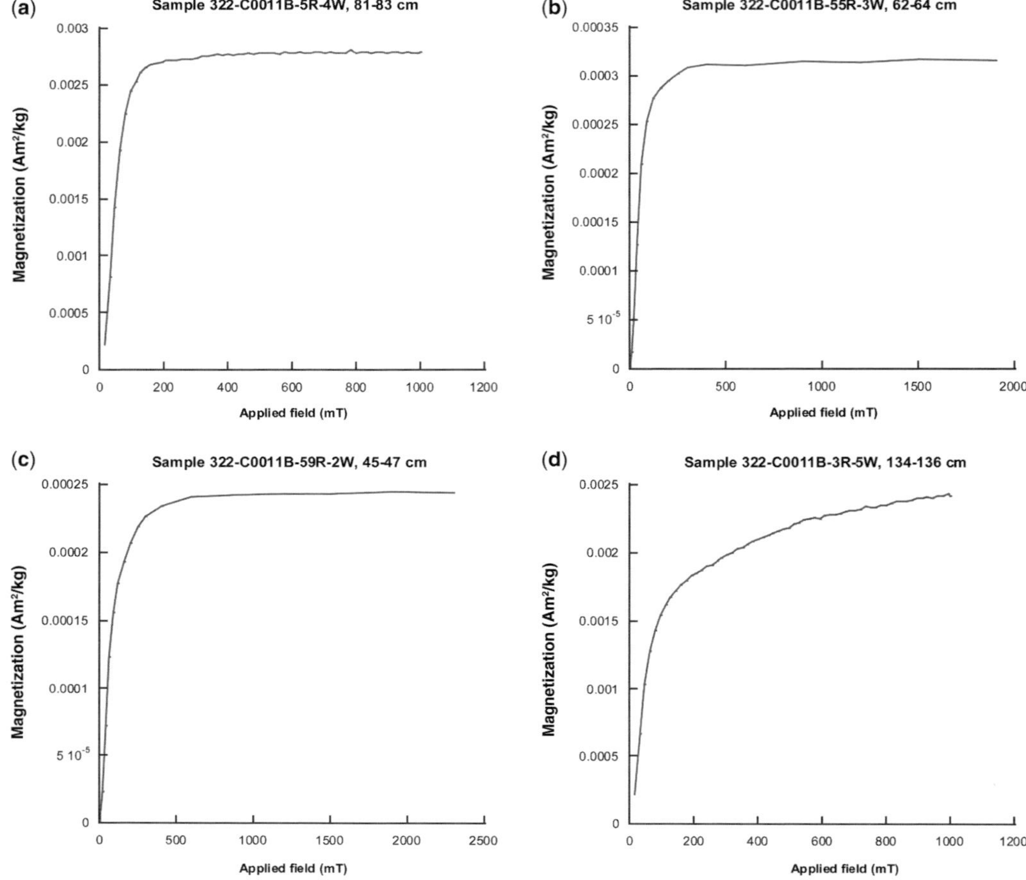

Fig. 9. Isothermal remanent magnetization acquisition curves for (**a**) Sample 322-C0011B-5R-4W, 81–83 cm; (**b**) Sample 322-C0011B-55R-3W, 62–64 cm; (**c**) Sample 322-C0011B-3R-5W, 134–136 cm; and (**d**) Sample 322-C0011B-3R-5W, 134–136 cm.

Overview of palaeomagnetic and rock magnetic data and stability tests

Stable magnetization components are observed throughout the majority of the recovered core in Holes C0011B and C0012A, after removal of a low-stability drilling-induced remanence. The inclination values of ChRM are moderate downward or upward, consistent with the expected inclination for the site and with the hypothesis that no significant inclination shallowing owing to compaction has occurred in these sediments. Similar inclination values in different lithologies and good agreement between the shipboard data and discrete sample inclinations from the same intervals in shore-based studies are observed. In addition, the mean normal and reversed directions for Units I–III of Site C0012 are not significantly different from antipodal

(Table 3), affording a positive reversal test and suggesting that the sedimentary cores from Nankai Trough have recorded the original depositional remanent magnetization acquired during the sediment deposition.

At each site, several relatively well-defined polarity intervals were identified in downhole magnetostratigraphic records. Using biostratigraphic and fission-track data (see next section for details), we were able to correlate patterns of the magnetic polarity interval recorded in the sediments against the standard geomagnetic reversal timescale. The fact that identified patterns of magnetic reversals can be matched with the established polarity time scale indicates that the ChRM is free of the secondary component of magnetization.

The rock magnetic results of core samples from Holes C0011B and C0012A corroborate the

Table 4. *Summary of rock magnetic properties of rock samples considered for this study*

Core, section, interval (cm)	Depth (m CSF)	T_c (°C) [T_{c1}, T_{c2}] (κ bridge)	H_c (mT)	H_{cr} (mT)	H_{cr}/H_c	J_r (mA m^2 kg^{-1})	J_s (mA m^2 kg^{-1})	J_r/J_s
C0011B-3R-5, 134–136	364.09		15.1	41.3	2.74	4.06	17.72	0.23
C0011B-4R-4, 3–5	367.29		6.5	29.1	4.48	15.46	189.78	0.08
C0011B-5R-1, 43–45	374.95	4 51 581						
C0011B-5R-4, 81–83	377.79		13.1	35.4	2.7	6.22	39.64	0.16
C0011B-13R-1, 34–36	450.86	575						
C0011B-19R-4, 113–115	494.07		17.4	44.7	2.57	22.2	80.4	0.28
C0011B-26R-1, 86.5–88.5	549.89		11.2	34.5	3.08	3.02	27.2	0.11
C0011B-26R-6, 17–19	555.28	588						
C0011B-37R-4, 31–33	646.84		2.3	63.5	27.61	0.78	21.2	0.04
C0011B-40R-2, 124–126	674.66	574						
C0011B-55R-3, 62–64	780.74	585						
C0012A-5R-3, 94.0–96.0	90.84	594	13	–	–	7.08	52.4	0.14
C0012A-8R-4, 5.0–7.0	115.79	581	12.4	–	–	3.98	27.6	0.14
C0012A-10R-6, 31.0–33.0	135.76		13.4	–	–	2.16	17.16	0.13
C0012A-11R-6, 42.0–44.0	145.94	590	7.4	–	–	31.8	345.8	0.09
C0012A-13R-4, 40.0–42.0	161.95	591	13.4	–	–	9.56	58.6	0.16
C0012A-17R-4, 29.0–31.0	199.41		13.8	–	–	2.42	18.5	0.13
C0012A-22R-3, 129.0–131.0	247.1		11.3	–	–	3.96	29.4	0.13
C0012A-34R-2, 9.0–11.0	358.44		0.57	–	–	0.28	1.66	0.17
C0012A-43R-5, 30.0–32.0	446.73	586	11.9	–	–	0.82	5.9	0.14
C0012A-45R-4, 14.0–16.0	464.94		19	–	–	2.26	9.5	0.24

Notes: T_c, Curie temperature (T_{c1}, T_{c2}, low and high Curie temperature, respectively).

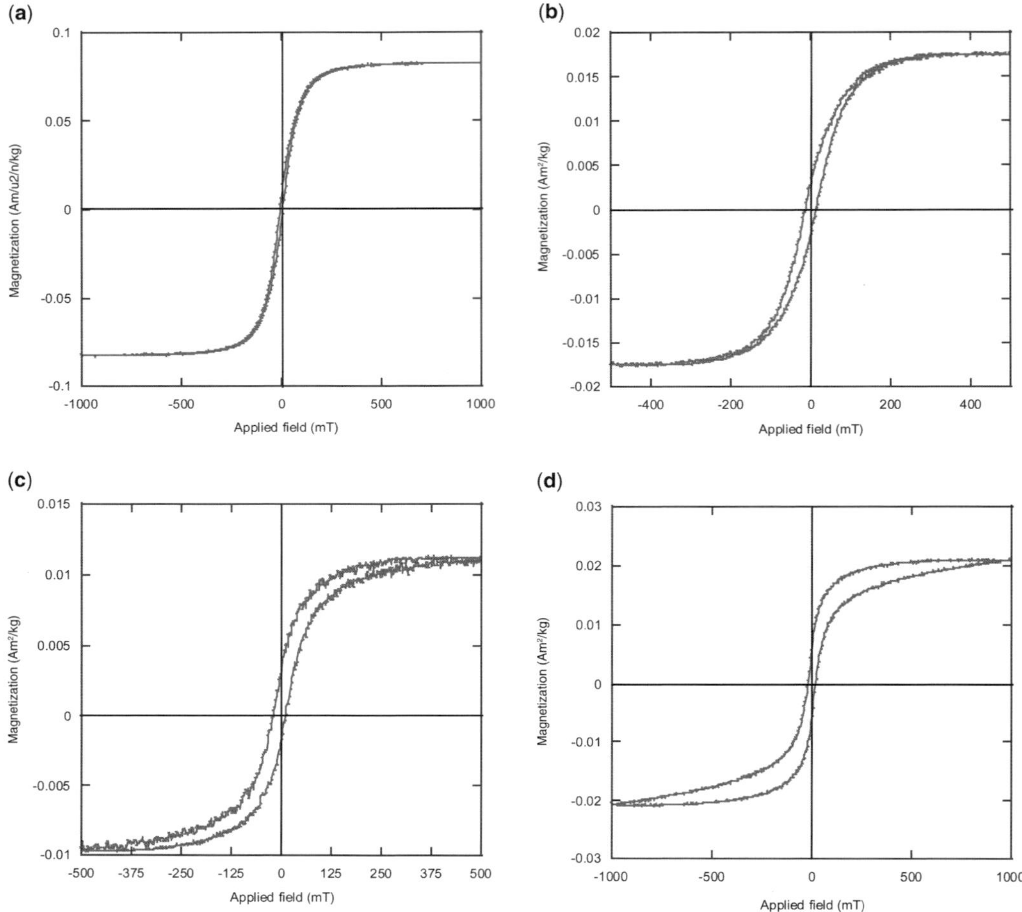

Fig. 10. Diagrams of room temperature hysteresis loops for representative rock samples in Site C0011. (**a**) Sample 322-C0011B-4R-4W, 3–5 cm; (**b**) Sample 322-C0011B-5R-4W, 81–83 cm; (**c**) Sample 322-C0011B-3R-5W, 134–136 cm; (**d**) 'wasp-waisted' Sample 322-C0011B-19R-4W, 113–115 cm. Horizontal axis is applied field. Vertical axis is normalized magnetization (after correction for slope).

demagnetization behaviour and give important information about the origin of remanence and the magnetic minerals present in the cores (Tables 4 and 5). Samples with Ti-poor titanomagnetite exhibit a strong Verwey transition in the vicinity of 120 K and thermomagnetic curves exhibit little difference between heating and cooling of the samples. These results are in good agreement with the hysteresis ratios. Therefore, these rocks are most likely good palaeomagnetic recorders and probably have preserved original and stable magnetic remanences. Several samples have more than one Curie temperature, thus suggesting the presence of multiple magnetic phases. The thermomagnetic signature indicates the inversion of titanomaghemite to a strongly magnetized magnetite, as shown by the irreversible cooling curve, and low-temperature curves do not show the Verwey transition clearly. The hysteresis ratios for rocks in this group still fall in the pseudo-single-domain region. Chemical remanent magnetization owing to oxidization of titanomagnetite and inversion of titanomaghemite has been shown to parallel the original thermoremance (Johnson & Merrill 1974; Hall 1977; Özdemir & Dunlop 1985; Dunlop & Özdemir 1997), plus the stable mean inclinations identified from these samples, and we believe that these rocks also retain stable remanence magnetization suitable for magnetostratigraphic and age–depth model studies (Tables 1 & 2).

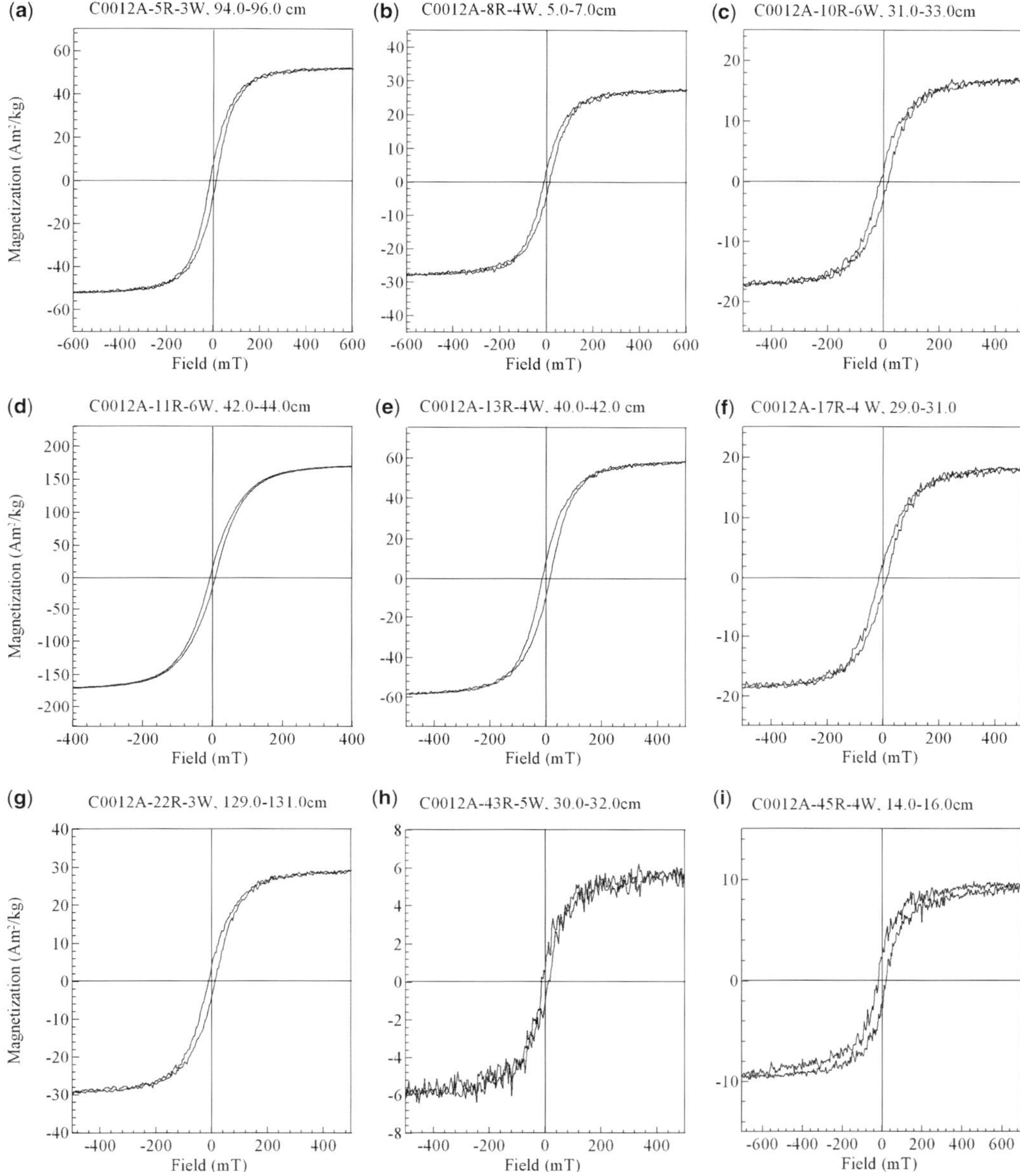

Fig. 11. Diagrams of room temperature hysteresis loops for representative rock samples in Site C0012. Horizontal axis is applied field. Vertical axis is normalized magnetization (after correction for slope).

Magnetic polarity sequence and integrated age–depth models

Magnetic polarity record for Hole C0011B

A number of clearly defined magnetic reversals can be discerned on the basis of changes in sign of inclinations of the cores from Hole C0011B. As with most applications of magnetostratigraphy, one of the greatest problems is correctly matching the observed sequence of magnetic polarity zones with the appropriate part of the GPTS.

One diagnostic feature in the palaeomagnetic data obtained for Hole C0011B is that some changes in magnetic polarity can be correlated with changes in biostratigraphic zonations. These

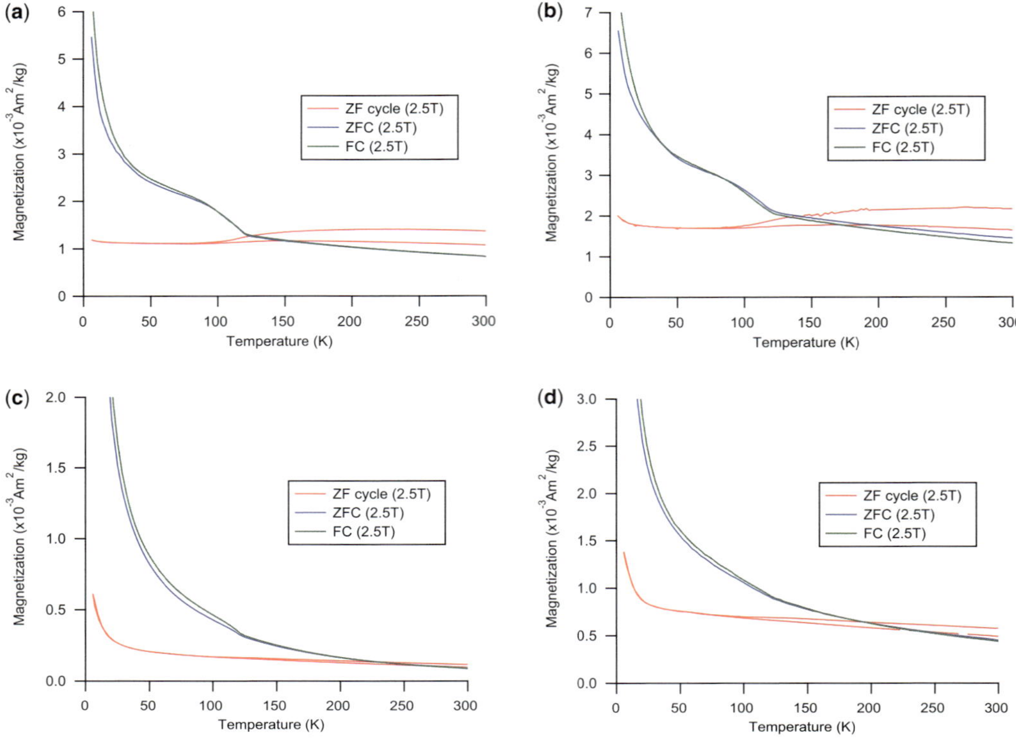

Fig. 12. Low-temperature variations of saturation isothermal remanence for several representative samples during zero-field and field (2.5 T) cooling from 300 to 6 K and warming back to 300 K, which display different behaviours of low-temperature magnetometry, see text. (**a**) Sample 322-C0011B-7R-5W, 65–67 cm; (**b**) Sample 322-C0012A-13R-4, 22–24 cm; (**c**) Sample 322-C0011B-32R-4W, 43–45 cm; and (**d**) Sample 322-C0011B-58R-4W, 138–140 cm. The directions of arrows indicate heating and cooling curves.

correlation points allow the determination of the magnetostratigraphy for Site C0011. For example, core sections between 394.69 and 452.54 m CSF (Sections 322-C0011B-7R-1, 118–120 cm, through 13R-2, 64 cm) show dominantly reversed polarity, indicating that a reversed chron was recorded in these sections. Biostratigraphic Zone NN11a/ NN10b with well-defined FO (first occurrence) *Discoaster berggrenii* events (Zone NN11a/NN10 boundary) is also placed at this interval (see Expedition 322 Scientists 2010*a*), suggesting that the reversed polarity should correlate with chron C4r (8.108–8.769 Ma; Fig. 13).

Biostratigraphic data also suggest that sediments within 495.49–557.73 m CSF are older than 9.61 Ma but younger than 10.81 Ma (see Expedition 322 Scientists 2010*a*). This information is in good agreement with palaeomagnetic observations and suggests that the observed long normal polarity interval between 507.90 and 564.89 m CSF (Sections 322-C00011B-21R-7, 81 cm, through 28R-1, 40 cm) should correspond to the normal polarity

chron C5n.2n (9.99–11.04 Ma). This match is definite because chron C5n.2n is the only long normal chron of this particular age. Furthermore, biostratigraphic data suggest that sediments between 653 and 667 m CSF in Unit IV may be 11.88–12.04 Ma in age (see Expedition 322 Scientists 2010*a*). This information suggests that the relatively well-defined record of the normal/reversed-polarity boundary at 695 m CSF should correspond to the end of chron C5An.2r (12.41 Ma).

Integrated age model and sedimentation rates for Hole C0011B

Palaeomagnetic and biostratigraphic datum events are summarized in Table 6, and the main features of the magnetostratigraphic interpretation along with the inferred biostratigraphic zones at Site C0011 are presented in Figure 13. It is clear from this compilation that the palaeomagnetic and palaeontological age determinations for the Late Miocene

Table 5. *Summary of low-temperature rock magnetic measurements on sediment samples considered for this study*

Core, section, interval (cm)	Depth (m CSF)	Verwey transition (K)						Comments
		First derivative (maximum slope)			Second derivative (maximum curvature)			
		ZFC	FC	Average	ZFC	FC	Average	
Hole C0011B								
C0011B-7R-5W, 65–67	397.495	112.4	116.2	114.3	120.5	120.3	120.4	Clear transition
C0011B-32R-4W, 43–45	600.685	118.0	118.0	118.0	121.0	121.0	121.0	Smeared transition
C0011B-37R-7W, 35–50	650.924	118.5	118.5	118.5	120.5	122.4	121.5	Subdued transition
C0011B-49R-5W, 63–65	734.200	116.8	110.0	113.4	123.5	123.8	123.7	Subdued transition
C0011B-57R-5W, 123–141	852.365	114.6	114.5	114.6	122.5	124.5	123.5	Subdued transition
C0011B-58R-4W, 138–140	860.276	–	114.6	114.6	123.1	121.5	122.3	Subdued transition
C0011B-58R-6W, 90–107	862.429	NA	114.5	114.5	NA	122.5	122.5	Smeared transition
Hole C0012A								
C0012A-13R-4W, 22–24	161.765	114.9	115.3	115.1	120.8	123.0	121.9	Clear transition
C0012A-27R-4W, 88–90	294.350	119.4	NA	119.4	121.7	NA	121.7	Subdued transition
C0012A-37R-1W, 22–24	418.325	120.7	NA	120.7	123.1	NA	123.1	Subdued transition
C0012A-40R-5W, 102.5–121	418.654	118.7	118.5	118.6	122.5	122.5	122.5	Smeared transition
C0012A-48R-2W, 0–18	490.565	110.5	NA	110.5	120.4	NA	120.4	Clear transition
C0012A-49R-1W, 29–31	499.800	114.1	NA	114.1	118.8	NA	118.8	Smeared transition

Note: NA, No data; –, not recognized.
Verwey transition temperature (T_v) was measured as temperatures corresponding to maximum slope or maximum curvature on ZFC (Zero-field cool) or FC (Field cool) warming curve measured by magnetic property measurement system. Average T_v was calculated based on the two curves.

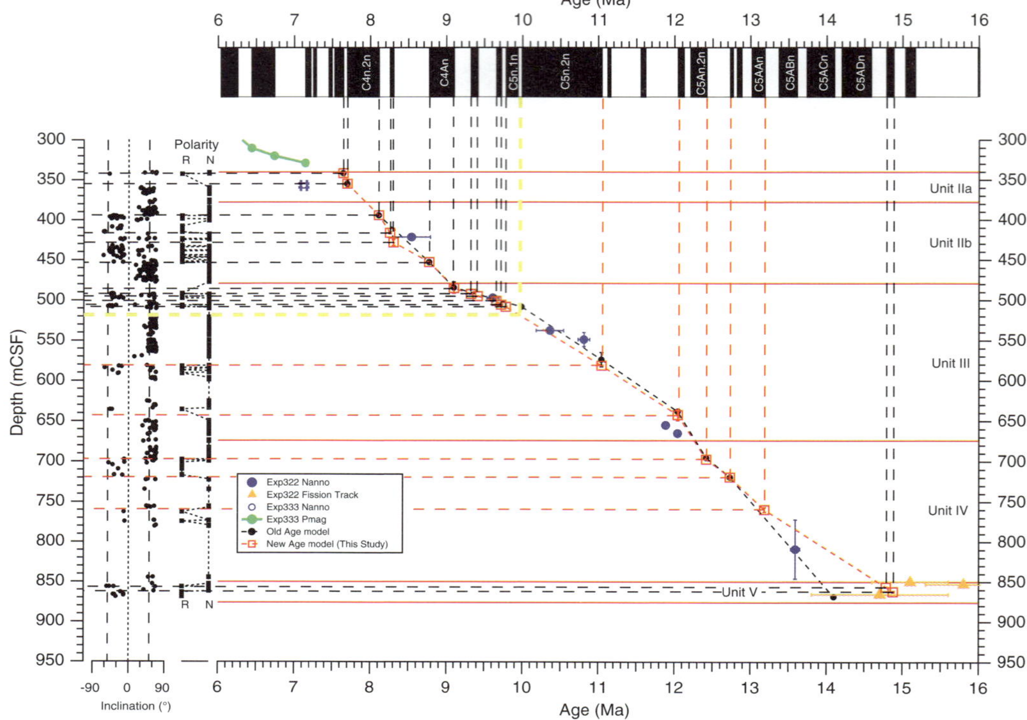

Fig. 13. Diagram of magnetostratigraphic interpretation at Site C0011. Black solid circles and red open rectangles are magnetostratigraphc age model based on only shipboard measurements (Saito *et al.* 2010) and on all the shipboard and shore-based palaeomagnetic measurements (this study). Each marker represent a chron boundary recognized based on the sequence of palaeomagnetic polarity interpretations (R or N) for individual specimens shown by solid rectangles on the left side that is correlated to a standard magnetostratigraphic timescale for the Neogene period (ATNTS2004; Lourens *et al.* 2004). Each correlation is shown by vertical and horizontal broken lines. Blue circles and orange triangles are nannofossil datums (Saito *et al.* 2010) and fission track age (Nakajima *et al.* 2011) for IODP Exp. 322 (Hole C0011B) with error bars. Nannofossil datums and palaeomagnetic results of shipboard measurements obtained for Exp. 333 (Holes C0011C&D; Expedition 333 Scientists 2011) are also shown as open blue circles and solid green circles with error bars. Vertical and horizontal yellow lines indicate the possible coring gap corresponding to C5n.1r. Vertical broken lines in the downhole inclination plot are the expected inclinations at the site.

sequence in Hole C0011B are compatible, but the resolution of the palaeomagnetic data is significantly greater. This allows a more precise determination of sedimentation rates and better definition of the times during which significant changes in sedimentation rate occurred.

The timescale of nannofossil biostratigraphy used for Exp. 322 and throughout the NanTroSEIZE is based on Raffi *et al.* (2006), which is calibrated relative to the astronomically tuned geomagnetic polarity timescale used for the magnetostratigraphy of Exp. 322 (Lourens *et al.* 2004). Thus, it is natural to conduct the combined analysis using both magnetostratigraphy and nannofossil stratigraphy on sedimentation rates and age models. As mentioned, Exp. 322 coring at Hole C0011B started at 340 m CSF. Exp. 333 reoccupied Hole C0011B and

cored through Lithologic Unit I. As a part of across-expedition collaboration within NanTroSEIZE, palaeomagnetic data of Unit I by Exp. 333 scientists (Henry *et al.* 2012) are also shown in Figure 13. In addition, Nakajima *et al.* (2011) reported four fission-track ages from tuff beds in the Unit V at Site C0011. These ages (15.1 ± 0.5, 16.1 ± 1.2, 15.8 ± 0.5 and 14.7 ± 0.9 Ma) are also shown in Fig. 13. The geometrical mean of these fission-track ages and their depth are 15.4 ± 0.8 Ma and 860 m CSF, respectively, which are consistent with the palaeomagnetic inferred age model. Thus, we have combined fission track, biostratigraphic, and palaeomagnetic data to create an integrated age model (orange-red curve in Fig. 13).

The combined data sets allow the determination of sedimentation accumulation rate values and

Table 6. *Palaeomagnetic and biostratigraphic age datums for Hole C0011B*

Magnetic datum (chron or subchron)	Normal or reversed	Boundary age (Ma)	Depth (m CSF)				Biostratigraphic datum (nannofossil)	Age (Ma)	Depth (m CSF)			
			Upper	Lower	Middle	Uncertainty			Upper	Lower	Middle	Uncertainty
C4n.1n												
		7.642	340.87	342.00	341.44	0.56						
C4n.1r												
		7.695	350.05	359.28	354.67	4.61						
C4n.2n												
		8.108	393.69	394.23	393.96	0.27						
C4r.1r												
		8.254	415.01	417.27	416.14	1.13						
C4r.1n												
		8.300	424.47	431.92	428.20	3.73						
C4r.2r							NN11a/NN10a	8.54 ± 0.24	417.67	425.12	421.39	3.72
		8.769	452.33	452.79	452.56	0.23						
C4An												
		9.098	478.69	491.33	485.01	6.32						
C4Ar.1r												
		9.321	492.06	492.40	492.23	0.17						
C4Ar.1n												
		9.409	494.67	495.36	495.02	0.34						
C4rAr.2r							NN10a/9	9.61 ± 0.08	495.49	498.51	497.00	1.51
		9.656	497.11	504.80	500.96	3.85						

(Continued)

Table 6. *Continued*

Magnetic datum (chron or subchron)	Normal or reversed	Boundary age (Ma)	Depth (m CSF) Upper	Lower	Middle	Uncertainty	Biostratigraphic datum (nannofossil)	Age (Ma)	Depth (m CSF) Upper	Lower	Middle	Uncertainty
C4Ar.2n	■											
		9.717	505.31	506.15	505.73	0.42						
C4Ar.3r												
		9.779	508.01	508.72	508.37	0.36						
C5n.1n												
		9.934										
C5n.1r	■											
		9.987										
C5n.2n	■						NN9/NN8	10.36 ± 0.18	535.46	539.94	537.70	2.24
							NN8/NN7	10.81 ± 0.08	539.94	557.55	548.74	8.80
		11.040	581.04	581.55	581.30	0.25						
C5r.1r												
		11.118										
C5r.1n	■											
		11.154										
C5r.2r C5r.2n	■											
							NN7/NN6	11.88 ± 0.02	653.60	656.20	654.90	1.30
C5r.3r												
							LO *C. floridanus*	12.04	662.20	667.76	664.98	2.78
		12.041	635.47	649.21	642.34	6.87						
C5An.1n	■											
		12.116										
C5An.1r												
		12.207										
C5An.2n	■											
		12.415	697.09	697.34	697.22	0.13						
C5Ar.1r												
		12.730	716.91	721.66	719.29	2.38						

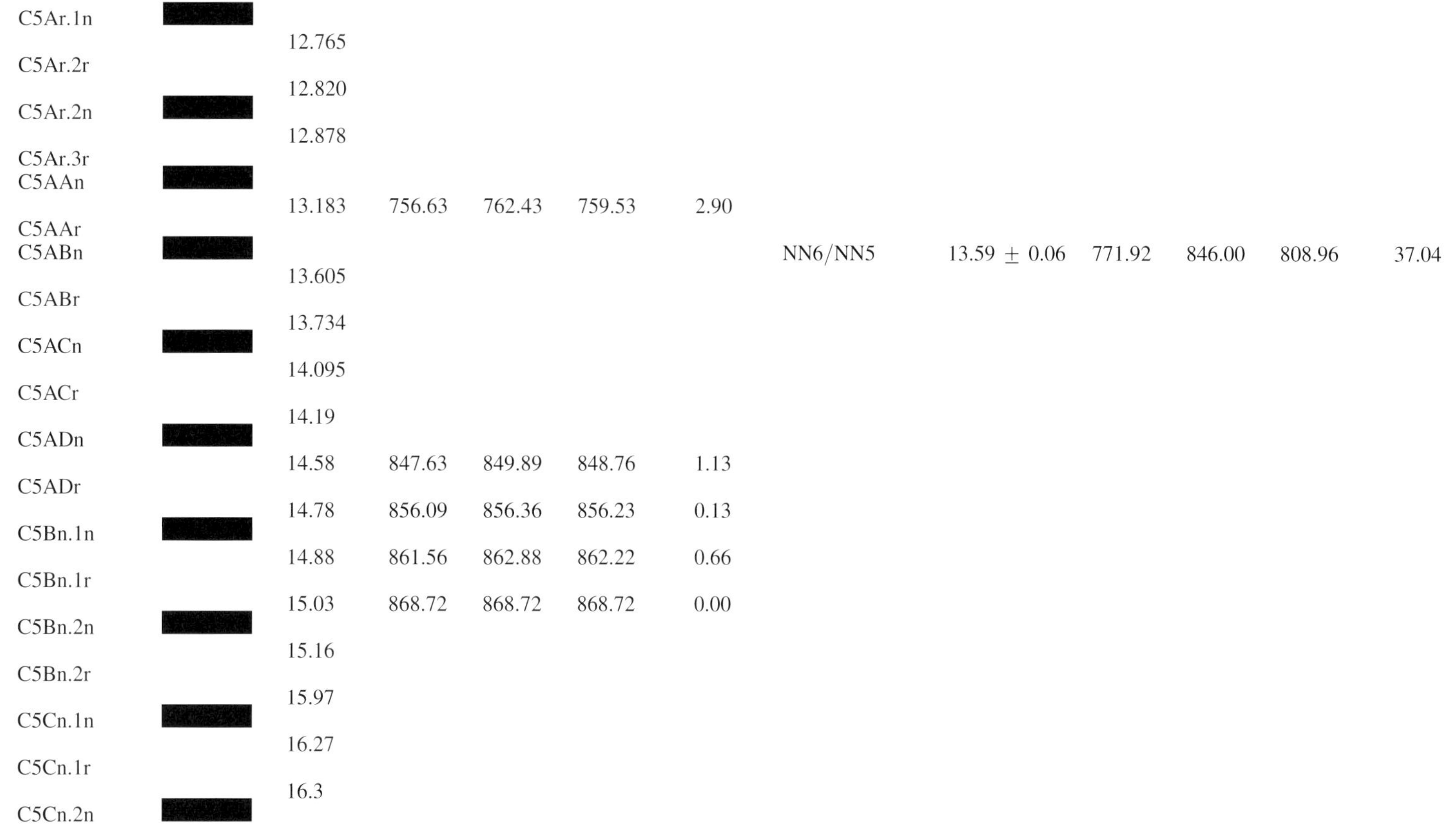
C5Ar.1n
12.765
C5Ar.2r
12.820
C5Ar.2n
12.878
C5Ar.3r
C5AAn
13.183   756.63   762.43   759.53   2.90
C5AAr
C5ABn
13.605   NN6/NN5   13.59 ± 0.06   771.92   846.00   808.96   37.04
C5ABr
13.734
C5ACn
14.095
C5ACr
14.19
C5ADn
14.58   847.63   849.89   848.76   1.13
C5ADr
14.78   856.09   856.36   856.23   0.13
C5Bn.1n
14.88   861.56   862.88   862.22   0.66
C5Bn.1r
15.03   868.72   868.72   868.72   0.00
C5Bn.2n
15.16
C5Bn.2r
15.97
C5Cn.1n
16.27
C5Cn.1r
16.3
C5Cn.2n

  X. ZHAO *ET AL.*

assignment of 'absolute' ages to the lithologic unit boundaries identified in Hole C0011B. The mean sediment accumulation rate is 9.5 cm/ka for Unit II, and 6.3 cm/ka for Unit III except for 480–540 m CSF, where the sedimentation rate is only 2.7 cm/ka Mean sediment accumulation rate for Unit IV is 11.0 cm/ka.

Magnetic polarity record for Hole C0012A

Similar to the magnetic records at Site C0011, several relatively well-defined polarity intervals have been identified in downhole magnetostratigraphic records. The polarity interval between 7.5 and 8.5 Ma could be well correlated with the standard GPTS unambiguously. For instance, shipboard biostratigraphic data suggest that sediments within 107.17–200.63 m CSF are older than 7.07 Ma but younger than 8.78 Ma (see Expedition 322 Scientists 2010b). This information suggests that the observed normal polarity interval between 142.11 and 169.39 m CSF (Sections 322-C0012A-11R-2, 86 cm, through 14R-1, 88 cm) should correspond to the normal polarity chron C4n.2n (7.695–8.108 Ma).

The interpretation of magneostratigraphy at Site C0012 strongly relies on the recognition of an apparently long normal interval between 216.84 and 238.04 m CSF as long normal chron C5n.2n (9.987–11.040 Ma), although there are considerable breaks between cores.

We have attempted two ways to correlate the successive sequence of relatively short polarity intervals found in the magnetostratigraphic record between 180 and 195 m CSF. The first is correlation of chrons C4Ar.2r to C4r.1r, which forms a straight line. A drawback of this correlation is the connection of the dominantly reversed-polarity interval between 197 and 207 m CSF with successive chrons between chrons C4Ar.3r and C5n.1r, where polarity is dominantly normal in the middle and only the margins are reversed. This can be compromised by attributing the normal chron (C5n.1n) to the breaks of the core, however. The second is imposing the normal sequence between 213 and 240 m CSF to between polarity chrons C4An and C5n.2n. The two relatively long reversed chrons C4Ar.1r and C4Ar.2r could be imposed on the possible hiatus close to the base of Unit II (c. 220 m CSF) suggested by nannofossil and sedimentological features (see Expedition 322 Scientists 2010b). Only the *hiatus* can explain the absence of reversed-polarity intervals and relatively slow sedimentation rate of c. 1.3 cm/ka for this period. One advantage of the second correlation is the fairly uniform sedimentation rate of c. 6 cm/ka in the depth range of 111–213 m CSF.

The polarity records for Units IV and V are less certain, especially for the lower part of the hole where core recovery was poor. A fission-track age of 13.2 ± 0.7 Ma was dated from volcaniclastic sandstone in the Unit V at Site C0012 (Nakajima *et al.* 2011), which is significantly younger than the other fission track dates in equivalent parts of the Site C0011. We emphasize that, although there are options for the selection of age models, the ChRM of rocks from Hole C0012A is stable and of high quality, as mentioned above. Thus, the possibility of misinterpretation of the polarity of the magnetozone is small compared with Hole C0011B, where interpretation of the magnetic polarity pattern itself was very difficult for the lower part of the hole. The missing intervals owing to poor core recovery and a possible hiatus leave room for the possibility of several interpretations. With the stability of ChRM, it is clear that possible future continuous piston coring at the site should unify the age model.

For the younger part of Hole C0012A, the available palaeomagnetic polarity records, biostratigraphic and tephra dating data from this site by Exp. 333 (Henry *et al.* 2012) have been used to assist in deciding which observed magnetic polarity zone or set of zones can be correlated with which magnetic chron on the GPTS (Table 7 and Fig. 14).

Integrated age model and sedimentation rates for Hole C0012A

Magnetostratigraphic and biostratigraphic (calcareous nannofossil) datum events are summarized in Table 6, and the main features of the magnetostratigraphic interpretation along with the inferred biostratigraphic zones and fission track dating at Site C0012 are presented in Fig. 14. The new data from this study as well as those by Expedition 333 Scientists (Henry *et al.* 2012) added significant anchoring points in the age model (red square curve in Fig. 14) and greatly improved the previous age models (Saito *et al.* 2010).

From the new age model we infer that a significant increase in sediment accumulation rate occurred in Unit I at c. 110 m CSF from c. 1 to c. 6 cm/ka for Model B (black dot curve in Fig. 14). The lithologic Unit I/II boundary can be assigned an age of 7.8 Ma. For Unit II, the new age model suggests successive changes in sediment accumulation rates from c. 6 to c. 1 cm/ka at c. 180 m CSF, from c. 1 to c. 7 cm/ka at c. 190 m CSF, and from c. 7 to c. 3 cm/ka at c. 210 m CSF. The new age model gives a Unit II/III boundary age of 10.2 Ma (Fig. 14). In the middle part of Unit III, the sedimentation rate increases slightly from c. 3 to 4 cm/ka at c. 250 m CSF and increases further to c. 6 cm/ka at c. 280 m CSF. The Unit III/IV boundary age can be estimated as 12.8 Ma

Table 7. *Palaeomagnetic and biostratigraphic age datums for Hole C0012A*

Magnetic datum (chron or subchron)	Normal or reversed	Boundary age (Ma)	Depth (m)				Biostratigraphic datum (nannofossil)	Age (Ma)	Depth (m CSF)			
			Upper	Lower	Middle	Uncertainty			Upper	Lower	Middle	Uncertainty
C3n.1r												
		4.49	64.25	80.19	72.22	7.97						
C3n.2n												
		4.63										
C3n.2r												
		4.8										
C3n.3n												
		4.9										
C3n.3r												
		5										
C3n.4n							NN13/NN12	5.0087 ± 0.033	80.79	93.05	86.92	6.13
		5.235	90.93	92.15	91.54	0.61						
C3r							NN12/NN11b	5.565 ± 0.025	80.79	93.05	86.92	6.13
		6.033										
C3An.1n												
		6.252										
C3An.1r												
		6.436	107.91	113.28	110.6	2.69						
C3An.2n												
		6.733										
C3Ar							PE R. pseudoumbilicus	7.122 ± 0.045	107.17	116.18	111.67	4.51
		7.140										
C3Bn												
		7.212	115.39	116.17	115.78	0.39						
C3Br.1r												
		7.251	116.3	116.58	116.44	0.14						
C3Br.1n												
		7.285	117.86	118.53	118.2	0.34						
C3Br.2r												
		7.454	128.44	128.80	128.62	0.18						
C3Br.2n												
		7.489	128.8	130.62	129.71	0.91						
C3Br.3r												
		7.528	132.09	133	132.55	0.45						
C4n.1n												
		7.642	137.75	140.15	138.95	1.2						

(*Continued*)

Table 7. *Continued*

Magnetic datum (chron or subchron)	Normal or reversed	Boundary age (Ma)	Depth (m)				Biostratigraphic datum (nannofossil)	Age (Ma)	Depth (m CSF)			
			Upper	Lower	Middle	Uncertainty			Upper	Lower	Middle	Uncertainty
C4n.1r												
C4n.2n		7.695	142.25	143.19	142.72	0.47						
C4r.1r		8.108	169.39	170.41	169.9	0.51						
C4r.1n		8.254	189.44	190.8	190.12	0.68						
		8.300	191.18	197.33	194.26	3.08	NN11a/NN10b	8.405 ± 0.115	197.42	200.63	199.02	1.6
C4r.2r		8.769	215.85	216.44	216.15	0.3	NN10a/NN10b	8.773 ± 0.012	200.63	206.91	203.77	3.14
C4An		9.098										
C4Ar.1r		9.321										
C4Ar.1n		9.409					LO D.hamatus	9.6085 ± 0.0785	216.03	235.77	225.9	9.87
C4rAr.2r		9.656										
C4Ar.2n		9.717										
C4Ar.3r		9.779										
C5n.1n		9.934										
C5n.1r		9.987					NN8/NN7	10.808 ± 0.078	216.03	235.77	225.9	9.87
C5n.2n		11.040	237.91	244.38	241.15	3.24						
C5r.1r		11.118	244.64	244.8	244.72	0.08						
C5r.1n		11.154	246.05	246.63	246.34	0.29						
C5r.2r		11.554	254.71	255.01	254.86	0.15						
C5r.2n		11.614	258.42	258.82	258.62	0.2						
C5r.3r		12.041					LO C. *floridanus*	12.037	284.86	293.04	288.95	4.09

Chron	Age				
C5An.1n	12.116				
C5An.1r	12.207	292.95	294.45	293.70	0.75
C5An.2n	12.415	304.38	312.01	308.20	3.82
C5Ar.1r	12.730				
C5Ar.1n	12.765				
C5Ar.2r	12.820	330.18	331.25	330.72	0.53
C5Ar.2n	12.878	332.91	333.24	333.08	0.16
C5Ar.3r	13.015	342.82	348.82	345.82	3.00
C5AAn	13.183	360.32	376.92	368.62	8.30
C5AAr	13.369				
C5ABn	13.605	427.12	428.00	427.56	0.44
C5ABr	13.734	433.58	435.12	434.35	0.77
C5ACn	14.095				
C5ACr	14.194				
C5ADn	14.581	473.44	474.37	473.91	0.47
C5ADr	14.784	500.93	509.80	505.37	4.44
C5Bn.1n	14.877				
C5Bn.1r	15.032				
C5Bn.2n	15.160				
C5Br	15.974				
C5Cn.1n					

Datum					
NN6/NN5	13.59 ± 0.06	415.92	418.26	417.09	1.17
NN5/NN4	14.914	491.52	529.59	510.55	19.04
NN3/NN2	18.921	491.52	529.59	510.55	19.04
O *H. ampliaperta*	19.657 ± 0.736	530.33	530.33	530.33	0

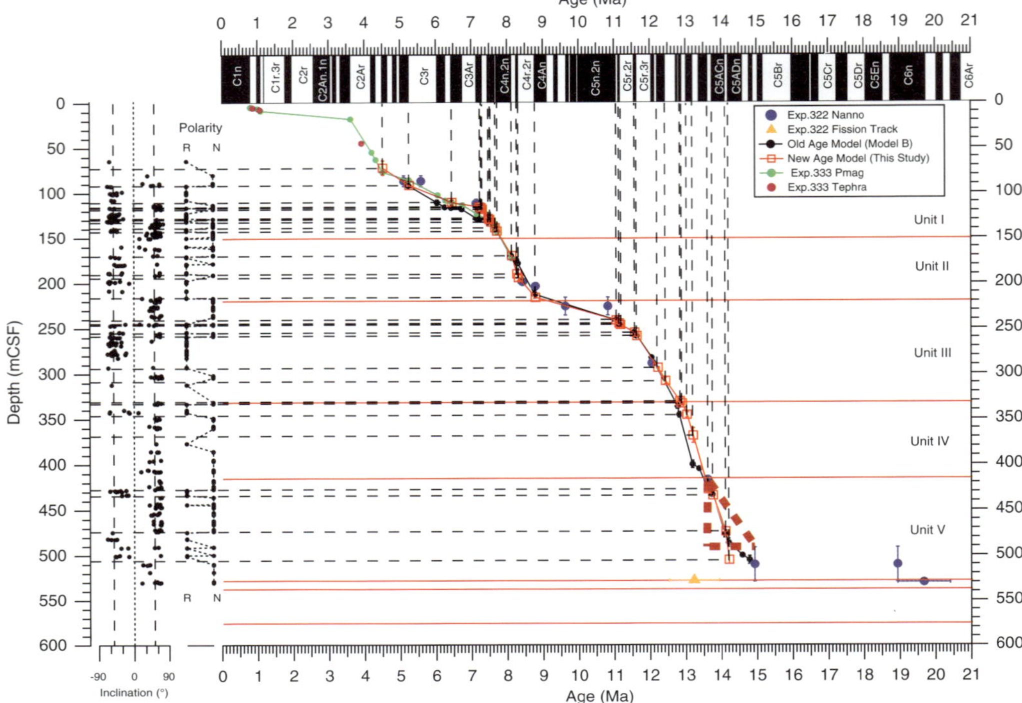

Fig. 14. Diagram of magnetostratigraphic interpretation at Site C0012. Black solid circles and red open rectangles are magnetostratigraphc age model based on only shipboard measurements (Model B; Saito *et al.* 2010) and on all the shipboard and shore-based palaeomagnetic measurements (this study). Each marker represent a chron boundary recognized based on the sequence of palaeomagnetic polarity interpretations (R or N) for individual specimens shown by solid rectangles on the left side that is correlated to a standard magnetostratigraphic timescale for the Neogene period (ATNTS2004; Lourens *et al.* 2004). Each correlation is shown by vertical and horizontal broken lines. Blue circles and orange triangles are nannofossil datums (Saito *et al.* 2010) and fission track age (Nakajima *et al.* 2011) for IODP Exp. 322 (Hole C0012A) with error bars. Tephra and palaeomagnetic results of shipboard measurements obtained for Exp. 333 (Holes C0012C&D; Expedition 333 Scientists 2011) are also shown as solid purple circles and solid green circles with error bars. Red triangle is a zone of arrowed age model based on the nannofossil datums at around 13.5 and 15 Ma. Vertical broken lines in the downhole inclination plot are the expected inclinations at the site.

according to the magnetostratigraphy, which is also consistent with the age model based on nannofossil datum. Although there is a considerably reliable nannofossil datum at 417 m CSF (Zone NN6/NN5 boundary), we tried to follow a straight line from Unit III down to the upper part of Unit V (430 m CSF) with a sedimentation rate of *c.* 6 cm/ka. The age model infers a Unit IV/V boundary age of 14.4 Ma. The sedimentation rate decreases to *c.* 8 cm/ka below 470 m CSF to 500 m CSF.

Our previous age Model B (Saito *et al.* 2010) favours an age of 13.5 Ma for the Unit IV/V boundary, which infers a smoother sedimentation rate for the base of Unit V, but a significantly increased sedimentation rate from Unit III (6.8 cm/ka) to Unit IV (10 cm/ka). According to this age model, the age of the thick sandstone units in the upper part of Unit V

(430~470 m CSF) is estimated to be 13.9 Ma. The age model for the bottom part (>435 m CSF) of the Hole C0012A remains to be improved owing to the apparently contradictory age of nannofossil and fission track dating (Fig. 14).

Discussion and concluding remarks

The ChRM directions observed at Holes C0011B and C0012A have both normal and reversed polarities. Fairly complete magnetostratigraphies were obtained from these two sites for the Late Miocene to Pleistocene.

As mentioned above, the positive results from the mean inclination reversal test (Arason & Levi 2010) on Units I–III of Site C0012 indicate that the ChRM is a record of the palaeomagnetic field close to the time of formation of these sediments.

Further evidence that the magnetization of these sediments was acquired close to their time of deposition is the fact that most of the observed magnetostratigraphy from the characteristic directions is in good agreement with that to be expected from the stratigraphic position of the sequence deduced from the biostratigraphic data. On these bases, it is argued that the characteristic directions of magnetization observed from these sediments are most likely the original primary magnetization acquired when the sediments were deposited.

Palaeomagnetic correlation within the Nankai Trough Cenozoic sequence is as promising as ever, especially now that a consistent magnetostratigraphy is emerging. The magnetostratigraphy of the Nankai section allows the determination of accurate sediment accumulation rates and better constraints of the timing of deformation. The composite age–depth model yields rates of sedimentation (uncorrected for either compaction or rapid event deposition by gravity flows) ranging from *c.* 4.0 cm/ka in the upper part of Unit III to 9.5 cm/ka in the lower part of lithologic Units III and IV. The average rate for lithologic Unit II is 9.4 cm/ka. The sedimentation rate within Unit III changed at *c.* 11 Ma.

Lithologic equivalents of Units II–V of the lower Shikoku Basin facies have been recovered in other drilling transects of the Nankai Trough area (Moore *et al.* 2001). For the Muroto transect, Sites 808 and 1174 (inside of the trench) and 1173 (outside of the trench) give sedimentation rates of 3.2, 3.5 and 2.7 cm/ka, respectively, for the lower Shikoku Basin facies deposits. These values are slightly lower than, but comparable to, the sedimentation rates we derived for the upper part of Unit III at Sites C0011 and C0012 (4.0 cm/ka). In general, they show significant changes in sedimentation rates at *c.* 11 Ma (Sites 808 and 1174) or *c.* 13 Ma (Site 1173; Moore *et al.* 2001). This can also be correlated with the change in sedimentation rate at 11 Ma for Site C0011. Site 1173, which is comparable to Site C0011, gives a slightly older age (*c.* 13 Ma). On the other hand, Site 1177, situated outside of the trench of the Ashizuri transect, gives a comparable sedimentation rate of 2.8 cm/ka.

In summary, the integrated age–depth model from biostratigraphy and magnetostratigraphy indicate that an abrupt decrease in the rate of sedimentation occurs in Unit III at about 11 Ma, which is also evident from lithofacies analyses showing that there is a substantial decrease in the rate of hemipelagic sedimentation at the Late Miocene (11 Ma) in the Shikoku Basin (Naruse *et al.* 2010). Changes in physical properties are also observed for the same depth interval at Site C0011 (Underwood *et al.* 2009). All of these resutls suggest that some fundamental palaeoenvironmental change in the Shikoku Basin and/or significant tectonic event may have occurred in Late Miocene.

We thank Prof L. Jovane and the two journal reviewers for the helpful reviews and constructive suggestions that greatly improved the original manuscript. This research was conducted with samples and data from the Integrated Ocean Drilling Program, an international marine research programme dedicated to advancing scientific understanding of the Earth, the deep biosphere, climate change and Earth processes by sampling and monitoring sub-seafloor environments. We thank shipboard scientists, the Marine Works Japan laboratory technicians and the crews of the *Chikyu* for their help and company during IODP Exp. 322. We also wish to express our appreciation to the shore-based IODP staff for all of their pre- and post-expedition efforts. Financial support for various parts of this research was provided by grants from the US Science Support Program of the Joint Oceanographic Institution Inc., US National Science Foundation (grant 0911331), National Science Foundation of China (grant 91128102), the Brazilian state of São Paulo FAPESP (grant 2012/07523-6) and Chinese Fundamental Research Funds for the Central Universities.

References

ANDO, M. 1975. Source mechanisms and tectonic significance of historical earthquake derived from tsunami data. *Physics of the Earth and Planetary Interiors*, **28**, 320–336.

ARASON, P. & LEVI, S. 2010. Maximum likelihood solution for inclination-only data in palaeomagnetism. *Geophysical Journal International*, **182**, 753–771, doi: 10.1111/j.1365-246X.2010.04671.x.

CANDE, S. C. & KENT, D. V. 1995. Revised calibration of the geomagnetic polarity time-scale for the Late Cretaceous and Cenozoic. *Journal of Geophysical Research*, **100**, 6093–6095.

DAY, R., FULLER, M. & SCHMIDT, V. A. 1977. Hysteresis properties of titanomagnetites: grain-size and compositional dependence. *Physics of the Earth and Planetary Interiors*, **13**, 260–267.

DUNLOP, D. J. 2002. Theory and application of the Day plot (M_{rs}/M_s v. Bc_r/Bc) 1. Theoretical curves and tests using titanomagnetite data. *Journal of Geophsical Research*, **107**, doi: 10.1029/2001JB000486.

DUNLOP, D. & ÖZDEMIR, Ö. 1997. *Rock-magnetism, Fundamentals and Frontiers*. Cambridge University Press, Cambridge.

ENKIN, R. J. 1994. *A Computer Program Package for Analysis and Presentation of Palaeomagnetic Data*. Pacific Geoscience Centre, Geologicalal Survey of Canada, Sidney, Canada.

EXPEDITION 322 SCIENTISTS. 2010*a*. Site C0011. *In*: SAITO, S., UNDERWOOD, M. B., KUBO, Y. & THE EXPEDITION 322 SCIENTISTS. (eds) *Proceedings of IODP*, **322**. Integrated Ocean Drilling Program Management International Inc., Tokyo, doi: 10.2204/iodp.proc.322. 103.2010.

EXPEDITION 322 SCIENTISTS. 2010b. Site C0012. In: SAITO, S., UNDERWOOD, M. B., KUBO, Y. & THE EXPEDITION 322 SCIENTISTS. (eds) Proceedings of IODP, **322**. Integrated Ocean Drilling Program Management International, Inc., Tokyo, doi: 10.2204/iodp.proc. 322.104.2010.

EXPEDITION 333 SCIENTISTS. 2011. NanTroSEIZE Stage 2: subduction inputs 2 and heat flow. IODP Preliminary Report, **333**, 295–305, doi:10.2204/iodp.pr.333.2011.

FISHER, R. A., 1953. Dispersion on a sphere. Proceedings of the Royal Society of London, **217**, 295–305.

GROMMÉ, C. S., WRIGHT, T. L. & PECK, D. L. 1969. Magnetic properties and oxidation of iron-titanium oxide minerals in Alae and Makaopuhi lava lakes, Hawaii. Journal of Geophysical Research, **74**, 5277–5293.

HALL, J. M. 1977. Does TRM occur in oceanic layer 2 basalts? Journal of Geomagnetism and Geoelectricity, **29**, 411–419.

HENRY, P., KANAMATSU, T., MOE, K. & THE EXPEDITION 322 SCIENTISTS. 2012. Proceedings of IODP, **333**. Integrated Ocean Drilling Program Management International, Inc., Tokyo, doi: 10.2204/iodp.proc.333. 2012.

JOHNSON, H. P. & MERRILL, R. T. 1974. Low temperature oxidation of a single-domain magnetite. Journal of Geophysical Research, **79**, 5533–5534.

KIMURA, G., KITAMURA, Y. ET AL. 2007. Transition of accretionary wedge structures around the up-dip limit of the seismogenic subduction zone. Earth and Planetary Science Letters, **225**, 471–484. doi:10.1016/j.epsl.2007.01.005.

KIRSCHVINK, J. L. 1980. The least-squares line and plane and the analysis of palaeomagnetic data. Geophysical Journal of the Royal Astronomical Society, **62**, 699–718.

KODAIRA, S., TAKAHASHI, N., NAKANISHI, A., MIURA, S. & KANEDA, Y. 2000. Subducted seamount imaged in the rupture zone of the 1946 Nankaido earthquake. Science, **289**, 104–106, doi: 10.1126/science.289.5476.104.

LAY, T., KANAMORI, H. ET AL. 2005. The great Sumatra-Andaman earthquake of 26 December 2004. Science, **308**, 1127–1133. doi:10.1126/science.1112250.

LOURENS, L. J., HILGEN, F. J., SHACKLETON, N. J., LASKAR, J. & WILSON, D. 2004. The Neogene period. In: GRADSTEIN, F. M., OGG, J. G. & SMITH, A. G. (eds) A Geologicalal Time Scale 2004. Cambridge University Press, Cambridge, 409–440.

MCFADDEN, P. L. & MCELHINNY, M. W. 1988. The combined analysis of remagnetization circles and direct observations in palaeomagnetism. Earth and Planetary Science Letters, **87**, 161–172.

MCFADDEN, P. L. & MCELHINNY, M. W. 1990. Classification of the reversal test in palaeomagnetism. Geophysical Journal International, **103**, 725–729.

MIYAZAKI, S. & HEKI, K. 2001. Crustal velocity field of southwest Japan: subduction and arc-arc collision. Journal of Geophysical Research, **106**, 4305–4326. doi:10.1029/2000JB900312.

MOORE, G. F., TAIRA, A., KLAUS, A. ET AL. 2001. Proc. ODP, Init. Repts., 190. College Station, TX. doi: 10.2973/odp.proc.ir.190.200.

MOSKOWITZ, B. M. 1981. Methods for estimating Curie temperatures of titanomaghemites from experimental J_s–T data. Earth and Planetary Science Letters, **53**, 84–88.

MOSKOWITZ, B. M., FRANKEL, R. B. & BAZYLINSKI, D. A. 1993. Rock magnetic criteria for the detection of biogenic magnetite. Earth and Planetary Science Letters, **120**, 283–300.

MOSKOWITZ, B. M., JACKSON, M. & KISSEL, C. 1998. Low-temperature magnetic behavior of titatnomagnetites. Earth and Planetary Science Letters, **157**, 141–149.

NAKAJIMA, T., NARUSE, H. ET AL. 2011. Sediment composition analysis and FT dating of the Shikoku Basin sediments drilled in the IODP Exp. 322, Abstract MIS022-15, Japan Geoscience Union Meeting.

NARUSE, H., PICKERING, K. ET AL. & IODP EXPEDITION 322 SHIPBOARD SCIENTIFIC PARTY. 2010. Abrupt change in the rate of hemipelagic sedimentation at the Late Miocene (~11 Ma) in the Shikoku Basin (the IODP Sites C0011 & C0012), Abstract for Japan Geological Union Meeting.

OKINO, K., SHIMAKAWA, Y. & NAGAOKA, S. 1994. Evolution of the Shikoku Basin. Journal of Geomagnetism and Geoelectricity, **46**, 463–479.

ÖZDEMIR, O. & DUNLOP, D. 1985. An experimental study of chemical remanent magnetization of synthetic monodomain titanomaghemites with initial thermoremanent magnetizations. Journal of Geophsical Research, **90**, 11 513–11 523.

RAFFI, I., BACKMAN, J., FORNACIARI, E., PÄLIKE, H., RIO, D., LOURENS, L. & HILGEN, F. 2006. A review of calcareous nannofossil astrobiochronology encompassing the past 25 million years. Quaternary Science Reviews, **25**, 3113–3137, doi: 10.1016/j.quascirev. 2006.07.007.

SAITO, S., UNDERWOOD, M. B., KUBO, Y. & THE EXPEDITION 322 SCIENTISTS. 2010. Proceedings of IODP, **322**. Integrated Ocean Drilling Program Management International Inc., Tokyo, doi: 10.2204/iodp.proc. 322.2010.

SENO, T., STEIN, S. & GRIPP, A. E. 1993. A model for the motion of the Philippine Sea Plate consistent with NUVEL-1 and geological data. Journal of Geophysical Research, **98**, 17941–17948, doi: 10.1029/93JB00782.

TAUXE, L. 1998. Palaeomagnetism, Principles and Practice. Kluwer Academic, Dordrecht.

TAUXE, L., MULLENDER, T. A. T. & PICKL, T. 1996. Potbellies, wasp-waists, and superparamagnetism in magnetic hysteresis. Journal of Geophysical Research, **101**, 571–583.

TOBIN, H. J. & KINOSHITA, M. 2006a. Investigations of seismogenesis at the Nankai Trough, Japan. IODP Publications, Scientific Prospectus, NanTroSEIZE Stage 1, doi: 10.2204/iodp.sp.nantroseize1.2006.

TOBIN, H. J. & KINOSHITA, M. 2006b. NanTroSEIZE: the IODP Nankai Trough Seismogenic Zone Experiment. Scientific Drilling, **2**, 23–27, doi: 10.2204/iodp.sd.2.06.2006.

TOBIN, H., KINOSHITA, M. ET AL. & THE EXPEDITION 314/315/316 SCIENTISTS. 2009. NanTroSEIZE Stage 1 expeditions: introduction and synthesis of key results. In: KINOSHITA, M., TOBIN, H. ET AL. & THE EXPEDITION 314/315/316 SCIENTISTS, Proc. IODP, 314/315/316. Integrated Ocean Drilling Program Management International, Inc., Washington, DC. doi:10.2204/iodp. proc.314315316.101.2009.

UNDERWOOD, M. B., SAITO, S., KUBO, Y. & THE EXPEDITION 322 SCIENTISTS. 2009. NanTroSEIZE Stage 2: subduction inputs. *IODP Preliminary Report*, **322**, doi: 10.2204/iodp.pr.322.2009.

VERWEY, E. J., HAAYMAN, P. W. & ROMEIJN, F. C. 1947. Physical properties and cation arrangements of oxides with spinal structure. *Journal of Chemical Physics*, **15**, 181–187.

ZIJDERVELD, J. D. A. 1967. A.C. demagnetization of rocks: analysis of results. *In*: COLLISION, D. W., CREER, K. M. & RUNCORN, S. K. (eds) *Methods in Palaeomagnetism*. Elsevier, New York, 254–286.

Evaluating Late Holocene radiocarbon-based chronologies by matching palaeomagnetic secular variations to geomagnetic field models: an example from Lake Kalimpaa (Sulawesi, Indonesia)

TORSTEN HABERZETTL[1]*, GUILLAUME ST-ONGE[2], HERMANN BEHLING[3] & WIEBKE KIRLEIS[4]

[1]Physical Geography, Institute of Geography, Friedrich-Schiller-University Jena, Löbdergraben 32, 07743 Jena, Germany

[2]Canada Research Chair in Marine Geology, Institut des sciences de la mer de Rimouski and GEOTOP Research Center, Université du Québec à Rimouski, 310, allée des Ursulines, Rimouski, Québec, Canada G5L 3A1

[3]Department of Palynology and Climate Dynamics, Albrecht-von-Haller-Institute for Plant Sciences, University of Göttingen, Untere Karspüle 2, 37073 Göttingen, Germany

[4]Graduate School Human Development in Landscapes/Institute for Prehistoric and Protohistoric Archaeology, Christian-Albrechts-University Kiel, Johanna-Mestorf-Strasse 2-6, 24118 Kiel, Germany

*Corresponding author (e-mail: torsten.haberzettl@uni-jena.de)

Abstract: Palaeomagnetic and palaeoenvironmental data from Indonesia and particularly from the Island of Sulawesi are scarce and exact dating has turned out to be a challenge in many archives from this region. Here we outline difficulties in radiocarbon dating of the palaeoenvironmental record from Lake Kalimpaa, Sulawesi, Indonesia. These difficulties demand the integration of additional parameters to obtain a reliable chronology for this record. Thus, we compare the palaeomagnetic secular variation data from this record with the CALS3k.4 spherical harmonic geomagnetic model of the 0–3 ka field (Korte & Constable 2011). The resulting age–depth model for the Lake Kalimpaa sequence provides a profound basis for further multi-proxy investigations on this record. For the first time, high-resolution palaeomagnetic secular variation data continuously spanning the past 1300 years are presented for this region, which complement existing records with lower temporal resolution or records missing the top-most sections.

The area between mainland Asia and Australia is of great significance for understanding modern climatic processes. Sea surface temperatures in the western Pacific warm pool drive the dynamics of El Nino Southern Oscillation and the Asian Monsoon (Oppo *et al.* 2009; Linsley *et al.* 2010). Influenced by water masses passing from the warm western Pacific to the Indian Ocean, this region acts as a considerable source of latent heat for the globe (Hope 2001). In particular, palaeorecords from the region help to trace the structures behind recurrent environmental changes. However, palaeoenvironmental information from Indonesia, and particularly from the Island of Sulawesi, is scarce. To complement the information derived from marine records and to improve the understanding of SE Asian–Pacific palaeoclimate, its amplitude and the exact timing of palaeoenvironmental variations, further records from terrestrial archives in this region are essential (Dam *et al.* 2001). Palaeoenvironmental studies have been carried out in Java (Stuijts *et al.* 1988; van der Kaars & Dam 1995; van der Kaars *et al.* 2001; Crausbay *et al.* 2006), Kalimantan (Morley 1981; Anshari *et al.* 2001, 2004; Page *et al.* 2004; Hope *et al.* 2005; Yulianto *et al.* 2005), Halmahera (Suparan *et al.* 2001) and New Guinea (Hope *et al.* 1988; Haberle 1998; Haberle & Ledru 2001; Haberle *et al.* 2001; Hope 2009; Fig. 1). Marine sediment cores are the only sources of palaeoenvironmental information in the intermediate area between the islands (van der Kaars *et al.* 2000; Hope 2001; Spooner *et al.* 2005; Newton *et al.* 2006; Brijker *et al.* 2007; Langton *et al.* 2008). Until now, on the Island of Sulawesi itself, which is the main area of interest for this contribution, palaeoenvironmental knowledge of the past two millennia is extremely limited. Palynological investigations have been carried out on Wanda Swamp (2°33′ S, 121°23′ E; Hope 2001), Danau (Lake) Tempe (4°5′S, 119°55′E; Gremmen 1990) and

From: JOVANE, L., HERRERO-BERVERA, E., HINNOV, L. A. & HOUSEN, B. A. (eds) 2013. *Magnetic Methods and the Timing of Geological Processes*. Geological Society, London, Special Publications, **373**, 245–259.
First published online August 14, 2012, http://dx.doi.org/10.1144/SP373.10 © The Geological Society of London 2013.
Publishing disclaimer: www.geolsoc.org.uk/pub_ethics

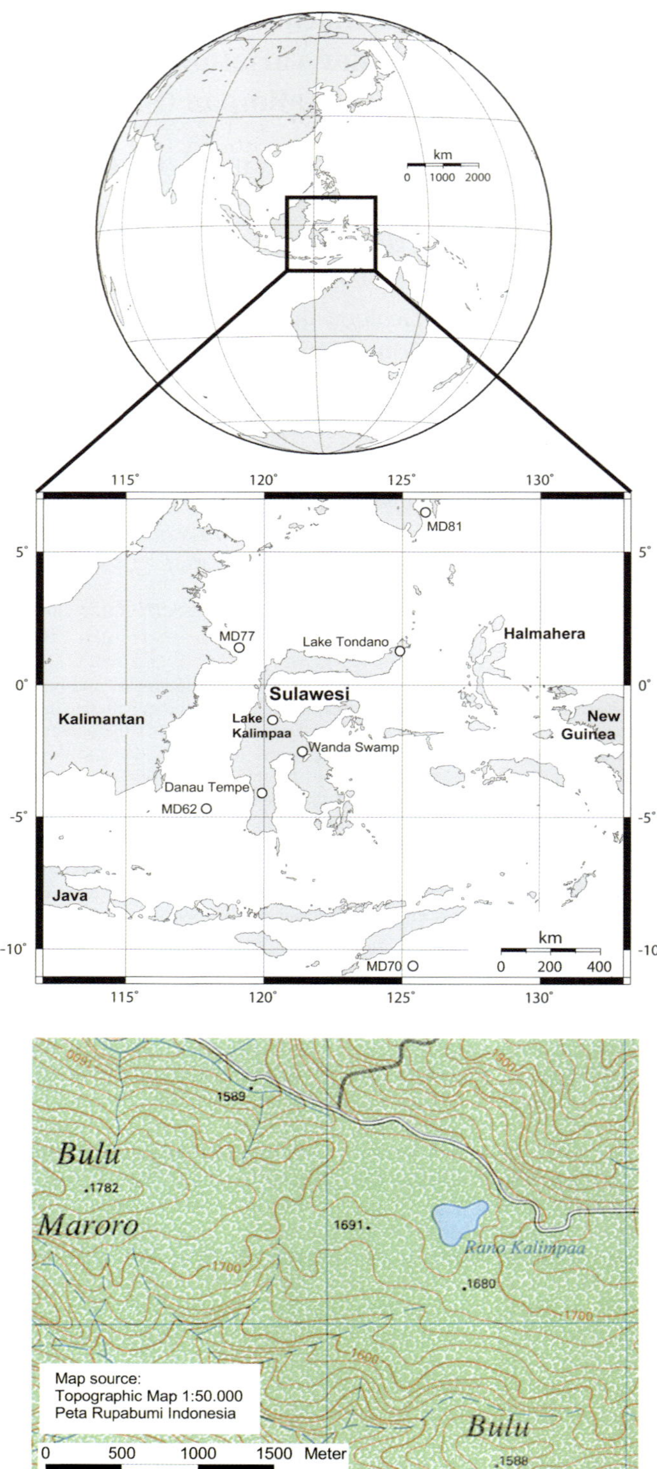
km
0 1000 2000
115°
120°
125°
130°
MD81
5°
5°
MD77
Lake Tondano
Halmahera
0°
Sulawesi
0°
Kalimantan
Lake
Kalimpaa
New
Guinea
Wanda Swamp
Danau Tempe
-5°
MD62
-5°
Java
km
-10°
-10°
MD70
0 200 400
115°
120°
125°
130°
Bulu
.1782
Maroro
1691.
Rano Kalimpaa
.1680
1700
1600
Bulu
.1588
Map source:
Topographic Map 1:50.000
Peta Rupabumi Indonesia
0 500 1000 1500 Meter

Lake Tondano (1°16′N, 124°54′E; Dam *et al.* 2001; Fig. 1). However, these studies focus on long-lasting variations like the Pleistocene–Holocene transition and are of low temporal resolution. Recently, a sequence focusing on Late Holocene developments in central Sulawesi was published (Kirleis *et al.* 2011).

As far as palaeomagnetic information is concerned, the lack of spatial coverage is even poorer. A number of deep-sea sediment cores (MD62, MD70, MD77 and MD81) in the vicinity of Indonesia and the Philippine Islands, which form a transect from 6.3°N to 10.6°S, were recovered (Lund *et al.* 2006; Fig. 1). These are the first high-resolution palaeomagnetic records of Holocene palaeomagnetic secular variation recovered within 17° of the Equator. As the focus of these records has been the entire Holocene, it appears that there is little information about the most recent palaeomagnetic history available in most records; for example, in MD70 (Fig. 1) the last *c.* 2000 years of sediment are missing (Lund *et al.* 2006). The same holds for a marine record recovered further north at ODP Leg 195 Hole 1202B (24°48.24′S, 122°30.00′E; Richter *et al.* 2006), for four marine cores obtained from the Ontong Java Plateau (1°24.1′N–3°36.3′S, 156°59.9′–158°41.6′E; Constable & Tauxe 1987) and for two records from northeastern Australian crater lakes (Constable 1985). Based on the age–depth model of the ODP site, it is possible that the last *c.* 1000 years are missing (Richter *et al.* 2006) and the Ontong Java Plateau study only provides relative palaeointensity data with a resolution of two to three points for the past two millennia (Constable & Tauxe 1987). In the northeastern Australian lakes, the top-most 1000–3000 years are missing completely (Constable 1985). Therefore, comparable archives are only available at a greater distance (e.g. southern Australia).

However, for eastern Asian lake records it was demonstrated that palaeosecular variation dating can be useful for stratigraphic correlation on a regional scale. The method could help to resolve ambiguities in the ages of some high-frequency climate events within different regions and to understand their significance in global change (Yang *et al.* 2009).

In the following, we compile an age control and present the first high-resolution palaeomagnetic data from Lake Kalimpaa, located on the Island of Sulawesi (Fig. 1). Radiocarbon dates in former studies on Sulawesi reveal that generating age–depth

models that provide the key to proper palaeoenvironmental reconstruction can be very difficult as they show age reversals, missing surfaces and/or unexpected modern dates (Dam *et al.* 2001; Hope 2001). We will illustrate the difficulties associated with dating at Lake Kalimpaa and, in combination with palaeomagnetic data, try to evaluate an age–depth model providing the basis for further multiproxy investigations. Until now, little attention has been given to the practical use of palaeomagnetic secular variations computed by the CALSxk model series as a dating tool (Barletta *et al.* 2010). Recently, good matches have been shown between the CALS3k.3 model (Donadini *et al.* 2009; Korte *et al.* 2009 – the model preceding the CALS3k.4 model (Korte & Constable 2011) used in this study) and marine sediments from the western (Barletta *et al.* 2010) and eastern (Ledu *et al.* 2010) Canadian Arctic, as well as between the model and lacustrine sediments from Lake Nam Co (Kasper *et al.* 2012), located on the Tibetan Plateau. For the location of lake Kalimpaa, some of the above-mentioned archives are contained in the CALS3k.4 model database, as well as higher-resolution records covering the relevant time span from further south (i.e. southern Australia) and further north (e.g. Lake Biwa; Ali *et al.* 1999; Donadini *et al.* 2009; Korte & Constable 2011).

Regional setting

Lake Kalimpaa (1°19′34.8″S, 120°18′31.9″E; Fig. 1), sometimes also referred to as Danau Tambing, is located close to the northeastern border of the protected area of the Lore Lindu National Park in Central Sulawesi, Indonesia. The national park has existed since 1993. However, parts of the park have been deforested (Merker 2003). A recreational area was recently constructed in the immediate surroundings of the lake and a small stream was redirected, bringing more water and sediment into the lake.

The lake itself is located at 1660 m asl. The surrounding geology consists of intrusive and metamorphic rocks (Villeneuve *et al.* 2002). Neighbouring crests reach altitudes of >2300 m asl, resulting in distinct topographic differences. The lake covers an area of 6.5 ha and has a maximum depth of 6.6 m. Reeds grow on the shore surrounding the lake. A reed belt is situated in the north and NE that fades to a swamp forest. About 200 m

Fig. 1. Map of the research area showing Sulawesi and adjacent islands as well as locations mentioned in the text. The basic map was created using Online Map Creation. A map of the lake including the immediate surrounding is shown at the bottom. This is based on a Topographic Map, 1:50 000 Peta Rupabumi Indonesia, edited by Badan Koordinasi Survey Dan Pemetaan Nasional (Bakosurtanal) in Cibinong–Bogor, Indonesia.

eastwards from the shoreline the main asphalt road through the National Park passes by (Fig. 1) and divides the lake area from the steeper mountainous area reaching to Mount Rorekatimbu (2300 m asl). Vegetation in the south and west of the lake consists of lower- to mid-mountain rainforest species. Inflow reaches the lake mainly through the swamp forest area in the NE, and an outflow is located to the SW.

Climatic patterns in this area are complex and poorly understood, with considerable variation over short distances (Hope 2001). The general pattern shows constant temperatures throughout the year ranging from 25 °C during the night up to 35 °C during the day. In the mountainous regions where Lake Kalimpaa is located, mean temperatures are around 20 °C. Annual precipitation is in the order of 2000–3000 mm, with less precipitation in the rain shadow of the mountains (Weber 2006).

Materials and methods

In 2006, three sediment cores (KAL1 to 3) were recovered from Lake Kalimpaa. Palaeomagnetic investigations were applied to the irregularly laminated, longest, best-preserved and hence most suitable core, KAL1 (211 cm). KAL2 was investigated by Indonesian counterparts using standard sedimentological analyses. KAL3 is only a short core of 25 cm in length and thus was not considered for further investigations. KAL1 consists of three overlapping sections (KAL1-1 to -3) that were recovered from a water depth of 6.5 m using a Livingstone piston corer. Core sections were recovered in a tube and transported to the laboratory of the Department of Palynology and Climate Dynamics, University of Göttingen, Germany, for further analyses. In the laboratory, cores were stored in darkness at 4 °C before being split, photographed and described lithologically. Sections were correlated using distinct macroscopic marker layers (Gutknecht 2008).

As no macro remains for radiocarbon dating were available, a first set of six bulk sediment samples was submitted to the accelerator mass spectrometry radiocarbon laboratory of the University of Erlangen, Germany (Table 1). Subsequently, three additional bulk sediment samples were sent to the Poznan Radiocarbon Laboratory, Poland (Table 1). Considering inconsistencies, all six samples previously dated in Erlangen were measured a second time (labelled as Erlangen 2 in Table 1). Most radiocarbon ages were calibrated using the Southern Hemisphere calibration curve (shcal04; McCormac *et al.* 2004) of the online CALIB 5.0.2 software (Stuiver & Reimer 1993). Calibrated ages are given as median ages with 2σ errors and are reported as cal BP (Table 1). Ages exceeding this curve were not calibrated and are displayed as BP

(Fig. 2). Modern ages were calibrated using CALI-Bomb (Reimer *et al.* 2004) and the year of the highest probability was chosen for age–depth modelling (Table 1). As the sediment–water interface was intact, it was used to represent the year of coring, which was 2006. Owing to very low sedimentary calcite contents (Gutknecht 2008), a hard water effect can be excluded for Lake Kalimpaa.

Sections were subsampled with u-channels and subsequently shipped to the Sedimentary Palaeomagnetism Laboratory at the Institut des sciences de la mer de Rimouski of the University of Québec at Rimouski, Canada. u-Channels of sections KAL1-2 and 3 were cut off at a very distinctive correlating marker layer and taped together. Small gaps in KAL1-2 were closed by gently pushing the sediment from the top, avoiding compaction of the sediment itself. This resulted in a reduction of the total length of the record by 3 cm (=208 cm total length). Using the many marker layers, the depths of the radiocarbon ages were adjusted to the new u-channel depth scale.

Palaeomagnetic data were acquired at 1 cm intervals on the u-channels using a 2 G Enterprises™ 760R cryogenic magnetometer and pulse magnetizer module (for isothermal remanent magnetization, IRM). Owing to the width of the response function of the magnetometer pick-up coils of *c.* 7 cm, some smoothing occurs. As the material in KAL1-2 and -3 seemed to be very different, the last 7 cm of KAL1-2 and the first 7 cm of KAL1-3 were excluded. It was not necessary to exclude the lower part of KAL1-1 or the upper part of KAL1-2 as these sections extended further down and up, respectively, from the main correlating marker layer. Measurements on these overlapping extensions provided identical results, so no data had to be removed.

The natural remanent magnetization (NRM) was measured first using stepwise alternating field (AF) demagnetization at peak fields of 0–75 mT with 5 mT increments. Inclination and declination of the characteristic remanent magnetization (ChRM) were calculated using an Excel spreadsheet developed for that purpose (Mazaud 2005) with AF demagnetization steps from 10 to 50 mT (nine steps). These steps were chosen in order to minimize maximum angular deviation (MAD) values obtained by principal component analysis (Kirschvink 1980) with the same macro. ChRM declinations were corrected for rotation at section breaks, that is, it was assumed that the lowermost data point of the upper section is identical to the uppermost data point of the lower section. Declination data are relative and centred at zero since the coring was not azimuthally oriented. An anhysteretic remanent magnetization (ARM) was then induced at peak AF of 100 mT with a 0.05 mT direct current (DC) biasing field and subsequently demagnetized and measured

Table 1. *Accelerator mass spectrometry radiocarbon dates from Lake Kalimpaa. Medians of calibrated ages refer to the 2σ ranges*

Sediment depth (cm)		Error	Median calibration age (cal BP)	Error to present (years)	Error to past (years)	Laboratory	Laboratory no.	Dataset	Remark	δ^{13}C
	^{14}C age (BP)									
44	1198	40	1050	85	125	Erlangen	Erl-11299	shcal04.14c		−29.1
73	1288	32	1160	95	100	Erlangen	Erl-10576	shcal04.14c		−34.0
94	1963	37	1850	125	95	Erlangen	Erl-11300	shcal04.14c		−30.5
119	1920	37	1800	90	90	Erlangen	Erl-11301	shcal04.14c		−32.5
151	1774	39	1630	95	90	Erlangen	Erl-11302	shcal04.14c		−28.6
204	17893	74				Erlangen	Erl-11303		>shcal04.14c	−25.3
73	1539	41	1370	70	145	Erlangen 2	Erl-12181	shcal04.14c		−34.8
94	1649	41	1470	95	120	Erlangen 2	Erl-12182	shcal04.14c		−28.4
119	1735	43	1590	170	120	Erlangen 2	Erl-12183	shcal04.14c		−31.4
151	1501	41	**1340**	50	65	Erlangen 2	Erl-12184	shcal04.14c		**−30.0**
204	13413	91				Erlangen 2	Erl-12185		>shcal04.14c	−26.5
136	1080	30	**940**	120	105	Poznan	Poz-24168	shcal04.14c		**−24.0**
203	4300	50	4760	180	195	Poznan	Poz-24169	shcal04.14c	0.4 mg C	−29.5
	Percentage modern carbon									
	105.75	0.37	**−46** −45 −43 −9 −8			Poznan	Poz-24225	SH1.14c	Highest probability printed in bold	**−29.3**
	102.62	0.52	−46 −45 −9 **−8**			Erlangen 2	Erl-12180	SH1.14c	Highest probability printed in bold	**−30.4**

Note: Ages labelled 'Erlangen 2' were re-dated and correspond to the same sample as the sample 'Erlangen' at the respective sediment depth. Ages used in the final chronology are printed in bold.

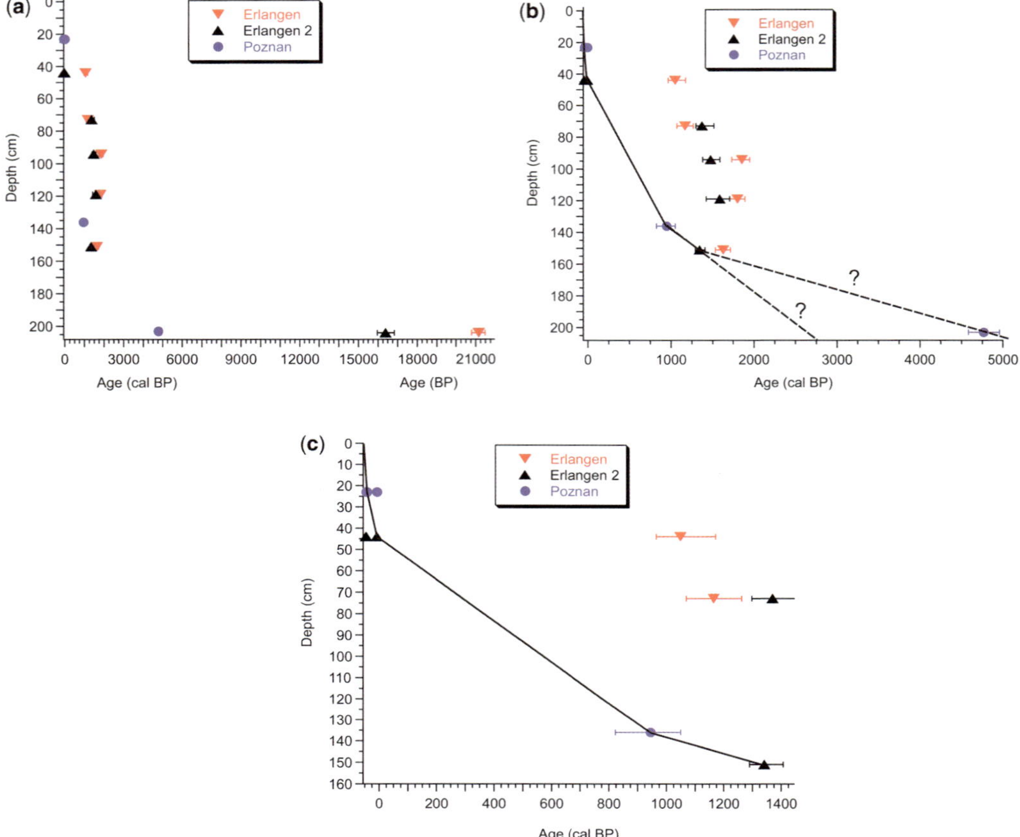

Fig. 2. Age–depth model of the Lake Kalimpaa record. Note different scales. (**a**) Ages presented as medians of calibrated ages referring to the 2σ ranges (except for two exceeding the Southern Hemisphere calibration curve). The age at 204 cm sediment depth is too old to be calibrated using the Southern Hemisphere calibration curve (McCormac *et al.* 2004). (**b**) All calibrated ages presented as medians referring to the 2σ ranges. Theoretical age–depth models for the lowermost part of the record are indicated by the dashed lines. (**c**) Final age–depth model for Lake Kalimpaa cut at 151 cm sediment depth.

from 0 to 60 mT with 5 mT steps as well as at 70 and 80 mT. An isothermal remanent magnetization (IRM) was imparted with a DC field of 0.3 T and subsequently demagnetized and measured using the same steps as with the ARM. Similarly, a second IRM (corresponding to a saturated isothermal remanent magnetization, SIRM) was imparted with a higher DC field of 0.95 T and subsequently demagnetized and measured at 0, 30, 50 and 100 mT.

Hysteresis loops along with remanence curves were measured on small subsamples taken from selected intervals of each u-channel using an alternating gradient force magnetometer (Princeton Measurements Corp., MicroMag model 2900). Magnetization saturation (Ms), coercive force (Hc), saturation remanence (Mrs) and coercivity of

remanence (Hcr) were extracted from this data to characterize the magnetic mineralogy and magnetic domain state, which can be an indicator of grain size (Day *et al.* 1977).

Results and interpretation

Chronology

Results of the first set of bulk subsamples sent for radiocarbon dating showed inconsistencies, that is, ages were not in stratigraphic order and distinct jumps in sedimentation rate occurred (Fig. 2a, Table 1). Three additional samples dated in a different laboratory revealed even more inconsistencies and would have led to the exclusion of additional

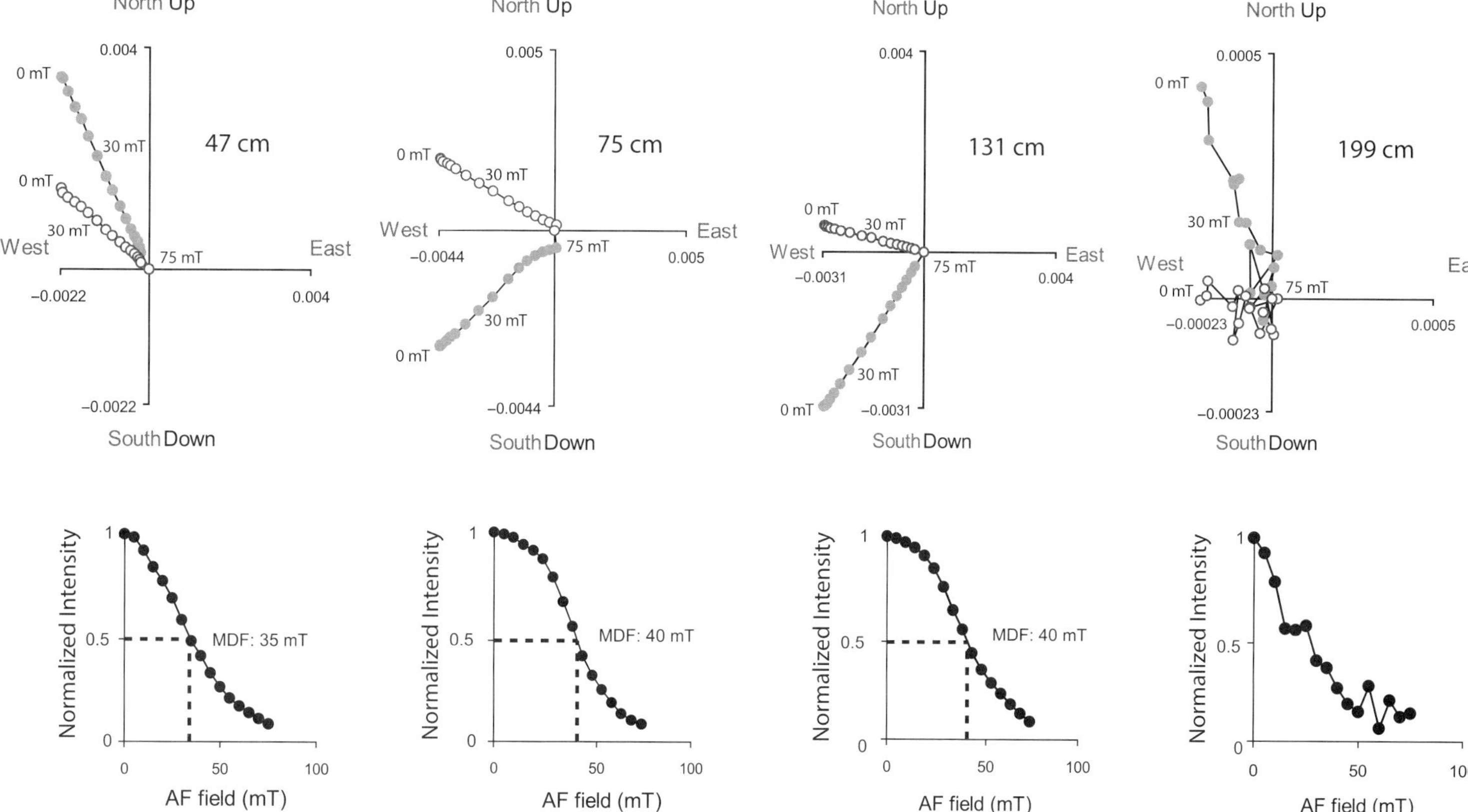

Fig. 3. Typical vector end-point orthogonal projection diagrams for the sediment record of Lake Kalimpaa. Solid dots refer to west–east whereas circles refer to up–down.

ages. For this reason, the first set of subsamples was re-dated and showed different ages from in the first attempt (Fig. 2, Table 1). As a hard-water effect can be excluded at Lake Kalimpaa, this leads to the assumption that there is a different kind of reservoir effect. Most likely, as often happens with bulk sediment samples (Haberzettl *et al.* 2005), not just the autochthonous carbon fraction was dated but also some allochthonous organic matter was incorporated into the samples. This probably led to the age offsets in either direction between the first and the second dating attempt of the first six samples (Fig. 2, Table 1). This hypothesis is supported by the low $\delta^{13}C$ values of the dated material ranging from -34.8 to $-24.0‰$, indicating non-algal organic matter contributions (Table 1; Haberzettl *et al.* 2005; Mayr *et al.* 2005). If this is the case, all ages are tentatively too old. In order to

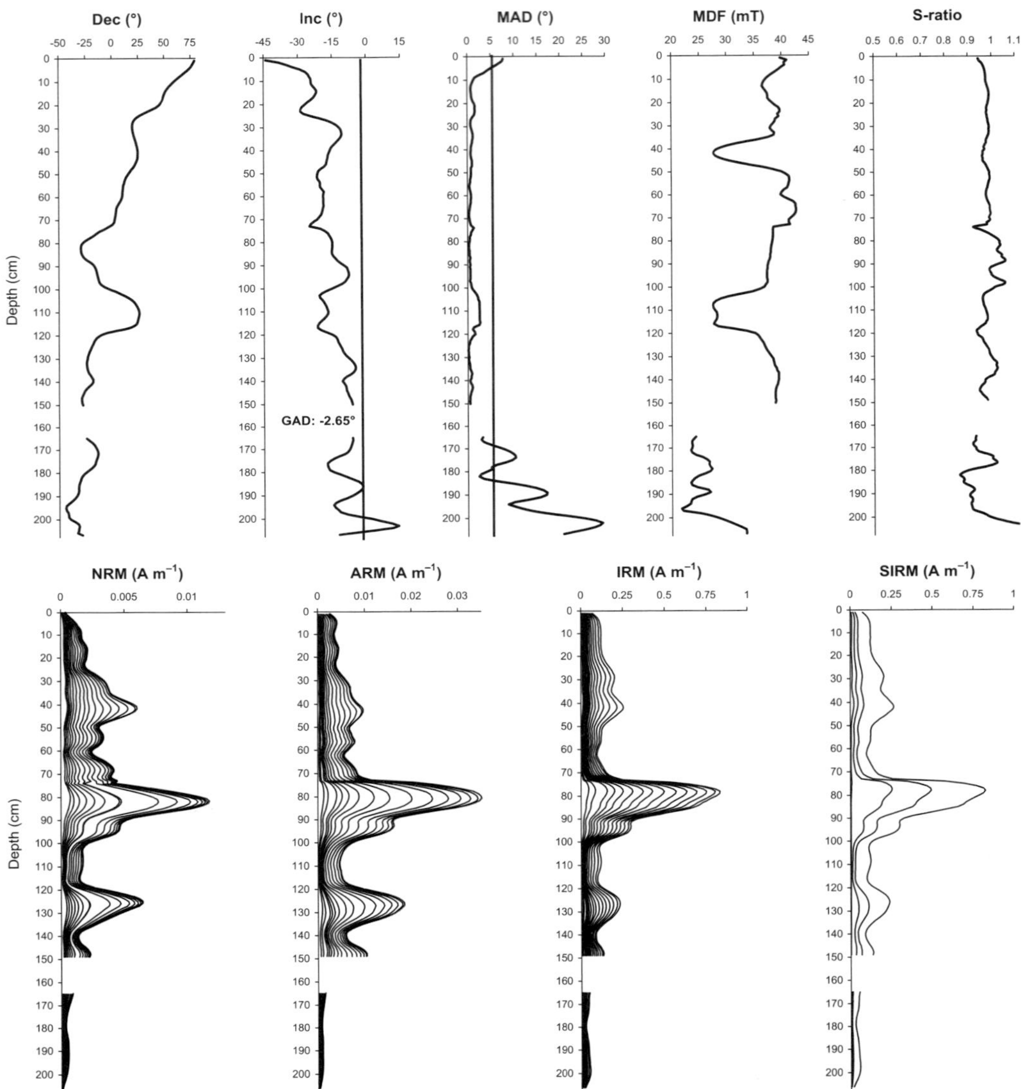

Fig. 4. Declination, inclination, maximum angular deviation, MDF, pseudo *S*-ratio ($IRM_{0mT}/SIRM_{0mT}$) as well as downcore variations of NRM, ARM, IRM and SIRM at all demagnetization steps from the Lake Kalimpaa record plotted on a depth scale. Declination and inclination are component directions calculated from the 20–50 mT demagnetization steps.

minimize this error, only the youngest dates were chosen for age–depth modelling and a linear interpolation was applied (Fig. 2). The two/three (one was re-dated) lowermost ages at 203 and 204 cm sediment depth show a clear offset to all ages above. Hence, it was not obvious whether it was suitable to include the youngest of the three in the age model or to do a linear extrapolation using the lowermost more reliable dates above (Fig. 2b). For this reason we decided to cut the record at the lowermost reliable date at 151 cm sediment depth when plotting data on an age scale. Accordingly, the age scale has a basal age of 1340 +65/−50 cal BP at 151 cm sediment depth (Fig. 2c).

The uppermost 44 cm (−8 cal BP to year of coring, 2006) of the record show an average sedimentation rate of an order of magnitude higher, passing from 0.8 mm a^{-1} to 9.2 mm a^{-1}. This significant increase in the most recent history can probably be explained by human disturbances in the immediate surroundings of the lake (e.g. campsite or road construction). In order to exclude any impact of this catchment activities, only data before 0 cal BP were used for comparisons later on in this study.

Palaeomagnetic data

A strong and stable characteristic remanent magnetization can be isolated between the AF demagnetization steps of 10 and 50 mT (Fig. 3). Declination data from c. 25–0 cm sediment depth trends from +79° to +25°. This seems to be a rather high rate of secular variation for a low-latitude site. We suspect that this is a coring artefact, as is often observed in the uppermost and water saturated part of sediment cores.

Inclination shows an increasing trend with depth (Fig. 4). Values are close to the expected geocentric axial dipole (GAD) (2.65°S for Lake Kalimpaa) showing an average offset of less than c. 12°. Inclination only reaches the expected GAD for the latitude of Lake Kalimpaa in the lowermost part of the record at c. 200 cm, which is in the part of the record that is not plotted on an age scale (Fig. 4).

Maximum angular deviation values are generally lower than 5° in the upper part of the record. MAD values increase to above 10° in the lower part, reaching 30° at the base of the record (Fig. 4). As MAD values can be used to help assess the quality of the directional record, intervals with MAD values above 5 should be considered suspect during times of stable polarity. High MAD values generally reflect a complex magnetization carried by unstable magnetic grains (Stoner & St-Onge 2007). For this reason and because of the uncertainties in the age–depth model in this part of the

record, no palaeomagnetic measurements of section KAL1-3 (165–208 cm sediment depth) were plotted on an age scale.

High MAD values between 208 and 165 cm sediment depth are probably related to the very low magnetic intensity at the base of the record (Fig. 4), resulting in a low signal-noise ratio, and to an increase in magnetic grain size as coarser sediment grains (sand) can be observed macroscopically in this part of the core (Fig. 5). The changes in magnetic grain size are supported by a jump in the median destructive field (MDF) from values oscillating around 40 mT to c. 25 mT (Fig. 4), as well as by samples plotting in the multi-domain range

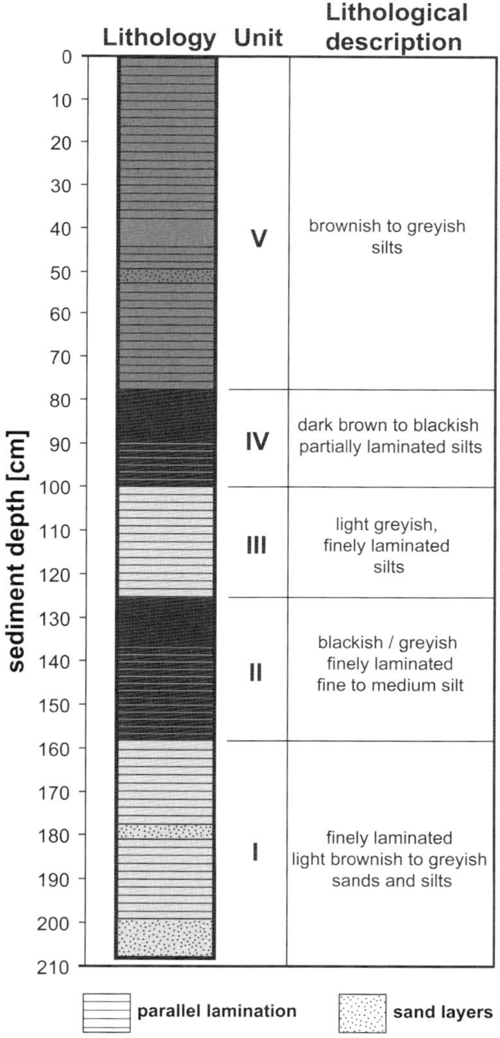

Fig. 5. Lithology of the investigated core.

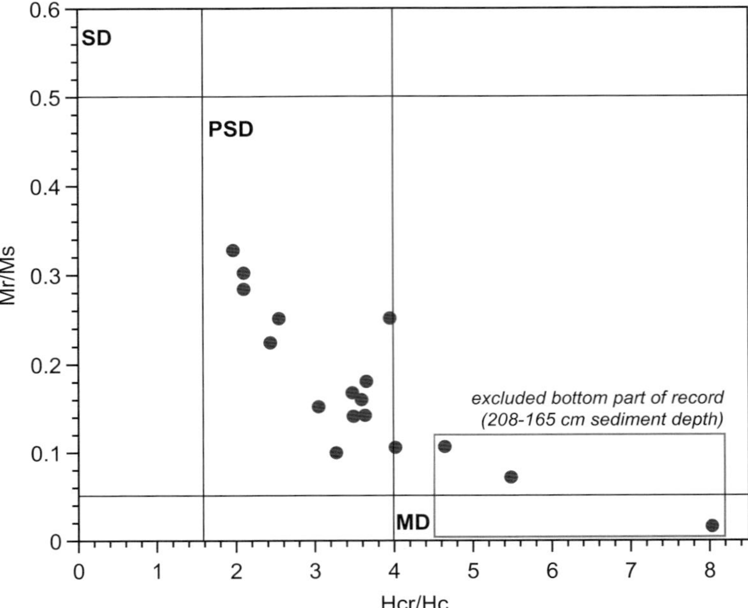

Fig. 6. Day plot using selected samples from the Lake Kalimpaa sediment record.

(Fig. 6). MDF is a grain-size as well as a coercivity-dependent parameter, and is also useful to estimate magnetic mineralogy (Dankers 1981). MDF values of the record indicate low-coercivity minerals in the upper part. This is in agreement with the pseudo *S*-ratio values ($IRM_{0mT}/SIRM_{0mT}$) close to 1, which are also indicative of low coercivity minerals such as magnetite (St-Onge *et al.* 2003). Apart from the excluded basal part of the record with coarser material, changes in magnetic concentration-dependant parameters (NRM, ARM, IRM) are within an order of magnitude (Fig. 4). Finally, most of the studied samples fall in the pseudo-single domain region for magnetite on a Day plot (Day *et al.* 1977). The only exception occurs at the excluded bottom of the record (208–165 cm sediment depth), where samples are in the multi-domain area (Fig. 6) and where MAD values are the highest.

These observations are consistent with a set of criteria (Tauxe 1993; Stoner & St-Onge 2007) proposed to be fulfilled for relative palaeointensity (RPI) determinations. Estimation of RPI of the geomagnetic field in sedimentary sequences is obtained by dividing the measured NRM by an appropriate normalizer in order to compensate for the variable concentration of ferrimagnetic minerals (Tauxe 1993). To construct our RPI proxy, we normalized the NRM_{30mT} by the ARM_{30mT}. No significant correlation between the RPI proxy and its normalizer

was observed ($R^2 = 0.046$), suggesting that ARM is suited for palaeointensity estimations. Additionally, no significant correlation was found between the RPI proxy and magnetic grain size proxies (e.g. ARM/IRM: $R^2 = 0.202$).

No obvious similarities are observed in the details between our RPI reconstruction and the CALS3k.4 (Korte & Constable 2011) spherical harmonic model output for the coring site. However, both graphs show the same trend and slope over the last 1000 cal BP. However, more detailed rock-magnetic work is needed to fully document potential changes in the magnetic carrier and its influence on the RPI estimation.

Discussion

Analogous conservative age–depth modelling approaches were used in other studies before, showing reasonable results when comparing proxy data with nearby other archives (Haberzettl *et al.* 2005; Mayr *et al.* 2005; Kasper *et al.* 2012). Although not to the same extent, similar age reversals observed in Lake Kalimpaa were also found in Wanda Swamp, and were attributed to the mixing of adjacent older sediments (Hope 2001). In Lake Tondano, a modern date was also found at 100–90 cm sediment depth (Dam *et al.* 2001), illustrating that dating lake records from Sulawesi have been similarly

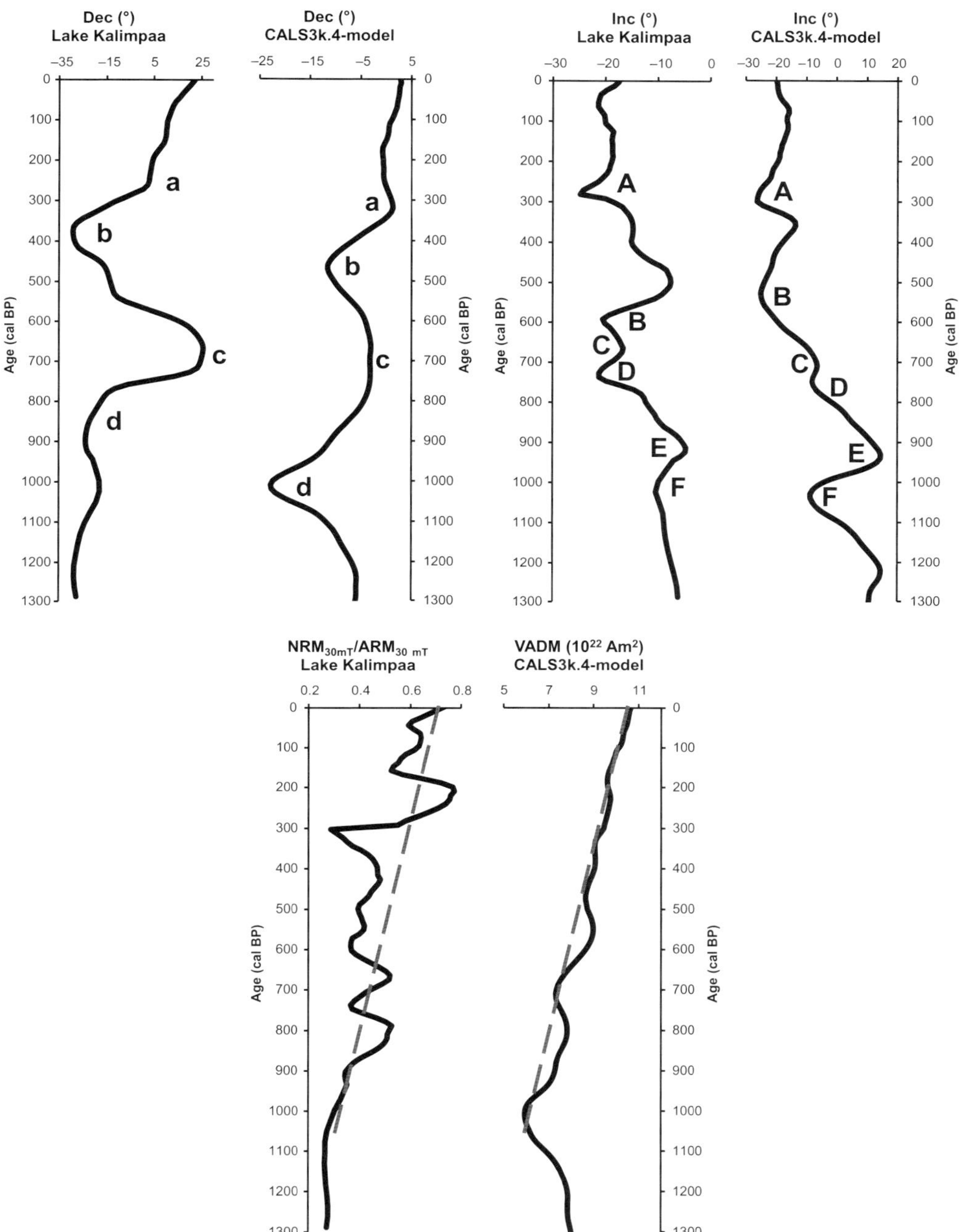

Fig. 7. Comparison of declination, inclination and palaeointensity from the Lake Kalimpaa record with the CALS3k.4 model (Korte & Constable 2011) output for the coordinates of Lake Kalimpaa.

complicated to those at Lake Kalimpaa. However, the authors of the latter do not attribute the young age to an increased sedimentation rate but to an admixture of recent soil organic matter. At Lake Kalimpaa this seems to be rather unlikely as two ages 21 cm apart show such young (modern) ages

(Fig. 2, Table 1). Hence, we have to assume that both ages are correct and attribute this to human influence in the catchment area.

Although it was challenging to construct an initial age model, there are many similarities in inclination and declination, as well as a similar trend in palaeointensity between the Lake Kalimpaa record and the CALS3k.4 model output (Korte & Constable 2011; Fig. 7). Comparison of the inclination from Lake Kalimpaa to the SE Australia inclination master curve (KBG) consisting of data from three Australian maars, that is, lakes Keilambete, Bullenmerri and Gnotuk in SW Victoria (Barton & Polach 1980; Barton & McElhinny 1981), also reveals similar trends (Fig. 8). This might be attributed to the fact that, after statistical treatment (Donadini *et al.* 2009), parts of the underlying three records are contained in the database of the CALS3k.4 model (Korte & Constable 2011). Correlating the remanence record of the SE Australia master curve is best done using inclination, which displays characteristic swings. The declination record is less suitable because it seems devoid of clear correlatable features between the Australian maars (Anker *et al.* 2001). However, chronologies of Lake Kalimpaa and the KGB are expected to differ distinctively. Comparisons of the KBG record with other data for SE Australia covering the past 2500 years (Barton & Barbetti 1982) suggested that KBG ages might be older by about 350 calendar years owing to a reservoir effect (Barton & Barbetti 1982; Anker *et al.* 2001). Because of these uncertainties, we consider the KBG record to support our measurements but do not use it to date the Kalimpaa record. Furthermore, the resolution of the KBG record is one order of magnitude lower than the Kalimpaa record (13 v. 150 data points, i.e. centennial v. decadal resolution). Despite this different resolution, the SE Australia records deviates from the GAD as in the Lake Kalimpaa record (Anker *et al.* 2001).

When developing the CALS3k.3 model (Korte *et al.* 2009), several inconsistencies either within records or when compared with other nearby records had already been observed for many records (Donadini *et al.* 2009). The authors tried to overcome these problems using a statistical rejection of outliers. This led to a rejection of 34% of the data (Donadini *et al.* 2009) and a resolution of *c.* 100 years (Korte & Constable 2003).

The similarities of the Lake Kalimpaa full vector palaeomagnetic data to the model (Fig. 7) indicate that the age model is a good first-order approximation. Differences lie within the error in the chronology determined by age-modelling artefacts owing to linear interpolation and the error in the radiocarbon dating method itself, which according to accepted ages in this study ranges between 50 and

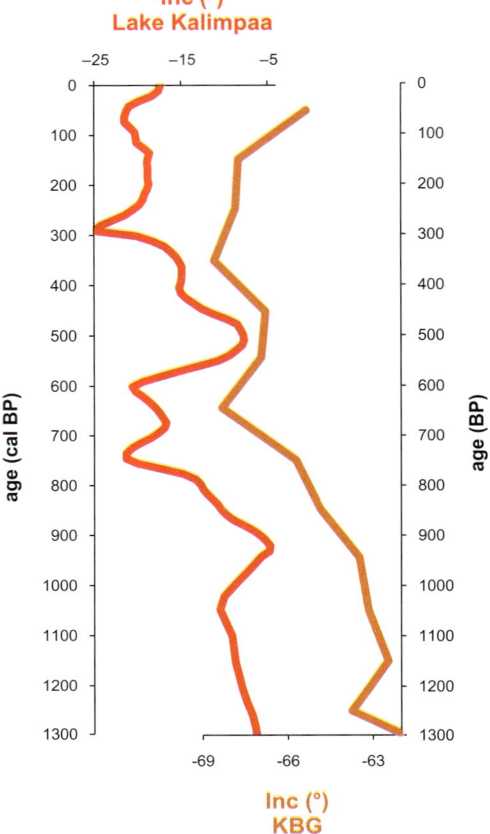

Fig. 8. Comparison of inclination from Lake Kalimpaa with inclination from the SE Australia inclination master curve (KBG; Barton & Polach 1980; Barton & McElhinny 1981) digitized from Anker *et al.* (2001). Note that Lake Kalimpaa data is given in cal BP whereas the KBG data is plotted on a BP age scale. Because of age and reservoir-effect uncertainties (Barton & Barbetti 1982; Anker *et al.* 2001), we consider the KBG record to support our measurements but do not use it to date the Kalimpaa record.

120 years (Table 1, Fig. 7). With the established age model, it is now possible for upcoming studies to set multi-proxy analyses in a time frame. This provides a reliable chronology for this record as further radiocarbon dates would not improve the chronology since they might also be influenced by remobilization of old organic matter. This is very important as the time span covered by the Lake Kalimpaa record is missing in the surrounding archives (Hope 2001).

As far as the palaeomagnetic data are concerned, this paper is the first high-resolution study in this area. Sedimentation rates are up to three times higher than in the available sediment records from the marine domain surrounding Sulawesi. These marine records normally cover the entire Holocene but often yield only a little information in the top-most part (Lund *et al.* 2006). Our data thus complement the published marine dataset.

The value of the Lake Kalimpaa record is increased further as it has good coverage of the past 400 years. There are few archives available for modelling palaeosecular variations after the year 1600, which can be attributed to frequent loss or disturbance of material near the sediment–water interface during coring (Donadini *et al.* 2009). Owing to the high sedimentation rate at the top of the Lake Kalimpaa record, even the top-most part (which might be influenced by coring artefacts) can be removed without great loss of information.

Conclusions

Although the radiocarbon-derived age–depth model of Lake Kalimpaa is conservative, a comparison of the full vector palaeomagnetic data with the CALS3k.4 geomagnetic spherical harmonic model confirms this approach. This leads to the conclusion that the presented age model is well suited for further multi-proxy palaeoenvironmental investigations, which are necessary as many archives from Indonesia miss the top-most part of the records. Finally, the palaeomagnetic data from this paper constitute the first continuous high-resolution terrestrial record from Indonesia, yielding a detailed full-vector pal aeomagnetic record from 1300 cal BP up to the present.

We would like to acknowledge the helping hands of C. Brunschön and J. Labrie during u-channel sub-sampling and laboratory work as well as comments from Monika Korte on an earlier version of this manuscript. M. Markussen is thanked for background information about the surrounding of Lake Kalimpaa, and M. Wündsch for assistance with graphs. T. Haberzettl was supported by a post-doctoral fellowship scholarship provided by Le Fonds québécois de la recherche sur la nature et les technologies (FQRNT). Guillaume St-Onge was supported by a Natural Sciences and Engineering Research Council of Canada discovery grant and Canadian Foundation for Innovation grants for the acquisition and operation of the cryogenic magnetometer. Sediment coring/research was carried out as part of subproject C7 of the Collaborative Research Centre SFB 552 'Stability in Rainforest Margins', at the University of Göttingen, and was funded by the DFG, Germany. We gratefully acknowledge logistic support from STORMA's Indonesian partner universities in Bogor and Palu, Institut Pertanian Bogor and Universitas Tadulako, the Ministry of Education in Jakarta, the Indonesian Institute of Sciences, and the authorities of the Lore Lindu National Park. Finally, we would like to thank Mr Iksam, State Museum Central Sulawesi, Palu, Indonesia, for his support with sediment coring.

References

ALI, M., ODA, H., HAYASHIDA, A., TAKEMURA, K. & TORII, M. 1999. Holocene palaeomagnetic secular variation at Lake Biwa, central Japan. *Geophysical Journal International*, **136**, 218–228.

ANKER, S. A., COLHOUN, E. A., BARTON, C. E., PETERSON, M. & BARBETTI, M. 2001. Holocene vegetation and paleoclimatic and paleomagnetic history from lake Johnston, Tasmania. *Quaternary Research*, **56**, 264–274.

ANSHARI, G., PETER KERSHAW, A. & VAN DER KAARS, S. 2001. A Late Pleistocene and Holocene pollen and charcoal record from peat swamp forest, Lake Sentarum Wildlife Reserve, West Kalimantan, Indonesia. *Palaeogeography, Palaeoclimatology, Palaeoecology*, **171**, 213–228.

ANSHARI, G., KERSHAW, A. P., VAN DER KAARS, S. & JACOBSEN, G. 2004. Environmental change and peatland forest dynamics in the Lake Sentarum area, West Kalimantan, Indonesia. *Journal of Quaternary Science*, **19**, 637–655.

BARLETTA, F., ST-ONGE, G., CHANNELL, J. E. T. & ROCHON, A. 2010. Dating of Holocene western Canadian Arctic sediments by matching paleomagnetic secular variation to a geomagnetic field model. *Quaternary Science Reviews*, **29**, 2315–2324.

BARTON, C. E. & BARBETTI, M. 1982. Geomagnetic secular variation from recent lake sediments, ancient fireplaces and historical measurements in southeastern Australia. *Earth and Planetary Science Letters*, **59**, 375–387.

BARTON, C. E. & MCELHINNY, M. W. 1981. A 10 000 yr geomagnetic secular variation record from three Australian maars. *Geophysical Journal International*, **67**, 465–485.

BARTON, C. E. & POLACH, H. A. 1980. ^{14}C ages and magnetic stratigraphy in 3 Australian maars. *Radiocarbon*, **22**, 728–739.

BRIJKER, J. M., JUNG, S. J. A., GANSSEN, G. M., BICKERT, T. & KROON, D. 2007. ENSO related decadal scale climate variability from the Indo-Pacific Warm Pool. *Earth and Planetary Science Letters*, **253**, 67–82.

CONSTABLE, C. G. 1985. Eastern Australian geomagnetic field intensity over the past 14000 yr. *Geophysical Journal of the Royal Astronomical Society*, **81**, 121–130.

CONSTABLE, C. G. & TAUXE, L. 1987. Palaeointensity in the pelagic realm: marine sediment data compared with archaeomagnetic and lake sediment records. *Geophysical Journal of the Royal Astronomical Society*, **90**, 43–59.

CRAUSBAY, S., RUSSELL, J. & SCHNURRENBERGER, D. 2006. A ca. 800-year lithologic record of drought from sub-annually laminated lake sediment, East Java. *Journal of Paleolimnology*, **35**, 641–659.

DAM, R. A. C., FLUIN, J., SUPARAN, P. & VAN DER KAARS, S. 2001. Palaeoenvironmental developments in the Lake Tondano area (N. Sulawesi, Indonesia) since 33 000 yr B.P. *Palaeogeography, Palaeoclimatology, Palaeoecology*, **171**, 147–183.

DANKERS, P. 1981. Relationship between median destructive field and remanent coercive forces for dispersed natural magnetite, titanomagnetite and hematite. *Geophysical Journal of the Royal Astronomical Society*, **64**, 447–461.

DAY, R., FULLER, M. & SCHMIDT, V. A. 1977. Hysteresis properties of titanomagnetites: grain-size and compositional dependence. *Physics of The Earth and Planetary Interiors*, **13**, 260–267.

DONADINI, F., KORTE, M. & CONSTABLE, C. G. 2009. Geomagnetic field for 0–3 ka: 1. New data sets for global modeling. *Geochemistry, Geophysics and Geosystems*, **10**, Q06007, http://dx.doi.org/10.1029/2008GC002295.

GREMMEN, W. H. E. 1990. Palynological investigations in the Danau Tempe depression, southwest Sulawesi (Celebes) Indonesia. *Modern Quaternary Research in Southeast Asia*, **11**, 123–134.

GUTKNECHT, B. 2008. *Organische Geochemie (CNS-Analytik) an einem Sedimentkern des Kalimpaa Sees (Indonesien)*. BSc thesis, University of Göttingen.

HABERLE, S. G. 1998. Late Quaternary vegetation change in the Tari Basin, Papua New Guinea. *Palaeogeography, Palaeoclimatology, Palaeoecology*, **137**, 1–24.

HABERLE, S. G. & LEDRU, M. P. 2001. Correlations among charcoal records of fires from the past 16 000 years in Indonesia, Papua New Guinea, and Central and South America. *Quaternary Research*, **55**, 97–104.

HABERLE, S. G., HOPE, G. S. & VAN DER KAARS, S. 2001. Biomass burning in Indonesia and Papua New Guinea: natural and human induced fire events in the fossil record. *Palaeogeography, Palaeoclimatology, Palaeoecology*, **171**, 259–268.

HABERZETTL, T., FEY, M. ET AL. 2005. Climatically induced lake level changes during the last two millennia as reflected in sediments of Laguna Potrok Aike, southern Patagonia (Santa Cruz, Argentina). *Journal of Paleolimnology*, **33**, 283–302.

HOPE, G. 2001. Environmental change in the Late Pleistocene and later Holocene at Wanda site, Soroako, South Sulawesi, Indonesia. *Palaeogeography, Palaeoclimatology, Palaeoecology*, **171**, 129–145.

HOPE, G. 2009. Environmental change and fire in the Owen Stanley Ranges, Papua New Guinea. *Quaternary Science Reviews*, **28**, 2261–2276.

HOPE, G., GILLIESON, D. & HEAD, J. 1988. A comparison of sedimentation and environmental change in New Guinea Shallow Lakes. *Journal of Biogeography*, **15**, 603–618.

HOPE, G., CHOKKALINGAM, U. & ANWAR, S. 2005. The stratigraphy and fire history of the Kutai Peatlands, Kalimantan, Indonesia. *Quaternary Research*, **64**, 407–417.

KASPER, T., HABERZETTL, T. ET AL. 2012. Indian Ocean Summer Monsoon (IOSM)-dynamics within the past 4 ka in the sediments of Lake Nam Co, central Tibetan Plateau. *Quaternary Science Reviews*, **39**, 73–85.

KIRLEIS, W., PILLAR, V. & BEHLING, H. 2011. Human–environment interactions in mountain rainforests: archaeobotanical evidence from central Sulawesi, Indonesia. *Vegetation History and Archaeobotany*, **20**, 165–179, http://dx.doi.org/10.1007/s00334-010-0272-0.

KIRSCHVINK, J. L. 1980. The least-squares line and plane and the analysis of palaeomagnetic data. *Geophysical Journal of the Royal Astronomical Society*, **62**, 699–718.

KORTE, M. & CONSTABLE, C. 2003. Continuous global geomagnetic field models for the past 3000 years.

Physics of the Earth and Planetary Interiors, **140**, 73–89.

KORTE, M. & CONSTABLE, C. 2011. Improving geomagnetic field reconstructions for 0–3 ka. *Physics of the Earth and Planetary Interiors*, **188**, 247–259.

KORTE, M., DONADINI, F. & CONSTABLE, C. G. 2009. Geomagnetic field for 0–3 ka: 2. A new series of time-varying global models. *Geochemistry, Geophysics and Geosystems*, **10**.

LANGTON, S. J., LINSLEY, B. K. ET AL. 2008. 3500 yr record of centennial-scale climate variability from the Western Pacific Warm Pool. *Geology*, **36**, 795–798.

LEDU, D., ROCHON, A., DE VERNAL, A. & ST-ONGE, G. 2010. Holocene paleoceanography of the northwest passage, Canadian Arctic Archipelago. *Quaternary Science Reviews*, **29**, 3468–3488.

LINSLEY, B. K., ROSENTHAL, Y. & OPPO, D. W. 2010. Holocene evolution of the Indonesian throughflow and the western Pacific warm pool. *Nature Geoscience*, **3**, 578–583.

LUND, S. P., STOTT, L., SCHWARTZ, M., THUNELL, R. & CHEN, A. 2006. Holocene paleomagnetic secular variation records from the western Equatorial Pacific Ocean. *Earth and Planetary Science Letters*, **246**, 381–392.

MAYR, C., FEY, M. ET AL. 2005. Palaeoenvironmental changes in southern Patagonia during the last millennium recorded in lake sediments from Laguna Azul (Argentina). *Palaeogeography, Palaeoclimatology, Palaeoecology*, **228**, 203–227.

MAZAUD, A. 2005. User-friendly software for vector analysis of the magnetization of long sediment cores. *Geochemistry, Geophysics and Geosystems*, **6**, Q12006, http://dx.doi.org/10.1029/2005GC001036

MCCORMAC, F., HOGG, A., BLACKWELL, P., BUCK, C., HIGHAM, T. & REIMER, P. 2004. SHCal104 Southern Hemisphere calibration 0–11.0 cal kyr BP. *Radiocarbon*, **46**, 1087–1092.

MERKER, S. 2003. *Vom Aussterben bedroht oder anpassungsfähig? – Der Koboldmaki Tarsius dianae in den Regenwäldern Sulawesis*. PhD thesis, University of Göttingen.

MORLEY, R. J. 1981. Development and vegetation dynamics of a lowland Ombrogenous peat swamp in Kalimantan Tengah, Indonesia. *Journal of Biogeography*, **8**, 383–404.

NEWTON, A., THUNELL, R. & STOTT, L. 2006. Climate and hydrographic variability in the Indo-Pacific Warm Pool during the last millennium. *Geophysics Research Letters*, **33**, L19710.

OPPO, D. W., ROSENTHAL, Y. & LINSLEY, B. K. 2009. 2000-Year-long temperature and hydrology reconstructions from the Indo-Pacific warm pool. *Nature*, **460**, 1113–1116.

PAGE, S. E., WÜST, R. A. J., WEISS, D., RIELEY, J. O., SHOTYK, W. & LIMIN, S. H. 2004. A record of Late Pleistocene and Holocene carbon accumulation and climate change from an equatorial peat bog (Kalimantan, Indonesia): implications for past, present and future carbon dynamics. *Journal of Quaternary Science*, **19**, 625–635.

REIMER, P. J., BROWN, T. A. & REIMER, R. W. 2004. Discussion: reporting and calibration of post-bomb ^{14}C data. *Radiocarbon*, **46**, 1299–1304.

RICHTER, C., VENUTI, A., VEROSUB, K. L. & WEI, K. Y. 2006. Variations of the geomagnetic field during the Holocene: Relative paleointensity and inclination record from the West Pacific (ODP Hole 1202B). *Physics of the Earth and Planetary Interiors*, **156**, 179–193.

SPOONER, M. I., BARROWS, T. T., DE DECKKER, P. & PATERNE, M. 2005. Palaeoceanography of the Banda Sea, and Late Pleistocene initiation of the Northwest Monsoon. *Global and Planetary Change*, **49**, 28–46.

STONER, J. S. & ST-ONGE, G. 2007. Magnetic stratigraphy in paleoceanography: reversals, excursions, paleointensity, and secular variation. *In: Developments in Marine Geology*. Elsevier, Amsterdam, **1**, 99–138.

ST-ONGE, G., STONER, J. S. & HILLAIRE-MARCEL, C. 2003. Holocene paleomagnetic records from the St. Lawrence Estuary, eastern Canada: centennial- to millennial-scale geomagnetic modulation of cosmogenic isotopes. *Earth and Planetary Science Letters*, **209**, 113–130.

STUIJTS, I., NEWSOME, J. C. & FLENLEY, J. R. 1988. Evidence for late quaternary vegetational change in the Sumatran and Javan highlands. *Review of Palaeobotany and Palynology*, **55**, 207–216.

STUIVER, M. & REIMER, P. 1993. Extended ^{14}C database and revised CALIB radiocarbon calibration program. *Radiocarbon*, **35**, 215–230.

SUPARAN, P., DAM, R. A. C., VAN DER KAARS, S. & WONG, T. E. 2001. Late Quaternary tropical lowland environments on Halmahera, Indonesia. *Palaeogeography, Palaeoclimatology, Palaeoecology*, **171**, 229–258.

TAUXE, L. 1993. Sedimentary records of relative paleointensity of the geomagnetic field: theory and practice. *Reviews in Geophysics*, **31**, 319–354.

VAN DER KAARS, W. A. & DAM, M. A. C. 1995. A 135 000-year record of vegetational and climatic change from the Bandung area, West-Java, Indonesia. *Palaeogeography, Palaeoclimatology, Palaeoecology*, **117**, 55–72.

VAN DER KAARS, S., WANG, X., KERSHAW, P., GUICHARD, F. & ARIFIN SETIABUDI, D. 2000. A Late Quaternary palaeoecological record from the Banda Sea, Indonesia: patterns of vegetation, climate and biomass burning in Indonesia and northern Australia. *Palaeogeography, Palaeoclimatology, Palaeoecology*, **155**, 135–153.

VAN DER KAARS, S., PENNY, D., TIBBY, J., FLUIN, J., DAM, R. A. C. & SUPARAN, P. 2001. Late Quaternary palaeoecology, palynology and palaeolimnology of a tropical lowland swamp: Rawa Danau, West-Java, Indonesia. *Palaeogeography, Palaeoclimatology, Palaeoecology*, **171**, 185–212.

VILLENEUVE, M., GUNAWAN, W., CORNEE, J. J. & VIDAL, O. 2002. Geology of the central Sulawesi belt (eastern Indonesia): constraints for geodynamic models. *International Journal of Earth Sciences*, **91**, 524–537.

WEBER, R. 2006. Kulturlandschaftswandel in Zentralsulawesi Historisch-geographische Analyse einer indonesischen Bergregenwaldregion. *Pazifik Forum Band*, **12**, 42–49.

YANG, X., HELLER, F., YANG, J. & SU, Z. 2009. Paleosecular variations since $\sim$9000 yr BP as recorded by sediments from maar lake Shuangchiling, Hainan, South China. *Earth and Planetary Science Letters*, **288**, 1–9.

YULIANTO, E., RAHARDJO, A. T., NOERADI, D., SIREGAR, D. A. & HIRAKAWA, K. 2005. A Holocene pollen record of vegetation and coastal environmental changes in the coastal swamp forest at Batulicin, South Kalimantan, Indonesia. *Journal of Asian Earth Sciences*, **25**, 1–8.

On the directional geomagnetic signature of the Pringle Falls excursion recorded at Pringle Falls, Oregon, USA

EMILIO HERRERO-BERVERA[1] & EDGARDO CAÑÓN-TAPIA[2]

[1]*Paleomagnetics and Petrofabrics Laboratory, SOEST-HIGP,
University of Hawaii at Manoa, 1680 East West Road, Honolulu, HI 96822, USA*

[2]*Division of Earth Sciences, CICESE, PO Box 434843, San Diego, CA 92143, USA*

**Corresponding author (e-mail: herrero@soest.hawaii.edu)*

Abstract: We studied the detailed characteristics of the Pringle Falls excursion from samples at the original site recovered from four profiles drilled along the Deschutes River, Oregon. We drilled 827 samples spaced along 5 km for their detailed directional study. The profiles registered a high-resolution (>10 cm/ka) palaeomagnetic record of the excursion ($c.$ 211 $\pm$ 13 ka) recorded by diatomaceous lacustrine sediments. We conducted palaeomagnetic and rock magnetic studies to investigate the reproducibility of the signal throughout the profiles. We performed low-field susceptibility v. temperature analysis that indicated that the main magnetic carrier is pure magnetite (Curie point 575 °C). The magnetic grain size also indicated single domain–multi domain (SD–MD) magnetite. The demagnetization was performed by alternating field experiments and the mean directions were determined by principal component analyses. The detailed behaviour of the palaeosignal is highly consistent since they are rapidly deposited sediments providing a detailed representation of the palaeofield. The dissected virtual geomagnetic pole paths in three different phases are highly internally consistent and are defined by clockwise and anticlockwise loops travelling from high northern latitudes over eastern North America and the North Atlantic to South America and then to high southern latitudes; then they return to high northern latitudes through the Pacific and over to Kamchatka.

An improved understanding of the origin of the geomagnetic field, and the process by which it reverses its polarity, is a longstanding goal in the Earth Sciences. One of the means by which this goal can be achieved is to examine, in detail, the temporal variations in the geomagnetic field direction and intensity associated with polarity reversals and short polarity excursions or events. One of the more fruitful recent efforts in this field is the utilization of geodynamo models (e.g. Glatzmaier & Roberts 1995; Camps & Prevot 1996; Glatzmaier *et al.* 1999; Coe *et al.* 2000; Hoffman & Singer 2008; Valet *et al.* 2008*a, b*) to test and refine models for geomagnetic field origin and reversal mechanisms, with comparisons made between spatially and temporally robust sets of full-vector (direction and palaeointensity) geomagnetic field data. One consequence of these efforts has been the recognition that there is a need for many high-quality records of geomagnetic field behaviour from a well-distributed set of site locations for a given time interval of geomagnetic field evolution. In the case of short polarity events or excursions, a number of high-quality records of palaeomagnetic directions and absolute palaeointensities from a variety of locations are needed to gain a full understanding of the origin of these

geomagnetic field phenomena and how their behaviour compares with the process by which full reversals of the geomagnetic field occur. Such data could ultimately be used to determine if polarity excursions and reversals are similar geomagnetic field phenomena. In particular, two questions are still pending: are excursions 'aborted reversals' (e.g. Merrill & McFadden 1999; Hoffman & Singer 2004), or are they possibly manifestations of a different process (e.g. Gubbins 1999; Valet *et al.* 2008*a, b*)? Finding answers to these questions certainly would improve the knowledge and understanding of geodynamo processes in general.

While there are many excellent studies of geomagnetic field polarity reversals (Valet *et al.* 1986, 1988*a, b*, 1989, 1992; Clement 1991; Clement & Kent 1991; Laj *et al.* 1991, 1992*a, b*; Tric *et al.* 1991*a, b*; Clement & Martinson 1992; Herrero-Bervera & Khan 1992; van Hoof & Langereis 1992*a, b*; McFadden *et al.* 1993; Zhu *et al.* 1994; Cisowski 1995; Channell & Lehman 1997; Herrero-Bervera & Coe 1999; Clement 2004; Herrero-Bervera & Valet 2005; Herrero-Bervera *et al.* 2007; Valet, J. P. & Herrero-Bervera, E. 2010; and others), there are very few full-vector studies of short-lived polarity events or excursions. The majority of such

From: JOVANE, L., HERRERO-BERVERA, E., HINNOV, L. A. & HOUSEN, B. A. (eds) 2013. *Magnetic Methods and the Timing of Geological Processes.* Geological Society, London, Special Publications, **373**, 261–278. First published online December 7, 2012, http://dx.doi.org/10.1144/SP373.12 © The Geological Society of London 2013.
Publishing disclaimer: www.geolsoc.org.uk/pub_ethics

data for excursions are from sedimentary rocks, for which only relative palaeointensity values can be inferred (Levi & Banerjee 1976; Meynadier *et al.* 1992). Studies of relative palaeointensity derived from sedimentary rocks typically find correlations between relative palaeointensity lows and geomagnetic field reversals or excursions (e.g. Meynadier *et al.* 1992, 1994; Valet & Meynadier 1993; Verosub *et al.* 1996; Channell 1999; Guyodo & Valet 1999; Lund *et al.* 2001; Channell *et al.* 2002; Laj *et al.* 2006; Channell 2006; Fuller 2006). While these types of studies are of great value too, today the interpretation of relative palaeointensity data recovered from sediments and sedimentary rocks characterized by high sedimentation rates (e.g. high-resolution records) is relatively simple. Certainly, if the directional data associated with these records is available (i.e. data are obtained from azimuthally oriented marine and lacustrine sediment cores), the information thus obtained augments its value. Unfortunately, owing to the relatively short duration of the geomagnetic field reversal/event process, volcanic and sediments/sedimentary records of geomagnetic field behaviour during reversals or excursions are relatively rare. There are only a handful of directional and palaeointensity records from polarity excursions (Coe *et al.* 1984; Roperch *et al.* 1988; Chauvin *et al.* 1989, 1991; Carlut *et al.* 1999; Tanaka & Kobayashi 2003; Riisager *et al.* 2004; Herrero-Bervera & Valet 2005; Cassata *et al.* 2008). Most of these studies document separate excursions (ranging in age from the Eocene to the Pliestocene), so there is clearly a need for additional studies of particular excursions recorded in several widely-spaced geographic areas (Merrill & McFadden 2005).

Of the short polarity excursions, the Blake and Pringle Falls polarity episodes are among the better recognized and studied, but they remain somewhat enigmatic. The age of the Blake event is generally reported as 110 ka (Tric *et al.* 1991*a*, *b*) based on studies of sediment cores from many parts of the world. In igneous rocks, partial records of the Blake event (usually a single cooling unit) have been recorded in China (Zhu *et al.* 2000) and in tephras from Japan (Takai *et al.* 2002). Of these studies, K–Ar ages provide age estimates of 123 ± 7 ka (Zhu *et al.* 2000), and an older age of 141 ka for the Aso-2 tephra (Takai *et al.* 2002). The nature of the geomagnetic field during the Blake event is complex and somewhat controversial. Some studies have documented two excursion 'pulses' (Herrero-Bervera *et al.* 1989; Tric *et al.* 1991*a*, *b*) or three (Zhu *et al.* 1994), spanning 10–50 ky, while in some sedimentary sequences the Blake event is absent in the directional records, and is at times only manifest by a low in relative palaeointensity values (see e.g. Channell 2006; Stoner & St-Onge 2007). Thus,

excursions are a very important part of the geodynamo behaviour over geological time scales that, in addition to reversals of the main dipole field, need to be fully studied and definitely considered when evaluating and assessing whether theoretical and numerical simulation of the dynamo produces Earth-like results (e.g. Herrero-Bervera & Helsley 1993; Herrero-Bervera & Runcorn 1997; Coe *et al.* 2000).

In this paper we present the directional behaviour of the geomagnetic signature of four successful records and their associated palaeomagnetic and rock magnetic characteristics of a short-duration geomagnetic event recorded at Pringle Falls, Oregon. This event took place at *c.* 211 ± 13 ka, based on a variety of age determinations from either tephras in sedimentary rocks, or from a limited number of locations where volcanic rocks record this event. This event can be considered to represent a very well-documented aborted reversal, particularly in terms of its age, directions and anisotropy of magnetic susceptibility (AMS) correlations (Herrero-Bervera *et al.* 1994; Singer *et al.* 2008; Canon-Tapia & Herrero-Bervera 2009). One of the greatest advantages of the record presented here is that the sediment cores have been azimuthally oriented, therefore allowing us to characterize the directional behaviour of the palaeofield with a high degree of confidence. Since our Pringle Falls record is characterized by an extremely high temporal resolution, this study will improve the knowledge of short excursions, which is extremely important now that excursions are regarded as two successive reversals bracketing an aborted polarity interval (Valet *et al.* 2008*a*, *b*).

Pringle Falls, Oregon

As described by MacLeod *et al.* (1982), the sedimentary lake sequence sampled originally for palaeomagnetic studies by Herrero-Bervera *et al.* (1989, 1994) as well as the subsequent re-visited sites sampled for this study, including the two additional profiles, are part of an extensive pre-historic fluvial and lacustrine system formed during the last million years located east of the Cascade mountains. The most consistent interpretation of the local geology suggests that the lake in question is the result of a late Pliocene/Pleistocene increase of the base level east of the sampling area close to the western margin of the Basin and Range structural province, intimately related to the development of the extensive volcanism associated with the Newberry volcano.

At the sampling sites, the lacustrine sequence is overlain by glacial sand and gravel. The unconformity has *c.* 30 m of local relief. It is unlikely that we have sampled the youngest portion of the sequence, as glacial erosion has removed at least 20 m of section at this particular site. The age of the lacustrine sequence is constrained by three events. These are (from the oldest to the youngest): (a) according to MacLeod *et al.* (1982) the onset of volcanism for the Newberry volcano, which is presumed to be younger than 700 ka, as all portions of the volcano are <500 ka; (b) a single diatom age on a sample taken at 195 m in a deep well that penetrates into 410 m of lake sediment which was dated as 600–200 ka based upon the similarity to the diatoms at Clear Lake; and (c) an upper-boundary date of 18–15 ka, based on the age of the latest glaciation in the area.

If one assumes that the basin has an age of no more than 1 Ma, and that the 410 m of section drilled near Bend, Oregon represents a typical maximum thickness, then the youngest sediment can be no younger than 68 ka (20 m times 2500 year m^{-1}, average sedimentation rate plus age of unconformity of 18 ka). Thus, we have sampled the Pringle Falls excursion even though the samples that contain the Pringle Falls geomagnetic polarity episode record come from the very top of our exposed section (see Herrero-Bervera *et al.* 1989, 1994). We have expanded the study of the original directional study (Herrero-Bervera *et al.* 1989, 1994) and radiometric dating (Singer *et al.* 2008) of the original Pringle Falls site. We have, thus far, drilled five separate sites (see Figs 1 & 2) that are currently under investigation and are presented and discussed in this paper.

Palaeomagnetic field and laboratory work

Sampling procedure

We drilled a total of 827 samples, and we studied for our palaeomagnetic analyses 1397 specimens (i.e. by means of a gasoline-powered portable drill and orientation using a magnetic compass). The lengths of the recovered cores ranged from 5 to 15 cm in length and were drilled with inclination angles from 3° to *c.* 60° with respect to the horizontal plane, allowing us to have a continuous sampling of the sediments in question. In some instances, this sampling technique provided an independent observation of the same stratigraphic level, as was the case for the three profiles (i.e. Profiles named 1989 La Pine and 1, 2 and 3) published by Herrero-Bervera *et al.* (1989, 1994) and Canon-Tapia & Herrero-Bervera 2009 (see Figs 1 and 2). The two additional profiles 1 and 3 were also drilled from the uppermost portions of the diatomaceous sediments in the lacustrine sedimentary sequences adjacent to Pringle Falls (43.7°N, 238°E) along the Deschutes River in Oregon, as was the case for the profiles published already. Profile 4 shown in Figures 1 and 2 is a very short segment of sediments that did not record the Pringle Falls excursion. Therefore such segment of sediments has been omitted from the description and interpretation of the excursion.

The four widely separated localities are excellent exposures owing to recent erosion near the falls, and

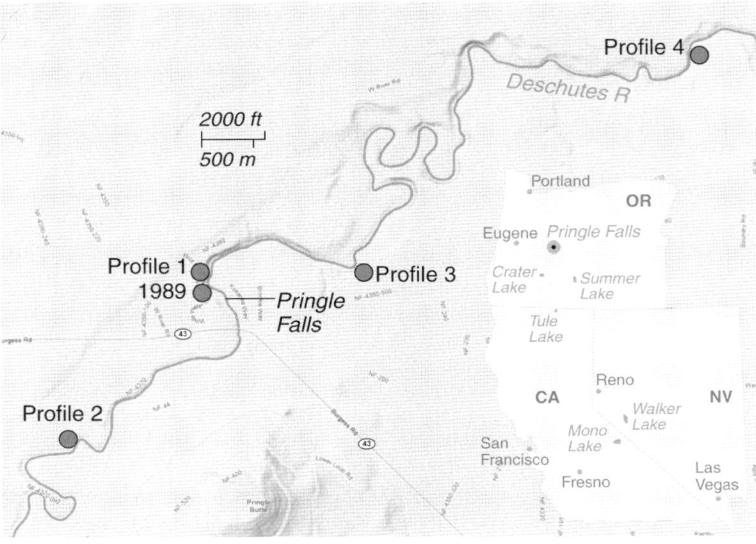

Fig. 1. Map of the southeast portion of Pringle Falls. Outcrops of the five sampling sites of our study. Profile 4 depicted in the figure is a site that was sampled but the sediments did not show a record of the Pringle Falls excursion.

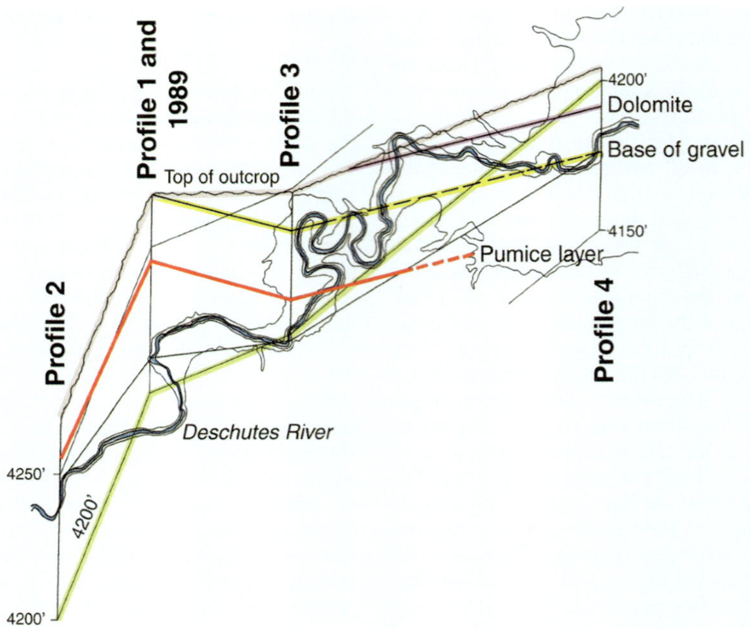

Fig. 2. Map of the sampling sites of our current study of the palaeosignal of Pringle Falls. The distance between Profile 2 and Profile 4 is 5 km. Stratigraphic correlation of the expanded study of the Pringle Falls episode along the Deschutes River in Oregon. Notice the Pumice Layer used as a marker horizon to correlate the sites. Profile 4 depicted in the figure is a site that was sampled but the sediments did not show a record of the Pringle Falls excursion.

also by the fact that the sections were drilled from the continuous exposures of flat-lying sediments. All the collected samples were cut into c. 2.5 cm-long cylinders and were measured on JR-5 spinner, ScT and 2G 755 SRM cryogenic magnetometers at the SOEST-HIGP Paleomagnetics and Petrofabrics Laboratory.

Laboratory experiments

Rock magnetic experiments

Magnetic properties were analysed to identify the magnetic carriers of the natural remanent magnetization (NRM) and to investigate the origin of the NRM. Studies of magnetic mineralogy were performed first using at least one sample from each of the five profiles under study. Low-field suscepti-bility v. temperature $(k–T)$ experiments were conducted in air using a KYL2 instrument in order to determine the Curie temperature of the samples. Thirteen specimens were progressively heated from room temperature up to 700 °C and subsequently cooled down using a KLY2-CS3 apparatus (Hrouda 1994; Hrouda *et al.* 1997) located at the SOEST-HIGP Petrofabrics and Paleomagnetics Labo-ratory. Several typical diagrams of susceptibility

v. temperature $(k–T)$ are shown in Figure 3. The curves have very similar heating and cooling pat-terns. Both show the presence of the Curie tempera-ture of pure magnetite.

We found that all of the specimens studied had reversible heating and cooling results, with single inflection points, indicating Curie temperatures between 550 and 565 °C. We interpreted these data to indicate the presence of low-Ti magnetite to pure magnetite as the primary magnetic minerals in these samples. The most probable sources of the magnetic carriers are therefore wind-blown Ti-low magnetites from the Newberry and Three Sisters volcanoes located close by the sampling sites.

Magnetic granulometry from hysteresis experiments

Magnetic hysteresis measurements were performed on c. 200 mg of powder from the diatomaceous lacustrine silt sediments using a variable field trans-lation balance up to 1.2 T. Saturation remanent mag-netization (M_r), saturation magnetization (M_s), and coercive force (H_c) were calculated after removing the paramagnetic contribution. We determined the hysteresis loops and the back-field demagnetiza-tion curve of the saturation isothermal remanent

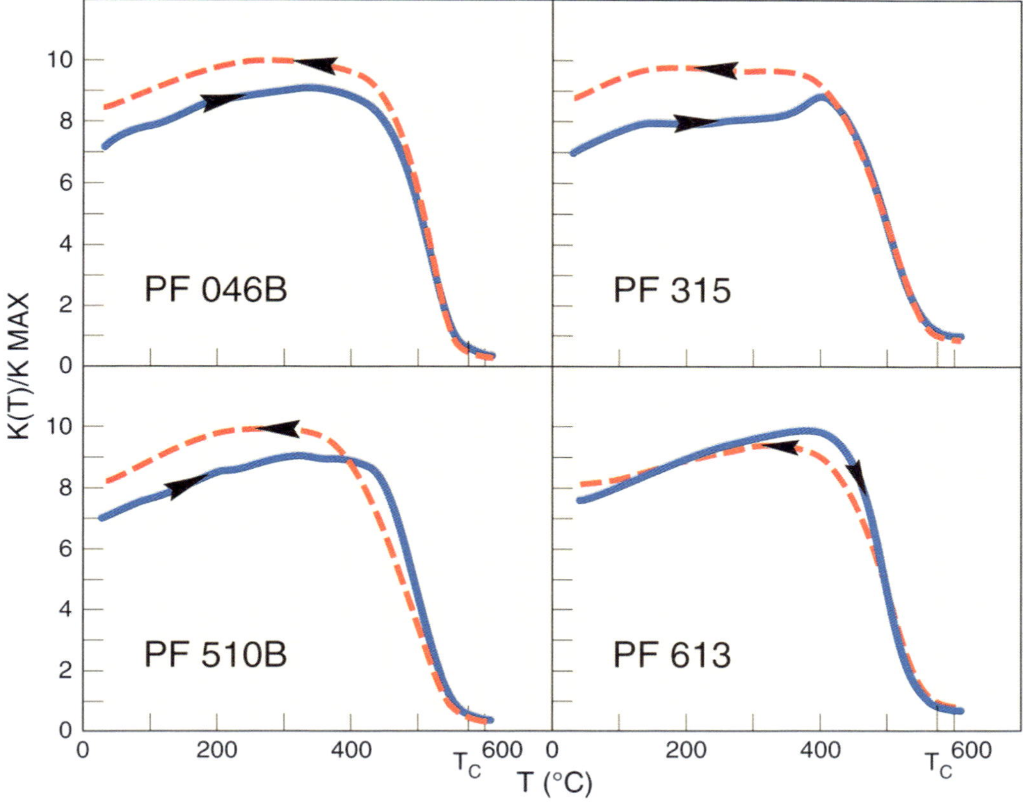

Fig. 3. Example thermomagnetic curves for lacustrine sediments. Data obtained from the HIGP KLY-2 low-field v. temperature (k–T) susceptibility instrument. These samples, as well as other samples analysed, have simple thermomagnetic behaviour, with Curie temperatures ranging from 550 to 565 °C. The samples also have very good reversibility upon cooling, indicating little alteration of the magnetic mineralogy during heating and cooling experiments. Blue curve is heating and red indicates cooling.

magnetization. The variable field translation balance instrument has a measuring range of 10^{-8}–10^{-2} A m^2. The coercivity of remanence (H_{cr}) suggests that the isothermal remanent magnetization is carried by low-coercivity grains. The ratios of hysteresis parameters were plotted in Figure 4 as a Day diagram (Day *et al.* 1977) following recent modifications by Dunlop (2002) for type curves and regions that have been defined for pure magnetite. Most grain sizes are tightly clustered within the pseudo-single domain range for the Pringle Falls profiles. It is striking that the distribution of coercivities is related to the unblocking temperatures. If we do not consider the influence of grain sizes, it is interesting to associate this distribution with the magnetic phases that are present in the sample. The hysteresis parameters have been initially defined from synthetic crystals of magnetite (Dunlop & Özdemir 1997, 2000). This would

probably explain why most studies deal with samples that are actually found within the pseudo-single domain range (e.g. Herrero-Bervera & Valet 2003, 2009).

NRM and demagnetization experiments

The remanent magnetization was measured with a 2G 755 Superconducting Rock Magnetometer (SRM) housed in the shielded room of the SOEST-HIGP Petrofabrics and Paleomagnetics Laboratory of the University of Hawaii. The samples under question were stepwise demagnetized by alternating fields (a.f.) from 5 to 60 mT. Typical demagnetization diagrams obtained are shown in Figure 5. The stable direction of the characteristic remanent magnetization (ChRM) was determined with no ambiguity. The ChRM was calculated using principal component analysis for the demagnetization

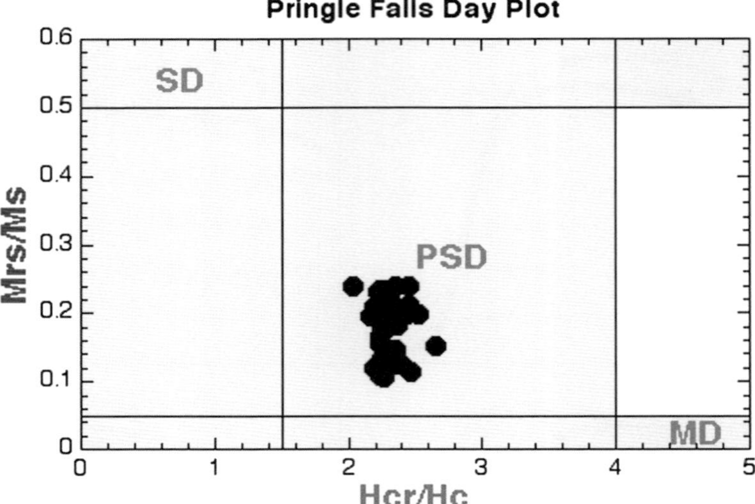

Fig. 4. Day plot showing magnetic grain size variations of Pringle Falls diatomaceous lacustrine silt sediments. Diagram represent the Hysteresis parameters, M_r/M_s v. H_{cr}/H_c plotted in the manner of Day *et al.* (1977) corrected according to Dunlop (2002). As is depicted in this figure we have found that the great majority of the specimens studied cover a range of magnetic sizes located in the pseudo single domain (PSD) area of the Day plot.

diagrams with a well-defined component trending to the origin.

The samples either showed consistent directions of magnetization to within 5° of the initial value for a.f. demagnetization up to 60 mT or showed a removal of a low-coercivity component and then univectorial behaviour (see Figs 5 & 6). The vectorial diagrams show that a characteristic component of magnetization was isolated by *c.* 15 mT and that a univectorial component (Zijderveld 1967) was isolated for the typical samples shown in Figure 6. No bias or systematic departure from the origin was accepted and in all cases the ChRM relies on a minimum of seven successive and up to 16 directions isolated during a.f. stepwise demagnetization. Complete demagnetization of the samples was obtained with a.f. demagnetization techniques. The demagnetization curves confirm that magnetite carries the NRM with a soft resistance to alternating fields.

Natural remanent magnetization results

The magnetostratigraphic results obtained previously and published by Herrero-Bervera *et al.* (1989) of the so-called La Pine profile are shown in Figure 7a. The magnetostratigraphic results of Profile 2 are also shown in Figure 7b and have already been published by Herrero-Bervera *et al.* (1994). The separation between the two sites is approximately 2 km, as can be seen in Figure 1, and both were drilled along the Deschutes River as depicted

in Figures 1 and 2. The demagnetization results of both profiles show that the directional characteristics in terms of the declination, inclination and intensity of magnetization of the drilled samples are repeated arguing for an excellent intrabasinal correlation of the two sites since the geomagnetic inclinations features A, B and C are conspicuous in both records. The second important characteristic of the La Pine record is the well-defined normal polarity from the base of the section (20 m from the top) to *c.* 7 m from the top.

The inclination from 7 to 20 m shows a close similarity to that of the present ambient field (62°) at the site from which the samples were obtained. The observed inclination values (61 ± 2°; $\alpha_{95} =$ 2.6°) were not significantly different from each other or from the inclination of the axial dipole field at the site, which was 62° as reported by Herrero-Bervera *et al.* (1989). This agreement between the observed inclination and that given by the geocentric axial dipole formula can be taken as evidence for the lack of an inclination error, especially as the data are averaged over a temporal and spatial extent that is satisfactory. More details of the demagnetized results can be found in the original paper published by Herrero-Bervera *et al.* (1989), particularly those related to the 20 m intensity record of the La Pine profile.

In order to correlate the Pringle Falls palaeosignal and verify the reproducibility of the directional magnetic signature of the excursion, we have studied two additional profiles. Figure 8 shows

Pringle Falls Excursion

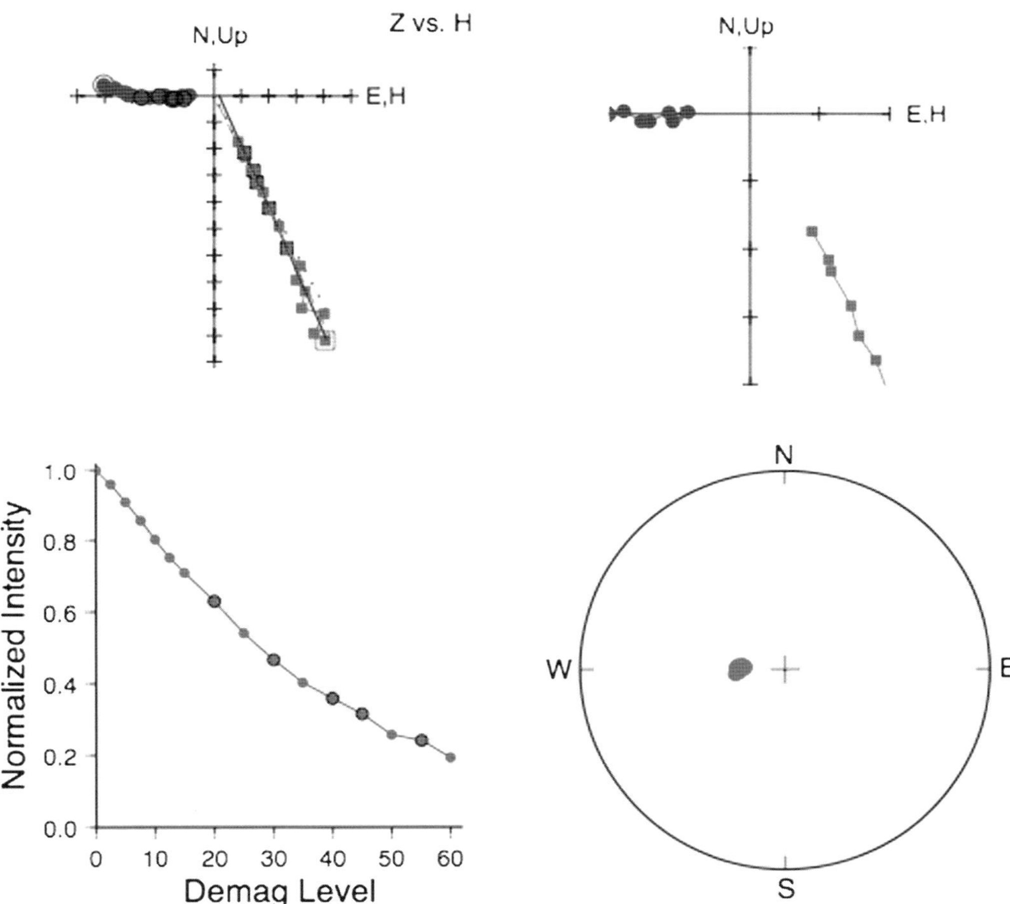

Fig. 5. Alternating field demagnetization results from an 'excursional' Pringle Falls sample. The top diagram shows vector end points of palaeomagnetic directions on orthogonal demagnetization diagrams or modified Zijderveld (1967) plots (squares are inclinations and circles are declinations depicting the z-axis v. the horizontal plane). The second orthogonal diagram on the right is an enlargement of the plot on the left side of the figure. The best-fit lines from principal component analysis (PCA) are shown for the Free PCA option (solid black line) and Anchored PCA option (dashed green line) for the vertical component only. The lower diagram shows the normalized intensity variation with progressive demagnetization as well as the stereographic projections of the declinations and inclinations.

the directional results, that is, inclination, declination and intensity of magnetization of Profile 1 (Fig. 8a) located adjacent (about 5 m apart) to the so-called La Pine profile, and also an additional fourth profile named Profile 3 (see Fig. 8b). Profile 1 is approximately 8.5 m thick whereas Profile 3 is approximately *c.* 18.4 m thick, despite the fact that such profile (i.e. Profile 3) is located about *c.* 1.7 km east from Profiles 1 and the original La Pine 1989 site (see Figs 1 & 2). The red arrows of Figure 8a,

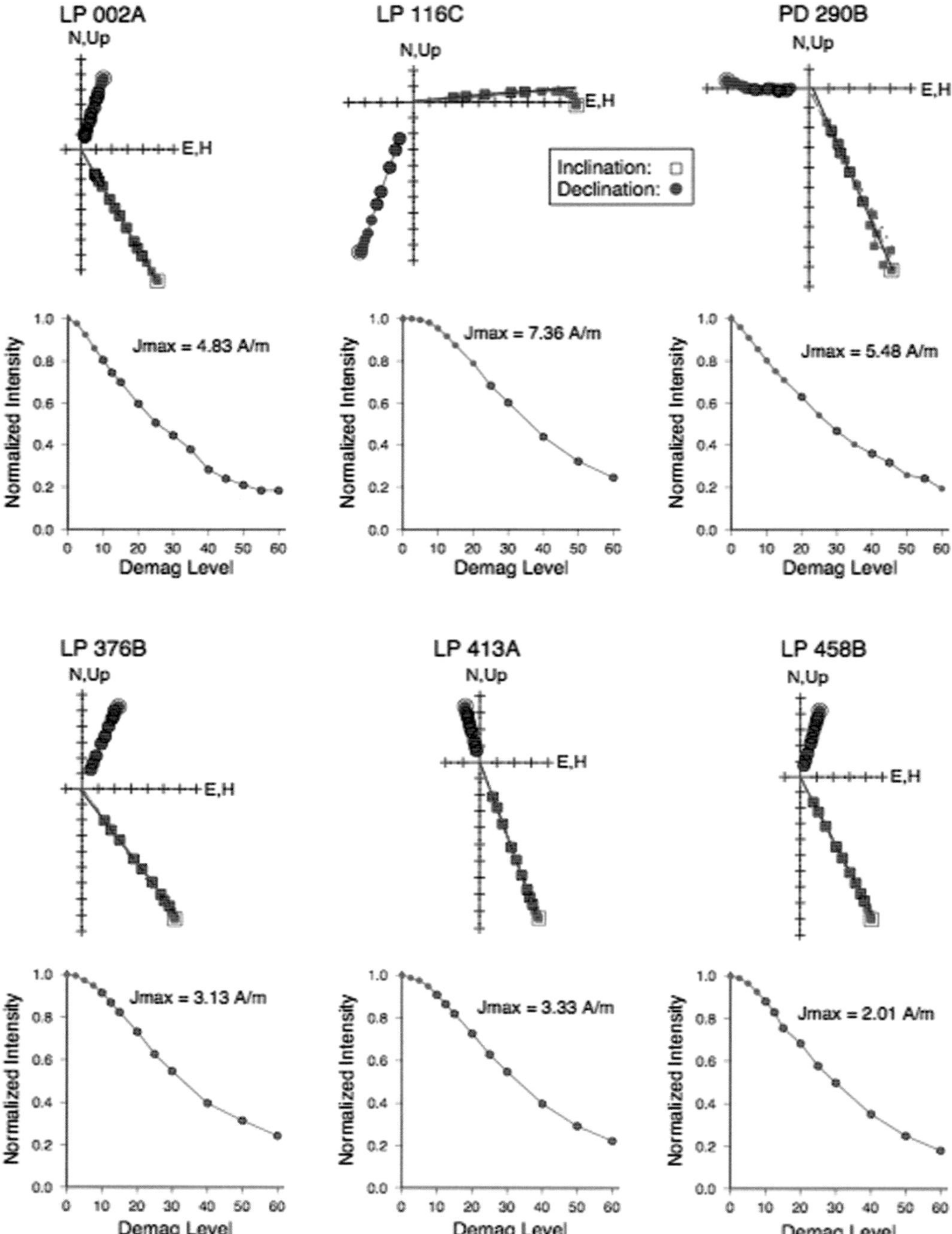

Fig. 6. Alternating field demagnetization results from six Pringle Falls samples. For each sample, the top diagram shows vector end points of palaeomagnetic directions on orthogonal demagnetization diagrams or modified Zijderveld plots (Zijderveld 1967) (squares are inclinations and circles are declinations). The best-fit lines from principal component analysis (PCA) are shown for the FREE PCA option (solid black line) and ANCHORED PCA option (dashed green line) for the vertical component only. For each sample, the lower diagram shows the normalized intensity variation with progressive demagnetization.

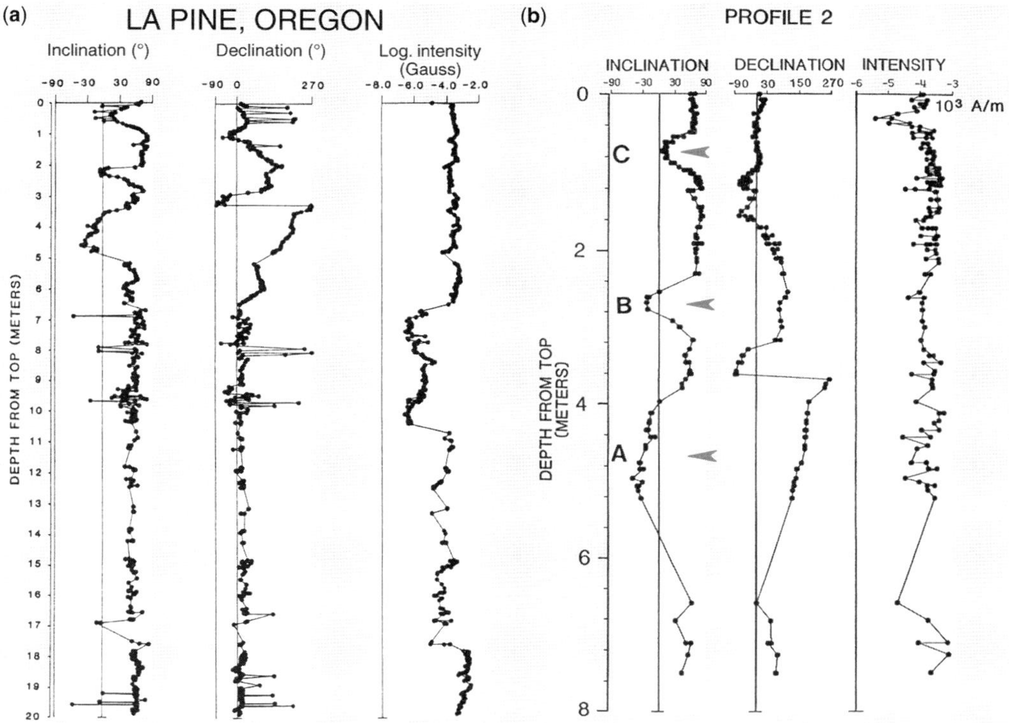

Fig. 7. (a) Magnetostratigraphic plot of declination, inclination and intensity of magnetization of the Pringle Falls Profile, originally named 'La Pine, Oregon', showing the characteristic excursional features that represent the magnetic signature of the Pringle Falls polarity episode (from Herrero-Bervera *et al.* 1989). (b) Magnetostratigraphic plot of declination, inclination and intensity (D, I, J) record of Pringle Falls Profile 2 depicting the characteristic features A, B and C as indicated by the red arrows. The distance between the two records is approximately 1.5 km (from Herrero-Bervera *et al.* 1994).

b indicate the three characteristic geomagnetic features of the Pringle Falls excursion, namely A, B and C. Figures 8b and 9 do not show the place where feature C should be recorded, and we argue that it is a gap where we do not have samples owing to a possible lack of suitable material for the study of such geomagnetic features.

It is also very important to mention that Figure 2 shows Profile 4 which is higher in the section than the other profiles (i.e. Profiles 1989, 1, 2 and 3) but we are not showing the magnetostratigraphic results owing to the lack of a record of the Pringle Falls excursional features, namely A, B and C, recorded in the other profiles under question.

The reproducibility of the palaeosignal can be easily shown by putting together four inclination records where the geomagnetic characteristic features A, B and C are conspicuous and where a correlation can be attained amongst the four sites in question. Figure 9 shows such intra-basinal correlation.

Virtual geomagnetic pole data and paths

The declination and inclination data presented in Figures 7 and 8 can be converted to virtual geomagnetic pole (VGP) plots representing the apparent motion of the pole during this event (Fig. 10). The directional data obtained at Pringle Falls (from four sites 1.5 and 3.0 km apart) are very detailed with a great number of transitional directions, from which we have calculated successive VGP positions. Figures 7 and 8 depict the directional records (e.g. D, I and J) from the four localities and Figure 10 shows the totality of the VGPs obtained from the four profiles in question. It can be seen that the transitional VGPs and directions that make up the polarity episode are represented from the bottom to the top of the inclination records (see Figs 7–9) in both cases.

The descriptive characteristics of the Pringle Falls VGPs of the three profiles published so far, that is, La Pine (Herrero-Bervera *et al.* 1989),

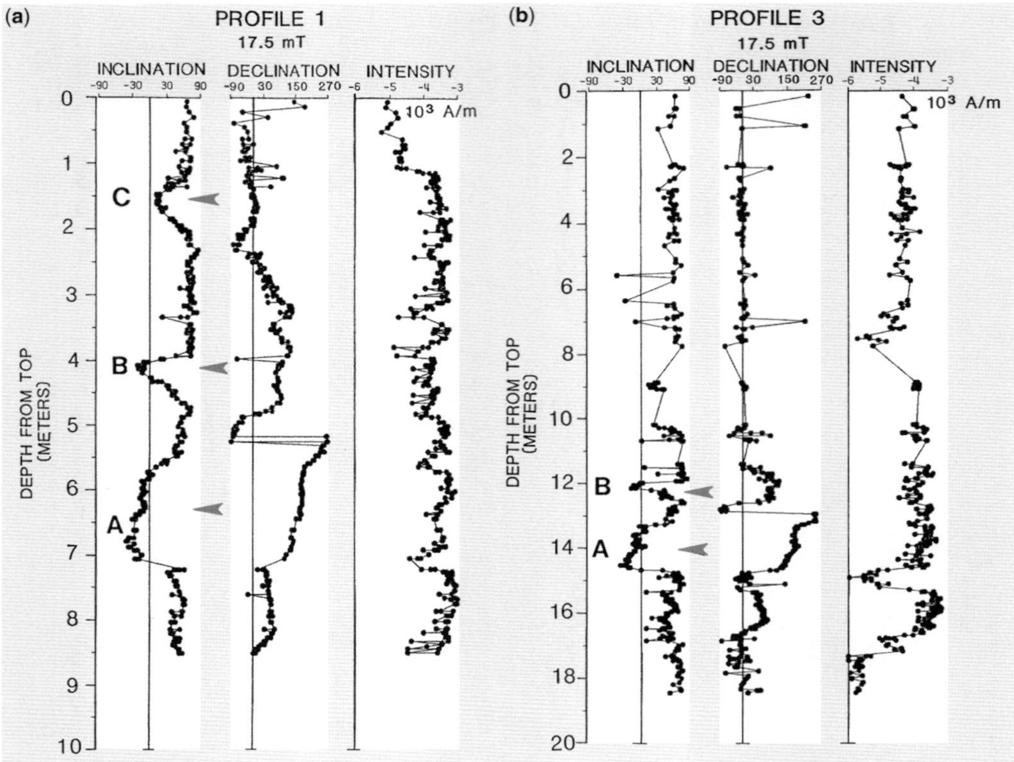

Fig. 8. (**a**) Magnetostatigraphic plot of declination, inclination and intensity of magnetization (D, I, J) Pringle Falls Profile 1, showing the characteristic excursional features, A, B and C as depicted by the red arrows that represent the magnetic signature of the Pringle Falls polarity episode. (**b**) Magnetostratigraphic plot of declination, inclination and intensity record of Pringle Falls Profile 3, depicting the characteristic features A and B, as indicated by the red arrows; geomagnetic feature C was not recorded at Profile 3. The distance between the two records is approximately 3.0 km.

Profiles 1 and 2 (Herrero-Bervera *et al.* 1989, 1994; Herrero-Bervera and Helsley 1993; Valet *et al.* 2008*a*, *b*), have been analysed from the view point of a geometrical configuration including the totality of the individual VGPs plotted up together. The two additional profiles that have not been published so far, that is, Profiles 1 and 3, also show the same general geometric characteristics as La Pine and Profile 2. The unique geomagnetic characteristics of the poles depicted in Figure 10 include rapid directional changes and slower periods or stand-still intervals. Also shown is the preferred behaviour of the VGP path(s) over the Americas and over Asia (Laj *et al.* 1991, 1992*a*, *b*). In addition to the travel of the VGPs through the Pacific Ocean, there is the antipodal behaviour of the path travelled between the eastern part of South America, the central Pacific and the final lingering of the poles to the Kamchatka Peninsula.

The composite VGP loop in Figure 10 suggests that this excursion may represent an aborted reversal

(e.g. Hoffman 1981) because the data include directions that are virtually antipodal to the present field. The overlapped VGP paths of the four studied profiles show a high degree of reproducibility of the palaeosignal, as was demonstrated when we analysed the directional records depicted in Figures 7–9. These overlapping VGP paths shown at the bottom of Figure 10 also attest to the intra-basinal reproducibility of the Pringle Falls palaeosignal of the excursion or 'aborted reversal' at the original Pringle Falls site sampled along the Deschutes River in Oregon.

The identification of the Pringle Falls geomagnetic signature by means of VGP paths

In order to have a clear visualization of the different stages of the evolution of the Pringle Falls excursional palaeofield we have broken down the VGP paths into three separate phases that encompass

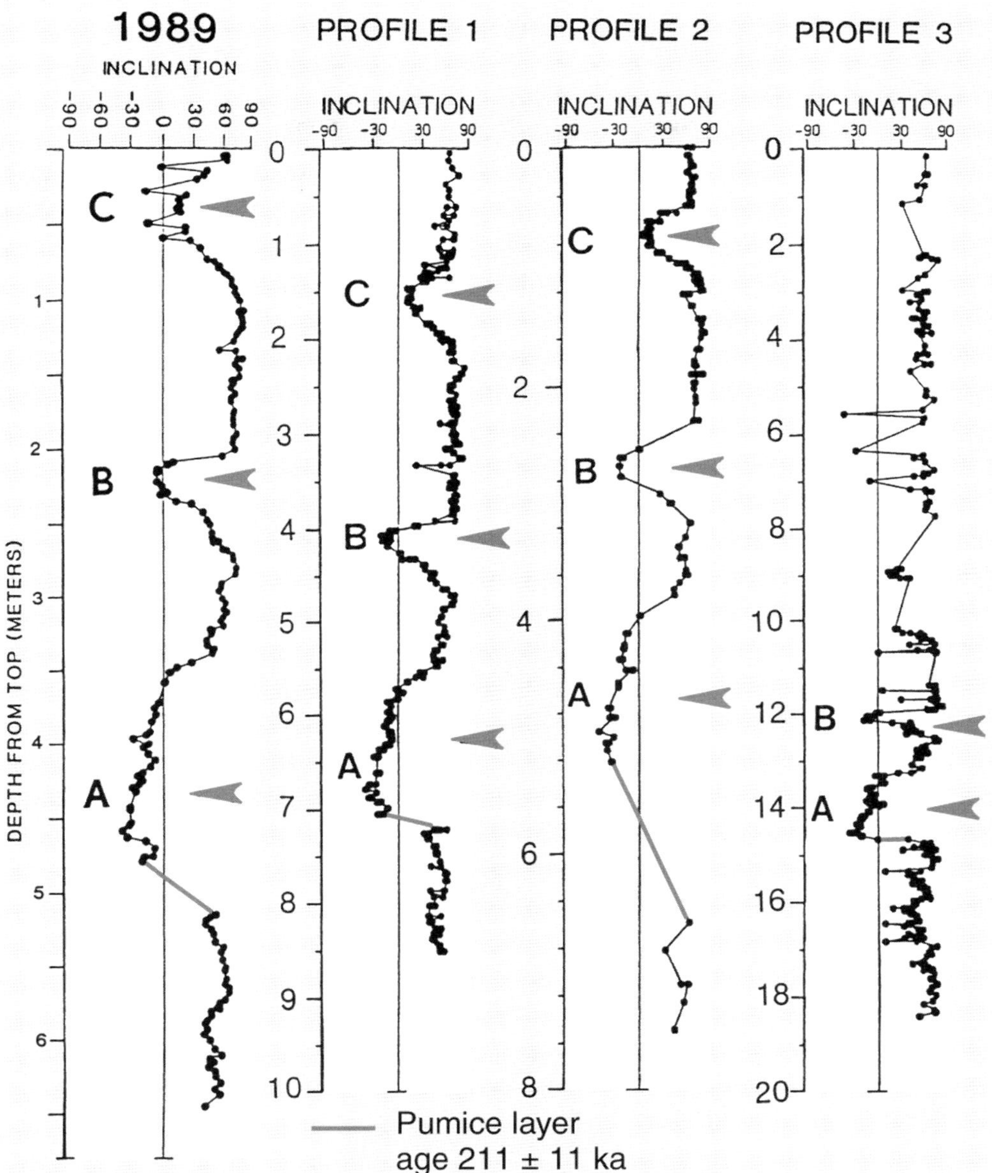

Fig. 9. Intra-basinal correlation of the inclination results of four profiles from Pringle Falls, showing features A, B and C (except for Profile 3). Notice the reproducibility of the three geomagnetic characteristic features of the excursion as indicated by the red arrows and above the recently already radiometrically dated pumice layer 'D' by $^{40}Ar/^{39}Ar$ methods, yielding an age of 211 ± 11 ka.

the three geomagnetic features that define the signature of the aborted reversal, namely A, B and C. Figure 11 shows the behaviour of the transitional field that we have called Phase A corresponding to the early part of the excursion (i.e. feature A, see Figs 7 & 8), followed by Phase B corresponding to the middle of the excursion path (i.e. feature B) and then Phase C correlated to the late part of the excursion and corresponding to feature C of the excursion. It is easy to observe that, once

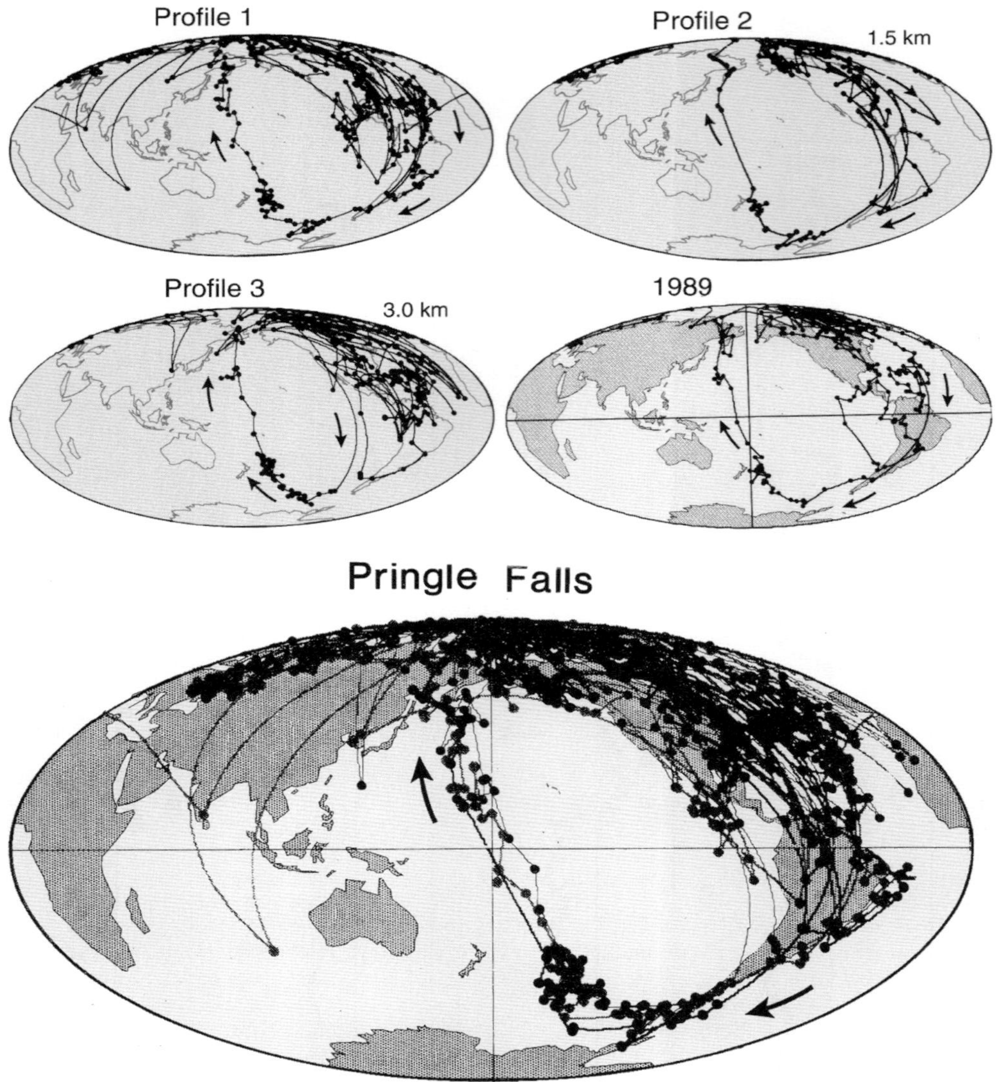

Fig. 10. Virtual geomagnetic pole paths of four profiles recovered along the Deschutes River Oregon. Profile labelled as 1989 and Profile 2 have already been published by Herrero-Bervera *et al.* (1994) and the distance between them is *c.* 1.5 km. Profiles 1 and 3 were recently obtained and studied for this study. The distance between Profile 3 and Profile 2 is *c.* 3.0 km. The lower VGP path named Pringle Falls is the overlapping of the four paths in question, slightly centred to the east, and therefore showing more VGPs on the eastern part of Brazil. All of these paths show the extraordinary intra-basinal correlation between and amongst the four drilled profiles.

the entire excursion has been dissected, the complex transitional evolution of the behaviour of the palaeofield emerges and is identifiable.

The dissected Pringle Falls paths show that the youngest feature A is characterized by an initial cluster of excursional VGPs, then a clockwise travel to South America followed by a departure of the field from the Patagonia area through the middle of the Pacific Ocean with a tendency towards the eastern part of the Pacific (i.e. Phase A, early part of the VGP path, see Fig. 11).

Subsequently, there is a lingering of the VGPs from the northeast part of the American continent and Greenland, travelling south to the eastern part

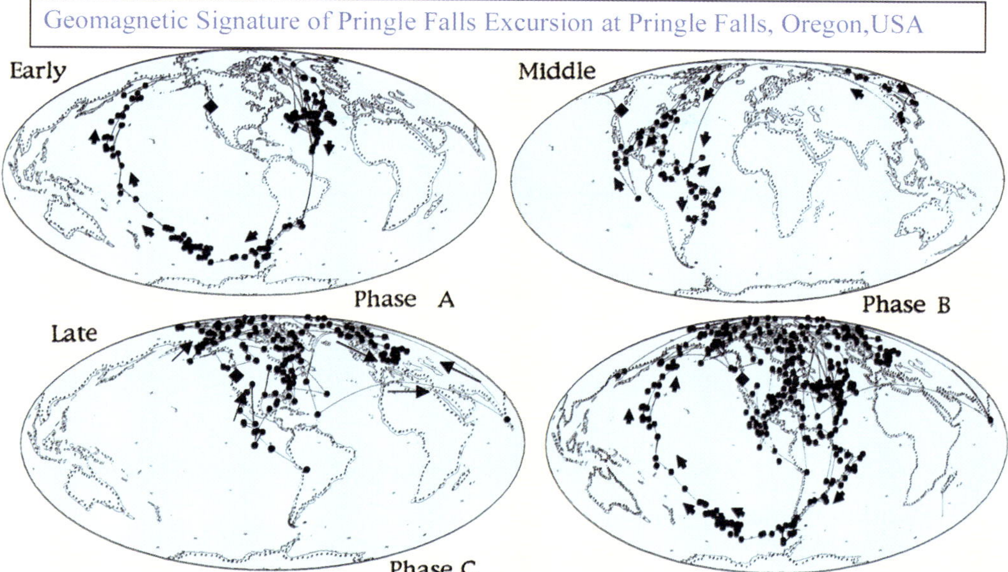

Fig. 11. Virtual geomagnetic pole paths of the dissected portions of the Pringle Falls geomagnetic signature. The data (i.e. the VGPS) used for the plot correspond to only one of the sampled profiles and not an integration of the entire data set of the four profiles in question. The upper left corner path represents the Early part of the excursional path corresponding to Phase A (i.e. geomagnetic feature A); the upper right corner path corresponds to the Middle part of the excursional travel and correlates to the Phase B (geomagnetic feature B). The lower left corner of the diagram represents the Late portion of the path corresponding to Phase C (i.e. geomagnetic feature C). The lower right corner of the diagram represents the overlapping or composite diagram of all the discrete VGPs of the entire excursion under question. The black arrows represent the direction of motion of the path from younger to older; notice the characteristic clockwise loop through the Pacific that identifies one of the signatures of the Pringle Falls excursion. Notice that the program used to plot the VGPs of this figure is different from the program used to plot the VGPs of Figure 10 and there is slight shift of the data to the east.

of the continent through the Gulf of Mexico on to the southern part of Baja California and then relatively fast travel to East Asia in the vicinity of Kamchatka, Japan and Korea with a sudden displacement of the VGPs to the eastern part of South America between Colombia–Venezuela to the northeast part of Brazil corresponding to feature B (i.e. Phase B, middle part of the VGP path, see Fig. 11).

The last part of the evolution of the excursional field is shown on the third dissected VGP path of Figure 11 and corresponds to feature C (i.e. Phase C, late part of the path). Such a path is characterized by the clustering of VGPs lingering between Western Europe and the entire North American continent with an initial departure from South America.

Correlation with other excursional records within the Brunhes Chron

We compared the excursional VGP paths that have been published recently belonging to the Brunhes

Chron with those of the recently obtained Pringle Falls recorded at the original site of Pringle Falls. Figure 12 shows that, if one makes a visual comparison of such VGP paths, one finds that all of them display remarkably different VGP paths (Valet *et al.* 2008*a*, *b*). The detailed description of the characteristics of the ages and VGP behaviour has recently been published by Valet *et al.* (2008*a*, *b*). It is important here to point out that there are two excursions that have been reported to be the same in terms of their age at the 195–200 ka level, but very recently such excursions have been identified with a certain degree of certainty with respect to the ages and they have been identified as two separate excursions. These two excursions are the Iceland basin excursion reported by Channell (2006), and Lund *et al.* (2006) with a recognized age date of 180–188 ka and the Pringle Falls aborted reversal dated twice at the Pringle Falls site yielding an age of 218 ± 10 ka (Herrero-Bervera *et al.* 1994) and 211 ± 12.8 ka (Singer *et al.* 2008). These two excursions are different with their respective VGP paths

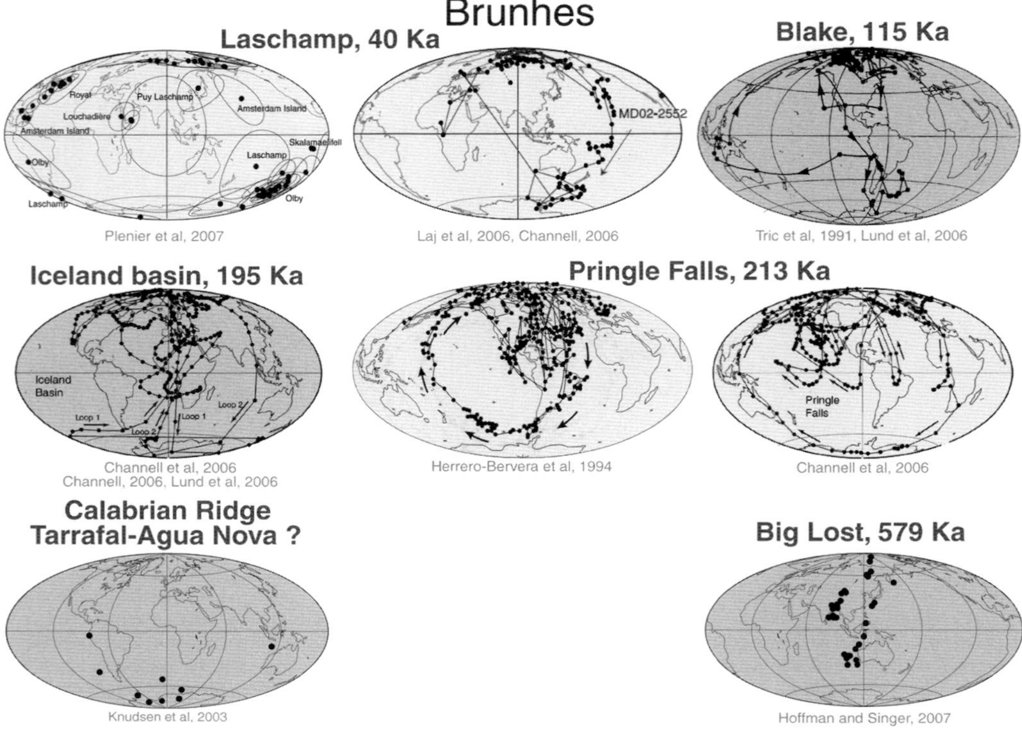

Fig. 12. Virtual geomagnetic pole paths of the most detailed records of excursions during the Brunhes Chron. After Valet *et al.* (2008*a, b*).

and their unique radiometric age determinations since the radiometric difference is of the order of 23 000 years (e.g. Valet *et al.* 2008*a, b*), as shown in Figure 12.

Discussion

It is known by now that geomagnetic excursions represent short episodes of a few thousand years at most during which the field considerably exceeds its normal range of variability during a polarity state. Palaeomagnetic records such as the Pringle Falls described here have now been obtained with extremely high temporal resolutions that have improved our knowledge of these short events.

Up until today, excursions have been defined as a single process linked to a large diminution of the dipole field and a large departure of the field from its initial polarity followed by a recovery to the initial state. According to this interpretation, excursions could be interpreted as manifestations of large-amplitude secular variations caused by the 'failure' and/or weakening of the dipole.

The presence of a period of a reversed polarity with a partially restored dipole even for a short time

suggests a different scenario. We already know that by definition, a reversal is restricted to the period over which the pole flips from one polarity to the other where pole positions are usually considered as transitional or excursional only when they are 60° apart in terms of latitude. Therefore, the onset and termination phases of excursions are actual field reversals, as they show the flip of the dipole between the two polarities, and the very short episode of the partial restoration of the dipole in the opposite polarity that can be seen as an 'aborted polarity state'.

Excursions as published by Valet *et al.* (2008*a, b*) then should be described 'as two successive reversals separated by an aborted polarity interval'. Note that this scheme is inconsistent with the hypothesis that excursions would represent a single clockwise (or anticlockwise rotation) of the dipole field. Owing to its large decrease the dipole becomes too weak to be dominant and the transitions are necessarily controlled by the secular variation of the non-dipole field (see Valet *et al.* 2008*a, b*). As can be seen from Figures 7, 8, 10 and 11, the quadruplicate directional profiles as well as the VGP paths show that the directions also reach opposite polarity, in addition, southern pole positions are also observed. Other Pringle Falls records, such as the one published by Channell (2006) from ODP Site 919 in the northern Atlantic, show reversed directions at all sites, which is an indication that such directions are associated with a dipolar field geometry.

Conclusions

In order to study the detailed excursional characteristics of the Pringle Falls polarity episode recorded at the original sampling site, we have studied four records of such aborted reversal. The Pringle Falls excursion is one of the five to six globally identified polarity episodes between 579 and 30 ka that have been well documented in several marine sediment cores or dated using $^{40}Ar/^{39}Ar$ methods on terrestrial rocks. Together with at least five other well-dated excursions between 730 and 520 ka, there are 10 excursions that define the Geomagnetic Instability Time Scale (e.g. Singer *et al.* 2008) for the Brunhes Chron. The Pringle Falls excursion recorded at the original site satisfies the minimum conditions to characterize it as a global excursion, namely, the Pringle Falls excursion has been dated by radiometric methods such as $^{40}Ar/^{39}Ar$ from two different lithologies such as igneous rocks (Singer *et al.* 2008) and lake and deep-sea sediments (e.g. Herrero-Bervera *et al.* 1989; 1994; Channell 2006) as well as from far removed localities from the Northern and Southern Hemispheres (e.g. McWilliams 2001).

The study of the detailed excursional characteristics of the Pringle Falls polarity episode recorded at the original sampling site was undertaken by drilling a total of 827 samples recovered from four widely spaced profiles sampled along the Deschutes River Oregon. The four profiles sampled documented a high-resolution palaeomagnetic excursion of the Pringle Falls magnetic polarity episode (*c.* 211 ± 13 ka, Singer *et al.* 2008), recorded by diatomaceous lacustrine sediments.

This sedimentary sequence was sampled as part of an extensive prehistoric fluvial and lacustrine complex that formed east of the Cascade Mountains during the last 1.0 Ma. The lake appears to have resulted from a late Pliocene/Pleistocene rise in the base level to the east of the sampling area near the western margin of the Basin and Range structural province and is related to the development of the extensive volcanism associated with the Newberry volcano.

We conducted palaeomagnetic and rock magnetic studies in order to investigate the reproducibility of the palaeomagnetic signal throughout the 5 km of the sampling of the four profiles that documented the Pringle Falls excursion. We performed low-field susceptibility v. temperature analysis to determine the magnetic carriers of the sediments and we found that the main magnetic carriers are pure and Ti-poor magnetite with Curie temperature ranging from 550 to 565 °C. Most grain sizes are tightly clustered within the pseudo-single domain range for the Pringle Falls profiles. The demagnetization of the sediments was done by means of alternating field methods, and the determination of the mean directions of Profiles 1 and 3 was done by principal component analyses. The level of detail of the palaeosignal of these four records is highly consistent since they are characterized by rapidly deposited sediments (greater than 10 cm/ka) that provide a detailed representation of field behaviour during the excursion.

The VGP paths are highly internally consistent and are defined by a clockwise loop travelling from high northern latitudes over the eastern part of North America and the North Atlantic to South America and then to high southern latitudes, then a return to high northern latitudes through the Pacific and over Kamchatka associated with the initial phase of the excursion, which corresponds to geomagnetic feature A. The other two geomagnetic features, namely B and C, corresponding to the middle and late stages of the evolution of the excursional field, have their own looping indicating a complex non-dipolar behaviour of the field. The initial or early phase Pringle Falls clockwise loop is characteristic of other recently found excursions like the Iceland Basin excursion (188 ka). The published age of the Pringle Falls excursion (*c.* 218 ± 14 ka; Herrero-Bervera *et al.* 1994) and the most recent radiometric ages at the Pringle Falls site (weighted mean 211 ± 13 ka; Singer *et al.* 2008), indicate that the dominance of such VGP paths (i.e. clockwise looping of the Pringle Falls, the Iceland Basin excursion and other excursions of the same age) shows that the excursional palaeofield had a relatively simple geometric characteristic. A corollary of the latter option is that palaeomagnetic polarity episodes of different ages may have similar transition and excursional polar paths, a conclusion implying that a common mechanism of the generation of the palaeofield was involved (e.g. see Herrero-Bervera *et al.* 1994).

We are very grateful to J. Lau for his field and laboratory assistance. Also we would like to express our thanks to the two anonymous referees for their very constructive and very helpful criticisms of our research work. Financial support to E. Herrero-Bervera was provided by SOEST-HIGP and the National Science Foundation grants EAR-9909206, EAR-INT-9906221, EAR-0207787, EAR-0213441, EAR-0510061, EAR-0710571, EAR-1015328 and EAR12-1215070, and the NSF EPSCoR program. This is SOEST contribution number 8701 and HIGP contribution number 1980.

References

CAMPS, P. & PREVOT, M. 1996. A statistical model of the fluctuations in the geomagnetic field from paleosecular variation to reversal. *Science*, **273**, 776–779.

CANON-TAPIA, E. & HERRERO-BERVERA, E. 2009. Is the Pringle Falls excursion a product of geomagnetic field behavior or an artefact of sedimentation processes? Insights from anisotropy of magnetic susceptibility (AMS) analyses. *Geophysical Journal International*, **178**, 702–712, http://dx.doi.org/10.1111/j.1365-246 X.2009.04223.x.

CARLUT, J., VALET, J-P., QUIDELLUR, X., COURTILLOT, V., KIDANE, T., GALLET, Y. & GILLOT, P-Y. 1999. Paleointensity across the Reunion event in Ethiopia. *Earth and Planetary Science Letters*, **170**, 17–34.

CASSATA, W., SINGER, B. S. & CASSIDY, J. 2008. Laschamp and Mono Lake geomagnetic excursions recorded in New Zealand. *Earth and Planetary Science Letters*, **268**, 76–88.

CHANNELL, J. E. T. 1999. Geomagnetic paleointensity and directional secular variation at Ocean Drilling Program (ODP) Site 984 (Bjorn Drift) since 500 ka: comparisons with ODP site 983 (Gardar Drift). *Journal of Geophysical Research*, **104**, 22937–22951.

CHANNELL, J. E. T. 2006. Late Brunhes polarity excursions (Mono Lake, Lachamp, Iceland Basin and Pringle Falls) recorded at ODP Site 919 (Irminger Basin). *Earth and Planetary Science Letters*, **244**, 378–393.

CHANNELL, J. & LEHMAN, B. 1997. The last two geomagnetic polarity reversals recorded in high-deposition-rate sediment drifts. *Nature*, **389**, 712–715.

CHANNELL, J. E. T., MAZAUD, A., SULLIVAN, P., TURNER, S. & RAYMO, M. E. 2002. Geomagnetic excursions and paleointensities in the Matuyama Chron at Ocean Drilling Program Sites 983 and 984 (Iceland Basin). *Journal of Geophysical Research*, **107**, 2001JB00 0491.

CHAUVIN, A., DUNCAN, R. A., BONHOMMET, N. & LEVI, S. 1989. Paleointensity of the earth's magnetic field and K–Ar dating of the Louchadiere volcanic flow (central France): new evidence for the Laschamp excursion. *Geophysics Research Letters*, **16**, 1189–1192.

CHAUVIN, A., GILLOT, P.-Y. & BONHOMMET, N. 1991. Paleointensity of the Earth's magnetic field recorded by two late Quaternary volcanic sequences at the Island of La Reunion (Indian Ocean). *Journal Geophysical Research*, **96**, 1981–2006.

CISOWSKI, S. 1995. Geomagnetic intensity and field direction changes associated with the Matuyama/Brunhes polarity reversal, as recorded in a sediment core from the North Pacific. *Proceedings of the Ocean Drilling Program – Scientific Results*, **145**, 475–482.

CLEMENT, B. 1991. Geographical distribution of transitional VGPs: evidence for nonzonal equatorial symmetry during the Brunhes–Matuyama geomagnetic reversal. *Earth and Planetary Science Letters*, **104**, 48–58.

CLEMENT, B. M. 2004. Dependence of the duration of geomagnetic polarity reversal on site latitude. *Nature*, **428**, 637–640.

CLEMENT, B. & KENT, D. V. 1991. A southern hemisphere record of the Matuyama–Brunhes polarity reversal. *Geophysics Research Letters*, **18**, 81–84.

CLEMENT, B. & MARTINSON, D. 1992. A quantitative comparison of two paleomagnetic records of the Cobb Mountain subchron from North Atlantic deep-sea sediments. *Journal of Geophysical Research*, **97**, 1735–1752.

COE, R. S., GROMMÉ, C. S. & MANKINEN, E. A. 1984. Geomagnetic paleointensities from excursion sequences in lavas on Oahu, Hazaii. *Journal of Geophysical Research*, **89**, 1059–1069.

COE, R. S., HONGRE, L. & GLATZMAIER, G. A. 2000. An examination of simulated geomagnetic reversals from a paleomagnetic perspective. *Philosophical Transactions of the Royal Society of London Series A*, **357**, 1787–1813.

DAY, R., FULLER, M. & SCHMIDT, V. A. 1977. Hysteresis properties of titanomagnetites: grain-size and compositional dependence. *Physics of the Earth and Planetary Interiors*, **13**, 260–266.

DUNLOP, D. J. 2002. Theory and application of the Day plot (M_{rs}/M_s versus H_{cr}/H_c) 1. Theoretical curves and tests using titanomagnetite. *Journal of Geophysical Research*, **107**, IB3, EPM 4 1-22.

DUNLOP, D. & ÖZDEMIR, Ö. 1997. *Rock Magnetism: Fundamentals and Frontiers*. Cambridge University Press, New York.

DUNLOP, D. & ÖZDEMIR, Ö. 2000. Effect of grain size and domain state on thermal demagnetization tails. *Geophysics Research Letters*, **27**, 1311–1314.

FULLER, M. 2006. Geomagnetic field intensity, excursions, reversals, and the 41 000 yr obliquity signal. *Earth and Planetary Science Letters*, **245**, 605–615.

GLATZMAIER, G. A. & ROBERTS, P. H. 1995. A three-dimensional convective dynamo solution with rotating and finitely conducting inner core and mantle. *Physics of the Earth and Planetary Interiors*, **91**, 63–75.

GLATZMAIER, G. A., COE, R. S., HONGRE, L. & ROBERTS, P. H. 1999. The role of the Earth's mantle in controlling the frequency of geomagnetic reversals. *Nature*, **401**, 885–890.

GUBBINS, D. 1999. The distinction between geomagnetic excursions and reversals. *Geophysical Journal International*, **137**, 1–3.

GUYODO, Y. & VALET, J. P. 1999. Global changes in geomagnetic intensity during the past 800 thousand years. *Nature*, **399**, 249–252.

HERRERO-BERVERA, E. & KHAN, M. A. 1992. Olduvai termination; Detailed palaeomagnetic analysis of a north central Pacific core. *Geophysical Journal International*, **108**, 535–545.

HERRERO-BERVERA, E. & HELSLEY, C. E. 1993. Global paleomagnetic correlation of the Blake geomagnetic polarity episode. *In*: AISSAOUI, D. M., MCNEILL, D. F. & HURLEY, N. F. (eds) *Application of Paleomagnetism to Sedimentary Geology*. SEPM Special Publications, **49**, 71–82.

HERRERO-BERVERA, E. & RUNCORN, S. K. 1997. Trasition fields during geomagnetic reversals and their geodynamic significance. *Philosophical Transactions of the Royal Society of London*, **355**, 1713–1742.

HERRERO-BERVERA, E. & COE, R. S. 1999. Transitional field behavior during the Gilbert-Gauss and Lower Mammoth reversals recorded in lavas from the Wai'anae volcano, O'ahu, Hawaii. *Journal of Geophysical Research*, **104**, 29,157–29,173.

HERRERO-BERVERA, E. & VALET, J.-P. 2005. Absolute paleointensity and reversal records from the Wai'anae sequence (O'ahu, Hawaii USA). *Earth and Planetary Science Letters*, **234**, 279–296.

HERRERO-BERVERA, E. & VALET, J.-P. 2009. Testing determinations of absolute paleointensity from the 1955 and 1960 Hawaiian flows. *Earth and Planetary Science Letters*, **287**, 420–433.

HERRERO-BERVERA, E., HELSLEY, C. E., HAMMOND, S. R. & CHITWOOD, L. A. 1989. A possible lacustrine paleomagnetic record of the Blake episode from Pringle Falls, Oregon, U.S.A. *Physics of the Earth and Planetary Interiors*, **56**, 112–123.

HERRERO-BERVERA, E., HELSLEY, C. E. ET AL. 1994. Age and correlation of a paleomagnetic episode in the western United States by $^{40}AR/^{39}Ar$ dating tephrochronology: the Jamaica, Blake, or a new polarity episode? *Journal of Geophysical Research*, **99**, 24 091–24 103.

HERRERO-BERVERA, E., BROWNE, E. J., VALET, J-P., SINGER, B. S. & JICHA, B. R. 2007. Cryptochron C2r.2r-1 recorded 2.51 Ma in the Koolau Volcano at Halawa, Oahu, Hawaii, USA: Paleomagnetic and $^{40}Ar/^{39}Ar$ evidence. *Earth and Planetary Science Letters*, **254**, 256–271.

HOFFMAN, K. A. 1981. Paleomagnetic excursions, aborted reversals and transitional fields, *Nature*, **294**, 67–68.

HOFFMAN, K. A. & SINGER, B. S. 2004. Regionally recurrent paleomagnetic transitional fields and mantle processes. *In: Timescales of the Paleomagnetic Field.* American Geophysical Union, Washington, DC. AGU Monographs, **145**, 233–244.

HOFFMAN, K. A. & SINGER, B. S. 2008. Magnetic source separation in Earth's outer core. *Science*, **321**, 1800.

HROUDA, F. 1994. A technique for the measurement of thermal changes of magnetic susceptibility of weakly magnetic rocks by the CS-2 apparatus and KLY-2 Kappabridge. *Geophysical Journal International*, **118**, 604–612.

HROUDA, F., JELINEK, V. & ZAPLETAL, K. 1997. Refined technique for susceptibility resolution into ferrimagnetic and paramagnetic components based on susceptibility temperature variation measurements. *Geophysical Journal International*, **129**, 715–719.

LAJ, C., MAZAUD, A., WEEKS, R., FULLER, M. D. & HERRERO-BERVERA, E. 1991. Geomagnetic reversal paths. *Nature*, **351**, 447.

LAJ, C., MAZAUD, A., WEEKS, R., FULLER, M. & HERRERO-BERVERA, E. 1992a. Geomagnetic reversal paths. [Discussion.] *Nature*, **359**, 111–112.

LAJ, C., MAZAUD, A., WEEKS, R., FULLER, M. & HERRERO-BERVERA, E. 1992b. Statistical assessement of the preferred longitudinal bands for recent geomagnetic field reversal records. *Geophysical Research Letters*, **19**, 2003–2006.

LAJ, C., KISSEL, C. & ROBERTS, A. P. 2006. Geomagnetic field behavior during the Iceland Basin and Laschamp geomagnetic excursions: A simple transitional field geometry? *Geochemistry Geophysics, Geosystems*, **7**, Q03004, http://dx.doi.org/10.1029/2005GC001122.

LEVI, S. & BANERJEE, S. K. 1976. On the possibility of obtaining relative paleointensities from lake sediments. *Earth and Planetary Science Letters.*, **29**, 219–226.

LUND, S. P., ACTON, G., CLEMENT, B., OKADA, M. & WILLIAMS, T. 2001. Brunhes Chron magnetic field excursions recovered from Leg 172 sediments. *In*: KEIGWIN, L. D., RIO, D., ACTON, G. D. & ARNOLD, E. (eds) *Proceedings of the Ocean Drilling Program, Scientific Results*, **172**. Ocean Drilling Program, Texas A & M University, College Station, TX, 1–20.

LUND, S., STONER, J. S., CHANNELL, J. E. T. & ACTON, G. 2006. A summary of Brunhes paleomagnetic field variability recorded in Ocean Drilling Program cores. *Physics of the Earth and Planetary Interiors*, **156**, 194–204.

MACLEOD, N. S., SHERROD, D. R. & CHITWOOD, L. 1982. *Geologic map of Newberry Volcano, Deschutes, Klamath and Lake Counties, Oregon*. U.S. Geological Survey Open File Report **82-847**.

MEYNADIER, L., VALET, J. P., WEEKS, R., SHACKLETON, N. J. & HAGEE, V. L. 1992. Relative geomagnetic intensity of the field during the last 140 ka. *Earth and Planetary Science Letters.*, **114**, 39–57.

MEYNADIER, L., VALET, J. P., BASSINOT, F. C., SHACKLETON, N. J. & GUYODO, Y. 1994. Asymmetrical sawtooth pattern of the geomagnetic field from equatorial sediments in the Pacific and Indian Oceans. *Earth and Planetary Science Letters*, **126**, 109–127.

MCFADDEN, P., BARTON, C. E. & MERRILL, R. T. 1993. Do virtual geomagnetic poles follow preferred paths during geomagnetic reversals? *Nature*, **361**, 342–344.

MCWILLIAMS, M. 2001. Global correlation of the 223 ka Pringle Falls event. *International Geology Review*, **43**, 191–195.

MERRILL, R. T. & MCFADDEN, P. L. 1999. Geomagnetic polarity transitions. *Reviews of Geophysics*, **37**, 201–226.

MERRILL, T. & MCFADDEN, P. L. 2005. The use of magnetic field excursions in stratigraphy. *Quaternary Research*, **63**, 232–237.

RIISAGER, J., RIISAGER, P., ZHAO, X., COE, R. S. & PEDERSEN, A. K. 2004. Paleointensity during a chron C26r excursion recorded in west Greenland lava flows. *Journal of Geophysical Research*, **109**, 2003JB002 887.

ROPERCH, P., BONHOMMET, N. & LEVI, S. 1988. Paleointensity of the Earth's magnetic field during the Laschamp event and its geomagnetic implications. *Earth and Planetary Science Letters*, **88**, 209–219.

SINGER, B. L., JICHA, B. R., KIRBY, B. T., ZHANG, X., GEISSMAN, J. W. & HERRERO-BERVERA, E. 2008. $^{40}Ar/^{39}Ar$ dating links Albuquerque Volcanoes to the Pringle Falls and the Geomagnetic Instability Time Scale. *Earth and Planetary Science Letters*, **267**, 584–595.

STONER, J. S. & ST-ONGE. 2007. Magnetic stratigraphy: reversals, excursions, paleointensity and secular variation. *In*: HILLAIRE-MARCEL, C. & DE VERNAL, A. (eds) *Development in Marine Geology: Volume 1, Proxies in Late Cenozoic Paleoceanography*. Elsevier, Amsterdam, 99–138.

TAKAI, A., SHIBUYA, H., YOSHIHARA, A. & HAMANO, Y. 2002. Paleointensity measurements of pyroclastic flow deposits co-born with widespread tephras in Kyushu Island, Japan. *Physics of the Earth and Planetary Interiors*, **133**, 159–179.

TANAKA, H. & KOBAYASHI, T. 2003. Paleomagnetism of the late Quaternary Ontake Volcano, Japan: directions, intensities, and excursions. *Earth Planets and Space*, **55**, 189–202.

TRIC, E., LAJ, C., JEHANNO, C., VALET, J-P., KISSEL, C., MAZAUD, A. & IACCARINO, S. 1991a. High-resolution record of the upper Olduvai transition from Po Valley (Italy) sediments; support for dipolar transition geometry? *Physics of the Earth and Planetary Interiors*, **65**, 319–336.

TRIC, E., LAJ, C., VALET, J.-P., TUCHOLKA, P., PATERNE, M. & GUICHARD, F. 1991b. The Blake geomagnetic event; transition geometry, dynamical characteristics and geomagnetic significance. *Earth and Planetary Science Letters*, **102**, 1–13.

VALET, J. P. & HERRERO-BERVERA, E. 2010. A few characteristic features of the geomagnetic field during reversals. *In*: PETROVSKY, E., HERRERO-BERVERA, E., HARINARAYANA, T. & IVERS, D. (eds) *The Earth's Magnetic Interior*. IAGA Special Sopron Book Series 1, Springer Science + Business Media B.V., 139–151, http://dx.doi.org/10.1007/978-94-007-0323-0_10.

VALET, J. P. & MEYNADIER, L. L. 1993. Geomagnetic field intensity and reversals during the past four million years. *Nature*, **366**, 234–238.

VALET, J.-P., LAJ, C. & TUCHOLKA, P. 1986. High-resolution sedimentary record of a geomagnetic reversal. *Nature*, **322**, 27–32.

VALET, J.-P., LAJ, C. & LANGEREIS, C. G. 1988a. Sequential geomagnetic reversals recorded in upper Tortonian marine clays in western Crete (Greece). *Journal of Geophysical Research*, **93**, 1131–1151.

VALET, J.-P., TAUXE, L. & CLARK, D. 1988b. The Matuyama–Brunhes transition recorded in Lake Tecopa sediments (California). *Earth and Planetary Science Letters*, **87**, 463–472.

VALET, J.-P., TAUXE, L. & CLEMENT, B. 1989. Equatorial and midlatitude records of the last geomagnetic reversal from the Atlantic Ocean. *Earth and Planetary Science*, **94**, 371–384.

VALET, J.-P., TUCHOLKA, P., COURTILLOT, V. & MEYNADIER, L. 1992. Paleomagnetic constraints on the geometry of the geomagnetic field during reversals. *Nature*, **356**, 400–407.

VALET, J. P., HERRERO-BERVERA, E., LeMOUEL, J. L. & PLENIER, G. 2008a. Secular variation of the geomagnetic dipole during the past 200 years. *Geochemistry Geophysics Geosystems*, **9**, Q01008, http://dx.doi.org/10.1029/2007GC001728.

VALET, J. P., PLENIER, G. & HERRERO-BERVERA, E. 2008b. Geomagnetic excursions reflect an aborted polarity state. *Earth and Planetary Science Letters*, **274**, 472–478, http://dx.doi.org/10.1016/j.epsl.07.056.

VAN HOOF, A. A. M. & LANGEREIS, C. G. 1992a. The upper Kaena sedimentary geomagnetic reversal record from Sicily. *Journal of Geophysical Research*, **97**, 6941–6957.

VAN HOOF, A. A. M. & LANGEREIS, C. G. 1992b.The upper and lower Thvera sedimentary geomagnetic reversal record from southern Sicily. *Earth and Planetary Science Letters*, **114**, 59–75.

VEROSUB, K. L., HERRERO-BERVERA, E. & ROBERTS, A. P. 1996. Relative geomagnetic paleointensity acrosws the Jaramillo subchron and the Matuyama/Brubnhes boundary. *Geophysical Research Letters*, **23**, 467–470.

ZHU, R., LAJ, C. & MAZAUD, A. 1994. The Matuyama–Brunhes and upper Jaramillo transitions recorded in a loess section at Weinan, north-central China. *Earth and Planetary Science Letters*, **125**, 143–158.

ZHU, R., PAN, Y. & COE, R. S. 2000. Paleointensity studies of a lava succession from Jilin Province, northeastern China: evidence for the Blake event. *Journal of Geophysical Research*, **105**, 8305–8317.

ZIJDERVELD, J. D. A. 1967. A.C. demagnetization of rocks: analysis of results. *In*: COLLINSON, D. W., CREER, K. M. & & RUNCORN, S. K. (eds) *Methods in Palaeomagnetism*. Elsevier, New York, 254–286.

On the palaeomagnetic and rock magnetic constraints regarding the age of IODP 325 Hole M0058A

EMILIO HERRERO-BERVERA[1]* & LUIGI JOVANE[2,3]

[1]*Hawaii Institute of Geophysics and Planetology, University of Hawaii, Honolulu, HI 96822, USA*

[2]*Department of Geology, Western Washington University, Bellingham, WA 98225, USA*

[3]*Instituto Ocenográfico, Universidade de São Paulo, São Paulo 05508-120, Brazil*

**Corresponding author (e-mail: herrero@soest.hawaii.edu)*

Abstract: We have studied the rock magnetic and palaeomagnetic properties of a 41 m long core (Hole M0058A) recovering calcareous sediments located seaward of Noggin Reef, offshore Queensland, Australia to decipher the magnetostratigrapy of the site. We deployed 1 cm^3 samples at every 10 cm down-core and subsampled the core by means of U-channels in order to obtain a continuous record. Stepwise alternating field demagnetization from natural remanent magnetization to 80 mT showed that the characteristic remanent magnetization was isolated at low demagnetization fields between 0 and 15 mT. We conducted magnetic granulometry analyses and Curie point determinations. The low-field v. temperature analyses indicate the presence of Ti-poor magnetite with Curie points from 560 to 563 °C. Hysteresis loop experiments were performed. The results show M_{rs}/M_s and H_{cr}/H_c ratios corresponding to single domain to multi-domain and super-paramagnetic to single domain ranges. Both discrete and continuous inclination results indicate a remarkable correlation of three excursional inclinations occurring during intervals of low intensity of magnetization closely corresponding to the Laschamp (c. 41 ka), Skálamælifell (c. 94 ka) and Blake (c. 115–120 ka) 'aborted reversals,' and indicate that the base of the core is much older than the Blake excursion.

One of the most remarkable features of the Earth's magnetic field is how it reverses polarity. Polarity reversals have been recorded in a variety of rocks (deep-sea sediments, continental sedimentary rocks, intrusive and extrusive rocks) in which the physico-chemical processes of formation and of magnetization are diverse. The Geomagnetic Polarity Time Scale, which compiles the time sequence of reversals for the past 542 Ma, includes a number of excursions (i.e. 'aborted reversals'), which took place over very short time intervals, geologically speaking, perhaps in a few thousand years. It has only been in recent years that the remarkable advance of determining the directions and intensity of the geomagnetic field during some polarity transitions and excursions has been made (e.g. Valet & Meynadier 1993; Herrero-Bervera & Runcorn 1997; Guyodo & Valet 1999; Thouveny *et al.* 2004; Valet *et al.* 2005; Channell 2006; Singer 2007). At the same time, the study of transitional fields (e.g. reversals and excursions) that are preserved in palaeomagnetic records is recognized as an essential step in understanding the instability of the core dynamo and deciphering the magnetic 'stratigraphy' of many volcanic and sedimentary sequences. Here, we present the palaeomagnetic and rock magnetic study of a calcareous sequence of approximately 41 m of sediments recovered during Integrated Ocean Drilling Program (IODP) Expedition 325 in order to further understand the correlation between the recorded excursions and radiometric dates determined by IODP Expedition 325, and the general stratigraphy of the site (i.e. M0058A).

Geological setting

Great Barrier Reef

The main aim of IODP Expedition 325 was to investigate the magnitude and nature of sea-level changes related to the Last Glacial Maximum. A transect of holes was drilled perpendicular to the coastline to reconstruct the sea-level history and shoreward migration of the Great Barrier Reef (GBR) as sea-levels rose c. 120 m over the most recent 15 kyr. It is estimated that the GBR began at the Mid Pleistocene Climate transition (Marine isotope stage MIS 11; 0.4–0.5 Ma). The drill-cores sampled marine sediments that are presently submerged from 40 to 167 m. The recovery of this transect is around 20%, but Hole M0058A had a recovery of 82% because the operation drilled soft muds instead of hard corals.

From: Jovane, L., Herrero-Bervera, E., Hinnov, L. A. & Housen, B. A. (eds) 2013. *Magnetic Methods and the Timing of Geological Processes*. Geological Society, London, Special Publications, **373**, 279–291. First published online April 25, 2013, http://dx.doi.org/10.1144/SP373.19 © The Geological Society of London 2013. Publishing disclaimer: www.geolsoc.org.uk/pub_ethics

IODP 325 Hole M0058A

Hole M0058A (146°35.357′E, 17°5.8356′S) is the deepest hole of IODP Expedition 325 in a current water depth of 167 m, drilled during the spring of 2010 along a seven-drill-site transect SE of Cairns, offshore of the Great Barrier Reef (Fig. 1; Expedition 325 Scientists 2011). This site is presently situated on the forereef slope, but was probably much shallower (*c.* 40 m) during glacial periods.

The 41.4 m long core of Hole M0058A samples a sedimentary sequence composed of three unconsolidated green mud sections separated by two sandy layers. The mud sediments have locally abundant planktonic foraminifera, scattered benthic foraminifera and mollusc shell fragments, and are characterized by relatively high reflectance (lighter colours) and low magnetic susceptibility. The upper, *c.* 7 m thick sandy layer from 7.2 m to 14.6 m is composed of fine to medium sand with intraclasts up to cobble size, and cemented grainstones with fragments of molluscs, bryozoa, halimeda, coralline algae, echinoids, larger benthic foraminifera and serpulids. The lower, 3 m thick sandy layer from 29.4 to 32.4 m is well defined by colour reflectance and magnetic susceptibility.

The mud–sand cycling in Hole M0058A is thought to represent deepening (fining) upward sequences corresponding to glacial–interglacial cycles from Marine Isotope Stage MIS-7 to MIS-1. During glacial intervals, Last Glacial Maximum and MIS-6, a live coralgal reef was established in close proximity to Hole M0058A, where the water depth was approximately 40 m during those times, with coarse neritic material being shed from shallower environments towards the site. This interpretation, of course, is contingent on the core's chronology (Expedition 325 Scientists 2011).

Chronology

Radiocarbon dating yielded an age of 48 030 ± 530 y BP at 8.6–8.8 m (Core 4X-CC, 0–2 cm), within the upper sandy layer. This result is actually beyond the radiocarbon dating limit, and indicates that the dated material is 'radiocarbon-dead'. Taken at face value, this result implies an age consistent with MIS 3 at the top of the upper sand layer that is interpreted as a low-stand. Owing to the absence of a geomagnetic reversal sequence in this young interval, only magnetic inclination stratigraphy might be used for chronology.

Methods

For most of the cores recovered by IODP Expedition 325, standard palaeomagnetic samples were obtained as 2.5 cm (1 inch) cores drilled perpendicular to the split face of the rock cores. Samples were spaced at irregular intervals in the rubble material sections, with the object of collecting at least one sample per section. Samples were collected at a spacing interval of *c.* 1 m depending on core recovery.

In Core M0058A, which had a relatively complete recovery, we collected discrete 1 cm^3 palaeomagnetic mini-boxes with a spacing interval of 10 cm and continuous U-channels, which are open-sided 2 × 2 cm square plastic containers with a length of up to 1.5 m. This allowed us to examine the magnetic properties and the complete magnetostratigraphy of these carbonate-rich margin sediments.

Rock magnetic experiments

Magnetic susceptibility. Low-field magnetic and mass-specific susceptibility were measured using a KLY-2 (AGICO) Kappabridge with an operating frequency of 920 Hz, and a magnetic induction of 0.4 mT (noise level of 2×10^{-10} m^3 kg^{-1}). Low-field magnetic susceptibility data were recorded as corrected mass specific units (χ, 10^{-6} m^3 kg^{-1}). The accuracy of the Kappabridge was checked using a calibration standard with a bulk susceptibility of 1153×10^{-6} SI. This calibration piece was centred in the Kappabridge at one calibration run per six core sections. Oriented palaeomagnetic samples were recovered when possible from all core intervals where up–down orientation was preserved. Unfortunately, some intervals were composed of rubble coral materials. Thus, the azimuthal orientation of each individual core section was random, and shorter intervals within each core section were obviously rotated relative to each other. In such cases, any potential palaeomagnetic analysis will be limited to inclination and relative palaeointensity determinations. The instrument used for the experiment is in the laboratory of the Bremen IODP Repository. The low-field bulk magnetic susceptibility values have been iteratively checked with the Multisensor Core Logger (MSCL) data and with reflectance in order to compare the measurements with other continuous proxies (e.g. density, volume-specific magnetic susceptibility and resistivity) and colour of the sediments. The U-channels as well as the 1 cm^3 mini-boxes were measured.

Curie point determinations and magnetic granulometry. In addition to the magnetic susceptibility experiments we conducted Curie point determinations as well magnetic grain size experiments. We analysed the field-dependent and temperature-dependent magnetic behaviour of five discrete samples (*c.* 200 mg) from the working half of the

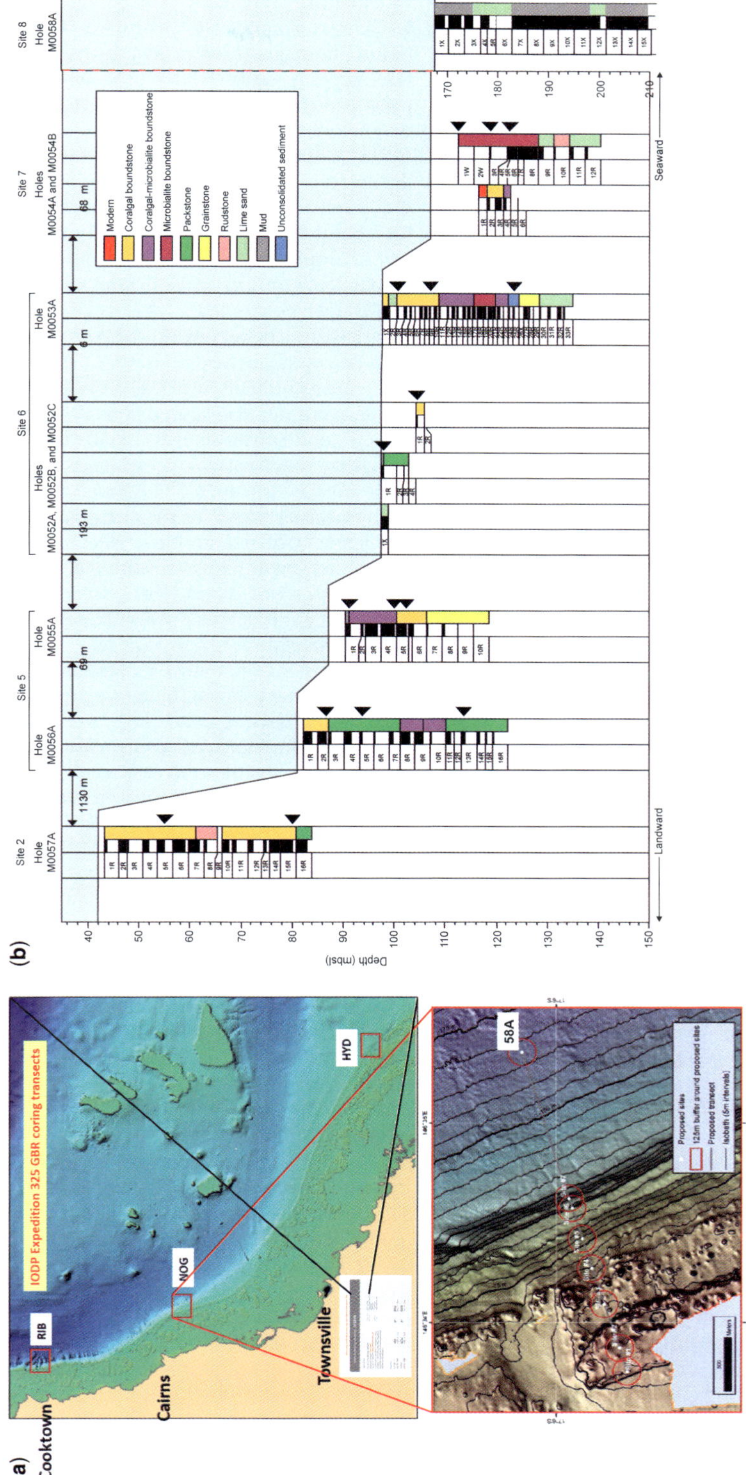

Fig. 1. (**a**) Location of the sampling site (M0058A) and its relation to the rest of the drilling sites during IODP Expedition 325. The inset shows the relative location of the Great Barrier Reef (GBR). (**b**) Lithostratigraphic summary of holes on the Transect NOG-01 (sites from landward to seaward: M0057A, M0056A, M0055B, M0052A, M0052B, M0052C, M0053A, M0054A, M0054B and M0058A). Modified from Expedition 325 Scientists (2011).

core, sampled from the top down to the end of the core at five different stratigraphic levels as follows: from the top (specimens 103 and 156), from the middle (specimen 197) and from the bottom (specimens 201 and 252) of the core. All analyses were conducted at the SOEST-HIGP Paleomagnetics and Petrofabrics Laboratory of the University of Hawaii at Manoa. The field-dependent measurements included determination of the low-field susceptibility v. temperature ($\kappa-T$) and hysteresis loops carried out at room temperature. Temperature-dependent measurements involved determination of the Curie temperature. Heating and cooling rates were close to 20 °C min^{-1}, with a maximum temperature of 700 °C using a KLY2-CS3 apparatus (Hrouda 1994; Hrouda *et al.* 1997). The low-field temperature v. susceptibility analyses of the five samples were conducted to determine the magnetic carrier of the natural remanent magnetization (NRM). These Curie point determinations were carried out on small amounts of powder and analysed in an air and Argon atmosphere.

We also performed hysteresis loop experiments for the determination of the magnetic grain size that provides insight into the intrinsic magnetic characteristics of the carbonate sediments and their possible subsequent alteration.

Magnetic hysteresis measurements were performed using a Petersen variable field translation balance (VFTB) capable of generating fields of up to 1.2 T located at the SOEST-HIGP Paleomagnetics and Petrofabrics Laboratory of the University of Hawaii at Manoa. For the VFTB measurements, sediment samples were exposed to a maximum field of 1 T. The hysteresis curve and the back-field demagnetization of SIRM were determined. The VFTB has a measurement range of 10^{-8}–10^{-2} A m^2. Saturation remanent magnetization (M_{rs}), saturation magnetization (M_s) and coercive force (H_c) were calculated after removing the paramagnetic contribution. Figure 3 shows the results of the hysteresis loops, which provide information about the main magnetic properties related to the amount, dimension and quality of the magnetic grains.

U-Channel methods. In addition to the discrete 1 cm^3 mini-cube sample measurements, we performed continuous measurements of the entire *c.* 41 m long core (Hole M0058A). Continuous U-channel samples were collected from the archival half of the core. U-Channel samples were enclosed in a plastic container with the same length as the core section (usually 150 cm) and a 2 × 2 cm square cross-section. One of the sides comprised a clip-on plastic lid allowing the sample to be sealed to inhibit dehydration and other chemical/physical changes.

Natural remanent magnetization of 29 U-channel samples was measured at 1 cm spacings before demagnetization and at 13 demagnetization steps in the 0–80 mT peak field range, in the pass-through 2G Enterprises automated cryogenic magnetometer (model 755R) equipped with three DC SQUID sensors (noise level $\leq 1 \times 10^{-9}$ emu) designed to measure U-channel samples housed in the shielded room of the Paleomagnetic Laboratory of the National Oceanography Centre, University of Southampton. NRM measurements and the demagnetization of several pilot discrete standard IODP palaeomagnetic plastic boxes (2 × 2 × 2 cm) were conducted at the University of Bremen palaeomagnetics laboratory. The first and last four measurements of the U-channel measurements were not considered reliable owing to edge effects and consequently were discarded. The response function of the magnetometer pick-up coils has a width, at half-height, of *c.* 4.5 cm; therefore, measurements at the 1 cm spacing are not independent of one another. Deconvolution of the U-channel data (Oda & Shibuya 1996; Guyodo *et al.* 2002) was carried out on parts of the NRM record in order to improve the resolution of the measurements.

Analysis of the data for component magnetization directions was computed for all NRM data, as well as for deconvolved data, using the standard principal component analysis (Kirschvink 1980). The demagnetization interval used to calculate the characterisitc remanent magnetization (ChRM) component was generally between 5 and 80 mT.

Results

Composite plots of the magnetic susceptibility experiments conducted on the soft sediments of site M0058A depict the correlation amongst the three different experiments (Fig. 2). The results indicate a good correlation of the features, for example, positive magnetic susceptibilities throughout the core ranging from 4.50 to 277.78 × 10^{-8} m^3 kg^{-1} with an arithmetic mean value of 38.24 × 10^{-8} m^3 kg^{-1}. There are two zones of high susceptibility with respect to the rest of the base values. The first zone occurs between 8.63 and 14.80 m below sea-floor (mbsf) and the second one from 27.00 to 32.60 mbsf with a maximum value of 277.78 × 10^{-8} m^3 kg^{-1}. These positive susceptibilities indicate the presence of ferromagnetic minerals repeated on the U-channel and MSCL records. Differences in sampling resolution and different instrumental and calibration methodologies between the U-channel, MSCL and discrete mini-boxes reveal dissimilar apparent responses in some intervals.

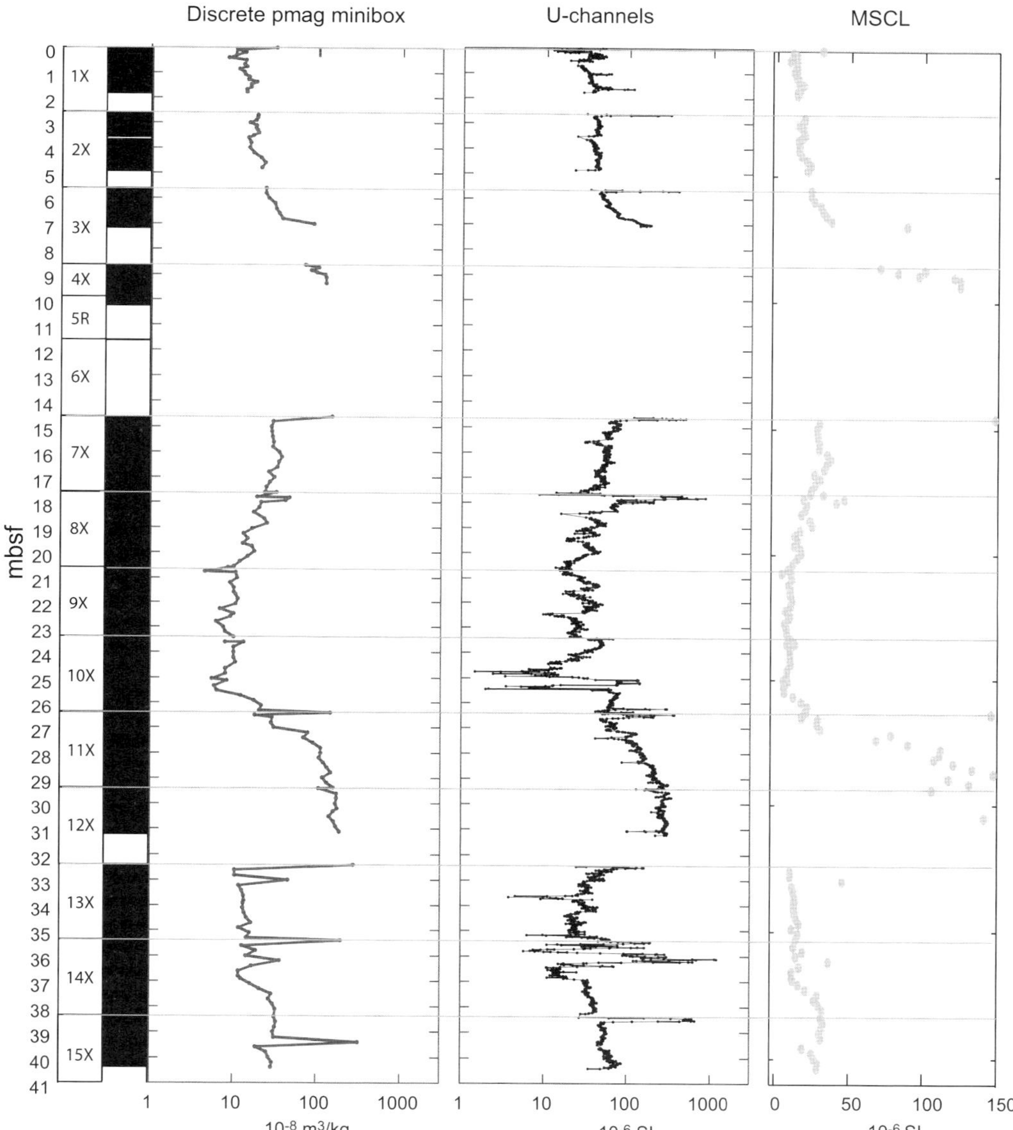

Fig. 2. Composite plot of the magnetic susceptibility experiments conducted on the soft sediments of core M0058A, depicting the correlation among the three different experiments, including the 1 cm^3 discrete samples, U-channels and *in-situ* multisensory core logger (i.e. MSCL). The boxes labelled 1X, 2X etc. represent the different sections of the core; the *y*-axis of the core is given in meters below sea-floor (mbsf). Notice the difference in density of samples between the discrete palaeomagnetic miniboxes taken every 10 cm and the U-channel sampling density at every 2 cm. The U-channel sampling gives a much better definition of the susceptibility signal.

Magnetic grain sizes can be indicative of the behaviour of the samples upon demagnetization and thus the stability of magnetization. The coercivity of remanence (H_{cr}) suggests that low-coercivity grains carry the NRM. The ratios of the hysteresis parameters show that most grain sizes are scattered within the multi-domain (MD) range. Samples were studied throughout the core. Specimen 103 ($M_{rs}/M_s = 0.10156$ and $H_{cr}/H_c = 10.5289$) fell along the super-paramagnetic (SP)–single domain (SD)

E. HERRERO-BERVERA & L. JOVANE

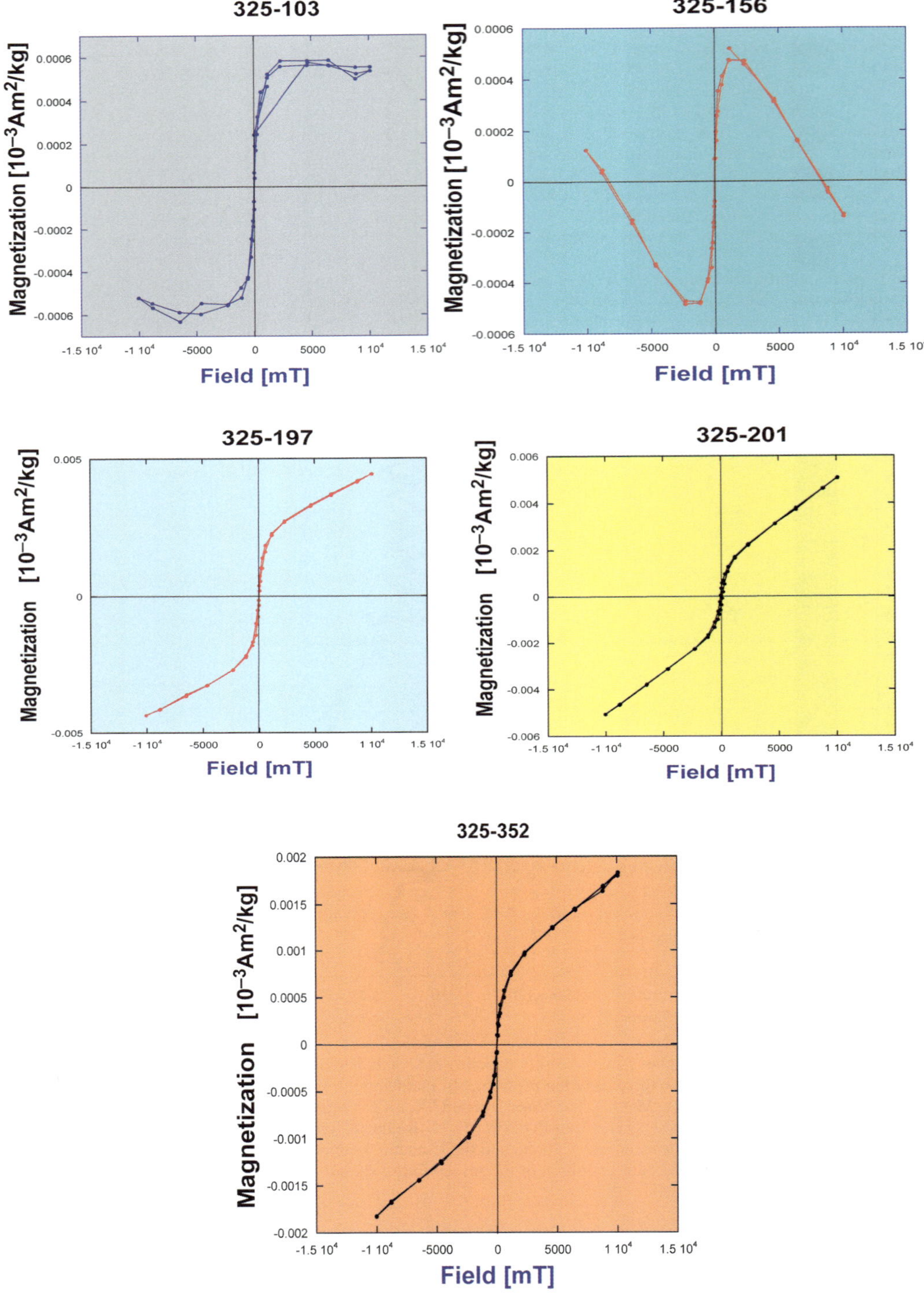

Fig. 3. Results of the hysteresis loops experiments performed on five specimens of soft sediment showing diverse M_{rs}/M_s and H_{cr}/H_c parameters.

mixing curve (10 nm); specimen 156 ($M_{rs}/M_s =$ 0.04072 and $H_{cr}/H_c = 8.2128$) was close to the SD–MD mixing curves; specimen 197 ($M_{rs}/M_s =$ 0.09399 and $H_{cr}/H_c = 9.2848$) fell in between the SP–SD and SD–MD mixing curves; specimen 201 ($M_{rs}/M_s = 0.21567$ and $H_{cr}/H_c = 2.6767$) fell on the SD–MD mixing curves; and specimen 252 ($M_{rs}/M_s = 0.00446$ and $H_{cr}/H_c = 5.9271$) fell

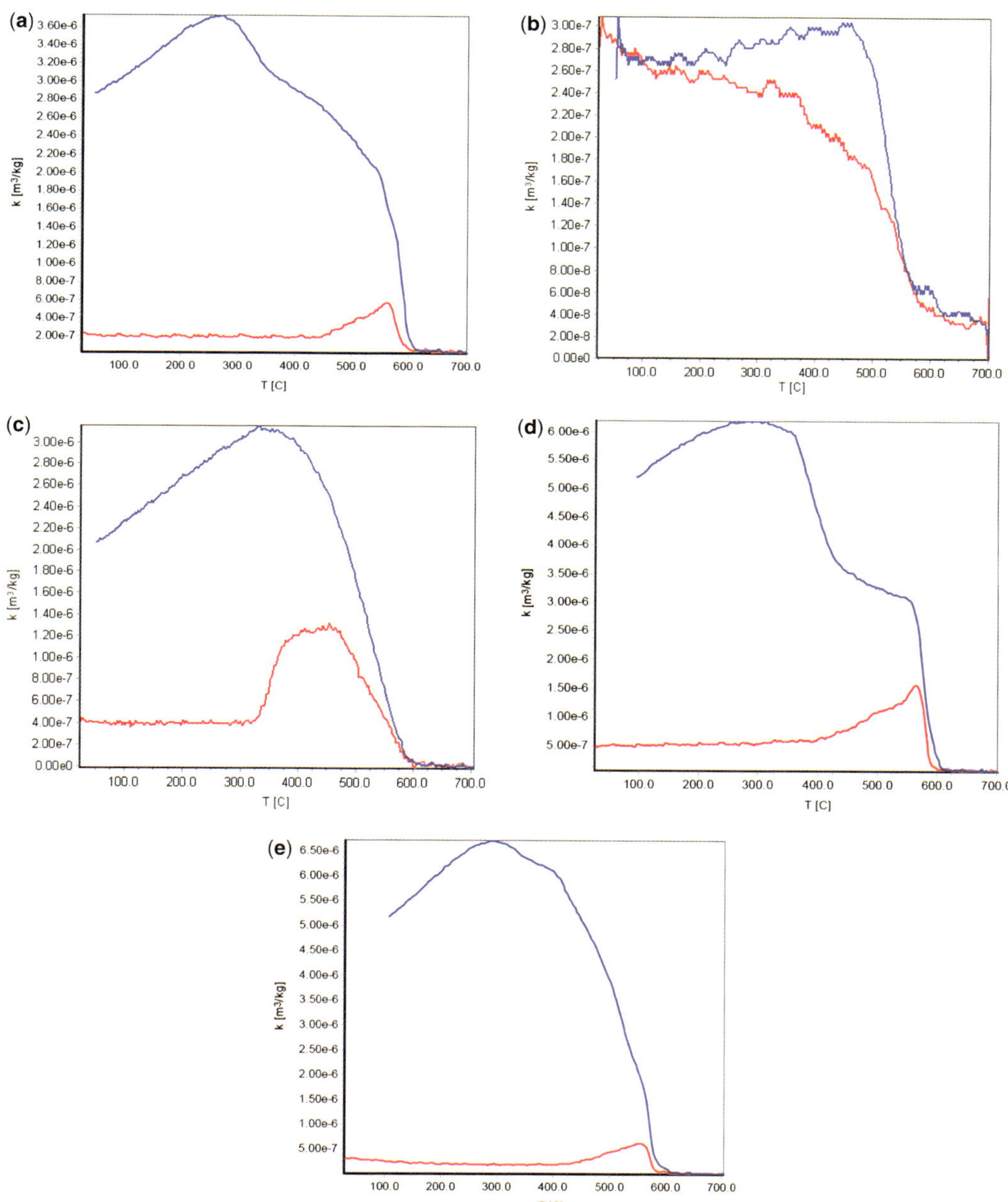

Fig. 4. Low-field susceptibility v. temperature (κ–T) Curie plots of five soft sediment specimens from Site M0058A. Samples **a**, **d** and **e** were run in air whereas specimens **b** and **c** were run in an argon gas atmosphere. The red colour indicates heating and the blue shows the cooling cycle of the experiment. The samples were taken at different stratigraphic levels from the top the middle and bottom of the *c.* 41 m long soft sediment core.

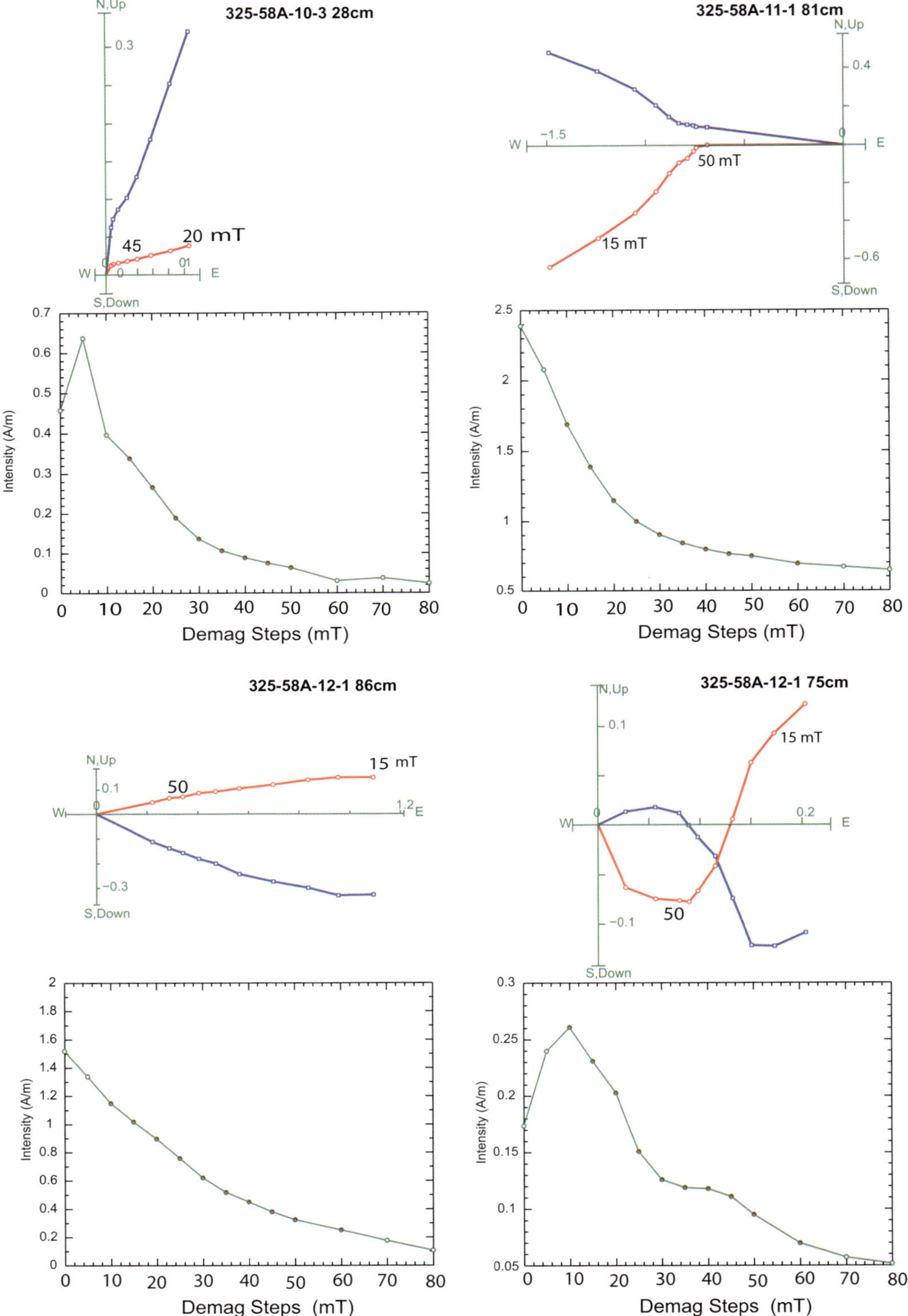
325-58A-10-3 28cm
N,Up
0.3
W
E
S,Down
45
20 mT
0
01
0.7
0.6
0.5
0.4
0.3
0.2
0.1
0
Intensity (A/m)
0 10 20 30 40 50 60 70 80
Demag Steps (mT)

325-58A-11-1 81cm
N,Up
0.4
W
-1.5
E
0
50 mT
15 mT
-0.6
S,Down
2.5
2
1.5
1
0.5
Intensity (A/m)
0 10 20 30 40 50 60 70 80
Demag Steps (mT)

325-58A-12-1 86cm
N,Up
0.1
15 mT
50
W
0
1,2 E
-0.3
S,Down
2
1.8
1.6
1.4
1.2
1
0.8
0.6
0.4
0.2
0
Intensity (A/m)
0 10 20 30 40 50 60 70 80
Demag Steps (mT)

325-58A-12-1 75cm
N,Up
0.1
15 mT
W
0
0.2 E
50
-0.1
S,Down
0.3
0.25
0.2
0.15
0.1
0.05
Intensity (A/m)
0 10 20 30 40 50 60 70 80
Demag Steps (mT)

along the SD–MD mixing theoretical curves of Dunlop (2002), suggesting that the magnetic carriers contain SP, SD and MD Ti-poor magnetite (see hysteresis curves shown in Fig. 3).

After measurement of the hysteresis loops and the back-field demagnetization curve of the SIRM, we carried out a low-temperature v. susceptibility analysis ($\kappa-T$) on the same five samples per stratigraphic level that were used in the magnetic granulometry experiments to determine the magnetic carriers of the NRM. These Curie point determinations were carried out on small amounts of powder and analysed in an air atmosphere. These experiments identified basically one distinct group (see Fig. 4), with the five specimens showing an irreversible complex $\kappa-T$ magnetic behaviour. All of the experiments showed a progressive increase in susceptibility up to $<560\,°C$, showing a Hopkinson peak. The heating curves indicate the presence of low-titanium to almost pure magnetite with Curie point temperatures ranging between 560–575 °C. The vast majority showed irreversible cooling curves indicative of the creation of oxidized magnetite during the cooling process.

Palaeomagnetic results (NRM)

The remanent magnetization of the entire set of 1 cm[3] mini-cubes was measured for 180 specimens, which were stepwise demagnetized by means of alternating fields from NRM, 5, 10, 15, 20, 25 30, 35, 40, 45, 50, 60, 70 and 80 mT. Typical demagnetization diagrams obtained from these analyses are displayed in Figure 5. The characteristic direction of each specimen was obtained by principal component analysis of a well-defined trend towards the origin on the demagnetization diagrams. Every principal component analysis result was based on at least five steps of the 180 analysed specimens, accepting only directions passing through the origin. In every case, the ChRM was isolated at relatively low alternating fields typically between 0 and 25 mT, indicative of SP magnetite as the primary carrier. Figure 5 shows the variety of results of the stepwise alternating field demagnetization experiments from NRM up to 80 mT. The excursional directions are characterized by high stability upon demagnetization at fields of 5–15 mT. Most of the soft secondary components are removed at fields <15 mT (Figs 6 & 7), as is the case for deep-sea sediments (e.g. Herrero-Bervera *et al.* 1987; Herrero-Bervera & Khan 1992).

Inclination and intensity of magnetization results

The magnetostratigraphic results of the analyses described above in terms of inclination and intensity of magnetization of the discrete samples are shown in Figure 6. The declinations, inclinations and intensity of magnetization along the core at 5 and 15 mT steps, respectively, are shown in Figure 7. The results of the two experiments (i.e. discrete and U-channel magnetostratigraphies) indicate that, upon alternating field demagnetization, the isolation of the ChRM components reveals three well-correlated excursions or 'aborted reversals' of the geomagnetic field. The demagnetized soft-sediment results of the declination and inclination indicate synchronous excursional changes at several stratigraphic levels. It is important to point out that there are very few soft sediment experiments where both discrete 1 cm[3] and U-channel (e.g. Kanamatsu *et al.* 2002) are correlated to check the internal consistency of the directions in question. Our combined results (i.e. discrete v. continuous direction) show a remarkable correspondence between the two methods, discussed further below.

Discussion

The rock magnetic parameters measured along the studied IODP Expedition 325 core site M0058A demonstrate that the sampled reef carbonate sequence contains a fairly concentrated ferromagnetic fraction. This fraction is primarily composed of Ti-poor magnetite with a widely ranging superparamagnetic distribution, and produced by the alteration and erosion of a range of rocks from continental Eastern Australia.

The Ti-poor magnetite is the main magnetic carrier of the carbonate sequence. It seems likely that the remanence was acquired through detrital and pre-diagenetical processes, although diagenetic and/or biogenic crystal growth cannot be ruled out. Palaeomagnetic analysis of sediment samples from ODP Leg 133, Site 820, located *c.* 10 km from the outer edge of the GBR, shows evidence for biogenic magnetite (Barton *et al.* 1993*a, b*). Chains of SD-sized crystals of pure magnetite are characteristic of biogenically produced (bacterial) magnetite (Blakemore & Frankel 1981) and are being found increasingly in marine environments (e.g. Yamazaki *et al.* 1991; Jovane *et al.* 2012). Caution is

Fig. 5. Orthogonal vectorial diagrams of stepwise demagnetization experiments of the continuous U-channel measurements. The blue vectorial lines indicate a projection onto the horizontal plane, whereas the red lines indicate projection onto the east–west oriented vertical plane. Numbers on the vectorial diagrams indicate alternating field peak field (in mT) used in alternating field demagnetization. The intensity of magnetization of the specimens are given in A m^{-1}.

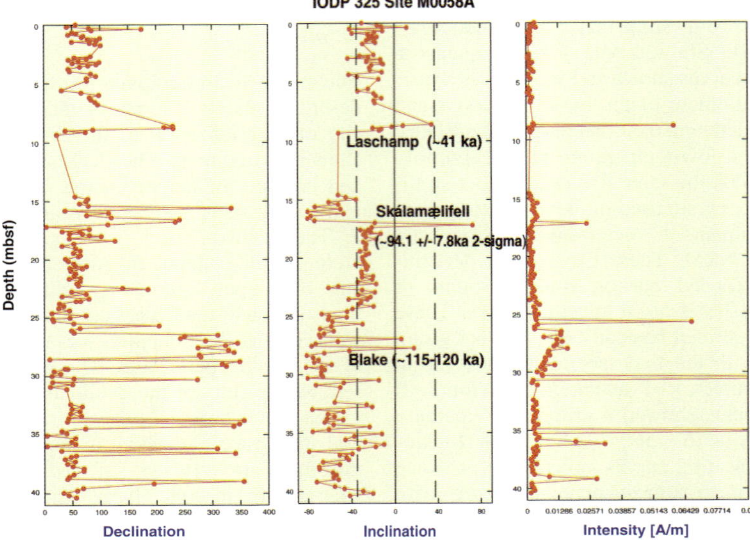

Fig. 6. Magnetostratigraphic results of the discrete specimens of (**a**) the declination, (**b**) the inclination and (**c**) intensity of the alternating field demagnetized results at 5 mT.

needed when interpreting such evidence, because (1) biogenic magnetite clustering into chains might be produced during the magnetic extraction process; (2) the sizes of the magnetic grains can be influenced by dissolution; and (3) dissolution favours the formation of isometric particles that will resemble biogenic magnetite crystals (Maher & Thompson 1999).

The studied GBR carbonates have a relatively strong and stable magnetization, with inclination values compatible with the GAD (Geocentric Axial Dipole) model at the site (GAD = $c.$ $-32.1°$). This suggests that they accurately recorded the geomagnetic field vector at the time of their formation and that they contain potentially reliable magnetostratigraphic information as depicted by the discrete and U-channel palaeomagnetic results (Figs 6 & 7). Similar results have also been reported recently for other sites, for example a late Pleistocene reef sequence in Tahiti that is part of a multidisciplinary study of the Holocene and Pleistocene reef sequences carried out by the IODP Expedition 310 (Menabreaz *et al.* 2010).

The excursional NRM directions (declination and inclination) at Site IODP 325 Site M0058A occur during intervals of low intensity of magnetization of the sediments from 6.0–9.35 mbsf (Excursion 1), 16.0–17.6 mbsf (Excursion 2), 26.5–27.4 mbsf (Excursion 3) and 27.6–28.40 mbsf (Excursion 4). Excursions 3 and 4 correspond to the double signal that characterizes the Blake excursion. There is a fair reproducibility of these four

excursions between the 1 cm³ discrete specimens results (Fig. 6; both of the NRM and at 5 mT) and the U-channel continuous measurements of the demagnetized data at 5 and 15 mT (Fig. 7).

A key factor for the correlation of these excursional inclinations to the Brunhes Chron excursions is the radiocarbon date of $c.$ 45 ka located 8 m from the top of the core. A second set of dates have been obtained from uranium–thorium methods from 27.88 m down to 35.80 m, ranging in age from as young as 133.482 ka to as old as 167.616 ka BP (A. Thomas, pers. comm. 2011).

These excursional inclinations have been well studied and reported in the literature, in deep-sea sediments as well as in igneous rocks (e.g. Valet *et al.* 2008). For instance, the Laschamp Excursion is now well established and has been firmly correlated to isotopic and ice-core models at 40–41 ka (Valet *et al.* 2008). This corresponds to our inclination Excursion 1. Our second inclination Excursion 2 corresponds to recent $^{40}Ar/^{39}Ar$ and $^{238}U–^{230}Th$ isochrons determined for two lava flows, yielding a weighted mean age of 96.0 ± 10.0 ka (2σ) (Valet *et al.* 2008). The weighted mean of the $^{40}Ar/^{39}Ar$ and $^{238}U–^{230}Th$ ages is 94.1 ± 7.8 ka (2σ), for lavas from Skálamælifell Hill, on the Reykjanes Peninsula in SW Iceland that recorded excursional geomagnetic field directions that correspond to virtual geomagnetic poles grouped at 12°S, 250°E in the eastern Pacific Ocean (Jicha *et al.* 2011). Our Excursions 3 and 4 correspond to the double geomagnetic features of

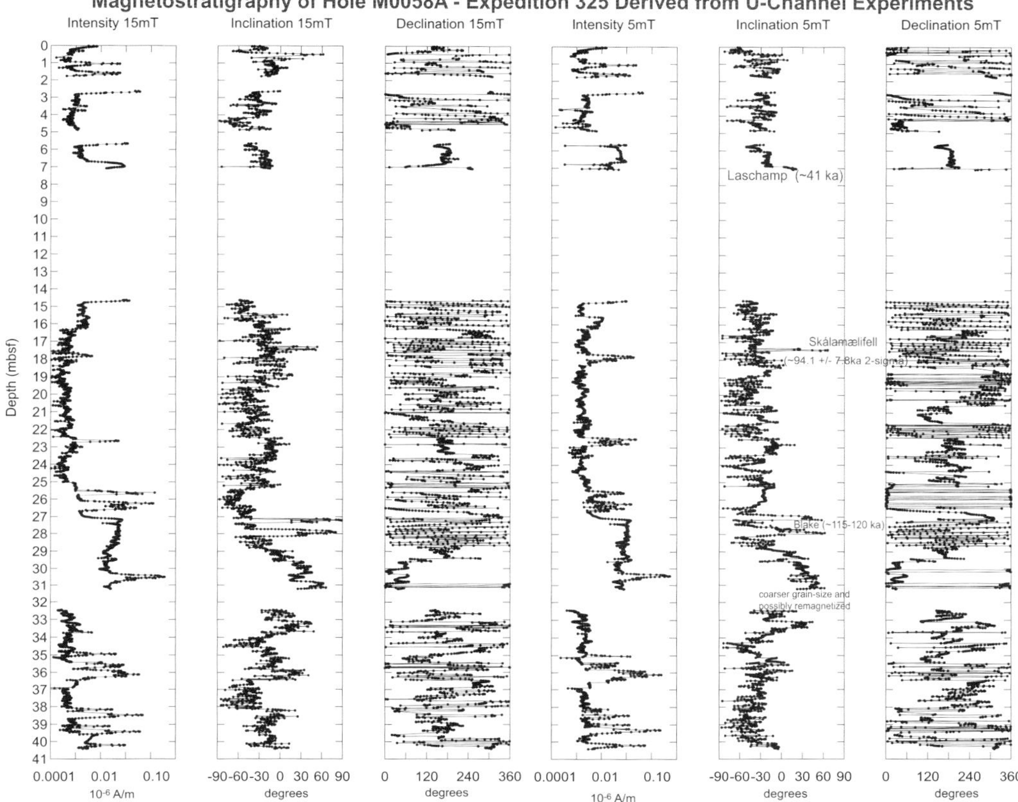

Fig. 7. Magnetostratigraphic results of the declination (**a**), inclination (**b**) and intensity (**c**) magnetization at two demagnetization steps at 5 and 15 mT of the *c*. 41 m long core from Site M0058A obtained from U-channel data down-core.

the Blake Excursion as mentioned above. The Blake Excursion has been documented in both igneous rocks and marine sediments, at *c*. 115–120 ka (e.g. Channell 2006; Valet *et al*. 2008).

Conclusions

We studied a *c*. 41 m long soft-sediment calcareous mud core from Expedition IODP 325 at Site M0058A. The core was sampled using both discrete 1 cm³ mini-cubes and U-channels in order to decipher the magnetostratigraphy in a continuous manner. The results are as follows:

- Rock magnetic properties such as low-field mass-specific magnetic susceptibility (χ), low-field susceptibility v. temperature (κ–T, Curie point determinations) and magnetic grain size analyses indicate that the main magnetic carriers are Ti-poor magnetites characterized by Curie

points ranging between 560 and 575 °C with magnetic grain sizes of low coercivities ranging between SP to admixtures of SP and SD and SD–MD grains of magnetites.
- Stepwise alternating field experiments isolated a ChRM at low fields (i.e. 0–15 mT) that revealed a set of three excursional inclinations and intensities. These inclinations and intensity characteristics make it difficult to discount these records as simply a product of multi-component magnetizations caused by smoothing as the sediments pass through a thick lock-in zone (Hyodo 1984) or as the product of antiparallel components that could generate a 'pseudo-excursion' record of the earth's magnetic field. We therefore interpret from this record that, during the excursional field, the field intensity diminished to about 20% of its initial or recovery value.
- The estimated times of the observed four excursions are consistent with those of the Laschamp,

Skálamælifell and the (double-event) Blake excursions.

- It follows that the base of the core M0058A, below Excursion 4, is older than the Blake Excursion, that is, older than 126 ka.

We would like to thank Mr J. Lau for his help during the on-shore laboratory phase of this project. We also thank the members of IODP Expedition 325 for all their help during the drilling and the core repository work in Bremen. Special thanks are given to the members of the Paleomagnetics Laboratory of the University of Bremen, who allowed us to run the discrete and U-channels samples. We also acknowledge the financial help of our respective institutions and USSSAC-Ocean Leadership grants to achieve the goals of this project. This is SOEST contribution no. 8827 and HIGP contribution no. 1999.

References

BARTON, C. E., LACKIE, M. & PEERDEMAN, F. M. 1993a. Environmental control of magnetic properties of upper slope sediments near the Great Barrier Reef: results from Leg 133, Site 820. *In*: MCKENZIE, J. A., DAVIES, P. J., PALMER-JULSON, A. *ET AL*. 1993. *Proceedings of the Ocean Drilling Program, Scientific Results*, **133**, Ocean Drilling Program, Texas A & M University, College Station, TX, 543–562.

BARTON, C. E., OMARZAI, S. K., ALEXANDER, I., PEERDER-MAN, F. & MCNEIL, D. 1993b. Magnetic stratigraphy and characterization of sediments from the Northeast margin of Australia and their relationship to environmental change during the Quaternary. *In*: MCKENZIE, J. A., DAVIES, P. J., PALMER-JULSON, *ET AL*. 1993. *Proceedings of the Ocean Drilling Program, Scientific Results*, **133**, *Ocean Drilling Program*, Texas A & M University, College Station, TX, 723–747.

BLAKEMORE, R. P. & FRANKEL, R. B. 1981. Magnetic navigation in bacteria. *Scientific American*, **245**, 58–65.

CHANNELL, J. E. T. 2006. Late Brunhes polarity excursions (Mono Lake, Laschamp, Iceland Basin and Pringle Falls) recorded at ODP Site 919 (Irminger Basin). *Earth and Planetary Science Letters*, **244**, 378–393.

DUNLOP, D. J. 2002. Theory and application of the Day plot (M_{rs}/M_s versus H_{cr}/H_c) 2. Application to data for rocks, sediments and soils. *Journal of Geophysical Research*, **107**, 5–15.

EXPEDITION 325 SCIENTISTS. 2011. Transect NOG- 01B. *In*: WEBSTER, J. M., YOKOYAMA, Y., COTTERILL, C. & THE EXPEDITION 325 SCIENTISTS, *Proceedings of the International Ocean Drilling Program*. Tokyo (Integrated Ocean Drilling Program Management International, Inc.), **325**, doi: 10.2204/iodp.proc.325. 106.2011.

GUYODO, Y. & VALET, J.-P. 1999. Global changes in geomagnetic intensity during the past 800 thousand years. *Nature*, **399**, 249–252.

GUYODO, Y., CHANNELL, J. E. T. & THOMAS, R. 2002. Deconvolution of U-channel paleomagnetic data near geomagnetic reversals and short events. *Geophysical Research Letters*, **29**, 1845, doi: 10.1029/2002GL014963.

HERRERO-BERVERA, E. & KHAN, M. A. 1992. Olduvai termination: detailed paleomagnetic analysis of a north central Pacific core. *Geophysical Journal International*, **108**, 535–545.

HERRERO-BERVERA, E. & RUNCORN, S. K. 1997. Transition fields during geomagnetic reversals and their geodynamic significance. *Philosophical Transactions of the Royal Society London*, **355**, 1713–1742.

HERRERO-BERVERA, E., THEYER, F. & HELSLEY, C. E. 1987. Olduvai onset polarity transition: two detailed paleomagnetic records from North Central Pacific sediments. *Physics of the Earth and Planetary Interiors*, **49**, 325–342.

HROUDA, F. 1994. A technique for the measurement of thermal changes of magnetic susceptibility of weakly magnetic rocks by the CS-2 apparatus and KLY-2 Kappabridge. *Geophysical Journal International*, **118**, 604–612.

HROUDA, F., JELINEK, V. & ZAPLETAL, K. 1997. Refined technique for susceptibility resolution into ferrimagnetic and paramagnetic components based on susceptibility temperature variation measurements. *Geophysical Journal International*, **129**, 715–719.

HYODO, M. 1984. Possibility of reconstruction of the past geomagnetic field from homogeneous sediments. *Journal of Geomagnetism and Geoelectricity*, **36**, 45–62.

JICHA, B. R., KRISTJANSSON, L., BROWN, M. C., SINGER, B. S., BEARD, B. L. & JOHNSON, C. 2011. New age for the Skálamælifell excursion and identification of a global geomagnetic event in the Late Brunhes Chron, *AGU Fall Meeting, GP23A-102*. EOS, San Francisco, CA.

JOVANE, L., FLORINDO, F., BAZYLINSKI, D. A. & LINS, U. 2012. Prismatic magnetite magnetosomes from cultivated Magnetovibrio blakemorei strain MV-1: a magnetic fingerprint in marine sediments? *Environmental Microbiology Reports*, **4**, 664–668, doi: 10.1111/1758-2229.12000.

KANAMATSU, T., HERRERO-BERVERA, E. & MCMURTRY, G. M. 2002. Magnetostratigraphy of Deep-Sea Sediments from Piston Cores Adjacent to the Hawaiian Islands: implication for ages of turbidites derived from submarine landslides. *In*: *Hawaiian Volcanoes: Deep Underwater Perspectives*. Geophysical Monograph **128**, American Geophysical Union, Washington, DC, 51–63.

KIRSCHVINK, J. L. 1980. The least squares lines and plane analyses of paleomagnetic data. *Geophysical Journal of the Royal Astronomical Society*, **62**, 699–718.

MAHER, B. & THOMPSON, R. 1999. *Quaternary Climates, Environments and Magnetism, Introduction*. Cambridge University Press, Cambridge.

MENABREAZ, L., THOUVENY, N., CAMOIN, G. & LUND, S. 2010. Paleomagnetic record of the late Pleistocene reef sequence of Tahiti (French Polynesia): a contribution to the chronology of the deposits. *Earth and Planetary Science Letters*, **294**, 58–68.

ODA, H. & SHIBUYA, H. 1996. Deconvolution of long core paleomagnetic data of Ocean Drilling Program by Akaike's Bayesian criterion minimization. *Journal of Geophysical Research*, **101**, 2815–2834.

SINGER, B. S. 2007. Polarity transitions radioisotopic dating. *In*: HERRERO-BERVERA, E. H. & GUBBINS, D.

(eds) *Encyclopedia of Geomagnetism and Paleomagnetism*. Springer, New York, 834–839.

THOUVENY, N., CARCAILLET, J., MORENO, E., LEDUC, G. & NERINI, D. 2004. Geomagnetic moment variation and paleomagnetic excursions since 400 kyr BP: a stacked record from sedimentary sequences of the Portuguese margin. *Earth and Planetary Science Letters*, **219**, 377–396.

VALET, J. P. & MEYNADIER, L. 1993. Geomagnetic field intensity and reversals during the last four million years. *Nature*, **366**, 234–238.

VALET, J.-P., MEYNADIER, L. & GUYODO, Y. 2005. Geomagnetic field strength and reversal rate over the past 2 million years. *Nature*, **435**, 802–805.

VALET, J. P., PLENIER, G. & HERRERO-BERVERA, E. 2008. Geomagnetic excursions reflect an aborted polarity state. *Earth and Planetary Science Letters*, **274**, 472–478.

YAMAZAKI, T., KATSURA, I. & MARUMO, K. 1991. Origin of stable remanent magnetization of siliceous sediments in the central equatorial Pacific. *Earth and Planetary Science Letters*, **105**, 81–93.

Characteristic wavelengths in VGP trajectories from magnetostratigraphic data of the Early Cretaceous Serra Geral lava piles, southern Brazil

GEORGE CAMINHA-MACIEL[1,2]* & MARCIA ERNESTO[2]

[1]*Present address: Universidade Federal do Pampa, Caçapava do Sul, RS, Brazil*

[2]*Department of Geophysics, University of São Paulo, São Paulo, SP, Brazil*

**Corresponding author (e-mail: caminha.maciel@unipampa.edu.br)*

Abstract: The virtual geomagnetic pole (VGP) trajectories during some geomagnetic polarity reversals of different ages are marked by anisotropic behaviour. This recurrent phenomenon may be reflected in the paleomagnetic data, even if the transitional field was not completely recorded. As the long-scale geomagnetic variations have a confined oscillatory character, the VGP paths from stratigraphically controlled sequences may be described on the basis of sine and cosine functions, even if time is not the independent variable. Here we considered longitude (or space) as the independent variable which had to be 'unrolled' to overcome the 360° repetitions as the VGPs moved around the geographic pole.

Sixteen VGP series from the Early Cretaceous Serra Geral lava flows of southern Brazil were analysed using a modified version of the periodogram for uneven data series, and a combination of information approach. The combination of all the spectra, as in a stacking procedure, reduces noise and results in a smooth curve highlighting features of interest. We found a set of highest correlation wavelengths of approximately 167, 190, 209, 257, 277 and 368°. Phase analyses using two different methods revealed strikingly good coherence for some of these wavelengths, indicating that they are not only artefacts of the spectral analysis. Similar analysis of magnetostratigraphic data from the Icelandic Magmatic Province indicated that the two datasets may have wavelengths of approximately 165 and 270° in common. These results suggest quasi-periodic behaviour, possibly with sub-harmonic instabilities owing to the modulating effect of inner Earth's anisotropies influencing the pole trajectory.

Understanding of the long-term variations of the Earth's magnetic field requires a full description of the spatial–temporal characteristics of the field. Analyses of secular variation have been mainly based on directional dispersion as a function of latitude in recent times (Tauxe & Kent 2004 and references therein), and very few analyses refer to older fields (e.g. Smirnov & Tarduno 2004; Biggin *et al.* 2008). Long-term geomagnetic variations are better observed in sedimentary rocks, as they may continuously record the behaviour of the paleomagnetic field, although knowledge of the time of the remanence acquisition is usually imprecise. Furthermore, depending on depositional rate, the geomagnetic record may be naturally smoothed. Lava flow sequences tend to produce a more faithful magnetic record. However, the intermittent character of volcanism does not allow continuous records, and the time gap between two consecutive flows is difficult to ascertain precisely. The character of the magmatic activity is such that adjacent volcanic sections may exhibit distinct paleomagnetic records of the same event (Valet & Herrero-Bervera 2003). This behaviour may prevent the identification of some features of the geomagnetic field. However, multiple magnetostratigraphic records may display complementary information about certain geomagnetic characteristics that can be assessed if the information is adequately combined.

Magnetostratigraphic records allow the observation of virtual geomagnetic pole (VGP) trajectories throughout the time interval spanned by a given geological formation. One interesting issue resulting from the comparison of VGP trajectories during recent geomagnetic reversals is that transitional VGPs may follow preferred paths along some longitudinal bands (Clement 1991; Laj *et al.* 1991; Hoffman 1992; Gubbins & Love 1998; Love 2000). This type of observation made on sedimentary sequences has been viewed with scepticism by some authors (Barton & McFadden 1995; Quidelleur *et al.* 1995), who claim that the phenomenon is an artefact of the magnetization process in those rocks. However, some volcanic rocks also show geographical VGP confinement, as pointed out by Love (1998), Valet & Herrero-Bervera (2003), Coe & Glen (2004), and others.

From: Jovane, L., Herrero-Bervera, E., Hinnov, L. A. & Housen, B. A. (eds) 2013. *Magnetic Methods and the Timing of Geological Processes*. Geological Society, London, Special Publications, **373**, 293–307.
First published online March 25, 2013, http://dx.doi.org/10.1144/SP373.15 © The Geological Society of London 2013.
Publishing disclaimer: www.geolsoc.org.uk/pub_ethics

Most of the observations of transitional VGP confinement are for reversals from the last 5 Ma, but older records (Late Permian through Early Cretaceous; Vizán *et al.* 1994) may show the same characteristics, at least on a statistical basis. Simulated reversals using the Glatzmaier–Roberts geodynamo model may also support (Coe *et al.* 2000) the preferred band hypothesis. Reversals may differ from one another, and the same reversal observed in different places may show distinctive records owing to the geomagnetic non-dipolar components that prevail at the sampling sites. However, the repetition of some observed patterns in long-term geomagnetic variations suggests the action of some regulatory phenomena (Clement 1991; Laj *et al.* 1991; Hoffman 1992; Gubbins & Love 1998; Coe *et al.* 2000; Costin & Buffet 2004; Leonhardt & Fabian 2007; Vizán & Van Zele 2008). Models for VGP paths (Kuznetsov 1999; Leonhardt & Fabian 2007) suggest that the trajectories may be conditioned by field anomalies superposing the dipolar field during reversals, leading to longitude confinements.

Unfortunately, many magnetostratigraphic datasets from igneous provinces are unsuitable for the investigation of transitional VGP confinement owing to their fragmentary character. They may show various polarity intervals but few or practically no transitional directions. However, they may carry valuable information on the behaviour of the magnetic field that may help better constrain geodynamo modelling. Volcanic magnetostratigraphic sequences are still underexplored, but advances may occur if new methods of analysis are developed and the problem is examined from a new perspective. This is the main objective of the present study.

In this study we apply a spectral analysis to the magnetostratigraphic sequences of the Lower Cretaceous Serra Geral Formation (Paraná Magmatic Province, southern Brazil). VGP sequences from several Serra Geral sections (Fig. 1) have already been obtained (Ernesto *et al.* 1990; Alva-Valdivia *et al.* 2003; Mena *et al.* 2006), but no complete reversal has been described despite the various polarity intervals recorded in the series (Fig. 2). However, if the VGP trajectories during polarity reversals or excursions are marked by anisotropic behaviour, it is possible that the datasets contain some recurrent features. In this way, if we approximate the VGP trajectory as a sum of oscillatory modes around a line (equator) in a 2D (latitude v. longitude) plane, then spectral properties (or correlations with sines and cosines) can be investigated. Because geochronologic control of the geomagnetic events is unavailable, we must proceed in a way that does not require an assumption about elapsed time. Here we address the problem by considering the spatial rather than the time behaviour of the VGP paths, taking longitudes as the independent variable.

The dataset

We analyse the Serra Geral Formation data published by Ernesto *et al.* (1990). These data come from 16 stratigraphically ordered lava piles from the southern part of the Paraná Magmatic Province (Fig. 1) where the best rock exposures are concentrated. In a recent re-evaluation of the available ^{40}Ar/^{39}Ar data, Thiede & Vasconcelos (2010) proposed an age interval of 134 ± 1 to 133 ± 1 Ma for the magmatic activity in that area, with younger ages towards the north. This short time interval of about 1 Ma for the emplacement of the magma flows, and the small differences in latitude (less than $4°$) between lava flow sequences, are good indicators that the rocks were magnetized by a relatively homogeneous paleomagnetic field. Most of the sequences record more than one polarity interval, although transitional directions are rare, as can be seen on the declination/inclination plots of Figure 2.

For the purpose of present analysis, the VGPs coordinates must be corrected of the effects of the South American plate displacement that has occurred since the Early Cretaceous. Hence, a correction of $5°$ to the whole dataset was applied taking into account the reference paleomagnetic pole given by Font *et al.* (2009) for the Early Cretaceous. The VGPs describe irregular loops around the geographic poles (north or south depending on the polarity state), causing longitudes to repeat. Therefore, the VGP trajectories must be 'unrolled', eliminating the longitude repetitions by adding $360°$ each time that data values moved backwards, as seen on the examples of Figure 3. This procedure does not affect the distribution of VGPs, and does not modify the spectral content in the series. This unwrapped sequence of longitude–latitude pairs can now be treated as an unevenly sampled 'time' series. Figure 4 shows the plots of the 16 unwrapped VGP sequences used in this work.

Methods

We seek to detect the long-term features in the VGP trajectories. Therefore, we have to model a smooth curve that fits the main path described during the secular variation cycles or complete polarity changes, regardless of the short-wavelength features that are responsible for the stationary features causing the VGPs to cluster at some latitudes. Once the VGPs in each series are arranged in order of increasing longitudes (Fig. 4), the various

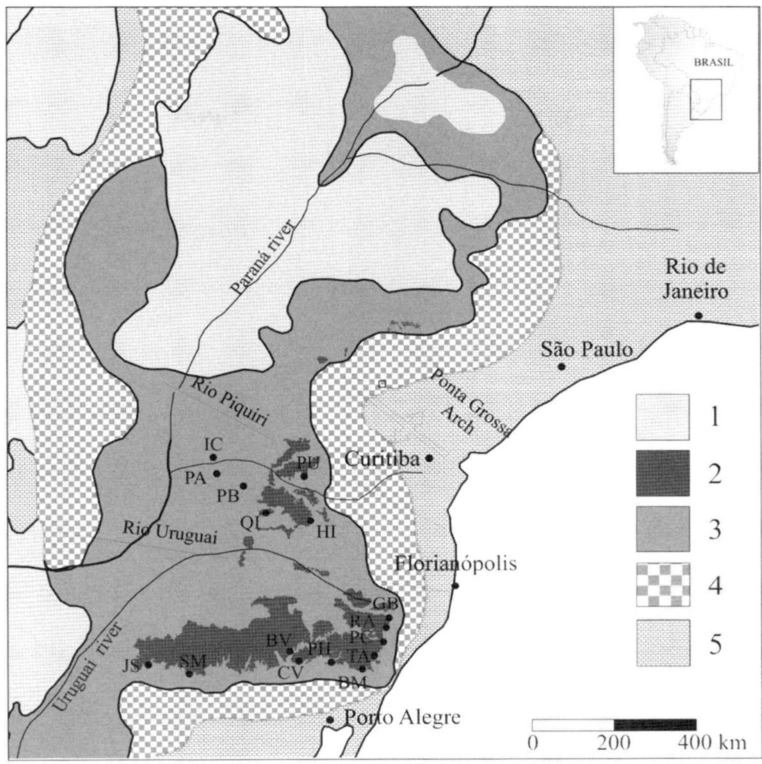

Fig. 1. Simplified map of the Paraná Magmatic Province showing the location of the Serra Geral sections sampled for the paleomagnetic work. Legend for the map: 1, post-volcanic sediments; 2, acid rocks; 3, tholeiitic rocks; 4, pre-volcanic sediments; 5, crystalline basement. Section codes and reference localities are according to Ernesto & Pacca (1988): JS, Jaguari-Santiago; SM, Santa Maria-Júlio de Castilhos; BV, Bento Gonçalves-Veranópolis; CV, São Sebastião do Caí-Carlos Barbosa; BM, Barra do Ouro-Morrinhos; PH, Nova Petrópolis-Novo Hamburgo; TA, Terra de Areia-Aratinga; PC, Praia Grande-Cambará do Sul; RA, Turvo-Timbé do Sul; GB, Guatá-Bom Jardim da Serra; IC, Rio Iguaçu-Cascavel; PA, Francisco Rio Iguaçu; PB, Pato Branco-Francisco Beltrão (section B); PU, Porto União-Guarapuava; QI, Quilombo-Iraní HI, Horizonte-Iraní.

data sequences can be treated as multiple independent samples of the same trajectory of a particle over an infinite 2D plane (latitude v. longitude). We want to recover the original function that gave rise to those samples. In a simplified picture, this function describes the trajectory of a particle with a general trend to the east/west, oscillating around the 90° parallels most of the time, and eventually crossing the equator. We have only randomly spaced sampled series for defining this function, but we can draw attention to three main constraints of the original function: (1) it is smooth (infinitely differentiable) – we are only interested in long period behaviour; (2) it is limited – all data must be contained in a limited space delimited by a finite number of parallels; and (3) it oscillates most of the time. The third condition suggests using Fourier basis to model the data, and the first two conditions allow us to do so mathematically.

The problem of estimating the original series from which we have only the unevenly sampled series may be solved by invoking the Popper–Bayes paradigm (Tarantola 2006): the data should not be used to choose a 'best' solution but instead to 'falsify' solutions over the parametric space. By 'falsifying' the author means setting a probability distribution over a subset of the parameter and improving that distribution by adding (combining) new experimental (or theoretical) information. In this framework, a Bayesian inversion scheme can be thought of as a combination of theoretical and experimental information, or as a combination of independent experimental information (Tarantola & Valette 1982). Therefore, the inversion of the spectral state of information functions we are dealing with can be the combination of the multiple records of the same events. These state of information functions are freely normalizable probability

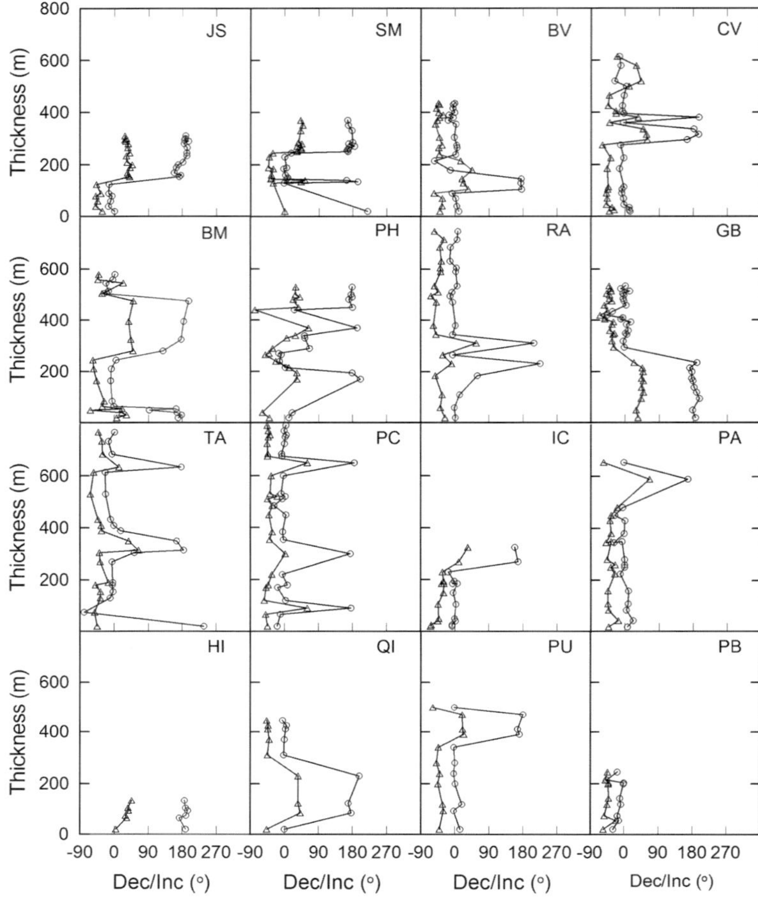

Fig. 2. Magnetic declination and inclination variations along the magma flow sequences showed in Figure 1. Circles are declinations and diamonds are inclinations.

distributions over a finite discrete set of points in the frequency domain. For each frequency we associate a probability value given by its spectral power estimate – the periodogram ordinate. For this, we made use of the optimal least-squares fitting properties of the Lomb–Scargle (LS) periodogram (Lomb 1976; Scargle 1982) for decomposing the data series variance using sinusoidal bases.

The periodogram is a suitable tool for spectral analysis of unevenly sampled time series, although it produces very noisy estimates even when the sample data are only slightly noisy (Scargle 1982; Hernandez 1999; Zechmeister & KürSter 2009). A solution could be to smooth the periodogram estimates by enlarging the main lobe or applying filters (windows), therefore reducing resolution, and consequently reducing the main lobe power (and detection efficiency). As a result, the power leakage from an existing spectral component to

neighbour frequencies is also reduced. This is the basis of the so-called 'smoothed periodogram method'.

As in the case of evenly spaced data, we need to define the frequency set for which power will be estimated. Usually the maximum frequency corresponds to the 'Nyquist frequency', which is the inverse of the half the minimum distance between any two consecutive points, the minimum frequency being the inverse of the total time interval covered by the data series. The frequency increment is taken as one and a half of the inverse of the whole series length. When the investigated phenomenon is sampled in an almost regular grid containing only small gaps or a few missing points, a characteristic sampling interval may be estimated by the average interval or by more elaborate estimates of the number of independent points in frequency domain (Horne & Baliunas 1986).

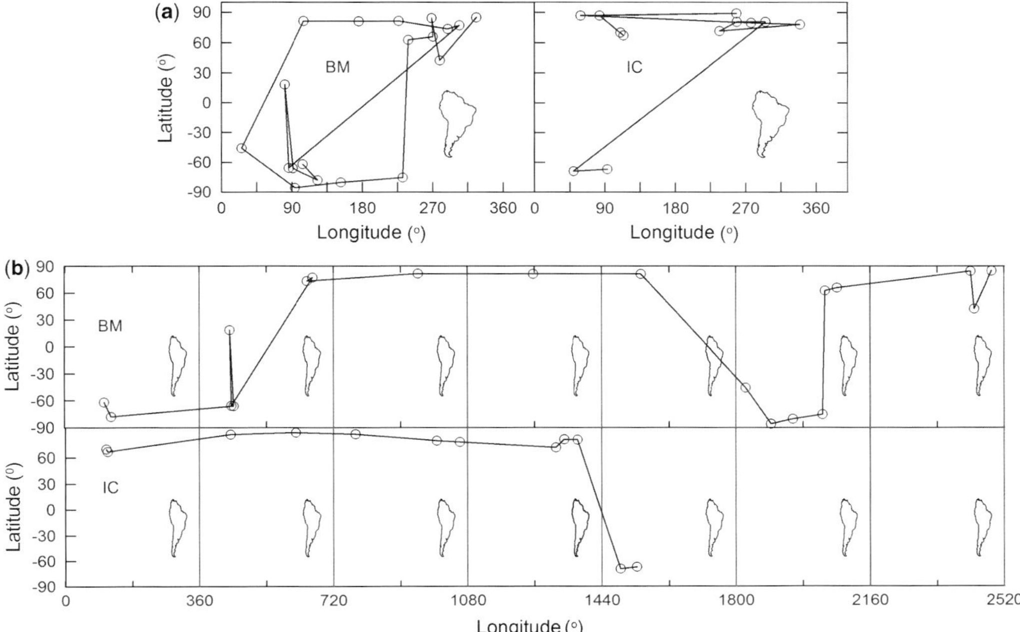

Fig. 3. Examples of latitude v. longitude plots of the Serra Geral VGP sequences with (**a**) original longitudes and (**b**) 'continued' longitudes.

In the present case the longitude difference between two consecutive points is extremely variable, and a new procedure is required for the estimate of the characteristic sampling interval as well as the Nyquist frequency. First, we set the frequency limits of the estimated spectra numerically in order to combine all information available in the series. The maximum available frequency range was computed as follows: from the inverse of the largest series length to the inverse of the shortest spacing between two consecutive points in any of the investigated data series (a kind of Nyquist frequency). Next, we reduced those limits in order to eliminate regions with highly incoherent signal from the estimated spectra (by comparison of all series), and regions with no resolution in the parameters estimation. With respect to spacing, we started from the smallest frequency spacing (the inverse of the greatest length), and progressively enlarged the spacing until we achieved smoother estimates for all individual series.

The challenge is to smooth those estimates, and this can be achieved through the combination of the information contained in the various data sequences. The obtained spectra can be compared and combined after being properly normalized. The adopted normalization parameter is the apparent total power, which corresponds to the power over the total investigated bandwidth. Each spectrum is normalized by the total area in such a way that the area under the curve is equal to unity; the most important aspect in this procedure is that the signal/noise ratio is preserved.

In the Bayesian approach we model our knowledge about a parameter rather than the parameter itself. Therefore, the determination of a restricted space solution does not necessarily mean that we know the parameter value with absolute precision (Tarantola 2006). We might interpret our obtained values (probability over frequencies) based on the data sample uncertainties and on the quantity (and quality) of the series. In this case, each sampled series represents an experiment of the spectral measurement or of the 'roughness' of the original series. In other words, if the distribution of the combined variable (estimation) is concentrated, it could be just because the combining distributions have small intersections or few common models, despite its low information content. In this approach we do not extract any central estimator (mean or variance), which according to Tarantola & Valette (1982) 'would certainly lead to a loss of information'. These authors indicate the use of the discrete probability distribution function (state of information function) itself as an elementary datum.

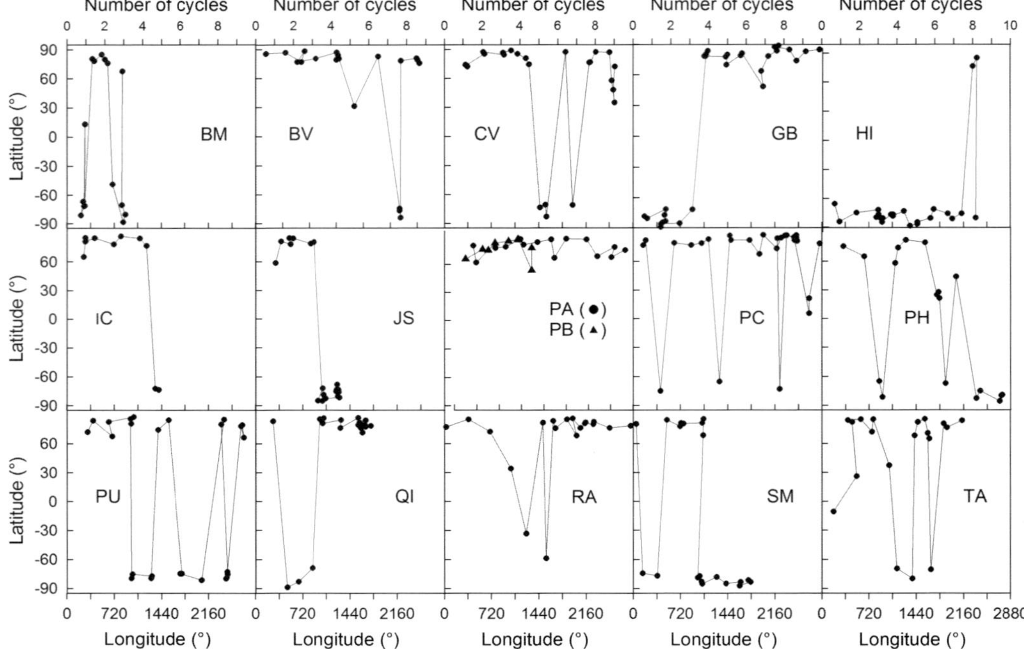

Fig. 4. Latitude v. 'continued' longitude plots of VGPs of the Serra Geral sequences used in this work.

The combination of the sampling functions furnishes two envelopes in the parametric space – one is larger and inclusive, and the other is smaller and exclusive. These envelopes respectively represent the union and intersection of the various states of information (as in usual set theory), and they correspond to the 'OR' and 'AND' operations (Tarantola & Mosegaard 2000). Mathematically, these operations can be performed as the sum (OR) and the product (AND) of the different estimates for each frequency. For the 'AND' operation we divide the resulting product by the null state of information function as defined by Tarantola & Valette (1982). We chose this function as a constant function over the interval, which is satisfactory in most cases as stated by those authors. The resulting combined estimate (combination of the multiple spectra) is also normalized to the total unit area.

Results

Periodogram analysis

In order to calculate the individual LS periodograms for the Serra Geral VGP sequences we had to define the sampling density as discussed above. The appropriate sampling density was then chosen empirically observing the following aspects: (1) as the number of points in the frequency domain increases more

noise is introduced; (2) fewer points in the frequency domain imply more 'scalloping error', that is, the error that arises when the real signal is between two observing points in the frequency domain; and (3) the chosen spacing interval should allow observation of the common spectral features in most of the spectra.

The LS periodograms for the VGP sequences are given in Figure 5. The upper part of the figure (a–c) shows the spectrum for each of the investigated sequences. As expected, the spectra are extremely noisy, especially in the shorter wavelength range, and few sections give clear indications of real wavelengths. However, it is easily seen on the individual curves that some common features appear in all spectra, although sometimes as a weak signal (below 20% of the maximum peak height).

Normalization and information states

The mean of all information in the individual spectra after normalization corresponds to the 'OR' curve (Fig. 5). Although still noisy, this is a smoother spectrum, mainly for long wavelengths, where probable real 'models' can be envisaged. The 'AND' curve (bottom of Fig. 5) represents a non-linear cascade of amplifiers, one for each data series furnishing one gain factor with encapsulated signal/noise ratio information. This curve combines

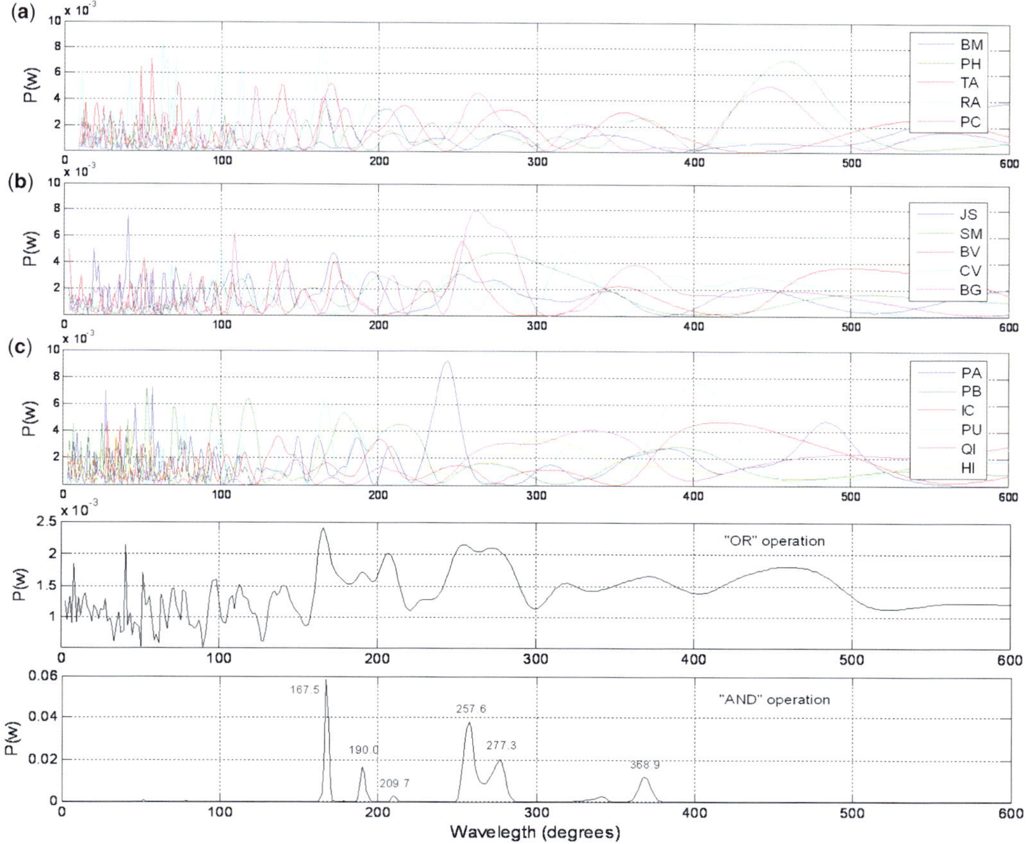

Fig. 5. LS spectra for the VGP sequences organized into three subsets (**a**–**c**) according to geographical proximity. The resulting spectra from the 'OR' and 'AND' operations combine the results from all series. $P(w)$ in arbitrary units.

only the common features in all individual spectra, and shows a surprisingly small number of discrete peaks, considering the large noise level and very low density of points relative to the length of the individual series. As a result we obtained a measurement of the probability over a subset of the frequency space that indicates which models fit better. In short, we used the information content in each series to falsify possible solutions in a subset of the parametric space. Therefore, we combined the general available information for all series in relation to the probability that a particular frequency can be a true Fourier component of the original trajectory.

Any spectral analysis is ultimately a correlation analysis of a finite, discretely sampled series with an infinite function basis (as in the LS periodogram), and so is subject to power leakage and aliasing. Therefore, we are interested not only in the maximum ordinate values of the resulting 'spectra', but we also have to investigate the local behaviour of the correlation maxima (neighbour frequencies), for which the continuity and the smoothness of the final distribution is important.

We found a set of highest correlation wavelengths of approximately 167, 190, 209, 257, 277 and 368°. The frequencies on the peaks $\lambda_1 = c.\ 167°$, $\lambda_2 = c.\ 257°$ are respectively 4 and 5 times greater than that at $\lambda_3 = c.\ 368°$. Then we infer a sub-harmonic relationship between λ_1 and λ_2, namely $\lambda_1/\lambda_2 = c.\ 2/3$, that is, λ_1 and λ_2 seem to be the second and third harmonic frequencies of a process for which $\lambda_f = c.\ 514°$ would be the fundamental one.

Performance of the method with simulated data

Synthetic data series possessing known harmonic composition, and with similar sampling characteristics to the real data studied in this work, were used to test the method. We considered a time series resulting from the superposition of sinusoidal components with known periods of 350, 480, 1100 and 4000 and amplitudes equal to 4, 6, 7 and 6, respectively. Noise proportional (5%) to the amplitude was added to each point. The resulting time series was then randomly sampled and uniformly distributed producing six different series as follows: three samplings of the whole series (21 cycles) with 25, 20 and 18 points (datasets 1–3); one sampling from the first two-thirds of the series with 15 points; one sampling from the last two-thirds of the series with 15 points; and one sampling with eight points from the middle third of the series. Figure 6 shows the resulting time series and sampled datasets.

The combined spectra for the six synthetic series are shown in Figure 7. The upper window (a) shows the spectrum for each series separately; the other windows show the result of the union and intersections of the state of information functions (operations 'OR' and 'AND'). The spectral combinations are able to filter most of the noise or spurious signals, and they highlight the signal in all the series ('AND' operation) for which the probability is higher. Despite the poor quality of the generated data series, the procedure presented here shows

excellent results considering the limited possible solutions in the parametric space. There are high-correlation regions in the OR and AND spectra, and the imposed periodicities (480, 1100 and 4000) in the original time series were satisfactorily recovered (487, 863 and 4207). In the investigated samples, no solution was found for frequency 350 based on the states of information used here.

The above results demonstrate that the method performs quite well for unevenly spaced time series affected by noise and with low density of points.

Discussion

The wavelength correlation suggested in the Serra Geral periodogram of Figure 5 needs confirmation; it is not possible to distinguish between actual components and spurious spectral effects based only on this analysis. Alternatively, we may suspect a 'coupling' effect of the spatial propagation modes or a spatially periodic variability in the media. These results seem to reinforce the initial guess that there could be periodic or quasi-periodic components in the VGP trajectories, and although these trajectories are extremely complex, their long-term behaviour

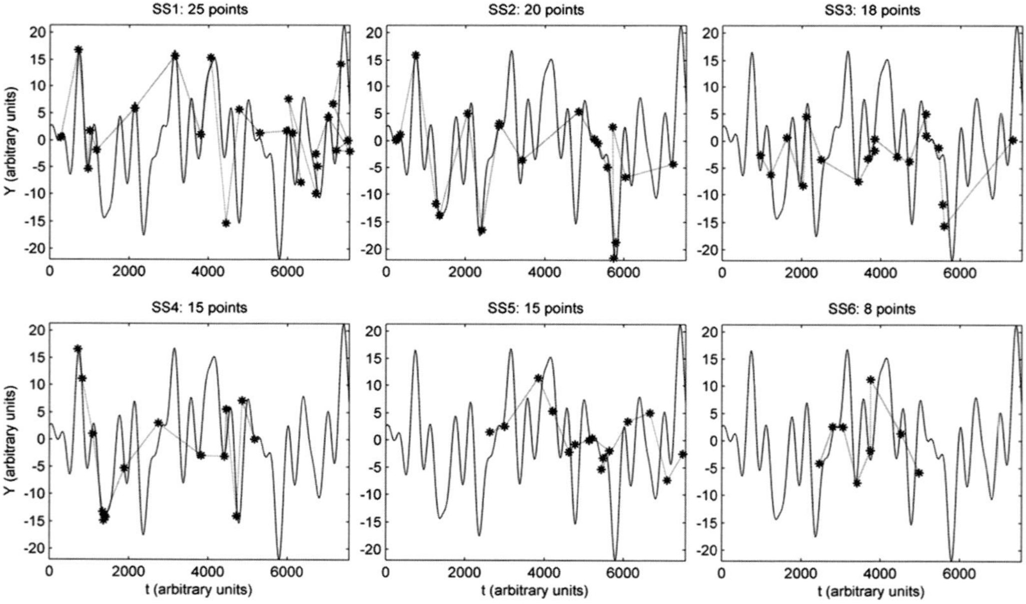

Fig. 6. Unevenly spaced datasets (stars and dashed lines) sampled from a time series (continuous line) corresponding to a superposition of four sine waves (periods 350, 480, 1100 and 4000). Sampled datasets are as follows: SS1 = 25 points, SS2 = 20 points, SS3 = 18 points from the whole synthetic series (21 cycles); SS4 = 15 points from the first two-thirds of the series; SS5 = 15 points from the last two-thirds of the series; SS6 = 8 points from the middle third of the series.

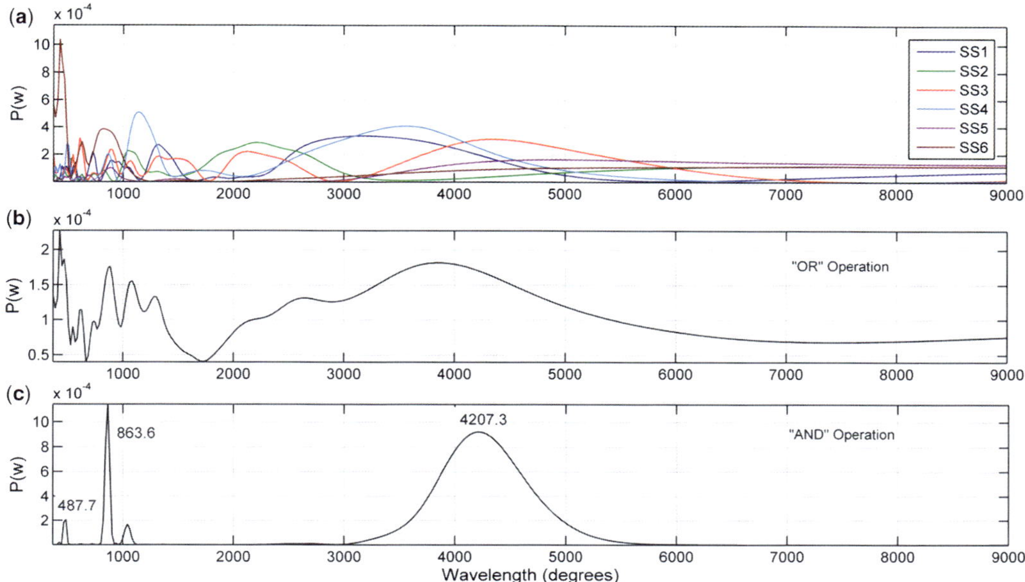

Fig. 7. (**a**) LS spectra for the unevenly spaced sampled datasets (SS1–SS6) from the synthetic series of Figure 6, and the subsequent (**b**) 'OR' and (**c**) 'AND' operations with the individual spectra. $P(w)$ in arbitrary units.

can be described by a relatively small number of sinusoidal components.

Spectra stability

In the case of data series sampled on a regular grid, the application of spectral windows to reduce lateral lobes owing to spectral leakage is a common procedure. This practice is not recommended in the irregular case (Hernandez 1999) because the spectral windows will be asymmetrical and non-homogeneous owing to irregularities of the original sampling. Furthermore, adapting the windows in the case of data distribution with strong asymmetries in spacing, as is the present case, is not trivial. Therefore, we applied two variations of the rectangular window: (1) we removed the extreme points of each series; and (2) we gradually attenuated the three first and last points of the series to levels of 20, 30 and 70% of the original values from the extremity inward. In both cases, the objective was to modify the structure of the lateral lobes in each spectrum, altering their combination. We intended to attenuate side lobes relative to the main lobes in each case; in doing so, we tried to separate those features generated exclusively by spectral leakage from the existing components in the original series. The results are given in Figure 8. The two procedures preserved some characteristics of the former spectrum, but considering the shortness of the original series, the first adopted procedure

apparently implied a considerable loss of information. Even so, peaks c. 255–270° still persisted, and peaks c. 450°–496°, which are close to the $\lambda_f = c$. 514° discussed above, are now evident.

The stability of the spectrum was also tested by disturbing the points of the original data series with uniformly distributed noise of the form ($x_i = x_i + K*rand$), with $K = 5$, 10, 15, 20 and 30°. This kind of filter is widely used in image processing and acts as a low-pass or 'anti-aliasing' filter. In Figure 9 the stability of the spectrum is noticeable as the main features were maintained. This indicates that the spectral characteristics can really represent long-term components of the trajectory, although some power transfer among peaks still occurred for increasing values of K, probably owing to aliasing.

Phase determination

Phases were determined by two methods: the classical Fourier phase estimator and the Hocke's (1998) estimator based on the LS periodogram. Despite the poorness of the sampling series, there is good agreement among the estimated phases for each data series, as seen in the rose diagrams of Figure 10. Apparently, the results for the wavelengths of 167, 277 and 368° cluster better, and this can be checked in Figure 11, where the results for each data sequence can be visualized. The fact that one or more sinusoidal components are in phase

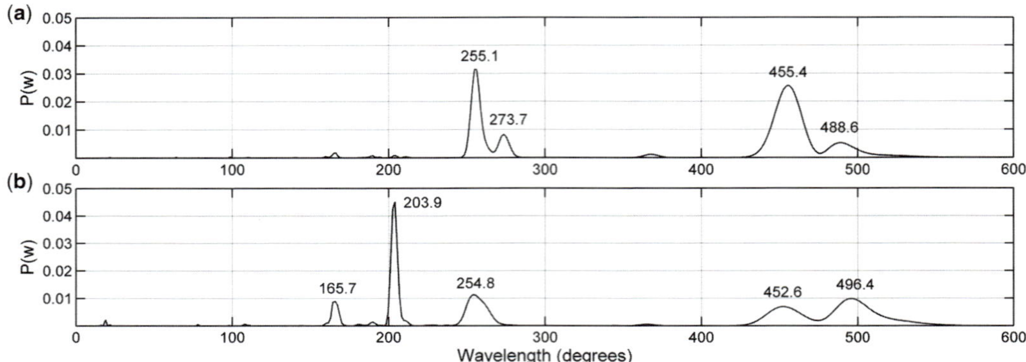

Fig. 8. The 'AND' spectra for the Serra Geral series after (**a**) removing the extreme points from each series, and (**b**) after progressively attenuating the three first and last points of the series. $P(w)$ in arbitrary units.

cannot be seen as a chance result, indicating that at least one of the components might be real. Furthermore, in the tests of spectrum stability, a large amount of power was transferred to the $c.$ 167 and $c.$ 277° peaks as the original data were becoming noisier (Fig. 9), indicating that probably they are not spurious effects.

Comparison with Icelandic data

At this point, it is instructive to investigate whether the characteristic wavelengths found in the Serra Geral spectra also represent a geometric feature in the VGP paths of other datasets. The Icelandic Magmatic Province is suitable for a comparative analysis, as the available paleomagnetic data (Kristjansson *et al.* 2004) show similar characteristics in

the record of secular variation and reversals. The data come from sections exposed in the northern part of Iceland, and the ages are within the range of 9–5 Ma. The individual sections (coded GR, KG, AF, AS, GL, SG and VE; Kristjansson *et al.* 2004) are partly superposed, and the composite section displays 17 reversals and excursions, but each one appears in one or two flows. To perform the spectral analysis in the same bandwidth as the Serra Geral spectra, the longer profiles GR and KG were subdivided, and only the upper parts were analysed. The VGP longitudes were 'continued' (Fig. 12) according to the same criteria used in the Serra Geral data.

The 'AND' spectrum of the Iceland data (Fig. 13) is smooth, with pronounced peaks at 169.1 and 265.2°. The latter is predominant, which

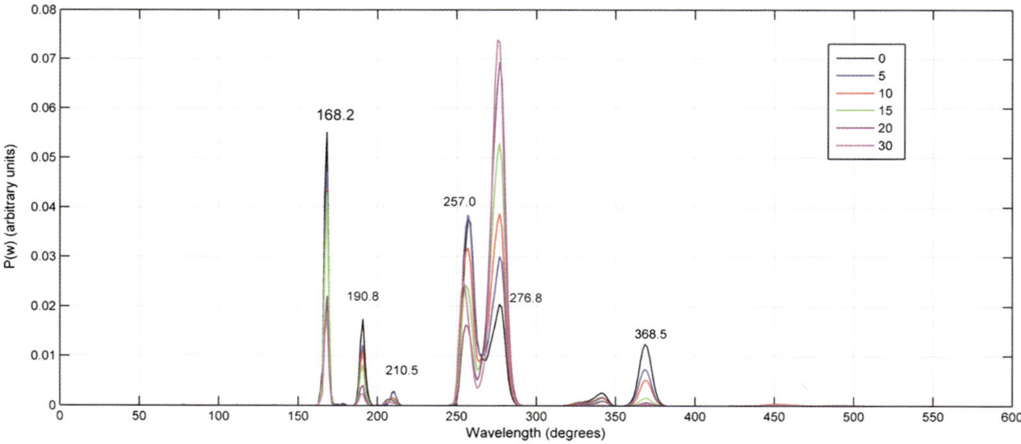

Fig. 9. The 'AND' spectrum for the Serra Geral series after disturbing the data with uniformly distributed noise of the form $(x_i = x_i + K*rand)$ with $K = $ 5, 10, 15, 20 and 30°. $P(w)$ in arbitrary units.

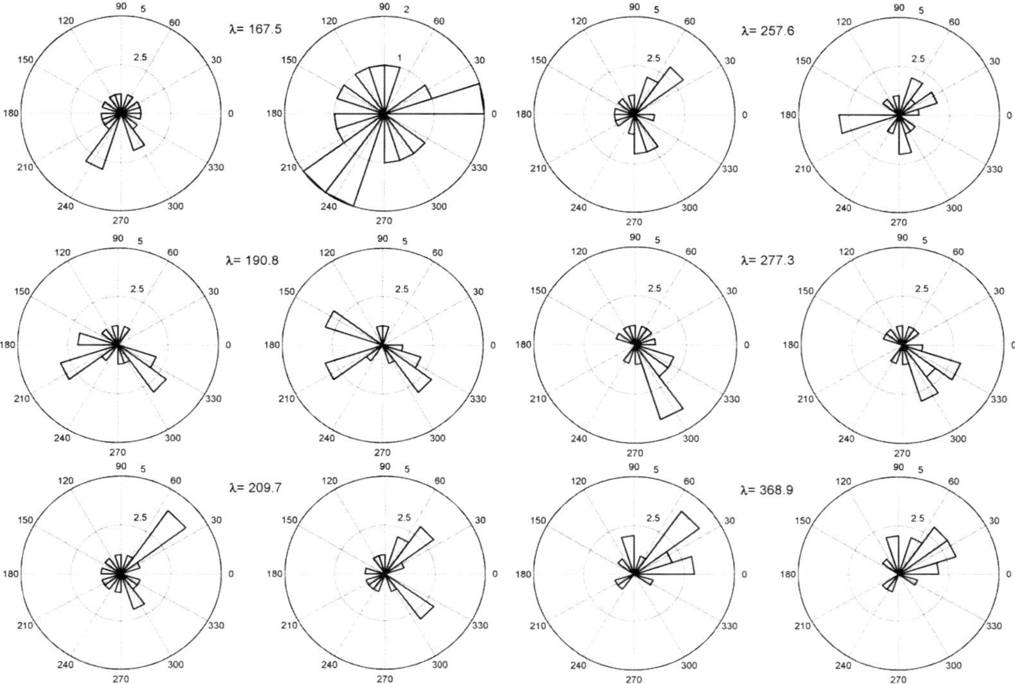

Fig. 10. Phase rose diagrams. For each frequency in the Serra Geral 'AND' spectrum phase was determined by the classical (left) and Hocke's (right) methods.

means that it is the most coherent feature in all the Iceland series. In the Serra Geral spectra, the highest peak is at *c.* 168°, although the doublet 257.6–277.3 also corresponds to a considerable part of the spectrum's energy (Fig. 5). This similarity between Iceland and the Serra Geral is somewhat surprising, as the two datasets refer to very different ages and geographic locations, and continuity in the geomagnetic field's behaviour is not necessarily expected.

Spatial behaviour of the VGP paths

The Serra Geral magnetostratigraphic sequences are appropriate for this type of analysis despite the low density sampling of the long-term geomagnetic variations; the high number of available sections allows a considerable smoothing of the individual curves. Considering the performance with synthetic data and the quality of the real data spectra, it is reasonable to say that we have successfully detected some characteristic wavelengths in the VGP trajectories. Moreover, independent datasets like those from the Icelandic magmatism have indications of similar geometric characteristics in the VGP paths.

It is important to stress that both the Serra Geral and Icelandic data record several geomagnetic polarity reversals, in both the N → R and R → N senses, and the record of the transitions themselves are scarce. Therefore, as a conclusion we may say that, although not fully observed, the secular variation cycles and/or polarity reversals show common features that are persistent for the time interval covered by the datasets, and that the general geomagnetic field behaviour may be similar during the Lower Cretaceous and younger ages.

The fact that the VGP trajectories may be modelled by the combination of some sinusoidal components does not mean that periodic phenomena were identified. We may only say that we have detected some possible wavelengths that are recurrent in all the investigated data sequences, and they clearly point to the existence of anisotropy in the VGP paths during secular variation cycles or polarity changes. Models for VGP paths during recent reversals were proposed by Leonhardt & Fabian (2007) and Kuznetsov (1999), among others, for which the trajectories may be conditioned by field anomalies superposing the dipolar field during polarity reversals, and leading to longitude confinements.

G. CAMINHA-MACIEL & M. ERNESTO

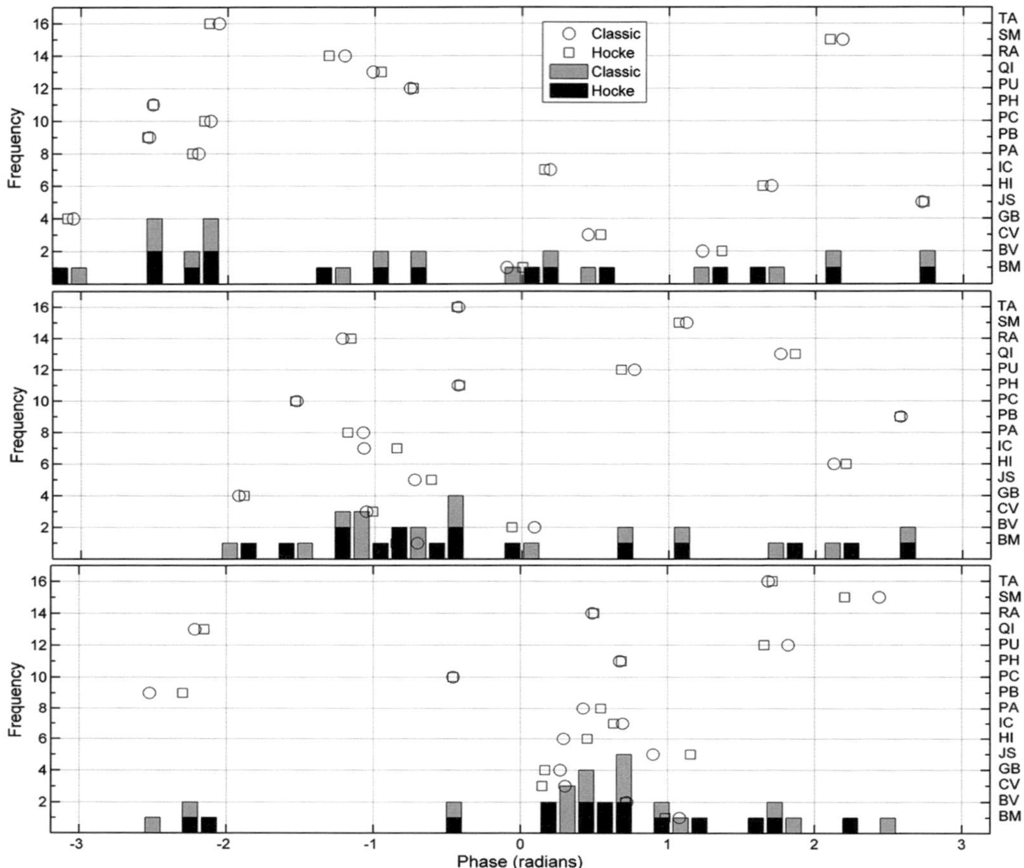

Fig. 11. Frequency histograms of the phase values for wavelengths 167.5, 277.3 and 368.9° (from top to bottom, respectively) determined by both the classical and Hocke's methods for each individual series. The horizontal lines discriminate the individual series. Legend on the top figure is the same for all figures.

In their analysis of Jurassic to Early Cretaceous intermediate VGPs, Vizán & Van Zele (2008) found data concentrations mainly at longitude bands 300–330°E and 120–150°E, indicating a strong asymmetry of the deep-earth anomalies causing the deviation of the VGP trajectories. The angular differences between these two bands are in the interval 150–210°, and may be easily associated with the *c.* 168–270° wavelengths found in this work. One possible interpretation is that this is the range of longitudes over which VGPs undergo changes in latitude as they describe loops of secular variation or move from one polarity state to the other. Therefore, these results might represent some of the components of the oscillatory VGP trajectories. The recovery of the complete trajectory should be possible in principle by the summation of all components if the respective amplitudes and phase distribution are known.

Conclusions

We have attempted to identify sinusoidal components in VGP trajectories recorded in magnetostratigraphic data of volcanic sequences using a combination of information approach. The starting point was the LS periodogram for unevenly sampled data series, although we do not consider the usual power spectrum, that is, in this case the independent variable is space and not time.

Without any characteristic sampling interval, and because data are too sparse in face of the phenomena of interest, the individual spectra did not give reliable information (periodicity or wavelength), or they exhibited a very weak signal. In the latter case, the problem could be by-passed, making use of some properties of the spectral estimator to suppress the spectral leakage. Power spreading across the (wavelength) frequency domain helped

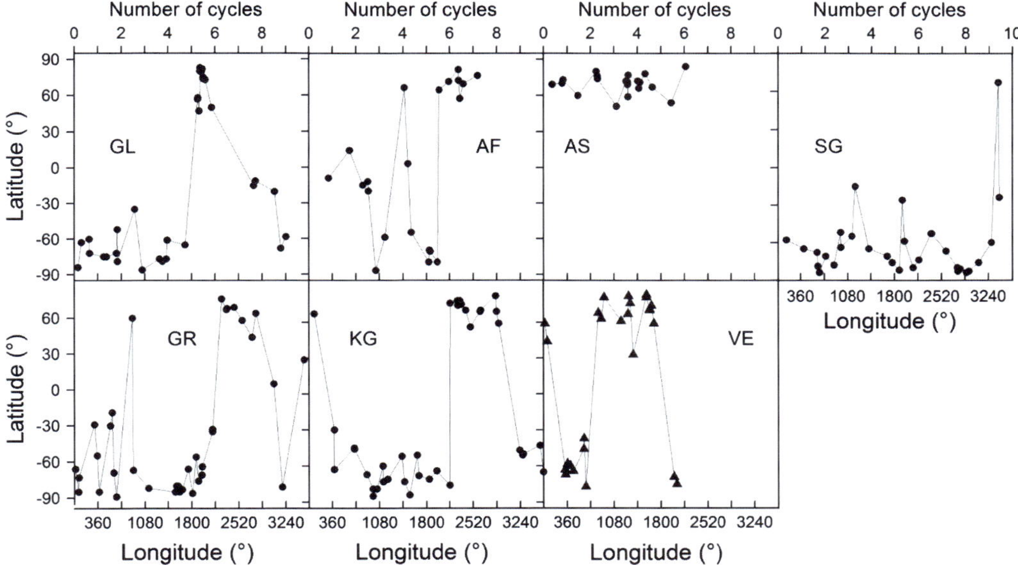

Fig. 12. The Iceland VGP series (Kristjansson *et al.* 2004) with continued longitudes.

in the signal identification as the addition of more serial information (other data series) enhanced the common signal.

As in seismic signal stacking, the combination of multiple spectra reduces noise and highlights the common features even if the signal is weak. The solution in the parametric space is delimited in terms of possible existing highly correlated trigonometric components. As a result, we obtained a smooth curve highlighting the features of interest.

The great advantage is that this method does not require any data pre-processing (e.g. de-trending,

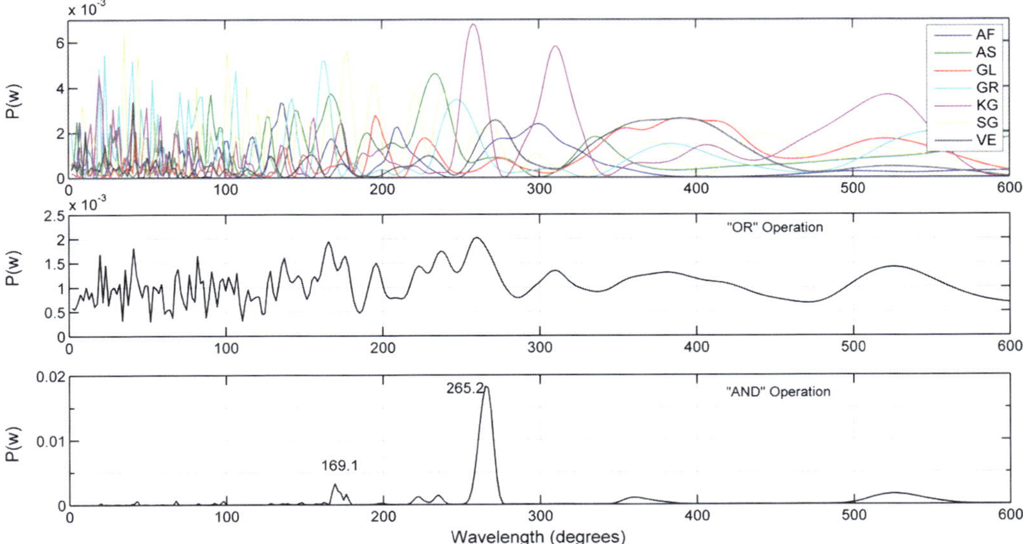

Fig. 13. The LS spectra for the Iceland series, and the subsequent 'OR' and 'AND' combinations. $P(w)$ in arbitrary units.

filtering or pre-whitening) that would modify the power spectrum and possibly cause the displacement of spectral peaks or modification in power (Hernandez 1999).

We found a set of highest correlation wavelengths of approximately 167, 190, 209, 257, 277 and 368°. Phase analyses using two different methods revealed strikingly good coherence for some of the wavelengths, indicating that they might not be artefacts of the spectral analysis. Similar analysis of magnetostratigraphic data from the Icelandic Magmatic Province indicates that the two datasets have two wavelengths of approximately 165 and 270° in common. These results suggest quasi-periodic behaviour, possibly with sub-harmonic instabilities owing to the modulating effect of inner Earth's anisotropies influencing the pole trajectory.

This work was funded by the Brazilian funding agency FAPESP (2004/05363-5). Thanks are due to W. Shukowsky and L.A. Diogo for valuable comments on the methodology. The paper was greatly improved by constructive comments and suggestions from X. Zhao, another unknown reviewer, and the special editor for this volume, L. Hinnov.

References

ALVA-VALDIVIA, L. M., GOGUITCHAICHVILI, A., URRUTIA-FUCUGAUCHI, J., RIISAGER, J., RIISAGER, P. & LOPES, O. F. 2003. Paleomagnetic poles and paleosecular variation of basaltsfrom Paraná Magmatic Province, Brazil: geomagneticand geodynamic implications. *Physics of the Earth and Planetary Interiors*, **138**, 183–196.

BARTON, C. & MCFADDEN, P. L. 1995. Inclination shallowing and preferred transitional VGP paths. *Earth and Planetary Science Letters*, **140**, 147–157.

BIGGIN, A. J., VAN HINSBERGENA, D. J. J., LANGEREIS, C. G., STRAATHOFA, G. B. & MARTIJN DEENEN, M. H. L. 2008. Geomagnetic secular variation in the Cretaceous Normal Superchron and in the Jurassic. *Physics of the Earth and Planetary Interiors*, **169**, 3–19.

CLEMENT, B. M. 1991. Geographical distribution of transitional VGPs: evidence for non-zonal equatorial symmetry during the Matuyama–Brunhes geomagnetic reversal. *Earth and Planetary Science Letters*, **104**, 48–58.

COE, R. S. & GLEN, J. M. G. 2004. The complexity of reversals. *In*: CHANNELL, J. E. T., KENT, D., LOWRIE, W. & MEERT, J. G. (eds) *Timescales of the Paleomagnetic Field*, AGU, Washington DC, Geophysical Monograph Series, **145**, 221–232.

COE, R. S., HONGRE, L. & GLATZMAIER, A. 2000. An examination of simulated geomagnetic reversals from a palaeomagnetic perspective. *Philosophical Transactions of the Royal Society B*, **A358**, 1141–1170.

COSTIN, S. O. & BUFFET, B. A. 2004. Preferred reversal paths caused by a heterogenous conducting layer at the base of the mantle. *Journal of Geophysical Research*, **109**, B06101.

ERNESTO, M. & PACCA, I. G. 1988. Paleomagnetism of the Paraná Basin flood volcanics, southern Brazil. *In*: MELFI, A. J. & PICCIRILLO, E. M. (eds) *The Mesozoic Flood Volcanism of the Paraná Basin: Petrogenetic and Geophysical Aspects*. IAG/USP, Brazil, 229–255.

ERNESTO, M., PACCA, I. G., HIODO, F. Y. & NARDY, A. J. R. 1990. Paleomagnetism of the Mesozoic Serra Geral Formation, southern Brazil. *Physics of the Earth and Planetary Interiors*, **64**, 153–175.

FONT, E. C., ERNESTO, M., SILVA, P. F., CORREIA, P. B. & NASCIMENTO, M. A. L. 2009. Paleomagnetism, rock magnetism, and AMS of the Cabo Magmatic Province, NE Brazil, and the opening of South Atlantic. *Geophysical Journal International*, **179**, 905–922.

GUBBINS, D. & LOVE, J. J. 1998. Preferred VGP paths during geomagnetic polarity reversals: symmetry considerations. *Geophysical Research Letters*, **25**, 1079–1082.

HERNANDEZ, G. 1999. Time series, periodograms, and significance. *Journal of Geophysical Research*, **104**, 10 355–10 368.

HOCKE, K. 1998. Phase estimation with the Lomb-Scargle periodogram method. *Annales Geophysicae*, **16**, 356–358.

HOFFMAN, K. A. 1992. Dipolar reversal state of the geomagnetic field and core-mantle dynamics. *Nature*, **359**, 789–794.

HORNE, J. H. & BALIUNAS, S. L. 1986. A prescription for period analysis of unevenly sampled time series. *The Astrophysical Journal*, **302**, 757–763.

KRISTJANSSON, L., GUDMUNDSSON, A. & HARDARSON, B. S. 2004. Stratigraphy and Paleomagnetism of a 2.9-km composite lava section in Eyjafjördur, Northern Iceland: a reconnaissance study. *International Journal of Earth Sciences (Geologische Rundschau)*, **93**, 582–595.

KUZNETSOV, V. V. 1999. A model of virtual geomagnetic pole motion during reversals. *Physics of the Earth and Planetary Interiors*, **115**, 173–179.

LAJ, C. A., MAZAUD, M., WEEKS, M., FULLER, M. & HERRERO-BERVERA, E. 1991. Geomagnetic reversals paths. *Nature*, **351**, 447.

LEONHARDT, R. & FABIAN, K. 2007. Paleomagnetic reconstruction of the global geomagnetic field evolution during the Matuyama/Brunhes transition: iterative Bayensian inversion and independent verification. *Earth and Planetary Science Letters*, **253**, 172–195.

LOMB, N. R. 1976. Least-squares frequency analysis of unequally spaced data. *Astrophysics and Space Science*, **39**, 447–462.

LOVE, J. J. 1998. Paleomagnetic volcanic data and geometric regularity of reversals and excursions. *Journal of Geophysical Research*, **103**, 12435–12452.

LOVE, J. J. 2000. On the anisotropy of secular variation deduced from paleomagnetic volcanic data. *Journal of Geophysical Research*, **105**, 5799–5816.

MENA, M., ORGEIRA, M. J. & LAGORIO, S. 2006. Paleomagnetism, rock-magnetism and geochemical aspects of early Cretaceous basalts of the Paraná Magmatic Province, Misiones, Argentina. *Earth Planets Space*, **58**, 1283–1293.

QUIDELLEUR, X., HOLT, J. & VALET, J. P. 1995. Confounding influence of magnetic fabric on sedimentary records of a field reversal. *Nature*, **374**, 246–249.

SCARGLE, J. D. 1982. Studies in astronomical time series analysis, II, Statistical aspects of spectral analysis of unevenly spaced data. *Astrophysical Journal*, **263**, 835–853.

SMIRNOV, A. V. & TARDUNO, J. A. 2004. Secular variation of the Late Archean–Early Proterozoic geodynamo. *Geophysical Research Letters*, **31**, L16607.

TARANTOLA, A. 2006. Popper, Bayes and the inverse problem. *Nature Physics*, **2**, 492–494.

TARANTOLA, A. & MOSEGAARD, K. 2000. *Mathematical Basis for Physical Inference*. Cornell University Library, New York, arXiv:math-ph/0009029v1

TARANTOLA, A. & VALETTE, B. 1982. Inverse problem = quest for information. *Journal of Geophysics*, **50**, 159–170.

TAUXE, L. & KENT, D. V. 2004. A simplified statistical model for the geomagnetic field and the detection of shallow bias in paleomagnetic inclinations: was the ancient magnetic field dipolar? *In*: CHANNELL, J. E. T., KENT, D., LOWRIE, W. & MEERT, J. G. (eds) *Timescales of the Paleomagnetic Field*. AGU, Washington DC, Geophysical Monograph Series, **145**, 101–115.

THIEDE, D. S. & VASCONCELOS, P. M. 2010. Paraná flood basalts: rapid extrusion hypothesis by new ^{40}Ar/^{39}Ar results. *Geology*, **38**, 747–750.

VALET, J. P. & HERRERO-BERVERA, E. 2003. Some characteristics of geomagnetic reversals inferred from detailed volcanic records. *Comptes Rendue Geosciences*, **2003**, 79–90.

VIZÁN, H. & VAN ZELE, M. A. 2008. Jurassic–Early Cretaceous intermediate virtual geomagnetic poles and Pangaean subduction zones. *Earth and Planetary Science Letters*, **266**, 1–13.

VIZÁN, H., SOMOZA, R., ORGEIRA, M. J., VÁSQUEZ, C. A., MENA, M. & VILAS, J. F. 1994. Late Palaeozoic–Mesozoic geomagnetic reversal paths and core-mantle boundary. *Geophysical Journal International*, **117**, 819–826.

ZECHMEISTER, M. & KÜRSTER, M. 2009. The generalized Lomb–Scargle periodogram: a new formalism for the floating for the floating-mean and Keplerian periodograms. *Astronomy & Astrophysics*, **496**, 577–584, arXiv:0901. 2573v1 [astro-ph.IM] 16jan2009.

Rock-magnetic cyclostratigraphy for the Late Pliocene–Early Pleistocene Stirone section, Northern Apennine mountain front, Italy

KELLEN L. GUNDERSON*, KENNETH P. KODAMA,
DAVID J. ANASTASIO & FRANK J. PAZZAGLIA

*Department of Earth and Environmental Sciences, Lehigh University,
1 West Packer Ave, Bethlehem, PA 18015, USA*

Corresponding author (e-mail: kellen.gunderson@lehigh.edu)

Abstract: Lithostratigraphic, magnetostratigraphic and rock-magnetic cyclostratigraphic data were combined to create a high-resolution age model for 342 m of Late Pliocene–Middle Pleistocene marine deposits exposed in the Stirone River, northern Italy. Magnetostratigraphic analysis of 74 oriented samples at 21 stratigraphic horizons recognized five polarity zones between *c.* 3.0 and 1.0 Ma. Unoriented samples were collected every metre between 0 and 311 m and low-field magnetic susceptibility (χ) was measured for cyclostratigraphic analysis. The χ data series was tied to absolute time using the magnetostratigraphy and subjected to multi-taper method spectral analysis. The resultant power spectra revealed significant frequency peaks that are aligned with eccentricity, obliquity and precession Milankovitch orbital cycles. The χ data, correlated to the 41 ka obliquity and the 23 ka/19 ka precession cycles and anchored to a well-established biostratigraphic horizon, were used to create a high-resolution age model for the Stirone section between 2.99 and 1.81 Ma, where stratigraphic positions of magnetic reversals were previously poorly defined. This cyclostratigraphic age model reveals that the length of an important depositional hiatus at the base of the C2An.1n subchron is 200 ka shorter than previously determined. We link the precession-aligned variability in χ to global mid-latitude, insolation-induced variability in runoff and ocean circulation.

Assigning absolute time to stratigraphic sections can be approached using several methods, each having its own unique temporal resolution, technical challenges and uncertainties. Conventional methods of geochronology, lithostratigraphy, biostratigraphy and magnetostratigraphy are typically used, but these methods do not provide the temporal resolution required for many tectonic, palaeoclimatic or sedimentological problems. The emergence of cyclostratigraphic correlation to accurate models of orbital variability (i.e. eccentricity, obliquity, precession) has provided a high-resolution metronome of Earth time that can be applied to stratigraphic sections (Hinnov 2000; Hinnov *et al.* 2004).

Cyclostratigraphy aims at recovering variations in rock textural, compositional or geochemical characteristics that serve as direct proxies of climatic variability resulting from orbital forcing functions (Hays *et al.* 1976). Cyclostratigraphic studies have employed a wide range of climatic proxies such as grain size, carbonate content (Shackleton *et al.* 1995), facies (Olsen 1986), biogenic silica (Williams *et al.* 1997) and stable isotopes (Hays *et al.* 1976), among others. An alternative approach is to use rock-magnetic measurements as climate proxies. Low-field magnetic susceptibility (χ) (Bloemendal & deMenocal 1989; Shackleton *et al.* 1999; Jovane *et al.* 2006; Ellwood *et al.* 2008; Jovane *et al.* 2010),

natural remanent magnetization (NRM; Kruiver *et al.* 2002) and anhysteretic remanent magnetization (ARM; Latta *et al.* 2006; Kodama *et al.* 2010) are rock-magnetic measurements that were demonstrated to vary at frequencies associated with astronomically forced cycles. These rock-magnetic measurements record changes in magnetic parameters such as magnetic mineralogy, grain size and magnetic mineral concentration. The variability in these parameters was shown to be influenced by climatically controlled processes, including glacial/interglacial soil production (Heller & Evans 1995), aeolian dust flux (Latta *et al.* 2006), and changes in wet season runoff (Kodama *et al.* 2010). Rock-magnetic measurements are useful palaeoclimatic proxies because the measurements are objective, non-destructive and fast. They also reveal variability that is not otherwise observable in lithologically homogenous sections (Bloemendal *et al.* 1988).

In this study we demonstrate how rock-magnetic cyclostratigraphy can be used to improve biostratigraphic and magnetostratigraphic correlations by adding high-resolution time control in an important stratigraphic section that shows no major observable lithological cyclicity. We present the results of a combined lithostratigraphic, magnetostratigraphic and cyclostratigraphic interpretation of the Pliocene–Early Pleistocene Stirone section

From: JOVANE, L., HERRERO-BERVERA, E., HINNOV, L. A. & HOUSEN, B. A. (eds) 2013. *Magnetic Methods and the Timing of Geological Processes*. Geological Society, London, Special Publications, **373**, 309–323.
First published online August 14, 2012, http://dx.doi.org/10.1144/SP373.8 © The Geological Society of London 2013.
Publishing disclaimer: www.geolsoc.org.uk/pub_ethics

deposited at the Northern Apennine mountain front adjacent to the Po Plain, Italy. We used low-field magnetic susceptibility (χ) to correlate the Stirone section to the theoretical 41 ka obliquity and 23/19 ka precession orbital models (Laskar *et al.* 2004) during the Late Pliocene–Early Pleistocene time. This time interval contains widely spaced magnetic polarity reversal boundaries that were poorly constrained by previous magnetostratigraphic studies in the Stirone section (Mary *et al.* 1993). We use this high-resolution age model to adjust the timing of an important biostratigraphically determined unconformity and calculate high-resolution sediment accumulation rates. The high-resolution age model we present can be used in subsequent studies to measure deformation rates along the Apennine

mountain front, determine the timing of biostratigraphic horizons or investigate the sequence stratigraphic framework of the Po plain sediments.

Geological setting

The Stirone River banks expose over 600 m of Messinian to recent foreland-dipping marine and continental synorogenic growth strata that are in close proximity to the original Piacenzian Stage (Late Pliocene) unit strato-type, located in the nearby Castell'Aquarto Basin (Fig. 1; Rio *et al.* 1990*b*, 1991; Mary *et al.* 1993; Channell *et al.* 1994; Artoni *et al.* 2007). The Stirone section is a coarsening upward succession of growth strata exposed on the

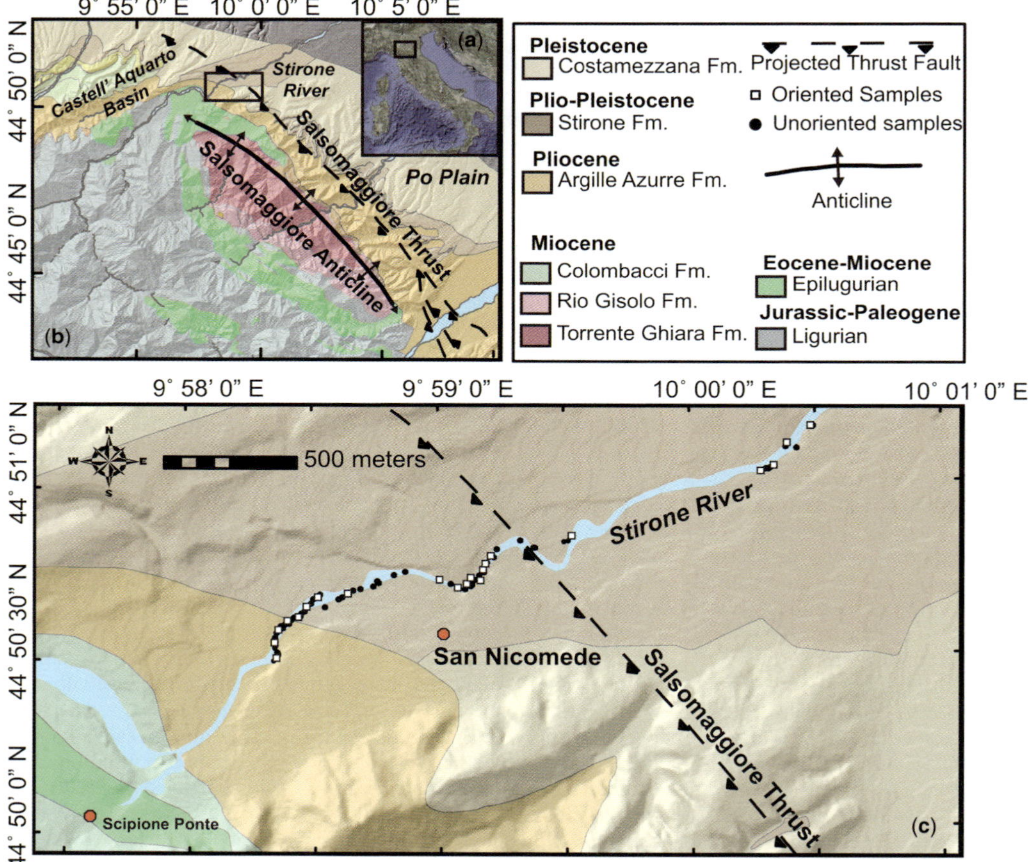

Fig. 1. Map of the study area in the Northern Apennines, Italy. (**a**) Index map inset. (**b**) Bedrock geological map of the Northern Apennine mountain front in the vicinity of the Salsomaggiore anticline and the adjacent Po Plain. The Messinian–Pleistocene Stirone section is exposed in the Stirone River valley. (**c**) Bedrock geological map of the Stirone River study area showing our sample locations. The white squares represent palaeomagnetic sites, where oriented samples were collected. Black circles represent the location of every tenth rock-magnetic sample. Geological maps modified from Di Dio (2005).

forelimb of the Salsomaggiore anticline that forms the structural front bordering the Po foreland basin (Artoni *et al.* 2004). The Late Miocene basal unit consists of marginal marine siltstones and sandstones called the Colombacci Formation (Fm). The Colombacci Fm is overlain by blue-grey Pliocene marine mudstones of the Argille Azzurre Fm (Mary *et al.* 1993; Channell *et al.* 1994; Amorosi *et al.* 1998*a*), which grades up-section into the Plio-Pleistocene fossiliferous silty muds containing several calcarenite layers of the Stirone Fm (Dominici 2001; Di Dio 2005). The fossils of the Stirone Fm were used to calibrate Pleistocene biostratigraphic horizons (Dondi 1961; Papani & Pelosio 1963). At the top of the section is a unit consisting of yellow littoral sands variously known as the Sabbie di Imola, Sabbie Gialle or Costamezzana Fm (Amorosi *et al.*

1998*b*; Di Dio 2005). Two separate palaeomagnetic studies conducted in the newly exposed Argille Azzurre Fm at the Stirone (Mary *et al.* 1993; Channell *et al.* 1994) document several magnetic reversals spanning the Early Pliocene to Early Pleistocene. Over the last two decades, the Stirone River exposures have improved and lengthened as incision of the river accelerated. These improved and continuous exposures provide the opportunity to refine the previous magnetostratigraphic age model and produce a high-resolution age model for the section using cyclostratigraphy. This study focuses on the upper 342 m of the section beginning at the base of the Stirone Fm (Fig. 2). The investigated section begins directly above a prominent metre-thick fossil chemoherm oriented steeply to bedding. Chemoherms like this are ubiquitous in

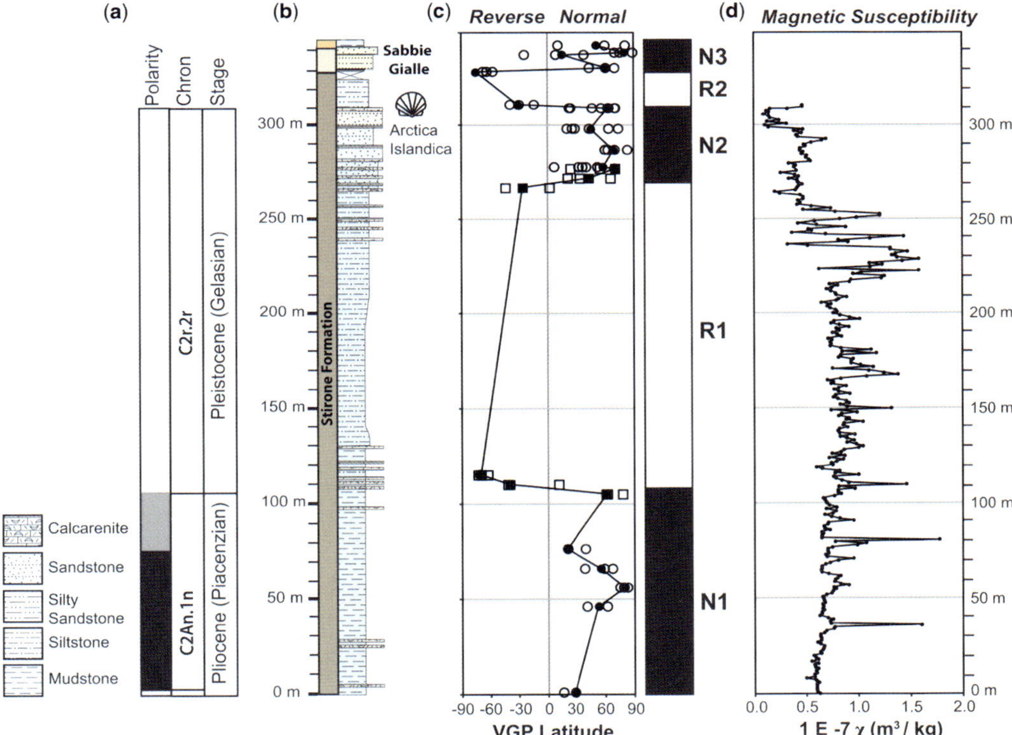

Fig. 2. (**a**) Shows the previous magnetostratigraphy for the upper Stirone section by Mary *et al.* (1993) in which two polarity zones were defined. These two zones were separated by a large zone of uncertain polarity. (**b**) Lithology of the Stirone section. The section is a coarsening upward succession of mud, silt, sand and calcarenites that displays no major lithological cyclicity. The stratigraphic level of the first occurrence of the mollusc *Arctica islandica* is also shown. (**c**) Virtual geomagnetic pole (VGP) latitude of oriented samples in the Stirone section. Solid symbols represent mean VGP for each horizon and open symbols show the individual sample VGPs for the same horizon. The circles represent sites that were collected during our first field season and the squares are samples collected during our second field season. Using the average VGP latitudes we define five polarity zones in the upper Stirone section. (**d**) Low-field magnetic susceptibility (χ) of the upper Stirone section. The solid circles show the values for each sample collected at 1 m spacing.

Neogene marine rocks along the northern Apennine front (Conti & Fontana 1999). They are caused by methane expulsion from the ocean floor and are analogous to the cold seeps and methanogenic carbonates generated along the Cascadia accretionary prism (Carson *et al.* 1990). The measured stratigraphic section (Fig. 2) is separated into two major lithostratigraphic units that exhibit bedding orientations that become progressively shallower up-section. Between 0 and 330 m, the section consists of the coarsening upwards, blue-grey mud and silt, with bundles of 0.25–1 m thick fossiliferous calcarenites of the Late Pliocene–Middle Pleistocene Stirone Formation (Bertolani Marchetti *et al.* 1979). A short 20 m interval containing sapropels is present at the base of the section (Channell *et al.* 1994). At 311 m the first occurrence of the mollusc *Arctica islandica* occurs, which approximates the end of the Gelasian stage of the Early Pleistocene (*c.* 1.81; Raffi 1986). Above the Stirone Fm (330–340 m) a medium, thin-bedded, well-sorted, tan to yellow cross-bedded sand is exposed. This sand represents a littoral facies and is the Costamezzana Fm (or Sabbie Gialle) (Amorosi *et al.* 1998*b*; Di Dio 2005). The contact between the Costamezzana and Stirone Fm is gradational over 10–15 m. The upper-most unit (340–342 m) is a blue freshwater mud known as the AEI (Lower Emilia synthem), which lies unconformably over the littoral sands.

Methods

Magnetostratigraphic analysis

We measured 342 m of section beginning just above the fossil chemoherm at the base of the Stirone Fm (Figs 1 & 2) and described the lithology every 1 m. Oriented samples were collected at 21 target horizons. A previous magnetostratigraphic study by Mary *et al.* (1993) recognized a single magnetic polarity reversal in the lower part of the section (Fig. 2), but acknowledged an *c.* 30 m interval of uncertain polarity. Additionally, Mary *et al.* (1993) did not identify any additional reversals at the top of the section. Because our magnetostratigraphy was meant to refine the previous reversal stratigraphy by Mary *et al.* (1993) and gain an absolute time reference for our cyclostratigraphic analysis, our target horizons were not equally spaced throughout the section. Instead, we sampled densely near the uncertain polarity zone observed by Mary *et al.* (1993) (Fig. 2). We also sampled densely between 260 and 315 m in order to identify the presence of the Olduvai normal chron, which was not observed in the previous study by Mary *et al.* (1993). The first occurrence of *Arctica islandica* (*c.* 1.81 Ma) at 311 m predicts the normal Olduvai chron to be located stratigraphically below that horizon. Finally,

we collected oriented samples from five sites in the upper 15 m of the section from the Sabbie Gialle sands, from which no previous magnetostratigraphic samples had been collected. At least three samples were collected from each horizon and 74 oriented samples were collected in total. Oriented samples were collected by carving small pedestals in the outcrop and by orienting 8 cm^3 plastic boxes on the pedestals. Four sites were located in well-indurated calcarenite units where 2.54 cm diameter oriented palaeomagnetic cores were collected using a Pomeroy EZ Core drill. From each site that was drilled we obtained four or five cores, from which we procured one to two samples per core.

Oriented samples were subjected to alternating-field and thermal demagnetization. Alternating-field (AF) demagnetization was conducted from 10–100 mT in 10 mT steps using a 2 G Enterprises superconducting magnetometer at Lehigh University. Samples subjected to progressive thermal demagnetization were heated to 600 °C in 50 °C steps, using an ASC-TD-48 thermal demagnetizer. Principal component analysis (PCA) was conducted to determine the characteristic remanent magnetization (ChRM) of each sample (Kirschvink 1980). Mean remanent declinations and inclinations were calculated for each horizon (site) using Fisher statistics (Fisher 1953). Virtual geomagnetic poles (VGPs) were calculated from the site mean directions. The VGP latitude was used to determine the polarity of a horizon.

Detailed rock-magnetic measurements

Partial anhysteretic remanent magnetization (pARM), isothermal remanent magnetization (IRM) and low-temperature (-196 °C) magnetic susceptibility (χ) were measured for representative samples to determine the magnetic mineralogy of the Stirone Fm. A pARM spectrum from 0 to 100 mT was obtained by applying a pARM in 5 mT steps with a DC bias field of 97 μT. IRM acquisition experiments were conducted using an ASC Impulse Magnetizer. Samples were subjected to a stepwise increase in field from 0 to 1 T and the magnetization was measured after each of the 21 steps. Modelling of the IRM acquisition curves was conducted using the software developed by Kruiver *et al.* (2001). Low-temperature χ was determined by measuring χ using a KLY-3 s Kappa bridge immediately after immersing the sample in liquid nitrogen (-196 °C) for 1 min.

Rock-magnetic cyclostratigraphy

Unoriented rock magnetic samples were collected every 1 m between 0 and 311 m. All rock- magnetic samples were collected in pre-weighed 8 cm^3 plastic boxes. Low-field magnetic susceptibility (χ) was measured using a KLY-3S Kappa bridge at room temperature and was normalized by sample mass. The χ data series was used to investigate magnetic mineral concentration variations in the Stirone section for possible Milankovitch cyclicity. Cyclostratigraphic analysis of the rock-magnetic measurements was focused on the 0–311 m part of the section because the two magnetic reversals in that interval had uncertain stratigraphic positions and no major biostratigraphic hiatuses were previously recognized there.

The multi-taper method (MTM) (Thompson 1982, 1990) was used to determine the power spectrum of the χ data series. In order to remove undersampled low frequency variability, the data series was de-trended by calculating the residuals of a best-fit second-order polynomial. The χ data series was re-scaled using the new magnetostratigraphic correlations from this study and re-sampled in even intervals with simple linear interpolation using the program Analyseries 2.0.4.2 (Paillard *et al.* 1996). MTM power spectra were calculated using the SSA-MTM program (Ghil *et al.* 2002), and 95 and 99% confidence intervals were determined for a robust red noise model (Mann & Lees 1996). Spectral results were band-passed using Gaussian filters where peaks coincided with Milankovitch frequencies. The filtered time series was used to aid the correlation to the theoretical orbital series (Laskar *et al.* 2004) by matching the peaks in the filtered series to the peaks in the theoretical model (see the Discussion).

Results

Magnetic reversal stratigraphy

Thermal and AF demagnetization techniques were shown to be similarly successful in isolating the ChRM of the oriented samples (Fig. 3); 70 our of 74 samples yielded clustered ChRM directions after PCA was conducted. Vector endpoint diagrams (Zijderveld 1967) show representative samples with normal and reversed ChRMs (Fig. 3). Progressive demagnetization curves reveal that some thermally demagnetized samples exhibit a decrease in NRM intensity near 300 °C, while others show an intensity decrease around 550 °C. The primary magnetization is represented by the low-temperature component for samples that exhibit a 300 °C decrease in NRM intensity. A few samples that were demagnetized using the AF technique revealed an unstable magnetization at intermediate to high coercivities (50–100 mT).

The VGP latitudes of oriented samples reveal four geomagnetic field reversals that define five polarity zones in the Stirone River section (Fig. 2).

The polarity zones are defined by consecutive sites exhibiting a similar polarity. The polarity for each site was determined using the VGP latitude calculated from the mean remanence direction for the site. Most sites show a consistent polarity within the stratigraphic horizon with the exception of two samples that show a polarity opposite the site average. Because the sampling strategy was designed to refine the position of previously established polarity zone boundaries, oriented samples were not collected uniformly throughout the section. The base of the section exhibits a normal polarity. A reversal is observed between the sites located at 105 and 110 m, which is within the uncertain polarity zone of Mary *et al.* (1993). The reversed interval ends between the 267 and 272 m sites. The next reversal is located directly above the uppermost calcarenite at 311 m, coincident with the first occurrence of the mollusc *Arctica islandica*, which occurs approximately at the end of the Gelasian stage in the Early Pleistocene (Raffi 1986; Dominici 2001). The final reversal occurs at 330 m, at the base of the yellow littoral Sabbie Gialle sands.

Magnetic mineralogy

Results of the low-temperature χ experiment indicate a magnetic mineralogy that contains both ferromagnetic and paramagnetic components (Fig. 4). Paramagnetic χ is temperature dependent; samples in which χ is completely dominated by paramagnetic components are predicted to exhibit a *c.* 400% increase in χ at -196 °C (Richter & van der Pluijm 1994). Ferromagnetic susceptibility, however, is not temperature-dependent and samples with a ferromagnetic mineralogy should display no increase in χ at low temperature. The Stirone samples displayed an increase in χ at low temperature; however, that increase was not large enough (125–200% increase) to conclude that either paramagnetic or ferromagnetic components alone dominate the susceptibility, but that paramagnetics and ferromagnetics both contribute to χ.

The pARM acquisition experiments (Fig. 5) exhibit two characteristic spectra: first, spectra that contain a single peak indicating the sample is dominated by a single low-coercivity (*c.* 25 mT) ferromagnetic component of magnetization; and second, spectra with multiple peaks indicating multiple ferromagnetic components of magnetization with slightly overlapping coercivity ranges (*c.* 30 and *c.* 85 mT; Jackson *et al.* 1988). IRM acquisition modelling results (Fig. 6) also suggest two dominant ferromagnetic components of magnetization. The high coercivity component, ranging from 67 to 83 mT, is shown to comprise *c.* 84% of the total magnetization while the low coercivity component, ranging from 17 to 25 mT, comprises *c.* 15% of the

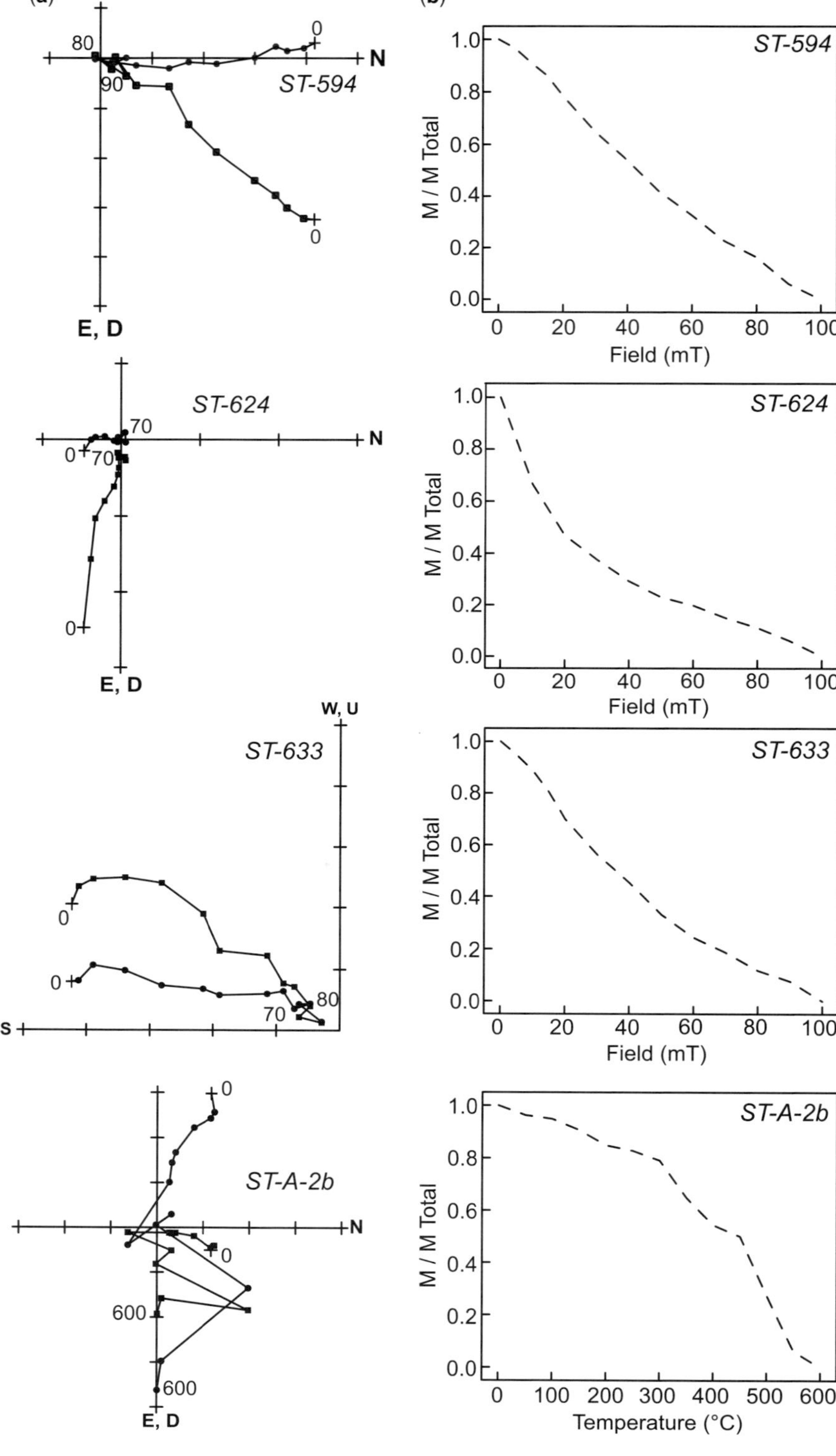
(a)
(b)
80
0
90
ST-594
N
E, D
ST-594
1.0
0.8
0.6
0.4
0.2
0.0
M / M Total
0   20   40   60   80   100
Field (mT)
ST-624
70
0   70
0
E, D
N
ST-624
1.0
0.8
0.6
0.4
0.2
0.0
M / M Total
0   20   40   60   80   100
Field (mT)
W, U
ST-633
0
0
70
80
S
ST-633
1.0
0.8
0.6
0.4
0.2
0.0
M / M Total
0   20   40   60   80   100
Field (mT)
0
ST-A-2b
0
600
600
E, D
N
ST-A-2b
1.0
0.8
0.6
0.4
0.2
0.0
M / M Total
0   100   200   300   400   500   600
Temperature (°C)

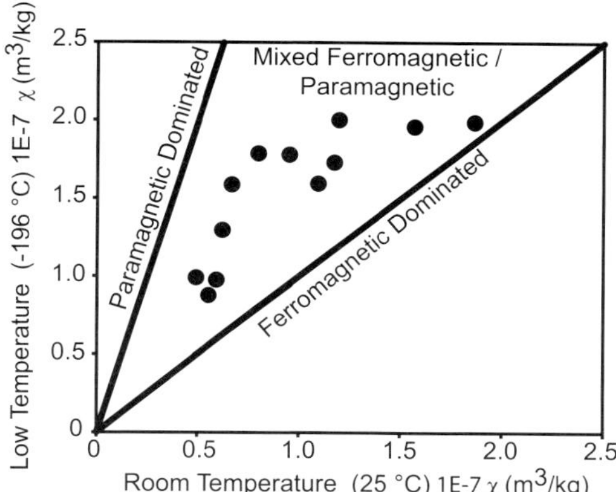

Fig. 4. Results of our low-temperature (-196 °C) χ experiment delineate zones showing the relative contributions of paramagnetic and ferromagnetic minerals to the total χ. The ratio of low-temperature χ to room-temperature χ is expected to be 4:1 for samples with a paramagnetic mineralogy. Samples with ferromagnetic mineralogy are expected to show little variation in χ at low temperature.

total magnetization. Minor tertiary component(s) comprise the remaining c. 1% of the magnetization.

Cyclostratigraphy

The χ data series (Fig. 2) exhibits multi-hierarchical, high frequency variability. The χ data displays little variation in the basal c. 30 m with large swings in amplitude occurring between 220 and 260 m, coincident with the deposition of the upper calcarenite beds and a general coarsening of the lithology. Our MTM spectral analysis of the un-tuned, residual data series reveals significant spectral peaks with ratios that suggest the presence of Milankovitch cyclicity (Fig. 7). The χ power spectrum exhibits significant peaks at frequencies of $1/85.3$, $1/10.2$, $1/6.3$ and $1/5.8$ m. To help identify Milankovitch forcing, we used the ratios between the observed periodicities. The predicted ratios of the long eccentricity and obliquity periods to the 23 and 19 ka precession peaks are 17.6:1 and 21.3:1 for the 405 ka eccentricity cycle and 1.8:1 and 2.1:1 for the 41 ka obliquity cycle. Assuming that the highest frequency peaks in the χ data series represent precession cycles,

the two low-frequency cycles approximate the predicted ratios for long eccentricity and obliquity cycles (13.5:1 and 14.7:1 for the long eccentricity cycle; 1.6:1 and 1.8:1 for the obliquity cycle).

Discussion

Rock magnetics

The high-coercivity component of magnetization that is observed in the pARM spectra and in the IRM acquisition modelling experiments is interpreted to represent secondary iron sulfide minerals (i.e. pyrrhotite or greigite) and the low-coercivity component is most likely detrital magnetite. The coercivity ranges exhibited in the pARM and IRM experiments are consistent with the documented coercivity ranges for these minerals (Peters & Dekkers 2003). Iron sulfide minerals, especially greigite, are common in Pliocene–Pleistocene mudstones in the region (Sagnotti & Winkler 1999). The presence of these two components is further supported by the thermal and AF demagnetization behaviour shown in Figure 3. The drop in magnetization at 300–350 °C in the thermal demagnetization curves is most likely caused by iron sulfide minerals that become magnetically unstable at those temperatures. In contrast, the decrease in magnetization above 550 °C is consistent with the behaviour of detrital magnetite. The presence of a mixed magnetite/iron-sulfide mineralogy is in agreement with the detailed rock magnetic work of Mary *et al.* (1993) that also concluded a mixed ferromagnetic mineralogy consisting of detrital magnetite and diagenetic iron sulfides. They also noted that the relative abundance of magnetite and sulfide minerals is variable throughout the Stirone section. Mary *et al.* (1993) described a zone that is dominated by iron sulfide ferromagnetic mineralogy in the upper part of the section above c. 130 m. They demonstrated that the remainder of the section is dominated by magnetite with iron sulfides as a minor secondary component.

Correlation to orbital cycles

We correlated our reversals to the Gradstein *et al.* (2004) geomagnetic polarity time scale (Fig. 8). We correlated the long normal polarity zone at the base of the section (N1) to the C2An.1n subchron.

Fig. 3. (**a**) Vector endpoint diagrams for representative samples. Sample ST-A-2-B shows endpoint vectors for a representative sample subject to thermal demagnetization. ST-594, ST-624 and ST-633 show the endpoint vectors for representative samples subject to AF demagnetization. Square symbols represent the vertical component of magnetization and circular symbols represent the horizontal component. (**b**) Progressive demagnetization curves for the samples shown in (a). Curves show the sample magnetization remaining after each demagnetization step, normalized to the natural remanent magnetization.

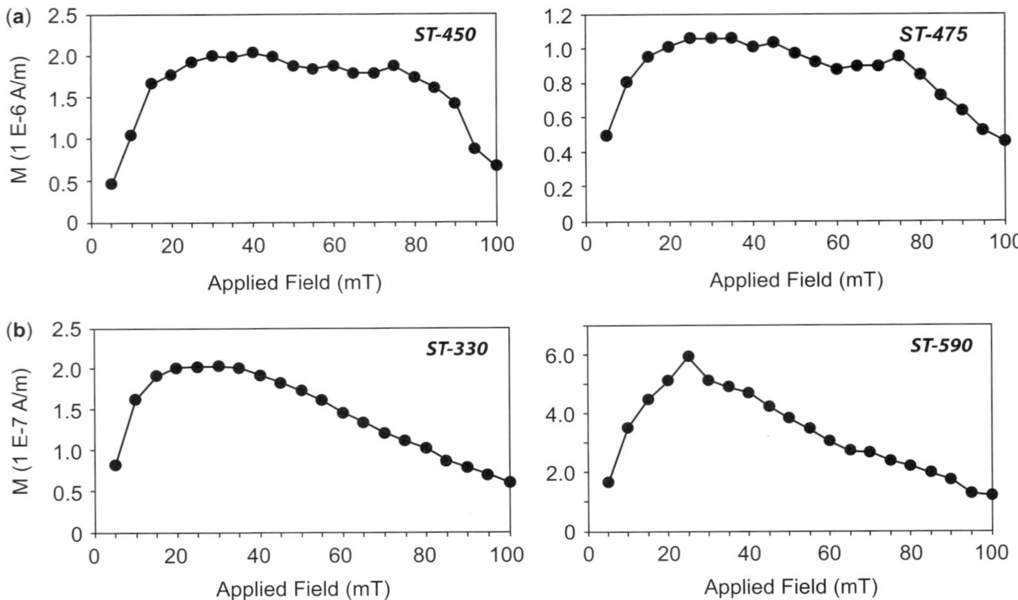

Fig. 5. Partial anhysteretic remanent magnetization (pARM) spectra for representative unoriented samples from the Stirone section. Solid symbols represent magnetization at each ARM step. A peak in the ARM spectrum represents the mean coercivity of the magnetization component. Spectra with two peaks have two components of magnetization.

This correlation is supported by the previous magnetostratigraphic studies conducted by Mary *et al.* (1993) and Channell *et al.* (1994). However, the age at the base of our section is not resolvable using the magnetostratigraphy alone; the biostratigraphic data of Channell *et al.* (1994) suggest that *c.* 200 ka of

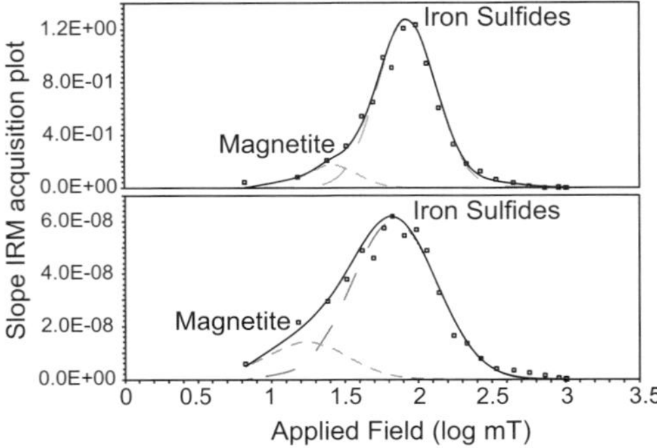

Fig. 6. Results from IRM acquisition modelling for representative samples indicate two major components of magnetization in the Stirone section. The squares represent the gradient of the IRM acquisition curve at each magnetization step. Solid black line indicates a best-fit curve to the acquisition data and represents a sum of the dashed lines, which represent the individual mineral components.

time is missing at the base of the C2An.1n subchron at the Stirone section. The R1 interval at the Stirone-section corresponds to the C2r.2r subchron. The magnetostratigraphy of Mary *et al.* (1993) interpreted this reversed subchron as extending all the way through the calcarenites between 270 and 310 m in the section. Our magnetostratigraphy observed several normal polarity sites within the calcarenites in this interval that we correlate to the Olduvai subchron (C2n). Our magnetostratigraphy failed to resolve the short-lived Réunion normal subchron that occurs at *c.* 2.14 Ma because our sample spacing was not close enough to resolve the 10 ka event. The top of the Olduvai subchron corresponds to the interval directly above the uppermost calcarenite in the Stirone section. This correlation is strengthened by the first occurrence of the mollusc *Arctica islandica*, which occurs at the same horizon. The C1r.2r subchron (R2 zone) is represented by the R2 polarity zone and appears to be a condensed stratigraphic section at the Stirone. The final reversal, at the base of the yellow littoral sands, corresponds to the base of the Jaramillo subchron (C1r.1n) at *c.* 1.07 Ma. Our age determination for these sands is in good agreement with previous magnetostratigraphic and cosmogenic studies further east along the mountain front that determined an *c.* 1 Ma depositional age for the Sabbie Gialle (Marabini *et al.* 1995; Cyr & Granger 2008).

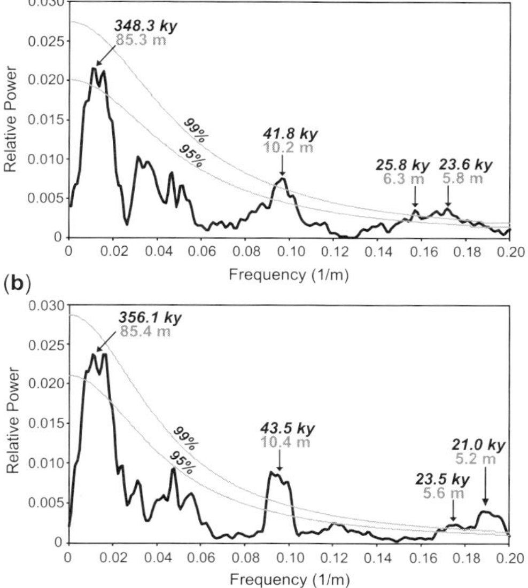

Fig. 7. Multi-taper method power spectrum of the (**a**) un-tuned and the (**b**) obliquity-tuned χ data series show significant cycles aligned with Milankovitch frequencies. The solid grey lines represent the 95 and 99% confidence intervals above a robust linear red-noise model (Mann & Lees 1996). The bandwidth is 0.0096 m^{-1}. Spectral power at each frequency has been normalized to the total power of the spectrum. The temporal periodicities of each of the significant peaks were calculated using the average sedimentation rate between of 0.24 mm a^{-1} as determined by the magnetostratigraphy. The improved alignment in the precession band in the obliquity-tuned power spectrum (b) demonstrates the efficacy of our obliquity correlation (Fig. 10).

Because the contact between the yellow sands and the overlying AEI unit is unconformable at the Stirone section and both units exhibit a normal polarity, it is impossible to tell if the contact lies within the Jaramillo subchron or spans several reversals.

We used our magnetostratigraphic correlations (Fig. 8) to assign absolute time to the section and confirm the presence of orbitally forced cyclicity in the χ data series. We fixed the timing of each magnetic reversal at the stratigraphic horizon where the reversal was observed and assumed constant sediment accumulation rates between the reversal boundaries. We found the significant spectral peaks in the χ power spectrum occurred at 348.3 ka (1/85.3 m), 41.8 ka (1/10.2 m), 25.8 ka (1/6.3 m) and 23.6 ka (1/5.8 m) (Fig. 7). The spectral peak that is best aligned with its predicted orbital frequency is the 41 ka obliquity peak. The

two precession cycles are longer than predicted by the orbital model while the long eccentricity cycle is shorter than expected. The slightly misaligned peaks are probably due to variations in sediment accumulation rate that cannot be resolved at the magnetostratigraphic scale.

The recognition of significant climate cycles in the χ data series allowed us to correlate the χ variability to theoretical orbital models (Fig. 9). We used three magnetic polarity reversals recognized in our magnetostratigraphy at 2.58, 1.95 and 1.81 Ma as tie points for our correlation. Beginning at these tie points, we correlated the major peaks in the χ data series, spaced approximately 10 m apart, to corresponding periods of low obliquity using the orbital model proposed by Laskar *et al.* (2004). We correlated peak χ values to low obliquity, rather than high obliquity, because the magnetic reversal we recognized in our section at 2.58 Ma occurs during a time of low obliquity (Laskar *et al.* 2004), but is coincident with a peak in our χ data series (Fig. 9). By following this convention, we allowed the magnetostratigraphy to guide our correlation instead of any assumptions regarding the orbital forcing mechanism. The validity of this correlation is demonstrated by the improved alignment of the χ power spectrum in the precessional bandwidth following our correlation (Fig. 7). The correlation procedure adjusted the χ data series for variations in sediment accumulation rate that caused the precession peaks in the un-tuned power spectrum to become misaligned.

In between each interpreted obliquity peak there exists smaller, high-frequency peaks representing precession cycles. In order to aid the correlation of χ peaks to the Laskar *et al.* (2004) precession orbital model, we applied a Gaussian filter centred at 0.047 with a bandwidth of 0.004 to the data and used the filtered data series in our correlation (Fig. 9). This allowed us to remove the low-frequency eccentricity and obliquity-related variability to better recognize the high-frequency peaks associated with precession cycles. Many of the interpreted precession peaks were coincident with the interpreted obliquity-related χ peaks. This is simply because the precession cycles (23 ka, 19 ka) are approximately half as long as the obliquity cycle (41 ka), and we therefore expect every other precession peak to coincide with an obliquity forced peak.

Stirone section age model

We used both the obliquity and precession correlations shown in Figure 9 to create a high-resolution age model for the Stirone section (Fig. 10; Table 1). The age model obtained by correlation to the obliquity cycles differs slightly from the model obtained

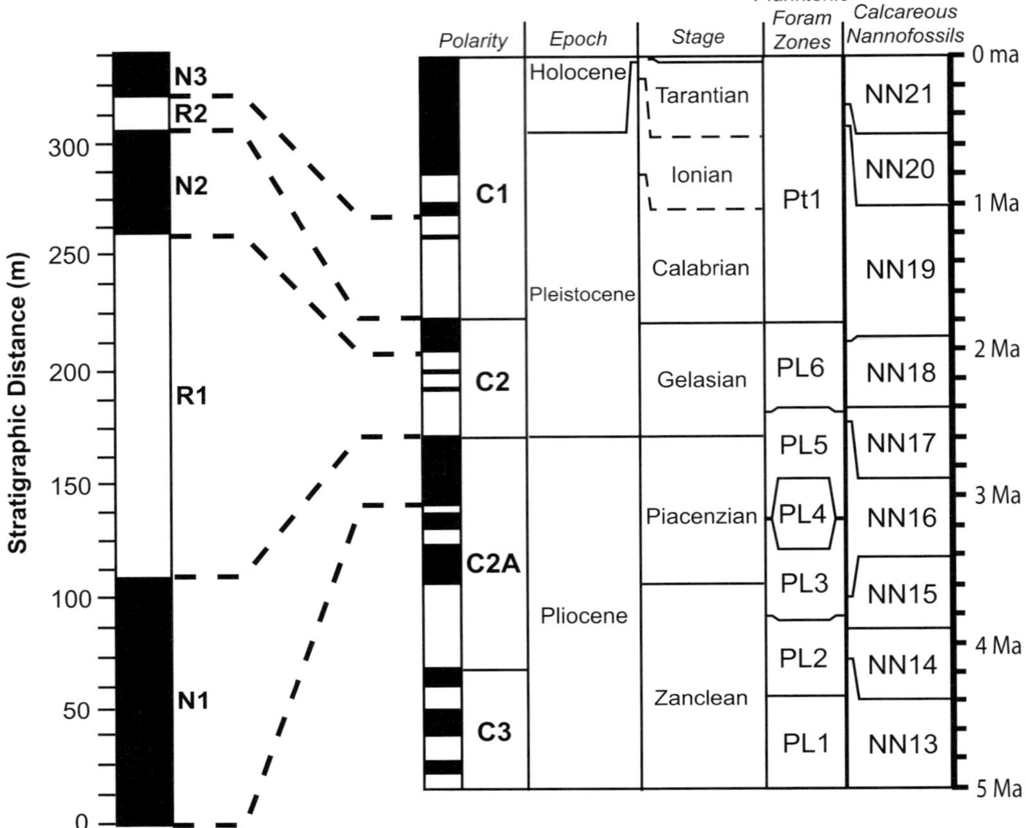

Fig. 8. The Stirone section's magnetostratigraphic correlation to the Gradstein *et al.* (2004) geomagnetic polarity timescale is indicated by the dashed lines. The correlation at the base of the N1 polarity zone was in accordance with the previous magnetostratigraphic interpretations of Mary *et al.* (1993) and Channell *et al.* (1994). The N2/R2 boundary is coincident with the first occurrence of the mollusc *Arctica islandica* occurring at *c.* 1.81 Ma. The R2/N3 boundary occurs at the base of the Sabbie Gialle sands.

by correlation to precession. Both of these models, however, show variability in sediment accumulation rate that cannot be resolved using the magnetostratigraphic time scale alone. The parts of the section where the obliquity age model converges to the precession age model coincide with time periods when the theoretical obliquity and precession cycles are in phase. This suggests that the variability in the χ data series is most likely controlled by both of the forcing parameters. The short time period when the two age models diverge, between *c.* 2.2 and 2.0 Ma, coincides with large amplitude swings in the χ data series and a coarsening of the Stirone section lithology (Fig. 2). This suggests that subtle changes in lithology and magnetic mineralogy in this interval mask the encoding of one or more of the orbital forcing parameters.

Our new high-resolution age model requires an adjustment to the timing of a hiatus at the base of our measured section. Channell *et al.* (1994) reported a stratigraphic hiatus that ended directly above the chemoherm bed at the base of our section. Based on the last occurrence (LO) of calcareous nannofossil *D. tamalis* and the nannofossil chronology of Rio *et al.* (1990*a*), Channell *et al.* (1994) report an age of 2.77 Ma for the end of the hiatus. Our cyclostratigraphic correlation requires the LO of *D. tamalis* to be 2.99 Ma in the Stirone section. This is not surprising as the LO of *D. tamalis* has already been adjusted several times. Rio *et al.* (1990*a*, 1991) assigned the LO of *D. tamalis* to occur at 2.60 Ma, just prior to the Plio-Pleistocene boundary at 2.58 Ma. Berggren *et al.* (1995) reported a 2.78 Ma age for the LO of *D. tamalis* in their astronomically tuned Neogene

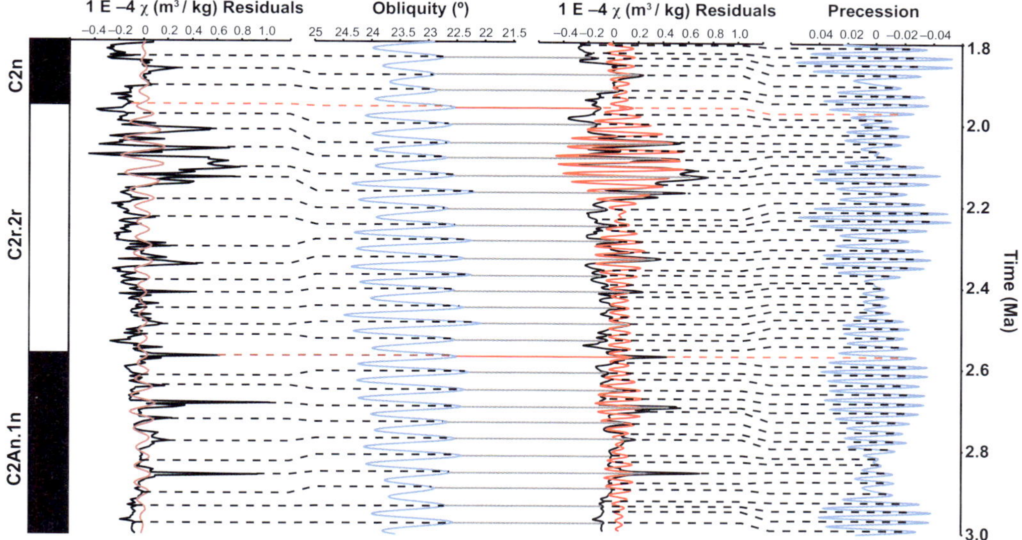

Fig. 9. Correlation of the de-trended (residual of second-order polynomial) low-field χ data series to the obliquity and precession theoretical orbital models (Laskar *et al.* 2004). Before the correlation, the data series was tied to absolute time using our magnetostratigraphic correlations (Fig. 8). The red dashed lines represent intervals where our correlation is anchored by the magnetostratigraphic reversals and the black dashed lines show our correlation to the theoretical orbital models. After correlating to obliquity, we applied a Gaussian filter centred at 0.047 with a bandwidth of 0.004 and used the filtered χ data series in our final correlation to the precessional orbital model. The filtered χ data series (red) and unfiltered χ data series (black) have the same scale.

timescale, an estimate which has subsequently been adjusted to 2.87–2.88 Ma (Di Stefano 1998; Raffi *et al.* 2006). Raffi *et al.*'s (2006) review of calcareous nannofossil chronology evaluated each Neogene biostratigraphic horizon and concluded that the LO of *D. tamalis* is diachronous worldwide, meaning that the age variations of the LO in individual sections are larger than *c.* 100 ka. Our tuned timescale places the LO of *D. tamalis* within 120 ka of the most recent adjustment and is consistent with the diachronous nature of *D. tamalis*.

A detailed biostratigraphic and magnetostratigraphic study of *c.* 30 m of section directly above the chemoherm layer at the base of our section, that was conducted as part of a thesis by Cau (2007) at the University of Parma, provides insight into the strength and limitations of our cyclostratigraphic approach. Cau (2007) observed a 4 m-thick reversed interval beginning 10 m above the chemoherm, and interpreted there to be a shorter hiatus than Channell *et al.* (1994) reported at the base of this zone, spanning 3.31–3.57 Ma. Additionally, a short, limited occurrence of *D. tamalis* was observed 20 m above the base of our section, which supports our interpretation that the base of our section is older than initially proposed by Channell *et al.* (1994). Cau (2007) then used a cyclostratigraphic analysis

of the sapropels in this short interval to tune this section to the Laskar *et al.* (2004) insolation curve. These data support our cyclostratigraphic correlation; by tuning our section to the theoretical precession model we assign the horizon at 21 m, near the top of Cau's (2007) section, as 2.93 Ma. Cau's (2007) own age determination for this horizon differs from ours by only 10 ka. Cau (2007), however, does assign the base of the section to be *c.* 3.31 Ma, in contrast to the 2.99 Ma age required by our age model. The apparent age discrepancy between the two age models only occurs in the first 20 m of section, where the amplitude of the χ data series is very small, indicating possible aliasing of the precession signal in this short interval caused by very slow sediment accumulation rates. Our age model requires the basal 20 m span 2.99–2.93 Ma, while Cau (2007) suggests the timing to be between 3.31 and 2.94 Ma.

Our age model also confirms the timing of regional calcarenite deposition. The Stirone Fm facies display a shallowing-upward progression probably caused by the combined effects of fault-related folding of the Salsomaggiore anticline, regional uplift of the Apennine range and the progradation of the Po Delta. Superimposed on the shallowing-upward trend is the record of climatically driven sea-level fluctuations, represented by the

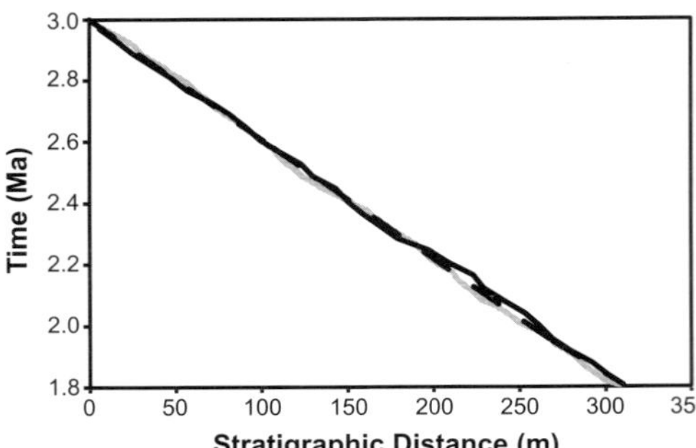

Fig. 10. Age models for the Stirone section determined from the magnetostratigraphy (black dashed line), obliquity correlation (black sold line) and precession correlation (grey solid line). All three models agree remarkably well and display little variation in sediment accumulation rate over a 1.2 Ma period of time.

occurrence of calcarenite beds. Previous studies have linked the deposition of calcarenites in the Northern Apennines to orbital forcing parameters; for instance, Roveri & Taviani's (2003) cyclostratigraphic investigation of Plio-Pleistocene calcarenites in the nearby Castell'Aquarto Basin (Fig. 1) determined that bundles of calcarenites in Mediterranean basins are controlled by 100 and 400 ka eccentricity cycles during this period. Roveri and Taviani (2003) observed calcarenite bundles centred at *c.* 3.0, 2.6, 2.2 and 1.8 Ma. The high-resolution age model we developed using the χ data series at the Stirone section (Fig. 10) show that calcarenite bundles are similarly timed and centred at *c.* 3.0, 2.6 and 1.9 Ma.

Encoding climate cycles

Variability in χ in the Stirone section is related to precessional and obliquity orbital forcing. Precession-controlled cyclicity in the formation of sapropels in Mediterranean marine sections is already well established (Rossignol-Strick *et al.* 1982; Rossignol-Strick 1985; Rohling & Hilgen 1991). Sapropel formation occurs during insolation maxima, when strengthened African monsoons increase runoff in North African rivers. This increased runoff contributes to the formation of sapropels in the Mediterranean by delivering more organic material to the ocean and creating a low-salinity lid that inhibits ocean circulation. Larrasoaña *et al.* (2003, 2006) also related enhancement in certain rock magnetic parameters that measure fine-grained ferromagnetic minerals to the same processes responsible for the deposition of sapropels, namely a strengthened African monsoon causing increased runoff and restricted ocean circulation. Our rock magnetic experiments show that χ is

Table 1. *Cyclostratigraphic age models for the Stirone section*

Stratigraphic distance (m)	Obliquity age model (Ma)	Stratigraphic distance (m)	Precession age model (Ma)
6	2.970	6	2.990
16	2.931	13	2.969
25	2.890	16	2.947
36	2.850	20	2.925
47	2.811	27	2.904
57	2.768	30	2.874
71	2.728	37	2.853
81	2.689	44	2.831
91	2.649	48	2.815
99	2.606	58	2.796
110	2.568	61	2.775
123	2.528	68	2.753
130	2.488	76	2.728
143	2.448	78	2.705
150	2.406	84	2.682
158	2.368	89	2.659
168	2.328	96	2.637
179	2.285	100	2.613
197	2.247	106	2.590
209	2.205	110	2.569
223	2.164	114	2.545
229	2.124	120	2.520
241	2.081	124	2.497
253	2.042	130	2.479
261	2.002	135	2.458
269	1.958	141	2.439
279	1.918	148	2.420
292	1.878	152	2.398
301	1.837	155	2.387
310	1.806	161	2.368
		165	2.347
		172	2.325
		176	2.304
		183	2.279
		190	2.254
		193	2.232
		197	2.211
		204	2.188
		214	2.165
		219	2.140
		224	2.118
		228	2.096
		235	2.075
		240	2.061
		245	2.044
		249	2.023
		256	2.003
		266	1.987
		271	1.968
		275	1.947
		279	1.925
		288	1.905
		294	1.875
		298	1.854
		301	1.832
		310	1.811

controlled by both ferromagnetic and paramagnetic components, and the relative contribution of these components creates the variability observed in the χ data series. It is during times of increased runoff and restricted Mediterranean circulation that the Stirone section displays peaks in χ. Since χ represents paramagnetic, iron sulfide and magnetite components, the variability in all of these components must be explained using the same encoding process. Enhanced runoff probably increased the delivery of detrital magnetite and Fe-rich detrital clays while the restricted Mediterranean circulation caused by the low salinity lid during high runoff times increased the productivity of diagenetic sulfide minerals.

The presence of significant obliquity cycles in the χ data series is more difficult to explain since obliquity exerts little control on insolation and runoff at middle and low latitudes. Orbital obliquity, however, is a major control on sea-level during the Late Pliocene–Early Pleistocene time (Lisiecki & Raymo 2005) with phases of low obliquity corresponding to an increased build-up in high latitude ice sheets and subsequently to lower sea-levels. Using the constraints imposed by our magnetostratigraphy, we correlated the large peaks in χ to obliquity minima, which coincide with sea-level lows. It is possible that the lower sea-level during obliquity minima contributed to the restriction of deep-water ventilation in the Mediterranean, thereby intensifying the anoxic conditions that contributed to the increased productions of iron-sulfides minerals. This sea-level effect, however, is probably secondary compared with the precession controlled runoff and ocean circulation variability.

An intriguing outcome of our spectral results for the Stirone section is that extensive diagenetic alteration of magnetite to iron sulfide minerals did not obscure the palaeomagnetic or cyclostratigraphic signals in the Stirone section. This is true despite the fact that different parts of the section have experienced different intensities of reduction diagenesis of magnetite (Mary *et al.* 1993). This is probably because the anoxic conditions prevalent during the deposition of sapropels and magnetic enhancement are ideal for the formation of iron-sulfide minerals such as pyrrhotite and greigite. The relatively fast and nearly constant sedimentation rate at the Stirone during this time (0.24 m/ka) meant that the magnetic minerals probably spent little time in the upper 1–2 m critical zone where reduction diagenesis occurred, ideal for the preservation of the anoxic signal.

Conclusion

We used rock-magnetic cyclostratigraphy to generate a high-resolution age model for the Stirone Fm, an important Plio-Pleistocene marine stratigraphic section that exhibits no lithological cyclicity. Our timescale correlation improved the timing of the LO of *D. tamalis* in this section, adjusted the length of a sedimentological disconformity, and determined high-resolution sedimentation rates that can be used for tectonic, sequence stratigraphic or sedimentological studies. The efficacy of rock magnetic cyclostratigraphic correlations is dependent on our ability to independently determine absolute time, estimate sediment accumulations rates and recognize depositional hiatuses. These constraints can be overcome when cyclostratigraphic analyses are combined with detailed lithostratigraphic, biostratigraphic and magnetostratigraphic analysis.

At the Stirone River section, we linked cyclic variations in χ to precessional scale changes in runoff and ocean circulation that control the concentrations of fine-grained ferromagnetic magnetite and iron sulfides and paramagnetic clays. The additional presence of obliquity cycles in the χ data series provides an insight regarding the ability of global ice volume-induced sea-level changes to subtly affect the precession-dominated variability by contributing to the restriction of deep water Mediterranean circulation.

This work was supported by a National Science Foundation grant (EAR-0809722) to D. J. Anastasio, K. P. Kodama and F. J. Pazzaglia, and a Geological Society of America student research grant to K. L. Gunderson. We thank Parco Fluviale Regionale dello Stirone for facilitating access to the section. S. Brachfeld, B. Ellwood and two anonymous reviewers contributed helpful reviews. We thank the editors for their suggestions and consideration of this manuscript. We thank D. Smith, A. Ponza and R. Gunderson for field assistance. A. Artoni and V. Picotti contributed their insight regarding the local geology.

References

AMOROSI, A., BARBIERI, M. *ET AL.* 1998*a*. Sedimentology, micropalaeontology, and strontium-isotope dating of a lower-Middle Pleistocene marine succession (Argille Azzurre) in the Romagna Apennines, northern Italy. *Bollettino della Società Geologica Italiana*, **117**, 789–806.

AMOROSI, A., CAPORALE, L. *ET AL.* 1998*b*. The Pleistocene littoral deposits (Imola Sands) of the northern Apennines foothills. *Giornale di Geologia*, **60**, 83–118.

ARTONI, A., PAPANI, G. *ET AL.* 2004. The Salsomaggiore structure (northwestern Apennine foothills, Italy); a Messinian mountain front shaped by mass-wasting products. *Geo-Acta*, **3**, 107–127.

ARTONI, A., RIZZINI, F. *ET AL.* 2007. Tectonic and climatic controls on sedimentation in late Miocene Cortemaggiore wedge-top basin (northwestern Apennines, Italy). *In*: LACOMBE, O., LAVE, J., ROURE, F. M. & VERGES, J. (eds) *Thrust Belts and Foreland Basins;*

from Fold Kinematics to Hydrocarbon Systems. Springer, Berlin, 431–456, http://dx.doi.org/10.1007/978-3-540-69426-7

BERGGREN, W. A., HILGEN, F. J. *ET AL.* 1995. Late Neogene chronology: new perspectives in high-resolution stratigraphy. *Geological Society of America Bulletin*, **107**, 1272–1287, http://dx.doi.org/10.1130/0016-7606(1995)1072.3.CO;2

BERTOLANI MARCHETTI, D., ACCORSI, C. A., PELOSIO, G. & RAFFI, S. 1979. Palynology and stratigraphy of the Plio-Pleistocene sequence of the Stirone River (Northern Italy). *Pollen et spores*, **21**, 150–167.

BLOEMENDAL, J. & DE MENOCAL, P. 1989. Evidence for a change in the periodicity of tropical climate cycles at 2.4 Myr from whole-core magnetic susceptibility measurements. *Nature (London)*, **342**, 897–900.

BLOEMENDAL, J., LAMB, B. & KING, J. W. 1988. Paleoenvironmental implications of rock-magnetic properties of late Quaternary sediment cores from the eastern Equatorial Atlantic. *Paleoceanography*, **3**, 61–87, http://dx.doi.org/10.1029/PA003i001p00061

CARSON, B., SUESS, E., STRASSER, J. C., LANGSETH, M. G. & MOORE, J. C. 1990. Fluid flow and mass flux determinations at vent sites on the Cascadia margin accretionary prism; Special section on the Role of fluids in sediment accretion, deformation, diagenesis, and metamorphism in subduction zones. *Journal of Geophysical Research*, **95**, 8891–8897, http://dx.doi.org/10.1029/JB095iB06p08891

CAU, S. 2007. *Paleoecologia delle comunita chemiosintetiche al passaggio Zancleano-Piacenziano nel torrente Stirone.* M.S. thesis, Università DegliStudi di Parma.

CHANNELL, J. E. T., POLI, M. S., RIO, D., SPROVIERI, R. & VILLA, G. 1994. Magnetic stratigraphy and biostratigraphy of Pliocene 'argilleazzurre' (Northern Apennines, Italy). *Palaeogeography, Palaeoclimatology, Palaeoecology*, **110**, 83–102.

CONTI, S. & FONTANA, D. 1999. Miocene chemoherms of the northern Apennines, Italy. *Geology*, **27**, 927–930, http://dx.doi.org/10.1130/0091-7613(1999)0272.3.CO;2

CYR, A. J. & GRANGER, D. E. 2008. Dynamic equilibrium among erosion, river incision, and coastal uplift in the northern and central Apennines, Italy. *Geology*, **36**, 103, http://dx.doi.org/10.1130/G24003A.1

DI DIO, G. 2005. *Cartageologica d'Italia alla scala 1:50 000.* Regione Emilia-Romagna, ServizioGeologico, Sismico e dei Suoli, Firenze.

DI STEFANO, E. 1998. Calcareous nannofossilquanitative biostratigraphy of holes 969E and 963B (Eastern Mediterranean). *Proceedings of the Ocean Drilling Program, Scientific Results*, **160**, 99–112, http://dx.doi.org/10.2973/odp.proc.sr.160.009.1998

DOMINICI, S. 2001. Taphonomy and paleoecology of shallow marine macrofossil assemblages in a collisional setting (late Pliocene–early Pleistocene, western Emilia, (Italy). *Palaios, United States*, **16**, 336–353.

DONDI, L. 1961. Nota paleontologico-stratigrafica sul Pedeappennino padano. *Bollettino della Società GeologicaItaliana*, **81**, 113–245.

ELLWOOD, B. B., TOMKIN, J. H., FEBO, L. A. & STUART, C. N. 2008. Time series analysis of magnetic susceptibility variations in deep marine sedimentary rocks: a test using the Upper Danian–Lower Selandian proposed GSSP, Spain. *Palaeogeography, Palaeoclimatology,*

Palaeoecology, **261**, 270–279, http://dx.doi.org/10.1016/j.palaeo.2008.01.022

FISHER, R. A. 1953. Introductions to statistical methods applied to directional data. *Proceedings of the Royal Society of London*, **A217**, 295–305.

GHIL, M., ALLEN, M. R. *ET AL.* 2002. Advanced spectral methods for climatic time series. *Reviews of Geophysics*, **40**, 1–41, http://dx.doi.org/10.1029/2000RG000092

GRADSTEIN, F. M., OGG, J. G. & SMITH, A. G. 2004. *A Geologic Time Scale 2004.* Cambridge University Press, Cambridge, 409–471.

HAYS, J. D., IMBRIE, J. & SHACKLETON, N. J. 1976. Variations in the Earth's orbit; pacemaker of the ice ages. *Science*, **194**, 1121–1132.

HELLER, F. & EVANS, M. E. 1995. Loess magnetism. *Reviews of Geophysics*, **33**, 211–240, http://dx.doi.org/10.1029/95RG00579

HINNOV, L. A. 2000. New perspectives on orbitally forced stratigraphy. *Annual Reviews of Earth Planetary Sciences*, **28**, 419–475.

HINNOV, L. A., GRADSTEIN, F. M., OGG, J. G. & SMITH, A. G. 2004. *Earth's Orbital Parameters and Cycle Stratigraphy.* Cambridge University Press, Cambridge, 55–62.

JACKSON, M., GRUBER, W., MARVIN, J. & BANERJEE, S. 1988. Partial anhysteretic remanence and its anisotropy: applications and grain size dependence. *Geophysical Research Letters*, **15**, 440–443.

JOVANE, L., FLORINDO, F., SPROVIERI, M. & PALIKE, H. 2006. Astronomic calibration of the late Eocene/ early Oligocene Massignano section (central Italy). *Geochemistry, Geophysics, Geosystems*, **7**, 1–10, http://dx.doi.org/10.1029/2005GC001195

JOVANE, L., SPROVIERI, M., COCCIONI, R., FLORINDO, F. & MARSILI, A. 2010. Astronomical calibration of the middle Eocene Contessa Highway section (Gubbio, Italy). *Earth and Planetary Science Letters, Netherlands*, **298**, 77–88, http://dx.doi.org/10.1016/j.epsl.2010.07.027

KIRSCHVINK, J. 1980. The least-squares line and plane and the analysis of paleomagnetic data. *Geophysical Journal. Royal Astronomical Society*, **62**, 699–718.

KODAMA, K. P., ANASTASIO, D. J., NEWTON, M. L., PARES, J. M. & HINNOV, L. A. 2010. High-resolution rock magnetic cyclostratigraphy in an Eocene flysch, Spanish Pyrenees. *Geochemistry, Geophysics, Geosystems*, **11**, 1–22, http://dx.doi.org/10.1029/2010GC003069

KRUIVER, P. P., DEKKERS, M. J. & HESLOP, D. 2001. Quantification of magnetic coercivity components by the analysis of acquisition curves of isothermal remanent magnetisation. *Earth and Planetary Science Letters*, **189**, 269–276.

KRUIVER, P. P., KRIJGSMAN, W., LANGEREIS, C. G. & DEKKERS, M. J. 2002. Cyclostratigraphy and rockmagnetic investigation of the NRM signal in late Miocene palustrine–alluvial deposits of the Librilla section (SE Spain). *Journal of Geophysical Research*, **107**, 1–18, http://dx.doi.org/10.1029/2001JB000945

LARRASOAÑA, J. C., ROBERTS, A. P. *ET AL.* 2003. A new proxy for bottom-water ventilation in the eastern Mediterranean based on diagenetically controlled magnetic properties of sapropel-bearing sediments; Paleoclimatic and paleoceanographic records in Mediterranean sapropels and Mesozoic black shales. *Palaeogeography, Palaeoclimatology, Palaeoecology*, **190**, 221–242.

LARRASOAÑA, J. C., ROBERTS, A. P. ET AL. 2006. Detecting missing beats in the Mediterranean climate rhythm from magnetic identification of oxidized sapropels (Ocean Drilling Program Leg 160); ODP contributions to paleomagnetism. *Physics of the Earth and Planetary Interiors*, **156**, 283–293, http://dx.doi.org/10.1016/j.pepi.2005.04.017

LASKAR, J., ROBUTEL, P., JOUTEL, F., GASTINEAU, M., CORREIA, A. C. M. & LEVRARD, B. 2004. A long-term numerical solution for the insolation quantities of the Earth. *Astronomy Astrophysics*, **428**, 261–285, http://dx.doi.org/10.1051/0004-6361:20041335

LATTA, D. K., ANASTASIO, D. J., HINNOV, L. A., ELRICK, M. & KODAMA, K. P. 2006. Magnetic record of Milankovitch rhythms in lithologically noncyclic marine carbonates. *Geology*, **34**, 29–32, http://dx.doi.org/10.1130/G21918.1

LISIECKI, L. E. & RAYMO, M. E. 2005. A Pliocene–Pleistocene stack of 57 globally disturbed benthic δ18 O records. *Paleoceanography*, **20**, 17, http://dx.doi.org/10.1029/2004PA001071

MANN, M. & LEES, J. 1996. Robust estimation of background noise and signal detection in climatic time series. *Climate Change*, **33**, 409–445, http://dx.doi.org/10.1007/BF00142586

MARABINI, S., TAVIANI, M., VAI, G. B. & VIGLIOTTI, L. 1995. Yellow sand facies with *Arctica islandica*: low-stand signature in an early Pleistocene Front–Apennine Basin. *Giornale di Geologia*, **57**, 259–275.

MARY, C., IACCARINO, S., COURTILLOT, V., BESSE, J. & AISSAOUI, D. M. 1993. Magnetostratigraphy of pliocene sediments from the Stirone River (Po Valley). *Geophysical Journal International*, **112**, 359–380, http://dx.doi.org/10.1111/j.1365-246X.1993.tb01175.x

OLSEN, P. E. 1986. A 40-million-year lake record of early Mesozoic orbital climatic forcing. *Science*, **234**, 842–848.

PAILLARD, D., LABEYRIE, L. & YIOU, P. 1996. Macintosh program performs time-series analysis. *Eos Transactions*, **77**, 379–379.

PAPANI, G. & PELOSIO, G. 1963. La serieplio-pleistocenicadel T. Stirone (Parmense occidentale). *Bollettino della Società Geologica Italiana*, **81**, 293–335.

PETERS, C. & DEKKERS, M. J. 2003. Selected room temperature magnetic parameters as a function of mineralogy, concentration and grain size. *Physics and Chemistry of the Earth*, **28**, 659–667, http://dx.doi.org/10.1016/S1474–7065(03)00120-7

RAFFI, S. 1986. The significance of marine boreal molluscs in the early Pleistocene faunas of the Mediterranean area. *Palaeogeography, Palaeoclimatology, Palaeoecology*, **52**, 267–289.

RAFFI, I., BACKMAN, J. ET AL. 2006. A review of calcareous nannofossilastrobiochronology encompassing the past 25 million years. *Quaternary Science Reviews*, **25**, 3113–3137, http://dx.doi.org/10.1016/j.quascirev.2006.07.007

RICHTER, C. & VAN DER PLUIJM, B. A. 1994. Separation of paramagnetic and ferrimagnetic susceptibilities using low temperature magnetic susceptibilities and comparison with high field methods. *Physics of the Earth and Planetary Interiors*, **82**, 113–123.

RIO, D., RAFFI, I. & VILLA, G. 1990*a*. 32. Pliocene–Pleistocene calcareous nannofossil distribution patterns in the western Mediterranean. *Proceedings of the Ocean Drilling Program, Scientific Results*, **107**, 513–533.

RIO, D., SPROVIERI, R. & CHANNELL, J. 1990*b*. Pliocene–Early Pleistocene Chronostratigraphy and the Tyrrhenian deep-sea record from Site 653. *Proceedings of the Ocean Drilling Program, Scientific Results*, **107**, 705–714, http://dx.doi.org/10.2973/odp.proc.sr.107.185.1990

RIO, D., SPROVIERI, R. & THUNELL, R. 1991. Pliocene–lower Pleistocene chronostratigraphy; a re-evaluation of Mediterranean type sections. *Geological Society of America Bulletin*, **103**, 1049–1058, http://dx.doi.org/10.1130/0016-7606(1991)1032.3.CO;2

ROHLING, E. J. & HILGEN, F. J. 1991. The eastern Mediterranean climate at times of sapropel formation: a review. *Geologie en Mijnbouw (Netherlands Journal of Geosciences)*, **70**, 253–264.

ROSSIGNOL-STRICK, M. 1985. Mediterranean Quaternary sapropels, and immediate response of the African monsoon to variation of insolation. *Palaeogeography, Palaeoclimatology, Palaeoecology*, **49**, 237–263.

ROSSIGNOL-STRICK, M., NESTEROFF, W., OLIVE, P. & VERGNAUD-GRAZZINI, C. 1982. After the deluge; Mediterranean stagnation and sapropel formation. *Nature (London)*, **295**, 105–110.

ROVERI, M. & TAVIANI, M. 2003. Calcarenite and sapropel deposition in the Mediterranean Pliocene: shallow- and deep-water record of astronomically driven climatic events. *Terra Nova*, **15**, 279–286, http://dx.doi.org/10.1046/j.1365-3121.2003.00492.x

SAGNOTTI, L. & WINKLER, A. 1999. Rock magnetism and palaeomagnetism of greigite-bearing mudstones in the Italian peninsula. *Earth and Planetary Science Letters*, **165**, 67–80.

SHACKLETON, N., CROWHURST, S., HAGELBERG, T., PISIAS, N. G. & SCHNEIDER, D. A. 1995. A new Late Neogene time scale: application to Leg 138 sites. *Proceedings of the Ocean Drilling Program, Scientific Results*, **138**, 74–101.

SHACKLETON, N., CROWHURST, S., WEEDON, G. & LASKAR, J. 1999. Astronomical calibration of Oligocene–Miocene time. *Philosophical Transactions: Mathematical, Physical and Engineering Sciences*, **357**, 1907–1929.

THOMPSON, D. J. 1982. Spectrum estimation and harmonic analysis. *Proceedings of the IEEE*, **70**, 1055–1096.

THOMPSON, D. J. 1990. Time series analysis of Holocene climate data. *Philosophical Transactions of the Royal Society of London. Series A, Mathematical and Physical Sciences*, **330**, 601–616.

WILLIAMS, D. F., PECK, J. ET AL. 1997. Lake Baikal record of continental climate response to orbital insolation during the past 5 million years. *Science*, **278**, 1114–1117.

ZIJDERVELD, J. D. A. 1967. A.C. demagnetization of rocks: analysis of results. *In*: COLLINSON, D. W., CREER, K. M. & RUNCORN, S. K. (eds) *Methods in Paleomagnetism*. Elsevier, Amsterdam, 254–286.

Global Milankovitch cycles recorded in rock magnetism of the shallow marine lower Cretaceous Cupido Formation, northeastern Mexico

LINDA A. HINNOV[1], KENNETH P. KODAMA[2]*, DAVID J. ANASTASIO[2], MAYA ELRICK[3] & DIANA K. LATTA[2]

[1]*Department of Earth and Planetary Sciences, Johns Hopkins University, Baltimore, MD 21218, USA*

[2]*Department of Earth and Environmental Sciences, Lehigh University, Bethlehem, PA 18015, USA*

[3]*Department of Earth and Planetary Sciences, University of New Mexico, Albuquerque, NM 87131, USA*

**Corresponding author (e-mail: kpkodama@gmail.com)*

Abstract: Rock magnetic cyclostratigraphy was measured in the Barremian–Aptian Cupido ('Cupidito') Formation, northeastern Mexico. The goal was to develop an objective evaluation of palaeo-environmental variability recorded in the formation that is independent of facies analysis and interpretation. Anhysteretic remanent magnetization (ARM) was used to estimate magnetic mineral concentration variations for the upper 143 m of the formation, which is characterized by metre-scale carbonate cycles representative of inner- and middle-shelf marine environments. Isothermal remanent magnetization acquisition experiments and scanning electron microscope (SEM) examination indicate that micron-sized detrital magnetite from eolian dust carries the ARM signal. At the sampled sections from Garcia and Chico canyons, 25 km apart, ARM records a synchronous 30–35 m oscillation with maxima coinciding with fourth-order sequence boundaries, superimposed with prominent high-frequency variability. Calibrating the 30–35 m oscillation to a 405 kyr period (long eccentricity cycle) focuses the high frequencies into short eccentricity, obliquity and precession index bands; the precession-band signal modulates with an eccentricity signature. The ARM signal is correlated between sections, but decoupled from the interpreted fifth-order depositional cycles. ARM amplitudes diminish up-section with facies suggesting deepening conditions that diluted magnetite concentration. This probably signals a warming, increasingly humid climate, changing global circulation and/or greater dispersal of magnetite grains.

Earth's ancient shallow marine carbonate platforms are important archives of past global change. Their *in situ* biological genesis makes them sensitive indicators of physical conditions in shallow marine environments, including temperature, circulation and sea level (Wilson 1975). The accommodation space afforded by water depth is thought to govern the amount and type of sediment that accumulates at a given location; water depth in turn is controlled by a combination of tectonic subsidence, near-shore deposition and global sea level, which vary significantly through time (Schlager 2005).

The commonly observed stacking of metre-scale shallowing upward cycles in carbonate platform deposits has led to proposals of allocyclic v. autocyclic forcing mechanisms. The most widely cited allocyclic hypothesis is astronomically controlled eustasy (Fischer 1964; Hardie *et al.* 1986); autocyclic mechanisms involve infilling and oceanward progradation of sedimentation without eustatic change (Ginsburg 1971). To this day, the allocycle v. autocycle debate remains unresolved in understanding the origin of 10^4–10^5 year scale lithologic cyclicity in shallow marine carbonates (Schwarzacher 2000; Burgess 2006; Schlager 2010). Full shallowing to sea level may not develop in response to every sea-level oscillation (subtidal 'missed beats'), resulting in amalgamated cycles; conversely, prolonged episodes of subaerial exposure may occur in inner platform settings that are only briefly or never flooded with every sea-level oscillation, leading to condensed or peritidal 'missed beats' (Goldhammer *et al.* 1990).

Ultimately facies analysis of shallow marine sections is limited in teasing apart the interplay of eustasy owing to climate change, and subsidence arising from local tectonics. This is in large part owing to the relatively subjective interpretation of lithofacies that is required from the stratigrapher. Therefore we need to develop facies-independent

From: Jovane, L., Herrero-Bervera, E., Hinnov, L. A. & Housen, B. A. (eds) 2013. *Magnetic Methods and the Timing of Geological Processes*. Geological Society, London, Special Publications, **373**, 325–340.
First published online April 25, 2013, http://dx.doi.org/10.1144/SP373.20 © The Geological Society of London 2013.
Publishing disclaimer: www.geolsoc.org.uk/pub_ethics

tools to isolate the climate record from sedimentary strata. Rock magnetic properties indicate the concentration of magnetic mineral particles and hence variations in detrital mineral input, for example, far-field eolian dust influx. Thus, rock magnetic parameters have the potential to provide palaeo-environmental records that are not directly linked to facies.

Anhysteretic remanent magnetization (ARM) has emerged as an important proxy for palaeoclimate change in carbonate rocks (Kodama 2012). ARM measures the concentration of ferromagnetic minerals and is not sensitive to the diamagnetism of the carbonate content; thus it is an ideal tool for evaluating stratigraphic cyclicity independent of facies interpretations. Lean & McCave (1998) demonstrated that glacial–interglacial changes are recorded by ARM variations in Pleistocene sediments on the Chatham Rise, SW Pacific Ocean. In those sediments ARM increases during interglacial periods in association with decreased productivity, leading to an increased oxic zone in the sediments and hence increased production and concentration of single-domain magnetite grains produced by magnetotactic bacteria. Results from Quaternary marine sediments in the northern Atlantic Ocean show that ARM data can vary on astronomical time scales (Kruiver *et al.* 1999). ARM has also been used to detect Milankovitch climate cycles in basinal carbonates showing no appreciable facies changes, in deposits coeval to those investigated in the present study (Latta *et al.* 2006).

Our study focus is on carbonate cycles in the Lower Cretaceous Cupido Formation as exposed in the Garcia and Chico canyons, northeastern Mexico (Fig. 1). These cyclic platform deposits have been the subject of numerous facies analysis investigations (Goldhammer *et al.* 1991; Lehmann *et al.* 1998, 1999; Goldhammer 1999; Foster 2003; Latta 2005; Altobi 2007). However, there is no consensus on the origins of the prominent sedimentary cyclicity. Stacking patterns suggest an obliquity forcing origin (Goldhammer *et al.* 1991). Spectral analysis of multiple sections in the region suggest a combined eccentricity and obliquity forcing origin (Altobi 2007), but the absence of the precession index forcing that would be necessary for recording of the eccentricity is an unsolved problem.

Here we have approached this problem using objective rock magnetic data. We measured high-resolution ARM stratigraphy in the upper Cupido Formation ('Cupidito') in Garcia and Chico canyons, and characterized the magnetic mineralogy responsible for the ARM variations. The results led to the discovery of strong astronomical signal in the ARM record that can be attributed to climate forcing of atmospheric dust deposition to the region. This ARM signal is largely disconnected

from the sedimentary facies in which it occurs, but it is common to both the Garcia and Chico sections. Thus, for this Mexico case study ARM serves the dual purpose of a climate proxy and high-resolution correlation tool. The clear implication is that ARM is a transformative method that can elucidate detailed palaeoclimate information from carbonate stratigraphy with complex origins.

Geological setting

Regional geology

Passive margin development along the Gulf Coast began in the Late Triassic following continental rifting associated with the opening of the Gulf of Mexico (e.g. Pindell 1993; Dickinson & Lawton 2001). Uplifted Palaeozoic basement blocks and intervening grabens controlled Mesozoic depositional patterns along the Mexican Gulf Coast (e.g. Wilson 1999; Dickinson & Lawton 2001). In northeastern Mexico, the Coahuila block, presently bounded by the Sabinas and Parras Basins, served as a basement high and a proximal source of clastic material, which influenced evaporite and subsequent carbonate deposition. By the earliest Cretaceous, extensive carbonate platforms developed around the entire Gulf of Mexico (Wilson 1999). In NE Mexico the shelf was shallow and rimmed with a stable carbonate platform represented by the late Barremian–Albian Cupido and Aurora Formations (Wilson 1999, Lehmann *et al.* 1999, 2000). The Cupido Formation (940 m thick) is conformable with respect to the underlying shales and limestones of the Barremian to early Aptian Taraises Formation (490 m thick), as well as with the overlying the deeper-water shales and lime-mudstones of the La Peña Formation (*c.* 20–100 m thick).

The Cupido Formation

The Cupido Formation is characterized by stacked, metre-scale (1–8 m thick), upward-shallowing peritidal cycles composed of subtidal facies overlain by intertidal to supratidal facies, which in some places are capped by subaerial exposure features (e.g. karst, palaeosols, rubble breccias; Goldhammer *et al.* 1991; Elrick 1995; Foster 2003). Facies patterns in the upper *c.* 150 m of the formation ('Cupidito') indicates gradual deepening which is followed by widespread platform drowning represented by the La Peña Formation (Wall *et al.* 1961; Tinker 1982; Goldhammer *et al.* 1991; Lehmann *et al.* 1998, 1999).

Our study focuses on two sections on the Cupido platform: Garcia Canyon, composed of

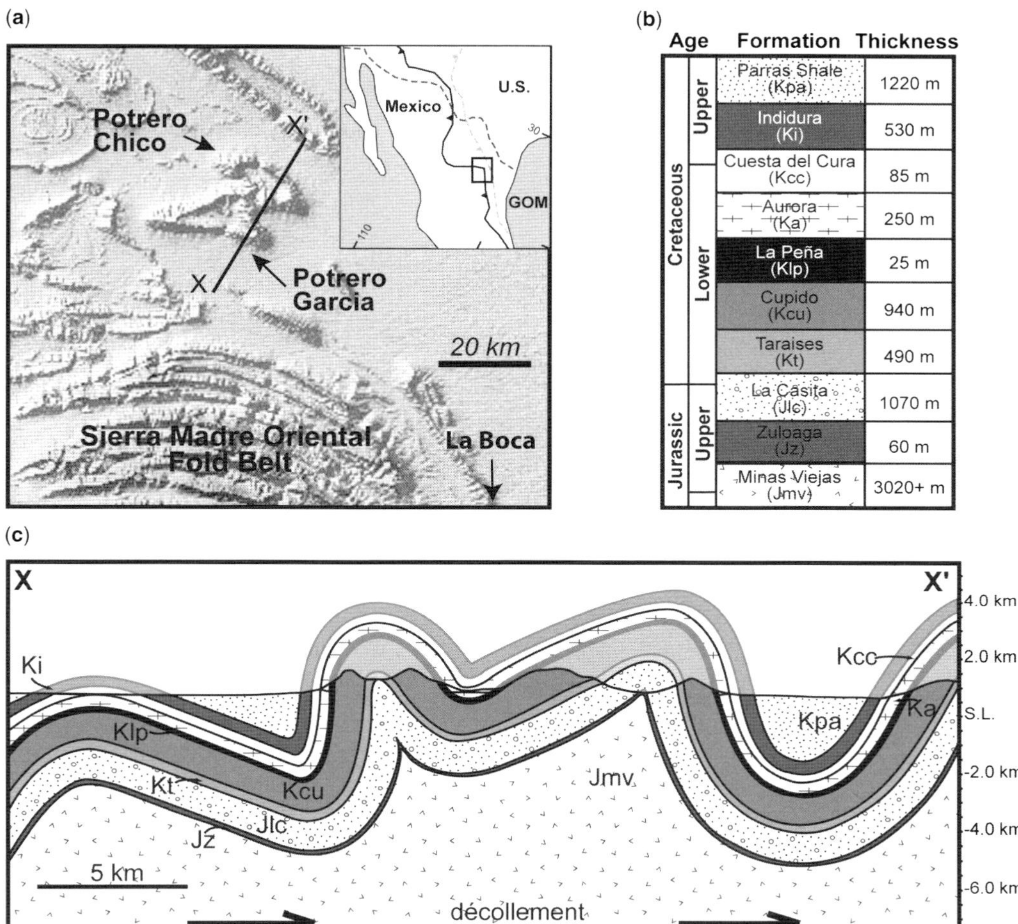

Fig. 1. Location and stratigraphic framework of the Lower Cretaceous inner (Garcia) and middle (Chico) carbonate platform sections in the foreland of the Sierra Madre Oriental fold belt, NE Mexico. (**a**) Index and hillshade maps locating the Garcia, Chico and La Boca sections in the foreland of the Sierra Madre Oriental fold belt, NE Mexico. (**b**) Regional stratigraphy of northeastern Mexico. The studied section is the upper Cupido ('Cupidito') Formation of latest Barremian–early Aptian age. (**c**) Cross section through the Sierra del Fraile anticlinorium drawn perpendicular to the strike of the average fold axes (see a). From Latta & Anastasio (2007).

peritidal-dominated inner-shelf facies; and Chico Canyon, dominated by shallow subtidal middle-shelf deposits (Fig. 2). The inner- and middle-shelf deposits were originally separated by *c.* 25 km, according to cross section restorations of Late Cretaceous Sevier–Laramide deformation (Latta & Anastasio 2007). The inner-shelf deposits are characterized by thin-bedded, peritidal and subtidal facies, including evaporite, laminated mudstone, and burrowed skeletal packstone units. The middle-shelf deposits contain similar peritidal and subtidal facies, but exhibit less frequent (less cyclic) facies changes, resulting in thicker-bedded units. In the

inner-shelf deposits at Garcia Canyon, Goldhammer *et al.* (1991) identified 78 upward-shallowing cycles within a *c.* 270 m section. Biostratigraphic age control indicates a late Barremian to early Aptian age (Alfonso-Zwanzinger 1978). Goldhammer *et al.* (1991) interpreted the cycles (average thickness 3.44 m) to represent 40 kyr obliquity forcing based on age, number of cycles and cycle bundling ratios between sequence boundaries.

The inner- and middle-shelf deposits of the upper Cupido Formation at Garcia and Chico Canyons are overlain by the relatively isochronous basal La Peña Formation (Goldhammer 1999;

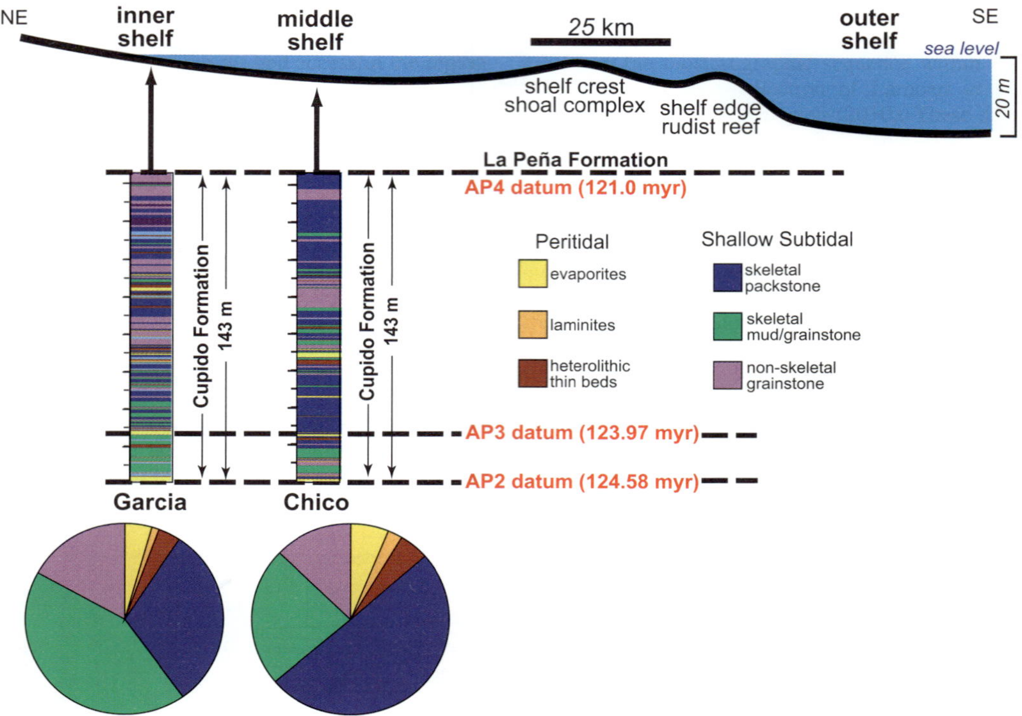

Fig. 2. Coeval inner- and middle-shelf deposits of the upper Cupido ('Cupidito') Formation. Regionally correlative datums (AP2, AP3 and AP4, see text) provide chronostratigraphic constraints, and exposure facies allow lithostratigraphic correlation. Pie charts illustrate facies distribution across the platform and indicate that the inner-shelf deposits at Garcia are dominated by peritidal facies and the middle-shelf deposits at Chico are dominated by subtidal facies.

Lehrmann & Goldhammer 1999; Fig. 2). The biostratigraphic controls on the La Peña Formation are ambiguous. At Santa Rosa Canyon (SW of the study area), the La Peña starts in the upper *Globigerinelloides blowi* planktonic foraminifer zone (Bralower *et al.* 1999), which places it below Oceanic Anoxic Event 1a (OAE1a, the 'Selli' event) described by Schlanger & Jenkyns (1976), Coccioni *et al.* (1987, 1989, 2012), and many others. However, in La Huasteca Canyon (south of the study area), the La Peña starts in the *Dufrenoyia furcata* (=*Dufrenoyia justinae* Zone in Mexico) ammonite zone (Barragan & Maurrasse 2008), which places it above OAE1a.

The Garcia and Chico sections have three chronostratigraphically significant sequence boundaries in the upper Cupido Formation (Fig. 3). These regionally correlative datums were originally assumed to correspond to global sequence boundaries for AP2, AP3 and AP4 (Goldhammer 1999), for which dates have been recently updated from 120.6 to 124.58 Ma, from 120.0 to 123.97 Ma

and from 117.07 to 121.0 Ma, respectively (Hardenbol *et al.* 1998; updated by Snedden & Liu 2010). These sequence boundaries can be correlated to biostratigraphically constrained sequence boundaries in the northern Gulf of Mexico (Goldhammer 1999), guyots in the Western Pacific (Röhl & Ogg 1998) and western European sedimentary basins (Jacquin *et al.* 1998). Thus these datums appear to represent global sea level changes.

However, new astrochronology of the Italian Piobbico core indicates that OAE1a starts at 124.55 Ma, and ends at 123.16 Ma (Huang *et al.* 2010). Thus, if the basal La Peña pre-dates OAE1a (i.e. is older than 124.55 Ma, as indicated by foraminifera stratigraphy) then these datums may need to be reassigned to older sequence boundaries, for example, Barr6 (126.2 Ma), AP1 (125.0 Ma) and AP2 (124.58 Ma). On the other hand, if the basal La Peña post-dates OAE1a (i.e. is younger than 123.16 Ma, as indicated by ammonite stratigraphy), then the presently assumed global sequence boundary assignments may stand.

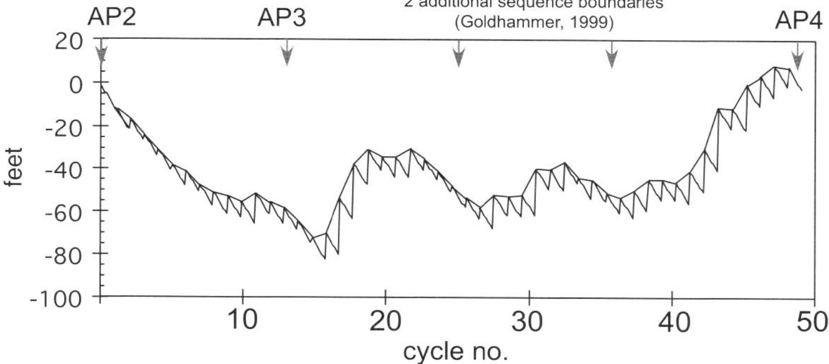

Fig. 3. Fischer plot of the Cupidito carbonate cycle succession at Garcia Canyon. Arrows indicate sequence boundaries as interpreted by Goldhammer (1999). The section duration is 3.58 or 1.62 myr (see text), which in turn indicates a cycle duration of 73 or 33 kyr, and sequence durations of *c.* 730–900 or 330–407 kyrs. There are 10–12 cycles per sequence, that is, no sustained 5:1 cycle bundling.

Data and methods

Sample collection

Samples were collected from the upper Cupido Formation at Garcia (inner-shelf) and Chico (middle-shelf) canyons. The sample spacing was maintained between 20 and 50 cm in order to sample each identified carbonate cycle at least four or five times. This resulted in a total of 297 samples at Garcia and 283 samples at Chico.

Rock magnetic analysis

Individual, unoriented samples were collected and hand crushed to 2–4 mm size pieces and weighed in individual, pre-weighed, 8 cm^3 plastic sample

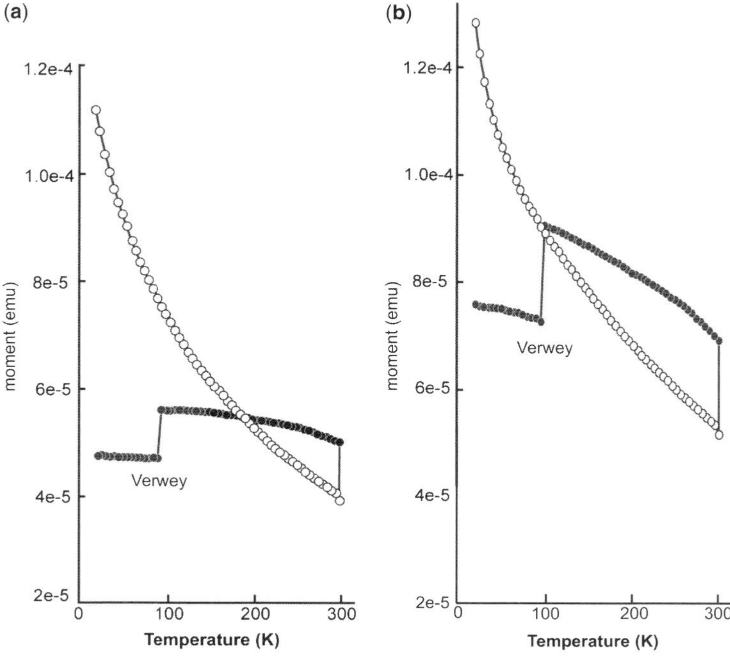

Fig. 4. Representative low-temperature magnetic measurements for inner (**a**) and middle (**b**), shelf deposits, indicating a sharp transition (*c.* 100 K) upon cooling (black circles), suggesting Verwey transitions, and that the ARM signal is probably controlled by ferromagnetic magnetite (e.g. Muxworthy 1999). From Latta (2005).

boxes. Low-field bulk magnetic susceptibility was measured on a Agico KLY-3s Kappabridge, while ARM was acquired in a peak alternating field of 100 mT with a DC field of 0.1 mT and measured on a 2G Enterprises, Inc. superconducting magnetometer. Both measurements were normalized for mass and all were conducted at Lehigh University.

Variations in ARM can be caused by fluctuations in primary ferromagnetic mineral concentrations or from diagenetic processes. In order to further characterize magnetic grain size and composition, rock magnetic experiments were undertaken at the Institute for Rock Magnetism, University of Minnesota. Ferromagnetic and paramagnetic grains were characterized in 24 samples (11 from the inner-shelf Garcia section and 13 from the middle-shelf Chico section) using the superconducting susceptometer (MPMS). After applying a 2.5 T field, the magnetic moment of each sample was measured in a 0 T field at 5 K increments while heated from 20 to 300 K. A 2.5 T field was then again applied at 300 K

and the magnetic moment of each sample was again measured in a 0 T field at 5 K increments while cooled back to 20 K.

The relative populations of magnetic grains, based on variations in grain size and the mineralogy of the ferromagnetic grains, were determined from 32 samples (16 from inner-shelf Garcia, and 16 from middle-shelf Chico) using the vibrating sample magnetometer (VSM1) where the field varied between 1.0 and −1.0 T. In an additional experiment to determine the magnetic mineralogy, isothermal remanences (IRMs; Lowrie 1990) were imparted to a single outer-shelf sample from La Boca Canyon that displayed relatively high ferromagnetic mineral concentrations and were subsequently thermally demagnetized (Latta *et al.* 2006). Finally, scanning electron microscope/energy dispersive X-ray microscopy images of magnetic grains collected from HCl-insoluble residues of samples that display hysteresis parameters characteristic of all samples, were used to help elucidate ferromagnetic grain morphology and size.

Time series methods

To study the frequency content of the ARM records we used the multitaper spectral analysis function in freeware *Analyseries* (Thomson 1982; Paillard *et al.* 1996). The ARM series were resampled to uniform spacing using the linear interpolation in *Analyseries*. The Garcia facies rank series was interpolated to a uniform spacing using the 'nearest neighbor' option in Matlab's interp1 function. Taner bandpass filtering (Taner 2003) was applied in Matlab to isolate the precession index signal in the 405 kyr tuned Garcia ARM record. Amplitude envelope analysis was performed in Matlab using Hilbert transformation (Hinnov 2000).

Results

Magnetic mineralogy

The MPMS results indicate the presence of magnetite as the dominant magnetic mineralogy in the Cupido Formation carbonates. The solid circles in Figure 4 indicate the cooling cycle in the MPMS measurements. The abrupt decrease in magnetic intensity at about 100 K in the cooling curve is interpreted to be the Verwey transition of magnetite. The presence of magnetite is important since it indicates a primary depositional magnetic mineral carrying the rock magnetic cyclostratigraphy in the Cupido Formation.

The hysteresis curves (Fig. 5) for the two sampled localities show a wasp-waisted loop for the inner shelf (Garcia) and a narrow loop for

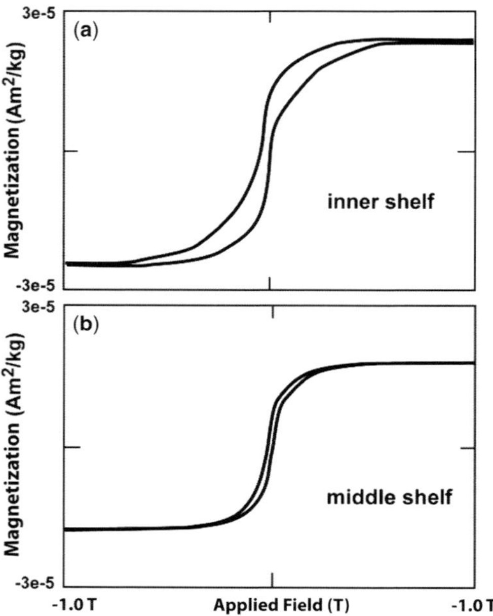

Fig. 5. Experimental results showing hysteresis curves of representative samples from each measured section, with the diamagnetic component removed. (a) Inner-shelf deposits exhibit two different grain size distributions of the magnetic mineral, most likely magnetite based on the low-temperature results, as shown by the wasp-waisted hysteresis curves, indicating a mixture of single domain and superparamagnetic grains. (b) Middle-shelf deposits, indicates the presence of a single low-coercivity ferromagnetic mineral population, probably in the pseudo-single domain grain size range.

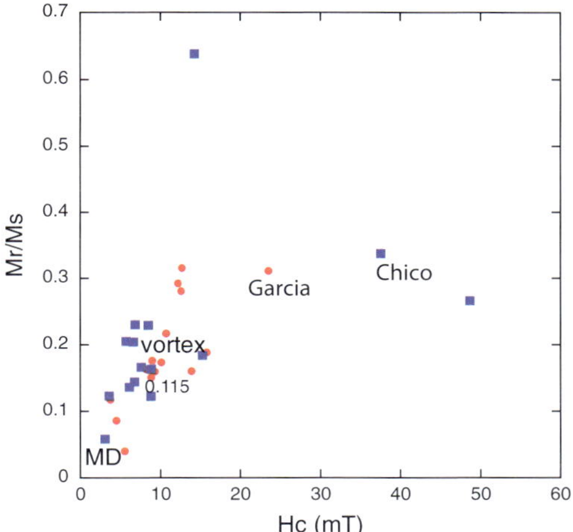

Fig. 6. Magnetic granulometry at Garcia and Chico canyons. This squareness diagram indicates that most of the samples measured from Garcia (red circles) and Chico (blue squares) are in the vortex to MD state, suggesting grain sizes of 0.1–1 μm.

the middle shelf (Chico). The wasp-waisted loop indicates a mixture of single domain and superparamagnetic ferromagnetic grains in the rocks (Tauxe *et al.* 1996), suggesting a range of magnetic particle sizes in the sub-micron range. The narrow hysteresis loop from the middle shelf indicates pseudosingle domain grains. The squareness (M_r/M_s) v. coercivity (H_c) plot of the hysteresis parameters

(Fig. 6) suggests that the magnetic particle size for the magnetite in the Cupido rocks is the vortex magnetic state (Tauxe *et al.* 2002), suggesting pseudo single domain grains with a grain size of *c.* 0.115 μm, with slightly larger particles in the outer shelf (Chico). Finally, anhysteretic remanent magnetization/saturation isothermal remanence magnetization values of 0.05–0.2 and *Raf* values of 0.35–0.4, where *Raf* is the crossover point between the IRM acquisition curve and the *af* demagnetization curve of the SIRM, suggest that no magnetosomes are present in the Cupido carbonates. Magnetosomes are produced by magnetotactic bacteria (Bazylinski *et al.* 1988).

The rock magnetic results, therefore, indicate that the dominant magnetic mineral carrying the rock magnetic cyclostratigraphy is primary, depositional inorganic magnetite grains. SEM/EDX reveals numerous iron oxide grains ranging in size from 1.5 to 12 μm, where nearly 50% of the grains fall between 3.0 and 4.0 μm (Fig. 7). SEM analyses further support the interpretation that the ferromagnetic grains are detrital. The iron oxide grains are observed to have quartz coatings, as expected from weathered bedrock, and an absence of textures consistent with diagenetic processes.

ARM cyclostratigraphy

The raw ARM measurements show a strong decrease in magnitude upsection (Figs 8 & 9). This trend corroborates facies observations of increasing water depths (Fig. 2) and flooding of the

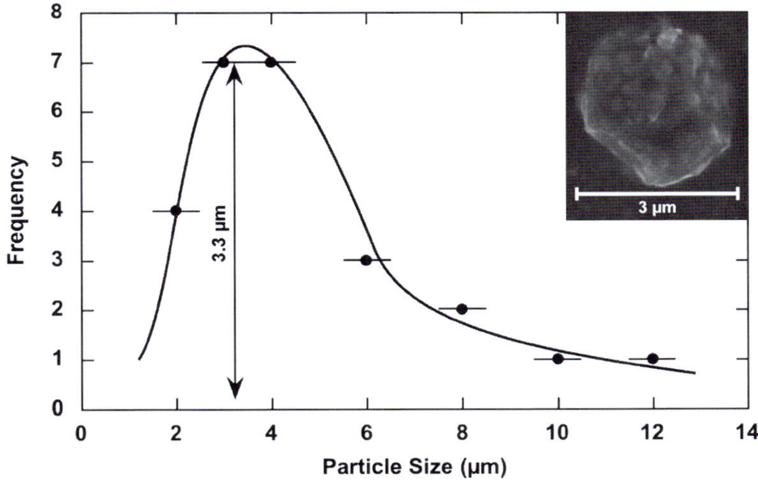

Fig. 7. Magnetite grain sizes analysed from HCl-insoluble residues by SEM/EDX. Average grain size 3.3 μm (±1 μm) is consistent with that of far-travelled atmospheric dust particles. Inset background electron image is of an iron-oxide grain from the Garcia section with quartz coatings, suggestive of original quartz cement and detrital origin. From Latta (2005).

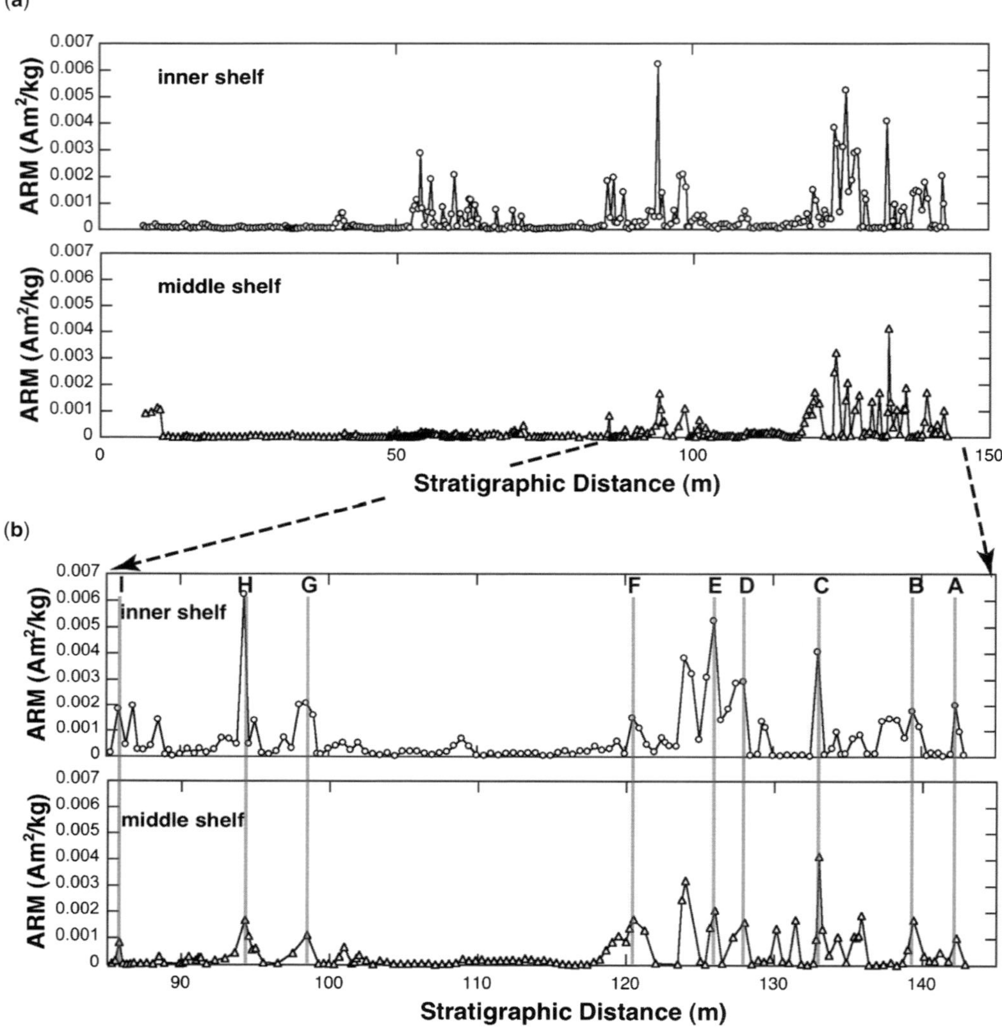

Fig. 8. (a) Comparison of ARM values between inner- (Garcia) and middle- (Chico) shelf measured sections in the upper Cupido Formation. The length of the sampled Garcia section is 136 m (7–143 m), and for the Chico section is 131 m (6–137 m). Up is to the left. (b) Detailed view of the upper 60 m; note the close correlation of ARM values marked by representative peaks (A–I) From Latta (2005).

Cupido Platform and subsequent deposition of the deep marine La Peña Formation. The Garcia and Chico records share numerous ARM peaks in stratigraphic order (labelled A–I in Fig. 8b), which attest to a remarkable consistency in magnetic mineral input between the two localities, which were separated by *c.* 25 km and had generally different water depths.

Log transforming the ARM ('$\log_{10}$ARM') dramatically improves our view of the ARM signal, revealing strong ARM cyclicity at multiple scales

in both sections (Fig. 9). The fitted linear trends show that ARM declines at the same rate in both sections; the offset of the trends shows that Garcia has a higher overall ARM throughout, presumably owing to its shallower depositional site. Most spectacularly the sections share a strong, phase-locked *c.* 30 m cycle. This cycle is also captured in the Fischer plot of the Garcia section (Fig. 3). This Fischer plot was used by Goldhammer (1999) to interpret fourth-order sequence boundaries, which coincide with the ARM maxima of the *c.* 30 m

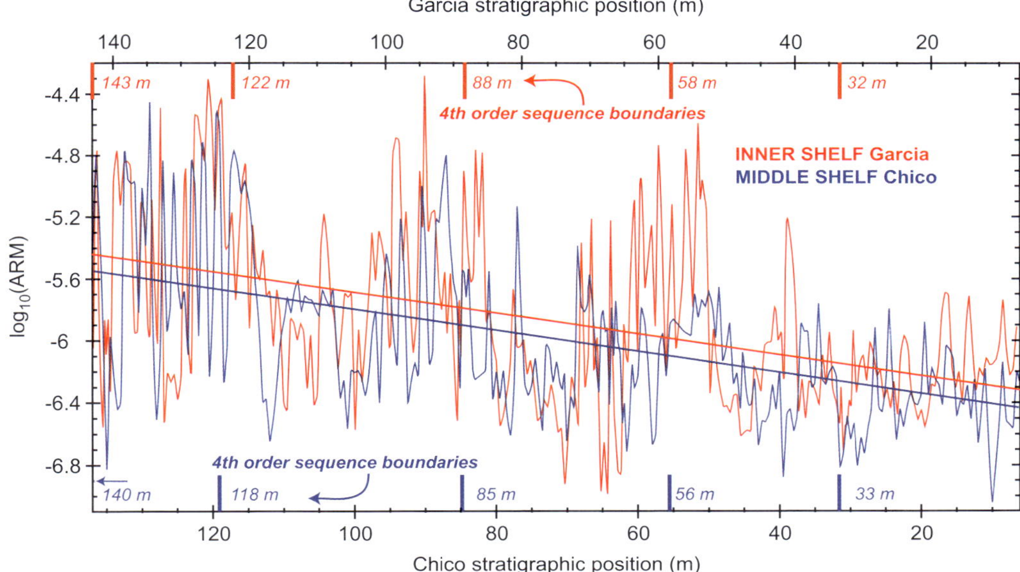

Fig. 9. Log$_{10}$ transformed ARM record for Garcia (red) and Chico (blue). Up is to the right. Fourth-order sequence boundaries are from Goldhammer (1999). The linear trend shows identical decline in ARM in both sections because of the deepening depositional environment upsection. The Garcia trend is offset from Chico owing to Garcia's deposition in a shallower setting. The cross-correlation coefficient between the two records, assuming that the bases and tops (basal contact of the La Peña Formation) of the sections are time-correlative points, is 0.68.

cycle (Fig. 9). Finally, superimposed on this dominant cycle are high-frequency variations that persist through both sections.

Spectral analysis

Power spectral analysis of the Garcia and Chico log$_{10}$ARM records in the stratigraphic domain (Fig. 10) confirm that the two records share important spectral characteristics. The 33.3 m cycle dominates both spectra; the Chico spectrum has high-power peaks at 14.9 and 11.4 m, whereas at Garcia there is only one, low-power peak at 13.5 m. Power declines rapidly in both spectra at higher frequencies, where spectral peaks occur at different frequencies. These mismatched peaks are due to different sedimentation rates affecting the two depositional environments. The Garcia spectrum, with its greater loss of power at higher frequencies, may reflect some abbreviation (condensation) of the depositional record owing to the higher incidence of subaerial exposure.

Discussion

Multiple sedimentological investigations have been undertaken on the Garcia section (Goldhammer

et al. 1991; Goldhammer 1999; Foster 2003; Latta 2005) that recognize pronounced cyclicity and intervals of subaerial exposure, thus providing evidence for sea-level oscillations. Therefore, in this discussion we focus on analysing the ARM record at Garcia, seeking evidence for astronomical forcing, and comparing it with the upward shallowing carbonate cycles.

ARM as eolian dust proxy

The SEM/EDX analysis reveals that iron oxide shapes probably represent detrital grains (Fig. 7). This, together with the small sizes of the analysed grains, is consistent with a far-travelled atmospheric dust origin (e.g. Pye 1987). Early Cretaceous atmospheric circulation indicates easterly trade winds affecting ancestral Mexico (e.g. Hasegawa *et al.* 2012). This could allow transport of eolian material from ancestral Africa across the widening Atlantic Ocean (Fig. 11), similar to what occurs today (Goudie & Middleton 2006). In support of our ARM evidence in Mexico, Deep Sea Drilling Project Sites 387, 391C, and 367 from the North Atlantic Basin contain Lower Cretaceous eolian dust transported from northwestern Africa (Dean & Arthur 1999). This eolian flux to

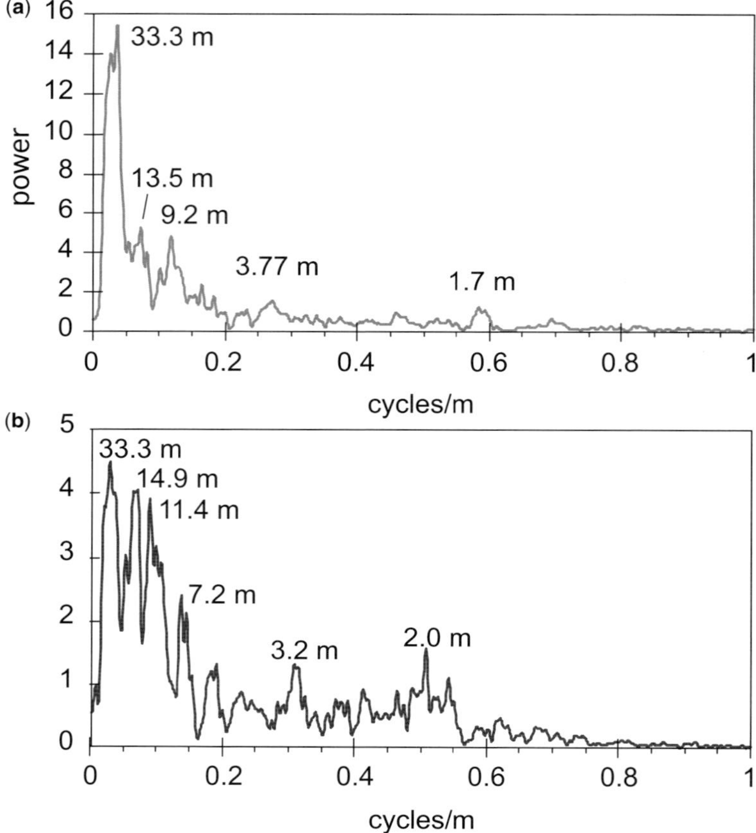

Fig. 10. 2π Power spectrum of the Garcia $\log_{10}$ARM (**a**) and Chico $\log_{10}$ARM (**b**) records in the stratigraphic domain. Labels indicate peak cyclicities in metres.

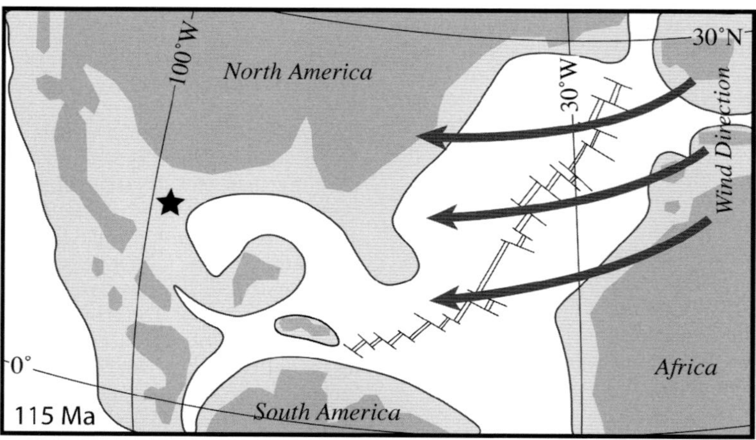

Fig. 11. Lower Cretaceous (115 Ma) palaeogeography, modified from Scotese *et al.* (1989), showing possible routes of eolian dust from Africa to the depositional site in Mexico (black star). Dark grey, continents; light grey, shallow shelves; white, ocean. From Latta *et al.* (2006).

Mexico would therefore be expected to reflect climatic forcing that affected low latitude atmospheric circulation. At the $10^4 - 10^5$ year timescales resolved by the Cupido carbonate cycles, evidence for astronomical forcing of eolian processes should be in sharp focus.

The ARM astronomical signal at Garcia Canyon

The Garcia $\log_{10}$ARM spectral peaks (Fig. 10a) have a frequency distribution that is suggestive of astronomical forcing (33.3:13.5:9.2:3.77:1.7 m is close in ratio to 405:130:97:c. 40:20 kyr). There is elevated power in the frequency band corresponding to the precession index (1.7 m), but strong peaks do not occur, possibly owing to variable sedimentation rates changing the thickness of precession scale cycles through the section.

At present, there is not an accurate chronology for these sections. Biostratigraphic evidence cannot establish whether the top of the Cupidito (basal La Peña) is pre-OAE1a or post-OAE1a in age; no geochronology is directly tied to either section. The global sequences AP2, AP3 and AP4 interpreted by Goldhammer (1999) (Figs 2 & 3) indicate a duration of 3.58 myr for the section. Alternatively, if these global sequences are instead Barr6, AP1 and

AP2, then the duration is 1.62 myr. Both of these estimated durations have unknown certainties that are probably in the half-million year range based on radioisotope dating constraints (Ogg & Hinnov 2012).

We tested the hypothesis that the 33.3 m $\log_{10}$ARM cycles were caused by 405 kyr orbital eccentricity forcing by tuning the former to the latter and assessing periodicity in the tuned spectrum. This timing corresponds to the alternative Barr6-AP2 assignment discussed above. The results (Fig. 12) show focusing of power in the 100, 44 and 20 kyr bands, as might be expected for an astronomical signal. Strong forcing by the orbital eccentricity, however, is a misnomer. The actual source of eccentricity power in stratigraphy is from precession index forcing. The stratigraphic recording process transfers power from precession index cycling into the precession modulator, that is, the eccentricity, in an effect known as 'rectification' (Weedon 2003). To examine whether there is a direct connection between the recorded precession index and eccentricity, we filtered the precession band (see Data and Methods section) and compared the amplitude envelope of the filtered signal with the eccentricity as recorded in the original tuned $\log_{10}$ARM signal. Filtered signal amplitudes are high when 405 kyr cycles reach their maxima, and low when the cycles are at

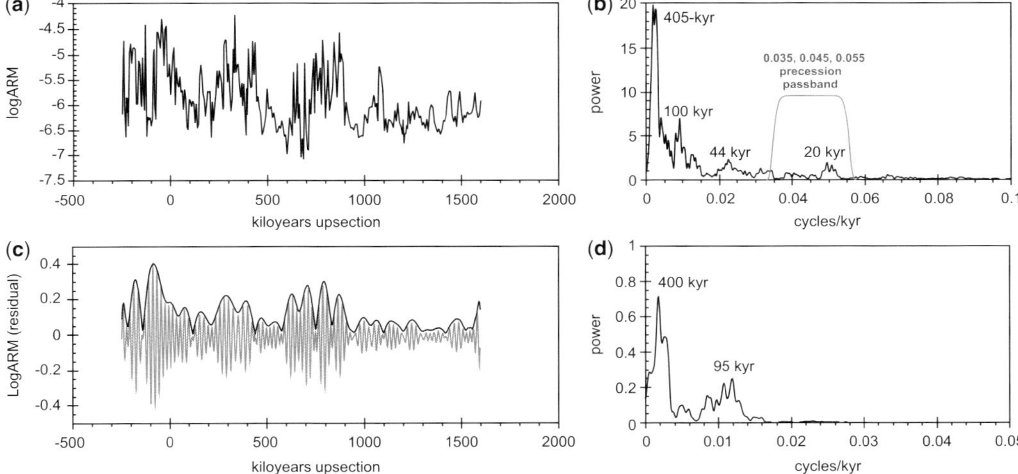

Fig. 12. Garcia $\log_{10}$ARM record calibrated to 405 kyr cycles. (**a**) 405 kyr calibrated Garcia logARM time series. The 405 kyr intervals were defined at 122, 88, 58 and 32 m (sequence boundaries of Goldhammer 1999) in the Garcia section. The total length of this time series is 1.84556 myr. (**b**) 2π Power spectrum of the 405 ky-calibrated Garcia $\log_{10}$ARM time series. The passband of the Taner filter (Taner 2003) applied to extract a precession index signal is shown, where the cut-off frequencies are 0.035 and 0.055 cycles kyr^{-1}, and the centre frequency of the band is 0.045 cycles kyr^{-1}. (**c**) Taner-filtered precession index signal (red line, grey in print version; passband shown in b) and amplitude envelope (black line) obtained through Hilbert transformation (Hinnov 2000). (**d**) 2π Power spectrum of the amplitude envelope shown in (c).

their minima (compare Fig. 12a, c), supporting the hypothesis that the *c.* 1.7 m scale ARM cycles were precession-driven.

ARM record v. depositional cyclicity

Comparison of the ARM signal and carbonate cycles at Garcia reveals some surprises. The maxima of the ARM 33.3 m cycles correlate with thinner and more subaerially exposed intervals (Fig. 13a). This suggests that, when sea level is low, high ARM associated with increased windiness and atmospheric dustiness is recorded in the Cupido Platform. When sea level is high, low ARM is recorded as water depths are greater, resulting in greater dispersion of eolian materials and dilution by increased carbonate accumulation.

However, the individual carbonate cycles are not matched by any of the high-frequency ARM cycles. Instead, there are two or more ARM cycles per carbonate cycle (Fig. 13b). Thus there is a decoupling between the sea-level oscillations forcing carbonate cycle development and the eolian influx producing the ARM signal. The strong astronomical signature in the ARM signal is not matched by a comparable one in carbonate cycling. The power spectra of the two Garcia records (Fig. 14) show that they share a common 405 kyr cycle, but at higher frequencies,

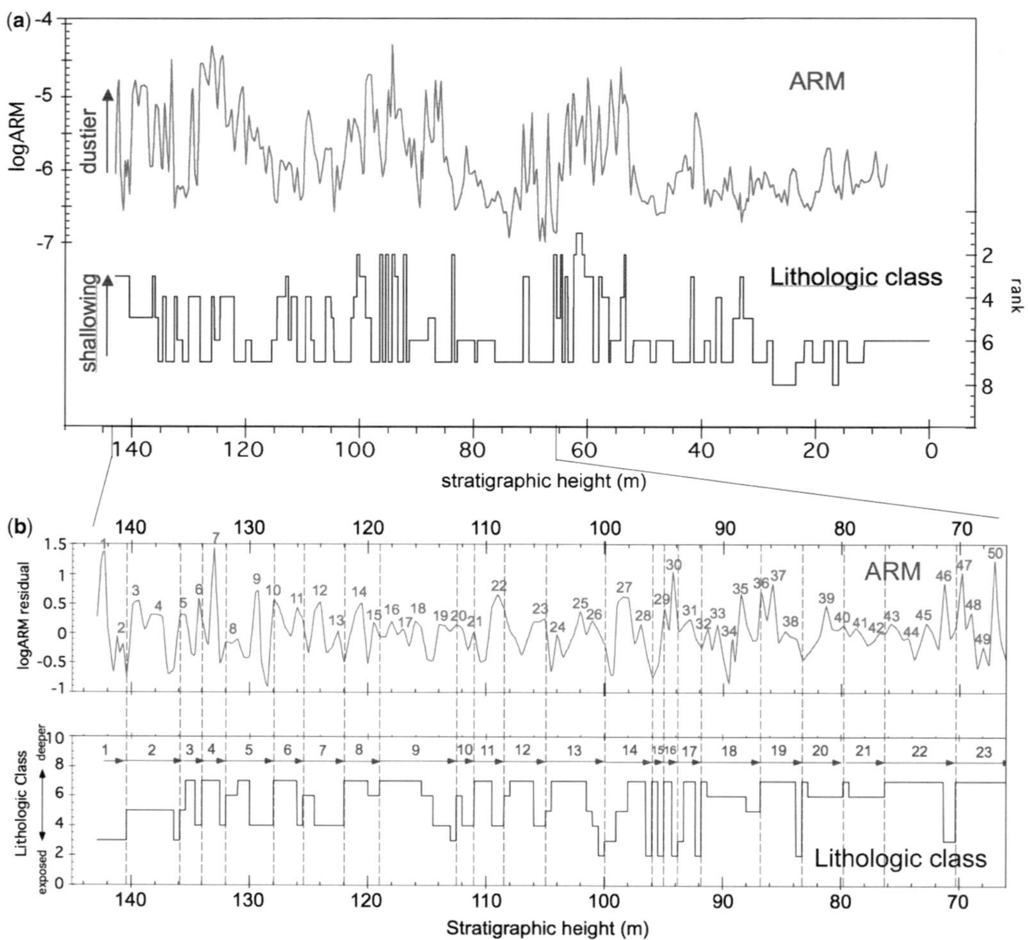

Fig. 13. Decoupling of the $\log_{10}$ARM record v. depositional cycles at Garcia Canyon. (**a**) The entire section. (**b**) Close-up of 70–140 m, where the $\log_{10}$ARM record has been prewhitened to remove the linear trend and 33 m cycle. Lithologic classes: 1, subaqueous gypsum; 2, laminites; 3, heterolithic thin beds; 4, thalassanoides burrowed wackestones; 5, non-skeletal grainstones; 6, bedded requienids and chrondrodonts; 7, skeletal packstones; 8, rudistid bioherms. (Adapted from Foster 2003.)

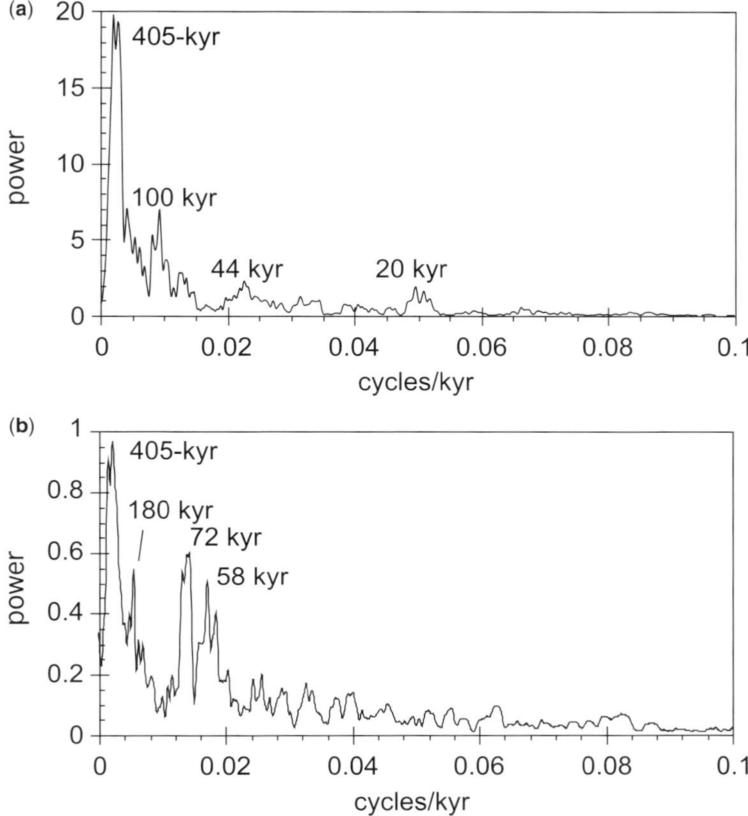

Fig. 14. 2π Power spectra of the 405 kyr calibrated ARM time series (**a**) and facies rank series (**b**) shown in Figure 13a.

where the ARM displays astronomical frequencies (spectral peaks at periods of 100, *c.* 44 and 20 kyr), the facies rank series spectrum has strong spectral peaks at periods of 72 and 58 kyr, neither of which is recognizable as an astronomical period. However, since the two records share a strong 405 kyr cycle, the 72 and 58 kyr periods in the facies spectrum may caused by of 'missed beats' in response to sea-level oscillations originating from high latitude astronomically forced ice sheets.

Cupidito sequence stratigraphy

The fourth-order sequences (Fig. 3) are part of a 'supersequence' encompassing the entire Cupidito section that progresses from a lowstand to a transgressive systems tract (Goldhammer 1999). This designation accounts for deepening trends in the upper Cupido Formation ('Cupidito') and the La Peña drowning. The base of the Cupidito (the lowstand) was originally assigned an age of 112 Ma; the top of the Cupidito was assigned an age of

110 Ma. Later, the sequences were correlated into the Hardenbol *et al.* (1998) global sequence framework with AP2 at the base of the Cupidito and AP4 at the top (Lehmann *et al.* 2000). AP3 is assigned to the sequence boundary within the Cupidito associated with the most intense features of subaerial exposure (Fig. 2). This framework, however, does not account for two additional fourth-order sequences within the Cupidito indicated by Goldhammer (1999) (Fig. 3). These sequence boundaries coincide with ARM maxima of the 33.3 m cycles, except at 32 m (Garcia) and 33 m (Chico) (Fig. 9).

The evidence presented above for a 405 kyr (orbital eccentricity) timing of the Goldhammer fourth-order sequences suggests that the Cupidito section is 1.846 myr in duration (Fig. 12). This duration is close to the original time frame indicated by Goldhammer (1999), and not the longer chronology (3 myr) of the Hardenbol *et al.* (1998) global sequence framework, or the updated chronology (3.58 myr) by Snedden & Liu (2010). The ambiguity imposed by biostratigraphy (La Peña either pre- or post-OAE1a) allows that the top of the Cupidito may be pre-OAE1a, which could indicate a AP2 global sequence boundary with an age of 124.58 Ma. Proceeding downward for 1.846 myr to 126.42 Ma brings us close to the Barr6 sequence boundary age of 126.2 Ma. However, this still leaves us with two extra sequences that are not described in the global sequence framework. To solve this problem we need new chronostratigraphic data. Specifically, carbon isotope stratigraphy, which has a distinctive evolution through the OAE1a interval (Li *et al.* 2008; Malinverno *et al.* 2010), may help determine whether the Cupidito carbonates are pre- or post-OAE1a in age.

Conclusions

ARM of the upper Cupido ('Cupidito') Formation, northeastern Mexico, was measured at two key sections at Garcia and Chico canyons to investigate the origin of the metre-scale carbonate cyclicity. The results uncovered an unexpectedly strong astronomical signal hidden in the ARM stratigraphy. This signal is common to both of the studied sections, which were separated by 25 km, and accumulated in different water depth conditions. ARM proved to be a powerful tool for correlating these carbonate sections at extremely high resolution (within a few thousand years).

The source of the ARM in the carbonates was traced to fine-grained detrital magnetite consistent with eolian dust. Early Cretaceous models of atmospheric circulation indicates that the most probable source of this dust would have been from northeastern Africa; easterly trade winds would have transported the dust across the North Atlantic Basin to the depositional site in Mexico. Astronomical forcing controlled the intensity and/or direction of the winds, and/or aridity on the African continent, modulating the amount of dust arriving at the Cupido Platform.

While the astronomical signal in the Cupidito ARM record is strong, the cyclic carbonate stratigraphy lacks a clear astronomical signal, although it has some attributes associated with 405 kyr orbital eccentricity forcing. This corroborates long-standing suspicions that shallow marine carbonates are imperfect recorders of astronomical forcing; at the same time, the rock magnetism approach offers a solution to this problem. Since the platform recorded sea-level variations that had at least a prominent 405 kyr cycling at Garcia, we can infer that the sea-level oscillations had attributes of insolation forcing that included a strong precessional component.

In sum, in this study the carbonate facies cyclicity is an 'oceanic' proxy, and the eolian dust stored in the carbonates is an 'atmospheric' proxy. Both are far-field proxies: sea-level changes originated from the dynamics of continental ice sheets and other water reservoirs, as well as from local tectonics, and eolian dust flux was controlled by changes in continental aridity and global wind patterns. We anticipate that, in the future, the development of similar multiple proxies in shallow marine carbonate sections such as the Cupido Formation will transform our understanding of the local depositional conditions as well as the processes that controlled deposition. Rock magnetic properties of carbonates in particular provide the additional, very powerful, proxies for such studies.

L. H. was supported by US National Science Foundation grant EAR-0718905. We thank M. Jackson and P. Solheid for analytical assistance and advice during D. L.'s fellowship at the Institute of Rock Magnetism, University of Minnesota. K. K., D. A. and M. E. were supported by US National Science Foundation grant EAR-0230053; D.L. was also supported by a student grant from the Geological Society of America. This manuscript is an extension of D. L.'s PhD dissertation at Lehigh University.

References

ALFONSO-ZWANZINGER, J. 1978. Geologia regional de sistema sedimentario Cupido. *Boletin de la Asociacion Mexicanna de Geologos Petroleros*, **30**, 1–56.

ALTOBI, Y. K. 2007. *Milankovitch orbital forcing control on shallow-water carbonate cyclicity and early dolomitization: insights from the Lower Cretaceous Cupido Platform, NE Mexico*. PhD dissertation, University of Texas at Austin, Austin, TX.

BARRAGAN, R. & MAURRASSE, F. J.-M. R. 2008. Lower Aptian (Lower Cretaceous) ammonites from the La Peña Formation of Nuevo Leon State, northeast Mexico: chronostratigraphic implications. *Revista Mexicana de Ciencias Geologicas*, **25**, 145–157.

BRALOWER, T. J., COBABE, E., CLEMENT, B., SLITER, W. V., OSBURN, C. L. & LONGORIA, J. 1999. The Record of global change in mid-Cretaceous (Barremian-Albian) sections from the Sierra Madre, northeastern Mexico. *Journal of Foraminiferal Research*, **29**, 418–437.

BURGESS, P. M. 2006. The signal and the noise: forward modeling of allocyclic and autocyclic processes influencing peritidal carbonate stacking patterns. *Journal of Sedimentary Research*, **76**, 962–977.

COCCIONI, R., NESCI, O., TRAMONTANA, M., WEZEL, F. C. & MORETTI, E. 1987. Descrizione di un livello–guida 'radiolaritico–bituminoso–ittiolitico' alla base delle Marne a Fucoidi nell'Appennino Umbro–Marchigiano. *Bollettino della Società Geologica Italiana*, **106**, 183–192.

COCCIONI, R., FRANCHI, R., NESCI, O., WEZEL, F. C., BATTISTINI, F. & PALLECCHI, P. 1989. Stratigraphy and mineralogy of the Selli Level (Early Aptian) at the base of the Marne a Fucoidi in the Umbro-Marchean Apennines, Italy. *In*: WIEDMANN, J. (ed.) *Cretaceous of the Western Tethys. Proceedings 3rd International*

Cretaceous Symposium. E. Schweizerbart'sche Verlagsbuchhandlung, Stuttgart, 563–584.

COCCIONI, R., JOVANE, L. *ET AL.* 2012. Umbria-Marche Basin, Central Italy: a reference section for the Aptian-Albian interval at low latitudes. *Scientific Drilling*, **13**, 42–46, doi:10.2204/iodp.sd.13.07.2011.

DEAN, W. E. & ARTHUR, M. A. 1999. Sensitivity of the North Atlantic basin to cyclic climatic forcing during the Early Cretaceous. *Journal of Foraminiferal Research*, **29**, 465–486.

DICKINSON, W. R. & LAWTON, T. F. 2001. Carboniferous to Cretaceous assembly and fragmentation of Mexico. *Geological Society of America Bulletin*, **113**, 1142–1160.

ELRICK, M. 1995. Cyclostratigraphy and sequence stratigraphy of the Lower Cretaceous Cupido Formation, northeastern Mexico. *Annual Meeting Abstracts – American Association of Petroleum Geologists and Society of Economic Paleontologists and Mineralogists*, **4**, 26.

FISCHER, A. G. 1964. The Lofer cyclothems of the Alpine Triassic. *In*: MERRIAM, D. F. (ed.) *Symposium on Cyclic Sedimentation*, Lawrence, KS. Kansas Geological Survey Bulletin, **169**, 107–149.

FOSTER, T. R. 2003. *The evolution of a Lower Cretaceous carbonate platform within a divergent margin setting: The Cupido Formation, northeastern Mexico.* MSc thesis, The University of Texas at Austin, Austin, TX.

GINSBURG, R. N. 1971. Landward movement of carbonate mud: a new model for regressive cycles in carbonates (abstract). *American Association of Petroleum Geologists, Annual Conference*, Program Abstracts, **55**, 340.

GOLDHAMMER, R. K. 1999. Mesozoic sequence stratigraphy and paleogeographic evolution of northeast Mexico. *In*: BARTOLINI, C., WILSON, J. L. & LAWTON, T. F. (eds) *Mesozoic Sedimentary and Tectonic History of North-Central Mexico.* Geological Society of America, Boulder, CO, Special Papers, **340**, 1–58.

GOLDHAMMER, R. K., DUNN, P. A. & HARDIE, L. A. 1990. Depositional cycles, composite sea-level changes, cyclic stacking patterns, and the hierarchy of stratigraphic forcing: examples from Alpine Triassic platform carbonates. *Geological Society of America Bulletin*, **102**, 535–562.

GOLDHAMMER, R. K., LEHMANN, P. J., TODD, R. G., WILSON, J. L., WARD, W. C. & JOHNSON, C. R. 1991. *Sequence Stratigraphy and Cyclostratigraphy of the Mesozoic of the Sierra Madre Oriental, Northeast Mexico, a Field Guidebook.* Gulf Coast Section, Society of Economic Paleontologists and Mineralogists, Tulsa, OK.

GOUDIE, A. S. & MIDDLETON, N. J. 2006. *Desert Dust in the Global System.* Springer, Berlin.

HARDENBOL, J., THIERRY, J., FARLEY, M. B., JACQUIN, T., DE GRACIANSKY, P.-C. & VAIL, P. R. 1998. Chart 4: Mesozoic and Cenozoic sequence chronostratigraphic framework of European basins. *In*: DE GRACIANSKY, P.-C., HARDENBOL, J., JACQUIN, T. & VAIL, P. R. (eds) *Mesozoic and Cenozoic Sequence Stratigraphy of European Basins.* Society for Sedimentary Geology, Tulsa, OK, Special Publications, **60**, 763–781.

HARDIE, L. A., BOSELLINI, A. & GOLDHAMMER, R. K. 1986. Repeated subaerial exposure of subtidal carbonate platforms, Triassic, northern Italy: evidence for high frequency sea level oscillations on a 10^4 year scale. *Paleoceanography*, **1**, 447–457.

HASEGAWA, H., TADA, R. *ET AL.* 2012. Drastic shrinking of the Hadley circulation during the mid-Cretaceous Supergreenhouse. *Climates of the Past*, **8**, 1323–1337.

HINNOV, L. A. 2000. New perspectives on orbitally forced stratigraphy. *Annual Review of Earth and Planetary Sciences*, **28**, 419–475.

HUANG, C., HINNOV, L. A., FISCHER, A. G., GRIPPO, A. & HERBERT, T. 2010. Astronomical tuning of the Aptian stage from Italian reference sections. *Geology*, **238**, 899–903.

JACQUIN, T., RUSCIADELLI, G., AMEDRO, F., DE GRACIANSKY, P.-C. & MAGNIEZ-JANNIN, F. 1998. The North Atlantic cycle; an overview of 2nd-order transgressive/regressive facies cycles in the Lower Cretaceous of Western Europe. *In*: DE GRACIANSKY, P.-C., HARDENBOL, J., JACQUIN, T. & VAIL, P. R. (eds) *Mesozoic and Cenozoic Sequence Stratigraphy of European Basins.* Society for Sedimentary Geology, Tulsa, OK, Special Publications, **60**, 397–409.

KODAMA, K. P. 2012. *Paleomagnetism of Sediments and Sedimentary Rocks: Process and Interpretation.* Wiley, Chichester.

KRUIVER, P. P., KOK, Y. S., DEKKERS, M. J., LANGEREIS, C. G. & LAJ, C. 1999. A pseudo-Thellier relative paleointensity record, and rock magnetic and geochemical parameters in relation to climate during the last 276 kyr in the Azores region. *Geophysical Journal International*, **136**, 757–770.

LATTA, D. K. 2005. *Structural, Lithotectonic, and Rock Magnetic Studies of Decollement Folding, Coahuila Marginal Folded Province, NE Mexico.* PhD dissertation, Lehigh University, Bethlehem, PA.

LATTA, D. K. & ANASTASIO, D. J. 2007. Multiple scales of mechanical stratification and decollement fold kinematics, Sierra Madre Oriental foreland, northeast Mexico. *Journal of Structural Geology*, **29**, 1241–1255.

LATTA, D. K., ANASTASIO, D. J., HINNOV, L. A., ELRICK, M. & KODAMA, K. P. 2006. A magnetic record of Milankovitch rhythms in lithologically non-cyclic marine carbonates. *Geology*, **34**, 29–32.

LEAN, C. M. B. & McCAVE, I. N. 1998. Glacial to interglacial mineral magnetic and palaeoceanographic changes at Chatham Rise, SW Pacific Ocean. *Earth and Planetary Science Letters*, **163**, 247–260.

LEHMANN, C., OSLEGER, D. A. & MONTANEZ, I. P. 1998. Controls on cyclostratigraphy of Lower Cretaceous carbonates and evaporites, Cupido and Coahuila platforms, northeastern Mexico. *Journal of Sedimentary Research*, **68**, 1109–1130.

LEHMANN, C., OSLEGER, D. A., MONTAÑEZ, I. P., SLITER, W., ARNAUD-VANNEAU, A. & BANNER, J. 1999. Evolution of Cupido and Coahuila carbonate platforms, Early Cretaceous, northeastern Mexico. *Geological Society of America Bulletin*, **111**, 1010–1029.

LEHMANN, C., OSLEGER, D. A. & MONTANEZ, I. P. 2000. Sequence stratigraphy of Lower Cretaceous (Barremian-Albian) carbonate platforms of northeastern Mexico: regional and global correlations. *Journal of Sedimentary Research*, **70**, 373–391.

LEHRMANN, D. J. & GOLDHAMMER, R. K. 1999. Secular variation in parasequence and facies stacking patterns of platform carbonates: a guide to application of stacking-patterns analysis in strata of diverse ages and settings. *In*: HARRIS, P. M., SALLER, A. H. & SIMO, J. A. (eds) *Advances in Carbonate Sequence Stratigraphy; Application to Reservoirs, Outcrops and Models.* Society for Sedimentary Geology, Tulsa, OK, Special Publications, **63**, 187–225.

LI, Y. X., BRALOWER, T. J. *ET AL.* 2008. Toward an orbital chronology for the early Aptian Oceanic Anoxic Event (OAE1a, 120 Ma). *Earth and Planetary Science Letters*, **271**, 88–100.

LOWRIE, W. 1990. Identification of ferromagnetic minerals in a rock by coercivity and unblocking temperature properties. *Geophysical Research Letters*, **17**, 159–162.

MALINVERNO, A., ERBA, E. & HERBERT, T. D. 2010. Orbital tuning as an inverse problem: chronology of the early Aptian oceanic anoxic event 1a (Selli Level) in the Cismon APTICORE. *Paleoceanography*, **25**, PA2203, doi: 10.1029/2009PA001769.

MUXWORTHY, A. R. 1999. Low-temperature susceptibility and hysteresis of magnetite. *Earth and Planetary Science Letters*, **169**, 51–58.

OGG, J. & HINNOV, L. A. 2012. The Cretaceous Period. *In*: GRADSTEIN, F., OGG, J., OGG, G. & SMITH, D. (eds) *A Geologic Time Scale 2012.* Elsevier, Oxford, 793–853.

PAILLARD, D., LABEYRIE, L. & YIOU, P. 1996. Macintosh program performs time-series analysis. *Eos Transactions, American Geophysical Union*, **77**, 379.

PINDELL, J. L. 1993. Regional synopsis of Gulf of Mexico and Caribbean evolution. *In*: PINDELL, J. L. & PERKINS, B. F. (eds) *Mesozoic and Early Cenozoic Development of the Gulf of Mexico and Caribbean Region, A Context for Hydrocarbon Exploration: Gulf Coast SEPM Foundation 13th Annual Research Conference Proceedings*, 251–274.

PYE, K. 1987. *Aeolian Dust and Dust Deposits.* Academic Press, London.

RÖHL, U. & OGG, J. G. 1998. *Aptian–Albian Eustatic Sea-Levels.* International Association of Sedimentologists, Wiley-Blackwell, London, Special Publications, **25**, 95–136.

SCHLAGER, W. 2005. *Carbonate Sedimentology and Sequence Stratigraphy.* Society of Economic Paleontologists and Mineralogists, Tulsa, OK. Concepts in Sedimentology and Paleontology, **8**, 200.

SCHLAGER, W. 2010. Ordered heiarchy versus scale invariance in sequence stratigraphy. *International Journal of Earth Sciences*, **99**(suppl 1), S139–S151.

SCHLANGER, S. O. & JENKYNS, H. C. 1976. Cretaceous oceanic anoxic events: causes and consequences. *Geologie en Mijnbouw*, **55**, 179–184.

SCHWARZACHER, W. 2000. Repetitions and cycles in stratigraphy. *Earth-Science Reviews*, **50**, 51–75.

SCOTESE, C. R., GAHAGAN, L. M. & LARSON, R. L. 1989. Plate tectonic reconstructions of the Cretaceous and Cenozoic ocean basins. *In*: SCOTESE, C. R. & SAGER, W. W. (eds) *Mesozoic and Cenozoic Plate Reconstructions.* Elsevier, Amsterdam, 27–48.

SNEDDEN, J. W. & LIU, C. 2010. Compilation of Phanerozoic sea-level change, coastal onlaps, and recommended sequence designations. *AAPG Search and Discovery*, article no. 40594, http://www.searchand discovery.com/documents/2010/40594snedden/ ndx_snedden.pdf

TANER, T. H. 2003. *Attributes revisited.* Technical publication/ Rock Solid Images, Inc., Houston, TX, http://rocksolidimages.com/pdf/attrib_revisited.htm

TAUXE, L., MULLENDER, T. A. T. & PICK, T. 1996. Potbellies, wasp-waists, and superparamagnetism in magnetic hysteresis. *Journal of Geophysical Research*, **101**, 571–583.

TAUXE, L., BERTRAM, H. & SEBERINO, C. 2002. Physical interpretation of hysteresis loops: micromagnetic modeling of fine particle magnetite. *Geochemistry, Geophysics, Geosystems*, **3**, doi: 10.1029/2001GC 000280.

THOMSON, D. 1982. Spectrum estimation and harmonic analysis. *Proceedings of the IEEE*, **70**, 1055–1096.

TINKER, S. W. 1982. *Lithostratigraphy and biostratigraphy of the Aptian La Peña Formation, northeast Mexico and southeast Texas, and the depositional setting of the Aptian Pearsall-La Peña Formations, Texas subsurface and northeast Mexico: Why is there not another Fairway Field?* MSc thesis. University of Michigan, Ann Arbor, MI.

WALL, J. R., MURRAY, G. E. & DIAZ, T. 1961. Geology of the Monterrey area. *Nuevo Leon, Mexico: Transactions–Gulf Coast Association of Geological Societies*, **11**, 57–71.

WEEDON, G. P. 2003. *Time-Series Analysis and Cyclostratigraphy: Examining Stratigraphic Records of Environmental Cyclicity.* Cambridge University Press, Cambridge.

WILSON, J. L. 1975. *Carbonate Facies in Geologic Time.* Springer, New York.

WILSON, J. L. 1999. Controls on the wandering path of the Cupido Reef trend in northeastern Mexico. *In*: BARTOLINI, C., WILSON, J. L. & LAWTON, T. F. (eds) *Mesozoic Sedimentary and Tectonic History of North-Central Mexico.* Geological Society of America, Boulder, CO, Special Papers, **340**, 135–143.

Visual identification and quantification of Milankovitch climate cycles in outcrop: an example from the Upper Ordovician Kope Formation, Northern Kentucky

BROOKS B. ELLWOOD[1]*, CARLTON E. BRETT[2], JONATHAN H. TOMKIN[3]
& WILLIAM D. MACDONALD[4]

[1]*Department of Geology and Geophysics, Louisiana State University, Baton Rouge, LA 70803, USA*

[2]*Department of Geology, University of Cincinnati, 500 Geology/Physics Building, Cincinnati, OH 45221, USA*

[3]*School of Earth, Society, and Environment, University of Illinois, 428 Natural History Building, 1301 W. Green Street, Urbana, IL 61801, USA*

[4]*Department of Geological Sciences, State University of New York, Binghamton, NY 13902, USA*

**Corresponding author (e-mail: ellwood@lsu.edu)*

Abstract: Applying time-series analyses using Fourier transform and multi-taper methods to low-field, mass-specific magnetic susceptibility (χ) measurements on marine samples from well-studied shale and limestone outcrops of the Upper Ordovician (Edenian Stage; Upper Katian) Kope Formation, northern Kentucky, corroborates direct visual identification in outcrops of Milankovitch eccentricity (*c.* 405 and 100 ka), obliquity and precessional climate cycles. Because individual outcrops were too short and deposition too chaotic to yield significant time-series results, it was necessary to build a *c.* 50 m thick composite sequence from three well-correlated outcrops to quantify the cyclicity. Time-series analysis was then performed using χ measured for 1004 closely spaced samples covering the section. Milankovitch bands are recorded in the time-series data from the composite. We tested this result by comparison of these bands to cyclic packages in outcrop, which correspond to thicknesses represented in the time-series datasets. This is particularly well defined for the eccentricity and obliquity cycles, with precessional bands being evident but as less well-defined packages of beds.

Cycles are often visually apparent as repeating patterns of facies in sedimentary outcrops. We often ask ourselves what these cycles represent. Is it climate, eustasy, a function of diagenesis or depositional setting, or is something else controlling what we are seeing? The interpretation of these lithic patterns in terms of climate-driven cycles is supported by isotopic studies of Cenozoic marine strata, using isotopic and lithic cycles as proxies for climatic cycles, presumably of global significance (e.g. Imbrie *et al.* 1984; Dinarès-Turell *et al.* 2007; Hilgen 2010). Tests of these hypotheses have shown that, in general, the basic climate-cycle assumptions are correct, and because of the short-term nature of many climate cycles, these datasets may provide much higher resolution timescales than are available using biostratigraphic information alone.

Bulk (initial) low-field, mass-specific magnetic susceptibility (χ) provides another proxy, which has proven to be very useful for correlating among marine (e.g. Crick *et al.* 1997; Hansen *et al.* 1999; Ellwood *et al.* 2000, 2001) and non-marine (e.g. Ellwood *et al.* 2004) sequences. Magnetic susceptibility can often be used as a climate proxy because regional and global processes that drive erosion, such as climate and eustasy, bring the detrital/aeolian components that are responsible for the χ signature into the marine and other environments where this record is preserved (Ellwood *et al.* 2000). These detrital/aeolian components dominate the χ observed in most marine sedimentary rocks. Examples of Milankovitch band cyclicity recorded in the χ signature of sedimentary rocks of diverse ages have been reported (e.g. Bloemendal & deMenocal 1989; deMenocal *et al.* 1991; Weedon *et al.* 1999; Shackleton *et al.* 1999*a*; Ellwood *et al.* 2000, 2007*b*, 2008*a*, *b*; Boulila *et al.* 2008), and it has been demonstrated that the cyclostratigraphy recorded in these sequences can be used for astronomical calibration of geological timescales (Mead *et al.* 1986; Weedon *et al.* 1997, 1999; Shackleton *et al.* 1999*b*; Crick *et al.* 2001).

From: Jovane, L., Herrero-Bervera, E., Hinnov, L. A. & Housen, B. A. (eds) 2013. *Magnetic Methods and the Timing of Geological Processes.* Geological Society, London, Special Publications, **373**, 341–353.
First published online August 14, 2012, http://dx.doi.org/10.1144/SP373.2 © The Geological Society of London 2013.
Publishing disclaimer: www.geolsoc.org.uk/pub_ethics

Time-series datasets have not been well documented for most of the Phanerozoic, however, because stratigraphic sequences are imperfect recorders of time owing to erosion, non-deposition or alteration modifying the primary climate-controlled signature in these materials. Short-term (high-frequency) Milankovitch bands (Earth's obliquity and precession) are not as well documented in older rocks as they are in younger sequences. Longer-term climate cycles have been found to be more readily identified in older rock sequences; in particular, the *c.* 405 ka eccentricity cycle has a robust, long-term palaeoclimatic stratigraphic signal that has been recorded in many ancient sediments (Shackleton *et al.* 1999*a*; Lasker *et al.* 2004). In fact, this cycle has been observed for Triassic lacustrine (Olsen & Kent 2000), Middle Carboniferous marine (Ellwood *et al.* 2007*b*), and Jurassic marine sediments (Boulila *et al.* 2008). Given that longer-term climate cycles are preserved in the sedimentary record, any robust method that can track these changes should provide useful proxies for climate.

One problem with applying climate proxies to the sedimentary record is that it requires reasonable biostratigraphic control to develop a chronostratigraphic framework where these data can be directly correlated with high precision among sections, because biostratigraphic uncertainties or disconformities may exist within sections (Ellwood *et al.* 2006, 2007*a*). The variable χ is useful in this regard because samples are relatively easy and fast to collect, allowing large numbers of samples to be obtained, thus producing a robust dataset.

The purpose of this project has been to use χ in marine sediments that clearly show cyclicity in outcrop. We chose the Upper Ordovician Kope Formation exposures in northern Kentucky (Fig. 1) because the sequence, cyclicity and biostratigraphy are well documented (Holland *et al.* 1997, 2001*a*, *b*; Brett & Algeo 2001*c*; Brett *et al.* 2003, 2008). Our initial attempts to correlate among outcrop sections were successful (Ellwood *et al.* 2007*a*). Local erosion caused some variability in sediment accumulation rate (SAR), however, so this work on single outcrops did not allow us to quantify the cyclicity, using time-series analysis, without first adjusting SARs in the sequence, something that introduces error in this type of analysis. To avoid this problem, we constructed a *c.* 50 m-thick composite section from well-documented and correlated outcrops, and applied time-series analysis techniques to these new data. That then allowed us to extract Milankovitch cyclic bands from the dataset and apply these results to individual exposed sequences to identify visual manifestations of these cycles in outcrop. Those data are reported here.

Geological setting and previous work

During the Late Ordovician the Ohio–Kentucky border region lay in the southern subtropics,

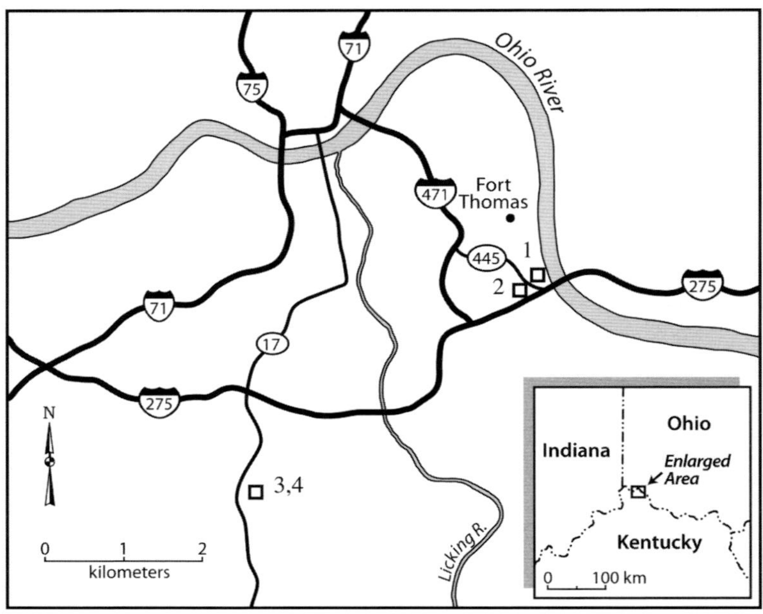

Fig. 1. Location of the studied sections (1–3; note that two sections were sampled at location 3).

approximately 25°S. Cyclic, thinly bedded limestone and mudstone bed accumulated in relatively shallow marine environments, below the fair weather wave base but generally above the storm wave base and in the dysphotic to shallow euphotic zone based on algal and cyanobacterial microendolith borings in shells (Vogel & Brett 2009). Tropical storms had a major impact on deposition, reworking and winnowing detrital sediments being carried to the site from eroded Taconic orogen source rocks to the east and NE (e.g. Ettensohn 1992; Jennette & Pryor 1993; Holland *et al.* 2001*a*). As a result, storm deposits are evident in the Cincinnatian, and metre-scale limestone mudstone cycles have previously been attributed to storm winnowing (Jennette & Pryor 1993; Holland *et al.* 2001*b*). According to this model, shell-rich limestones reflect periodic strong winnowing of the seafloor, because of either lowering of sea-level or increased frequency/intensity of storm activity. However, recently certain assumptions of this model have been challenged by new field and subsurface data and an alternative mechanism, involving periodic siliciclastic sediment-starvation proposed to explain the accumulation of *in situ* skeletal carbonates (Brett *et al.* 2003, 2008; Dattilo *et al.* 2008). Such starvation might result from sequestering of terrigenous sediments in coastal areas flooded by sea-level rise or possibly shifts to more arid conditions in source terranes.

The Kope Formation in the study area is *c.* 70–75 m thick (Brett & Algeo 2001*b*) and contains much of the Edenian Stage of the Upper Ordovician, Cincinnatian Series (Weiss & Sweet 1964). The Kope is composed primarily of cyclic alternations of shales/mudstones and bundles of thin (3–10 cm) beds of skeletal (brachiopod, bryozoan and crinoidal debris) packstones and grainstones, 0.5–10 m (Jennette & Pryor 1993; Holland *et al.* 1997). Lithostratigraphic and biostratigraphic studies of the Kope show detailed correlations among closely spaced sections (Tobin & Pryor 1981; Tobin 1982; Jennette & Pryor 1993; Holland *et al.* 1997; Brett & Algeo 2001*a*, *b*, *c*). Samples were collected for this study from the Economy Member, Brent Submember, Southgate Member, Pioneer Valley, Snag Creek, Alexandria and Grand View Submembers, within exposures of the Kope Formation (summarized by Brett & Algeo 2001*b*; Ellwood *et al.* 2007*a*).

Bulk, low-field, mass specific magnetic susceptibility (χ)

The bulk, low-field, mass specific magnetic susceptibility on samples is a measure of the induced magnetization of all materials when placed in a magnetic field (for details see Ellwood *et al.*, 2012). The magnetic susceptibility is the basis for developing magnetostratigraphy susceptibility (MSS), defined by Salvador (1994), and derives from magnetostratigraphy, a term applied to all types of magnetic measurements that are made in a stratigraphic context. For marine sedimentary materials, it is has been shown that χ is dominated by the detrital/aeolian component within samples (e.g. Bloemendal & deMenocal 1989; Ellwood *et al.* 2000). In the very low alternating, inducing magnetic fields that are generally applied, χ is largely a function of the mineralogy, concentration and grain morphology of minerals in measured samples.

Applications of bulk (initial) low-field mass-specific χ

Sample variability for χ results primarily from changes in the amount of iron, clay and ferromagnesian constituents. This variability in sedimentary successions may be primary, produced during deposition, or secondary, resulting from alteration of the deposited sequence. The main controlling factors on χ include rates of weathering, erosion, SAR and biological productivity. These factors are tied to climate, tectonic and volcanic processes, availability of nutrients and local (relative) or global sea-level changes. Secondary processes include pedogenesis, diagenesis and within-sediment redox effects driven by sulfate-reducing bacteria and other organisms.

Studies of marine sedimentary sequences

Marine sediments are composed of terrigenous, detrital and aeolian components, and a biogenic component of the carbonate and/or siliceous tests of marine organisms (e.g. Lisitzin 1972). There is also a component of sediment derived from marine weathering processes or chemical precipitation of silica or carbonate, but this is generally minor. The χ signature mainly represents contributions from the terrigenous and biogenic components. Because the dominant biogenic component is only weakly diamagnetic carbonate or silica, in general, the terrigenous component dominates χ. In cases where biogenic processes generate siderite or pyrite, both paramagnetic, this results in the iron being preserved in the sample and such changes usually do not significantly change the paramagnetic component's contribution to χ (Ellwood *et al.* 1988). Therefore, the processes controlling the influx into the marine environment of terrigenous components will usually account for χ variations observed. Thus, climate cycles that cause the erosion and transport of terrigenous material typically result in a cyclic signal for χ recorded in the deposited sediments. There are many factors that can modify this signature;

although the expected variations may be reduced in magnitude, the character of the cycles generally remains and can be extracted (i.e. Weedon *et al.* 1999). Superimposed on these cyclic variations are contributions from unique events, such as volcanic eruptions or meteorite impacts, which may be of local, regional or global significance.

Results from lithified marine sedimentary rock sequences

During diagenesis the χ of marine sediments decreases. This occurs because ferrimagnetic grains that dominate in unlithified sediments are generally converted to paramagnetic phases during burial and diagenesis, mainly by the action of sulfate-reducing bacteria. The result is that, within less than a metre of the sediment–water interface, new iron minerals, commonly pyrite and siderite, are formed and magnetite is lost (Karlin & Levi 1983, 1985; Ellwood *et al.* 1988). As a result, χ magnitudes are rapidly reduced from values observed in many unconsolidated near-surface sediments ($2–10 \times 10^{-7} \, \mathrm{m}^3 \, \mathrm{kg}^{-1}$; Ellwood *et al.* 2006) to χ values in lithified marine rocks, typically ranging from *c.* 1×10^{-9} to $2 \times 10^{-7} \, \mathrm{m}^3 \, \mathrm{kg}^{-1}$ (Ellwood *et al.* 2000). In general, however, this process does not remove iron from the system, so that total iron content is preserved. These new phases then contribute, along with the paramagnetic detrital components present and the remaining or authigenic ferrimagnetic constituents, to χ in lithified sediments. A test of χ variability in single limestone beds has demonstrated the non-varying nature of χ over distances up to 25 km (Ellwood *et al.* 1999). Identification of well-documented climate cyclicity in χ values indicates that primary climate cycles are preserved in these datasets (i.e. Bloemendal & deMenocal 1989; Weedon *et al.* 1999; Jovane *et al.* 2006; Ellwood *et al.* 2008*a, b*). Diagenesis also modifies some of the paramagnetic terrigenous constituents of marine sediments. In many cases clays dominate the detrital/aeolian component, and these are generally not destroyed but may be altered. The iron is still conserved in the secondary clays, as is the paramagnetic behaviour of these materials. Other common detrital components observed in marine rocks are the ferromagnesian minerals, including biotite, tourmaline and other compounds (see Ellwood *et al.* 2000 for a discussion).

Ultimately, following diagenesis, the χ observed for most marine sedimentary rocks can be used as a proxy for physical processes responsible for delivering the detrital/aeolian component of these sediments into marine sedimentary basins (e.g. Hladil *et al.* 2006). As observed for unlithified sediments, these processes in lithified sediments are both regional and global and include climate, volcanism, tectonic processes, relative sea-level changes and eustasy.

Climate cycles and other χ variations

The χ results, for almost all studies of marine sedimentary rocks, show many levels of cycles. Some of these cycles are very long-term and are interpreted as resulting from transgressive–regressive variations (T–R cycles) owing to sea-level rise and fall associated with eustasy and erosion along continental margins. Other cycles are much shorter and are interpreted to result from climatically driven processes, where cyclic detrital/aeolian erosional pulses are responsible for χ variations observed (Ellwood *et al.* 2001). Work by da Silva and Boulvain (2005) for Upper Devonian rocks in Belgium has shown a high correlation between fourth-order T–R cycles and χ variations. In addition to these cycles, there are distinct χ peaks that are correlated to maximum flooding surfaces, and to sudden influxes of detrital material into sedimentary basins by turbidity currents or other sediment suspensions.

Large-magnitude, rapid shifts in χ (represented by multiple data points) are interpreted to result from geological processes that reflect fundamental changes in the sediments being delivered to marine basins. For example, the sharp change from low χ in the Permian to high χ in the Triassic (Hansen *et al.* 1999) is argued to result from a sudden dramatic influx into ocean basins of detrital material that helped to cause the mass extinctions in the marine fauna at this time (Thoa *et al.* 2004). A similar pulse in χ is associated with the K–T boundary and is attributed to erosion following the meteorite impact at that time (Ellwood *et al.* 2003).

Methods

Field sampling

Our objective was to perform time-series analysis on a composite section of samples collected at close intervals from Kope exposures in the northern Kentucky (Fig. 1). The composite section constructed from these data represent a total of 1004 samples covering 50.1 m of section (Fig. 2). The sections making up the composite have been chosen for their excellent exposures that require minimal cleaning, thus allowing sampling at very close intervals. Stratigraphic position was measured and samples were collected at 5 cm continuous intervals within sections; samples are tied directly to the biostratigraphic sampling, published stratigraphy and bed notation. All lithologies encountered were collected, and include thinly interbedded limestone and carbonate-rich mudrock (marl) samples.

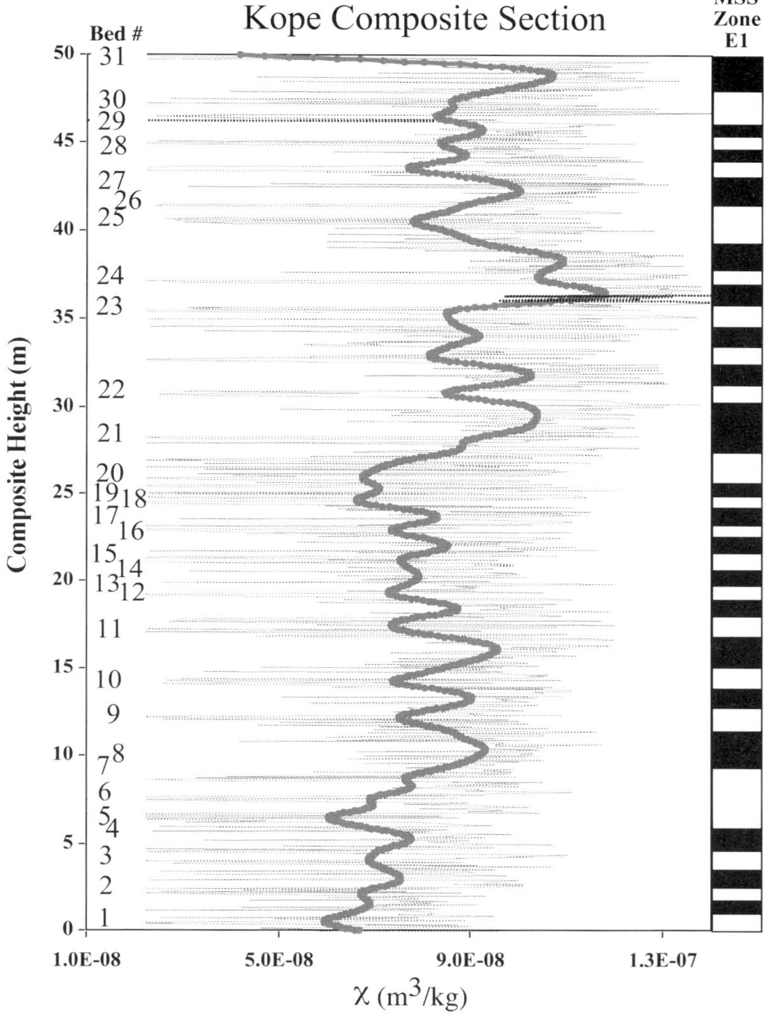

Fig. 2. Composite section (*c.* 50 m) from magnetic susceptibility (m³ kg⁻¹) data measured for Kope Formation samples collected from localities in Figure 1. Dotted lines connect initial χ data points for each sample; solid curve is smoothed χ data using splines to hold stratigraphic position (see text), and the MSS zone E1 bar-log was constructed to represent the χ smoothed trend at *c.* 100 ka. Labelled examples of outcrop expression of Milankovitch bands are indicated by arrows. Note that there is no lithology log because lithologies represented are thinly bedded limestone and marl beds, where individual beds at the scale of *c.* 50 m are thinner than the line drawn to represent them.

Samples were collected with a hammer (lithified materials such as limestone) or knife (unlithified materials such as friable marl) and unoriented. Approximately 10–20 g of material were placed in a small, pre-labelled plastic Zip-Lock (tr) type bag. These bags were then placed in a larger, labelled cloth bag and returned to the Louisiana State University (LSU) Rock Magnetism Laboratory for χ measurement.

Magnetic susceptibility measurements

All measurements reported in this paper were performed using the susceptibility bridge at LSU (constructed by Marshal Williams at the University of Georgia) that is calibrated relative to mass using standard salts reported by Swartzendruber (1992) and CRC tables. The low-field χ bridge at LSU can measure diamagnetic samples at least as low as

-4×10^{-9} m^3 kg^{-1}. This is illustrated by two relatively pure calcite samples collected from a core drilled from a standing speleothem in Carlsbad Caverns National Park, with values of $-3.37 \times 10^{-9} \pm 7.64 \times 10^{-11}$ and $-3.46 \times 10^{-9} \pm 8.69 \times 10^{-11}$ m^3 kg^{-1}. We report χ in terms of sample mass because it is much easier and faster to measure with high precision than volume, and it is now the standard for χ measurement. Bulk (initial) measurements of χ are often performed without consideration of the anisotropy of samples, and for most samples the bulk χ error that is due to this anisotropy is small. The anisotropy effect is further reduced when samples are crushed before the χ is measured.

To visually simplify and represent the χ cycles at varying frequencies, the bar-log format, similar to that previously established for polarity data presentations, is used here. Bar-logs are constructed from smoothed χ data and are accompanied by primary χ datasets (Fig. 2). As in remanent polarity studies, a point half-way between the highest and lowest points along a χ data trend toward higher or lower values is chosen as the boundary between MSS zones. This approach is somewhat subjective, but given that the primary data are presented in each case, the reader can judge if the boundary point chosen is reasonable. Here, primary χ data (dashed lines in Fig. 2) are smoothed using spline functions (solid curves in Fig. 2), to hold data points in stratigraphic position (splines were calculated using the JMP statistical software package from the SAS Institute Inc). Splines vary depending on the number of samples in each dataset being smoothed, and therefore using splines is a matter of choice, as are other smoothing techniques. Splines applied here were adjusted to produce an MSS zonation that conforms to a *c.* 100 ka eccentricity smoothing frequency (discussed below). Finer splines represent shorter cycles. The following bar-log plotting convention is used; if the χ cyclic trends increase or decrease and are represented by two or more data points, then this change is assumed to be significant and the highs and lows associated with these cycles are differentiated by filled (high χ values) or open (low χ values) bar-logs (shown in Fig. 2). This

method is best employed when high-resolution datasets are being analysed (large numbers of closely spaced samples) and helps to resolve variations associated with anomalous samples, such as variations owing to weathering effects, secondary alteration and metamorphism (Ellwood *et al.* 2000). Longer-term trends observed in such datasets can be due to factors such as plate-driven eustasy, as opposed to shorter-term climate cycles, or event sequences such as impacts and eruptions (Ellwood *et al.* 2003).

Results

Magnetic susceptibility for the Kope Formation samples seems to be lithologically dependent. That is, limestone generally has low χ values while mudstone exhibits higher χ values. This effect appears to be controlled exclusively by the amount of clay (identified as illite; Ellwood *et al.* 2007*a*) contained in the sample measured. Tobin and Pryor (1981), Brett and Algeo (2001*c*) and Brett *et al.* (2003, 2008) have argued that the cycles in the Kope were initiated by skeletal packstone and grainstone bundles that represent relative starvation owing to sea-level rise, with possible winnowing of clay during this phase of sediment accumulation. These beds are overlain by mudstone intervals that record increasingly rapid, episodic deposition of mud and minor silt, in part as storm deposits. Magnetic susceptibility results are consistent with this hypothesis, indicating that clay deposition, responsible for the paramagnetic signature, exhibits a series of T–R cycles. We infer that the cycles observed in the Kope datasets are the result of climate cyclicity driving differential erosion and deposition.

To test this, we performed time-series analyses using the primary χ data (independent of smoothing) for the *c.* 50 m-thick composite section (Fig. 2; for details on the time series methods used see Ellwood *et al.* 2012). The results are shown in Table 1 and Figure 3. Given that the P2 peak (*c.* 19.3 ka, a value derived from a Late Ordovician age of *c.* 450 Ma (Ogg *et al.* 2008) using the data from Berger *et al.* 1992) in Figure 3 exhibits a reasonably high

Table 1. *Kope formation time series results*

Cycle	Duration (ka)	Frequency (cycles m^{-1})	Confidence level	Harmonic number
E2	*c.* 405	*c.* 0.17	>90%	7
E1	*c.* 100	*c.* 0.40	Low	19
O2	*c.* 37.1	*c.* 1.10	>90%	54
O1	*c.* 30.5	*c.* 1.32	*c.* 99%	67
P2	*c.* 19.3	*c.* 2.10	*c.* 90%	105
P1	*c.* 16.4	*c.* 2.43	None	122

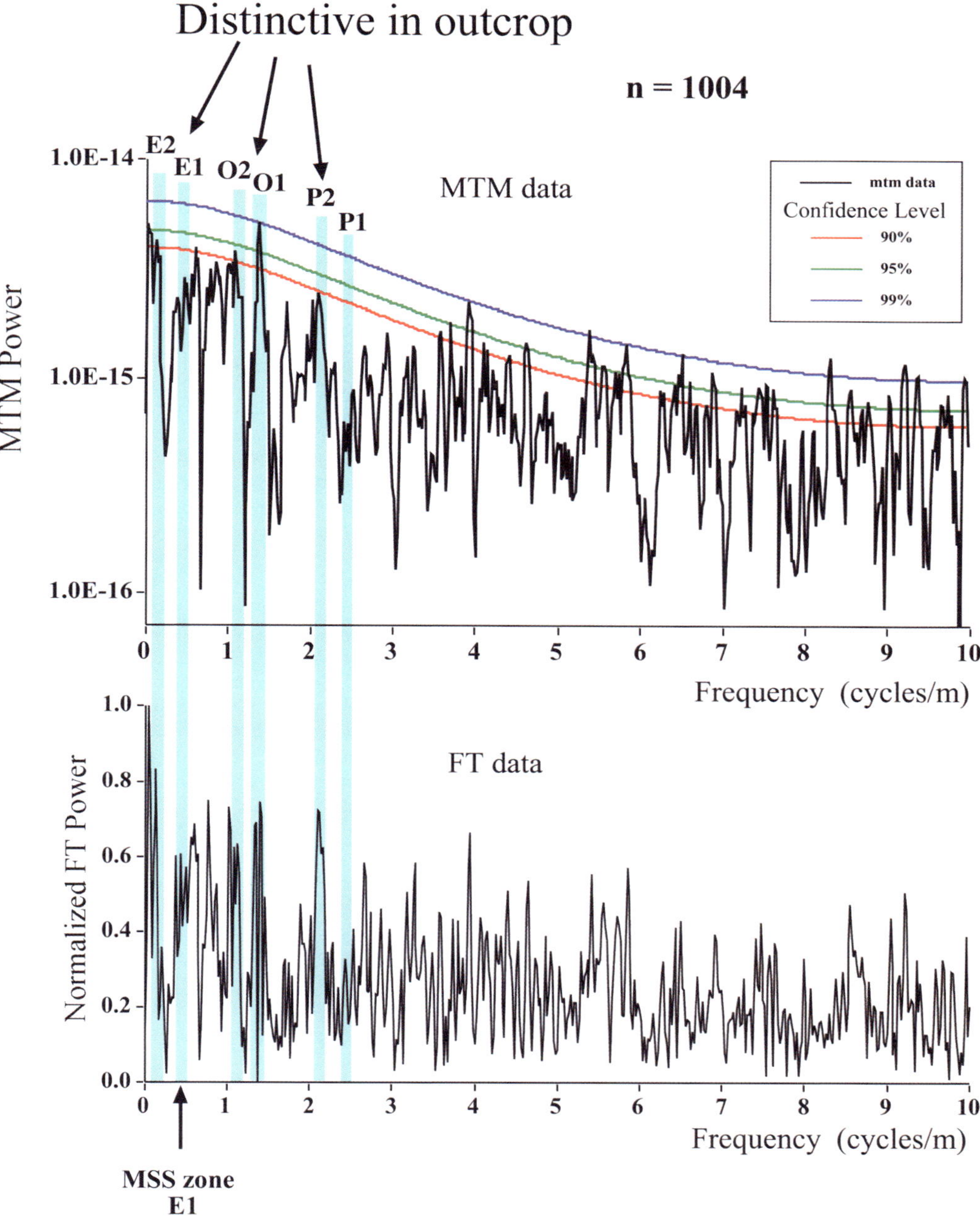

Fig. 3. Fourier transform (FT) and Multi-Taper (MTM) results for the primary χ data (before smoothing), reported in cycles m^{-1}. For this analysis, 1004 χ data points were used and six Milankovitch climate bands are illustrated, where E2 and E1 represent eccentricity at *c.* 405 and *c.* 100 ka, respectively, O1 and O2 represent obliquity at *c.* 30.5 and *c.* 37.1, respectively, and P1 and P2 represent precession at *c.* 16.4 and 19.3 ka, respectively. The obliquity and precessional bands are corrected for an age of *c.* 445 Ma (Upper Ordovician from Ogg *et al.* 2008) using corrections by Berger *et al.* (1992). Also given is the position of MSS zone E1 (from Fig. 2). Eccentricity, obliquity and precessional cycles are visually apparent in outcrops (Fig. 4).

multi-taper methods confidence level and Fourier transform (FT) power, and it represents a frequency distribution through the dataset of *c.* 2.10 cycles m^{-1} (harmonic number *c.* 105), then we can assign duration for the whole composite section of *c.* 2.027 Ma. In general, regions with high spectral power coincide with the presumed Milankovitch bands. The statistically significant spectral values at higher than orbital frequencies could be the result of the large sample size producing false positives, or given the slightly unusual structure of the FT, could indicate that a red noise null-hypothesis is not appropriate (Vaughan *et al.* 2011). Nevertheless, the spectral analysis is in general agreement with the other techniques. Note the correspondence between spectral power, χ zonation and relatively high-amplitude peaks in Figure 3 identified as E2 (*c.* 405 ka at *c.* 0.17 cycles m^{-1}; harmonic 7), E1 (*c.* 100 ka at *c.* 0.40 cycles m^{-1}; harmonic 19), O2 (*c.* 37.1 ka at *c.* 1.10 cycles m^{-1}; harmonic 54) and O1 (*c.* 30.5 ka at *c.* 1.32 cycles m^{-1}; harmonic 67; both obliquity bands derived from Berger *et al.* 1992). The MSS zone bar-log shown Figure 2 is adjusted to correspond to the E1 Milankovitch band, at *c.* 0.40 cycles m^{-1}, and this position is shown in Figure 3. Only the higher frequency P1 band appears not to have relatively strong power (again recalculated from Berger *et al.* 1992).

Discussion

Here we used 1004 samples from four different outcrops at three localities (Fig. 1) where χ has been previously measured and reported (Ellwood *et al.* 2007*a*). These sections had previously been well correlated to each other using litho- and biostratigraphy in combination with χ variations, so that at least some beds are coeval among sections. Using these data we developed a single composite reference section for the purpose of quantifying the cyclostratigraphy for the Upper Ordovician (Cincinnatian: Edenian) Kope Formation in northern Kentucky. This experiment was, in part, designed to allow us to quantify with reasonable certainty the visual cyclicity evident in many of the Kope exposures in this area. In addition, this work was designed to allow the calculation of SAR represented by cycles in outcrop, to communicate the sense of real time/thickness represented by the depositional record of an epicontinental sea during this time in Earth history. In turn this provides an understanding of the processes active during climate cycles of different magnitudes in the early Phanerozoic.

In the literature there is a large number of papers in which cycles are identified using quantitative techniques applied to geochemical or geophysical stratigraphic datasets, but very few studies relate these results to the visual record represented in the outcrops from which the data were obtained. In part this is the result of the fact that much of this work has been done on core recovered from wells, but also because it is difficult to take such datasets and extrapolate the quantitative result back to outcrop patterns observed. In this study we were able to do this with relatively high certainty, and the result is shown in Figure 4.

The roadcut shown in Figure 4 is typical for the exposures of the Kope in this area. Clearly seen are cyclic packages represented by ledges of relatively resistant limestone and siltstone alternating with the more gently sloped portions of the hillside represented by less resistant mudrock–marl units. These variations are visible close-up, after cleaning the section (Fig. 5), and a better sense of the cyclicity emerges (usually represented by limestone–shale packages), with notations added to indicate the magnitude of climate bands represented. However, the uncertainty in any one outcrop is seen by the clear differences in intervals between beds (Fig. 2). Condensation can be seen in some zones (i.e. beds 12–15) and expansion in others (i.e. beds 8–12). This condensation/expansion varies between outcrops, such that for any one outcrop the lack of uniformity prevents well-defined FT peaks. However, the effect is sufficiently small so that the 50 m composite FT result produces a result that exhibits power in five of the six Milankovitch band positions identified in Figure 3. Given that we have identified, for the whole dataset, the distribution in cycles m^{-1} for each Milankovitch band, and given that SARs are roughly uniform throughout the sections sampled (following from the excellent time-series results for the composite), then we can calculate where each band will fall within each individual outcrop and label these accordingly (Figs 4 & 5).

A very interesting observation from Figure 4 is the position of packages that represent the intersection (beginning) of precession, obliquity and 100 ka eccentricity bands within a single, longer-term *c.* 405 ka eccentricity cycle. Given that limestone dominates in these packages, we interpret these patterns to represent the focus during interglacial times of the temperature moderating effects of each of these cycles. Even though the limestones seem to record shallow-water faunas in many cases, the reinforcing effect of several cycles has produced a minor base level rise, thus starving the system more than usual of siliciclastics.

Sediment accumulation rate magnitude and variability

The MSS and time-series data developed for the composite section (Figs 2 & 3) can be used to

Fig. 4. Outcrop exposed at the Rt. 445 locality (Fig. 1) where we have identified the Milankovitch bands that are visually represented in the hillside. Note the pick-up truck for scale.

calculate an overall duration (see above) and SAR for the composite as well as incremental durations and SARs for the various beds and sections represented. Whereas the uncertainties for the overall composite are reasonably low, these increase as shorter lithological segments are evaluated. Given that the entire composite section of *c.* 50 m was deposited in *c.* 2.027 Ma, the SAR for the composite is *c.* 2.47 cm/ka.

The uncertainties associated with shorter depositional segments are larger. Here we use the MSS zone bar-logs (Fig. 2) to approximate the cyclicity through the section, where the MSS zone has been shown to correspond with the E1 (*c.* 100 ka) cyclicity (Fig. 3). Then we can assign SARs to segments within the larger composite sections using these MSS zones and the height represented within segments. This gives reasonable estimates of the uncertainties represented in short segments. Using the examples given above, for the expanded interval represented by beds 8–12, the thickness is *c.* 10.5 m through *c.* 3.75 MSS zones each *c.* 100 ka long,

Fig. 5. Two close-up views of the Rt. 445 locality (Fig. 1) where the section has been exposed through cleaning. Zones representing Milankovitch cycles are illustrated. Resistant beds are mainly limestone. The vertical bars represent a single obliquity cycle, a single eccentricity cycle and a range of precessional cycles.

resulting in an SAR for this segment of *c.* 2.8 cm/ ka. For the condensed segment represented by beds 12–15, the thickness is *c.* 3.3 m through only *c.* 1.5 MSS zones, resulting in an SAR of 2.2 cm/ka.

Conclusions

We have used 1004 samples collected from four different sections from the Upper Ordovician Kope Formation in Northern Kentucky to build a composite section *c.* 50 m thick. Samples represent marine sandstone, siltstone, limestone and carbonate-rich mudrock (marl) that were deposited in a relatively shallow epicontinental sea. Magnetic susceptibility data for these samples provide the basis for application of FT methods to the dataset with the goal of identifying climate-driven cyclic changes caused by delivery of detrital–aeolian compounds into the marine environment at this locality. Our results identified five, well-defined harmonic features that we attribute to Milankovitch eccentricity (*c.* 405 and 100 ka), obliquity (*c.* 30.5 and 37.1 ka) and precession (*c.* 19.3 ka) bands; obliquity and precession were recalculated using the proposed adjustments by Berger *et al.* (1992) for an approximate age for the sequence of 445 Ma (Ogg *et al.* 2008). These cycles were then visually identified in the outcrop. These data also provide the basis for estimates of SAR for the entire composite, and for intervals where condensation or expansion are apparent. MSS zonation also allows visual characterization of the composite, helping illustrate where anomalous SARs are impacting individual outcrops. These results demonstrate the utility of the MSS method in characterizing, visualizing and quantifying climate cycles, and the timing of these cycles recorded in marine sediments. Because these cycles exist in most such successions, MSS provides a very important tool that can be used in better understanding SAR and depositional processes active during deposition of important sequences preserved in the rock record.

Research by C. E. Brett was supported by a grant from the donors to the Petroleum Research Fund of the American Chemical Society. Partial funding for this project was provided by the Robey H. Clark endowment to LSU.

References

BERGER, A., LOUTRE, M. F. & LASKAR, J. 1992. Stability of the Astronomical frequencies over the Earth's history for paleoclimate studies. *Science*, **255**, 560–565.

BLOEMENDAL, J. & DEMENOCAL, P. 1989. Evidence for a change in the periodicity of tropical climate cycles at 2.4 Myr from whole-core magnetic susceptibility measurements. *Nature*, **342**, 897–900.

BOULILA, S., HINNOV, L. A. *ET AL.* 2008. Astronomical calibration of the Early Oxfordian (Vocontian and Paris basins, France): consequences of revising the Late Jurassic timescale. *Earth and Planetary Science Letters*, **276**, 40–51.

BRETT, C. E. & ALGEO, T. J. 2001*a*. Sequence stratigraphy of Upper Ordovician and Lower Silurian strata of the Cincinnati Arch region. *In*: ALGEO, T. J. & BRETT, C. E. (eds) *Sequence, Cycle & Event Stratigraphy of Upper Ordovician & Silurian Strata of the Cincinnati Arch Region. Field Trip Guidebook, 1999 Field Conference of the Great Lakes Section SEPM-SSG.* Kentucky Geological Survey Guidebook, Series, **12**. Kentucky Geological Survey, Lexington, KY, 34–46.

BRETT, C. E. & ALGEO, T. J. 2001*b*. Stratigraphy of the Upper Ordovician Kope Formation in its type area (Northern Kentucky) including a revised nomenclature. *In*: ALGEO, T. J. & BRETT, C. E. (eds) *Sequence, Cycle & Event Stratigraphy of Upper Ordovician & Silurian Strata of the Cincinnati Arch Region. Field Trip Guidebook, 1999 Field Conference of the Great Lakes Section SEPM-SSG.* Kentucky Geological Survey Guidebook, Series, **12**.Kentucky Geological Survey, Lexington, KY, 47–64.

BRETT, C. E. & ALGEO, T. J. 2001*c*. Event beds and small-scale cycles in Edenian to Lower Maysvillian strata (Upper Ordovician) of Northern Kentucky: identification, origin, and temporal constraints. *In*: ALGEO, T. J. & BRETT, C. E. (eds) *Sequence, Cycle & Event Stratigraphy of Upper Ordovician & Silurian Strata of the Cincinnati Arch Region. Field Trip Guidebook, 1999 Field Conference of the Great Lakes Section SEPM-SSG.* Kentucky Geological Survey Guidebook, Series, **12**. Kentucky Geological Survey, Lexington, KY, 65–92.

BRETT, C. E., ALGEO, T. J. & MCLAUGHLIN, P. I. 2003. Use of event beds and sedimentary cycles in high-resolution stratigraphic correlation of lithologically repetitive successions. *In*: HARRIES, P. J. (ed.) *High-Resolution Approaches in Stratigraphic Paleontology.* Kluwer Academic, Dordrecht, 315–350.

BRETT, C. E., KIRCHNER, B., TSUJITA, C. & DATTILO, B. 2008. Depositional dynamics recorded in mixed siliciclastic-carbonate marine successions: insights from the Upper Ordovician Kope Formation of Ohio and Kentucky, USA. *In*: PRATT, B. R. & HOLMDEN, C. (eds) *Dynamics of Epeiric Seas.* Geological Association of Canada, Toronto, Special Papers, **48**, 73–102.

CRICK, R. E., ELLWOOD, B. B., EL HASSANI, A., FEIST, R. & HLADIL, J. 1997. MagnetoSusceptibility event and cyclostratigraphy (MSEC) of the Eifelian–Givetian GSSP and associated boundary sequences in North Africa and Europe. *Episodes*, **20**, 167–175.

CRICK, R. E., ELLWOOD, B. B., EL HASSANI, A., HLADIL, J., HROUDA, F. & CHLUPAC, I. 2001. Magnetostratigraphy Susceptibility of the Pridoli-Lochkovian (Silurian–Devonian) GSSP (Klonk, Czech Republic) and a Coeval sequence in Anti-Atlas Morocco. *Palaeogeography, Palaeoclimatology, Palaeoecology*, **167**, 73–100.

DA SILVA, A.-C. & BOULVAIN, F. 2005. Upper Devonian carbonate platform correlations and sea level variations recorded in magnetic susceptibility. *Palaeogeography, Palaeoclimatology, Palaeoecology*, **240**, 373–388.

DATTILO, B., BRETT, C. E. & TSUJITA, C. S. 2008. Sediment supply v. storm winnowing in the development of muddy and shelly interbeds from the Upper Ordovician of the Cincinnati region, USA. *In*: TAPANILLA, L. (ed.) *Essays in Honor of Paul Copper. Canadian Journal of Earth Sciences*, **45**, 1–23.

DEMENOCAL, P., BLOEMENDAL, J. & KING, J. 1991. 22. A rock-magnetic record of monsoonal dust deposition to the Arabian Sea: Evidence for a shift in the mode of deposition at 2.4 Ma. *In*: PRELL, W. L., NIITSUMA, N. *ET AL.* (eds) *Proceedings of the Ocean Drilling Program, Scientific Results*, US Government Printing Office, Washington, **117**, 389–407.

DINARÈS-TURELL, J., BACETA, J. I., BERNAOLA, G., ORUE-ETXEBARRIA, X. & PUJALTE, V. 2007. Closing the Mid-Palaeocene gap: toward a complete astronomically tuned Palaeocene Epoch and Selandian and Thanetian GSSPs at Zumaia (Basque Basin, W Pyrenees). *Earth and Planetary Science Letters*, **262**, 450–467.

ELLWOOD, B. B., CHRZANOWSKI, T. H., HROUDA, F., LONG, G. J. & BUHL, M. L. 1988. Siderite formation in anoxic deep-sea sediments: a synergetic bacterially controlled process with important implications in paleomagnetism. *Geology*, **16**, 980–982.

ELLWOOD, B. B., CRICK, R. E. & EL HASSANI, A. 1999. The magnetosusceptibility event and cyclostratigraphy (MSEC) method used in geological correlation of devonian rocks from anti-atlas morocco. *AAPG Bulletin*, **83**, 1119–1134.

ELLWOOD, B. B., CRICK, R. E., EL HASSANI, A., BENOIST, S. & YOUNG, R. 2000. Magnetosusceptibility event and cyclostratigraphy (MSEC) in marine rocks and the question of detrital input v. carbonate productivity. *Geology*, **28**, 1135–1138.

ELLWOOD, B. B., CRICK, R. E. *ET AL.* 2001. Global correlation using magnetic susceptibility data from Lower Devonian rocks. *Geology*, **29**, 583–586.

ELLWOOD, B. B., BENOIST, S. L., EL HASSANI, A., WHEELER, C. & CRICK, R. E. 2003. Impact ejecta layer from the mid-devonian: possible connection to global mass extinctions. *Science*, **300**, 1734–1737.

ELLWOOD, B. B., HARROLD, F. B. *ET AL.* 2004. Magnetic susceptibility applied as an age-depth-climate relative dating technique using sediments from scladina cave, a late pleistocene cave site in Belgium. *Journal of Archaeological Science*, **31**, 283–293.

ELLWOOD, B. B., GARCÍA-ALCALDE, J. L. *ET AL.* 2006. Stratigraphy of the middle devonian boundary: formal definition of the susceptibility magnetostratotype in Germany with comparisons to sections in the Czech Republic, Morocco and Spain. *Tectonophysics*, **418**, 31–49.

ELLWOOD, B. B., BRETT, C. E. & MACDONALD, W. D. 2007a. Magnetosusceptibility stratigraphy of the upper ordovician kope formation, Northern Kentucky. *Palaeogeography, Palaeoclimatology, Palaeoecology*, **243**, 42–54.

ELLWOOD, B. B., TOMKIN, J., RICHARDS, B., BENOIST, S. L. & LAMBERT, L. L. 2007b. MSEC data sets record glacially driven cyclicity: examples from the arrow canyon Mississippian–Pennsylvanian GSSP and associated sections. *Palaeogeography, Palaeoclimatology, Palaeoecology*, **255**, 377–390, http://dx.doi.org/10.1016/j.palaeo.2007.08.006

ELLWOOD, B. B., TOMKIN, J. H., FEBO, L. A. & STUART, C. N., JR. 2008a. Time series analysis of magnetic susceptibility variations in deep marine sediments: a test using upper Danian-lower Selandian proposed GSSP, Spain. *Palaeogeography, Palaeoclimatology, Palaeoecology*, **261**, 270–279.

ELLWOOD, B. B., TOMKIN, J. H., RATCLIFFE, K. T., WRIGHT, M. & KAFAFY, A. M. 2008b. High resolution magnetic susceptibility and geochemistry for the Cenomanian/Turonian boundary GSSP with correlation to time equivalent core. *Palaeogeography, Palaeoclimatology, Palaeoecology*, **261**, 105–126.

ELLWOOD, B. B., LAMBERT, L. L., TOMKIN, J. H., BELL, G., NESTELL, M. K., NESTELL, G. P. & WARDLAW, B. R. 2012. Magnetostratigraphy susceptibility for the Guadalupian Series GSSPs (Middle Permian) in Guadalupe Mountains National Park and Adjacent Areas in West Texas. *In*: JOVANE, L., HERRERO-BERVERA, E., HINNOV, L. A. & HOUSEN, B. A. (eds) *Magnetic Methods and the Timing of Geological Processes*. Geological Society, London, Special Publications, **373**, http://dx.doi.org/10.1144/SP373.1.

ETTENSOHN, F. R. 1992. General Ordovician paleogeographic and tectonic framework for Kentucky. *In*: ETTENSOHN, F. R. (ed.) *Changing Interpretations of Kentucky Geology: Layer Cake, Facies, Flexure, and Eustasy*. Ohio Division of Geological Survey, Miscellaneous Reports, US Government Printing Office, Washington, **5**, 19–21.

HANSEN, H. J., LOJEN, S., TOFT, P., DOLENEC, T., YONG, J., MICHAELSEN, P. & SARKAR, A. 1999. Magnetic susceptibility of sediments across some marine and terrestrial Permo-Triassic boundaries. *Proceedings of the International Conference: Pangea and the Paleozoic–Mesozoic transaction*, China University of Geosciences, Hubei, China, 114–115.

HILGEN, F. J. 2010. Astronomical dating in the 19th century. *Earth-Science Reviews*, **98**, 65–80.

HLADIL, J., GERSL, M., STRNAD, L., FRANA, J., LANGROVA, A. & SPISIAK, J. 2006. Stratigraphic variation of complex impurities in platform limestones and possible significance of atmospheric dust: a study with emphasis on gamma-ray spectrometry and magnetic susceptibility outcrop logging (Eifelian–Frasnian, Moravia, Czech Republic). *International Journal of Earth Sciences (Geologische Rundschau)*, **95**, 703–723.

HOLLAND, S. M., MILLER, A. I., DATTILO, B. F., MEYER, D. L. & DIEKMEYER, S. L. 1997. Cycle anatomy and variability in the storm-dominated type Cincinnatian (Upper Ordovician): coming to grips with cycle delineation and genesis. *The Journal of Geology*, **105**, 135–152.

HOLLAND, S. M., MILLER, A. I. & MEYER, D. L. 2001a. Sequence stratigraphy of the Kope-Fairview interval (Upper Ordovician, Cincinnati, Ohio area). *In*: ALGEO, T. J. & BRETT, C. E. (eds) *Sequence, Cycle, and Event Stratigraphy of Upper Ordovician and Silurian Strata of the Cincinnati Arch Region*. Kentucky Geological Survey Guidebook **1**. Kentucky Geological Survey, Lexington, KY, 93–102.

HOLLAND, S. M., MILLER, A. I., MEYER, D. L. & DATILLO, B. F. 2001b. The detection and importance of subtle biofacies within a single lithofacies: the Upper

Ordovician Kope Formation of the Cincinnati, Ohio region. *Palaios*, **16**, 205–217.

IMBRIE, J., HAYS, J. D. *ET AL.* 1984. The orbital theory of pleistocene climate: support from a revised chronology of the marine delta 18O record. *In*: BERGER, A. L., IMBRIE, J., HAYS, J., KUKLA, G. & SALTZMAN, B. (eds) *Milankovitch and Climate. Part I*. Kluwer Academic, Dordrecht, 269–305.

JENNETTE, D. C. & PRYOR, W. A. 1993. Cyclic alternation of proximal and distal storm facies: kope and fairview formations (Upper Ordovician), Ohio and Kentucky. *Journal of Sedimentary Petrology*, **63**, 183–203.

JOVANE, L., FLORINO, F., SPROVIERI, M. & PÄLIKE, H. 2006. Astronomic calibration of the late Eocene/early Oligocene Massignano section (Central Italy). *Geochemistry, Geophysics, Geosystems*, **7**, Q07012, http://dx.doi.org/10.1029/2005GC001195

KARLIN, R. & LEVI, S. 1983. Diagenesis of magnetic minerals in recent haemipelagic sediments. *Nature*, **303**, 327–330.

KARLIN, R. & LEVI, S. 1985. Geochemical and sedimentological control of the magnetic properties of hemipelagic sediments. *Journal of Geophysical Research*, **90**, 10 373–10 392.

LASKAR, J., ROBUTEL, P., JOUTEL, F., GASTINEAU, M., CORREIA, A. & LEVRARD, B. 2004. A long-term numerical solution for the insolation quantities of the Earth. *Astronomy and Astrophysics*, **428**, 261–285.

LISITZIN, A. P. 1972. *Sedimentation in the World Ocean*. Society of Economic Paleontologists and Mineralogists, Tulsa, OK, Special Publications, **17**.

MEAD, G. A., YAUXE, L. & LABRECQUE, J. L. 1986. Oligocene paleoceanography of the South Atlantic: paleoclimate implications of sediment accumulation rates and magnetic susceptibility. *Paleoceanography*, **1**, 273–284.

OGG, J. G., OGG, G. & GRADSTEIN, F. M. 2008. *The Concise Geologic Timescale*. Cambridge University Press, Cambridge.

OLSEN, P. E. & KENT, D. V. 2000. High-resolution early Mesozoic Pangean climatic transect in lacustrine environments. *In*: BACHMAN, G. H. & LERCHE, I. (eds) *Zentralblatt fuer Geologie und Palaeontologie 1988*. Schweizerbart & Borntraeger Science Publishers, Stuttgart, 1475–1495.

SALVADOR, A. 1994. *International Stratigraphic Guide: A Guide to Stratigraphic Classification, Terminology, and Procedure*. The International Union of Geological Sciences and The Geological Society of America, Denver, CO.

SHACKLETON, N. J., McCAVE, I. N. & WEEDON, G. P. 1999a. Preface. *Philosophical Transactions of the Royal Society London, A*, **357**, 1733–1734.

SHACKLETON, N. J., CROWHURST, S. J., WEEDON, G. P. & LASKAR, J. 1999b. Astronomical calibration of Oligocene–Miocene time. *Philosophical Transactions of the Royal Society London, A*, **357**, 1907–1929.

SWARTZENDRUBER, L. J. 1992. Properties, units and constants in magnetism. *Journal of Magnetism and Magnetic Materials*, **100**, 573–575.

THOA, N. T. K., HUYEN, D. T., ELLWOOD, B. B., LAN, L. T. P. & TRUONG, D. N. 2004. Determination of Permian–Triassic boundary in limestone formations from Northeast of Vietnam by paleontological and MSEC methods. *Journal of Sciences of the Earth*, **26**, 222–232.

TOBIN, R. C. 1982. *A Model for Cyclic Deposition in the Cincinnatian Series of Southwestern Ohio, Northern Kentucky, and Southeastern Indiana*. Ph.D. dissertation, University of Cincinnati, Cincinnati, OH.

TOBIN, R. C. & PRYOR, W. A. 1981. Sedimentological interpretation of an Upper Ordovician carbonate-shale vertical sequence in Northern Kentucky. *In*: ROBERTS, T. G. (ed.) *Geological Society of America, 1981 Annual Meeting Field Trip Guidebooks. Vol. I: Stratigraphy and Sedimentology*. American Geological Institute, Falls Church, VA, 1–10.

VAUGHAN, S., BAILEY, R. J. & SMITH, D. G. 2011. Detecting cycles in stratigraphic data: spectral analysis in the presence of red noise. *Paleoceanography*, **26**, FA4211, http://dx.doi.org/10.1029/2011PA002195

VOGEL, K. & BRETT, C. E. 2009. Record of microendoliths in different facies of the Upper Ordovician in the Cincinnati Arch region USA: the early history of light-related microendolithic zonation. *Palaeogeography, Palaeoclimatology, Palaeoecology*, **281**, 1–24.

WEEDON, G. P., SHACKLETON, N. J. & PEARSON, P. N. 1997. The Oligocene timescale and cyclostratigraphy on the Ceara Rise, western equatorial Atlantic. *In*: SCHACKLETON, N. J., CURRY, W. B., RICHTER, C. & BRALOWER, T. J. (eds) *Proceedings of the Ocean Drilling Program, Scientific Results*. US Government Printing Office, Washington, **154**, 101–114.

WEEDON, G. P., JENKYNS, H. C., COE, A. L. & HESSELBO, S. P. 1999. Astronomical calibration of the Jurassic time-scale from cyclostratigraphy in British mudrock formations. *Philosophical Transactions of the Royal Society London A*, **357**, 1787–1813.

WEISS, M. P. & SWEET, W. C. 1964. Kope Formation (Upper Ordovician): Ohio and Kentucky. *Science*, **145**, 1296–1302.

Anisotropy of magnetic susceptibility and sedimentary cycle data from Permo-Carboniferous rhythmites (Paraná Basin, Brazil): a multiple proxy record of astronomical and millennial scale palaeoclimate change in a glacial setting

DANIEL R. FRANCO[1]* & LINDA A. HINNOV[2]

[1]*Department of Geophysics, National Observatory, R. Gal. José Cristino, 77, 20921-400 Rio de Janeiro, RJ, Brazil*

[2]*Department of Earth and Planetary Sciences, Johns Hopkins University, 3400 North Charles Street, Baltimore, MD 21218 USA*

**Corresponding author (e-mail: drfranco@on.br)*

Abstract: In this study we examine glaciogenic rhythmites from the Late Palaeozoic Itararé Group, Paraná Basin, Brazil. We conduct spectral analysis on lithological cycle ('couplet') thickness series, and declination of maximum axis of anisotropy of magnetic susceptibility ellipsoidal tensor (K1) data. We tested the efficiency of K1 as a palaeoclimatic proxy. To constrain the timescale of harmonic features in the data, we analysed the couplet thickness spectra, converting the spectra to the time domain using an astronomical calibration based on Milankovitch frequency ratios. Comparison of the two rhythmites provides insights into their sedimentation rate evolution and cyclicity. Millennial-scale mechanisms of climatic origin influenced the deposition of both rhythmites, generating the lithological couplets, and are consistent with millennial-scale variations recognized as triggers for large-scale climatic changes during the Late Pleistocene. The common harmonic features in the couplet thickness and K1 spectra support the view that the azimuth of the K1 axis in sedimentary fabric is a useful palaeoclimatic proxy, reflecting sedimentation processes that were directly influenced by flow-induced, sediment transport, which is linked to external climate factors.

Discussions about global warming are increasingly focused on the importance of improving our knowledge concerning palaeoclimatic change, and understanding the dynamics and amplitude of the present climatic system (Litt *et al.* 2001). It is well established from deep-sea stratigraphy that $10^4 - 10^5$ year scale insolation variations caused by Earth's astronomical parameters have given rise to quasi-periodic glacial episodes over the past 3 million years (e.g. Raymo & Huybers 2008). Accurate correlations have now been made between Milankovitch-forced climatic and stratigraphic cycles (Hilgen *et al.* 2012; Hinnov & Hilgen 2012). Thus, it is now increasingly 'routine' to study palaeoclimatic signals in the geological record and to relate these signals to recent climatic change.

The effects of climatic change on sedimentation in glacio-lacustrine and glacio-marine depositional systems are far less understood (Fielding *et al.* 2008). Low latitude glaciation, which does not occur today, is particularly poorly understood. In this context, the Itararé Group (IG; Paraná Basin, Brazil; Fig. 1) represents one of the main Palaeozoic glaciation sequences in Gondwana, characterized by laminated sedimentary rocks of glacial origin, consisting of alternating sandstone/siltstone and siltstone/claystone beds, forming lithological pairs or couplets. These rhythmites have been called 'varvitos' (varvites) (Leinz 1937; Rocha-Campos 1967) owing to their similarity to Pleistocene varvic clays. Past studies of the Itu rhythmites suggested an annual deposition for the constituent lithological couplets. Ernesto & Pacca (1981) performed spectral analysis on a series of couplet thicknesses from the Itu rhythmites, in an effort to characterize the variability of couplet deposition. The series was constructed with the assumption that each couplet corresponded to one year of sedimentation; the estimated spectrum indicated periodicities of 11 and 22 years, and other decadal cycles. This result reinforced the annual sedimentation hypothesis for the Itararé rhythmites, and set up an important geological precedence for Late Palaeozoic palaeomagnetic and palaeoenvironmental studies.

However, there are mounting contra-indications to the annual sedimentation hypothesis. Silva & Azambuja Filho (2005) concluded that couplet deposition was influenced by astronomical forcing and

From: Jovane, L., Herrero-Bervera, E., Hinnov, L. A. & Housen, B. A. (eds) 2013. *Magnetic Methods and the Timing of Geological Processes*. Geological Society, London, Special Publications, **373**, 355–374. First published online November 6, 2012, http://dx.doi.org/10.1144/SP373.11 © The Geological Society of London 2013. Publishing disclaimer: www.geolsoc.org.uk/pub_ethics

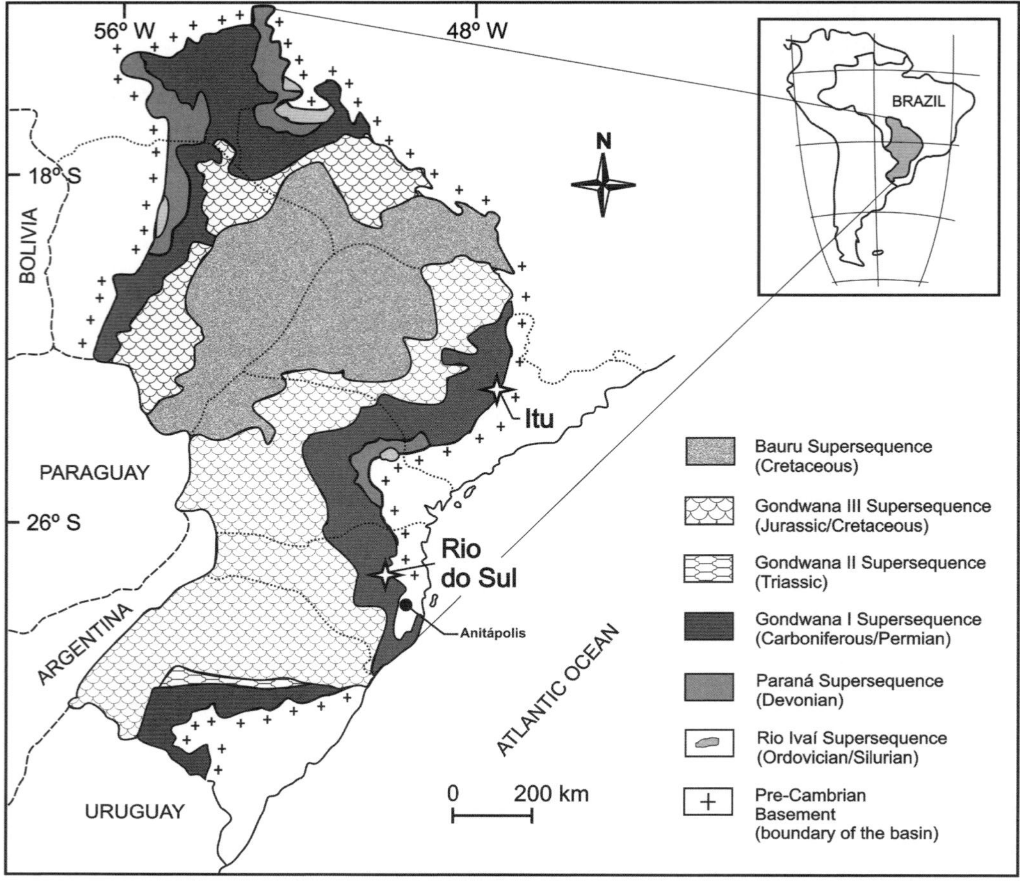

Fig. 1. Location and distribution of stratigraphic supersequences from Paraná Basin, southeastern Brazil. The Itararé Group is classified as a Gondwana I Supersequence. Locations for the Itu and Rio do Sul rhythmites outcrops are indicated, as well as the city of Anitápolis. A cross section showing the regional stratigraphy appears in Figure 1 of Franco *et al.* (2011). Modified after Souza *et al.* (2006).

millennial-scale phenomena in an analysis of a cored section of rhythmites near the city of Anitápolis (state of Santa Catarina; Fig. 1). They also investigated fine-scale lamination patterns using greyscale scans, documenting variations interpreted as decadal solar cycles and other quasiperiodicities related to solar activity. More recently, Franco *et al.* (2011) assessed the two IG rhythmite successions examined herein with analysis of lamination sequences within individual couplets and evaluation of directional variability in characteristic remanent magnetization (ChRM) along successive couplets. They found that millimetre-scale lamination sequences within couplets have a spectral content that is compatible with a decadal- to centennial-scale deposition of the couplets, and that there are distinct 'jumps' in ChRM declination of *c.* 20–50° from couplet to couplet,

suggesting significant missing time between couplets. Thus, individual couplets in the IG sequences must represent substantial time, much longer than the annual scale that was originally envisioned.

Here, we continue investigation of the depositional origins of these enigmatic Late Palaeozoic Gondwana rhythmites with two new assessments: (1) spectral analysis of couplet thickness series; and (2) spectral analysis of the maximum axis (K1) of anisotropy of magnetic susceptibility (AMS) ellipsoidal tensor data. Variations in the azimuth of K1 along the rhythmite successions provide information on the evolution of sediment transport, which can be linked to external climatic forcing. We will compare the spectral content of the data in order to examine the reliability of K1 as an efficient palaeoclimatic indicator.

Anisotropy of magnetic susceptibility: a tool for assessing sediment transport directionality and palaeoclimatic processes

Anisotropy of magnetic susceptibility is well established as a sensitive technique and efficient gauge of petro-fabric (e.g. Ising 1942; Graham 1954; Fuller 1963; Rees 1965). AMS can be determined through an easy and convenient set of procedures, and is capable of detecting very subtle anisotropies (Hrouda 1982; Borradaile 1988; Jackson 1991; Tarling & Hrouda 1993; Schmidt *et al.* 2007; Zhang *et al.* 2010). The method has been applied to a broad range of geological problems, from structural geology (e.g. Parés *et al.* 1999; Burmeister *et al.* 2009; Borradaile & Jackson 2010) to sediment deposition and palaeoenvironmental reconstruction (e.g. Kodama & Sun 1990; Lima *et al.* 2000; Parés *et al.* 2007; Bradák 2009).

In rocks, AMS is based on a quantitative evaluation of preferable crystallographic orientation, compositional layering distribution, size of microfractures and shape of grain fabrics (Parés *et al.* 2007; Schmidt *et al.* 2007). Measurements are made based on the fact that, by definition, AMS establishes a symmetrical, second-rank tensor with six independent matrix elements. The coordinate system is defined in terms of eigenvalues and eigenvectors, and the matrix may be described as an ellipsoid, with its semi-axes corresponding to the three main susceptibilities (maximum, intermediate and minimum susceptibilities, K1, K2 and K3, respectively; Hrouda 1982; Tarling & Hrouda 1993). The contribution of diamagnetic, paramagnetic and ferromagnetic (*sensu stricto*) mineral phases in a rock generates the full AMS.

From sedimentary rock fabric it is possible to recognize preferred grain orientation from which to infer palaeocurrent direction (Zhu *et al.* 2004; Chen *et al.* 2008; Bradák 2009). AMS is effective in providing information on flow directions in different sedimentary environments, and has proven useful in the study of transport mechanisms and sediment deposition (e.g. Kanamatsu 1996; Lagroix & Banerjee 2002; Veloso *et al.* 2007; Chen *et al.* 2008). The depositional surface in a sedimentary environment is indicated by the spatial distribution of AMS tensor axes (Tarling & Hrouda 1993). Typically, the magnetic foliation (i.e. the plane defined by the K1 and K3 axes) is parallel to the depositional surface, whereas magnetic lineation (i.e. the grouping of K1 axes), depending on the hydrodynamic regime, is parallel to the flow direction (Hrouda 1982; Rochette *et al.* 1992; Liu *et al.* 2001; Parés *et al.* 2007; Bradák 2009).

External climatic factors are typically responsible for current-induced fabric in sedimentary deposits (e.g. Reading 1986). Sediment is entrained by climate-controlled hydrodynamic flow (from rainstorms, floods, etc.), that is, by flowing water, and is rapidly and strongly structured in response to flow direction and intensity (Allen 1982). Thus the direction of the K1 axis in sedimentary rock fabric is a palaeocurrent indicator – and by extension, a palaeoclimatic proxy, if the directional variability of sedimentary deposition is climatically controlled, and mineralogical contributions are favourable. Consequently, study of K1 data from sedimentary rock successions has the potential to provide information on sediment transport intensity and variability, and by extension, palaeoclimatic change.

Geological setting

The Itararé Group, Paraná Basin, Brazil

The glaciogenic rhythmites investigated in this work are from outcropping sequences of undisturbed laminated rocks from the eastern belt and southern portion of the Permo-Carboniferous Itararé Group, Paraná Basin, Brazil (Fig. 1). The IG is related to the Gondwana I Supersequence – the major transgressive–regressive cycle recorded in Gondwana during the Late Palaeozoic (Ferreira 1997; Rocha-Campos *et al.* 1997; Milani & Zalán 1999; Weinschütz & Castro 2004; Souza *et al.* 2006; Vesely & Assine 2006).

The IG exhibits glaciolacustrine and/or brackish water palaeoenvironmental settings with an upwardly increasing marine influence (Gama *et al.* 1992; Eyles *et al.* 1993). It is composed of three major units (the Lagoa Azul, Campo Mourão and Taciba formations), each recording a renewed subsidence phase along structural NE–SW and NW–SE trending lineaments (França & Potter 1988; Eyles *et al.* 1993; França *et al.* 1996; Vesely & Assine 2004). Dominant facies from subaqueous gravity flows of sediments resulted in high sedimentation rates along steep, fault-bounded basin margins (Eyles *et al.* 1993). These aspects, along with the presence of roche moutonnée, striated surfaces, diamictites and rhythmites with dropstones, support a hypothesis for a repeated advance–retreat record of ice masses originating from a large South African ice sheet along a SE–NW direction (Rocha-Campos 1967; França & Potter 1991; Eyles *et al.* 1993; Santos *et al.* 1996; Rocha-Campos *et al.* 2000; Vesely & Assine 2002; Archanjo *et al.* 2006).

The eastern outcropping belt with the rhythmites discussed herein occurs within the Taciba Formation; the targeted Itu and Rio do Sul quarry outcrops (Fig. 1) occur within the Chapéu do Sol and Rio do Sul members, respectively, which occur more or less laterally with respect to each other. Recent chronostratigraphic data summarized in Franco *et al.* (2011)

indicate that the Itu rhythmites are from the Upper Carboniferous (Kasimovian–Gzhelian stages), and the Rio do Sul rhythmites are from the Lower Permian (Asselian–Sakmarian stages).

Itu (IT) rhythmites

Itu (IT) Quarry (23°16'S; 47°19'W; Fig. 2), near the city of Itu, state of São Paulo, is preserved as a national geological park ('Varvite Park') (Setti & Rocha-Campos 1999; Rocha-Campos 2002; http://sigep.cprm.gov.br/sitio062/sitio062english.htm). The outcropping rhythmites occur as a 15 m-thick succession, in a stack of 260 lithological pairs or couplets, each comprising a basal, centimetre- to decimetre-thick, light-coloured bed of fine sandstone/siltstone overlain by a dark-coloured millimetre-thick layer of siltstone/claystone (Ernesto & Pacca 1981; Sinito *et al.* 1981). The couplets thin upward, and couplet granulometry declines upsection. The upward thinning is due mainly to reductions in the light-coloured sandstone/siltstone members, from *c.* 20–50 cm at the bottom of the sequence to *c.* 1.5 cm at the top; the thickness of dark, siltstone/claystone members is almost constant along the sequence (*c.* 5 mm). The couplets are separated by sharp contacts, with abrupt but transitional contacts between members within couplets (Rocha-Campos *et al.* 1981; Setti & Rocha-Campos 1999). Abundant sedimentary structures are present in the sequence, mainly within the coarser, light members, which include flat and cross lamination, and normally graded laminae. Isolated clasts (millimetre–decimetre) composed of granite and quartzite occur as dropstones (Rocha-Campos & Sudaram 1981; Gama *et al.* 1992; Ferreira 1997; Setti & Rocha-Campos 1999; Caetano-Chang & Ferreira 2006).

The eastern margin of the Paraná Basin had a bay or indentation-like setting that opened toward the NW (Setti & Rocha-Campos 1999); palaeocurrent directions there are systematically oriented in a NW–SE direction (e.g. Gesicki *et al.* 2002; Vesely & Assine 2002). Dropstones and occasional intercalations of turbidites and other coarse-grained clastic deposits are indicative of a glacial influence on the depositional system; progressive glacier recession is suggested by the upward decreasing thicknesses of the couplets. Santos *et al.* (1996) proposed that rhythmites deposition along the eastern basin edge occurred in a restrictive marine environment in the presence of thawing and glacier recession as Gondwana moved northward from the southern polar region.

Rio do Sul rhythmites

The Rio do Sul (RS) rhythmites (27°10'S; 49°35'W; Fig. 2) have one of its best exposures at Itaú Quarry (city of Trombudo Central, state of Santa Catarina), in an outcrop of 9 m thickness, with a stack of 127 lithological couplets (Rocha-Campos *et al.* 1981). The basic RS couplet is composed of a light layer of siltstone overlain by a dark layer of claystone; the couplets vary in thickness from 1.2 to 32 cm. As with the IT couplets, the dark layer members remain very thin (2–9 mm) throughout the sequence (Rocha-Campos & Sundaram 1981; Rocha-Campos *et al.* 1981; Canuto 1993). There are sharp contacts between couplets, but gradational transitions between light and dark members within couplets. In general RS granulometry is finer than in the IT rhythmites (Franco 2007).

Sedimentary structures in the RS rhythmites include normal and reverse graded laminae, micro-cross lamination, clay partings within the lighter beds, and normal grading and fine sand partings in the darker layers (Rocha-Campos & Sundaram 1981; Rocha-Campos *et al.* 1981). There is a common occurrence of lenses of chips or pellets associated with shallow furrows on bedding planes, clasts (centimetre–decimetre scale), ichnofossils and a relatively diversified palynoflora (Canuto 1993). Canuto suggested that the RS rhythmites formed in a proximal glaciomarine depositional system; Eyles *et al.* (1993) proposed a low-energy depositional setting at some distance from major sediment input points. Santos *et al.* (1992) pointed to the clasts and spill deposits as evidence for glacial resurgence, and a large meltwater input.

Data and methods

Couplet thickness series

Couplet thickness data series were assembled from thickness measurements of the individual couplets along the rhythmite sections. This technique was devised by Schwarzacher (1964) to analyse long-period modulations of couplets in limestone–shale sequences. We assembled the thickness measurements in two ways. The first way simply assigns couplet thickness with respect to stratigraphic height, that is, to its centre position along the sequence; this is known as a 'stratigraphic' series (Herbert 1992). In this series, modulations in cycle thickness are reflected in the independent variable: modulations are recorded as distorted cycles along the thickness series. The second way assigns each couplet thickness to its couplet number in the sequence, and is known as a 'metronomic' series (Herbert 1992) to emphasize the underlying assumption that individual couplets represent the same amount of time. In this reference frame, sedimentation rate variations are confined to the dependent variable, and do not distort couplet modulation cycles (Hinnov 2000).

Fig. 2. Field photographs of the studied rhythmites. Itu Quarry (state of São Paulo). (**a**) Some of the thickest strata from Itu quarry (bottom of the sequence); abrupt contacts between couplets owing to intercalation of siltstone/mudstones are visible. (**b**) Detail showing couplet thickness variations and sedimentary structures, for example, ripple-drift cross lamination. (**c**) Dropstone between two pairs of thin lithologies (top of the sequence). Rio do Sul (state of Santa Catarina): (**d**) main outcrop at Itaú quarry. (**e**) Detail of contact between couplet lithologies.

There is no guarantee, however, that individual couplets will maintain the same time-scale throughout the series.

K1 data

The preparation of specimens for AMS measurements was as follows: oriented blocks were collected from each couplet; cylinders with a diameter of 2.2 cm were drilled through the entire thickness of each block. From these cylinders, 2.2 cm-long specimens were cut along the stratigraphically vertical direction. AMS measurements were made with a MINISEP device (Molspin Ltd, UK) from the Laboratory of Paleomagnetism, Institute of Astronomy, Geophysics and Atmospheric Sciences (University of São Paulo, Brazil). This instrument operates at low alternating induction magnetic field values (c. 10 kHz at a field value of 7×10^{-4} T). For a given stratigraphic height, at least three specimens were measured, depending on the availability of samples. Average K1 azimuth was determined by the software described in Lienert (1991), based on Monte Carlo perturbations in synthetic tensors generated for the estimate of Hext–Jelinek (H–J) elliptical confidence regions.

Time calibration and spectral analysis

Franco et al. (2011) argued for an annual scale for the millimetre-scale laminations in the IT and RS rhythmite sequences, and a multi-centennial to millennial scale for individual couplets. Analysis of ChRM directional series for both rhythmites in Franco et al. (2012a) led to specific low-frequency astronomical calibrations based on the method of frequency ratios. This method was employed by Kruiver et al. (2000), who posited that the frequency spacing of key spectral peaks of astronomically forced stratigraphic series, reported as 'wavelengths' in spatial units (e.g. centimetres), would have the same frequency separations as spectral peaks of the astronomical parameters, which can be reported in terms of frequency ratios. For example, short eccentricity:obliquity:long precession:short precession = 5.7:2.2:1.2:1 is predicted for Permian times (Berger et al. 1992).

To characterize harmonic features in the couplet thickness and K1 data series, spectral analysis was carried out using REDFIT software for unevenly spaced time series (Schulz & Mudelsee 2002). Significance tests for spectral peaks based on a theoretical first-order autoregressive process fitted to the data were also performed and reported with respect to the spectral confidence limits (80, 90, 95 and 99%). Finally, to seek evidence for astronomical-scale components, harmonic analysis (Schulz & Stategger 1997) was conducted on the very-low-frequency bands of the series.

Results

Itu rhythmites

The metronomic and stratigraphic couplet thickness series of the IT rhythmites reveal couplet thickness modulations throughout the outcrop. A total of 165 couplets were measured, starting above the high-energy strata, with an average thickness of 6.3 cm that progressively thin up-section (Fig. 3a, b). Couplets near the bottom of the sequence reach thicknesses of several tens of centimetres; here also is most of the sedimentary evidence for high-energy depositional conditions (Fig. 2).

The IT couplet thickness power spectra (Fig. 4a, b) show significant differences between the stratigraphic and metronomic scales. This indicates that the long-term modulations are correlated with couplet thickness, and cannot be pinpointed in the stratigraphic spectrum. On the other hand, modulation frequencies are clearly resolved in the metronomic spectrum. The broad 49.2 couplet 'bundling' modulations further decompose into three harmonics (inset of Fig. 4b), indicating bundling modulations of 56.8, 28.7 and 19.4 couplets, all visibly superimposed in the metronomic series (Fig. 3a).

K1 data were obtained from the lower 4.26 m of the Itu Quarry outcrop at an average sample spacing of 2.8 cm; each couplet was sampled at least three times. Examples of H–J results for two IT couplets are shown in Figure 5a, b, carried out with site-mean statistics and principal AMS tensor axes as described in Lienert (1991). A relatively high degree of anisotropy (1.11–1.15) is observed, and the K1 and K2 axes are lying on near-horizontal planes for the selected couplets (a detailed discussion on AMS behaviour throughout the succession appears in Franco et al. 2012b). The K1 stratigraphic series (Fig. 6a) displays highly visible cycling in contrast to subdued cycles in ChRM data from the same samples (Fig. 2 in Franco et al. 2012a). In the K1 stratigraphic spectrum (Fig. 7a), a high power cycle occurs at a wavelength of 81 mm, which is not related to couplet thickness (averaging 63 mm in this interval); a 921 mm cycle is also present, which is visible and persistent along the stratigraphic series (Fig. 3a).

Rio do Sul rhythmites

At the Rio do Sul outcrop, the first (lower) 94 couplets were measured out of the 127 total, with an average couplet thickness of 4.8 cm (Fig. 3c, d). The lower 75% of the sequence is characterized

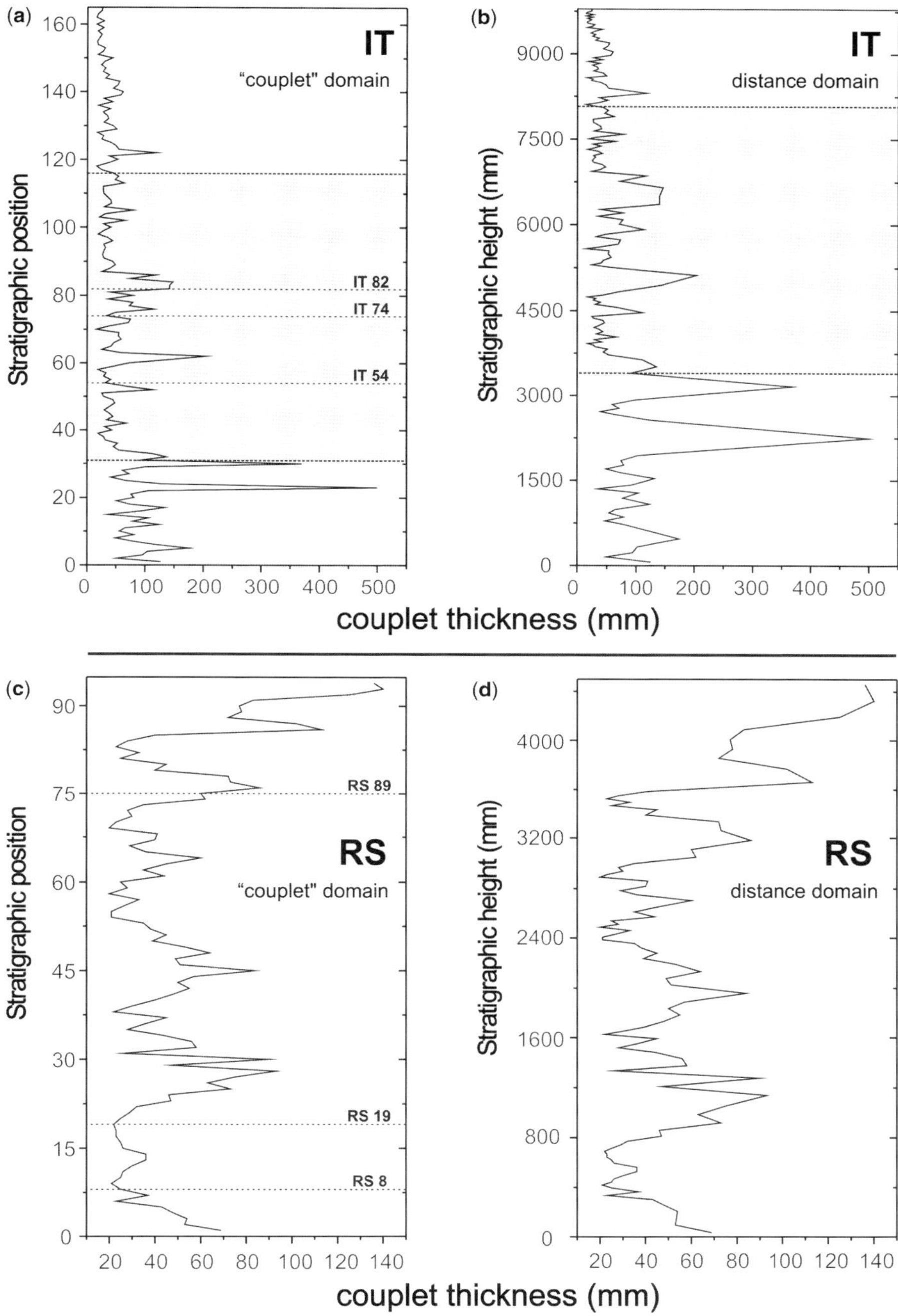

Fig. 3. Two assessments of couplet thickness series. (**a**) IT metronomic couplet thickness series (total number of couplets is 165); (**b**) IT stratigraphic couplet thickness series (total stratigraphic thickness is 9739 mm); (**c**) RS metronomic couplet thickness series (total number of couplets is 94); (**d**) RS stratigraphic couplet thickness series (total stratigraphic thickness is 4472 mm). Grey shading in (a) and (b) indicates the segment analysed for anisotropy of magnetic susceptibility (AMS) in this study. Positions of individual couplets with H–J diagrams in Figure 5a, c are shown – IT 54, IT 74, RS 8 and RS 19 – as well as the positions of couplets IT 82 and RS 89 analysed in Franco *et al.* (2011).

D. R. FRANCO & L. A. HINNOV

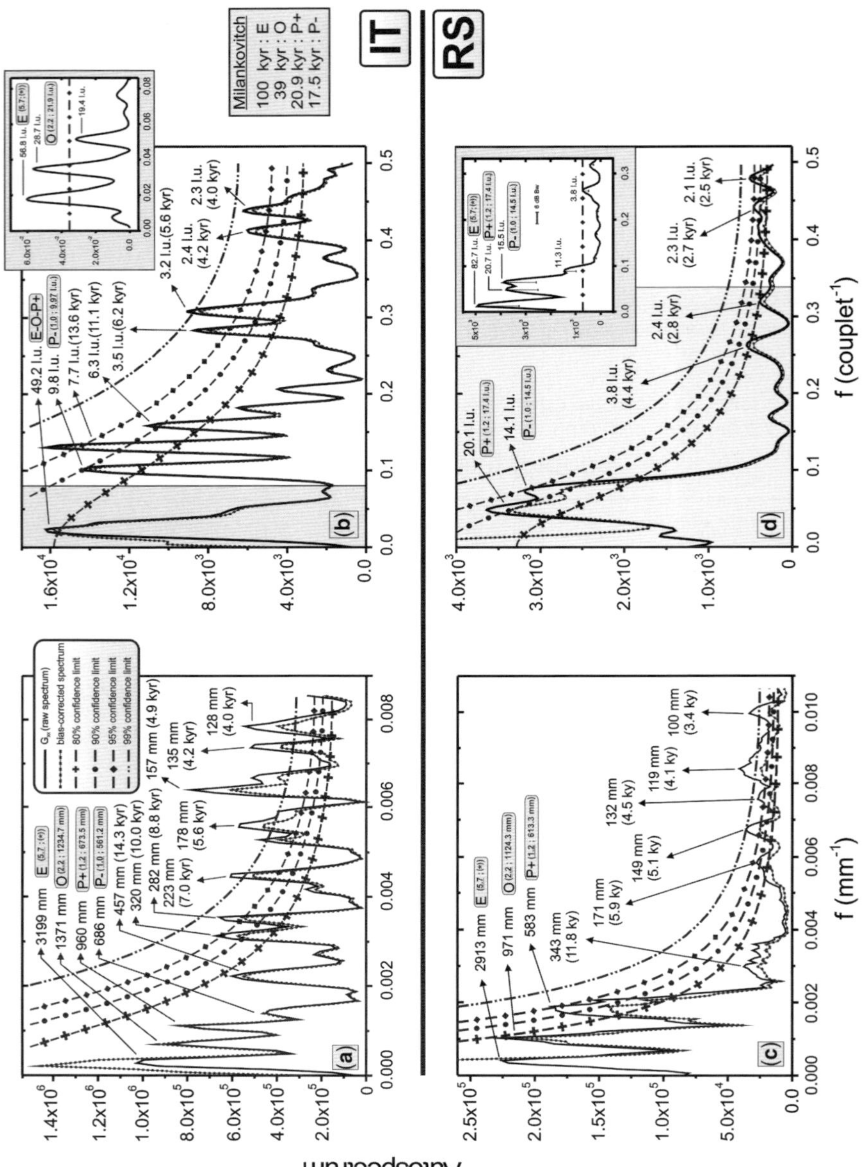

Fig. 4. Spectral analysis of the couplet thickness series. (**a**) Itu stratigraphic spectrum; (**b**) Itu metronomic spectrum; (**c**) RS stratigraphic spectrum; and (**d**) RS metronomic spectrum. Spectral peaks exceeding the confidence limits are labelled; time-periodicity values in ka are also given (see text for calibration details). Indications for Milankovitch-related components are represented by letters E (eccentricity), O (obliquity), P+ and P− (long and short precession, respectively); periods given in the grey legend are as predicted for the Permian (Berger *et al.* 1992). Inset figures in (b) and (d) are high-resolution harmonic analyses shown with 95% significance levels for null hypothesis tests for harmonic components (Schulz & Stattegger 1997), to isolate precession, obliquity and eccentricity frequencies. For all spectra: black line, raw spectrum; dashed line, bias-corrected spectrum; c, couplet.

by a regular cycling of thick couplets (thicknesses greater than 9.5 cm) with very thin ones (*c.* 2 cm). In the uppermost 25% of the sequence, there is a marked, upward thickening of couplets.

The RS couplet thickness spectra (Fig. 4c, d) are less complicated than the IT spectra. This is due in large part to the regularity of the thickness modulations through most of the succession. The metronomic spectrum (Fig. 4d) shows a strong bundling component modulating over 14–20 couplets. The harmonic analysis (inset in Fig. 4d) resolves a third component with a 82.7 couplet bundling cycle, although this is close to the length of the couplet thickness series (94 couplets) and does not represent a complete cycle.

K1 data were collected from the lowermost 4.46 m of the RS section; examples of H–J output are shown in Figure 5c, d. Similarly, RS results exhibit both K1 and K2 axes lying on horizontal planes with similar degrees of anisotropy, although with slightly smaller values (1.11–1.14) than those in the IT couplets. The data were collected at a median sample rate of 2.5 cm (there is a 28.5 cm gap just above the 1 m position; Fig. 6b). Thus, the K1 dataset in this case spans nearly the entire RS outcrop, and co-samples the RS couplet series. The K1 spectrum (Fig. 7b) has multiple spectral peaks, with none exceeding the 95% significance level, testament to the highly variable character of the series. However, it shares a 534 mm wavelength with the stratigraphic couplet spectrum (583 mm; Fig. 4c), which is thought to have captured a precession forcing signal, as discussed below.

Discussion

Time calibration to astronomical parameters

The time calibration for the IT and RS rhythmites, based on the method of ratios, is as follows: in the IT stratigraphic spectrum (Fig. 4a) the spectral peak with a wavelength of 3199 mm corresponds to the 100 ka-eccentricity cycle. Thus, the IT spectral peaks at wavelengths of 1371, 960 and 686 mm calibrate to 39 (obliquity), 20.9 and 17.5 ka (precession index) periods. In the RS stratigraphic spectrum (Fig. 4c) the peak at wavelength 583 mm corresponds to the 20.9 ka long precession cycle ('P+'); consequently the spectral peaks at wavelengths of 2913 and 971 mm calibrate to the 100 ka (short eccentricity) and 39 ka (obliquity) periods, respectively.

For the metronomic IT spectrum (Fig. 4b), astronomical frequencies indicated by the harmonic analysis include three components at 56.8 couplet units (100 ka, short eccentricity), 28.7 couplets (39 ka, obliquity) and a weak component at 19.4 couplets

(20.9 ka, long precession). The RS metronomic spectrum has spectral peaks at 82.7, 20.7, 15.5 and 14.1 couplet unit lengths, with spectral peak ratios (5.9:1.5:1.1:1) compared with theoretical Permian Milankovitch ratios (5.7:2.2:1.2:1), although the obliquity frequency ratio does not agree – 1.5 v. 2.2; Fig. 4d).

Interpretation of astronomical to millennial quasi-periodic harmonic features

Based on the astronomical calibration, the precession cycle played a predominant role in rhythmite deposition, especially for the RS rhythmites. Precessional forcing of climate and sedimentation has been widely discussed in the literature (e.g. Berger *et al.* 1993; Van der Zwan 2002). As discussed by Berger & Loutre (1997), harmonic components at *c.* 11.2–13.1 ka is compatible with hemi-precession; these harmonics may occur in the IT spectrum in peaks at 320 mm (10.0 ka; Fig. 4a), 6.3 couplets and 7.7 couplets (11.1 and 13.6 ka, respectively; Fig. 4b), as well as in the RS spectral peak at 343 mm (11.8 ka; Fig. 4c). Additionally, a short eccentricity-induced signal (100 ka) is a major spectral feature inferred from the IT couplet series. Heckel (2008) and Schmitz & Davydov (2012) reported that major glacial–interglacial fluctuations during the Late Carboniferous occurred with periodicities associated with the *c.* 400 ka (long) eccentricity cycle, and 100 ka (short) eccentricity and obliquity. However, it is not possible to infer a *c.* 400 ka signal from either of the Brazilian rhythmite successions. This is because, according to the time calibration (Franco *et al.*2012*a*), the IT series has an average sedimentation rate on the order of 3199 mm/ 100 ka = 31.99 mm/ka; the 11 419 mm long section therefore represents a duration of 11 419 mm/ (31.99 mm/ka) = 357 ka, that is, <400 ka. The RS succession is interpreted to have an average sedimentation rate of 583 mm/20.9 ka = 27.89 mm/ka; thus, the 4472 mm long section represents a duration of 4472 mm/(27.89 mm/ka) = 160 ka. (An expanded calibration analysis is given in Table 1, and is discussed further below.)

The couplet thickness spectra also show the influence of harmonic signals compatible with sub-Milankovitch quasi-periodicities, ranging from 8.8 to 4.0 ka (IT) and from 5.9 to 2.5 ka (RS). Contributions of non-linear effects related to Milankovitch cycles (e.g. harmonics, combination tones; King 1996; Kleiven *et al.* 2003) cannot be ruled out, although the extension of this influence on climate is questionable (Versteegh 2005). A number of processes for centennial-to-millennial climate change – which could also explain the harmonic features in these rhythmites – have been debated in the

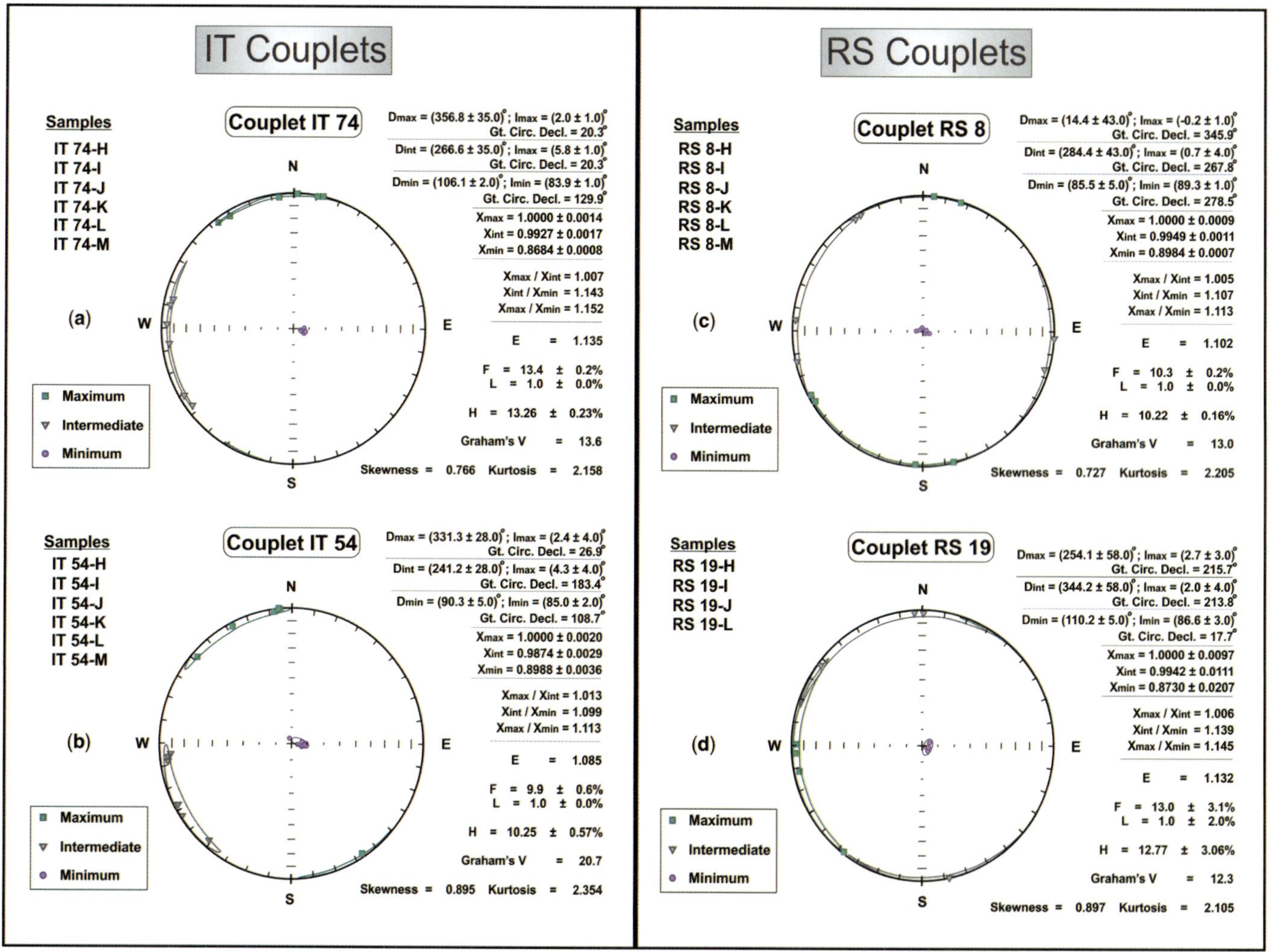

IT Couplets
RS Couplets

Couplet IT 74
Samples
IT 74-H
IT 74-I
IT 74-J
IT 74-K
IT 74-L
IT 74-M
(a)
N
W
E
S
Maximum
Intermediate
Minimum
Dmax = (356.8 ± 35.0)°; Imax = (2.0 ± 1.0)°
Gt. Circ. Decl. = 20.3°
Dint = (266.6 ± 35.0)°; Imax = (5.8 ± 1.0)°
Gt. Circ. Decl. = 20.3°
Dmin = (106.1 ± 2.0)°; Imin = (83.9 ± 1.0)°
Gt. Circ. Decl. = 129.9°
Xmax = 1.0000 ± 0.0014
Xint = 0.9927 ± 0.0017
Xmin = 0.8684 ± 0.0008
Xmax / Xint = 1.007
Xint / Xmin = 1.143
Xmax / Xmin = 1.152
E = 1.135
F = 13.4 ± 0.2%
L = 1.0 ± 0.0%
H = 13.26 ± 0.23%
Graham's V = 13.6
Skewness = 0.766   Kurtosis = 2.158

Couplet IT 54
Samples
IT 54-H
IT 54-I
IT 54-J
IT 54-K
IT 54-L
IT 54-M
(b)
N
W
E
S
Maximum
Intermediate
Minimum
Dmax = (331.3 ± 28.0)°; Imax = (2.4 ± 4.0)°
Gt. Circ. Decl. = 26.9°
Dint = (241.2 ± 28.0)°; Imax = (4.3 ± 4.0)°
Gt. Circ. Decl. = 183.4°
Dmin = (90.3 ± 5.0)°; Imin = (85.0 ± 2.0)°
Gt. Circ. Decl. = 108.7°
Xmax = 1.0000 ± 0.0020
Xint = 0.9874 ± 0.0029
Xmin = 0.8988 ± 0.0036
Xmax / Xint = 1.013
Xint / Xmin = 1.099
Xmax / Xmin = 1.113
E = 1.085
F = 9.9 ± 0.6%
L = 1.0 ± 0.0%
H = 10.25 ± 0.57%
Graham's V = 20.7
Skewness = 0.895   Kurtosis = 2.354

Couplet RS 8
Samples
RS 8-H
RS 8-I
RS 8-J
RS 8-K
RS 8-L
RS 8-M
(c)
N
W
E
S
Maximum
Intermediate
Minimum
Dmax = (14.4 ± 43.0)°; Imax = (-0.2 ± 1.0)°
Gt. Circ. Decl. = 345.9°
Dint = (284.4 ± 43.0)°; Imax = (0.7 ± 4.0)°
Gt. Circ. Decl. = 267.8°
Dmin = (85.5 ± 5.0)°; Imin = (89.3 ± 1.0)°
Gt. Circ. Decl. = 278.5°
Xmax = 1.0000 ± 0.0009
Xint = 0.9949 ± 0.0011
Xmin = 0.8984 ± 0.0007
Xmax / Xint = 1.005
Xint / Xmin = 1.107
Xmax / Xmin = 1.113
E = 1.102
F = 10.3 ± 0.2%
L = 1.0 ± 0.0%
H = 10.22 ± 0.16%
Graham's V = 13.0
Skewness = 0.727   Kurtosis = 2.205

Couplet RS 19
Samples
RS 19-H
RS 19-I
RS 19-J
RS 19-L
(d)
N
W
E
S
Maximum
Intermediate
Minimum
Dmax = (254.1 ± 58.0)°; Imax = (2.7 ± 3.0)°
Gt. Circ. Decl. = 215.7°
Dint = (344.2 ± 58.0)°; Imax = (2.0 ± 4.0)°
Gt. Circ. Decl. = 213.8°
Dmin = (110.2 ± 5.0)°; Imin = (86.6 ± 3.0)°
Gt. Circ. Decl. = 17.7°
Xmax = 1.0000 ± 0.0097
Xint = 0.9942 ± 0.0111
Xmin = 0.8730 ± 0.0207
Xmax / Xint = 1.006
Xint / Xmin = 1.139
Xmax / Xmin = 1.145
E = 1.132
F = 13.0 ± 3.1%
L = 1.0 ± 2.0%
H = 12.77 ± 3.06%
Graham's V = 12.3
Skewness = 0.897   Kurtosis = 2.105

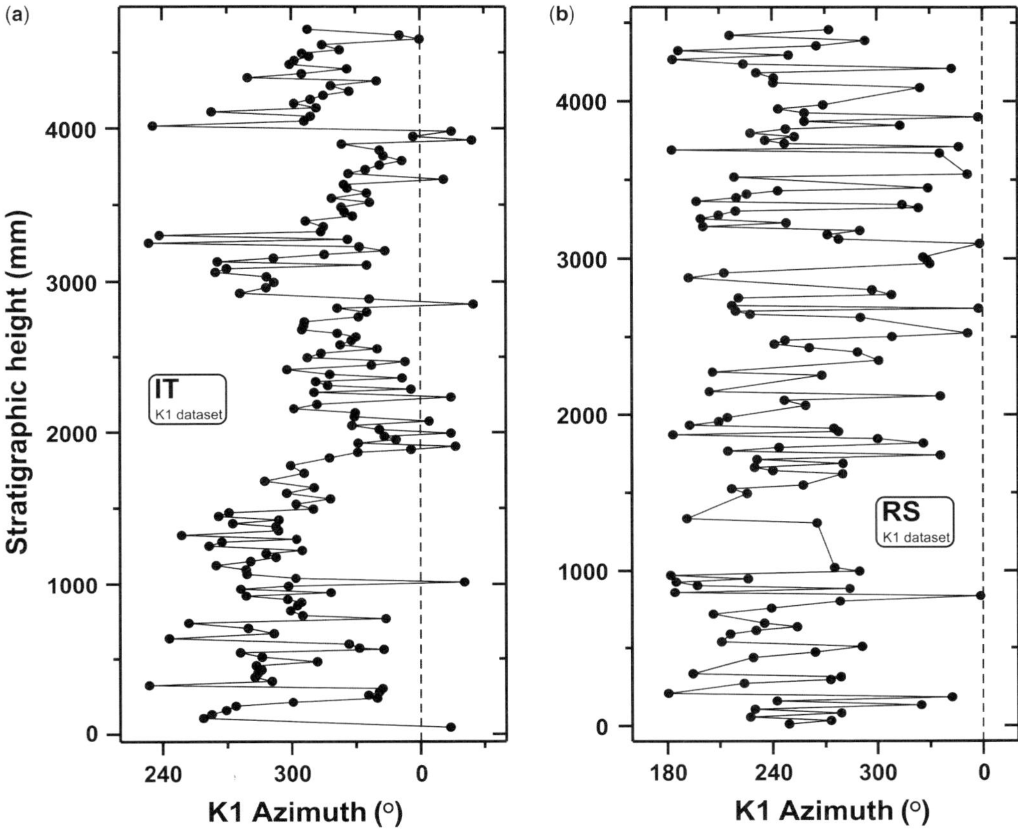

Fig. 6. K1 stratigraphic series for (**a**) IT rhythmites and (**b**) RS rhythmites. The lower portions of the outcrops were sampled for AMS measurements: 4.65 m for IT rhythmites; and 4.46 m for RS rhythmites.

literature (e.g. solar activity, oceanic–atmospheric feedbacks), most in the context of palaeoclimatic records from the last Ice Age (0–100 ka). Some changes occurred on a worldwide scale, and triggered abrupt climatic events (Stocker 1999; Luckman 2000; Paillard 2001; Bard 2002; Hinnov *et al.* 2002; Sharma 2002; Schulz *et al.* 2004; Gallet *et al.* 2005; Witt & Schumann 2005).

Significant periodicities between 2.8 and 2.4 ka in the metronomic RS spectrum are compatible with the *c.* 2.0–2.4 ka Hallstatt solar cycle, related to the long-period modulation of heliomagnetic activity (Ogurtsov *et al.* 2002; Clilverd *et al.* 2003). This harmonic feature, which has been linked to ^{14}C production variability and is recorded in Holocene tree rings (e.g. Thomson 1990; van Geel *et al.* 1999),

Fig. 5. Examples of site-mean statistics and principal AMS tensor axes from Monte-Carlo simulation and Hext–Jelinek statistical analysis (Lienert 1991), displayed in Schmidt equal-area projection, lower hemisphere. Results are indicated with respect to 95% elliptical cones of confidence for four representative IT and RS couplets: IT 74 (**a**), IT 54 (**b**), RS 8 (**c**) and RS 19 (**d**). (Stratigraphic positions are noted in Figure 3a, c.) It is possible to see the directions related to the individual samples from mean-tensor orientation. X_{max}, X_{int} and X_{min} are, respectively, maximum, intermediate and minimum susceptibility eigenvalues; $(D_{max}; I_{max})$, $(D_{int}; I_{int})$ and $(D_{min}; I_{min})$ are respectively mean declination and inclination values related to the maximum, intermediate and minimum susceptibility eigenvectors; X_{max}/X_{int}, X_{int}/X_{min} and X_{max}/X_{min} are, respectively, lineation, foliation and degree of anisotropy; F and L are respectively the planar and linear magnetic anisotropy; H is the total anisotropy; E is the length difference of the major and minor susceptibility axes.

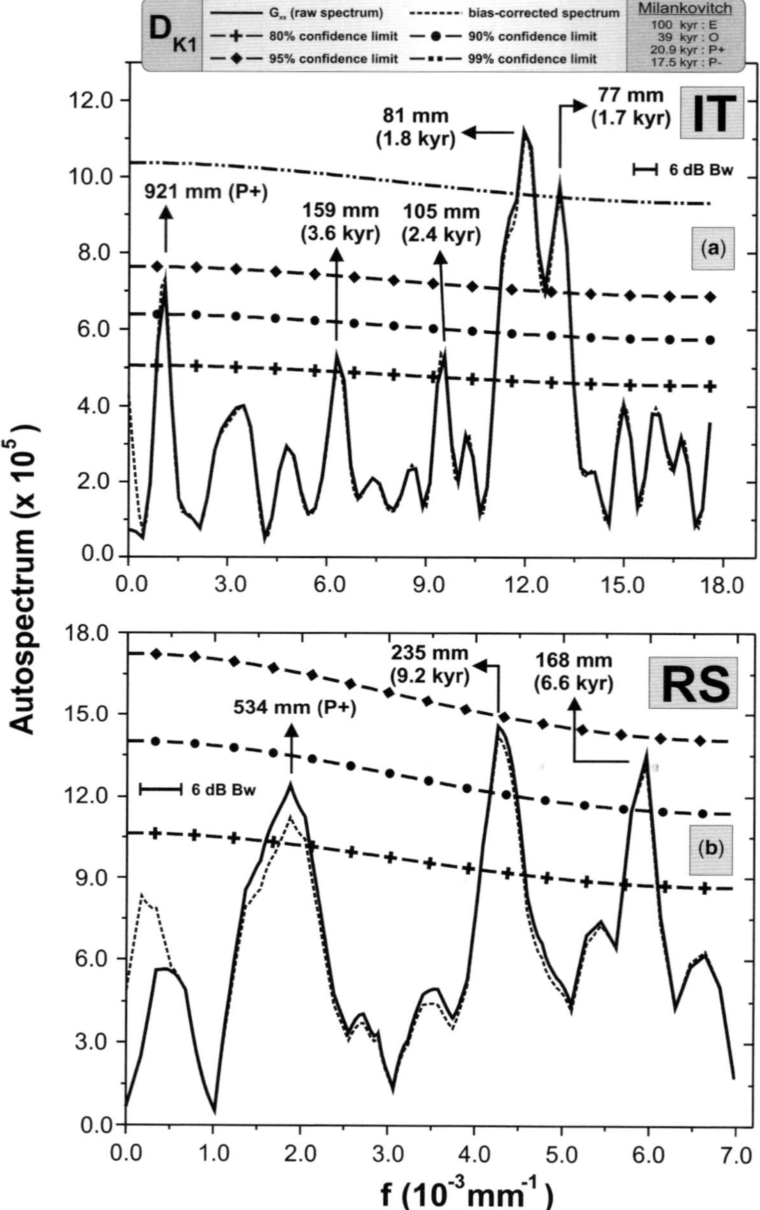

Fig. 7. Spectral analysis of the K1 stratigraphic series: (**a**) Itu and (**b**) Rio do Sul rhythmites. Spectral peaks exceeding the confidence limits are labelled; time-periodicity values in ka are also given (see the text for details). Indications for Milankovitch-related spectral peaks are represented by P+ (long precession). Milankovitch periods are given in the grey legend as predicted for the Permian (Berger *et al.* 1992). For all spectra: black line, raw spectrum; dashed line, bias-corrected spectrum.

was recently suggested to be generated by amplitude modulation of the Gleissberg (*c.* 210 year) solar cycle (Usoskin & Mursula 2003) and attributable to solar inertial motion (Charvátová 2000). The Hallstatt cycle is now well documented in Holocene climate records. There is also evidence from deep

Table 1. *Estimates of average sedimentation rates for Itu and Rio do Sul successions based on Milankovitch cycle periods and corresponding wavelengths in stratigraphic spectra*

IT	$T_e \equiv 100$ kyr	$T_O \equiv 39$ kyr	$T_{P+} \equiv 20.9$ kyr	$T_{P-} \equiv 17.5$ kyr
T_i (mm) 3199	$R_e = 31.99$			
1371		$R_o = 35.15$		
960			$R_{P+} = 45.93$	
$921^{(*)}$			$R_{P+}^{(*)} = 44.07$	
686				$R_{P-} = 39.20$
R_{IT}		39.3 $\pm$ 2.6 mm/ky		
RS	$T_e \equiv 100$ kyr	$T_O \equiv 39$ kyr	$T_{P+} \equiv 20.9$ kyr	$T_{P-} \equiv 17.5$ kyr
T_i (mm) 2913	$R_e = 29.13$			
971		$R_o = 24.90$		
583			$R_{P+} = 27.89$	
$534^{(*)}$			$R_{P+}^{(*)} = 25.55$	
–				–
R_{RS}		26.9 $\pm$ 1.0 mm/kyr		

T_e, T_O, T_{P+} and T_{P-} : Respectively, 100-kyr eccentricity, 39.0-kyr obliquity, 20.9-kyr and 17.5-kyr precession cycles for Permian time (Berger *et al.*, 1992); T_i: corresponding spectral peak wavelength, related to the i-Milankovitch cycle (i = e (eccentricity); o (obliquity); P+ (20.9-kyr precession) and P− (17.5-kyr precession) provided from stratigraphic spectra of IT and RS data series; (*) spectral peak wavelength from stratigraphic spectra of K1 data; R_{IT} and R_{RS}: average sedimentation rates respectively for Itu and Rio do Sul successions.

time for this cycle, notably the low-latitude Pangean evaporites of the Late Permian (Early Changhxingian) Castile Formation, southwestern USA, which exhibit strong hot-house palaeoclimatic fluctuations with a 2.3 ka period over an extended time interval of 200 ka (Anderson 2011). Here, the IT and RS rhythmites provide evidence that millennial-scale heliomagnetic activity was present 40–50 Ma earlier during the Permo-Carboniferous ice-house.

The IT and RS couplet thickness spectra are very similar to the characteristic remanent magnetization and bulk magnetic susceptibility spectra reported for these same rhythmites (Franco *et al.* 2012*a*). The ratios of the very low frequency components were proposed to represent Milankovitch periodicities, with spectral peaks in the stratigraphic domain spectra compatible with those found in this study. These palaeomagnetic data contain millennial-scale variations reminiscent of abrupt climatic oscillations during Late Quaternary times, that is, the so-called 'Bond cycles', a quasi-periodic interplay of successive 1470 year Dansgaard–Oeschger (DO) cycles packaged into 3.0–8.0 ka Heinrich warming events (Bond *et al.* 1993, 2001; Stocker 1999; Bard 2002; Berger & von Rad 2002; Rind 2002), as well as the 2.4 ka Hallstatt solar cycle. Eyles *et al.* (1997) reported that ice-rafted layers from the Australian Early Permian could be analogous to records generated by Heinrich events, which are themselves indicative of warming events.

This suggests that the same millennial forcing mechanisms have been prevalent throughout geological time.

Interpretation of the K1 power spectra

Both K1 spectra (Fig. 7) show evidence for cyclicity that can be assigned to the 20.9 ka precessional cycle. Based on the astronomical calibration, the spectral peaks in the IT K1 spectrum at 81 and 77 mm have 1.8 and 1.7 ka periods, close to the DO period. The peak at 159 mm converts to 3.6 ka, and may be evidence for Heinrich events (3.0–8.0 ka periodicity), although it is a multiple of 1.8 ka. The 2.4 ka period (at 105 mm) could reflect the *c.* 2.3–2.4 ka Hallstatt solar cycle. The RS K1 spectrum exhibits two spectral peaks at 168 and 235 mm, which convert to 6.6 and 9.2 ka. The shorter periodicity could be linked to Heinrich events; the 9.2 ka cycle could be a harmonic of the 17.5 ka short precession cycle.

The dia- and paramagnetic sediments of the IT rhythmites in particular appear to be controlled by bottom currents, resulting in strong inclination flattening in the ChRM (Franco *et al.* 2012*b*). The currents are in turn linked to large-scale climatic processes, as suggested by the spectral analysis. The IT rhythmites indicate a NW–SE directional pattern, whereas the RS rhythmites have almost a radial distribution of directions. The RS depositional

Table 2. *Spectral peaks from astronomical calibration of the couplet thickness spectra and maximum axis of anisotropy of magnetic susceptibility (AMS) tensor (K1) datasets (stratigraphic and metronomic) from Itu (IT) and Rio do Sul (RS) rhythmites, associated with suggested millennial time-scales and corresponding statistical significance*

Millennial time-scale range	IT: converted spectral peaks (in ka)		RS: converted spectral peaks (in ka)		IT: converted K1 spectral peaks(*) (in ka)	RS: converted K1 spectral peaks(*) (in ka)
	Stratigraphic	Metronomic	Stratigraphic	Metronomic	Stratigraphic	Stratigraphic
Milankovitch cycles (Berger *et al.* 1992)	100 (E; >80%*)	100 (E*)	100 (E; <80%*)	100 (E;* >95%)	—	—
	39 (O; <80%)	39 (O; <80%)	39 (O; >80%)	—	—	—
	20.9 (P+; <80%)		20.9 (P+;* >95%)	20.9 (P+; >5%)	20.9 (P+; >95%*)	20.9 (P+; >80%*)
	17.5 (P − ; <80%)	17.5 (P − ; >90%)	—	17.5 (P−; >95%)	—	—
Sub-Milankovitch range	14.3 (>80%)	13.6 (>90%)	—	—	—	—
	10.0 (>95%)	11.1 (>90%)	11.8 (<80%)	—	—	9.2 (>90%)
	8.8 (>95%)	—	—	—	—	—
	7.0 (>99%)	—	—	—	—	—
	—	6.2 (>95%)	—	—	—	6.6 (>90%)
2.3 ka Hallstattzeit cycle (?)	5.6 (>95%)	5.6 (>99%)	5.9 (>90%)	—	—	—
	4.9 (>99%)	—	5.1 (>99%)	—	—	—
H events (?)	—	—	—	—	—	—
	4.2 (>99%)	4.2 (>95%)	4.5 (>95%)	4.4 (>80%)	—	—
	4.0 (>99%)	4.0 (>95%)	4.1 (>99%)	—	—	—
DO events (?)	—	—	3.4 (>99%)	—	3.6 (>80%)	—
	—	—	—	—	—	—
	—	—	—	2.8 (>80%)	—	—
c. 2.3 ka Hallstattzeit cycle (?)	—	—	—	2.7 (>95%)	—	—
	—	—	—	2.5 (>95%)	2.4 (>80%)	—
DO events range (?)	—	—	—	—	1.8 (>99%)	—
	—	—	—	—	1.7 (>99%)	—

E, O, P+ and P−: Respectively, 100 ka-eccentricity, 39.0 ka-obliquity, 20.9 ka and 17.5 ka-precession cycles (Berger *et al.* 1992); *: defined spectral peak for astronomical calibration; H-events: Heinrich events; DO events: Dansgaard-Oeschger events.

Table 3. *Average deposition periods per couplet for Itu and Rio do Sul successions, based on Milankovitch cycle periods and its corresponding wavelengths (number of couplets) in metronomic spectra*

IT		$T_e \equiv 100$ ka	$T_O \equiv 39$ ka	$T_{P+} \equiv 20.9$ ka	$T_P - \equiv 17.5$ ka
T_i (number of couplets)	$56.8_{(e)}$	$T_e = 1.76$ ka/ couplet			
	$28.7_{(O)}$		$T_O = 1.36$ ka/ couplet		
	$9.8_{(P-)}$				$T_P = 1.79$ ka/ couplet
$\bar{T}_{IT}$			1.64 ± 0.14 ka/couplet		
RS		$T_e \equiv 100$ ka	$T_O \equiv 39$ ka	$T_{P+} \equiv 20.9$ ka	$T_P \equiv 17.5$ ka
T_i (number of couplets)	$82.7_{(e)}$	$T_e = 1.21$ ka/ couplet			
	$20.7_{(P+)}$			$T_{P-} = 1.01$ ka/ couplet	
	$15.5_{(P-)}$				$T_P = 1.13$ ka/ couplet
T_{RS}			1.12 ± 0.10 ka/couplet		

T_e, T_O, T_{P+} and T_{P-}, Respectively, 100 ka eccentricity, 39.0 ka obliquity, 20.9 and 17.5 ka precession cycles (Berger *et al.*, 1992); T_i, corresponding spectral peak wavelength, related to the *i*-Milankovitch cycle ($i = $ e (eccentricity); o (obliquity); P+ (20.9 ka precession) and P− (17.5 ka precession) from metronomic spectra of IT and RS data series; T_{IT} and T_{RS}, average depositional periods per couplet for Itu and Rio do Sul successions, respectively.

site was far from major sedimentary sources, as testified by a fine granulometry and the absence of traction load sedimentary structures.

Average sedimentation rates, and the time scale of the rhythmite couplets

The spectral peaks compatible with Milankovitch cycles were recorded at shorter wavelengths and over lower numbers of couplets in the IT rhythmites compared with those of the RS rhythmites. This can be interpreted as evidence for a difference in average sedimentation rate between the two rhythmites. We estimated average sedimentation rates for both successions through astronomical calibration of the stratigraphic couplet thickness and K1 spectra (Table1). For each succession, cycle wavelengths (Figs 4a, c & 7a, b) were divided by their related Milankovitch periods as identified through the method of ratios. This provided different thickness to Milankovitch cycle period ratios (*Ri*, where $i \equiv$ corresponding Milankovitch cycle) in millimetres per thousand years. The average sedimentation rate values are 39.3 ± 2.6 mm/ka and 26.9 ± 1.0 mm/ka for the IT and RS successions, respectively.

These results are compatible with the maximum and minimum range of values for Permian sedimentation rates as discussed in Kukal (1990): 2–22 cm/ ka, with average rate of *c.* 10 cm/ka. Through an independent spectral analysis and interpretation of well log data and high-resolution scanning of Itararé Group rhythmites recovered by drill cores at Anitápolis (Fig. 1), Silva & Azambuja Filho (2005) inferred sedimentation rates between 2 and 20 cm/ka. Also noteworthy is comparison of our results with sedimentation rates estimated for glacial and interglacial deposits in the Northern Hemisphere. Elverhøi & Henrich (1996) indicate sedimentation rates between 1 and 5 cm/ka for marine and shelf sediments during interstadial periods, and higher values >10 cm/ka during stadials.

We also developed new information for insights into possible palaeoenvironmental processes, which could explain the remarkable repetition of lithologies at these outcrops – and as an alternative for the hypothesis annual deposition of lithological couplets. We divided the values of the spectral peaks related to Milankovitch forcing in the metronomic couplet thickness spectra (Fig. 4b, d and Table 2) by their inferred Milankovitch cycles. It resulted, for both successions, in a set of estimates for 'ka/couplet' (Table 3). The average values, 1.64 ± 0.14 and 1.12 ± 0.10 ka/couplet for IT and RS successions, respectively, are remarkably compatible, considering that the time calibrations that were used are independent for the two successions. Together, the estimates indicate an average value of 1.38 ka/couplet, which is close to the *c.* 1.5 ka timing of the Pleistocene DO events.

Conclusions

The procedures carried out in this study are summarized as follows:

- Sedimentary cycle ('couplet') thicknesses and the maximum axis (K1) of AMS ellipsoidal tensor were measured along the outcrops of two glacio-lacustrine rhythmite sequences in the Itararé Group (Paraná Basin, Brazil), from the Late Carboniferous Itu rhythmites and Early Permian Rio do Sul rhythmites.
- Analysis of couplet thickness sequences revealed thickness modulation patterns that, through the method of frequency ratios and in the absence of independent geochronology, can be linked to Milankovitch cycles.
- Astronomical calibration of the couplet thickness and K1 spectra provided average sedimentation rate estimates for the successions: 39.3 ± 2.6 mm/ka for the Itu rhythmites and 26.9 ± 1.0 mm/ka for the Rio do Sul rhythmites.
- The astronomical calibration also indicated an average 1.38 ka duration for the couplets that characterize both successions, reminiscent of the 1.5 ka timing of Late Pleistocene DO cycles.
- Spectral peaks with periods of 8.8–4.0 ka (IT thickness spectra), 5.9–3.4 ka (RS thickness spectra) and 6.6 ka (RS K1 spectrum) are compatible with the timing of Late Pleistocene Heinrich events; spectral peaks with 1.8–1.7 ka periods for the high-resolution RS K1 spectrum is reminiscent of the c. 1.5 ka cycle related to DO events.
- For both outcrops, wavelengths of K1 spectral peaks inferred as P+ (long precession cycle) have matching counterparts in the couplet thickness spectra. This suggests that the maximum axis (K1) of the AMS ellipsoidal tensor can be exploited as a palaeoclimatic proxy in depositional systems that are influenced by currents linked to climatic forcing.
- Spectral peaks with periods ranging from 2.8 to 2.5 ka in the RS metronomic couplet thickness spectrum and the spectral peak at 2.4 ka in the K1 IT spectrum could indicate the presence of a (2.4 ka) Hallstatt solar cycle.

Our interpretation is in opposition to the traditional hypothesis of a varvic depositional character of the rhythmite couplets. Franco et al. (2011) analysed high-resolution greyscale scans of sediment cores cut from representative couplets of these successions, which supported an annual timescale for the pervasive millimetre-scale laminations found within couplets. A similar conclusion was reached by Silva & Azambuja Filho (2005).

What kind of perpetual harmonic phenomenon could cause the observed, repeated couplet lithologies – the almost constant, discrete thickness of dark lamina separating couplets – and the similarity in the harmonic process involved in the deposition of these rhythmites, separated by such large distances (c. 1000 km) and times (c. 10 million years)? For these same successions, Franco et al. (2012a) reported harmonic features from a spectral analysis of bulk magnetic susceptibility and ChRM data series, for which they found evidence for a quasi-periodic c. 2.3–2.4 ka Halstatt cycle, as well as periodicities suggestive of DO and H events. This evidence – which is reinforced by the results of the present study – indicates the prevalence of these harmonic features in the palaeoclimate system throughout geological time; hence they are not restricted only to the Quaternary Period.

The thickness variability of sedimentary cycles, as well as bulk susceptibility K_Z – a palaeoclimatic proxy indicating ferromagnetic mineral concentration (Verosub & Roberts 1995) – in rhythmic sedimentary successions, indicate strong influence from external climatic factors. For these Permo-Carboniferous rhythmites, the similarities between the couplet thickness and K1 spectra, indicating common millennial-scale patterns, are also in accordance with those in K_Z spectra for the same successions (Franco et al. 2012a). The compatibility of harmonic features in the rhythmite couplet thickness sequences and K1 data support the view that K1 is an effective palaeoclimatic proxy for sedimentation influenced by sediment transport, which in turn is forced by climatic processes. This result opens a new prospect for a palaeoclimatic 'marker' based on the directional variability of sedimentation, which could provide information on major, large-scale climatic processes linked to the time evolution of changing factors in depositional and transport mechanisms.

The authors are grateful to M. Ernesto, I. Pacca (IAG-USP) and A. Rocha-Campos (IGc-USP) for valuable discussions that contributed to the development of the present study. We thank B. Ellwood and A. Sinito for thoughtful and thorough reviews that significantly improved the paper. This work was supported by Brazilian agencies FAPESP (grant 02/06480-0) and CAPES (grant 2603-07-1).

References

ALLEN, J. R. L. 1982. *Sedimentary Structures: Their Character and Physical Basis*, **1** and **2**, Elsevier, Amsterdam.

ANDERSON, R. Y. 2011. Enhanced climate variability in the tropics: a 200 000 yr annual record of monsoon variability from Pangea's equator. *Climate of the Past*, **7**, 757–770.

ARCHANJO, C. J., SILVA, M. G., CASTRO, J. C., LAUNEAU, P., TRINDADE, R. I. F. & MACEDO, J. W. P. 2006. AMS and grain shape fabric of the Late Palaeozoic diamictites of the Southeastern Paraná Basin, Brazil. *Journal of Geological Society*, **163**, 95–106.

BARD, E. 2002. Climate shock: abrupt changes over millennial time scales. *Physics Today*, **55**, 32–38.

BERGER, A. & LOUTRE, M.-F. 1997. Intertropical latitudes and precessional and half-precessional cycles. *Science*, **278**, 1476–1478.

BERGER, W. H. & VON RAD, U. 2002. Decadal to millennial cyclicity in varves and turbidites from the Arabian Sea: hypothesis of tidal origin. *Global and Planetary Change*, **34**, 313–325.

BERGER, A., LOUTRE, M. F. & LASKAR, J. 1992. Stability of the astronomical frequencies over the Earth's history for paleoclimate studies. *Science*, **255**, 560–566.

BERGER, A., LOUTRE, M.-F. & TRICOT, C. 1993. Insolation and earth's orbital periods. *Journal of Geophysical Research*, **98**, 10341e11036.

BOND, G., BROECKER, W., JOHNSEN, S., MCMANUS, J., LABEYRIE, L., JOUZEL, J. & BONANI, G. 1993. Correlations between climate records from North Atlantic sediments and Greenland ice. *Nature*, **365**, 143–147.

BOND, G., KROMER, B. *ET AL.* 2001. Persistent solar influence on North Atlantic climate during the Holocene. *Science*, **294**, 2130–2136.

BORRADAILE, G. J. 1988. Magnetic susceptibility, petrofabrics, and strain, a review. *Tectonophysics*, **156**, 1–20.

BORRADAILE, G. J. & JACKSON, M. 2010. Structural geology, petrofabrics and magnetic fabrics (AMS, AARM, AIRM). *Journal of Structural Geology*, **32**, 1519–1551.

BRADÁK, B. 2009. Application of anisotropy of magnetic susceptibility (AMS) for the determination of paleo-wind directions and paleo-environment during the accumulation period of Bag Tephra, Hungary. *Quaternary International*, **198**, 77–84.

BURMEISTER, K. C., HARRISON, M. J., MARSHAK, S., FERRÉ, E. C., BANNISTER, R. A. & KODAMA, K. P. 2009. Comparison of Fry strain ellipse and AMS ellipsoid trends to tectonic fabric trends in very low-strain sandstone of the Appalachian fold–thrust belt. *Journal of Structural Geology*, **31**, 1028–1038.

CAETANO-CHANG, M. R. & FERREIRA, S. M. 2006. Ritmitos de Itu: Petrografia e Considerações Paleodeposicionais *Geociências (São Paulo)*, **25**, 345–358.

CANUTO, J. R. 1993. *Fácies e Ambientes de Sedimentação da Formação Rio do Sul (Permiano), Bacia do Paraná, na Região de Rio do Sul, Estado de Santa Catarina.* PhD thesis, IGC-USP, Brazil.

CHARVÁTOVÁ, I. 2000. Can origin of the 2400-year cycle of solar activity be caused by solar inertial motion? *Annales de Geophysique*, **18**, 399–405.

CHEN, Q., LI, C., LI, P., LIU, B. & SUN, H. 2008. Late Quaternary palaeosols in the Yangtze Delta, China, and their palaeoenvironmental implications. *Geomorphology*, **100**, 465–483.

CLILVERD, M. A., CLARKE, E., RISHBETH, H., CLARK, T. D. G. & ULICH, T. 2003. Solar cycle: solar activity levels in 2100. *Astronomy and Geophysics*, **44**, 5.20–5.22.

ELVERHØI, A. & HENRICH, R. 1996. Glaciomarine environments 'ancient glaciomarine sediments. *In*: MENZIES, J. (ed.) *Past Glacial Environments: Sediments, Forms and Techniques, Glacial Environments*, **2**. Butterworth-Heinemann, Oxford, 179–211.

ERNESTO, M. & PACCA, I. G. 1981. Spectral analysis of Permocarboniferous geomagnetic variation data from glacial rhythmites. *Journal of the Royal Astronomical Society*, **67**, 641–647.

EYLES, C. H., EYLES, N. & FRANÇA, A. B. 1993. Glaciation and tectonics in an intracratonic basin: the Late Palaeozoic Itararé Group, Paraná Basin, Brazil. *Sedimentology*, **40**, 1–25.

EYLES, N., EYLES, C. H. & GOSTIN, V. A. 1997. Iceberg rafting and scouring in the Early Permian Shoalhaven Group of New South Wales, Australia: evidence of Heinrich-like events? *Palaeogeography, Palaeoclimatology and Palaeoecology*, **136**, 1–17.

FERREIRA, S. M. 1997. *Ritmitos Várvicos do subgrupo Itararé – O Exemplo da Pedreira de Varvitos de Itu*, M.Sc. Dissertation, IGCE-UNESP, Brazil.

FIELDING, C. R., FRANK, T. D. & ISBELL, J. 2008. The late Paleozoic ice age – a review of current understanding. *In*: FIELDING, C. R., FRANK, T. D. & ISBELL, J. (eds) *Resolving the Late Paleozoic Ice Age in Time and Space*. Geological Society of America, Boulder, CO, Special Papers, **441**, 343–354.

FRANÇA, A. B. & POTTER, P. E. 1988. Estratigrafia, ambiente deposicional e análise de reservatório do Grupo Itararé (Permocarbonífero), Bacia do Paraná (parte 1). *Boletim de Geociências da Petrobrás*, **2**, 147–191.

FRANÇA, A. B. & POTTER, P. E. 1991. Stratigraphy and reservoir potential of glacial deposits of the Itararé group (Carboniferous–Permian), Paraná Basin, Brazil. *Bulletin of the American Association Petroleum Geology*, **75**, 62–85.

FRANÇA, A. B., WINTER, W. R. & ASSINE, M. L. 1996. Arenitos Lapa–Vila Velha: um modelo de trato de sistemas subaquosos canal-lobos sob influência glacial, Grupo Itararé (C-P), Bacia do Paraná. *Revista Brasileira de Geociências*, **26**, 43–56.

FRANCO, D. R. 2007. *Magnetoestratigrafia e Análise Espectral de Ritmitos Permocarboníferos da Bacia do Paraná: Influências dos Ciclos Orbitais no Regime Deposicional*. PhD thesis, IAG-USP, Brazil.

FRANCO, D. R., HINNOV, L. A. & ERNESTO, M. 2011. Spectral analysis and modeling of microcyclostratigraphy in Late Paleozoic glaciogenic rhythmites (Paraná Basin, Brazil). *Geochemistry, Geophysics and Geosystems*, **12**, Q09003, http://dx.doi.org/10.1029/2011GC003602

FRANCO, D. R., HINNOV, L. A. & ERNESTO, M. 2012a. Millennial-scale climate cycles in Permian-Carboniferous rhythmites: Permanent feature throughout geologic time? *Geology*, **40**, 19–22.

FRANCO, D. R., ERNESTO, M., PONTE-NETO, C. F., HINNOV, L. A., BERQUÓ, T. S., FABRIS, J. D. & ROSIÈRE, C. A. 2012b. Magnetostratigraphy and mid-paleolatitude VGP dispersion during the Permo-Carboniferous Superchron: results from Paraná Basin (southern Brazil) Rhythmites. *Geophysical Journal International*, http://onlinelibrary.wiley.com/doi/10.1111/j.1365-246X.2012.05670.x/abstract.

FULLER, M. 1963. Magnetic anisotropy and paleomagnetism. *Journal of Geophysical Research*, **68**, 293–309.

GALLET, Y., GENEVEY, A. & FLUTEAU, F. 2005. Does Earth's magnetic field secular variation control centennial climate change? *Earth and Planetary Science Letters*, **236**, 339–347.

GAMA, E. G. JR, PERINOTTO, J. A. J., RIBEIRO, H. J. P. S. & PADULA, E. K. 1992. Contribuição ao estudo da ressedimentação no subgrupo Itararé: Tratos de fácies e hidrodinâmica deposicional. *Revista Brasileira de Geociências.*, **22**, 228–236.

GESICKI, A. L. D., RICCOMINI, C. & BOGGIANI, P. C. 2002. Ice flow direction during late Paleozoic glaciation in western Paraná Basin, Brazil. *Journal of the South American Earth Sciences*, **14**, 933–939.

GRAHAM, J. W. 1954. Magnetic susceptibility anisotropy, an unexploited petrofabric element. *Geological Society of American Bulletin*, **65**, 1257–1258.

HECKEL, P. H. 2008. Pennsylvanian cyclothems in Midcontinent North America as far-field effects of waxing and waning of Gondwana ice sheets. *In*: FIELDING, C. R., FRANK, T. D. & ISBELL, J. L. (eds) *Resolving the Late Paleozoic Ice Age in Time and Space.* Geological Society of America, Boulder, CO, Special Papers, **441**, 275–289.

HERBERT, T. D. 1992. Paleomagnetic calibration of Milankovitch cyclicity in Lower Cretaceous sediments. *Earth and Planet Science Letters*, **112**, 15–28.

HILGEN, F. J., LOURENS, L. J. & VAN DAM, J. 2012. Chapter 29: the neogene period. *In*: GRADSTEIN, F., OGG, J., OGG, G. & SMITH, D. (eds) *A Geologic Time Scale 2012.* Elsevier, Amsterdam, 947–1002.

HINNOV, L. A. 2000. New perspectives on orbitally forced stratigraphy. *Annual Review of Earth and Planetary Sciences*, **28**, 419–475.

HINNOV, L. A. & HILGEN, F. J. 2012. cyclostratigraphy and astrochronology. *In*: GRADSTEIN, F., OGG, J., OGG, G. & SMITH, D. (eds) *A Geologic Time Scale 2012.* Elsevier, Amsterdam, 63–83.

HINNOV, L. A., SCHULZ, M. & YIOU, P. 2002. Interhemispheric space-time attributes of the Dansgaard–Oeschger oscillations between 0–100 ka. *Quaternary Science Reviews*, **21**, 1213–1228.

HROUDA, F. 1982. Magnetic anisotropy of rocks and its application in geology and geophysics. *Geophysical Surveys*, **5**, 37–82.

ISING, G. 1942. On the magnetic properties of varved clay. *Arkiv för Matematik Astronomi och Fysik*, **29A**, 1–37.

JACKSON, M. 1991. Anisotropy of magnetic remanence: a brief review of mineralogical sources, physical origins, and geological applications, and comparison with susceptibility anisotropy. *Pure and Applied Geophysics*, **136**, 1–28.

KANAMATSU, T. 1996. Magnetic fabric analysis of fine-grained sediments, Iberia Abyssal Plain. *In*: WHITMARSH, R. B., SAWYER, D. S., KLAUS, A. & MASSON, D. G. (eds) *Proceedings of the Ocean Drilling Program, Part B, Scientific Results*, **149**. Ocean Drilling Program, Texas A & M University, College Station, TX, 335–337.

KING, T. 1996. Quantifying nonlinearity and geometry in time series of climate. *Quaternary Science Review*, **15**, 247–266.

KLEIVEN, H. F., JANSEN, E., CURRY, W. B., HODELL, D. A. & VENZ, K. 2003. Atlantic Ocean thermohaline circulation changes on orbital to suborbital timescales during the mid-Pleistocene. *Paleoceanography*, **18**, 1008.

KODAMA, K. P. & SUN, W. W. 1990. SEM and magnetic fabric study of a compacting sediment. *Geophysical Research Letters*, **17**, 795–798.

KRUIVER, P. P., DEKKERS, M. J. & LANGEREIS, C. G. 2000. Secular variation in Permian red beds from Dôme de Barrot, SE France. *Earth and Planetary Science Letters*, **179**, 205–217.

KUKAL, Z. 1990. The rate of geological processes. *Earth Science Reviews*, **28** 1–284.

LAGROIX, F. & BANERJEE, S. K. 2002. Paleowind directions from the magnetic fabric of loess profiles in central Alaska. *Earth and Planetary Science Letters*, **195**, 99–112.

LEINZ, V. 1937. *Estudos sobre a Glaciação Permo-Carbonífera do Sul do Brasil*, DNPM/DFPM, bol. 21.

LIENERT, B. R. 1991. Monte Carlo simulation of errors in the anisotropy of magnetic susceptibility: a second-rank symmetric tensor. *Journal of Geophysical Research*, **96**, 19539–19544.

LIMA, R. G., ARCHANJO, C. J., MACEDO, J. W. P. & MELO, G. JR 2000. Anomalias de suscetibilidade magnética no batólito granítico de Teixeira (Província da Borborema, nordeste do Brasil) e sua relação com a zona de cisalhamento de Itapetim. *Revista Brasileira de Geociências*, **30**, 685–692.

LITT, T., BRAUER, A. *ET AL.* 2001. Correlation and synchronisation of late glacial continental sequences in northern central Europe based on annually laminated lacustrine sediments. *Quaternary Science Reviews*, **20**, 1233–1249.

LIU, B., SAITO, Y., YAMAZAKI, T., ABDELDAYEM, A., ODA, H., HORI, K. & ZHAO, Q. 2001. Paleocurrent analysis for the Late Pleistocene-Holocene incised valley fill of the Yangtze delta, China by using anisotropy of magnetic susceptibility data. *Marine Geology*, **176**, 175–189.

LUCKMAN, B. H. 2000. The Little Ice Age in the Canadian Rockies. *Geomorphology*, **2**, 357–384.

MILANI, E. J. & ZALÁN, P. V. 1999. An outline of the geology and petroleum systems of the Paleozoic interior basins of South America. *Episodes*, **22**, 199–205.

OGURTSOV, M. G., NAGOVITSYN, Y. A., KOCHAROV, G. E. & JUNGNER, H. 2002. Long-period cycles of the sun's activity recorded in direct solar data and proxies. *Solar Physics*, **211**, 371–394.

PAILLARD, D. 2001. Glacial cycles: toward a new paradigm. *Reviews of Geophysics*, **39**, 325–346.

PARÉS, J. M., VAN DER PLUIJM, B. A. & DINARÈS-TURELL, J. 1999. Evolution of magnetic fabrics during incipient deformation of mudrocks (Pyrenees, northern Spain). *Tectonophysics*, **307**, 1–14.

PARÉS, J. M., HASSOLD, N. J. C., REA, D. K. & VAN DER PLUIJM, B. A. 2007. Paleocurrent directions from paleomagnetic reorientation of magnetic fabrics in deep-sea sediments at the Antarctic Peninsula Pacific margin (ODP Sites 1095, 1101). *Marine Geology*, **242**, 261–269.

RAYMO, M. E. & HUYBERS, P. 2008. Unlocking the mysteries of the ice ages. *Nature*, **451**, 284–285.

READING, H. G. 1986. *Sedimentary Environments and Facies*, 2nd edn. Blackwell, Oxford.

REES, A. I. 1965. The use of anisotropy of magnetic susceptibility in the estimation of sedimentary fabric. *Sedimentology*, **4**, 257–271.

RIND, D. 2002. The Sun's role in climate variations. *Science*, **296**, 673–677.

Rocha-Campos, A. C. 1967. The Tubarão Group in the Brazilian portion of the Paraná Basin. *In*: Bigarella, J. J., Becker, R. D. & Pinto, I. D. (eds) *Problems in Brazilian Gondwana Geology*. Instituto de Geologia, Curitiba, 27–102.

Rocha-Campos, A. C. 2002. Varvito de Itu, SP. Registro clássico da glaciação neopaleozóica. *In*: Schobbenhaus, C., Campos, D. A., Queiroz, E. T., Winge, M. & Berbert-Born, M. (eds) *Sítios Geológicos e Paleontológicos do Brasil I*. DNPM/CPRM/SIGEP, Brasília, 147–154.

Rocha-Campos, A. C. & Sundaram, D. 1981. Geological and palynological observations on Late Paleozoic varvites from the Itararé Subgroup, Paraná Basin, Brazil. *In*: Il Congresso Latino-Americano de Paleontologia, Actas, **1**, 275–275.

Rocha-Campos, A. C., Ernesto, M. & Sundaram, D. 1981. Geological, palynological and paleomagnetic investigations on Late Paleozoic Varvites from the Paraná Basin, Brazil. *In*: 3rd Simpósio Regional de Geologia, Actas, **2**, 162–175.

Rocha-Campos, A. C., Santos, P. R. & Canuto, J. R. 1997. Distribuição e Padrões da Glaciação Neopaleozóica na Bacia do Paraná, Brasil. *In*: 3rd Simpósio sobre Cronoestratigrafia da Bacia do Paraná, Barra das Garças (MT), Brasil, Actas, 1–32.

Rocha-Campos, A. C., Canuto, J. R. & Santos, P. R. 2000. Late Paleozoic glaciotectonic structures in northern Paraná Basin, Brazil. *Sedimentary Geology*, **130**, 131–143.

Rochette, P., Jackson, M. & Aubourg, C. 1992. Rock magnetism and the interpretation of anisotropy of magnetic susceptibility. *Reviews of Geophysics*, **30**, 209–226.

Santos, P. R., Rocha-Campos, A. C. & Canuto, J. R. 1992. Estruturas de arrasto de icebergs em ritmitos do Subgrupo Itararé (Neopaleozóico), Trombudo Central, SC, *Bol. IG-USP, Série Científica*, **23**, 1–18.

Santos, P. R., Rocha-Campos, A. C. & Canuto, J. R. 1996. Patterns of Late Paleozoic deglaciation in the Paraná Basin, Brazil. *Palaeogeography, Palaeoclimatology and Palaeoecology*, **125**, 165–184.

Schmidt, V., Hirt, A. M., Rosselli, P. & Martín-Hernández, F. 2007. Separation of diamagnetic and paramagnetic anisotropy by high-field, low-temperature torque measurements. *Geophysical Journal International*, **168**, 40–47.

Schmitz, M. D. & Davydov, V. I. 2012. Quantitative radiometric and biostratigraphic calibration of the Pennsylvanian–Early Permian (Cisuralian) time scale and pan-Euramerican chronostratigraphic correlation. *Geological Society of America Bulletin*, **124**, 549–577.

Schulz, M. & Mudelsee, M. 2002. REDFIT: estimating red-noise spectra directly from unevenly spaced paleoclimatic time series. *Computers and Geosciences*, **28**, 421–426.

Schulz, M. & Stattegger, K. 1997. SPECTRUM: spectral analysis of unevenly spaced paleoclimatic time series. *Computers and Geosciences*, **23**, 929–945.

Schulz, M., Paul, A. & Timmermann, A. 2004. Glacial–interglacial contrast in climate variability at centennial-to-millennial timescales: observations and conceptual model. *Quaternary Science Reviews*, **23**, 2219–2230.

Schwarzacher, W. 1964. An application of statistical time-series analysis of a limestone/shale sequence. *Journal of Geology*, **72**, 195–213.

Setti, G. C. X. & Rocha-Campos, A. C. 1999. Facies and environment of deposition of varvite and associated rocks (Itararé Subgroup, Late Paleozoic) from Itú, SP. *Anais da Academia Brasileira de Ciencias*, **71**, 836–837.

Sharma, M. 2002. Variations in solar magnetic activity during the last 200,000 years: is there a Sun–climate connection? *Earth and Planetary Science Letters*, **199**, 459–472.

Silva, J. G. R. & Azambuja Filho, N. C. 2005. Cicloestratigrafia do Eopermiano – Estudo de caso no Grupo Itararé, Bacia do Paraná (parte 2): Evidências de indução astronômica (orbital e solar) no clima e na sedimentação. *Revista Brasileira de Geociéncias*, **35**, 77–106.

Sinito, A. M., Valencio, D. A., Ernesto, M. & Pacca, I. G. 1981. Palaeomagnetic study of Permocarboniferous glacial varves from the Itararé subgroup, southern Brazil. *Geophysical Journal of the Royal Astronomical Society*, **67**, 635–640.

Souza, P. A., Amaral, P. G. C. & Bernardes de Oliveira, M. E. C. 2006. A Late carboniferous palynoflora from the Itararé subgroup (Paraná Basin) in Campinas, São Paulo State, Brazil. *Revue de Micropaléontologie*, **49**, 105–115.

Stocker, T. F. 1999. Abrupt climate changes: from past to the future – a review. *International Journal of Earth Sciences*, **88**, 365–374.

Tarling, D. H. & Hrouda, F. 1993. *The Magnetic Anisotropy of Rocks*. Chapman & Hall, London.

Thomson, D. J. 1990. Time series analysis of Holocene climate data. *Philosophical Transactions of the Royal Society London A*, **330**, 601–616.

Usoskin, I. G. & Mursula, K. 2003. Long-term solar cycle evolution: review of recent developments. *Solar Physics*, **218**, 319–343.

Van der Zwan, C. J. 2002. The impact of Milankovitch-scale climatic forcing on sediment supply. *Sedimentary Geology*, **147**, 271–294.

van Geel, B., Raspopov, O. M., Renssen, H., van der Plicht, J., Dergachev, V. A. & Meijer, H. A. J. 1999. The role of solar forcing upon climate change. *Quaternary Science Reviews*, **18**, 331–338.

Veloso, E. E., Anma, R., Ota, T., Komiya, T., Kagashima, S.-i. & Yamazaki, T. 2007. Paleocurrent patterns of the sedimentary sequence of the Taitao ophiolite constrained by anisotropy of magnetic susceptibility and paleomagnetic analyses. *Sedimentary Geology* **201**, 446–460.

Verosub, K. L. & Roberts, A. P. 1995. Environmental magnetism: past, present, and future. *Journal of Geophysical Research*, **100**, 2175–2192, http://dx.doi.org/10.1029/94JB02713

Versteegh, G. J. M. 2005. Solar forcing of climate. 2: Evidence from the past. *Space Science Review*, **120**, 243–286.

Vesely, F. F. & Assine, M. L. 2002. Superfícies estriadas em arenitos do Grupo Itararé produzidas por gelo flutuante, sudeste do estado do Paraná. *Revista Brasileira de Geociéncias*, **32**, 587–594.

VESELY, F. F. & ASSINE, M. L. 2004. Seqüências e tratos de sistemas deposicionais do Grupo Itararé, norte do Estado do Paraná. *Revista Brasileira de Geociéncias*, **34**, 219–230.

VESELY, F. F. & ASSINE, M. L. 2006. Deglaciation sequences in the Permo-Carboniferous Itararé Group, Paraná Basin, southern Brazil. *Journal of the South American Earth Sciences*, **22**, 156–168.

WEINSCHÜTZ, L. C. & CASTRO, J. C. 2004. Arcabouço cronoestratigráfico da Formação Mafra (intervalo médio) na região de Rio Negro/PR – Mafra/SC, borda leste da Bacia do Paraná. *Revista Brasileira de Minas, Ouro Preto*, **57**, 151–156.

WITT, A. & SCHUMANN, A. Y. 2005. Holocene climate variability on millennial scales recorded in Greenland ice cores. *Nonlinear Processes in Geophysics*, **12**, 345–352.

ZHANG, R., KRAVCHINSKY, V. A., ZHU, R. & YUE, L. 2010. Paleomonsoon route reconstruction along a W–E transect in the Chinese Loess Plateau using the anisotropy of magnetic susceptibility: Summer monsoon model. *Earth and Planetary Science Letters*, **299**, 436–446.

ZHU, R., LIU, Q. & JACKSON, M. J. 2004. Paleoenvironmental signi¢cance of the magnetic fabrics in Chinese loess-paleosols since the last interglacial (<130 ka). *Earth and Planetary Science Letters*, **221**, 55–69.

Magnetostratigraphy susceptibility for the Guadalupian series GSSPs (Middle Permian) in Guadalupe Mountains National Park and adjacent areas in West Texas

BROOKS B. ELLWOOD[1]*, LANCE L. LAMBERT[2],
JONATHAN H. TOMKIN[3], GORDEN L. BELL[4], MERLYND K. NESTELL[5],
GALINA P. NESTELL[5] & BRUCE R. WARDLAW[6]

[1]*Department of Geology and Geophysics, Louisiana State University,
Baton Rouge, LA 70803, USA*

[2]*Department of Geological Sciences, University of Texas at San Antonio,
San Antonio, TX 78249, USA*

[3]*School of Earth, Society, and Environment, University of Illinois, 428 Natural History
Building, 1301 W. Green Street, Urbana, IL 61801, USA*

[4]*Great Basin National Park, Baker, NV 89311, USA*

[5]*Department of Earth and Environmental Sciences, University of Texas at Arlington,
Arlington, TX 76019–0049, USA*

[6]*United States Geological Survey, Reston, VA 20192, USA*

**Corresponding author (e-mail: ellwood@lsu.edu)*

Abstract: Here we establish a magnetostratigraphy susceptibility zonation for the three Middle Permian Global boundary Stratotype Sections and Points (GSSPs) that have recently been defined, located in Guadalupe Mountains National Park, West Texas, USA. These GSSPs, all within the Middle Permian Guadalupian Series, define (1) the base of the Roadian Stage (base of the Guadalupian Series), (2) the base of the Wordian Stage and (3) the base of the Capitanian Stage. Data from two additional stratigraphic successions in the region, equivalent in age to the Kungurian–Roadian and Wordian–Capitanian boundary intervals, are also reported. Based on low-field, mass specific magnetic susceptibility (χ) measurements of 706 closely spaced samples from these stratigraphic sections and time-series analysis of one of these sections, we (1) define the magnetostratigraphy susceptibility zonation for the three Guadalupian Series Global boundary Stratotype Sections and Points; (2) demonstrate that χ datasets provide a proxy for climate cyclicity; (3) give quantitative estimates of the time it took for some of these sediments to accumulate; (4) give the rates at which sediments were accumulated; (5) allow more precise correlation to equivalent sections in the region; (6) identify anomalous stratigraphic horizons; and (7) give estimates for timing and duration of geological events within sections.

Guadalupe Mountains National Park contains three Global boundary Stratotype Sections and Points (GSSPs) that define the base of the Guadalupian Series of the Middle Permian (Ogg *et al.* 2008). These three GSSPs were ratified in 2001 by the International Union of Geological Sciences, International Commission on Stratigraphy (see http://stratigraphy.science.purdue.edu/gssp/ for a current listing of GSSPs), and are, in ascending stratigraphic order, the Kungurian–Roadian (K–R; beginning of the Roadian Stage) in Stratotype Canyon, the Roadian–Wordian (R–W; beginning of the Wordian Stage) on Getaway Ledge, and the Wordian–Capitanian (W–C; beginning of the Capitanian Stage) near the crest of Nipple Hill, all within the National Park (Fig. 1).

The Guadalupian Series is one of the best-studied stratigraphic units in the world, and after careful evaluation of all the proposed sites for these GSSPs, the West Texas successions in Guadalupe Mountains National Park were chosen as the location for the GSSPs that define the Middle Permian. Exposures in the Park include shelf through basin depositional environments, although the GSSP localities mainly represent basin and lower slope carbonates, and where they were sampled by

From: JOVANE, L., HERRERO-BERVERA, E., HINNOV, L. A. & HOUSEN, B. A. (eds) 2013. *Magnetic Methods and the Timing of Geological Processes*. Geological Society, London, Special Publications, **373**, 375–394.
First published online August 14, 2012, http://dx.doi.org/10.1144/SP373.1 © The Geological Society of London 2013.
Publishing disclaimer: www.geolsoc.org.uk/pub_ethics

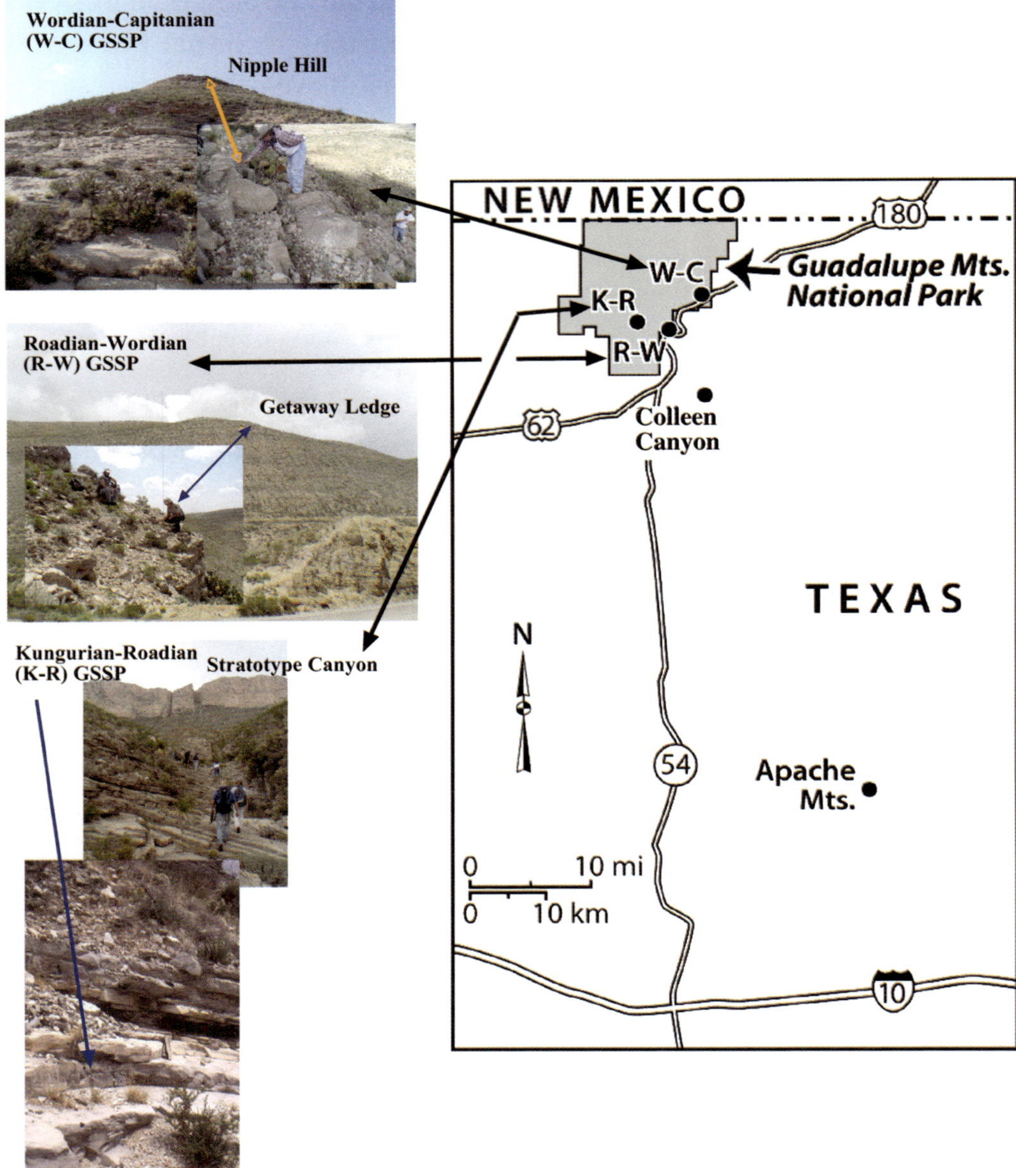

Fig. 1. Location map of the studied sections in West Texas. Photos show positions within Guadalupe Mountains National Park of the three Guadalupian Series GSSPs analysed: in stratigraphic bottom-up order, the Kungurian–Roadian (K–R) GSSP in Stratotype Canyon, also defining the base of the Guadalupian Series; the Roadian–Wordian GSSP on Getaway Ledge; and the Wordian–Capitanian GSSP at the crest of Nipple Hill.

us they contained some interbedded shale to marl lithologies. This setting provides a wide range of litho- and biofacies for study, containing ammonoid, fusulinid and conodont occurrences (Glenister *et al.* 1992, 1999, 2004; Lambert *et al.* 1995, 2000).

The evolutionary events that define the GSSPs occur within the lineage of the conodont genus *Jinogondolella*. The Kungurian–Roadian GSSP, marking the beginning of both the Guadalupian Series and the Roadian Stage, is defined by the cladogenic appearance of *Jinogondolella nankingenis*, which evolved from *Mesogondolella lamberti* (Glenister *et al.* 1999; Lambert *et al.* 2007; Ogg *et al.* 2008). The Roadian–Wordian GSSP, which marks the beginning of the Wordian Stage, is defined by the first appearance of *Jinogondolella aserrata* (Glenister *et al.* 1999; Ogg *et al.* 2008). The Wordian–Capitanian GSSP, marking the beginning of the Capitanian Stage, is defined by the first appearance of *Jinogondolella postserrata* (Glenister *et al.* 1999; Ogg *et al.* 2008).

We have been working for many years in the West Texas region on a number of Permian sections because of the outstanding exposures and well-defined biostratigraphy. Now, because of the close proximity of three GSSPs and the possibilities for precise correlation among exposed sections of Middle and Late Permian age strata, we decided to develop the magnetostratigraphy susceptibility (MSS) for these boundary sections, using low-field, mass-specific, bulk magnetic susceptibility (χ) measurements on outcrop samples from the GSSP sections within the Park. The required permissions were applied for, the work authorized by the US National Park Service, and we began collecting the GSSP sections in 1998 with the Kungurian–Roadian Stage/Guadalupian Series boundary in Stratotype Canyon (Fig. 1). Subsequently, we also collected coeval sequences in the Delaware and Apache Mountains of West Texas, and the results for part of our completed work are presented here.

Previous work

Global stratigraphy

Our understanding of the timing of events in Earth's history has been seriously hampered by age uncertainties for most of geological time. This problem becomes critically important in resolving the timing of significant events on a global scale. It is possible that age uncertainties have actually 'created' distinctive changes in the interpretation of Earth's history, through circular logic (e.g. recognizing a widespread extinction event when the relevant time correlations are based on extinctions), or the importance of some changes have been completely missed because we have not realized that abrupt changes identified within globally distributed stratigraphic successions, that are thought to be of different ages, are in fact coeval.

The aforementioned complications are exacerbated by the traditional patchwork of regional time-scale terms used in different parts of the world. A pertinent example is illustrated by the changing correlations reported for the Middle Permian over the last 30 years, as stratigraphic specialists have tried to more precisely correlate the leading regional standards. In 1978, widely cited Middle Permian terminology included four successive stages – the Kungurian, Kazanian, Punjabian and Djulfian (Dzhulfian) – and the age for the base of the Middle Permian was *c.* 260 Ma (Waterhouse 1978). At times in the past there was no formal Middle Permian Series recognized, only a lower and upper series (e.g. Jin *et al.* 1994*a*, *b*). Now the international community formally recognizes three stages in a reinstated Middle Permian Series: the Kungurian Stage has been moved into the Lower Permian Series (Cisuralian), and the Middle Permian Series (Guadalupian) now comprises the Roadian, Wordian and Capitanian stages (Jin *et al.* 1997). The age for the Lower–Middle Permian boundary is now recognized at *c.* 270.6 Ma (Ogg *et al.* 2008), older by almost 10 Ma than previously recognized. Such variations directly affect papers that compile data for time-synthesis, and that are not using current or converted time scales. This change can result in reported ages for significant events in Earth's history that can vary substantially from present usage owing to large age uncertainties. As a result, work synthesizing long-term changes in the geological record, such as estimates of mass extinctions (Sepkoski 1996) or atmospheric CO_2 concentration (Berner 1990, 2001), requires excellent biostratigraphic age control and each age reference must be adjusted to some base-level for overall age-control estimates. This adjustment requires both (1) knowing the specific time scale that was used in each regional reference, and (2) having high-resolution correlation among datasets. If age uncertainties are not resolved, the number of misinterpretations for global geological datasets rises dramatically.

In an effort to establish better biostratigraphic precision and reduce age uncertainties, the International Union of Geological Sciences and International Commission on Stratigraphy have been working towards establishing uniform standards to represent geological time globally by defining a system of GSSPs for all geological stage boundaries within the Phanerozoic. These GSSPs should be placed in stratigraphic successions that provide reference standards to which other sections of equivalent age can be compared. Once a stage boundary has been formally established through the GSSP process, that stage can then be modelled using any technique whose datasets can be tied directly to the bounding GSSPs, and thus eventually provides an age framework that hopefully will allow marine and non-marine successions all over the world to be linked. As part of the International Commission on Stratigraphy mandated work, it is stressed that abiotic methods should be used in conjunction with biostratigraphy for boundary correlation purposes. The χ of rock strata is one such abiotic method that has great potential for reducing age uncertainties. The method can also be calibrated against other stratigraphic datasets.

Magnetostratigraphic susceptibility, general comments

All materials are 'susceptible' to becoming magnetized in the presence of an external magnetic field, and χ is an indicator of the strength of this transient magnetism. The basic unit of measurement in developing magnetostratigraphy susceptibility (MSS) datasets in stratigraphy (as defined in the International Stratigraphic Guide; Salvador 1994) is χ. Magnetostratigraphy is a term applied to all types of magnetic measurements made in a stratigraphic context, as magnetostratigraphic units or 'magnetozones'. Examples are MSS, magnetostratigraphy polarity and magnetostratigraphy anhysteretic remanent magnetism. In marine sediments and sedimentary rocks, geochemical and experimental studies have clearly shown that χ is dominated by the detrital/aeolian component within samples (e.g. Bloemendal & deMenocal 1989; Ellwood *et al.* 2000). It is very different from remanent magnetism (RM), the intrinsic magnetization that accounts for the polarity of materials. In marine stratigraphic successions, χ is generally considered to be an indicator of detrital iron-containing paramagnetic and ferrimagnetic grains, mainly ferromagnesian and clay minerals (Ellwood *et al.* 2000; 2008*b*; da Silva *et al.* 2009), and can be quickly and easily measured on small samples. In the very low inducing magnetic fields that are generally applied, χ is largely a function of the concentration and composition of the magnetizable material in a sample, and can be measured on small, irregular lithic fragments and on highly friable material that is difficult to sample for RM measurement. An important aspect of χ datasets is that they most often exhibit cycles, and it has been shown that these cycles exhibit climate cyclicity or eustasy (e.g. Bloemendal & deMenocal 1989; Weedon *et al.* 1999) because changes in climate cause changes in weathering and erosion rates, resulting in changes in detrital/ aeolian delivery rates in the marine environment, causing changes in marine sediment χ.

Magnetic susceptibility in marine sedimentary rocks

Ferrimagnetic iron oxide minerals deposited in marine sediments are observed to be rapidly chemically reduced by nitrate or sulfate-reducing bacterial organisms. The result is that, within less than a metre of the sediment water interface, new iron minerals, commonly pyrite and siderite, are formed and magnetite is lost (Karlin & Levi 1983, 1985; Ellwood

et al. 1988). As a result, χ magnitudes are rapidly reduced from values observed in unconsolidated near-surface sediments (Ellwood *et al.* 2006*a*), to the χ values typically ranging from *c.* 1×10^{-9} to 2×10^{-7} m^3 kg^{-1} (Ellwood *et al.* 2000) found in marine sedimentary rocks. Even where very high sediment accumulation rates are observed, these processes are known to occur (Karlin & Levi 1985).

Many workers use RM measurements to identify minerals contributing to the χ signature in samples. One difficulty with using such techniques to evaluate if paramagnetic or ferrimagnetic minerals dominate χ is that most remanent techniques only identify ferrimagnetic components in samples, not paramagnetic or diamagnetic components. Because ferrimagnetic constituents are ubiquitous in most natural sedimentary systems, such RM methods will usually enhance the RM of ferrimagnetic grains while almost always missing the contributions of either paramagnetic or diamagnetic components. An example of this problem was shown for surface sediments from the Gulf of Mexico by Ellwood *et al.* (2006*a*). In their experiment, the authors collected core-top and grab samples from over 200 sites and measured both the anhysteretic remanent magnetism (ARM) and χ for these samples. The pattern for the ARM dataset showed a symmetrical eastern and western Gulf ARM pattern. Magnetic susceptibility showed a very asymmetrical pattern, with relatively high χ in the west, reflecting fluvial sediment input and current redistribution, and diamagnetic values associated with the relatively clean carbonate and quartz-dominated sands along the eastern and southern margins of the Gulf. Because there are extremely fine ferrimagnetic particles in all of these samples, the ARM showed high values, whereas the χ was dominated by diamagnetic components.

A second problem with the argument that most marine sediments are dominated by ferrimagnetic components is shown in other experiments by our group (Ellwood *et al.* 2000). When large blocks of limestone were dissolved and the residue weighed and plotted against χ values of these samples measured before dissolution, the result showed a direct relationship between natural χ values and the mass of the residue (detrital/aeolian component) in each sample. This relationship was replicated experimentally using paramagnetic mineral powders diluted in reagent-grade calcium carbonate acting as the 'rock matrix'. However, when powders of rocks containing a small amount of ferrimagnetic minerals, like magnetite, were used, the χ values were observed to lie outside the range of normal marine χ values. Magnetic susceptibility in magnetite is extremely high, even in very small amounts. Therefore, to produce normal χ values for lithified marine sedimentary rocks containing abundant diamagnetic and paramagnetic compounds, magnetite content must be extremely low. Certainly there are ferrimagnetic constituents in these rocks, but experiments using χ methods, such as thermomagnetic susceptibility measurement, as opposed to RM measurement techniques, best characterize the relative contributions of materials to the χ. When such measurements are performed, the χ has been shown most often to be dominated by paramagnetic components, with a small ferrimagnetic component that is present but usually does not dominate the χ (Ellwood *et al.* 2007*a*; 2008*b*). In almost all of these measurements, however, the detrital/ aeolian components are controlling the χ signal. To characterize the χ signature in our samples, we use thermomagnetic measurements.

Climate

Isotopic studies for Cenozoic marine sediments/ rocks rely on the assumption that isotopic (and cyclic lithological) changes are a proxy for climatic cycles that are assumed to be global (Imbrie *et al.* 1984; Dinarès-Turell *et al.* 2007). Tests of these hypotheses have shown that in general these basic assumptions are correct, and that these datasets can provide a much higher resolution for time scales than are available using biostratigraphic information alone. Such datasets have not been documented for most of the Phanerozoic, however, because stratigraphic successions are imperfect recorders of time owing to erosion, non-deposition, bioturbation, alteration and other processes. As a consequence of these processes, short-term Milankovitch bands (Earth's orbital obliquity and precession) are not as well developed in older rocks as they are in younger rocks. It is therefore expected that longer-term climate cycles are more readily identified in older rock sequences, and that the *c.* 405 ka eccentricity cycle has a robust, long-term palaeoclimatic stratigraphic signal that should be recorded (Shackleton *et al.* 1999*a*; Lasker *et al.* 2004). This band has been observed for Triassic lacustrine (Olsen & Kent 2000), Middle Carboniferous marine (Ellwood *et al.* 2007*b*) and Jurassic marine sediments (Boulila *et al.* 2008). Given that longer-term climate cycles are preserved in the sedimentary record, any robust method that can track these changes, such as χ, should provide useful proxies for climate.

The χ method has been shown to track climate cyclicity for both unlithified and lithified marine sediments (e.g. Bloemendal & deMenocal 1989; Weedon *et al.* 1999). As such, the cyclostratigraphy recorded in these sequences can be used for astronomical calibration of geological time scales (Mead *et al.* 1986; Hartl *et al.* 1995; Weedon *et al.* 1997; Shackleton *et al.* 1999*b*; Crick *et al.* 2001).

In addition to its use in palaeoclimatic studies, MSS can be used for high-resolution correlation among marine sedimentary rocks of broadly differing facies with regional and global extent (Crick *et al.* 1997, 2000; Ellwood *et al.* 1999, 2000, 2007*a, b*; Whalen & Day 2008). MSS, when used as a correlation method, provides a robust dataset to independently evaluate and adjust stratigraphic position among geological sections. It requires reasonable biostratigraphic control to initially develop a chronostratigraphic framework where distinctive MSS zones can be directly correlated with high precision among sections, even when biostratigraphic uncertainties or slight unconformities are known to exist within these sections (Ellwood *et al.* 2006*b*, 2007*a*). The method is particularly useful for independent age control because it can extract data from sections that are not amenable to other magnetostratigraphic techniques, such as remanent magnetization (Berggren *et al.* 1995; Gradstein *et al.* 2004), as it does not require that the rock or sediment analysed be oriented. MSS works as a climate proxy because regional and global processes that drive erosion, including climate and eustasy, bring the detrital components responsible for the χ signature into the marine environment where its stratigraphy is often preserved (Ellwood *et al.* 2000).

Methods

Field sampling

Five stratigraphic sections located in West Texas were collected for this study. Section lithologies are limestone-dominated with sandstone and siltstone interbeds. The collections include the following: 227 samples from the Kungurian–Roadian GSSP interval, covering 11.35 m of section; 87 samples from the Roadian–Wordian GSSP, covering 4.30 m; 86 samples from the Wordian–Capitanian GSSP, covering 4.25 m; 190 samples from the Colleen Canyon (Delaware Mountains) Kungurian– Roadian section, covering 18.91 m; and 116 samples from the Apache Mountains B Section (Lambert *et al.* 2002), covering 5.83 m, including the Wordian–Capitanian boundary interval. In the field we first cleaned each section using scrapers and brushes, so that all beds and lithologies were exposed. Highly weathered zones were cleaned by digging, chipping and brushing to expose relatively fresh surfaces and remove, as much as possible, pedogenic effects. Samples were collected for χ measurement at 5 cm intervals, with the exception of the Colleen Canyon section, which was collected at 10 cm intervals because of its relative lithostratigraphic thickness. All samples were returned to the Rock Magnetism Laboratory at Louisiana State University (LSU) for study.

Laboratory measurement

In the laboratory samples were cleaned and crushed. Anomalous materials (secondary calcite veins, pyrite nodules, etc.) were removed, and then *c.* 10 g of sample was used to make the χ measurements reported here. Each sample was measured three times using the low-field χ bridge at LSU, and the mean of these measurements is reported. The standard deviation is not shown because it is *c.* 1 order of magnitude lower than the χ value measured, it is difficult to represent for both paramagnetic and diamagnetic samples, and so many samples are represented in our diagrams that, if shown, clarity would be lost. Primarily we use the standard deviation observed for each sample to evaluate the quality of the data. If the standard deviation is large (rarely found), we then examine the result for error in measurement, or in data calculation, and look for contamination in samples. The instrument is calibrated, relative to mass, using standard salts reported by Swartzendruber (1992) and the CRC tables. We report χ in terms of sample mass because it is much easier and faster to measure with high precision than volume, and it is now the standard for χ measurement.

Thermomagnetic susceptibility measurements

Thermomagnetic susceptibility measurements, using the KLY-3S Kappa Bridge at LSU, were performed on selected limestone and shale–marl samples. This measurement involves heating a sample from room temperature to 700 °C while measuring low-field, volume magnetic susceptibility (k) as the temperature rises. At low temperatures, paramagnetic minerals exhibit a parabolic-shaped k decay during this process because the k in paramagnetic minerals is inversely proportional to temperature of measurement (Hrouda 1994). Ferrimagnetic minerals, on the other hand, usually show an increase in k up to a point where the k decays towards the Curie temperature of the minerals responsible for that component of k. For magnetite this temperature is *c.* 580 °C. In some cases, as the temperature increases, unstable minerals in the samples break down and ferrimagnetic minerals form. Therefore, the initial k begins to convert to a ferrimagnetic phase, usually at temperatures above 250 °C. This phase is often magnetite and has a *c.* 580 °C Curie point. This behaviour is often observed when illite is present in samples (e.g. Ellwood *et al.* 2007*a*).

Magnetic susceptibility: general comments

Magnetic susceptibility data for most marine sedimentary rocks (>99.9%) range from 1×10^{-9} to 2×10^{-7} m^3 kg^{-1}. For presentation purposes and inter-dataset comparisons, the bar-log format, similar to that previously established for RM polarity data presentations, is used here. Bar logs are constructed from smoothed χ data and where presented are accompanied by primary χ datasets. As in polarity studies, a point half-way between the highest and lowest points in a trend is chosen as the boundary between MSS zones. This approach has always been somewhat subjective, but given that the primary data are also presented in each case, the reader can judge if the boundary point chosen is reasonable. Here, primary χ data are smoothed using splines to hold data points in stratigraphic position (splines were calculated using the JMP statistical software package from the SAS Institute Inc). Splines vary depending on the number of samples in each dataset being smoothed, and therefore using splines is a matter of choice, as are other smoothing techniques. Splines applied here were adjusted in increments of 10 to produce an MSS zonation that conforms to the desired smoothing frequency (discussed below). Finer splines represent shorter cycles.

Because χ datasets are cyclic, for the purpose of correlating among geological sequences we use a bar-log plotting convention, so that if a χ cyclic trend is represented by two or more data points, then this trend is assumed to be significant and the highs and lows associated with these cycles are differentiated by filled (high χ values) or open (low χ values) bar-logs. This method is best employed when high-resolution datasets are being analysed (large numbers of closely spaced samples as in this paper). High-resolution datasets help resolve χ variations associated with anomalous samples. Such variations may be due to weathering effects, secondary alteration and metamorphism, or to longer-term trends owing to factors such as eustasy (Ellwood *et al.* 2008*a*). In addition, variations in detrital input between localities or a change in detrital sediment source is resolved by developing and comparing bar-logs between different localities.

Time-series analysis using Fourier transform and multi-taper spectral estimation

We performed a time-series analysis of the primary χ data (independent of smoothing) for the sections sampled. To avoid introducing error, studied sections were selected where covered intervals could either be avoided or be sampled after careful cleaning of the section. We then collected uniformly spaced samples and assumed that this uniform spacing is linear relative to time (i.e. Δx is proportional to Δt), so that harmonic analysis methods could be used. The less this assumption is true, whether owing to variations in sediment accumulation rate, differential diagenesis or other factors, the more noise is produced in the spectral graph and the lower the power in spectral peaks will be.

The spectral power for χ datasets was obtained using both the multi-taper (MTM) and Fourier transform (FT) methods after the data were both detrended and subjected to a Hanning window to reduce spectral leakage and increase the dynamic range (Jenkins & Watts 1968; Thompson 1982). Incidences of statistically significant peaks (at the 90, 95 and 99% confidence limits) in the resulting spectra were determined by employing MTM (Ghil *et al.* 2002), as calculated with the SSA-MTM toolkit (Dettinger *et al.* 1995). A null hypothesis of red noise was assumed (low frequency, high

power in the spectrum, sloping towards lower values at high frequencies), using a three-taper model. As this method is prone to producing false positives, we limited the use of statistical significance here to its role in supporting (or not) the positions of multiple Milankovitch bands within the dataset. The positions of these bands were fixed relative to each other, and so a climate forcing mechanism was supported by the spectral analysis when the Milankovitch frequencies were also frequencies of high spectral power. The MTM and FT methods, unlike the spline-smoothed χ data and bar-logs derived from these datasets, are capable of resolving high-frequency features in the data.

The FT and MTM methods, when compared with the spline-smoothed χ data and bar-logs derived

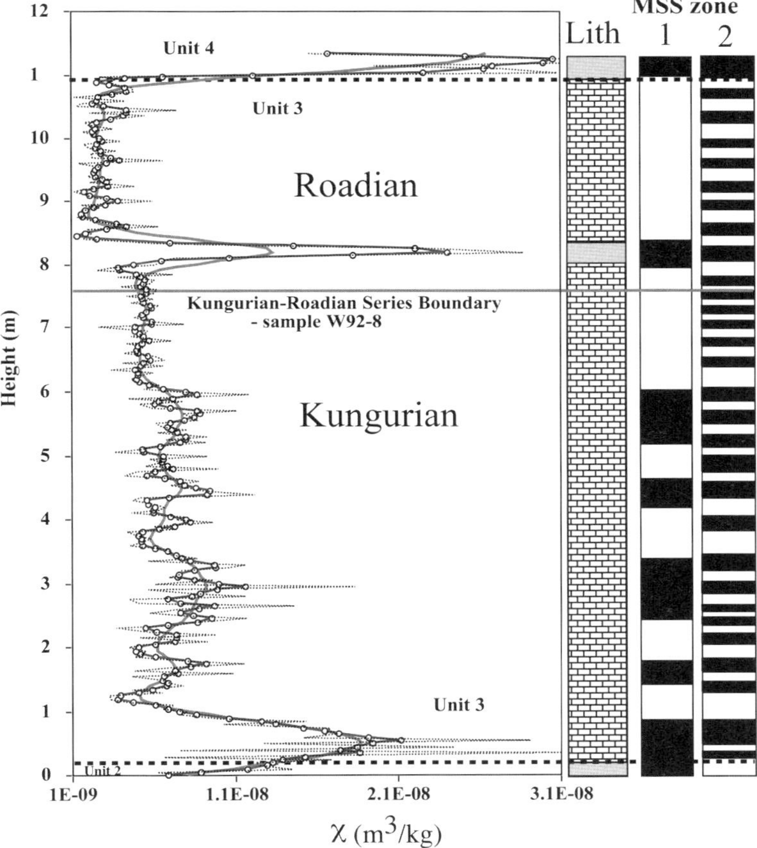

Fig. 2. Magnetic susceptibility data for the Kungurian–Roadian GSSP (Fig. 1). Raw (dotted curve) and smoothed (solid curves) χ data using splines to hold stratigraphic position are given. MSS zone bar-logs are developed from the smoothed χ data (see text) and represent (1) coarse and (2) fine smoothed datasets. Stratigraphic units 2–4 constitute the El Centro and Shumard Members of the Cutoff Formation (Wardlaw *et al.*, in press) and the boundaries between units are represented by horizontal dotted lines. The Kungurian–Roadian boundary position is given by the solid horizontal line. Lith represents limestone and shale-marl intervals. Units 3 and 4 are black, high-organic material.

from these datasets, allow both lower- and higher-frequency high-amplitude peaks (peaks with power) to be visualized. This comparison then can be used to evaluate how uniform the dataset is, that is, where there are clear variations in sediment accumulation rates (SARs), as well as to characterize the general quality of χ cyclic datasets.

Results

In this paper we define the MSS for all the Guadalupian Series GSSPs, now ratified for localities within Guadalupe Mountains National Park. To do this, 706 samples were collected from five sections (reported above) and the χ measured for each of those samples. The MSS character is reasonably clear for all sections, even though the magnitude of the χ signature is very low relative to most marine rocks.

Magnetic susceptibility values for the Kungurian–Roadian sections measured (Figs 2 & 3) were mainly low, with most values less than 1×10^{-8} m^3 kg^{-1}. Lithology is mainly bedded limestone, and shale and limestone Units 3 and 4 are black. Unit 3, containing the Kungurian–Roadian boundary, is the major unit sampled in both sections. In the Stratotype Canyon GSSP, the measured and collected interval begins with low χ values in Unit 2, increasing towards the end of Unit 2 and into the base of Unit 3. These relatively large χ values (Fig. 2) then drop in χ and exhibit a relatively

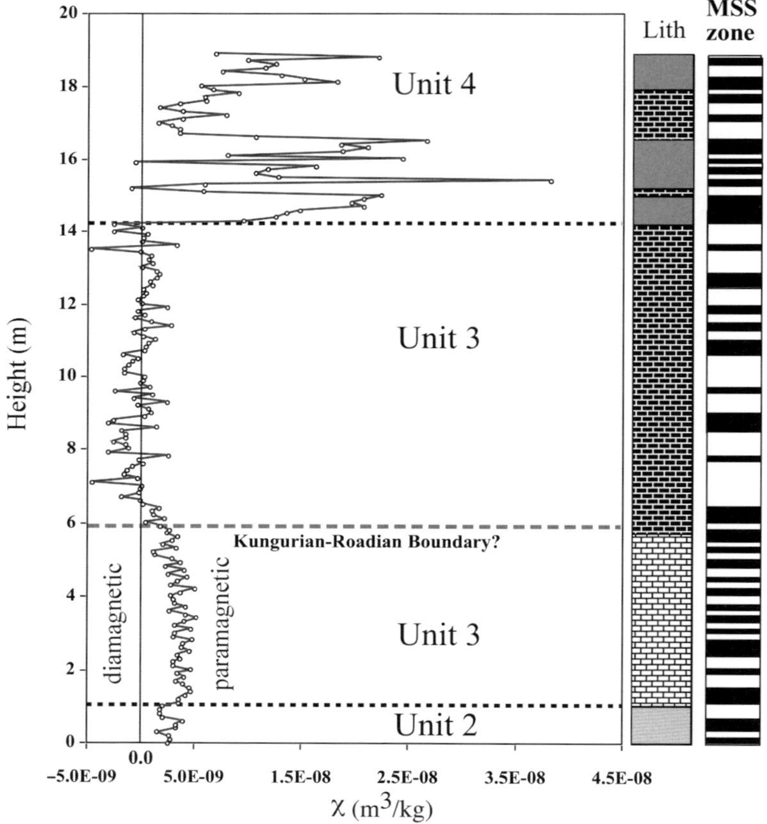

Fig. 3. Magnetic susceptibility data for the Kungurian–Roadian boundary interval exposed at Colleen Canyon in the Delaware Mountains, West Texas (Fig. 1). Only raw (dotted curve) χ data are given because splines cannot be performed on $-\chi$ data. MSS zone bar-log is developed from the χ data (see text). Correlated stratigraphic units 2–4 are given and the boundaries between these units are represented by horizontal dotted lines. The Kungurian–Roadian boundary position is labelled and given by a dashed horizontal line. Samples lie within the Cutoff Formation (Fig. 2). Lith, limestone and shale–marl intervals. Units 3 and 4 are black, highly organic material. Diamagnetic samples are $-\chi$; paramagnetic samples are low $+\chi$.

stable cyclicity through most of Unit 3. Just above the boundary, there is a distinctive high χ that is associated with the shale unit at that level. Above this, χ falls to the lowest values in the section, but it still exhibits clear cyclicity. A drop in χ above the boundary in Unit 3 is also observed in the Colleen Canyon section (Fig. 3). However, the χ peak and shale observed just above the GSSP boundary point in the GSSP section is missing at the Colleen Canyon locality. Through Units 2 and 3 in the Colleen Canyon section, χ values are quite low ($<5 \times 10^{-9}\,\mathrm{m}^3\,\mathrm{kg}^{-1}$; Fig. 3), with a number of samples in the upper part of Unit 3 being diamagnetic. Whereas there is reasonable cyclicity, it is poorly defined in the upper part of Unit 3, although still evident. In both sections there is strong χ recovery in overlying Unit 4 samples (Figs 2 & 3), indicating greater detrital flux into these sections at that time.

In the GSSP we define two MSS zones, (1) a low-frequency and (2) a high-frequency component, that were developed into bar-logs. These MSS zones were used for correlation among the GSSP and the Colleen Canyon section (Fig. 3), as discussed below.

Magnetic susceptibility values for the Roadian–Wordian GSSP, located along Getaway Ledge (Fig. 1), are presented in Figure 4. Here, χ is very low ($<5 \times 10^{-9}\,\mathrm{m}^3\,\mathrm{kg}^{-1}$) in the lower part of the section (below the boundary) and shows muted but reasonably clear cyclicity. Susceptibility values exhibit an overall increase throughout the section, and then larger variability and significantly higher χ values above the boundary. Lithologies through this trend change from limestone at the base to marl–shale in the upper part of the succession. These trends are indicative of high-frequency T–R cycles superimposed on a relatively long-term

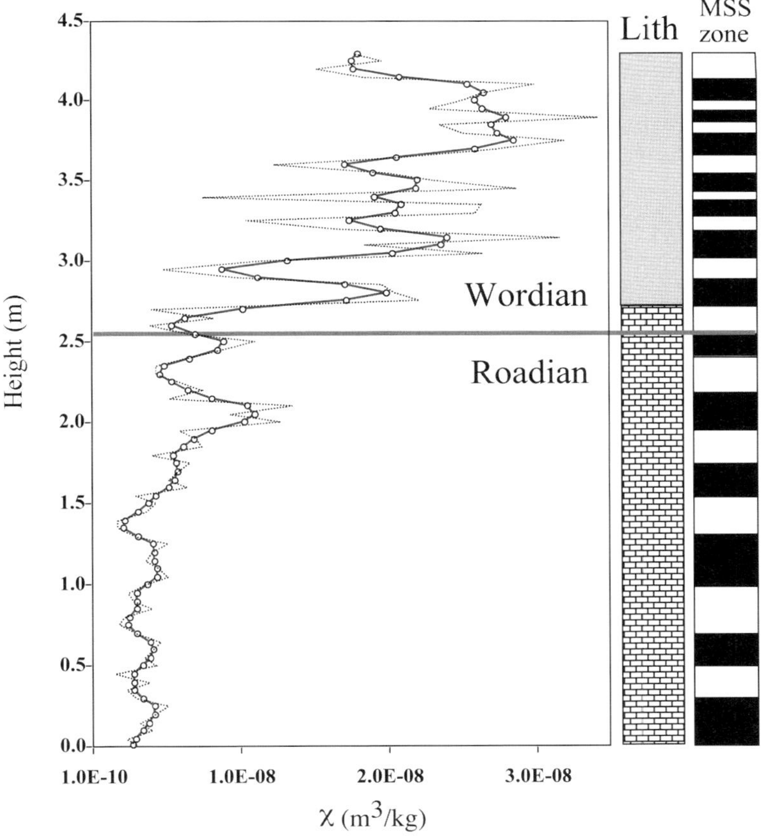

Fig. 4. Magnetic susceptibility data for the Roadian–Wordian GSSP (Fig. 1). Raw (dotted curve) and smoothed χ data using splines to hold stratigraphic position are given. MSS zone bar-logs are developed from the smoothed χ data (see text) and represent a fine smoothed dataset. The Roadian–Wordian boundary position is given by the solid horizontal line. Samples were collected within the Getaway Member of the Cherry Canyon Formation. Lith, limestone and shale–marl intervals.

regression through the boundary interval. There were no other biostratigraphically Roadian–Wordian sections to which we had access in the region, so at this time we have not been able to test the cyclicity observed for this GSSP against other time-equivalent sections.

Susceptibility values and lithologies for the Wordian–Capitanian GSSP, located at the top of Nipple Hill within the National Park (Fig. 1) are given in Figure 5. Lithologies through the interval sampled are represented by indurated, bedded limestone. Susceptibility values are very low throughout the section, with most samples $<5 \times 10^{-9}\,\mathrm{m}^3\,\mathrm{kg}^{-1}$, and a significant number of samples exhibiting diamagnetism. One serious problem with this GSSP is that the section is truncated less than 1 m above the stage boundary point, at the top of Nipple Hill. To make up for this problem, we were able to obtain samples that extend somewhat higher

(*c.* 2.5 m) above the Wordian–Capitanian boundary interval at a locality *c.* 70 km to the SSE of the GSSP, in the biostratigraphic sequence within the Apache Mountains B Section of Lambert *et al.* (2002) and Wardlaw & Nestell (2010). These data are presented in Figure 6. Here, χ values are much higher than for the GSSP section, indicating significantly higher detrital/aeolian component fluxes into the site. In addition, good cyclicity is observed for the dataset. Whereas χ magnitudes are very different between the two sections, we observe that both sections exhibit a general up-section transgressive trend towards very slightly lower χ values.

To characterize the magnetic mineralogy responsible for the χ in our samples, we performed thermomagnetic susceptibility measurement on five randomly selected limestone and one shale samples covering the main lithologies and all of the successions sampled. These measurements are reported in

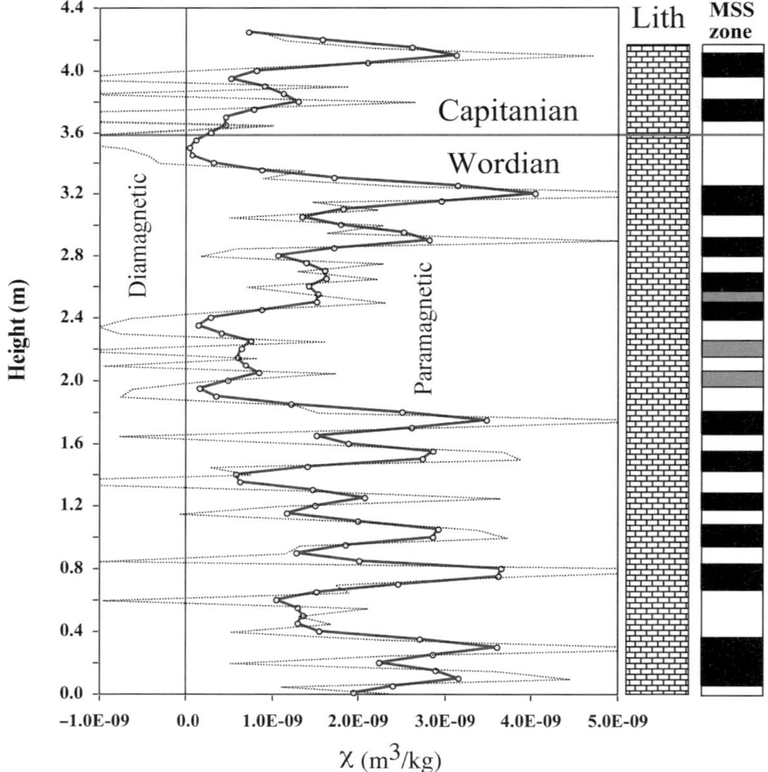

Fig. 5. Magnetic susceptibility data for the Wordian–Capitanian GSSP (Fig. 1). Raw (dotted curve) and smoothed χ data using splines to hold stratigraphic position are given. To calculate splines, $-\chi$ data were given a very low χ value of $1 \times 10^{-10}\,\mathrm{m}^3\,\mathrm{kg}^{-1}$. MSS zone bar-logs are developed from the smoothed χ data (see text) and represent a fine smoothed dataset. Grey bar-log elements represent poorly defined segments that were determined through iterative graphic comparison. The Wordian–Capitanian boundary position is given by the solid horizontal line. Samples all lie within the Pinery Limestone Member of the Bell Canyon Formation. Lith, limestone intervals.

Figure 7. Initial k values for all samples were very low, with the highest value reported for an Apache Mountain shale sample. A slight parabolic decay with heating is observed at low temperatures in all samples, indicating that the initial low-field k in these samples is dominated by a paramagnetic component. At higher temperatures for all samples, mainly above 300 °C, there are mineral breakdown effects resulting in new, ferrimagnetic mineral components that exhibit magnetite Curie points (*c.* 580 °C). Two samples, one from Stratotype Canyon and the other from Colleen Canyon (SC and CC in Fig. 7, respectively) exhibit a large breakdown peak that is typical for thermomagnetic breakdown of illite (Ellwood *et al.* 2007*a*) or siderite, which we have observed for similar high-magnitude k peaks at 500–550 °C.

Discussion

Graphic comparison

Graphic comparison was used for correlation among sections. The graphic comparison technique involves plotting the bottoms and tops of corresponding MSS zones and then drawing a Line of Correlation (LOC) through straight-line segments of these data points to evaluate how strong the correlation is. The process is similar to the graphic correlation method (Shaw 1964). In addition, we add 'tunnels' that enclose packages of relatively uniform LOC segments, and within these tunnels only minor variations in LOC segments occur, thus allowing estimates of relative SARs for sections being compared (Fig. 8).

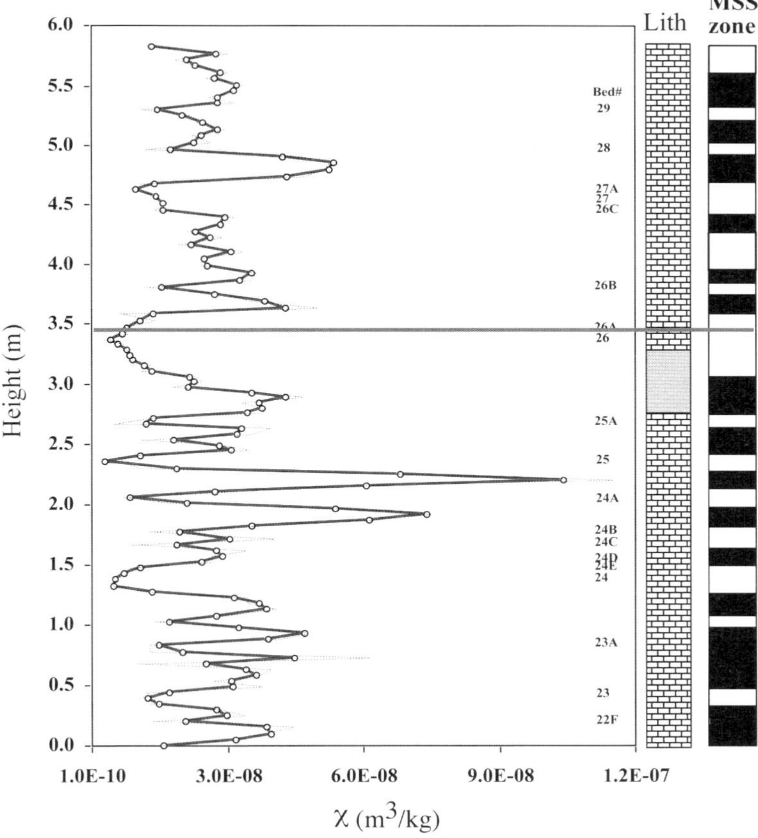

Fig. 6. Magnetic susceptibility data for the Apache Mountains B Section (Lambert *et al.* 2002; Wardlaw & Nestell 2010) that brackets the Wordian–Capitanian boundary exposed in the Apache Mountains, West Texas (Fig. 1). Raw (dotted curve) and smoothed χ data using splines to hold stratigraphic position are given. MSS zone bar-logs are developed from the smoothed χ data (see text). The Wordian–Capitanian boundary position is given by the solid horizontal line. Samples all lie within strata of biostratigraphically equivalent age to the Pinery Limestone Member of the Bell Canyon Formation. Lith, limestone and shale–marl intervals.

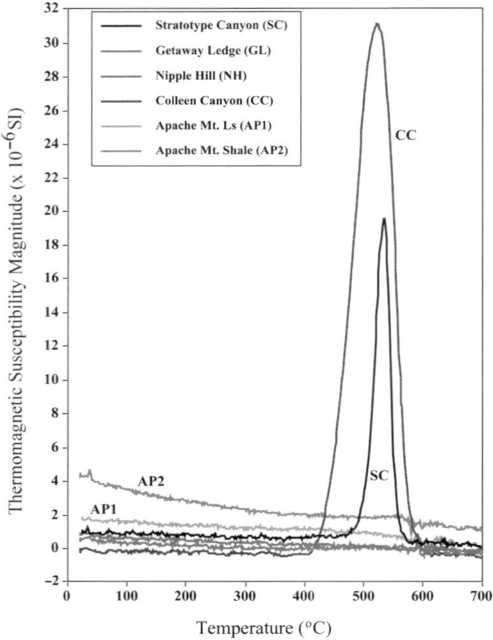

Fig. 7. Thermomagnetic susceptibility measurements for five randomly selected limestone and one shale–marl sample used in this study. A dominant paramagnetic behaviour is shown to control the susceptibility at low temperatures.

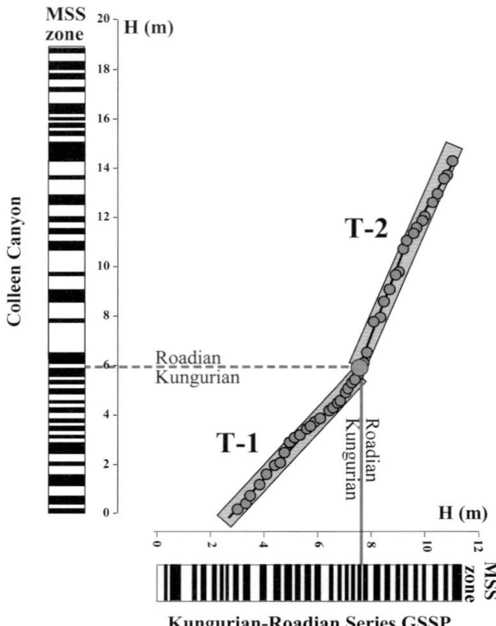

Fig. 8. Graphic comparison of the Kungurian–Roadian GSSP to the Colleen Canyon MSS datasets. Fine-stippled circles represent correlation points between the MSS zone 2 boundaries from Stratotype Canyon (Fig. 2) and the MSS zones from Colleen Canyon (Fig. 5). The boundary interval is identified for the GSSP (solid vertical line) and projected through the LOC into the Colleen Canyon dataset (dashed horizontal line). Solid lines through data points are Line of Correlation (LOC) segments fitted to the data. Two 'tunnels', T1 and T2, are fitted to the data and generally define the dataset trends and relative sediment accumulation rates.

The first comparison, the Kungurian–Roadian Stratotype Canyon GSSP, v. the equivalent Colleen Canyon section, is presented in Figure 8. Here the MSS zones from the GSSP and Colleen Canyon are compared. Because cycle frequency is relatively high in these datasets (stippled circles in Fig. 8), unrecognized trivial breaks in sedimentation tend to offset individual LOC segments, resulting in some scatter along individual LOCs. The MSS comparison shows excellent correspondence between these two datasets, with a clear change beginning just below the boundary, indicating a relative increase in SAR in the Colleen Canyon section above the K–R boundary. We enclose the individual MSS zone data points with two tunnels, T1 and T2 (Fig. 8) that visually represent this change in relative SAR between the two sections. The succession sampled at Colleen Canyon is almost twice as thick as that sampled at the GSSP in Stratotype Canyon, *c.* 15 km to the NW (Fig. 1), and the sample interval at Colleen Canyon is twice that at the GSSP. The adjusted sampling reflects the basinal thickening of the Cutoff Formation. Even with these differences, and the extremely low χ values observed, we find that the method produces

a reasonably defined dataset. Given the excellent correspondence between MSS zones (Fig. 8), we can project the K–R GSSP point from the Stratotype Canyon section into the Colleen Canyon section, and tentatively identify the position at *c.* 6 m height within the Colleen Canyon measured section as the beginning of the Roadian Stage at this locality. Although conodont recoveries are sparse in the Colleen Canyon section, this χ result is fully compatible with our biostratigraphic data.

Our second graphic comparison, the Wordian–Capitanian Nipple Hill GSSP, v. the Apache Mountains Wordian–Capitanian Section B, yields a very good correlation (Fig. 9), expected because both sections have excellent biostratigraphic control. When well-defined cyclicity is observed, we argue that the χ datasets allow much higher correlation precision than does biostratigraphy alone, indicating one reason for using χ data in this way. We were

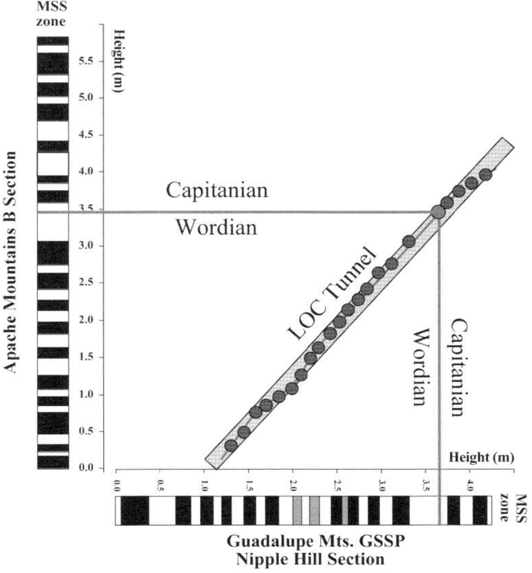

Fig. 9. Graphic comparison of the Wordian–Capitanian GSSP to the Apache Mountains B Section MSS datasets. Symbols as in Figure 8. The boundary interval is represented by vertical and horizontal solid lines.

able to fit a single, narrow tunnel to the W–C dataset because the MSS zone comparisons show relatively low variability, indicating similar SARs in both sections through the interval sampled, even though the two datasets are separated by *c.* 70 km (Fig. 1).

Fourier transform and multi-taper analyses

Given that most χ datasets show reasonable cyclicity, we have performed time-series analysis of the raw χ data (independent of smoothing) on two sampled sections, the Stratotype Canyon K–R GSSP and the Apache Mountains W–C Section B. The results for the χ data, measured for the Wordian–Capitanian boundary, Apache Mountains B Section (Fig. 6) are presented in Figure 10 and Table 1. There is statistical power at greater than the 95% confidence level for the peaks at frequencies of *c.* 0.2 and *c.* 3.7 cycles m^{-1}. These fit with the *c.* 405 ka eccentricity (E2) and *c.* 17.6 ka precessional (P1) Milankovitch bands, respectively (Fig. 10; Table 1; corrected (Berger *et al.* 1992) for an age for this boundary of *c.* 266 Ma (Ogg *et al.* 2008)). In addition, the position of these two peaks constrains the *c.* 34.9 ka (O1) obliquity band to fall coincident with an FT major peak with power in the spectrum, and a near 90% MTM confidence level peak (given in Fig. 10 at *c.* 2.0 cycles m^{-1}). Thus, three of the peaks in the spectrum are constrained to coincide with important Milankovitch bands. The MSS data closely approximate the O1

obliquity time series data. These results demonstrate that the MSS zonation is controlled by climatically driven detrital–aeolian fluxes into the marine environment.

A similar time series analysis of the Kungurian–Roadian GSSP at Stratotype Canyon produces an even better defined dataset (Fig. 11; Table 2), where there are several peaks with reasonable power that are identified. If we assign the peak at *c.* 6.6 cycles m^{-1} a value of *c.* 17.6 ka, the P1 precession position, with confidence levels at *c.* 99%, then the 0.35 cycles m^{-1} peak is equivalent to the Milankovitch eccentricity (E2) of *c.* 405 ka and exhibits a confidence level of >99% (Fig. 11; Table 2). In addition, both E1 eccentricity and O2 obliquity have strong FT power and nearly 90% confidence level, and the MSS zone data from Stratotype Canyon closely approximate the E1 cyclicity. Whereas this dataset is not ideal, it does allow close approximation to cyclicity for several Milankovitch peaks and, therefore, using P1 values for harmonic number and cyclic duration given in Table 2, it does allow an estimate for total time represented by the K–R GSSP measured section of *c.* 1.32 Ma.

A floating-point time scale model

Because χ data from the Wordian–Capitanian boundary interval from Apache Mountains Section B have been shown to represent Milankovitch climate cyclicity, the χ data can also be used in testing a floating-point time scale model for specific climate bands identified in the time series datasets. Here we use the MSS zone from Figure 6, shown to correlate with the O1 Milankovitch band in Figure 10, for comparison with the time-scale model in Figure 12. Here we build a uniform cyclicity model where each division represents a Permian O1 half-cycle of *c.* 20 ka. Then, depending on the absolute geological time scale used (e.g. Ogg *et al.* (2008) report a value of 265.8 Ma), it is possible to graphically correlate MSS zones to the uniform time-scale model and assign specific ages to each MSS zone boundary. This process facilitates estimation of biostratigraphic zonation timing, bio-events identified in the sequence, overall time represented by the section, and the absolute SAR. We have done this in Figure 12 by graphically comparing the MSS zonation from Figure 6, developed for the Apache Mountains Wordian–Capitanian Section B, with our uniform model for climate cyclicity.

The results in Figure 12 give this comparison and place the location of the boundary interval by projecting the boundary position from the Apache Mountains section into the floating point time-scale model. The comparison data points and low scatter

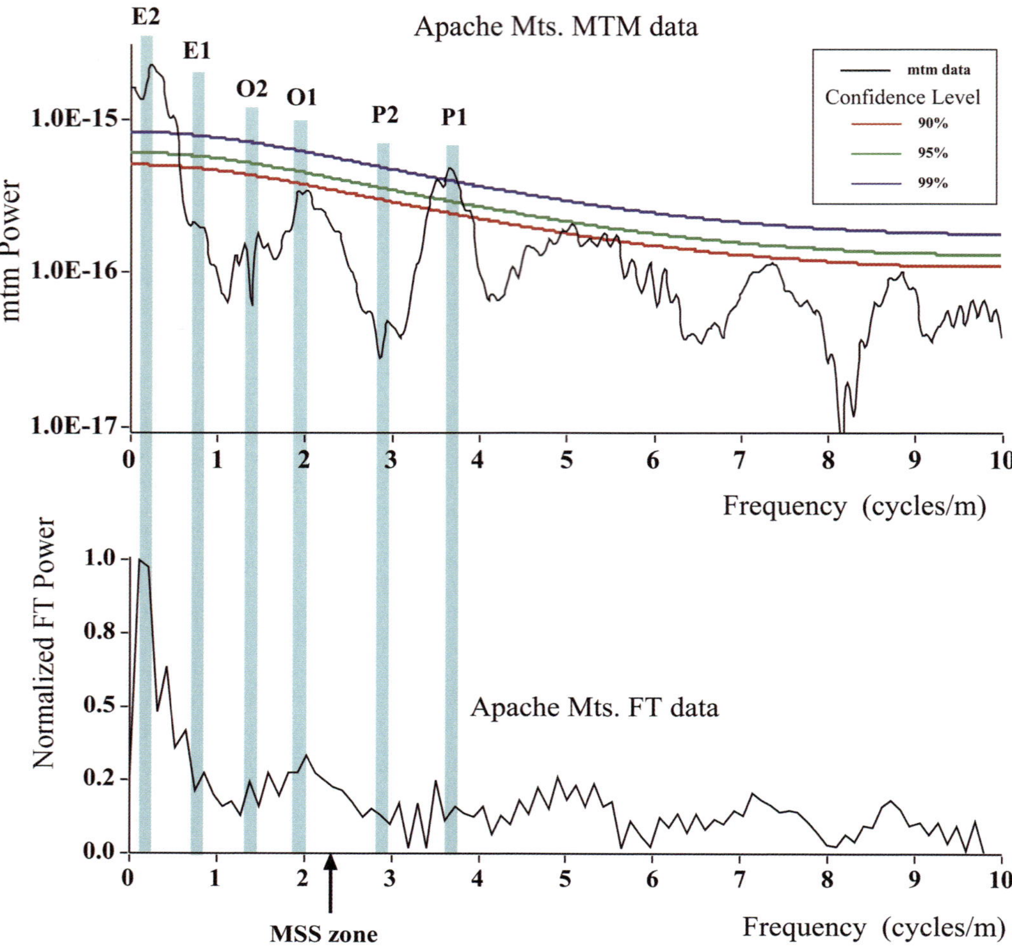

Fig. 10. Spectral analysis using the Fourier transform (FT) and multi-taper (MTM) methods on the raw χ data from the Apache Mountains B section, bracketing the Wordian–Capitanian boundary (Fig. 6). The vertical bars denote frequency (cycles m^{-1}) data for E2 (eccentricity at *c.* 405 ka), E1 (eccentricity at *c.* 100 ka), O2 (obliquity at *c.* 43.9 ka), O1 (obliquity at *c.* 34.9 ka), P2 (precession at *c.* 21.0 ka) and P1 (precession at *c.* 17.6 ka), where O2, O1, P2 and P1 are corrected for an age of *c.* 265.8 Ma given by Ogg *et al.* (2008) using corrected periods from Berger *et al.* (1992).

in LOC segment fits show low variability throughout the dataset and are defined by two main tunnels. Because the comparison is being made with a dataset representing uniform time, we can project time boundaries through LOC segments back into the Apache Mountains Section B and

Table 1. *Apache Mountains time series results*

Cycle	Duration (ka)	Frequency (cycles m^{-1})	Confidence level	Harmonic number
E2	*c.* 405	*c.* 0.2	>99	2
E1	*c.* 100	–	None	–
O2	*c.* 43.9	–	None	–
O1	*c.* 34.9	2.0	<90	19
P2	*c.* 21.0	–	None	–
P1	*c.* 17.6	*c.* 3.7	>99	35

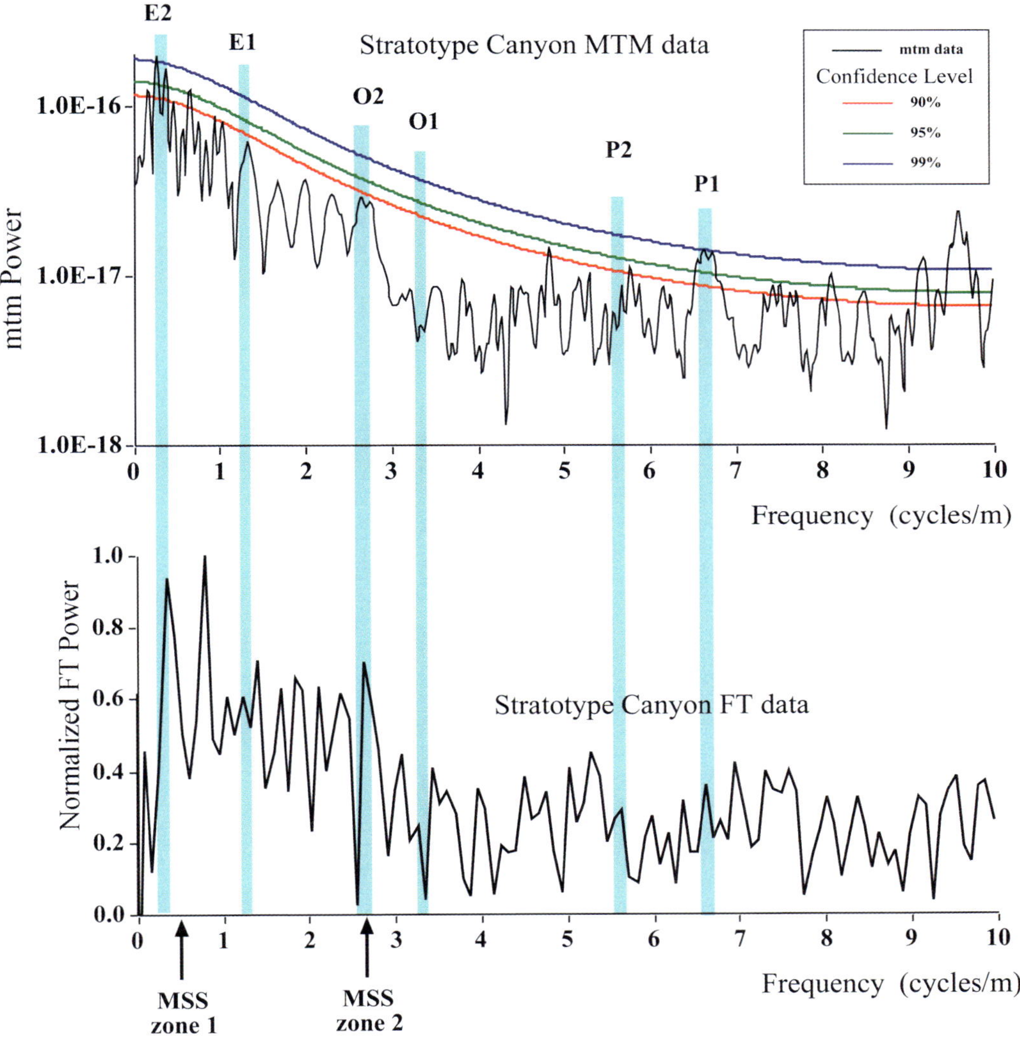

Fig. 11. Spectral analysis using the FT and MTM methods on the raw χ data from the Stratotype Canyon GSSP section (Fig. 6) bracketing the Kungurian–Roadian boundary. Vertical bars as in Fig. 10.

assign ages to those MSS zone boundaries. Using LOC segments we can then evaluate SAR variability throughout the sequence represented. For well-constrained tunnels, we can then estimate SARs, and for Apache Mountains Section B. This yields the following results: T1 *c.* 1.25 cm/ka; T2 *c.* 0.80 cm/ka; T3 *c.* 1.95 cm/ka and T4 *c.* 0.92 cm/ka (Fig. 12). The SARs for T2 and T4 are low for marine rocks, but not unreasonable. T2 represents an increase in detrital–aeolian component flux into the marine system, beginning *c.* 30–40 ka before the boundary, and then slowing to a similar

but slightly higher rate above the boundary than observed through most of T2 time.

The 'T3 Climate Event', as referred to here, represents a significant, lower frequency event than those occurring throughout the rest of the succession. This interval is dominated by decreasing detrital components, culminating in deposition of limestone Beds 26 and 26A (beds identified in Fig. 6), the interval that contains the first occurrence of the boundary identifying conodont, *Jinogondolella postserrata* (Wardlaw & Nestell 2010), the lowest χ values in the section, and the peak of a

Table 2. *Stratotype canyon time series results*

Cycle	Duration (ka)	Frequency (cycles m^{-1})	Confidence level	Harmonic number
E2	*c.* 405	*c.* 0.35	>99	3.5
E1	*c.* 100	*c.* 1.25	<90	14
O2	*c.* 43.9	*c.* 2.65	<90	31
O1	*c.* 34.9	–	None	–
P2	*c.* 21.0	*c.* 5.65	Low	64
P1	*c.* 17.6	*c.* 6.60	*c.* 99	75

highstand system track at that time. This distinctive boundary event is clearly identified in both the Nipple Hill W–C GSSP (Fig. 5) and the Apache Mountains B Section (Fig. 6). The estimated duration for this interval is *c.* 22 ka (Fig. 12). By comparing the Apache Mountains time series data with the model (Fig. 12), and then comparing the Apache Mountains MSS zones with the Nipple Hill MSS zones (Fig. 9), we have estimated the length of time represented for our measured sections; 540 ka for the Nipple Hill and *c.* 580 ka for the Apache Mountains Section B, respectively.

To plug into a specific time scale, an absolute age must be used. If different time scales are used, or the time scale changes, it is a simple matter to recalculate the new age for the MSS zone of interest. Timing of these zones can then be compared on a global scale with other sections using the MSS zonation identified for those successions, with comparison to the W–C climate model, to evaluate the global timing and consistency of the biostratigraphic dataset in question. Here (Fig. 12) we assign a value for the Wordian–Capitanian boundary of 265.8 Ma (Ogg *et al.* 2008). From that evaluation, fine-scale timing within the Apache Mountains B Section (and therefore to the Nipple Hill W–C GSSP) can be estimated from a half-cycle floating point time scale model (Fig. 12) by first projecting each small time segment into the Apache Mountains B MSS zonation in Figure 12. Then, using Figure 9 and our age assignments for Apache Mountains Section B, ages can also be projected into the Nipple Hill Wordian–Capitanian GSSP and assigned to the MSS zonation there.

We have also used the time series data from the Kungurian–Roadian GSSP in Stratotype Canyon (Fig. 11) to estimate the amount of time and SAR for deposition of the sampled portion of the Stratotype Canyon section (Fig. 13). Using MSS zone 2 (Fig. 2) and the time series results (Fig. 11) for the succession, we compared the MSS zone 2 data for the Kungurian–Roadian GSSP with a standard time model representing O2 (obliquity) cyclicity. LOC segments yield four tunnels that best fit the data. Two of these, T1 and T3, represent short segments yielding essentially identical SARs of

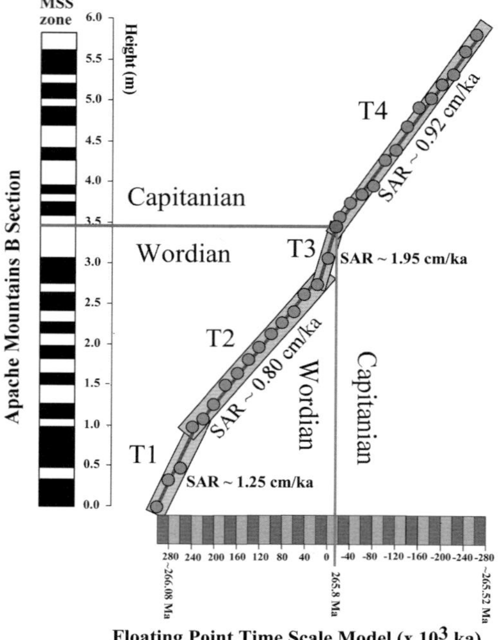

Floating Point Time Scale Model (x 10^3 ka)

Fig. 12. Graphic comparison of the Wordian–Capitanian MSS zonation to an obliquity floating-point time scale model, using a *c.* 40 ka cyclicity determined for the χ dataset presented in Figure 6, and for which the time series analysis yields results consistent with Milankovitch cyclicity (Fig. 10; Table 1). Stippled circles represent MSS zone comparison points to the model. Solid circle represents the Wordian–Capitanian boundary correlation point. Solid lines are lines of correlation (LOC) segments fitted to the data. Four reasonably well-constrained 'tunnels', T1–T4 are fitted through the LOC data points below and above the boundary, respectively. Note the change in LOC slope and high SAR (*c.* 1.36 cm/ka) just below and to the boundary interval. Changes in LOC segments illustrate correspondence/departure from a uniform varying cyclicity (see text). Sediment accumulation rates for T2, *c.* 0.80 cm/ka, and T4, *c.* 0.92 cm/ka, are given and represent most of the data.

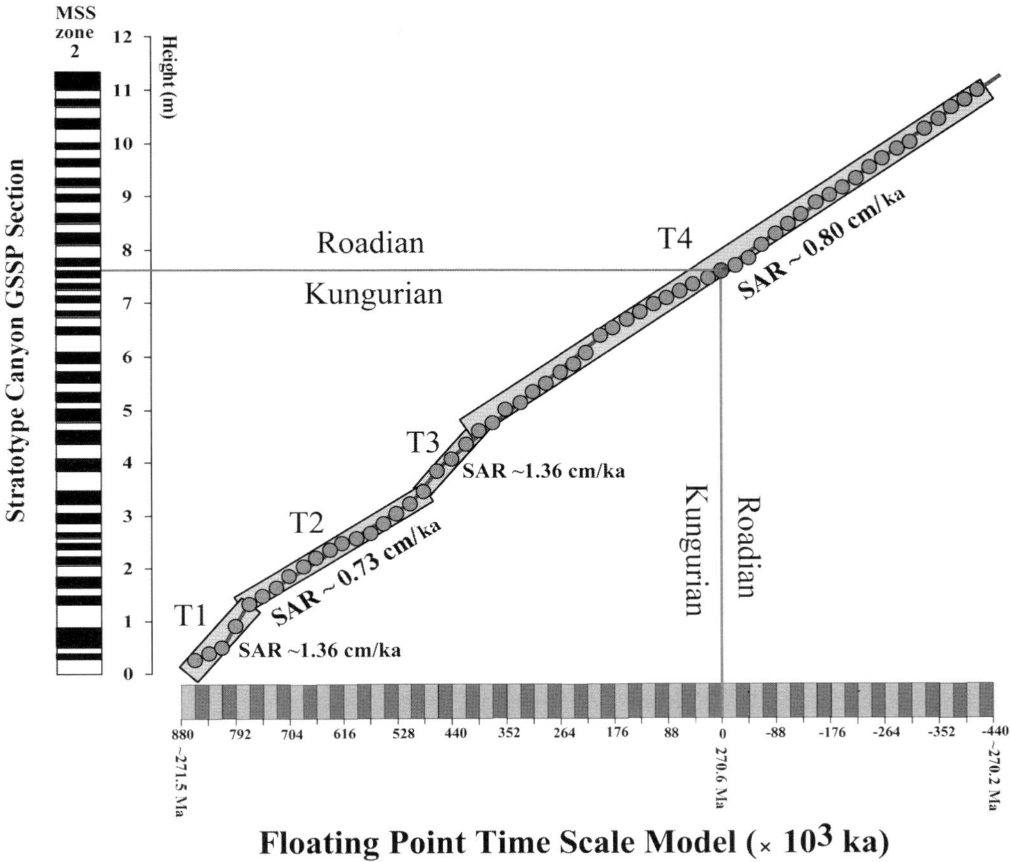

Floating Point Time Scale Model ($\times 10^3$ ka)

Fig. 13. Graphic comparison of the Kungurian–Roadian GSSP in Stratotype Canyon to an O2 floating-point time scale model, using a *c.* 44 ka climate cyclicity determined for the χ dataset presented in Figure 2, and for which the time series analysis yields high-confidence levels consistent with Milankovitch cyclicity (Fig. 11; Table 2). Stippled circles represent MSS zone 2 comparison points to the model. Solid circle represents the Kungurian–Roadian boundary correlation point. Solid lines are LOC segments fitted to the data. Four well-constrained 'tunnels', T1–T4 are fitted through the LOC data points below and above the boundary, respectively. T1 and T3 represent similar short changes in LOC slope exhibiting an SAR *c.* 1.36 cm/ka. Changes in LOC segments illustrate correspondence/departure from a uniform varying cyclicity (see text). Sediment accumulation rates for T2, *c.* 0.73 cm/ka, and T4, *c.* 0.80 cm/ka, are given.

c. 1.36 cm/ka. T2, a longer tunnel, represents a lower SAR at *c.* 0.73 cm/ka. Most of the section, *c.* 6.4 m, is enclosed within one tunnel that runs through the K–R boundary interval and yields an SAR of *c.* 0.80 cm km^{-1}. The total length of time represented for all LOC tunnels over the interval sampled in Stratotype Canyon through the K–R GSSP (including *c.* 7.5 m of Kungurian and only *c.* 4 m of Roadian) is *c.* 1.32 Ma. It is interesting to note that the SARs determined for both the Kungurian–Roadian and Wordian–Capitanian sequences in the Guadalupe Mountains exhibit very similar values for all LOC tunnels that include long data segments. The

low SAR values observed are consistent with similar modern marine environments.

Conclusions

We have collected samples from the three GSSP localities exposed in Guadalupe Mountains National Park, West Texas. These GSSPs define the base of the three stages, Roadian, Wordian and Capitanian, within the Middle Permian Guadalupian Series, including the base of the series, defined by the Kungurian–Roadian GSSP. Overall, this work (1) demonstrates that χ datasets provide a proxy for climate cyclicity, (2) defines the MSS character for the three Guadalupian Series GSSPs in Guadalupe Mountains National Park, (3) gives quantitative estimates for the time it took for these sediments to accumulate, (4) gives the rates at which these sediments were accumulating, (5) allows correlation to equivalent sections in the region, (6) identifies anomalous stratigraphic horizons within some of these sections, and (7) gives floating point time scales for the Kungurian–Roadian and Wordian–Capitanian GSSPs and provides time scale resolution at the *c.* 20 ka level for the intervals measured intervals.

We report the MSS for these GSSPs and for two equivalent sections that are also exposed in West Texas. The magnetic susceptibility was measured for a total of 706 closely spaced samples collected through the boundary intervals in each of the five sections. Magnetic susceptibility variability is cyclic and χ values are generally extremely low relative to most marine rocks. As a result, there is some scatter in some parts of these sections, and the MSS cyclicity is marginal in these zones.

Using the MSS cyclicity along with the known biostratigraphy in the GSSPs, we were able to make correlations with high precision among coeval sampled sections. These included: (1) correlating, using the graphic comparison method, the Kungurian–Roadian GSSP (the GSSP defining the base of the Guadalupian Series) exposed within Stratotype Canyon in the Park, with an equivalent succession in Colleen Canyon, located *c.* 14 km to the SSE in the Delaware Mountains outside the Park; and (2) correlating, using the graphic correlation method, the Wordian–Capitanian GSSP, exposed near the crest of Nipple Hill in the Park, with the coeval Apache Mountains Section B, located *c.* 70 km to the SSE of the GSSP. The Roadian–Wordian GSSP located along Getaway Ledge within the Park was also sampled, χ measured and the MSS zonation determined, but no equivalent section was available for sampling.

Time series analysis, using Fourier transform multi-taper methods, was performed on the raw χ data from two of the studied sections, the Kungurian–Roadian GSSP in Stratotype Canyon and the Wordian–Capitanian boundary contained within the Apache Mountains B Section. Time series results were reasonable for both sections and we were able to define a number of supported Milankovitch peaks, indicating that these datasets are controlled by climate fluxes of detrital/aeolian material into the marine environment during that time.

The time series work for the Kungurian–Roadian GSSP showed good cyclicity and a number of strong, supported Milankovitch peaks, including: two peaks with at least 99% confidence levels,

eccentricity (E2, *c.* 405 ka) at *c.* 0.35 cycles m^{-1}, and precession (P1, *c.* 17.6 ka) at *c.* 6.6 cycles m^{-1}; and two peaks with near 90 confidence levels, eccentricity (E1, *c.* 100 ka) at *c.* 1.25 cycles m^{-1}, and obliquity (O2, *c.* 43.9 ka) at *c.* 2.65 cycles m^{-1}. The Apache Mountains Section B time series data showed good cyclicity at $>99\%$ confidence levels for E2 at *c.* 0.2 cycles m^{-1} and P1 at *c.* 3.7 cycles m^{-1}.

We thank the National Park Service for permission to sample the GSSP sections in the park. S. Ellwood is acknowledged for her help in sampling all sections. Funding for this work was provided by small grants from the University of Texas at Arlington (to B.B.E., M.K.N. and G.P.N.) and Louisiana State University (to B.B.E.) and from CONOCO Inc. (to L.L.L.).

References

BERGER, A., LOUTRE, M. F. & LASKER, J. 1992. Stability of the astronomical frequencies over the Earth's history for paleoclimate studies. *Science*, **255**, 560–566.

BERGGREN, W. A., KENT, D. V., AUBRY, M-P. & HARDENBOL, J. 1995. *Geochronology, Time Scales and Global Stratigraphic Correlation.* SEPM Special Publication **54**. Society for Sedimentary Geology, Tulsa, OK.

BERNER, R. A. 1990. Atmospheric carbon dioxide levels over Phanerozoic time. *Science*, **249**, 1382–1386.

BERNER, R. A. 2001. The effect of the rise of land plants on atmospheric CO2 during the Paleozoic. *In:* GENSEL, P. G. & EDWARDS, D. (eds) *Plants Invade the Land: Evolutionary and Environmental Perspectives, Critical Moments and Perspectives in Earth History and Paleobiology.* Columbia University Press, New York, 173–178.

BLOEMENDAL, J. & DEMENOCAL, P. 1989. Evidence for a change in the periodicity of tropical climate cycles at 2.4 Myr from whole-core magnetic susceptibility measurements. *Nature*, **342**, 897–900.

BOULILA, S., HINNOV, L. A. *ET AL.* 2008. Astronomical calibration of the Early Oxfordian (Vocontian and Paris basins, France): consequences of revising the Late Jurassic time Scale. *Earth and Planetary Science Letters*, **276**, 40–51.

CRICK, R. E., ELLWOOD, B. B., EL HASSANI, A., FEIST, R. & HLADIL, J. 1997. Magnetosusceptibility event and cyclostratigraphy (MSEC) of the Eifelian–Givetian GSSP and associated boundary sequences in north Africa and Europe. *Episodes*, **20**, 167–175.

CRICK, R. E., ELLWOOD, B. B., EL HASSANI, A. & FEIST, R. 2000. Proposed magnetostratigraphy susceptibility magnetostratotype for the Eifelian–Givetian GSSP (Anti-Atlas Morocco). *Episodes*, **23**, 93–101.

CRICK, R. E., ELLWOOD, B. B., EL HASSANI, A., HLADIL, J., HROUDA, F. & CHLUPAC, I. 2001. Magnetostratigraphy susceptibility of the Pridoli-Lochkovian (Silurian–Devonian) GSSP (Klonk, Czech Republic) and a Coeval sequence in Anti-Atlas Morocco. *Palaeogeography, Palaeoclimatology, Palaeoecology*, **167**, 73–100.

DA SILVA, A-C., POTMA, K., WEISSENBERGER, J. A. W., WHALEN, M. T., HUMBLET, M., MABILLE, C. & BOULVAIN, F. 2009. Magnetic susceptibility evolution and sedimentary environments on carbonate platform sediments and atolls, comparison of the Frasnian from Belgium and Alberta, Canada. *Sedimentary Geology*, **214**, 3–18.

DETTINGER, M. D., GHIL, M., STRONG, C. M., WEIBEL, W. & YIOU, P. 1995. Software expedites singular-spectrum analysis of noisy time series. *Eos Transactions of the American Geophysical Union*, **76**, 12–21.

DINARÈS-TURELL, J., BACETA, J. I., BERNAOLA, G., ORUE-ETXEBARRIA, X. & PUJALTE, V. 2007. Closing the Mid-Palaeocene gap: toward a complete astronomically tuned Palaeocene Epoch and Selandian and Thanetian GSSPs at Zumaia (Basque Basin, W Pyrenees). *Earth and Planetary Science Letters*, **262**, 450–467.

ELLWOOD, B. B., CHRZANOWSKI, T. H., HROUDA, F., LONG, G. J. & BUHL, M. L. 1988. Siderite formation in anoxic deep-sea sediments: a synergetic bacterially controlled process with important implications in paleomagnetism. *Geology*, **16**, 980–982.

ELLWOOD, B. B., CRICK, R. E. & EL HASSANI, A. 1999. The magneto-susceptibility event and cyclostratigraphy (MSEC) method used in geological correlation of Devonian rocks from Anti-Atlas Morocco. *American Association of Petroleum Geologists Bulletin*, **83**, 1119–1134.

ELLWOOD, B. B., CRICK, R. E., EL HASSANI, A., BENOIST, S. L. & YOUNG, R. H. 2000. The magnetosusceptibility event and cyclostratigraphy (MSEC) method applied to marine rocks: detrital input v. carbonate productivity. *Geology*, **28**, 1134–1138.

ELLWOOD, B. B., BALSAM, W. L. & ROBERTS, H. H. 2006a. Gulf of Mexico sediment sources and sediment transport trends from magnetic susceptibility measurements of surface samples. *Marine Geology*, **230**, 237–248.

ELLWOOD, B. B., GARCÍA-ALCALDE, J. L. ET AL. 2006b. Stratigraphy of the middle devonian boundary: formal definition of the susceptibility magnetostratotype in Germany with comparisons to Sections in the Czech Republic, Morocco and Spain. *Tectonophysics*, **418**, 31–49.

ELLWOOD, B. B., BRETT, C. E. & MACDONALD, W. D. 2007a. Magnetosusceptibility stratigraphy of the upper ordovician kope formation, Northern Kentucky. *Palaeogeography, Palaeoclimatology, Palaeoecology*, **243**, 42–54.

ELLWOOD, B. B., TOMKIN, J., RICHARDS, B., BENOIST, S. L. & LAMBERT, L. L. 2007b. MSEC data sets record glacially driven cyclicity: examples from the arrow canyon Mississippian–Pennsylvanian GSSP and associated sections. *Palaeogeography, Palaeoclimatology, Palaeoecology*, **255**, 377–390, http://dx.doi.org/10.1016/j.palaeo.2007.08.006

ELLWOOD, B. B., TOMKIN, J. H., FEBO, L. A. & STUART, C. N. JR. 2008a. Time series analysis of magnetic susceptibility variations in deep marine sediments: a test using upper Danian–Lower Selandian proposed GSSP, Spain. *Palaeogeography, Palaeoclimatology, Palaeoecology*, **261**, 270–279.

ELLWOOD, B. B., TOMKIN, J. H., RATCLIFFE, K. T., WRIGHT, M. & KAFAFY, A. M. 2008b. High resolution magnetic susceptibility and geochemistry for the cenomanian/turonian boundary GSSP with correlation to time equivalent core. *Palaeogeography, Palaeoclimatology, Palaeoecology*, **261**, 105–126.

GHIL, M., ALLEN, R. M. ET AL. 2002. Advanced spectral methods for climatic time series. *Reviews in Geophysics*, **40**, 3.1–3.41, http://dx.doi.org/10.1029/2000RG000092

GLENISTER, B. F., BOYD, D. W. ET AL. 1992. The Guadalupian: proposed international standard for a Middle Permian Series. *International Geology Review*, **34**, 857–888.

GLENISTER, B. F., WARDLAW, B. R. ET AL. 1999. Proposal of Guadalupian and component Roadian, Wordian, and Capitanian stages as international standards for the Middle Permian Series. Permophiles (Newsletter of the Subcommission on Permian Stratigraphy. *International Union of Geological Sciences*), **34**, 3–11.

GLENISTER, B. F., WARDLAW, B. R. & LAMBERT, L. L. 2004. Guadalupian Series: International standard for Middle Permian time. *In*: ARMSTRONG, F. R. & KELLERLYNN, K. (eds) *The Guadalupe Mountains Symposium – Guadalupe Mountains National Park 25th Anniversary Conference, Research and Resource Management Proceedings*, Guadalupe Mountains National Park, Pine Springs, TX, 245–250.

GRADSTEIN, F. M., OGG, J. G. & SMITH, A. G. 2004. *A Geologic Time Scale 2004*. Cambridge University Press, Cambridge.

HARTL, P., TAUXE, L. & HERBERT, T. 1995. Earliest Oligocene increase in South Atlantic productivity as interpreted from 'rock magnetics' at Deep Sea drilling Site 522. *Paleoceanography*, **10**, 311–326.

HROUDA, F. 1994. A technique for the measurement of thermal changes of magnetic susceptibility of weakly magnetic rocks by the CS-2 apparatus and the KLY-2 Kappabridge. *Geophysical Journal International*, **118**, 604–612.

IMBRIE, J., HAYS, J. D. ET AL. 1984. The orbital theory of pleistocene climate: support from a revised chronology of the marine delta 18O record. *In*: BERGER, A. L., IMBRIE, J., HAYS, J., KUKLA, G. & SALTZMAN, B. (eds) *Milankovitch and Climate, Part I*. Kluwer Academic, Dordrecht, 269–305.

JENKINS, G. M. & WATTS, D. G. 1968. *Spectral Analysis and its Applications*. Holden-Day, San Francisco, CA.

JIN, Y., GLENISTER, B. F., KOTLYAR, G. V. & SHENG, J. 1994a. An operational scheme of Permian chronostratigraphy. *Palaeoworld*, **4**, 1–13.

JIN, Y., SHENG, J. ET AL. 1994b. Revised operational scheme of Permian chronostratigraphy. *Permophiles*, **25**, 12–15.

JIN, Y., WARDLAW, B. R., GLENISTER, B. F. & KOTLYAR, G. V. 1997. Permian chronostratigraphic subdivisions. *Episodes*, **20**, 10–15.

KARLIN, R. & LEVI, S. 1983. Diagenesis of magnetic minerals in recent haemipelagic sediments. *Nature*, **303**, 327–330.

KARLIN, R. & LEVI, S. 1985. Geochemical and sedimentological control of the magnetic properties of hemipelagic sediments. *Journal of Geophysical Research*, **90**, 10373–10392.

LAMBERT, L. L., WARDLAW, B. R. & GLENISTER, B. F. 1995. West Texas Permian Series as world chronostratigraphic reference standards – A status report. *In*: GARBER, R. A. & LINDSAY, R. F. (eds)

Wolfcampian–Leonardian Shelf Margin Facies of the Sierra Diablo: Seismic Scale Models for Subsurface Exploration. West Texas Geological Society, Midland, Texas, Publication 95–97, 123–133.

LAMBERT, L. L., LEHRMANN, D. J. & HARRIS, M. T. 2000. Correlation of the Road Canyon and Cutoff Formations, west Texas, and its relevance to establishing an international Middle Permian (Guadalupian) Series. *In*: WARDLAW, B. R., GRANT, R. E. & ROHR, D. M. (eds) *The Guadalupian Symposium*. Smithsonian Contributions to the Earth Sciences, Smithsonian Institution Press, Washington, **32**, 153–183.

LAMBERT, L. L., WARDLAW, B. R., NESTELL, M. K. & NESTELL, G. P. 2002. Latest Guadalupian (Middle Permian) conodonts and foraminifers from West Texas. *Micropaleontology*, **48**, 343–364.

LAMBERT, L. L., WARDLAW, B. R. & HENDERSON, C. M. 2007. Mesogondolella and Jinogondolella (Conodonta): multielement definition of the taxa that bracket the basal Guadalupian (Middle Permian Series) GSSP. *Palaeoworld*, **16**, 208–221.

LASKAR, J., ROBUTEL, P., JOUTEL, F., GASTINEAU, M., CORREIA, A. & LEVRARD, B. 2004. A long-term numerical solution for the insolation quantities of the Earth. *Astronomy and Astrophysics*, **428**, 261–285.

MEAD, G. A., YAUXE, L. & LABRECQUE, J. L. 1986. Oligocene paleoceanography of the South Atlantic: paleoclimate implications of sediment accumulation rates and magnetic susceptibility. *Paleoceanography*, **1**, 273–284.

OGG, J. G., OGG, G. & GRADSTEIN, F. M. 2008. *The Concise Geologic Time Scale*. Cambridge University Press, Cambridge.

OLSEN, P. E. & KENT, D. V. 2000. High-resolution early Mesozoic Pangean climatic transect in lacustrine environments. *In*: BACHMAN, G. H. & LERCHE, I. (eds) *Zentralblatt fuer Geologie und Palaeontologie, 1988.* Schweizerbart & Borntraeger, Stuttgart, 1475–1495.

SALVADOR, A. 1994. *International Stratigraphic Guide: A Guide to Stratigraphic Classification, Terminology, and Procedure.* The International Union of Geological Sciences and The Geological Society of America, Denver, CO.

SEPKOSKI, J. J. 1996. Patterns of Phanerozoic extinctions: a perspective from global data bases. *In*: WALLISER, O. H. (ed.) *Global Events and Event Stratigraphy in the Phanerozoic*. Springer, Berlin, 21–34.

SHACKLETON, N. J., MCCAVE, I. N. & WEEDON, G. P. 1999a. Preface. *Philosophical Transactions of the Royal Society London, A*, **357**, 1733–1734.

SHACKLETON, N. J., CROWHURST, S. J., WEEDON, G. P. & LASKAR, J. 1999b. Astronomical calibration of Oligocene–Miocene time. *Philosophical Transactions of the Royal Society London, A*, **357**, 1907–1929.

SHAW, A. B. 1964. *Time in Stratigraphy*. McGraw Hill, New York.

SWARTZENDRUBER, L. J. 1992. Properties, units and constants in magnetism. *Journal of Magnetic Materials*, **100**, 573–575.

THOMPSON, D. J. 1982. Epectrum estimation and harmonic analysis. *IEEE Proceedings*, **70**, 1055–1096.

WATERHOUSE, J. B. 1978. Chronostratigraphy for the world Permian. *In*: COHEE, G. V., GLAESSNER, M. F. & HEDBERG, H. D. (eds) *Contributions to The Geologic Time Scale*. American Association of Petroleum Geologists, Tulsa, OK, 299–322.

WEEDON, G. P., SHACKLETON, N. J. & PEARSON, P. N. 1997. The Oligocene time scale and cyclostratigraphy on the Ceara Rise, western equatorial Atlantic. *In*: SCHACKLETON, N. J., CURRY, W. B., RICHTER, C. & BRALOWER, T. J. (eds) *Proceedings of the Ocean Drilling Program, Scientific Results*. US Government Printing Office, Washington, **154**, 101–114.

WEEDON, G. P., JENKYNS, H. C., COE, A. L. & HESSELBO, S. P. 1999. Astronomical calibration of the Jurassic time-scale from cyclostratigraphy in British mudrock formations. *Philosophical Transactions of the Royal Society London, A*, **357**, 1787–1813.

WHALEN, M. T. & DAY, J. E. 2008. Magnetic susceptibility, biostratigraphy, and sequence stratigraphy: insights into Devonian carbonate platform development and basin infilling, Western Alberta. *In*: *Papers on Phanerozoic Reef Carbonates in Honor of Wolfgang Schlager*. Society for Sedimentary Geology, Tulsa, OK, Special Publications, **89**, 291–314.

Index

Page numbers in *italic* denote figures. Page numbers in **bold** denote tables.